Evolution Slam Dunk

Evolution Slam Dunk

Why the Reptile-Mammal Transition Proves *Macroevolution* & How Antievolutionists Ignore It

James Downard

Evolution Slam Dunk:
Why the Reptile-Mammal Transition Proves Macroevolution & How Antievolutionists Ignore It

Inquiries should be addressed to

Rulon James Downard
4033 N. Belt St.
Spokane, WA 99205

Evolution Slam-Dunk is part of the #TIP "Troubles in Paradise" project.

www.tortucan.wordpress.com& *www.tortucan.com*

First Printing: 2016
ISBN 978-1540736291

Contents

1. Introduction

Evolution.

The understanding that all life on Earth is related by natural branching common descent. It's the cornerstone of modern biological science, and yet the very word gets all too many people up in arms. It's only an unproven "theory," some say—misunderstanding that in science a Theory is at the top of the reasoning chain, a general well-attested explanation for a broad body of solid evidence, not some tentative hypothesis or guess.

So why is "evolution" a fighting word?

Unless you've lived in a cave, you've probably already noticed that the most vocal critics of evolution are objecting to it because it conflicts with their religious beliefs about how we humans originated and whether we serve some teleological purpose. Many of those will mention a lot of "evidence" along the way, of course--evidence which they typically got from a secondary source, and which they seldom fact-checked first before repeating it to others.

It's that *letting others do their "thinking" for them* aspect that lies at the root of why antievolutionism is so troubling and wrong, and studying that methodological aspect is a main object of this book.

I've been investigating the antievolution subculture and their underlying methodology for nigh on twenty years, and people can follow that end of the work at my #TIP "Troubles in Paradise" websites (*www.tortucan.com* & *www.tortucan.wordpress.com*). *Tortucan* is my take on the Latin word for turtle, a shorthand way to describe people who dwell under a stout mindshell that is largely impervious to contrary evidence. I'll be using "tortucan" occasionally along the way, as both noun and adjective, and will say more about the concept in Chapter 12

Although religious convictions underly so much of antievolution activity, this book is not about religious faith *per se*, or whether a belief in science can or cannot be accommodated with any particular religious doctrine. I'm not religious, but have no intrinsic problem with people who are.

But religion inevitably comes up along the way because so many of evolution's critics don't allow the science discussion to avoid it. For many, evolution is indeed a "God or matter" choose your side thing.

This book is about the scientific evidence, though—and there's a lot of it nowadays. I'll be covering many paleontological topics, including dinosaurs, and a lot about biological work that connect up to that story. But the core of it concerns a very specific segment of fossil evidence, the Reptile-Mammal Transition sequence that served to convince me that evolution is as well attested as any scientific concept could want. And it's about how antievolutionists *don't* deal with that evidence, with an obtuse consistency that in the end reveals an awful lot about what is (and isn't) going on in their heads.

Most modern antievolutionists freely acknowledge that "microevolution" (observable natural mutations in organisms) occurs, but firmly believe that "macroevolution" (really big changes in life, like us) lies beyond that microevolutionary boundary. Not only unsupported by the scientific evidence, but utterly refuted, or even theoretically impossible.

The Reptile-Mammal Transition says otherwise. It is macroevolution full blown, and only evolution has been able to account for it, applying the very microevolutioary processes antievolutionists supposedly accept.

The Reptile-Mammal Transition (or RMT for short) is a *Slam Dunk* for evolution.

Or a *Home Run*. Or *Touchdown*. Pick your sports metaphor. I'll be mixing mine all through this book, no apologies.

The evidence for the evolution of mammals from a particular group of reptiles (the synapsids) a quarter of a billion years ago represents a broad convergence of independent data, from fossils and developmental biology to genetics. You'll be reading about that work in this book. And other subjects, too, along the way, just to show that the bad method being seen in antievolutionist coverage of the Reptile-Mammal case isn't some isolated foible, but a fundamental pattern of behavior.

It's ultimately a quite simple issue: how do opponents of evolution explain all that evidence?

They don't.

And I mean it. *They don't.*

Even when they bring the subject up themselves—which they don't do very often, as we'll see.

Delving into the sources they rely on (and more importantly, the huge field of science work that they don't) goes a long way to explain how antievolutionists can sustain the quite flawed belief that there is no evidence for macroevolution.

And a nod of thanks at this point to Glenn Branch and the *National Center for Science Education* (NCSE) for making available to me a loaner copy of Michael Denton's 2016 book, so that my survey of the field could be as comprehensive as I could make it.

By the time you finish this book, I think you'll see something else besides the amazing gymnastic evasions of antievolutionists (as entertaining or frustrating as those may be). You'll discover how the macroevolutionary process isn't some polar opposite from microevolution, after all.

The macroevolution visible in the RMT is composed of nothing but microevolutionary increments. Muscles shifting as tiny changes morphed bone arrangements, or subtle shifts in gene regulation opening up new adaptive possibilities.

That's how big evolution works: tiny changes, none of which are by themselves a "macrovolutionary" leap, but taken together, piled up over a hundred million years or more (and that is a *really* long time), you can end up with a whole new vertebrate class, as happened with the Mammals.

The vertebrate class we humans belong to.

Which means the synapsids and their descendants are not strangers. They're *us*, along with all of our many nearer cousins, from humpback whales to the bats that pollinate the tequila cactus. Drink up.

The Reptile-Mammal Transition is *our* story.

Be proud of our lineage's history, and amazed. Because it is pretty amazing.

And it has the considerable added pleasure of actually being *true*.

Notes on Taxonomy and the *Map of Time*

Since they're going to play an important background role in the story, some benchmarks on taxonomical terms and chronology are in order. Get the nasty dull medicine out of the way, up front.

Despite the wrinkles of cladistic taxonomy (a method of analyzing the features of things to better understand their relationships, a topic I'll have more to say about in Chapter 6), the older Linnean classification system remains in use. Worked out by botanist Carolus Linnaeus (1707-1778) over a century before Darwinian evolution turned such relationships into a template of branching lineages, Linnaeus categorized all life in seven nesting categories, from the most general all the way through to the species level of interbreeding populations:

Kingdom / Phylum / Class / Order / Family / Genus / Species

In our case (and it's always about *us*, isn't it?), that would be:

Animalia / Chordata / Mammalia / Primate / Hominidae / Homo / sapiens

It's a feature of Linnean "binomial" classification that only the genus is capitalized, so it's *Homo sapiens* and *Tyrannosaurus rex*, not *Homo Sapiens* and *Tyrannosaurus Rex*. Those latter are goofs that can signal the scientific illiteracy of the author, as I discovered many years ago firsthand when making that mistake myself under the gaze of a knowledgable geologist who promptly set me straight.

Going by that, the capitalized and *italicized* critter names in this book will be refering to the animals' *genus* (plural, *genera*).

Now for the even Bigger Picture: the periods applied to geological history, and what dates belong to what. Some years ago I compiled a pdf for my #TIP project, "The Portable Map of Time", which put the whole shebang on a single sheet, from the Big Bang till now, with all the right number of zeros. It's free to download and share from the "Other Stuff" tab at *www.tortucan.wordpress.com*.

For those not wanting to go web-searching, here's a brief summary with approximate dates for the Earth end of things, so everybody can start with a common *Map of Time* (and yes, I know Young Earth creationists don't accept that sequence, but that's their problem and another topic).

With "bya" meaning billions of years ago, and "mya" millions of years ago, and riffing off the clever color-code scheme by geologist Robert Hazen (2012), here goes:

4.5 bya **Hadean Eon**, the *Black Earth* of volcanic basalt followed by the *Gray Earth* of continental granite formation.

3.6 bya **Archean Eon**, the *Blue Earth* of planetary oceans and early bacterial life.

2.5 bya **Proterozoic Eon**, the *Red Earth* as oxygen-spewing cyanobacteria rusted out the oceans, punctuated by a *White Earth* of planetary hyperglaciation, followed by the *Green Earth* of our complex multicellular animal life.

Things had got much busier by the **Phanerozoic Eon** (and those periods ending in **mass extinctions** are in **bold**):

mya		
540	Paleozoic Era	Cambrian Period began, lots of sea critters now.
500		**Ordovician** Period, even more sea critters.
440		Silurian, plants and animals colonized the land.
400		**Devonian**, tetrapods (our vertebrate group) appeared.
360		Carboniferous (includes Pennsylvanian), shelled eggs got laid.
310		**Permian**, the Reptile-Mammal Transition underway!
250	Mesozoic Era	**Triassic**, the first full mammals & dinosaurs.
215		Jurassic, the dinosaur-bird transition underway.
150		**Cretaceous**, dinos out at end, birds & mammals survived.
65	Cenozoic Era	Tertiary (AKA Paleogene/Paleocene), then 48-38 mya Eocene epoch.
30		Neogene (Miocene) Period, humans arrive, yay!?

This should give everybody a common reference frame for the journey ahead.

2. Meet the Furballs

What is a mammal, anyway?

We mammals are warmblooded furballs that give live birth instead of laying eggs. Except for the cetaceans (whales and dolphins) that lack surface hair, and the monotremes (the platypus and echidnas) that lay eggs. More complicated already.

But if you dive beneath the surface details, as generations of hard-working scientists have done, the features that unite our so disparate vertebrate class become clearer, along with the evolutionary reasons for why we have them.

Us being mammals, you don't have to look far. You can begin with your own body.

Four limbs to start—that's what the term *tetrapod* means. That goes back a long way, deep into our fishy roots, hundreds of millions of years ago.

There's been lots of sweat equity by paleontology on that end of things, filling in the fossil steps from purely aquatic fish to fully terrestrial animals.

Tetrapods emerged from a group that included two orders among the fishes: Dipnoi (lungfish), air-gulping since at least the Devonian, discussed by Alice Clement & John Long (2010), and the crossopterygians (the "living fossil" coelacanths and their extinct rhipidistian relatives, such as *Eusthenopteron*), grouped in the subclass Sarcopterygii that shared with the earliest amphibians (like *Ichthyostega* and *Acanthostega*) bony limbs, specialized vertebrae, and a two-part skull with internal nostrils and unique teeth.

There is no shortage of ready resources describing those critters and their features. Ones I'd gathered for my early "Troubles in Paradise" research included David Lambert & The Diagram Group (1985, 86-87, 90-93), Edwin Colbert & Michael Morales (1991, 64-69), Michael Benton (1993b, 79-83), Per Ahlberg *et al.* (1996), Patricia Rich *et al.* (1996, 367-371), Samantha Weinberg (2000, 98-102, 195-203) and Jennifer Clack *et al.* (2003).

Anyone with a Google instinct can discover plenty more online, but I'll be sticking to the scientific literature in this book (although a lot of that is obtainable online too, fortunately).

Comparing the structural layout of an advanced rhipidistian fish like *Eusthenopteron* with early amphibians, Leonard Radinsky (1987, 78-81) noted how relevant muscle transformations were traceable in the fossils. Paleontologists can tell this because muscle attachments leave telltale scrapes on the bones. Recent work suggests the aquatic *Eusthenopteron*'s bones may also have had marrow, Sophie Sanchez *et al.* (2014), putting that feature in the tetrapod kit bag long before any terrestrial bone growth adaptation.

Further fossils showed that the earliest amphibians retained internal fishlike gills, unlike the external ones of modern amphibians, and were exhibited in their tadpole phase, Michael Coates & Jennifer Clack (1991).

All this is important to remember, because the evolutionary process involves adapting and modifying inherited anatomy, every stage of which must be functioning biologically. It's an evolutionary parade that is also ever so gradual, with none of the novelties appearing *poof*, out of the middle of nowhere.

Such sequential appearance of traits as seen in the fossils prompted Kenneth Miller (1999, 40) to comment in his criticism of creationism (his *italics*): "*The first amphibians looked more like fish than any amphibian species that would follow them in the next 380 million years.*"

Work including Jennifer Clack (1998), Clack *et al.* (2003) and Gaël Clément *et al.* (2004) continued blazing the paleontology trail, leading up to the discovery of the celebrated "fish with legs" *Tiktaalik* in 2006. Recounted in the fine book by Neil Shubin (2008), *Your Inner Fish*, those preferring some of the ongoining primary source technical work may consult Edward Daeschler *et al.* (2006), Neil Shubin *et al.* (2006; 2014), Jason Downs *et al.* (2008), and Jason Anderson *et al.* (2015).

Daeschler *et al.* (2006, 757) described *Tiktaalik* as "a remarkable intermediate between *Panderichthys* and early tetrapods." You'll notice how that inclusion of *Panderichthys* marked the progress of paleontology over the years, turning up even more relevant critters to include on the *Map of Time* track, each telling a piece of the expanding scientific story, represented by Catherine Boisvert (2005) and Boisvert *et al.* (2008).

Consider *spiracles*, small holes behind the eyes of some fish that open onto the mouth. Jennifer Clack (1989) had already noted the robust stapes bone known from the late Devonian tetrapod *Acanthostega*, where "the stapes controlled palatal and spiracular movements in ventilation." So a bone we mammals now rely on exclusively for hearing started out with a very different function, related to breathing.

We'll hear more about those spiracles again in Chapter 7.

Want more evidence?

The well-preserved middle ear of *Panderichthys* further supported that spiracle breathing idea, evolving toward the tetrapod layout (advancing beyond the earlier configuration seen in its more distant cousin *Eusthenopteron*), but still showing the specializations indicating "that the middle ear of early tetrapods evolved initially as part of a spiracular breathing apparatus," Martin Brazeau & Per Ahlberg (2006). The polypterid fishes still rely on spiracular breathing, incidentally, offering an informative living analog for the situation found in those extinct forms, as Jeffrey Graham *et al.* (2014) have explained.

But there's something else about how *Tiktaalik* specifically was found that says something important about how science works (and why antievolutionist views are so unhelpful).

The paleontologists knew the constraints on *where* and *when* to look: shallow water deposits dating from the Devonian period. Dig in the wrong habitat (dry land or deep oceans), or at the wrong time (long before their evolution, or long after the modified descendants were already on the scene) and you'd miss the transitional players.

Only creationists have the weird idea that evolutionary ancestors have to still exist *now*, or that individual animals have to be doing the whole evolving thing right in front of you, from rock pile or pond scum to wildebeest or whale in one go.

No wonder they don't see "proof of evolution." They're looking for the wrong thing.

Paleontologists operate on a firmer playing field, paying attention to all the data, and working hard to get the sequence of events straight. That's why early tetrapods aren't ever going to show up in Cambrian deposits (way too early) or be hanging around completely unmodified in Cretaceous ones (way too late), even if some can occasionally persist long enough to be tagged a "living fossil" (though even then there will be derived features to distinguish them from their ancient counterparts).

But the paleontologists hunting for early tetrapods knew their field and once again were able to *anticipate the past*. Or that of the Designer?

If *Tiktaalik* was designed, in the way sports cars or cell phones are, we have to imagine that the Designer had a penchant for making hitherto unknown species that fit into a purely imaginary evolutionary pattern and then just coincidentally (or perversely) plopping them down in just the right contexts that would reward the hard work of paleontologists millions of years later (or a few thousand years on, if one leaves too many zeros out in the manner of Young Earth creationism), thus saving the paleontologists so much fruitless effort.

Have you ever tried digging on a cold bleak hillside in the north of Canada? It's hard work. You don't go after such things unless you've narrowed the field first.

So how thoughtful of the "Designer" to so reward Neil Shubin and colleagues, contriving things to make evolution look good once again. Or maybe it was a Fossil Genie, with time off from the bottle for good behavior.

Isn't there something drastically wrong with that picture?

That evolutionists are so proficient at this fossil finding game, turning up new animals that nonethless have the features evolutionists were expecting them to, is another signal that their success comes not from the convenient cooperation of a selectively dimwitted Designer (or compliant Fossil Genie), but from meticulous application of a theoretical evolutionary model that happens to be *true*.

Now antievolutionists nitpick the evidence for tetrapod origins, of course, along with anything that suggests an evolutionary reality that they want so not to be true. That's a fine subject for another book, but an example will suffice from one of the early stars of Creation Science apologetics, to illustrate the temper of their game.

In his book *Evolution: The Fossils STILL Say No!* Duane Gish (1995, 87) cited paleontologist Peter Forey (1988, 729) when Gish argued how "Evolutionists point to the presence of the fish-like tail fin as evidence that

Ichthyostega is a descendant of rhipidistian fish, but as Forey points out, such ideas are flawed because fish-like tails are characteristic of the general group of vertebrates with jaws (Gnathostomata)."

Gish was pulling a fast one.

Paleontologists try to find uniquely inherited features to trace specific ancestry. A character trait occurring early in the parade (like a fish's tail) wouldn't specify on its own which of the descendant lines were involved. But having that generally inherited feature certainly wouldn't be a surprise, or rule out ancestry. So it was a brave evasion for Gish to suggest early amphibians couldn't have evolved from any fish line because their tails were still too generally *fish-like*.

Incidentally, after Christopher McGowan (1984, 150-158) and Arthur Strahler (1987, 408-412) criticized some of the more confused bits on early amphibian anatomy in Gish (1978, 76-77), Gish (1995, 83-92) responded to their attentive scholarly prodding by *removing* those items from his new coverage (but *not* acknowledging the debt or the corrections). Gish's revision also skipped discussing parietal skull openings, a topic Joe Cain (1988, 102) had criticized Gish (1978, 86-87) on for the creationist's lopping off the rest of a quote that had significantly changed its meaning.

Tidier at least on those points, but overall Gish continued to follow the "quote mining" addiction of piling up selectively staged authorities in place of investigating the taxonomical features in detail. And the glib sidestepping of a Duane Gish may be compared to any of the regular science reporting on early tetrapod evolution during that same time, such as Oleg Lebedev (1997), John Wilford (1998, 111-114), Daeschler & Shubin (1998), Henry Gee (1999, 46-66) or Richard Ellis (2001, 130-138).

Kerri Westenberg (1999), Douglas Palmer (1999, 78-79), or *Prehistoric Life* (2009, 128-139) have some handy visuals showing the assorted players. Try looking at them without spotting how their very existence showed animals whose slow changes were tracking the evolutionary process.

Returning to our reptile-mammal trail, there's a few more technical benchmarks to get on the table before moving on to the fossil record of mammal origins and how antievolutionists don't deal with it.

Check out your hand

Five digits—we're *pentadactyl*, another feature that goes way back, but not as far as being a tetrapod, as the earliest land vetebrates started out with *more than five fingers*. And toes, you could count them too.

Having seven or eight fingers and toes apparently became a hassle on land, though, and constraints imposed by the developmental biology have locked in the model we're stuck with now, examined by Frietson Galis *et al.* (2001). Though even that isn't entirely rigid, as occasionally digit reductions have occurred, such as digit loss in archosaurs (a broad group running from crocodiles to dinosaurs and birds) covered by Michael Shapiro *et al.* (2003), Hans Larsson *et al.* (2010) and Merijn de Bakker *et al.* (2013), and comparable loss in mammals discussed by Bau-lin Huang & Susan Mackem (2014) surveying papers by Kimberly Cooper *et al.* (2014) and Javier Lopez-Rios *et al.* (2014).

Novel exceptions can occur, too, such as Christian Mitgutsch *et al.* (2012) on the unusual "thumb" that has evolved in a species of Spanish mole. And the detailed genetics and biogeography of such things can be quite involved, as Trip Lamb & David Beamer (2012) discovered regarding digit reduction and reaquisition among dwarf salamanders.

Though the pentadactyl format became the mode for tetrapods afterward, the genetic systems for more fingers and toes hadn't disappeared, and a late-blooming seven-digit model (*Nanchangosaurus*) is known from the Triassic, Xiao-Chun Wu *et al.* (2003). No coincidence that it was another marine animal, where having extra digits kicking back in wasn't so much of a problem as it would be were it scrambling on land, and Donald Prothero (2007, 239) noted this was a very primitive intermediate stage in the evolution of the ichthyosaurs (the dolphin-shaped marine reptiles of the Mesozoic).

Like those ancient marine reptiles, the mammalian cetaceans have undergone substantial limb loss and variation too. Fortunately there are living representatives of those to compare with their fossil counterparts, allowing more of the biology and genetics of their limb evolution to be worked out, such as by Lisa Cooper *et al.* (2007) and Zhe Wang *et al.* (2009).

"The evolution of fingers and toes is associated with changes in the timing and expression of the more ancient *Hox* genes that regulate development of the body axis and appendages," Timothy Rowe (2004, 399) noted, citing Neil Shubin *et al.* (1997) and Robert Carroll (2001). The actual pentadactyl state in vertebrates involved the disconnection of one of the homeobox genes (*Hoxa11*) during the deployment of *Hoxa13*, as recently discovered by Yacine Kherdjemil *et al.* (2016)

The evolution of appendages across the vertebrate range (from crocodiles to whales and birds) have turned out to involve regulatory genes beyond the *Hox* bunch, including *Gli3*, *Ptch1*, bone morphogenetic protein (*Bmp*) and sonic hedgehog (*Shh*)—yes, that was named for the videogame character. Who says scientists don't have a sense of humor? Some of the genetic relationships are surprisingly direct, such as repeats of the amino acid alanine occuring in the *Hoxd13* gene (more or less alanines, potentially more or less fingers or toes).

The progress of the field in determining how limbs, hands, wrists and digits have originated and evolved in vertebrates can be tracked in Paolo Sordino *et al.* (1995), József Zákány *et al.* (1997), Michael Coates & Martin Cohn (1998), Michel Laurin *et al.* (2000), Günter Wagner & Chia-Hua Chiu (2001), Marie Kmita *et al.* (2002), Ying Litingtung *et al.* (2002), Keiti Anan *et al.* (2007), Marcus Davis *et al.* (2007), Thomas Montavon *et al.* (2008; 2011), Alexander Vargas *et al.* (2008), Natalie Butterfield *et al.* (2009), Carolina Minguillón *et al.* (2009), Joost Woltering & Denis Duboule (2010), Igor Schneider *et al.* (2011), Renata Freitas *et al.* (2012), Gretchen Vogel (2012) on

Rushikesh Sheth *et al.* (2012), Guillaume Andrey *et al.* (2013), Kathryn Kavanagh *et al.* (2013), Takayuki Suzuki (2013), Tohru Yano & Koji Tamura (2013), J. Raspopovic *et al.* (2014), Woltering *et al.* (2014), Andrew Gehrke *et al.* (2015), Noritaka Adachi *et al.* (2016), Tetsuya Nakamura *et al.* (2016) and the aforementioned Kherdjemil *et al.* (2016).

That's a lot of continuing work.

And the lack of progress over in the antievolution literature may be symbolized by Intelligent Design advocate Michael Denton (2016a, 157-159; 2016b) arguing in his revised *Evolution: Still a Theory in Crisis* book and summarized for the *Discovery Institute*'s *Evolution News & Views* online newsletter that the pentadactyl limb was somehow *nonadaptive*, and thereby posed "An Existential Challenge to Darwinian Evolution." It wasn't as though he was unaware of relevant work on the evolution of limbs in fossils and the genetics of living models, as Denton (2016a, 156, 161, 163-164, 225, 271, 313-315n, 328n, 340n) cited the likes of Coates & Cohn (1998), Davis *et al.* (2007) Montavon *et al.* (2011), Lamb & Beamer (2012), Sheth *et al.* (2012), Kavanagh *et al.* (2013), Suzuki (2013), Yano & Tamura (2013) and Woltering *et al.* (2014). It's just that Denton's notion of "adaptation" (and how scientists were supposed to spot it) was excessively vague, which I'll have more to say about in Chapters 5, 6 and 11.

But back to our trail.

The upshot of the paleontology and recent biology findings show that the vertebrate body plan has remained fairly stable for a very long time, even as the parts morphed into new shapes, sometimes reduced, or even shut down. The same bones have been adapted to many new uses (including flight and novel ways of hearing), and the genetics of this process are being ferreted out—and by evolutionists, not their assorted antievolutionary critics.

For example, small variations can have sigificant effects over the long haul, such as slight shifts in the regulatory gene *Hoxc8* determining the number and placement of neck and back vertebrae in chickens (along the diapsid line that includes dinosaurs and birds along with modern reptiles) and mice (a descedant of the long synapsid line that led to us mammals), discussed by Heinz-Georg Belting *et al.* (1998).

Unlike dinosaurs, where extra neck vertebrae popped up repeatedly in sauropods, almost all mammals have only seven cervicals (sloths and manatees being the exceptions), Frietson Galis (1999) and Robert Asher *et al.* (2011). The range of natural vertebral variation in manatees has been examined by Emily Buchholtz *et al.* (2007), along with their adaptive and evolutionary implications (such as how an especially short neck can contribute to drag reduction in conjunction with flippers positioned farther back on some species). Pleiotropic effects (where changes in one gene or feature spur alteration in other body parts or processes linked to them developmentally) appear to be playing a part too, as do selective constraints against additional trunk vertebrae in species adapted to fast running, Galis *et al.* (2006; 2014). Comparable natural factors govern the famously long giraffe neck, Megu Gunji

& Hideki Endo (2016). Only recently have more of the cervical genetic players been identified there, such as the *T (brachyury)* gene, Andreas Kromik *et al.* (2015).

The big split in our taxonomy tale concerns the *diapsids* and *synapsids*, named for the number of openings at the back of the skull for major jaw muscle attachments. Diapsids have two while the synapsids had only one, the basics surveyed neatly by Michael Benton (1997). Apart from that, their earliest members started out looking a lot alike. It's only after a very long time that the two branches evolved to show distinctly different forms, such as reptiles, dinosaurs and birds on the diapsid side, and us mammals representing what's remained of the equally long synapsid line.

Now many antievolutionists spin all this as merely common *design*, similar genes being deployed by the Designer, not a reflection of common *ancestry*.

But all that looks less plausible the more data you include.

Take teeth.

They've been set in a dentary bone for hundreds of millions of years, yet how they look on a particular jaw depends on the special history of each lineage. In the case of tooth enamel, the core *amelogenin* gene traces back long before vertebrate jaws or teeth, all the way into the Precambrian, Sidney Delgado *et al.* (2001) and Qingming Qu *et al.* (2015), while the *enamelin family* underwent lots of specialization and variation in such derived models as frogs, lizards, crocodiles, and birds, Tiphaine Davit-Béal *et al.* (2009) and Nawfal Al-Hashimi *et al.* (2010).

Yes, *birds* with their toothless beaks have tooth enamel genes lurking as shut down pseudogenes. So do the toothless baleen whales, Robert Meredith *et al.* (2009; 2013) and Meredith & John Gatesy *et al.* (2011). In fact, *all* toothless mammals still retain the ancestral tooth enamel genes, baggage held over from the time when their ancestors made teeth.

Now one might wonder, if we're talking Intelligent Design (ID for short), just how smart a move was it to go out of the way to equip tooth enamel genes in toothless birds or mammals? Isn't that a bit like putting a gas tank on a Tesla?

Now, consider hair

That birds have keratin filament feathers while mammals are covered in keratin filament hair may represent one of the neater contingent splits in vertebrate history, reflecting what may have started out as the most trivial of differences between the diapsids and synapsids.

The tendency of the skin to pimple outwards (evagination, as it does in reptiles) for example, compared to dimpling inwards (invagination, which it does in mammals), spotted by G. Oster & P. Alberch (1982). Richard Prum & Alan Brush (2002, 289) noted how such contingent differences constrained the generation of evolutionary novelties like feathers, and Lorenzo Alibardi *et al.* (2009, 562-567) illustrated the varied developmental paths and gene interplay from basal sauropsid skin to the feathers and hair of birds and mammals.

Starting of course with the underlying genes, and how they've been modfied and coopted through evolution.

Deep at the level of axial symmetry (front and back, top and bottom, etc.) sonic hedgehog (*Shh*) and bone morphogenetic protein (*Bmp*) pop up again, along with *Pitx2, Wnt-7a* and retinoic acid, Cheng-Ming Chuong *et al.* (1992; 1996; 2003), Aimee Ryan *et al.* (1998), Chijen Lin *et al.* (1999), Concepcíon Rodríguez-Esteban *et al.* (1999), Dorothy Supp *et al.* (1999), Tohru Tsukui *et al.* (1999), Randall Widelitz *et al.* (1999), Cheng-Ming Chuong & Nipam Patel *et al.* (2000) and Thomas Schlange *et al.* (2002).

But it doesn't stop there. Those same gene groups play out farther downstream in the skin and what they're made of. In the case of the reptiles and their closer kin, Matthew Harris *et al.* (2002, 160) noted "that the anterior-posterior expression polarity of Sonic hedgehog (*Shh*) and Bone morphogenetic protein 2 (*Bmp2*) in the primordia of feathers, avian scales, and alligator scales is conserved and phylogenetically primitive to archosaurian integumentary appendages."

Said more informally: the same genes are helping to make their oh-so-varied *skin*.

Apply that set of genes (*BMP* and *Shh*) to variations in skin buckling, tossing in some more players like "transforming growth factor" TGF-β2 and the β-catenins, and you're building the substrate for feathers evolving on the reptile side and hair for mammals, explored in a continuing series of work by Sheree Ting-Berreth & Cheng-Ming Chuong (1996), Han-Sung Jung *et al.* (1998), Bruce Morgan *et al.* (1998), Kerstin Foitzik *et al.* (1999), Ting-Xin Jiang *et al.* (1999), Holger Kulessa *et al.* (2000) and Kentaro Suzuki *et al.* (2009).

For those looking for the experimental, it had been known since Danielle Dhouailly *et al.* (1980) that feathers could be induced on the scales of a chicken's foot by the common moderator of retinoic acid. Since then it was learned it was the underlying β-catenins triggering feather development, with particular mutations in them switching a bird's foot scales into feather bud mode, Selina Noramly *et al.* (1999) and Randall Widelitz *et al.* (2000).

All of that biology is before the keratins enter the picture, the stuff scales and feathers and hair are directly made of. Roger Sawyer *et al.* (2000; 2003; 2005), Sawyer & Loren Knapp (2003), Cheng Chang *et al.* (2009) and Ping Wu *et al.* (2015) reflect recent work plotting out how keratins are specifically deployed in reptile and bird skin.

Pull back to the Big Picture level, and scales, hair and feathers can be seen to have originated from interacting variations among the ancestral vertebrate skin placodes, surveyed by Ping Wu & Lianhai Hou *et al.* (2004), Alibardi *et al.* (2009), Cheng Chang *et al.* (2009), Dhouailly (2009), Kevin Painter *et al.* (2012) and Nicolas Di-Poï & Michel Milinkovitch (2016). In a related contingent split, the foregoing of standard reptilian scales in the mammalian lineage appears to have opened the adaptive options for sensory evolution still drawing on the shared ancestral nerve growth factor (NGF)-

Hoxd1 signaling pathways, Ting Guo *et al.* (2011).

Naturally, there's lots of hard work involved to gradually identify what all the pieces of the process are, and work out how they have evolved over time. Just on feather budding, there's work by Stephen Gould *et al.* (1995), Alexander Noveen *et al.* (1995), Chia-Wei Chen *et al.* (1997), Cheng-Ming Chuong & Rajas Chodankar *et al.* (2000), Paul Maderson & Lorenzo Alibardi (2000), Sharon Cahoon-Metzger *et al.* (2001), and Chih-Feng Chen *et al.* (2015).

And over on the mammal hair side of things, there's Denis Duboule (1998) commenting on Alan Godwin & Mario Capecchi (1998), Jiro Kishimoto *et al.* (2000), Satrajit Sinha & Elaine Fuchs (2001), Yuanxiang Zhao & Steven Potter (2001), Randall Widelitz *et al.* (2003; 2007) and Sung-Ho Huh *et al.* (2013).

Research on the human end of this trail tends to relate to hair loss and how to stop it, Elia Ben-Ari (2000), which means plenty of opportunity to discover yet more evolutionary variations in play, such as Lutz Langbein *et al.* (1999; 2001) on several variant forms of human hair keratins that have turned up.

Natural evolutionary variation never stops, nor does the interesting work on it by scientists fueled by curiosity and a sound method.

Incidentallly, there is direct evidence that the synapsid epidermis had been diverging from the ancestral amniote model for some time. The only trace of therapsid skin that has turned up comes from the Middle Permian dinocephalian *Estemmenosuchus*, one of many clunky looking animals knocking around in those days, such as its cousin *Moschops*, which looked to me in my childhood dino-loving days rather like an overgrown grinning frog, though modern depictions like the one in *Prehistoric Life* (2009, 190-191) are more dramatic. As *Estemmenosuchus* had short brow horns, you could think of it as something like an elk-frog.

As random samplings go, the skin fragment from *Estemmenosuchus* was a dandy: embedded in its smooth skin were signs of *glands*, something uncommon on the diapsid side of the fence, but entirely reasonable to expect from a group whose metabolism would eventually end up sweating through them for thermoregulation, Peter Chudinov (1965). Sylvia Czerkas & Stephen Czerkas (1991, 51, 54-55, 57) noted the gland matter along with illustrations of *Estemmenosuchus* and *Moschops*. Evidence suggests glandular secretions (including eventually mammary lactation) and a higher metabolic rate accompanied synapsid evolution, William Hillenius (1994), John Ruben & Terry Jones (2000, 588-590) and Olav Oftedal (2002).

On the diapsid side, incidentally, birds have developed a few glands of their own, especially for preening, Peter Stettenheim (2000, 467-468), though Gopinathan Menon & Jaishri Menon (2000) described how the avian lipid secretion system differs from the mammalian analogs.

Just how far-ranging these processes have been in vertebrate history is hinted at by the recent discovery that the extinct pterosaurs (another group of energetic vertebrate fliers, along with birds and bats) had fuzzy wing membranes, Xiaolin Wang *et al.* (2002) on *Jeholopterus*, with summaries by Michael Benton (2005, 226-228) and David Unwin (2006, 131-135), followed by further fossil confirmation by Alexander Kellner *et al.* (2010).

With no living examples to do genetic studies on, of course, the role of *Shh* or *BMP* on yet another variant lineage of integumentary fibers would seem beyond experiment. But for the mammal side of things you have living members, with more and more of their genes coming under detailed scrutiny, and new technologies to reveal seemingly irretrievable details.

Julien Benoit *et al.* (2016) have followed exactly that kind of trail, starting with the role the *Msx2* homeogene plays in the formation of mammary glands and hair, and how closely related those are to the nervous system and skin sensitivity. Move to facial sensitivity, such as a mammal's whiskers, and that correlates with a particular set of vertebrate skull openings, including the maxillary canal, "a bony tube which runs parallel to the tooth row in the maxilla and premaxilla bones." Knowing how those vary between reptiles and mammals, dozens of therapsid fossils were given CT scans for comparison, and what that implied about what their Msx2 gene must have been doing to produce such a configuration.

Benoit *et al.* suggested a "single mutation event" in the *Msx2* gene had arisen around 250 mya, with cascading effects for "some of the most prominent characters defining mammals, such as hair coverage, lactation, lateral cerebellar expansion and endothermy."

One question is whether design advocates would ever bother to connect those dots, or make such inferences. Stay tuned to Chapter 10, when we'll return to Benoit's paper.

Now that you've got a taste of the broad spread of technical data swirling around the issue of mammals, and how they (we) got to be the way they (we) are, it's time to look closer at the fossils.

3. "An Earful of Jaw"

As I pointed out in the previous chapter, vetebrates can be kind of "boring" in a lot of their anatomy, insofar as they use the same bones (often morphed around a lot) in the same general arrangement.

Not so the mammal skull and jaw. Here things get really interesting.

Consider your own jawbone.

Everybody who's seen a Halloween skeleton knows the layout, and you can easily feel it under your skin: that great big *dentary* bone running all the way from the chin to the hinge point on the main skull, which for mammals is the *squamosal* bone. And sticking up from the back of our one big dentary on either side: the *coronoid process*, which makes our jawbone look like an L laid on its side when viewed edge on.

This mammal jaw layout is very different from other vertebrates (fish, amphibians and reptiles), which by the time we get to the basal diapsid and synapsid split consists of seven bones in their jaw, with four prominant ones: the *dentary* in front, followed by the *surangular* and *angular* (one atop the other), and the *articular* in the rear. The articular is called that because it is the normal vertebrate jaw hinge articulation, which uses the *quadrate*, a bone resting *below* the squamosal skull bone.

So at first glance, the mammal layout seems entirely different from any immediate reptilian ancestor, and that's exactly how antievolutionists insist on seeing it, as we'll be exploring in later chapters.

But there were clues that something else may have been going on, concerning what happens in the embryological development of mammals, and this has been known for quite awhile.

In the early 1800s biologists began to notice the similarity between the early embryological development of mammals and reptiles, where the jaw layout started out *the same in both*. The reptilian configuration represented a very ancient default, with *all* the standard jaw bones appearing in cartilage form, and hinged on the *quadrate*. Only as the mammal embryo developed did the dentary *expand* until it linked up with the growing *squamosal* bone, at which point four of the jaw bones *moved on into the ear* developing in that same spot.

The articular and quadrate bones become the *malleus* and *incus* of our mammalian middle ear, while the angular supplies the *tympanic bone* that supports the mammal eardrum, and the fetal prearticular bone eventually fuses to the front of the malleus as the *goniale* bone.

Experiments by Karl Reichert (1811-1883) began to identify that bone migration in 1837, twenty years before Charles Darwin (1809-1882) rolled out his theory of evolution to eventually make sense of such relationships, a historical sequence noted by Neil Shubin (2008, 160-162), Masaki Takechi & Shigeru Kuratani (2010, 419) or Wolfgang Maier & Irina Ruf (2016, 270-271) in their discussions of the history of study on mammalian ear evolution.

This process is even more apparent among newborn marsupials (a group considered to retain some of the traits of the earliest pre-placental mammals), where the quadrate and articular only take up their auditory position *after* the dentary completes its post-natal growth. I must thank Frank Sonleitner (personal communication) for first calling my attention to the delayed character of marsupial jawbone growth.

Timothy Rowe (1996), Marcelo Sánchez-Villagra *et al.* (2002), Thomas Schmelzle *et al.* (2005; 2007) and Héctor Ramírez-Chaves *et al.* (2016) reviewed marsupial ear bones regarding early mammal evolution and more recent adaptions, including their varied relationship to the concurrently expanding brain neocortex, and Takechi & Kuratani (2010, 421) compared the lower jaw of ancestral mammals to embryonic and adult marsupials. The postnatal bone shift in marsupials is quite a developmental twist, by the way, depending on the lack of jaw movement while the newborn is occupied suckling, Maier & Ruf (2016, 277).

Which means, even after over a hundred million years of specialization "the condition of the marsupial neonate resembles that of the embryonic condition of mammalian ancestors," Sánchez-Villagra *et al.* (2002, 233).

There has been so much work done since those 19th century scientists first spotted that jaw/ear shift, identifying the genes involved and how they illuminate the developmental biology and the incoming paleontology. There are concise reviews to draw on, of course, like Michael Benton (1990, 228-231), Cynthia Kenyon (1994, 178) or Elizabeth Pennisi (1999, 577).

And no shortage of technical papers for those wanting that.

Zhe-Xi Luo *et al.* (1995) and Zhe-Xi Luo (2001) connected more of the paleontology, while Yasuyo Shigetani *et al.* (2002), Sang-Hwy Lee *et al.* (2004) and Daniel Meulemans Medeiros & Gage Crump (2012) studied the process at the larger scale of jaw formation. There are the *Hox* connections too, such as James Martin *et al.* (1995) and Sophie Creuzet *et al.* (2002), while other researchers have homed in on specific genetic players, like how *Dlx* plays out in the pharyngeal arch jaw-building game, the commentary by Georgy Koentges & Toshiyuki Matsuoka (2002) putting the work of Michael Depew *et al.* (2002; 2005) in perspective.

The jaw/ear transformation is only a part of what happens among the neural crest cells, which generate features from the dentine-forming cells of teeth and pigment cells in bird feathers and mammal hair, to the dorsal fins and gill arches of fish, noted in summary by general works like Werner Muller (1996, 237-238), or explored in more detail by Tatjana Sauka-Spengler *et al.* (2007) and Brian Hall (2009).

Ears have to do basic tasks. Sound waves have to be collected, converted from pressure waves to fluidic signals, and that sensory cascade eventually processed by some manner of brain. That overall formal structure is as true for insects as for mammals, as Ronald Hoy (2012) reminded regarding the convergence analysis of Fernando Montealegre-Z. *et al.* (2012). Among vertebrates, there are many common developmental elements that have come into play to form the various middle ears there, surveyed by Susan Chapman (2011).

Focussing in on just the mammalian side of things, our middle ears involve yet more activity by our old friend, *sonic hedgehog*, Sandrine Testaz *et al.* (2001), along with others like *Pax2, Hmx2, Hmx3, Bapx1, Bmp* and *Tgf*, Juan Represa *et al.* (2000) and Zhe-Xi Luo (2011, 372-373). One experimental "knockout" of the *MHox* gene in mice (where genes are selectively removed to learn more about what particular DNA does in the organism) had a relevant effect: "the incus fails to demarcate and the palatoquadrate is partially retained, forming an articulation with the malleus. Thus, the *MHox* mutation results in cranial skeletal components that are morphologically similar to those seen in phylogenetically more primitive animals such as reptiles," James Martin *et al.* (1995, 1246).

This is a direct experimental window that allows modern science to peek into the past for clues as to what mutations contributed to the evolution of our own hearing. Seen in the light of modern genetics, the evolution of the mammal middle ear involves yet another sequence of "changes in the developmental program of the ancient chordate pharyngeal arches," Takechi & Kuratani (2010, 417). The science work has continued to progress, of course, such as the review of middle ear dynamics by Neal Anthwal *et al.* (2013) pulling together a broad range of work, including more "knockout" experiments. All of that study continues to confirm that the scientists are on the right track.

Furthermore, if you're of the evolutionary persuasion, it shouldn't come as a surprise that the regulatory *Bapx1* gene that comes into play with the mammalian malleus and incus operate in their ancestral counterparts, the articular and quadrate bones in fish, reptiles and birds, Craig Miller *et al.* (2003), Joanne Wilson & Abigail Tucker (2004), Abigail Tucker *et al.* (2004), Frank Eames & Richard Schneider (2008) and Zhe-Xi Luo (2011, 372-373).

Next door to that, the vertebrate inner ear (involving balance and pressure sensation in the cochlea) has similarly built on a large network of interacting genes and systems, studied by Jinwoong Bok *et al.* (2007; 2011), Evan Braunstein *et al.* (2009) and Byron Hartman *et al.* (2010). The cavitation process forming the mammal inner ear epithelium membrane has developed in a specialized way compared to that in reptiles and birds, Hannah Thompson & Abigail Tucker (2013) with commentary by Donna Fekete & Drew Noden (2013).

The inner ear's sensory hairs have built on genetic systems tracing all the way back to the ciliated neurons that underlie the metazoan mechanosensory system, the *bHLH* gene family operating there long predating the appearance of vertebrates, Kurt Beisel & Bernd Fritzsch (2004), and Fritzsch *et al.* (2007; 2010). Move on down the line some hundreds of millions of years to the coelacanth (where we have living examples of those, fortunately, to study directly), and you've got inner ear dynamics hovering tantalizingly close to the tetrapod layout with a system becoming attuned to detecting pressure sensations, Fritzsch (1987; 2003) pertaining to technical points raised by Peter Bernstein (2003).

Regarding the cochlea, the Intelligent Design perspective of the Seattle-based antievolutionist *Discovery Institute* has a different view. We'll take a look at their version of things in Chapter 11.

In areas were the genetics haven't been been pinned down (yet), many unresolved questions remain, of course, such as the origin of the tympanic membrane or the specific mechanisms behind the shift to the secondary squamosal jaw attachment, all noted by Takechi & Kuratani (2010, 426-429). The Anthwal 2013 survey paper highlighted those issues too, and added ones like how secondary cartilage is formed, or what adaptive reasons may have been involved in favoring that shift to the secondary squamosal jaw joint so clearly shown in the biology and fossils.

It's worth a look at the background on that tympanic issue. Maier & Ruf (2016, 271-275) surveyed 19th and 20th century embryological work on the tympanic bone and membrane, but it really wasn't until the modern era of developmental genetics that more of the process could be teased out.

Although frogs, turtles, lizards, crocodiles, dinosaurs, birds and mammals all have tympanic middle ear systems, they turned out to be independent developments as each lineage adapted the ancient stapes bone to detecting airborne pressure waves, an evolutionary story worked out over many decades by investigators from John Bolt & Eric Lombard (1985) and Jakob Christensen-Dalsgaard & Catherine Carr (2008) to Taro Kitazawa *et al.* (2015). The first tympanic ears didn't appear in land tetrapods until the Triassic, and Christensen-Dalsgaard *et al.* (2011) explored the ways in which the hearing abilities of the living African lungfish offered clues to how pressure hearing had developed in tetrapods.

Farther down the evolutionary line, although birds and mammals have developed their own distinctive formats, both built on deep common roots, but diverged in part because of a happenstance of jaw arrangement (as initially trivial as the evagination/invagination split). Kitazawa *et al.* (2015) identified how "the relative positions of the primary jaw joint and the first pharyngeal pouch led to the coupling of tympanic membrane formation with the lower jaw in mammals, but with the upper jaw in diapsids." Maier & Ruf (2016, 276) handily illustrated those diapsid and mammal membrane locations with red and yellow circles overlaying the relevant skulls, making the relationships especially vivid.

While fossil finds for the miniscule tympanic bones are rare, its been possible more recently to do digital reconstructions of what has turned up so far, as Michael Laaß (2016) did of one Late Permian therapsid, *Pristerodon*, whose "cochlear cavity represents a connecting link between the primitive therapsid inner ear and the mammalian inner ear."

Scientists, you see, want to first identify how things actually work, then honestly try and explain the why of it, taking into account every scrap of information.

Another interesting unresolved issue is the molecular mechanism underlying how inner ear hair cells connect neuronally to be transformed into the perception of sound. Just during this writing, a new paper by Oscar Diaz-Horta *et al.* (2016) identified *ROR1* (receptor tyrosine kinase like orphan receptor 1) "as a previously unrecognized gene that is essential for the development of the inner ear and hearing in humans and mice."

Although antievolutionists haven't hit on these unsolved matters (yet), it wouldn't come as a surprise if any of them do. In Chapter 11 we'll see Michael Denton bumping very close to the tympanic matter in 2016, but his behavior was only what's been standard operating procedure in antievolutionist circles from the get-go, the pointing out of the "not yet known" to cast doubt on the rest of the data they never properly discuss. We'll see Duane Gish doing a *lot* of that next chapter.

Though doing so runs the risk of relying on dated work, long after the "mystery" has been solved by later effort they didn't know about or failed to fairly describe. That was what Gish was doing above when he selectively invoked that twenty year old Forey paper on fish tails.

But the main reason why most antievolutionists haven't started waving tympanic membranes at their audience is that so few of them get anywhere close enough to the primary source technical literature to have spotted them. We won't have to guess at this. I'll be reviewing all of them in the chapters of this book, along with examining the sources they relied on for their arguments.

That means *you'll* know the information, too, even if they don't.

On the fossil trail

Now all this genetic and developmental biology that I've mentioned so far involves observation and experiment on living examples. To see the earlier stages in that proces—the smoking gun of the RMT spooling out over a hundred million years—we have to look back into time, which means the fossil record.

I've hinted at pieces of it already, of course, but time now for the closer look.

During the 19th century, as paleontology turned from an amateur avocation to a scientific discipline, more and more fossils were uncovered. The most famous ones were dinosaurs (nothing brings the public into a museum like a brand new and big dinosaur). But paleontologists were also uncovering more of those synapsid (single skull opening) reptiles.

A word on systematics: sticklers like Kevin Padian & Kenneth Angielczyk (2007, 215-218) and Angielczyk (2009) rightly note that nowadays the taxonomical category of "reptile" does not include their cousins, the synapsids. In the modern view, reflected in such works as Timothy Rowe (2004, 401), the reptiles split off from the amniotic tetrapods *after* the synapsids. Charting that split, Donald Prothero (2007, 232, 271) insisted that "anyone who still uses the obsolete and misleading term mammal-like reptile clearly doesn't know much about the current understanding of vertebrate evolution.

But calling the topic of this book "the Synapsid-Mammal Transition" or "The Pan-Mammalia Transition" or "the Basal Amniotic Tetrapod-Mammal Transition" would leave the historic usage of "reptile" too far afield for my purposes. Even Prothero (2007, 271) noted the earliest synapsids "are almost indistinguishable from the earliest true reptiles in most features," and it is in that sense only that I'm drawing on and using the popular notion of "reptile" in this book, and not its strict scientific usage.

So, Padian, Angielczyk and Prothero, don't go all grumpy cat on me regarding *reptiles*.

Now, back to the trail.

Most kids who played with dinosaur models (as I did), or saw prehistoric thrillers (as I did), are aware of one member of the synapsid clan: *Dimetrodon*, that venerable finback predator of the Permian period. Look at its skull, though, and it's very typical of the early synapsids: a basic "reptile" jaw layout except with only one skull opening in the back, noted by *Prehistoric Life* (2009, 188).

Not that the synapsids were monotonous. As Kenneth Angielczyk (2009, 258) enthusiastically reminded:

> Synapsids existed for over 80 million years before the first mammals evolved, and during this time, non-mammalian synapsids evolved a fascinating array of shapes, sizes, and ways of life. These forms include large sabre-toothed carnivores, herbivores with turtle-like beaks, carnivores and herbivores with tall sails on their backs, specialized burrowers, small weasel-like carnivores, hippo-sized herbivores with thickened skulls that may have been used for head-butting, and even a carnivorous species that might have been venomous. This diversity is sometimes overlooked because much of it is peripheral to the evolution of mammals.

Prothero (2007, 274-275) nicely charted the basic players and the many diagnostic characters that developed from early synapsids to early mammals (including body variations in vertebrae and toes), while Angielczyk (2009, 268) noted the incidence of secondary palates in the skull (evolving three times independently in the synapsid line) along with variations in nostril location. For those wanting still more details and examples without diving into the underlying technical literature, Douglas Futuyma (1998, 145-152) and Michael Benton (2005, 289-316) are fine textbook examples.

Thus there's no shortage of meticulous science data to report, and evolutionists more than able to do it.

So the fact is, there are a lot of critters knocking around in the synapsid fossil closet, most of those not leading off into the mammals. But as that is our subject here, we need to check on what was happening in the *later* synapsid crew that led paleontologists to come up with that "mammal-like reptile" tag now so out of taxonomical fashion.

As the decades of fossil digging went on in the late 19th century and the examples accumulated, something very interesting was seen in the synapsid gang—and *only* in the synapsids.

The models had changed so much that there were new terms on the scene to describe them: the "wild arch" therapsids and, later still, a suborder among them, the "dog teeth" cynodonts.

Those cynodonts are especially relevant because of how their jaws were laid out. It was something already known from our embryonic development.

"The mammalian jaw apparatus passes through a fetal stage strikingly similar in morphology to adult advanced cynodonts," noted Edgar Allin (1975, 405).

An extinct group of almost mammals were showing as adult features a developmental form now only known in our embryonic stage. Keep those cynodonts in mind, we'll be encountering them again a lot as antievolutionists try to dismiss their transitional character.

Put the beasts in order of appearance (and the geological system we know was well developed by then to do that) and you could see how the dentary bone was getting *bigger* in the synapsids generally as you went along (suggesting of course underlying genetic variations not available for scientific investigation in the fossils themselves).

The teeth of these branches were changing too, in the way we'd recognize from our mammal perch a quarter billion years later, where we have specialized and permanent teeth (like molars), not just a continuously replaced row of largely the same type of teeth as seen typically in living diapsid reptiles (or in those extinct dinosaurs). Though even there, current reptile tooth shape and replacement patterns show a great variety among species that represent hundreds of millions of years of their own evolution, as Marcela Buchtová *et al.* (2013, 119) briefly summarized.

Differentiating teeth were occurring early in the synapsid lineage, including highly specialized forms like the Permian amonodont *Tiarajudens* from Brazil (260 mya), a herbivore showing occluding teeth for chomping vegetation, yet also sporting sabertooth fangs (which, may have functioned for threat displays to discourage rivals or predators, much as in the equally herbivorous living deer), Jörg Fröbisch (2011) and Jeremy Berlin (2012) on Juan Cisneros *et al.* (2011). Incidentally, *Answers in Genesis* (2011) implied that such herbivorous saber-tooths justified the Young Earth creationism (YEC for short) dogma that the carnivorous theropod dinosaurs had been "created vegetarian." Indeed, doctrinal YEC doesn't allow for any carnivorous animals until after the Fall of Man (Adam and Eve and the apple eating thing) or in conjunction with the Flood, though exactly when the flip occurred has eluded the speculations of Ken Ham (1998, 47-50) or Jonathan Sarfati (2014). None looked all that close at the finer points of tooth diagnostics (such as wear patterns and lack of seration) that distinguish herbivorous fangs from functional meat-slicing teeth.

By the time full mammals were on the scene, their developmental variants made even as seemingly basic a task as counting how many premolars there were in early mammals rather tricky, Richard Cifelli (2000). That turned out to be due to the plasticity of mammalian tooth development, where tooth buds could actively merge and alter the number of finished teeth, as Jan Prochazka *et al.* (2010) later explored.

Farther out on the jaw, that coronoid process was developing too, along with the expansion of the synapsid skull opening into the mammalian zygomatic arch. The zygomatic arch is something else easily seen in the Halloween skull. And *felt*: it's the edge of our pronounced cheekbone. All *very* mammal.

As the dentary bone was expanding among the synapsids, though, the angular, surangular and articulate bones were shrinking, eventually becoming relegated to a slot along the side of the massively enlarged dentary bone, even as the quadrate and squamosal bones were shifting in the skull above.

Stephen Jay Gould (1993b) covered the basics of this in his "An Earful of Jaw" article for *Natural History*. A pattern was emerging: the synapsids were coming to resemble mammals, which is why they got to be called the mammal-like reptiles. *Duh.*

Because this suite of changes (distinctive skull shifts along with the mammal-like teeth) wasn't appearing in any other vertebrate lineage, by the later 19th century it was looking like the synapsids might have been the very group from which mammals evolved. Indeed, the *only* group from which they could have evolved.

Clean sweep by Robert Broom (1866-1951)

Once again, I must offer a nod of thanks to Frank Sonleitner for drawing my attention to Broom's role in the story, recounted with illustrations by Richard Aulie (1974b, 25-27), one of a trio of articles in *The American Biology Teacher* on the mammal evolution saga, Aulie (1974a-b; 1975).

Though later far better known for his pioneering work on a particular branch of the mammal line (the australopithecines he'd help uncover starting in the 1920s, part of our human origins trail), in 1912 Broom was more of a general mammal guy, having been digging in the synapsid-rich Karoo deposit of South Africa for many years—an effort which brought the combative and antiauthoritarian Broom into "acrimonious" conflict with a succession of South African and British paleontologists and museums over priority and recognition, recounted by Jesse Richmond (2009).

A parenthetical note on philosophy and religion. Broom was no atheist.

"With an unusual knowledge of the Bible, he occasionally spoke before church gatherings and viewed evolution as the working out of a divine purpose," Aulie (1975, 25). Broom doubted natural selection as a driving mechanism, though, and lived before the DNA genetic side of things had been uncovered. So while the general concept of common descent had been accepted in science by the 1930s, the idea that the Darwinian natural selection process was the main explanation for that change had not gone down so well, and Lamarckian inheritance of acquired characteristics was still very popular, especially among older American scientists.

Seeing humans as the pinnacle of an evolutionary process that had *stopped*, though, Broom rejected both "Darwinian" and "Lamarckian" mechanisms in favor of a blurry theistic evolution guided by "some intelligent controlling power" that might yet be actively prodding along enlightened folk of his own day, such as Gandhi. As covered by Jesse Richmond (2009, 498-503), Broom's liberal Christian convictions (open to faith healing and spiritualist seances) resembled the mystical paleontology of Jesuit Pierre Teilhard de Chardin (1881-1955), but I was equally reminded of Alfred Wallace (1823-1913), the co-proponent along with Charles Darwin of the concept of evolution by natural selection.

While modern evolutionists see purely natural forces in play in the evolutionary process, Wallace came to invoke a Spiritualist "Mind" to account for the origin of life and human consciousness, while leaving the rest to natural selection. It is interesting that both Broom and Wallace were skeptical of Darwin's sexual selection mechanism (now as much a part of the evolutionary kitbag as natural selection).

Wallace's life and views are ably covered by biographer Ross Slotten (2004), while my own Downard (2015a, 14-19) surveyed recent efforts by the Intelligent Design movement to hijack the overtly Spiritualist Wallace for their cause.

Since Broom was neither atheist nor materialist, though, that philosophical issue can be set aside as a red herring as we focus on the fossil data, as Broom did in 1912.

You could see what the fossils were telling about the shifting synapsid skull layout, and the detailed biology of our mammal jaw was old hat knowledge by then (at least for regular scientists--we'll see later how antievolutionists manage to consistently miss it). So it only took an application of clear thinking to set down what would be needed to switch from the known synapsid configuration to the one found in mammals.

In Broom's assessment, there was only one way for the switchover to work from an evolutionary perspective, where every stage in the process remained fully functional and biologically plausible. The skull would need to morph in specific ways, putting the quadrate hinge right beside the eventual mammal one on the squamosal bone, allowing a double-jointed arrangement.

But it was more than just the bones that had to move. The muscle attachments had to fit, too, pulled along by the bone rearrangement to maintain the utility of the jaw (and remember, muscle attachments can be identified by their tellale indentations in the bone). The parameters Broom laid out were the *only way* he could see for the evolutionary transition to work.

The only way.

This is really important. No animal with this intermediate structure was known to exist in 1912, either as a living example or as a fossil. Broom was *predicting* the existence of a hitherto unknown form of life, exhibiting an intermediate jaw configuration unknown to science or nature. And he did all this by *applying* his evolutionary understanding of all the available evidence to fill in a gap in our knowledge.

Quite a gauntlet thrown.

So what happened?

When the little synapsid fossil *Diarthrognathus* ("two joint jaw") was found in 1932 (it got its present name in 1958 when Arthur Crompton did a full systematic description) its jaw layout exactly matched the prediction Robert Broom had made back in 1912, when he deduced what an intermediate jaw structure *had to have looked like* to link reptiles and mammals, and Crompton named it *Diarthrognathus broomi* in his honor.

Another even older fossil found later, *Probainognathus* ("progressive jaw"), had the same distinctive layout. Both were small critters, by the way, resembling mice by size and form.

Barbara Stahl (1985, 408-410, 445-446) illustrated the skulls of *Probainognathus* and *Diarthrognathus*, and coverage of these and related taxa and their phylogenetic context can be followed in summaries by McGowan (1984, 137-138), James Hopson (1987) or Colbert & Morales (1991, 229 231), and at more technical length by James Hopson & James Kitching (2001), Bruce Rubidge & Christian Sidor (2001, 457), Zhe-Xi Luo & Zofia Kielan-Jaworowska *et al.* (2002, 19-20), Agustín Martinelli & José Bonaparte (2011) and Martinelli *et al.* (2016).

For Colbert & Morales (1991, 228), *Probainognathus* represented a good example "of what the ultimate mammalian ancestor may have been like."

This discovery was no stray accident, then, any more than finding *Tiktaalik* in the right geological window was, almost a hundred years later. The predicted double-jaw alignment had occurred in more than one synapsid genus living way back then, some 230 million years ago. A specifically evolutionary prediction made—and fulfilled.

As Richard Aulie put it in his 1974 article, such a success "can be expected in evolutionary theory but not in the doctrine of special creation."

And the paleontology beat goes on

George Gaylord Simpson (1902-1984) offered the view of the fossils from the early 1950s, writing how "numerous lineages of therapsid reptiles were all changing adaptively in a mammal-like direction. Paleontologists use the arbitrary criterion that a reptile became a mammal when a dentary-squamosal joint developed and the functional jaw movement ceased to be on the articular-quadrate joint. This line was probably crossed separately by at least five different lineages (leading to monotremes, multituberculates, triconodonts, symmetrodonts, and pantotheres, although it is just possible that two or three of these early differentiated from a single crossing of the lines; there may have been some other late Triassic-early Jurassic crossings with early extinction)," Simpson (1953, 348).

The paleontology continued, of course, and newer fossils were integrated into the clarifying perspective. Barbara Stahl (1985, 397-398, 409-410) showed a lot of them, including some full skeletons, while *Prehistoric Life* (2009, 182) or Takechi & Kuratani (2010, 420) highlighted specific details with varied fossil examples.

Richard Cifelli (2001, 1214-1218) surveyed the shifts in interpretation of synapsid and early mammal phylogeny based on the improving fossil collection. That analysis may be compared to Bruce Rubidge & Christian Sidor (2001) or Zhe-Xi Luo *et al.* (2002) to see how the pattern was becoming more resolved at the beginning of our current century.

Maier & Ruf (2016, 278) pointed out that mammals like *Morganudocon* from the Triassic-Jurassic boundary retained a "mandibular middle ear" connected to the jaw, meaning "its translocation to the basicranium must have occurred later, i.e. during the Jurassic and Cretaceous. Fortunately enough, many excellently preserved fossils, mostly from China, document this important evolutionary step," citing seven notable papers.

(1) Jørn Hurum *et al.* (1996) and **(2)** Gulliermo Rougier *et al.* (1996) reported on the Late Cretaceous *Chulsanbataar* and *Kryptobaatar* from Mongolia, both from the now-extinct multituberculate lineage of mammals. Incidently, the subsequent find of a fine skeleton of *Rugosodon* from the Late Jurassic further rooted multituberculate origins, as Sid Perkins (2013c) explained regarding Chong-Xi Yuan *et al.* (2013). We'll return to *Rugosodon* in Chapter 10.

(3 & 4) Zhe-Xi Luo (2007; 2011) delved into how diverse the early mammal radiation had been, and the developmental processes for jaws and ears on display in living and fossil mammals, profusely illustrated in his 2011 monograph. It's interesting also to see David Krause *et al.* (2014, 3767) charting many of the newer fossil finds chronologically, and set against a changing paleobiogeographical background where shifting continental landmasses were isolating lineages regionally.

(5) Qiang Ji *et al.* (2009) covered the Early Cretaceous *Maotherium*, including its revealing preservation of ossified Meckel's cartilage (the bit that embryologically links the postdentary bones to the jaw) *as an adult*. The commentary by Thomas Martin & Irina Ruf (2009) highlighted how "Developmental heterochrony—that is, the differing timing of developmental processes during embryonic growth—can lead to a premature fixation of ancestral character states and the retention of embryonic patterns in the adult." Mary MacDonald & Brian Hall (2001) and Jonathan Jeffery *et al.* (2002) offer introductory surveys of heterochronic factors identified in vertebrate embryological development.

Taken together, the fossils and developmental biology suggested a heterochronic shift had happened in the evolution of the mammalian middle ear, and Luo (2011, 366-371) pictured numerous fossil examples along with noting the genes now known to govern the corresponding homologous features in living mammals, remarking specifically of the middle ear transformation, "This difference in fossils is now significantly attributable to these signaling pathways and genes," Luo (2011, 372).

(6) Jin Meng *et al.* (2011) described *Liaoconodon* from the Early Cretaceous of China. The commentary by Anne Weil (2011) further noted how this cumulative fossil evidence was suggesting the anchoring of the eardrum was a critical factor in allowing the middle ear to separate from the jaw and become devoted exclusively to hearing.

(7) Chang-Fu Zhou *et al.* (2013) described the Middle Jurassic *Megaconus* from Mongolia, which still had (no shock from an evolutionary perspective) the ancestral mandibular ear arrangement, along with evidence for something else rarely preserved: its *fur*. So not yet fully mammal in respect of the ear, but definitely so for its hair.

A companion paper not cited by Maier & Ruf is worth noting: Xiaoting Zheng *et al.* (2013) on the arboreal *Arboroharamiya* from the same geological deposit has stirred up the taxonomy of their early mammal group, which had very few fossils to go on before and required further finds to sort out fully, as indicated by commentaries by Richard Cifelli & Brian Davis (2013), and Sid Perkins (2013b).

So much hard work is needed for paleontology and systematics, yet equally a labor of love and great enthusiasm by those of compelling curiosity.

There are more examples even than those cited by Meier & Ruf, such as the preserved Meckel's cartilage in the Early Cretaceous *Repenomamus* from China, Yuanqing Wang *et al.* (2001). And an informative diagram by Qiang Ji *et al.* (2009, 280) located many of these new fossils on the line of the developing ear layout in these later stages of the RMT. *Morganucodon* comes earlier in the sequence, followed by *Hadrocodium* and *Yanoconodon*, and *Maotherium* just about at the full-blown mammalian jaw-ear layout.

Restudy of *Hadrocodium* had already found it had apparently lost the distinctive "postdentary trough" where the diminished old reptile jaw bones had nestled, suggesting its middle ear bones had fully separated from the jaw as an adult in that lineage, Zhe-Xi Luo & Alfred Crompton *et al.* (2001). Later work noted by Alexander Averianov & Alexey Lopatin (2014, 430) suggested the trough absence may have been an artifact of that fossil's limited preservation (it was an extremely tiny early mammal, barely an inch and a half long), and this point will bear on the argument of creationist David Coppedge that we'll get to in Chapter 10.

As for *Yanoconodon*, described by Zhe-Xi Luo & Peiji Chen *et al.* (2007), this early Cretaceous critter from China represented a time span over *fifty million years* after the probainognathids had arrived on the scene. And exactly as one might expect from a slowly evolving sequence, this little adult (only about five inches long) showed the connecting cartilage still found in monotreme newborns, illustrated by Takechi & Kuratani (2010, 421). *Yanoconodon*'s interesting vertebral variations will also be a point of contention for David Coppedge.

Modern technology is naturally being applied to older fossils, such as the high-resolution computed tomography (CT) review of the Late Triassic mammaliaform *Haramiyavia*, Zhe-Xi Luo & Stephen Gatesy *et al.* (2015), showing its retention of the "postdentary trough for mandibular attachment of the middle ear—a transitional condition of the predecessors to crown Mammalia."

All told, that's a *lot* of "missing links" found.

Not a series of stray flukes here—this is an evolving *pattern*, clear and unmistakable, for all who think to look at the full range of the evidence.

Now consider the non-evolutionary alternative.

If a Designer were responsible for all this, and not evolution, He/She/It/They seemed positively obsessive about "designing" the very animals evolutionists needed to fill in the stages, and perversely committed to making sure they all landed in the right geological sequence to match the pattern.

Or a Fossil Genie that really likes evolutionary paleontologists.

I, for one, don't find any of that very likely.

And it's not just the ears.

If you looked at all the evidence, what bones were involved and how muscles were attached to them, and how the new teeth and ear bones interacted as an evolving package from one example to the next, some interesting observations could be made.

Alfred Crompton (1963) realized that the new "molariform" teeth showing up in the advanced therapsids might have been the adaptive way to increase the bite force even as the evolving therapsid jaw relied on a less forceful muscle layout to cushion the sound sensing going on with the shrinking auditory bones. Alfred Crompton & Pamela Parker (1978) reviewed the evolution of muscle attachments over the range of the fossils known by then, stressing how the later stages of the process could only occur because of the evolution going on in the middle ear.

Things are connected, you see.

As Crompton & Parker (1978, 200) put it:

"Once the new articulation was established, the need to retain the old reptilian jaw joint as a jaw hinge was eliminated. The angular bone, which supported the mandibular tympanic membrane and some of the reduced postdentary bones and the quadrate, could separate from the jaw and become isolated in a middle-ear cavity (Fig. 9)," which compared the layouts in the Triassic mammal-like reptile *Thrinaxodon* with a modern opossom. "And the development of a new mammalian jaw joint that was not involved in sound conduction once again allowed the jaw joint to bear both vertically and medially directed forces. The relatively rapid development of different types of jaws and dental configuration during the Jurassic may have been in part related to the acquisition of this new jaw articulation."

This passage from Crompton & Parker from 1978, so concise and informative in its clarity and vision, is yet another example of why scientists have been able to work out how critters (living and extinct) are related, how they got to be that way, and what happened to them afterwards. They're paying attention to *all* the data.

Who's paying attention now?

I regard the RMT as a clear Slam Dunk for evolution, a chain of macroevolutionary evidence that deserves to be at least as widely known (and popular) as dinosaurs and their bird offshoots are. And yet it isn't always properly highlighted by defenders of evolution.

The first treatment I came across that made a big deal about it was an article by Robert Sloan (1983) in Peter Zetterberg's anthology, *Evolution versus Creationism.* Philip Kitcher (1982, 110-114) highlighted the RMT, Christopher McGowan (1984, 127-141) devoted a whole chapter to it, and the section in Arthur Strahler (1987, 413-420) was very extensive and detailed.

But an equal number of notable criticisms of creationism at the time failed to take note of the synapsid case, such as Niles Eldredge (1982), Laurie Godfrey (1983a), David Wilson (1983) and Robert Hanson (1986).

Joe Cain (1988, 94) complained that "Research into synapsid history and the synapsid-mammal transition is extensive, although little is translated to popular media," and that situation hasn't substantially changed a quarter century later.

Niles Eldredge (2000, 191n) slightly redressed the ommission in the revamp of his 1982 book, briefly noting how "Paleontologist James Hopson of the Field Museum of Natural History in Chicago has been especially eloquent in expressing the fruits of his research on the evolution of mammals from mammal-like reptiles, providing one of the best antidotes to the tired old creationist claim that the fossil record reveals no transitions between 'major kinds.'"

But the RMT deserved better billing than an *endnote*.

Moving on into the era of Intelligent Design, it still hasn't been discussed all that often.

Jerry Coyne (2009, 52-53, 258-259) mentioned the reptile-mammal case briefly in his *Why Evolution Is True*, but cited only James Hopson (1987) in his references.

Not to fault Hopson's summary. In fact, it's an excellent article by a leader of the field, and its contents will figure notably as a roadblock for both Duane Gish (Chapter 4) and several Intelligent Design advocates (Chapters 7 and 9). But it obviously is not of much help on the paleontology that has turned up in the many decades *since 1987*, or allow a targeted response to the assorted antievolution claims fielded on the RMT in the remaining apologetics of the 21st century.

More of the mixed bag: Neil Shubin (2008, 158-164) included the jaw-ear transformation in *Your Inner Fish* (which was expanded into a very fine episode when it was done as a three part television documentary), but Richard Dawkins (2009) inexplicably didn't mention the RMT at all in his lengthy volume, *The Greatest Show on Earth.*

For a collection of evidence as gobsmackingly impressive as the RMT, its underplaying in the popular coverage of evolution is simply unacceptable.

Well, my book hopes to remedy some of that, pulling together not only how rich and varied the physical evidence for the transition is, but also offering a detailed and comprehensive examination of how lame and evasive antievolutionists are when it comes to that evidence.

That side of things is what we'll be turning to in the next eight chapters.

4. When Creation Science Came to Bat, Mighty Gish Struck Out

Although it might seem like full blown "the Earth is only 6000 years old and there were dinosaurs on Noah's Ark" creationism has been around from the get-go, or at least since Bishop Ussher in the 17th century, the chronology says otherwise. As the cliché goes, centuries of theologians hadn't really bothered all that much over the age of rocks, being more concerned with the salvational Rock of Ages.

Geology was becoming established as a scientific discipline over those years, though, and began whittling away at the old Biblical scheme before dinosaurs were even discovered. By the time Charles Darwin's evolutionary idea appeared in 1859, the scientific community was knee deep in a very ancient Earth and cosmos that were thought to be at least many millions of years old.

The Young Earth Creationism brand we know today is a product of the *20th century*, when Seventh Day Adventist founder Ellen White (1827-1915) got a revelation that the Flood really happened, no matter what regular geology was by then saying. The late Henry Morris (1918-2006) dusted off White's "Flood Geology" and established the *Institute for Creation Research* (ICR) in the 1960s, while Ken Ham's rambunctuous *Answers in Genesis* didn't come along until the 1990s.

And while YEC antievolutionism is seeminly awash with detail, hammering away on a plethora of topics in their books, videos and website postings, there is considerably less there than meets the eye or ear. As I found in my #TIP "Troubles in Paradise" research, over half of the 7500-plus antievolution works I had catalogued came from only some *87* writers. Which meant there was an awful lot of repetition.

When it comes to the science work supporting evolution, antievolutionism is very much a *reactive* enterprise, more likely to respond to what they encounter in the popular culture, or in general education, than pouring through the technical science literature where the evidence for such things as the RMT are to be found.

When it came to our topic, 1960s YEC started out on a very slim footing. Henry Morris (1963) had little specifically to say about any fossils, and certainly missed noticing the RMT. Move ahead twenty years, and Morris' 1985 *Scientific Creationism* (which was intended as a school text back in the days when Creation Science was angling to get included in public school science classes) tiptoed just a smidge farther.

Morris (1985, 83-84) declared how "each of the various orders of amphibians, reptiles and mammals appears suddenly in the fossil record, without incipient forms leading up to it and without transitional forms between it and any other order," and that "the paleontologist George Gaylord

Simpson notes that each of the 32 orders of mammals in the classification system appears suddenly in the fossil record with all its distinct ordinal characteristics fully expressed."

He then quoted Simpson (1944, 106) on this point: "This regular absence of transitional forms is not confined to mammals, but is an almost universal phenomenon, as has long been noted by paleontologists."

If you Google that short text you'll find an extensive litany of creationist websites still using the quote, very likely having mined it from Henry Morris or some like-minded writer, such as Duane Gish (1980) who couldn't resist it, either. What they haven't done is look through the *context* of the quote, or whether it meant quite what Morris or Gish wanted it to.

Simpson's book was actually grappling with the fossil preservation and speciation rate issues that Stephen Jay Gould and Niles Eldredge had by 1985 clarified under the Punctuated Equilibrium ("Punc-Eek") concept. Simpson called it "quantum evolution" back in 1944, but the idea was the same: the recognition that the fossil appearance of new taxa was constrained by the fact that a lot of speciation events involved *geographic isolation* (Ernst Mayr's *allopatric* speciation, in the jargon, as opposed to "in one place" *sympatric* speciation that was the mode of thinking of a lot of evolutionists at the time). See William Provine (2004) if you want more on that background.

On the theoretical end, C. Newman *et al.* (1985) suggested punctuated patterns would emerge naturally under the Neo-Darwinian processes of mutation plus natural selection.

The originators of the idea were certainly not shy about discussing or defending it: Gould & Eldredge (1977; 1993), Eldredge (1991, 34-58; 2005, 176-182; 2008; 2015) and Gould (2002, 745-1024). As for the "controversy" swirling around the concept (which has certainly simmered down in the decades since), Frank Sonleitner (1987), Donald Prothero (1992), Jeffrey Schwartz (1999, 320-330), Michael Shermer (2001, 97-116) and Robert Asher (2012, 73-78) offer historical perspectives. Anastasia Thanukos (2008) illustrated how the Punc-Eek idea plays out in fossil contexts.

The upshot of all this is that, if most vertebrate speciation events are *allopatric* in nature, the fossil grabbag at any location would represent only narrow time slices, not nearly broad enough as a *spatial* range to be likely to catch much of individual speciation events sprawling across what can be regional or even continental landscapes.

Not a difficult idea to grasp, one might think, but it's a subject that antievolutionists have consistently failed to understand (including Henry Morris and Duane Gish), as I surveyed across the antievolution field in Downard (2016).

More to our point here, Simpson's 1944 book had *not* discussed the therapsid case, or dwelt on how one identified transitional features in specific examples, so it's possible Morris and Gish got too frisky with finding a congenial authority quote and turned their investigative brains off.

Such as, what does it mean to be an "order" among Class Mammalia?

Taxonomy involves laying out all the characteristics of an animal and deciding how to classify it within the rules. The Linnean system didn't really allow for "almost this" or "nearly that" forms. An animal had to be pegged under a single category, no overlaps.

You can see the problem if you try to make a list of movies, and require that any one film can be listed only once. So is the 1962 *The Music Man* a Comedy (it's really funny) or a musical (people are singing a lot)? Easiest to peg it as a Musical, since the songs are used to tell the story. But what about *The Great Race* from 1965? It's also very funny, but although there's a big musical number in it, and at one point Natalie Wood sings with offscreen orchestral accompaniment, few would classify it as a Musical.

How much singing would there need to be in a comedic film to justify its classification as a Musical (and no longer as a Comedy)? Would there not inevitably be a blurry point where it could go either way, depending on your mood? *A Funny Thing Happened on the Way to the Forum* maybe? But since there's no place in the classification scheme for *almosts*, any film you like would have to be pegged as one or the other. That's the taxonomical issue that Morris was stumbling past in his haste to buttress his creationist argument with the gloss of Simpson's rather dated authority.

So there are "32 orders of mammals." What does that mean?

Remember that *order* is the taxonomical category between *class* and *family* (recall, if needed, the listing on that in Chapter 1).

But are there 32 of them? All nicely fixed and immutable?

Well, by the 1980s (when Morris and Gish were writing) it seems the list had got bigger since 1944, without their noticing. That was reflected in the listing of mammal orders in a slightly later book by Colbert & Morales (1991, 434-437): 39 of them by then, with 18 living and 21 extinct.

Relying on a forty year old authority quote, the creationists were missing how older classifications had needed revision based on newer fossils and improved systematic measures. And these included several *ancestral* orders.

Eupantotheria was now seen as the base group for marsupials and placental mammals, for example, which included the extinct Dryolestida (close to that transition but known only from fragmentary fossils) that persisted down into the Miocene period with *Necrolestes*, Nicholás Chimento *et al.* (2012) and Guillermo Rougier *et al.* (2011; 2012).

Then there's the Acreodi for cetaceans (whales and dolphins). This order included the dog-sized Mesonychids, which turned out to be more close cousins to cetaceans than direct ancestors when a spectacular set of intermediate whale fossils (like *Pakicetus* and *Ambulocetus*) were found in the decades after Colbert & Morales (1991). We'll be bumping into whales now and then in further chapters.

Working as a duo, Henry Morris & Gary Parker (1987) never mentioned the RMT at all, even while repeatedly insisting how much a failure evolution supposedly was. The closest Morris & Parker (1987, 134) got was a blurry geology chart where a *Dimetrodon* was barely visible, which the creationists where using only to highlight the supposed problems in *plant* evolution.

Parker was the one who had written their chapter on fossils, and his section on "Vertebrates: Animals with Backbones," Morris & Parker (1987, 135-140), was devoted entirely to dismissing the *Archaeopteryx* bird case. So already Parker was whittling down a vast vertebrate fossil record to one exemplar, and even there repeatedly missed the points, parading selective quotation while he avoided the science details of even what he did bring up. For those interested in the grisly details, I examined all of Parker's arguments and the information he left out in Downard (2003b, 68-73).

Now if taxonomical ambiquity isn't your thing, then quote mining can be done in another way: changing the subject, as when the *Institute for Creation Research*'s William Bauer (1979) invoked French zoologist Pierre Grassé (1977, 67), "Even when he makes statements such as 'For instance, the genesis of mammals from reptiles is rather well known,' Grasse goes on to say, 'In paleontology, however, the discovery of a new fossil can considerasbly modify our views and make interpretations obsolete which were previously thought to be definitive.'"

Grassé is another popular quote mine target by creationists, by the way, which should by now come as no surprise.

But follow what Bauer tried to do here. Did he mean that scientists were *not* supposed to revise views based on new evidence, or that any of that new evidence challenged fundamental contentions, such as the track of the mammals? Bauer wasn't stopping long enough in his hop-skip-and-jump to discuss any examples.

Bauer obviously missed Grassé (1977, 35): "The shaping of the mammalian form, which lasted from 50 to 60 million years, occured in a smooth and gradual manner." And Grassé (1977, 34) had even charted "the distribution of the reptilian and mammalian characters" among the main lines and "twigs" leading to mammals.

But why bother looking at the mere factual details in your own source, when quotes can be bashed together to make the data go away, poof!

Duane Gish (1921-2013)

In the RMT-waffling department, a very different situation pertains to Duane Gish, the great detail-fiddler of creationist apologetics, who stood out for repeatedly touching on the RMT story, and extensively citing primary sources. And therein lay his doom, at least for his credibility as a scholar—an anchor that by connection dragged down any of the many other creationists who have freely drawn on Gish over the years, ending up as the ICR pundit of record for a whole generation of creationists, from Dave Nutting's *Alpha Omega Institute*, interviewing Gish (1993b), to Luke Randall (2013) in the online era.

Nutting's *Alpha Omega Institute* and Luke Randall's *Was Darwin Right?* website are typical of many grassroots creationists in how ecumenical their secondary parasitism is, as easily invoking Duane Gish as Intelligent Design, such as Mike Shaver *et al.* (1996) showing comparable enthusiasm about the retooled creationist textbook *Of Pandas and People as* Mary Jo Nutting (1997) and *Think & Believe* (1997; 1998) were concerning the Irreducible Complexity arguments emanating from non-YECer Michael Behe. There will be much more to be said of the claims on mammals and other subjects offered by *Pandas* and Behe in subsequent chapters.

Duane Gish (1981) laid out his main approach to the RMT in an ICR *Impact* (later known as *Acts & Facts*) article, and reprised his primary arguments in condensed form for kids in Gish (1990, 61-64).

The 1981 Gish play first generally acknowledged that mammal-like reptiles existed and possessed the features claimed by paleontologists, but then dismissed their relevance by suggesting that animals can appear that contain "some characteristics which are possessed by a second class of creatures" without their being transitional or intermediate. Gish offered the reptile-like amphibian *Seymouria* (noting it lived too late to be a direct transitional to reptiles, that had already appeared) and the monotreme platypus (bringing in its "duck bill" and webbed feet as more obvious red herrings).

Gish (1995, 149-150) repeated this in his revised version of his main work (*Evolution: The Fossils STILL Say No!*), and this *it's just a mosaic* argument would remain a common one for antievolutionists, as we'll be seeing down to the present.

Only the RMT doesn't turn on isolated features, but on the *whole* animal, and how lots of examples fall into a *temporal sequence*, both subjects that Gish studiously avoided in the 1981 article and his 1995 book. Instead, he indulged in a series of tactical evasive manuevers, trying to distract the reader away from the actual evidence by presenting obstacles to evolution entirely of his own contrivance.

Like this from the 1981 piece:

> Mammal-like reptiles appeared supposedly right at the start of the reptiles, gradually became more mammal-like through the Permian and Triassic, and finally culminated in the appearance of the first real mammals at the end of the Triassic. At this time the mammal-like reptiles essentially became extinct, even though earlier they had been amongst the most numerous of all reptiles, world-wide in distribution. Since evolution is supposed to have involved natural selection, in which the more highly adapted creatures reproduce in larger numbers and thus gradually replace the less fit, we would now expect the mammals, triumphant at last, to flourish in vast numbers and to dominate the world. A very strange thing happened, however. For all practical purposes, the mammals disappeared from the scene for the next 100 million years! During this supposed vast stretch of time, the 'reptile-like' reptiles, including dinosaurs and many other land-dwelling creatures, the marine reptiles, and the flying reptiles, swarmed over the earth. As far as the mammals were concerned, however, the 'fittest' that replaced the mammal-like reptiles, they were almost nowhere to be found. Most of the fossil remains of mammals recovered to date from the Jurassic and Cretaceous Periods, allegedly covering more than 100 million

> years, could be contained in two cupped hands. Most such mammals are represented by a few teeth. If evolution is supposed to involve survival of the fittest, and the fittest are defined as those that reproduce in larger numbers, the origin of mammals represents something very strange, indeed. Since they survived in very few numbers, evolution apparently occurred by survival of the unfit!

Gish (1995, 173-174) carried this farrago over into his updated revision.

What Gish was tap-dancing around was the Permian mass extinction around 250 mya, in which almost everything died, not just the almost-mammals, ably recounted in Michael Benton (2003). The moving continents were forming the Pangea supercontinent, with a *changing climate* that ultimately favored the new tropical-friendly dinosaurs over the temperate-adapted mammal furballs. Gish left that part out too.

Traditional Young Earth creationists rejected plate tectonics in those days, by the way, though more recent ones have "evolved" their views to try and coopt what they could no longer ignore, squeezing tectonics into their Flood Geology mandate. Another story for another book.

The big problem for Gish's argument was that mammals had *not* disappeared from the Mesozoic. This should have been plain enough from even a cursory inspection of the paleontology available in 1981, let alone 1995. Mammals were no longer the major players on the scene, as dinosaurs dominated the Mesozoic ecosystem until the mass extinction event 65 mya, but they had not vanished.

"Mesozoic mammals and their closest extinct relatives were tiny animals, and their fossils are notoriously difficult to collect," reminded Timothy Rowe (2004, 402). Douglas Palmer (1999, 106-107) illustrated a range of these diminutive early mammals, with the thumb-sized Late Triassic *Morganucodon* being especially small, *Prehistoric Life* (2009, 221), the skull shown to scale by Czerkas & Czerkas (1991, 118) being barely an inch long. Averaging mouse-sized, large swaths of these Mesozoic critters are known only by the pieces of them that stood the best chance of getting fossilized: their diagnostic *teeth*.

Often insectivorous, the early mammals had more of a nocturnal lifestyle that operated less under the feet of the day-dominating dinosaurs (though even some of them had fine night vision). Later on in the Cretaceous, a species of *Repenomamus* (recall them from last chapter regarding their preserved Meckel's cartilage) grew large enough to feed on younger dinosaurs, Anne Weil (2005) commenting on Yaoming Hu *et al.* (2005), but overall mammal size remained fairly modest. We'll bump into *Repenomamus* again in Chapter 10.

Christine Janis (1993, 171-172) described the adaptive advantages of small size in that dinosaur world. Exploiting the ecological niche of wide-eyed nocturnal miniature scampering insectivore carried mammals through the long dinosaur preeminence, and the hearing skills these smaller ones honed would eventually come in handy for some early mammal spin-offs, the acoustic-navigating bats and cetaceans.

More relevant still, in the years since Duane Gish wasn't thinking about it, Timothy Rowe (1996) laid out the part that a nocturnal and arboreal lifestyle played in the evolution of the distinctive mammalian neocortex. The mammal lineage certainly had a long wind-up for that nocturnal niche, by the way, as Kenneth Angielczyk & Lars Schmitz (2014) traced it deep into the synapsid line. And Jung-Woong Kim *et al.* (2016) looked into the recruitment of rod photoreceptors from cones that facilitated mammalian night vision during the long dinosaur dominance.

Amazingly, only a few pages later Gish (1995, 176-177) commented on those very Mesozoic mammals that he had mistakenly swept from the field, commenting on some Cretaceous finds that were being reported in the general scientific press in 1992.

This bit of Gish (1995, 177) really dropped my derived mammalian jaw, though: "Especially exciting is the discovery of mammalian fossil skulls with intact middle-ear ossicles and other important cranial features. Creationists expect that whatever is found there, each kind will be complete with no evidence of transitional features and thus are eagerly awaiting publication of the findings in a scientific journal."

That was definitely *not* the case, either for the fossils keeping to Gish's cue sheet, or that creationists were chomping at the bit to discuss them.

Falling within Gish's timeframe was one article he could have spotted but apparently didn't: the *Scientific American* article on the new Mongolian finds, Michael Novacek *et al.* (1994). Focusing mainly on the many dinosaurs found, they did mention (on page 66) *Deltatheridium*, a genus whose fossils "seem to straddle the line between marsupials and placentals."

Those more complete fossils eventually nudged *Deltatheridium* more on the marsupial side, Guillermo Rougier *et al.* (1998). Still other specimens represented the early eutherian (placental) mammals, and these still retained the epipubic bones (an abdominal projection from the pelvis) that have been lost in later mammals, Michael Novacek *et al.* (1997) and Michael Benton (2005, 306, 311, 315, 324). Still more diversity has turned up in later fossils found in that group, Gregory Wilson & Jeremy Riedel (2010), Brian Davis & Richard Cifelli (2011), Zhe-Xi Luo & Chong-Xi Yuan *et al.* (2011, 444), Shundong Bi *et al.* (2015) and Guillermo Rougier & Michael Novacek *et al.* (2015).

As for the mammals specifically with their "intact middle-ear ossicles," well that bunch was definitely not at all kind to the concept of creationist *kinds*, as covered in the last chapter, and creationists have not been stampeding to discuss them since. We'll bump into a few sad exceptions in later chapters.

Underlying Gish's faulty argument was more than just missing the players. Whether deliberately or by his own ignorance, Gish was seriously misunderstanding the concept of evolutionary *fitness*.

Fitness is not a fixed point, but a shifting one based on conditions. No matter how exceptional an animal may be, for however long a run, it's no guarantee that it will go on forever. In fact, it's about dead certain it won't, since 99% of all species are extinct. Sometimes the reasons are clear (volcanic eruptions or asteroid splats), other times not. And is that surprising, really, given we don't have all the available data in the fossil record?

But we still have a lot, and somehow Gish (and antievolutionists generally, not just YECers) manage to leave out that critical *biogeographical* context, the where and when of what has lived on Earth.

Even if he hadn't missed that, why would the relative lack of success of *later* mammals against the dinosaurs rule out their evolutionary origins, or make the evidence for that go away? It wouldn't, of course, and that Gish tried to imply that it somehow would suggested how easily the creationist was willing to grasp at irrelevant straws.

In his extensive *Science and Earth History* criticism of creationism, Arthur Strahler (1987, 412-413) took Gish to task for this weird argument, which the creationist had continued to field into the mid-1980s. Interestingly, Gish did not mention Strahler's criticism when he repeated the claim in his 1995 retooling. That he left in it at all, though, meant Gish really wanted it to be true, no matter what.

But Gish by then was certainly aware of Strahler's book, having selectively quote-mined Strahler (1987, 316, 405, 408) on fossil fish and Cambrian chordates in Gish (1993a, 79-80; 1995, 74-76). But apart from those hit-and-run jabs, Gish managed to skip addressing *any* of Strahler's many substantive criticisms of the creationist's work. I delved into Gish's misleading use of the Strahler quotes in Downard (2003b, 54-86; 2004c, 218-220).

Much of the chapter on mammal evolution in Gish's revised 1995 book was devoted to the RMT evidence, Gish (1995, 147-176), and half of his source citations consisted of new quotations mined from Thomas Kemp's 1982 book, *Mammal-like Reptiles and the Origin of Mammals*, a technical summary much closer to Gish's 1981 article than to the mid-1990s cutting edge of the homeobox genetic revolution.

Gish 1995 was like a kid in a candy store, relishing every opportunity to extract quotes which served his apologetic purpose. Gish (1995, 152-154) was especially revealing, as the creationist offered some artfully selective quotes from Tom Kemp and others to sow doubts about the dating and sequence of the fossil record, suggesting they were being dictated by "evolutionary expectations."

This was about the closest Gish got to uncovering his own hidden creationist ball, their dogmatic insistence that the fossil record was laid down in the Biblical Flood. The issue of creationist complaints about geological dating and the viability of Flood Geology have been tackled by many a critic. I took a crack at it in Downard (2004c).

That YEC context needs to be kept in mind when Gish (1995, 159) trotted out a quote from Kemp (1982a, 3): "Of course, there are many gaps in the synapsid fossil record, with intermediates found between the various known groups invariably unknown. However, the known groups have enough features in common, that it is possible to reconstruct hypothetical intermediate stage."

So what are these "known groups"? And how much *did* they differ? Is it possible to interpolate such intermediates? Remember, that is *exactly* what Robert Broom did with the probainognathids.

All these were substantive issues of taxonomy and comparative anatomy. But Gish was not about to discuss them. Instead, he continued with a quintessential presentation of a YEC argument (with some important tag phrases that I've put in **bold**):

> Kemp asserts that the reptile to mammal transition is the best documented case for evolution, but then must admit that hypothetical transitional forms must be constructed because intermediate forms between known groups are almost invariably unknown! The first mammal-like reptiles appear in the rocks of the Upper Pennsylvanian, **allegedly** about 350 million years ago, and **supposedly** became extinct at the end of the Triassic. Thus, evolutionists believe the mammal-like reptiles spent almost 200 million years evolving before reaching mammalian status. Countless billions of transitional forms would have lived and died during that vast stretch of time. Our museums should have many thousands of **actual transitional forms** on their shelves. Resorting to hypothetical intermediates would certaintly not be necessary if these creatures had actually evolved.

I highlighted that "allegedly" because it is a background reflection of the Young Earth creationism Gish believed in, but never quite got around to overtly defending or applying in his book. And the "supposedly" because it is part of the dogma of Flood Geology that all the *kinds* of animal life were preserved on Noah's Ark, and therefore *must have survived* into a time only 4500 years ago. Though in a friendlier venue (like being interviewed for Dave Nutting's *Alpha Omega Institute*), Gish (1993b) loosened up more on that, doubting the plain paleontological fact that "fishes existed before the amphibians."

But what about these "mammal-like reptiles"? Where were they now, since they would have trundled down the Ark gangway after dry land appeared, alongside familiar animals like cows and giraffes, and a much larger range of extinct life? Are those therapsids (and dinosaurs and trilobites) actually all only as old as the pyramids of Egypt, also some 4500 years old? Really? Could the known fossil record for them all be crammed into a YEC framework and make sense? If they couldn't, didn't that fatally compromise Gish's whole argument?

It may seem like a gracious thing for Gish to provisionally accept the standard geological framework when he didn't actually believe it, but it's also a way to sidle around the fact that Flood Geology was (and remains) an untenable mess, which would have become obvious if he'd tried to apply it to actual fossil examples, investigating their depositional context.

But we didn't get that fun argument. Instead we got that claim of *billions* and *thousands*.

Which still evades a big question: how much of the past has been preserved?

The Bermuda Triangle Defense

If you've ever caught one of the many videos on the subject, you know the drill. There is supposedly a zone of mystery in the Bermuda Triangle out in the Atlantic Ocean, and believers will discourse on the "mysterious" disappearance of the SS *Such-and-So* on the night of September X, but fail to mention that the worst hurricane in twenty years tore through the area on the night of September X. Do you think that might have had something to do with the "mysterious" ship disappearance?

Now think of fossils. Mightn't the absence of locations to dig in, or the physical size of small animals with fragile bones, have something to do with some of these evolutionary "fossil gaps"?

Because of the methodological similarity of the two arguments, in "Troubles in Paradise," in Downard (2004b, 94-95) I dubbed the creationist failure to appreciate the geological context of "fossil gaps" the *Bermuda Triangle Defense*.

Geologists and paleontologists know that the odds of any individual organism (or even a sampling of a whole population) alive at any given moment in time becoming fossilized and enduring down to the present is vanishing low. Most of all the life that has ever lived is not on view.

All we get are little time slices, snapshots going on at particular locations. The sediments may be laid down for centuries without preserving anything except hardy pollen grains or stray teeth. Other deposits may be great capture points for body parts, such as a bend in a river, but not everything alive upstream will end up conveniently washed there, and likely not all in one piece.

There's a whole scientific field devoted to the subject of how living things end up as the fossil bits they do, by the way: *taphonomy*.

George Simpson (1983, 14-26) and Anna Behrensmeyer (1984) described the general principles, and Derek Briggs (1995) showed its experimental side. Even gene usage can reveal clues on ancestral biology and lifestyles, Andreas Wagner (2000b).

Fortunately, a lot can be gleaned from even fossil fragments. As anthropologist Laurie Godfrey (1983b, 195) reminded, taphonomists "can recognize the telltale signs of postmortem gnawing, of trampling, of slow or rapid water transport, of oxidizing or reducing depositional environments, of physical and chemical weathering, of postdepositional deformation, and so on. They can tell you why no shoulder blades, vertebrae, hand bones, and foot bones may be represented in a deposit loaded with skulls, jaws, and occasional long bones of fossil vertebrates." More recently, digital technology has entered the field too, surveyed by Stephan Lautenschlager (2016).

Because taphonomy wrecks Flood Geology (which posits everything in the fossil record sloshing into sediments whose characteristocs have to be consistent with a catastrophic global inundation), creationists have tended to skirt the issue.

Gish (1993) totally bypassed it in his tactically worded rejoinder to Godfrey and other critics of creationism. With a straight face, Henry Morris & John Morris (1996, 263) authority-quoted of all people Behrensmeyer (1984, 560, 561) about how rapid preservation needed to be to protect fossils, as though this automatically meant the much more catastrophic Flood Geology conditions! Bernard Northrup (1997) was a rare example of a creationist venturing into this area, a meandering rationalization of evidence that precluded a single Flood, inspiring him to sound a lot like 18th and 19th century catastrophists by proposing several deluges plus three or more vaguely outlined non-aquatic cataclysms.

So it is that the presence of even a single stray tooth unique to a particular animal would testify to that genus' existence in that slice of time, and the taphonomy of its geological deposition could in turn suggest how and why that tooth ended up there. But unless you were really lucky and the relevant parts of the *body* were preserved, that blip would only tell you that the group apparently was around then, not what might have been going on in the rest of its skeleton that you didn't get a look at.

And if there were only selected time slices known, perhaps spread across continents and many millions of years, to make sense of why there might be "gaps" in the show would require paying attention to all that.

Yet at no point in Gish's book (not just the reptile-mammal section), or at any time in any of his apologetic writings that I know of, did the creationist stop to investigate the *geological context* of the supposed fossil gaps or the individual fossils.

And it's a lapse that *all* antivolutionists share, not just creationists like Duane Gish. I literally know of no counterexamples, and will be reminding you of that circumstance now and then by noting the recurring *Bermuda Triangle Defense* plays. Just for fun, I'll even keep count of new players as they show up on the field.

An especially clear example of what I'm talking about, which I'll lay down as our *Bermuda Triangle Defense* (Player #1), concerned the evolution of the iconic horned dinosaur *Triceratops*. By the mid-1990s it was clear that the cerotopsians had evolved from the smaller protoceratopsids, and those derived in turn from the bipedal psittacosaurids, covered variously by David Norman (1985, 128-145), Peter Dodson & Philip Currie (1990), Lambert & The Diagram Group (1990, 169-171), Dodson (1997a-b) and Paul Sereno (1990; 1997).

Gish (1995, 122) brazenly dismissed one of the pertinent characters thus: "If *Protoceratops* was the ancestor of the horned dinosaurs, it should be found in geological formations such as the Middle or Lower Cretaceous, presumed to be older than the Upper Cretaceous."

And where was one supposed to look for such deposits, Duane? Gish conveniently neglected to specify.

But this time the ommision was more than casually bold. If you look at a standard geology text, you'll find the geological periods broken down into subcategories. There's an Early, Middle, and Late Triassic, for example. And there's an Early, Middle, and Late Jurassic, too. But check on the *Cretaceous*, and it's divided into Early and Late Cretaceous only.

There is *no* "Middle Cretaceous" division.

Why? Because erosion was winning out over deposition during that time, leaving so few strata available from the swath around 100 mya that the geologists don't even bother tagging a distinct category for it.

Though creationist Charles Creager (2003) failed to get the memo on this, as he similarly fretted over the absence of "fossils from the mid Cretaceous" regarding the RMT. "Could it be that relevant fossils that are classified as the mid Cretaceous didn't fit their evolutionary assumptions?" That would be a "No" on that one, Charles. Chalk up *Bermuda Triangle Defense* (Player #2).

We'll bump into Creager's bouncy analysis again in Chapter 10.

Now the biogeographical distribution of fossil ceratopsids suggested they evolved from something like the aforementioned *Psittacosaurus*. Found in Asia starting around 120 mya, the sparcely known family was the first to have the distinctive beaklike ceratopsian version of the ornithschian predentary bone--they also turn out to have had filamentous bristles on their tail, Gerald Mayr *et al.* (2002) and shown in *Prehistoric Life* (2009, 346-347), a circumstance of relevance to the evolution of feathers over in their more distant saurischian cousins.

The fully quadrupedal protoceratopsids don't show up in the fossil record until around 100 mya (smack dab in what would have been the "Middle Cretaceous"), first in Asia but spreading over the land bridge into North America. With their rudimentary neck frill, they resembled miniature versions of the later ceratopsids that were around by 85 mya in North America. To pin down the transitions better would require locating more rocks of the *right age and location*—think the *Tiktaalik* case again.

The chart of main Asiatic Cretaceous sites in Colbert & Morales (1991, 211) showed only nine for the Late Cretaceous, and only *one* for the Early Cretaceous. A map in Peter Dodson (1996, 13) indicated *Psittacosaurus* came from exactly four spots in all Asia, two of them adjacent in Mongolia. Lambert & The Diagram Group (1990, 222) also showed the global distribution of Mesozoic deposits, which are concentrated primarily in the western United States, and David Weishampel (1990, 63-139) provided a similar but more comprehensive survey.

The physical instances of rare preservation guaranteed that the fossil trail would possess inevitable gaps, and creationists like Gish have been keen to focus on all the *gaps* instead off turning to the fossil evidence whose preservation were generating the "gaps." It was like not only missing the forest for the trees, but not noticing the trees either.

Gish's whole discussion of dinosaurs turned on such shell games, as I went into in Downard (2004c, 194-200). Though that Fossil Genie refused to sit idly by, including obligingly plopping down more blips on the psittacosaurid-protoceratopsid-ceratopsian field in the years since, tracked by Brenda Chinnery & David Weishampel (1998), Xing Xu *et al.* (2002; 2006) and Wenjie Zheng *et al.* (2015).

A comparable situation applies to the geology of the RMT, a lot of the evidence for that coming from only a few premier deposits, such as the Permian Period Karoo Supergroup in South Africa and the later Ischigualasto Formation of Argentina from the Triassic.

Such fundamentals haven't changed even after Gish's 1995 book.

Although of great significance for understanding Permian life, reflected in Michael Benton (2003) or Bruce Rubidge (2005), the relevant slices of the Karoo Basin represent only a small segment of the geological frame, explored at length by Octavian Catuneanu *et al.* (2005).

The Ischigualasto figures prominently in dinosaur paleontology too, as their earliest representatives are showing up at the same time as the first clear mammals, Raymond Rogers (1997, 372-374). But that site is also but a very small slice of Triassic time available in the region, as indicated by the stratigraphy chart in Fernando Novas (1997, 679).

The fossil representation of Jurassic mammals remains sparse, as John Flynn *et al.* (1999) were able to push back the evidence for mammal presence in Madagascar by 25 million years with a single fossil find (with implications for mammal dispersal throughout the southern Gondwana landmass), or Oliver Rauhut *et al.* (2002) concerning the first specimen found in all of South America (next door to Africa back then). That shouldn't be surprising, since mammals overall were small during that time and many got preserved only by their durable teeth (the same is true for the cartilaginous sharks). And given such limits to the dataset, it inevitably leaves unclear spots in working out when taxa initially evolved and from what, reflected in the systematic review by Tom Kemp (2009b) for example.

Ahem, the *Bermuda Triangle Defense*, anyone?

Duane Gish pulls a "Garrett Hardin"

A dedicated environmentalist, Garrett Hardin (1915-2003) is most popularly known for Hardin (1968) where he coined the phrase "the tragedy of the commons" to describe his sour view of 1960s public land policy, where environmental degradation and pollution could go on by being dumped in the public air and water. Wearing his evolutionist hat, though, a short 1980 piece of his was included in Ashley Montague's *Science and Creationism*, Hardin (1984), which was an OK generalized criticism of creationism, but otherwise was so insubstantial that I had no cause to cite it anywhere in my own *Troubles in Paradise*.

But Hardin held a slot in my research notes for a quite different reason, a methodological one: his 1974 polemic *Mandatory Motherhood*, which sought to show the deleterious effects of restrictive abortion laws, and whose shrill polemic tone may be contrasted with the more temperate contemporary William Nolan (1978). At one point Hardin (1974, 36, 105-133) cited a 1966

Swedish study for some "statistically significant" findings about the poorer health and antisocial behavior of those born unwanted compared to those born intentionally. This report so impressed Hardin that he reprinted it for the reader's education--an incautious move, since it flatly contradicted the claims Hardin was making about its conclusions.

So, to "pull a Garrett Hardin" is to go out of your way to call attention to information that blows your argument to smithereens.

Now Duane Gish (1995, 155-158) is not the worst offender in this field (wait till you see *Explore Evolution* down in Chapter 9), but the creationist sauntered into that mode with training wheels on when he offered more Kemp quotes on the many mammal features not preserved in the fossils, like internal organs, and the overall issues of the evolution of endothermy (warm-bloodedness) and hair. That too will be a recurring antievolutionary trope.

As we saw in Chapter 2, though, the genetics of these processes weren't known in 1982 for Kemp to discuss, but writing over a decade later, Gish (1995, 156) too confidently declared that "Hairs, as do feathers, develop from follicles, and thus have a mode of development completely different from that of scales. Evolutionists must believe that somehow, via random, accidental genetic mistakes, reptiles 'solved' the problem of converting reptilian scales into mammalian hair."

It's interesting to compare Gish's creationist certainties with the typological ones of Michael Denton (1985, 106), who declared that "no structures are known which can be considered in any sense transitional between hair and any other vertebrate dermal structure." Incidentally, the Nuttings' *Think & Believe* (1990a) creationist newsletter freely mixed antievolutionist sources by quoting Michael Denton's view on hair as readily as they did Duane Gish, as we'll see later on with the organ of Corti.

Viewed from decades on today, Denton and Gish's 1985/1995 convictions illustrate how not predictive or informative ID or creationism turned out to be.

Gish (1995, 158-159) barreled on ahead, though, to undermine his own argument, by quoting this from Kemp (1982a, 331) on the matter of *homeostasis* (the ability to maintain a stable biological condition amid a fluctuating environment): "It was noted that the fossil record supports the view that evolution towards mammalian levels of homeostasis involved practically all aspects of the organism simultaneously. No single structure or function could evolve very far without being accompanied by appropriate changes in all the other features."

Gish then opined that "Kemp goes on to explain that just as the internal changes required to maintain homeostasis during the conversion of a reptile into a mammal must be gradual and intimately correlated, so also must the morphological changes be carefully correlated and thus gradual."

And Gish resumed quoting Kemp:

> To take as an example dicynodonts, their herbivorous specialization requires the replacement of the teeth by horny tooth plates, reorientation of

> the jaw musculature, changes in the form of the jaw hinge, and an extensive remodeling of the shape of the skull and lower jaws. Also suitable locomotion and central nervous programming and behavior are needed. No one of these features has much adaptive value unless accompanied by others, and therefore the evolution of the dicynodont type of organism must have followed a correlated progression, each feature evolving gradually and in association with changes in all the other features.

Gish must have thought this had somehow made his case, since he offered no comment on it, and did not discuss anything about dicynodonts himself. It wasn't as though there weren't works for him to cite or discuss in 1995, such as Gillian King *et al.* (1989) on the evolution of that dicynodont feeding system. But bringing up a topic without exploring it at depth was Gish's preferred mode.

Gish's non-approach to the details of paleontological topics *he* brought up concerning Kemp may compared to the later Kemp (2006a; 2007a-b) on therapsid origins, making use of that correlated progression concept and offering numerous specific examples from fossils and biology (the physical evidence that Gish was intimating didn't exist), as well as relating that to broader conceptual issues of general evolution. Incidentally, one of Kemp's fossil witnesses, the cynodont *Chiniquodon* (AKA *Probelesodon*), will be popping up again in later chapters for other antievolutionists to trip over.

Now take that issue of mammalian endothermy. The general suspicion was that warmbloodedness developed in early mammals in conjunction with adapting to a new nocturnal niche, Alfred Crompton *et al.* (1978). But in surveying the competing theories of the origin of mammalian endothermy, Kemp (2006b) noted one of the problems hobbling the debate was their assumption that these processes were somehow mutually exclusive, that factors like ecological adaptation and improvements in thermoregulation couldn't have been going on at the same time in a complex selective environment.

That more inclusive approach was exactly where modern evolutionary thinking was headed, and where creationists (including Gish) resolutely failed to follow, trying to parse away a technical argument with tactical authority quoting, while never quite getting around to paying attention to the full data set known at the time. Or doing any of the hard work to turn up more data on their own, as creationists so regularly operate as nothing more than opportunistic scavengers.

Gish (1995, 160) took another shot at the geological preservation issue with this: "Kemp attributes the absence of transitional forms between genera in the notion that at the speces level evolution occurs rapidly in small populations. He accepts the punctuated equilbrium theory of evolution suggested by Niles Eldredge and Stephen Jay Gould. This idea will be discussed in a later chapter."

Except Gish (1995, 353-356) never got around to doing that, simply repeating his claims about Punc-Eek depending on "unknown mechanisms" without any recognition of Ernst Mayr's allopatric speciation contribution, or investigating any fossil or living applications of it.

Gish (1995, 161-166) continued to play authority quoting games with Kemp, and then drew on a still older paper on the early synapsid pelycosaurs (the group *Dimetrodon* belonged to) by Alfred Romer & Llewellyn Price (1940), which Gish quoted on the pattern of new forms emerging after extinction events.

Wow, 1940?

1940 is way before a lot of the newer synapsid finds, and long before cladistic systematics clarified the "fit in one box only" Linnean classification scheme that complicated tracing lineages. Romer & Price (1940) had even stressed how fragmentary the fossil material was at the time, but Gish was in no mood to stop and investigate how later paleontological work may have altered the understanding of the players since Romer & Price's view before the Second World War.

Incidentally, Romer reflected a particular view that the laying of land eggs was the defining characteristic of reptiles, and this bore on how fossils were interpreted during the period, as noted by Stahl (1985, 271-275), another source directly known to Gish.

That the 1995 Gish felt obliged to trawl back another *forty-two years* before Kemp's thirteen-year-old book for another authority quote underscored how little he wanted to address the primary source paleontology available by the mid-1990s (or the jaw developmental biology of the 1830s). Gish couldn't resist quoting with disapproval Romer & Price alluding to features relating the pelycosaurs and therapsids, but the creationist never went on to explain what these features were, or what they'd need to look like to earn Gish's approval.

Take Gish (1995, 164): "Dimetrodon, which had hugely elongated neural spines, creating a large sail-like structure, believed possibly to have functioned as a heat-exchanger, was one of the most numerous of the sphenacodonts (fig 10). There are no fossils whatsoever of any transitional forms showing a gradual evolutionary origin of these enormous spines. Spenacodon was a more conservative member of the spenaocodontids (fig 11). No suggestion is made concerning which creature is the specific ancestor of the therapsids." The two figures were to skeletons of the two animals.

OK, here's some paleontology.

Dorsal fins apparently originated several times independently in the pelycosaurs, suggesting common genetic processes that could be selected repeatedly in related lineages, though not without variation, such as in *Edaphosaurus* discussed by David Berman (1979). Unlike the *Dimetrodon* line, there were side prongs on *Edaphosaurus*' fin spines, illustrated by *Prehistoric Life* (2009, 191). Many examples from these groups have turned up since, but as so many are fragmentary, such as the one described by David Mazierski & Robert Reisz (2010), it's difficult to resolve the pelyosaur fin issue without more fin information.

More *Bermuda Triangle Defense*, anyone?

Moving closer to the *Dimetrodon* bunch, though, fins popped up several times there. The basal *Cutleria wilmarthi* is known by only one fragmentary specimen, but likely had no tail spines, Michel Laurin (1994). The later *Secodontosaurus obtusidens* (first found in the 1870s) and the more recent find of *Cryptovenator hirschbergeri* were finned though, Jörg Fröbisch *et al.* (2011). While *Dimetrodon* had tall spines, there was certainly a range among its close cousin the sphenacodonts, with *Sphenacodon* showing short vertebral spines, while *Ctenospondylus* (known since the 1930s) had notably taller ones, Spencer Lucas *et al.* (2007).

Wouldn't such variation in the taxa known to have existed by the 1990s have warranted a closer look in the transitional feature department? What was preventing Gish from doing any of that?

More recent work has identified the common developmental processes in *Dimetrodon* and *Sphenacodon*'s vertebral spines, and in the convergent sails among their cousins the edaphosaurids, suggesting thermoregulation may have been less a factor in the spines' evolution than sexual display, Adam Huttenlocker *et al.* (2010; 2011). This is exactly the sort of investigation that is naturally inspired by evolutionary thinking, while nothing like it goes on in the argumentative quagmire of creationism, where nothing ends up meaning much of anything.

By focusing exclusively on *Dimetrodon*, was Gish deliberately avoiding any of the related taxa that showed the presence of slightly extended vertebral spines? It would seem harder to do, given that he had a picture of *Sphenacodon* right in front of him (his "fig 11"), but somehow he couldn't spot the connection, how its spines were extended into a rudimentary dorsal ridge.

More significantly, Gish never explained how much of a change would constitute a transitional spine, or how that would differ from what was visible in its close cousin *Sphenacodon*. Nor did he ever explore the geological context of the fossils, to identify where would one go to look for such specimens, and evaluate how likely it would be to acquire one with the right parts preserved to illuminate fin evolution.

Gish couldn't let go of the *Bermuda Triangle Defense*.

Dimetrodon and that gang are very early in the synapsid parade. And it's revealing that Gish's book never illustrated any of the *later* fossils, the critters more relevant to that pivotal jaw evolution. Was it because he couldn't get rights to the pictures? Or was it because it would have been a lot easier to spot the holes in Gish's argument if you could see what he was not honestly talking about?

Morganucodon* and *Kuehneotherium

Gish's main argument had remained consistent from his 1981 article through to his 1995 book revision. Gish (1981) had taken aim at two early mammals: "Now let us consider the two creatures, *Morganucodon* and *Kuehneotherium* that supposedly represent the most definitive transitional forms between reptiles and mammals. These are the creatures that, it is claimed, possessed the mammal-type jaw-joint side by side with the reptile-type jaw-joint."

Oh, but Gish didn't like that *mammal* classification. Not one bit. Adopting a royal tone, "we emphatically reject the idea of calling *Morganucodon* a mammal." And that was solely (by Gish's decree) because that animal retained all those *reptilian* jaw bones (exactly what a transitional would need to do, remember) beside the now functional mammalian jaw articulation (a skeletal characteristic that defines an animal as being a mammal).

Gish's effort to dismiss the intermediate nature of *Morganucodon* and *Kuehneotherium* as early mammals retaining reptilian ancestral features could only be done by pulling them out of the mammal category, and pigeonholing them as merely peculiar reptiles that (just by coincidence?) happened to have the distinctive jaw configuration of mammals too. Which he wasn't going to describe (or illustrate) so that his readers might accidently spot the annoying facts on their own.

Gish then tried to pull a fresh rabbit out of a completely different hat: "while thousands of fossil reptiles have been found which possess a single ear bone and multiple jaw bones, and thousands of fossil mammals have been found which possess three ear bones and a single bone in the jaw, not a single fossil creature has ever been found which represents an intermediate stage, such as one possessing three bones in the jaw and two bones in the ear."

This was a distinctive anomaly for Gish, insofar as he was specifying what he would accept as an intermediate form. Or was he?

Being a mammal paleontologist, James Hopson (1987, 24) spotted Gish's trick (Hopson's *italics*): "intermediates such as *he* describes never did exist. But his argument is a 'red herring,' intended, it would seem, to mislead the uninformed. As we have seen, the four reptilian jaw bones were incorporated into the mammalian middle ear *as a unit*."

Hopson's *American Biology Teacher* article noted another of Gish's manuevers, this time involving the creationist's selective invocation of Kenneth Kermack *et al.* (1973). Gish cited that paper "as stating that the accessory jaw bones and reptilian jaw joint of *Morganucodon* were not reduced in size from that of much earlier cynodonts." Gish's point here being to reaffirm his mantra of how garden variety reptilian *Moganucodon* was compared to the predecessors.

Gish had taken pains to highlight how fragmentary the fossils of *Morganucodon* were, which should have raised a flag about how improving specimens could change the perspective. Work which Gish could have known about too, had he been attentive to the more current technical literature.

We can say all that, because Hopson noted that *later work* by Allin (1975) had shown the retained jawbones in question were indeed reduced in size, and that Kermack *et al.* (1981) *had come to agree with this new analysis*. Along with every other paleontologist since, such as Colbert & Morales (1991,

234) illustrating *Morganucodon*'s jaw with its sliver-like reptilian bones nestled along the large dentary, or Michael Benton (2005, 294) comparing *Morganucodon*'s jaw and ear layout with earlier cynodonts *Thrinaxodon* and *Probainognathus*.

And what was all that modified anatomy doing?

Maier & Ruf (2016, 275) noted that Kermack's 1981 work had demonstrated "that in this basal mammaliaform an almost perfect tympanic cavity must have existed underneath the ear ossicles, but that the whole structural complex of the middle ear was still attached to the angle of the dentary."

In other words, that early mammal *Morganucodon* still retained the connection between its ear bones and the enlarged dentary jaw that had moved the old reptile bones into that new position and hearing function.

For frosting on the transitional cake, Hopson reminded his 1987 readers how *Morganucodon*'s "tiny quadrate closely resembles the incus of primitive living mammals," adding that this included "the long process which contacts the very mammalian stapes."

In Hopson's judgment, in the *Morganucodon* case "we have a fossil that in its jaw and ear precisely straddles the boundary between two higher categories of traditional classification—the Reptilia and Mammalia."

We can make some interesting observations now on Gish's *scholarship*.

As Gish's article was dated December 1981, Allin's 1975 paper and the long Kermack monograph on *Morganucodon*'s skull (which came out in January of 1981) could have been known to the creationist.

But not only had Gish missed those papers in 1981, they were still off his scope when Gish (1995, 167-171) fielded the *Morganucodon* jaw argument again.

And yet ... in *Creation Scientists Answer Their Critics*, Gish (1993a, 91-92, 108) had cited that very 1975 Allin paper in a daisychain response to a 1985 debate Gish had with Phillip Kitcher, who "displayed a slide showing alleged reconstructions of a series of mammal-like reptiles, supposedly bridging the gap between reptiles and mammals" and had challenged Gish to say where the gaps were.

How then could Gish possibly claim to not having known about the Allin work when he tossed off his now-obsolete 1981 mammal claim?

Oh, but things are way worse than that. Gish's 1993 response had done much more than just notice the Allin paper.

"Never having seen that particular illustration or the article in which it was found, I was at a loss to do so," Gish complained. "I did state that there were obvious gaps, because every mammal, living or fossil, has three bones in the ear and a single bone in the jaw, while every reptile, living or fossil, has a single bone in the ear and multiple bones in the jaw, and there are no intermediates. I further challenged Kitcher to explain how the intermediates managed to hear and chew, while they were dragging two bones from the jaw up into the ear."

Evidently displeased with being caught behind the debating curve by Kitcher, Gish reported his subsequent investigation (my **bold**):

> After being tipped off that there was some skullduggery involved in the use of the illustration, I wrote to Kitcher for the documentation used for that particular slide. He graciously supplied the information, saying that he actually had obtained the slide from Kenneth Miller. I obtained a copy of the publication, and found the illustration on page 430. The text that accompanied the illustration revealed the fact that 1) two of the "intermediates" in the series were **totally** hypothetical, 2) hypothetical structures had been added to **some** of the "intermediates," 3) the "intermediates" were not arranged in a **true time** sequence, and 4) the "intermediates" were **not drawn to scale**. If a creationist had used an illustration which incorporated any of these doctorings of the facts, even a single one, and evolutionists came into possession of that evidence, the creationist would be thoroughly roasted for distorting science and for deliberate falsehood.

Gish had apparently previously fielded the same criticisms during a 1988 debate with Ken Saladin, after Saladin used the same slide in his presentation, *Talk.Origins Archive* (1988). And thus long before having written his 1995 book revisions.

So what about Gish's claims about the Allin chart?

First, the chart appeared on page 431. It was the captions that were on page 430. But that's a mere quibble. We have much more serious issues than that.

Allin showed eight jaw illustrations, running from (1) an early mammal of the group including *Amphitherium* on the top, (2) the more primitive early mammal *Morganucodon* next, followed by some cynodont fossils (3) *Probainognathus*, (4) *Thrinaxodon* and a still more primitive (5) *Leavachia*. Then the first of the two **hypotheticals**, (6) an "advanced therocephalian ancestral to cynodonts," followed by another actual fossil (7) *Dimetrodon*, and the other **hypothetical**, (8) an "ancestral sphenacodont in which the angular complex is not yet formed."

For Gish's criticism of Allin's hypotheticals to hold, the creationist would have needed to explain what about them was so implausible. This Gish did not do.

Start with the (6) case. Its jaw is laid out as virtually identical to the later (5) *Leavachia*, apart from one significant proposed ancestral variation, a change in the end of the angular bone. The overall shape of the proposed transitional angular was a variation on the one *already known to have existed* in the earlier (7) *Dimetrodon*. Along the inside, the "reflected lamina of angular" (RL) character was shown as having expanded. This referred to a depression in the bone where muscles were attached. On the outside, an indentation in the ancestral bone would have needed to extend, eventually becoming an exposed flange as seen in the later fossils, like (5) *Leavachia*. Did Gish wish to contend that this variation was somehow biologically impossible?

Based, perhaps, on the review of the relevant developmental material that Gish never got around to offering?

Because (6) was early in the evolutionary sequence, the change in the (RL) didn't even relate to Gish's point of contention, the eventual jaw-ear coopting. Even if you pulled (6) out of the sequence, you'd still have all those five known fossils after it, and (7) *Dimetrodon* before it.

Now what about the even earlier hypothetical (8)?

That, too, was a variation of the known fossil (7) *Dimetrodon* that came later. The same bones, but with only two major proposed ancestral variations: the longer and flatter keel of the angular bone (Ank) occupied what would be the later reflected lamina (RL), while the grooved articular bone depression was oriented *upwards*, rather than having twisted slightly to face *backwards* as it would in *Dimetrodon* later on.

So, how accurate was it for Gish to claim the two proposed intermediates were "totally hypothetical" or that any truly "hypothetical" structures were being added to either of them?

Ah, if only Gish had stopped to *discuss* any of this.

But what about their not being in a "true time sequence"?

Sorry, Duane, but that was just plain codswallop. If you put a time frame to the groups involved you got this:

(1) ~170 mya - *Amphitherium*, Middle Jurassic
(2) ~214 mya - *Morganucodon*, Late Triassic
(3) ~234 mya - *Probainognathus*, earlier Late Triassic
(4) ~250 mya - *Thrinaxodon*, Early Triassic
(5) ~260 mya - *Leavachia*, Late Permian
(7) ~280 mya - *Dimetrodon*, Early Permian

Does any of that look out of sequence? Doesn't look that way to me.

Did Gish offer any evidence that they did? Nope.

So what about that final complaint, that the fossils "were not drawn to scale"?

This is perhaps the most intriguing of Gish's accusations. Why should the *scale* matter?

Was he trying to argue that transitional configurations cease to be that way solely because one example might be bigger or smaller than the other? Like a miniature horse can't be related to its bigger ancestor solely because of *scale*?

Get real.

Dimetrodons were about six feet long, but the bulk of the reptile-mammal transitionals are much smaller, rodent sized. Dithering about "scale" was just one more dodge Gish did to step past all the data he never bothered with.

Keep this "not drawn to scale" episode in mind, though. We'll see some later antievolutionists trying the very same trick in Chapter 9.

Since Gish cited the Allin paper in a way that suggested he must have seen the full text, we can ask what the subject of the paper was, and whether Gish ever really dealt with that either.

The main point of Allin's 1975 paper was to propose a new conception of what was going on in the development of the mammalian ear. Previously, the assumption had been that the jaw shift had occurred independently of the hearing changes, building on the middle ear found in "typical" living reptiles, with the angular/tympanic bone not involved in sound transmission. His paper proved very influential, although Maier & Ruf (2016, 273) noted that Allin had "neglected evidence from early embryology, which is very difficult to integrate."

Allin had especially noted the living monotremes, where the tympanic support bone apparently vibrates with the articular/malleus bone. And he pointed out there was actually a lot more variation in living reptiles, too, so that deciding what was "typical" in ancestral ones had to be thought through far more carefully. Reviewing the available evidence, Allin (1975, 406) indicated that "the middle ear region of the earliest reptiles is not similar to that of typical living reptiles."

Working through the components, Allin (1975, 407) found for the stapes side of things that "primitive reptiles probably had a persistent hyo-stapedial connection as does the living *Sphenodon*. Such a connection is present in the embryo of all amniotes." Incidentally, that *Sphenodon* is the venerable "living fossil" Tuatara lizard, representing a lineage tracing back into the Triassic with forms like *Diphydontosaurus* living 205 mya, illustrated by *Prehistoric Life* (2009, 208-209), showing more diversity and dietary range over tens of millions of years than the more restricted living tuatara, Marc Jones *et al.* (2009; 2013) and Oliver Rauhut & Alexander Heyng *et al.* (2012).

An email from Frank Sonleitner some years back called my attention to a pertinent living example of reptilian jaw audition described by Carl Gans & Ernest Wever (1972). Although the legless amphisbaenid "worm lizards" lack external ears, they hear via a flap of skin on the lower jaw (acting as a tympanum) that transmits sound to the inner ear by a long cartilaginous extension of the stapes crossing the jaw joint.

Another relevant witness hops up from the amphibian side, where the tiny Seychelle frog lacks both a middle ear and tympanum, but manages to convey sounds to the inner ear by more direct *bone conduction* (of obvious relevance to comparable intermediate stages among the therapsids) combined with mouth resonance, Renaud Boistel *et al.* (2013). At least one Late Permian synapsid showed signs of that bone-conduction hearing, as revealed by the digital modeling of *Kawingasaurus* by Michael Laaß (2015).

Lots of variations in vertebrate audition, then.

But getting back to the cynodonts, Allin (1975, 416) summarized their hearing case this way: "From the nature of its junction with the stapes, the cynodont quadrate obviously took part in sound conduction, relaying mandibular vibrations. Substrate-carried sound would have efficiently passed by this route when the jaw was in contact with the ground. Cynodonts were surely sensitive to air-borne sound as well, since even snakes, with no tympanic membrane, perceive low-frequency aerial sound quite acutely."

The cynodonts had these bones available to function as sound transmitters, and Allin continued, "Only long after the postdentary unit had departed from the mandible in primitive mammals could the angular (typanic) element be stablized as a static (non-vibrating) drum-supporting entity; this never took place in monotremes."

There was thus a *lot* of information just in Allin's paper, yet *none* of it filtered up through Gish's creationist method, even though he certainly knew of its contents, having previously cited it, only to blatantly misrepresent it in 1993 regarding those "hypotheticals" and time sequence.

Additional fossils had continued to illustrate cynodont diversity in the hearing department, and which Gish could have known about if his object was exploring as much of the available literature as he could.

For example, there was *Procynosuchus*, a nearly complete fossil skeleton of a very early cynodont found in 1974, and which Kemp (1979, 114-116) had evaluated specifically in terms of Allin's work. If you're interested in a stark contrast of method, you need only compare Kemp to Gish on Allin's points. Or skip on down to Chapter 8 to see John Woodmorappe bounding past *Procynosuchus* in 2001. Or press on to Chapter 10 to watch Charles Creager flipping it off. Even if taken together, none of their writing has advanced the cause of science even as far as the length of *Procynosuchus*' diminutive jawbone.

Despite all this information under his nose, Gish (1995, 169-170) doubled down on his *Morganucodon* and *Kuehneotherium* argument (with my **bold**), insisting "that these creatures had a fully-developed, powerful reptilian jaw joint. The anatomy required for such a jaw-joint, including the arrangement and mode of attachment of musculature, the arrangement and location of blood vessels and nerves, etc., **must be quite different** from that required for a mammalian jaw-joint. How then could a powerful, fully functional reptilian jaw-joint be accommodated along with a mammalian jaw-joint?"

And which anatomist said these "must be quite different"? Gish offered no sources, even though he had Allin (1975) as something of a starter to set him on the factual trail. As for the musculature changes, Joe Cain (1988, 96) summarized the issues in his critique of creationist muddling of the RMT: "In cynodont ancestors, the lower jaw shows typically reptilian musculature, whereas early cynodonts show evidence of an insipient masseter muscle, which elsewhere is found only in mammals (Kemp, 1982). Indeed, among the cynodonts, the lower jaw shows two patterns: (1) jaw musculature expanded over the outer surface of the jaw bone, which is elsewhere unique to mammals, and (2) the musculature differentiated into several distinct parts, each inserting at different angles, as in mammals (Figure 5) (Crompton and Parker, 1978). This evidence is derived from studying microscopic markings on the surface of the fossilized jaws that mark points of muscle connection on the bone."

Since Gish had pointedly cited Kemp's 1982 book, in principle he could have known about at least that part of it Cain managed to notice. Kemp's analysis remained a benchmark, as Michael Benton (2005, 296) illustrated cyndodont jaw muscle evolution based on his work. Peter Ungar (2010, 91) illustrated the reorganization of the lower jaw adductor muscles in synapsids, which eventually separated into the *temporalis* and *masseter* muscles. Incidently, the early cynodont *Charassognathus* found by Jennifer Botha *et al.* (2007) showed the first traces of that specialized masseter arrangement.

As for Gish's rhetorical question on the jaw accommodation, this too Gish had been repeating, and Strahler (1987, 419) explicitly quoted Gish's own cited source of Kemp (1982a, 256) on this very point (Strahler's ellipsis): "The axes of the two jaw hinges, dentary-squamosal and articular-quadrate, coincide along a lateral-medial line, and therefore the double jaw articulation of the most advanced cynodonts is still present.... The secondary, dentary-squamosal jaw hinge had enlarged (in the morganucodontids) and took a greater proportion if not all of the stresses at the jaw articulation. The articular-quadrate hinge was free to function solely in sound conduction."

Remember, Strahler's book was yet another relevant source Gish *already knew about*.

Incidentally, *Kuehneotherium* has been tough to characterize due to its fragmentary fossil representation, Doris Kermack *et al.* (1968). Though enough has come to beknown to shed light on its still-primitive auditory arrangement, along with its many derived dental features, subjects of extensive study in works from Gulliermo Rougier *et al.* (1996, 28-29) and Pascal Godefroit & Denise Sigogneau-Russell (1999) to Pamela Gill *et al.* (2014) and Maxime Debuysschere (2016 in press).

Probainognathus* and *Diarthrognathus

After his double-down on *Morganucodon* and *Kuehneotherium*, what Gish (1995, 170) tried to do next was to finally mention (and seek to elbow aside) two of the relevant fossils his 1981 article had paid no attention to: *Probainognathus* and *Diarthrognathus*.

Since we've already introduced them, and know how their anatomical features had been predicted by Robert Broom, it's interesting to see what Gish wrote about them.

Insisting that the double jaw-joint in those two "have been questioned," Gish first invoked Kemp (1982a, 271), Gish's ellipsis: "A second much quoted feature of *Probainognathus* that relates it to mammals is the secondary contact between the dentary and the squamosal. In fact, there is some doubt whether there is actual contact between these bones (Crompton and Jenkins, 1979)...."

And then Gish authority quoted Chris Gow (1981, 15): "The ictidosaur, *Diarthrognathus*, from the Clarens Formation (Cave Sandstone) (Crompton, 1958) is generally held to exhibit the expected morphological grade intermediate between cynodonts and mammals; more specifically, it is thought to have both reptilian and mammalian jaw-joint. However, several of Crompton's interpretations of the morphology of the lower jaw and its articulation with the skull were wrong; some, but not all of these he has conceded in print (Crompton, 1972)."

Probainognathus dates about *30 million years* before *Morganucodon*'s fuller contact point (see my list above). So that much earlier fossil had, according to Alfred Crompton, not yet pushed the dentary and squamosal into "actual contact", while the same Crompton is selectively deemed wrong on the similar *Diarthrognathus*. Or, at least, regarding features Gish did not otherwise specify.

Joe Cain (1988, 100) spotted Gish's evasions on these peripheral issues too, particularly and especially the failure of the creationist "abrupt appearance model" to account for any of the detailed evidence. But by then Gish (1995, 170) was on a spiral of denial, opining: "The manner in which these creatures are reconstructed and their function is visualized is often critically affected by preconceived notions of what should be expected." Oh, the irony, as this so plainly described what Gish was up to in his dizzy mazurka of authority quoting.

The jaw features of the fossils of both animals were not being haphazardly "reconstructed." Paleontology is a discipline way more rigorous that that. Heck, I read the Allin paper!

The paleontological facts were that both fossils clearly showed the layout that Gish so much didn't want to be there (no wonder he didn't put in any illustrations of them). For him, animals have to remain fixed in their neatly created "kinds." There can be no intermediate zones. Ever.

Meanwhile, another source Gish (1995) was aware of, Colbert & Morales (1991, 235), had no trouble connecting the all too obvious dots: "Here we see examples of the gradual transition from reptile to mammal. *Diarthrognathus*, from the upper Triassic of South Africa, is on the reptilian side of the line because, although it had the double jaw joint, the quadrate-articular articulation was still dominant. *Morganucodon*, from the upper Triassic of Europe and Asia is on the mammalian side of the line because, although it too had the elements of both articulations, the squamosal-dentary joint was the dominant one."

The more recent depiction of the jaw articulation change in Luo (2011, 359) was especially clear, viewing the contact points more from the back. In this perspective, the almost meeting dentary-squamosal of *Probainognathus* progressed through the Late Triassic *Brasilodon* (where the contact had been lightly made) to *Morganucodon* where the articulation was massive and strong. Viewed from that angle, Gish's insistance that *Morganucodon* was somehow using its shrunken articular-quadrate as its primary jaw hinge was physically absurd.

Gish obviously couldn't know of Luo's 2011 illustration in 1995, but the bones had still been there, and hadn't moved in the meantime. The creationist could have worked through the pieces just as Luo had. But at every stage of his argument on the RMT, Duane Gish sliced and diced the data and fiddled with sources, rather than forthrightly tackling every scrap of available information to offer a persuasive accounting of the facts.

Just the sort of genuine curiosity and careful method so on display in the many meticulous works by the paleontologists (from Allin to Luo) that were so plainly missing from Gish's clumsy creationist tapdance.

The conclusion of Gish's argument pressed onto what seemed especially safe ground, aspects of mammal anatomy that were genuinely unresolved issues at the time.

Gish (1981) offered the workings of the mammal inner ear (once more, I've highlighted some of his wording in **bold**):

> Now the anatomical problems associated with such a postulated process are vastly greater than merely **imagining** how two bones precisely shaped to perform in a powerfully effective jaw-joint could detach themselves, **force** their way into the middle ear, reshape themselves into the malleus and incus, which are precisely engineered to function with a remodeled stapes in a vastly different auditory apparatus, while all at the same time the creature continues to chew and to hear! As **insuperable** as this problem appears to be, it pales into relative insignificance when we consider the fact that the essential organ of hearing in the mammal is the organ of Corti, **an organ not possessed by a single reptile, nor is there any evidence that would provide even a hint of where this organ came from.**

Knowing how much of the fossil and biological data were available to Gish in 1981, his dismissal of the issue as mere "imagining" was gutsy, while his use of "force" was pure evasion. No one was proposing that the bones were *invading* the ear. They were being pulled along, step by step. But then Gish didn't go into the developmental biology of the mammal jaw and ear (which, remember, had been known since the 1830s).

Take note of that ommission of the developmental biology of the mammal jaw and ear. It will happen again, almost without exception, with all other antievolutioists. And those exceptions (which we'll get to in Chapter 10) are of the sort that "proves the rule."

Gish added the mammalian diaphragm for good measure (again my **bold**). "There is no structure in a reptile that is in any way similar or homologous to the mammalian diaphragm. There is **no structure found in a reptile from which it could have been derived**. Again, a complicated structure had to be created de novo (and by **mistake**!) to perform a function that was already being very satisfactorily performed in a different manner in the assumed reptilian ancestor."

When Gish retooled his 1978 antievolution book, he included these new claims, and the Nuttings' creationist *Think & Believe* (1990a) newsletter quoted the Corti bit without critical comment from Gish (1985, 101). Nor was Gish (1995, 172-173) about to let go of either the organ of Corti or diaphragm claims.

With the organ of Corti and the diaphragm Gish was on his safest spot, but only because he had positioned his soapbox in as clear a field as possible.

Very little was known about the genetics behind either in 1981. Understandably so, since so many of the genes involved weren't even discovered until the 21st century.

Though in his criticism of creationists' claims about the RMT, Joe Cain (1988, 101) reminded Gish that back in 1956 Alfred Romer had suggested the organ of Corti was an elongated *basilar papilla*, and that while so fragile a piece of soft tissue was unlikely to be preserved in fossils, the Corti/papilla homology had embryological support.

While antievolutionists have shown no gumption to do work in this area themselves, evolutionary biologists have not been so hesitant, and bit by bit work has proceeded in teasing out how the Corti organ develops from the basilar papilla in what becomes the vertebrate cochlea, Bernd Fritzsch *et al.* (2011; 2013). The problem is especially difficult, involving functionally retroengineering a sequence whose contributing genes are still not yet identified (a working task that can take researchers *decades*).

There were already a lot of anatomical details to consider and fossils to illustrate aspects of them, though, as Zhe-Xi Luo *et al.* (1995) covered in their review of the cochlea and its placement in the skull.

What was needed were more fossils, and once again the Fossil Genie has come through: the Late Jurassic *Dryolestes* still had an uncoiled cochlea, Zhe-Xi Luo & Irina Ruf *et al.* (2011). Geoffrey Manley (2012, 735) charted the evolution of mammal cochlea, including the *Dryolestes* example. "Indeed, the coiled cochlea, considered by many to be archetypically mammalian, arose only in one of the three lineages, the therians, and only after 100 Ma of mammalian evolution in that lineage," Manley (2012, 736). "Therians" include marsupial and placental mammals.

"The early organ of Corti was likely to have been a low- to mid-range frequency receptor receiving input from an insensitive middle ear," Manley (2012. 737). There were already a pair of hair-cells in early mammals (as yet of uncertain function) which would become involved in the development of the modern mammal ear.

We'll be returning to the cochlea issue in Chapters 8 & 10, to see whether Intelligent Design fares any better than Creation Science in the investigative gumption department.

Much the same Coming Attraction problem clouded Gish's other example, working out the evolutionary origins of the diaphragm. Without the benefit of knowing the genetics of its developmental biology (only discovered later), researchers like John Ruben *et al.* (1987) focused on trying to parse out the selective reasons for the diaphragm evolving, particularly in relation to higher oxygen processing in sustained activity (even before reaching a warm-blooded endothermic condition).

The idea that the diaphragm needn't have started out for respiration continued to be worked on by researchers, such as Mark Pickering & James Jones (2002), Steven Perry & Martin Sander (2004), and Perry *et al.* (2010).

By the time of Allyson Merrell & Gabrielle Kardon (2013), much more was known about the developmental biology and genetics of the diaphragm. That it develops from a specific set of shoulder muscles fitted in with something visible in the fossils: an increase in cervical vertebrae number early in synapsid history which triggered off a multistage evolutionary cascade, where a duplicated *brachial plexis* played off a particular set of body connections (illustrated by our old pal *Dimetrodon*) and only later got coopted in the pulmonary system, Tatsuya Hirasawa & Shigeru Kuratani (2013).

And the Fossil Genie generously plopped something on the field: the Early Cretaceous *Spinolestes* from Spain, a splendidly preserved eutriconodont mammal with signs of a diaphragm along with its hair and stiffer spines like a rudimentary hedgehog, Thomas Martin *et al.* (2015).

Gish had passed on by this time, of course, leaving it to a new generation of creationists to ignore the constantly accumulating ongoing work.

Or let an Intelligent Design advocate pick up the ball to be dropped, such as Michael Denton (2016a, 120, 305n), who put the mammalian diaphragm on a list of other "classic examples" of things that supposedly could not be accounted for "in terms of cumulative selection," but which he "could have included had space allowed." Denton cited only Perry *et al.* (2010) and Hirasawa & Kuratani (2013) in his endnotes, papers which went a long way toward the very thing he insisted couldn't be done, but Denton did not reflect on any of their specific content.

Kind of the way Gish behaved with his sources, wasn't it?

We'll be delving into some of Denton's less truncated arguments in Chapters 5, 6 & 11.

Creationist grab bag

It is fair to say that there was no one in the creationist literature who tackled the RMT with more gusto than Gish. Compared to his efforts, the remaining authors of his generation have either tossed off generalized claims, or not noticed it at all, such as Donald Chittick (1984) or Old Earth creationist Alan Hayward (1985).

In the loose end department, armed with his diploma in theology, Australian A. W. (Bill) Mehlert (1988; 1993) penned two critiques of the RMT for creationist journals. While neither of these were available for my perusal online as of this writing, John Woodmorappe (2001) cited Mehlert at the end of a paragraph, declaring how "it soon becomes obvious that many of the anatomically-based evolutionistic claims, when analyzed, turn out to be questionable." Woodmorappe also cited Mehlert on a crocodile tooth. No other antievolutionists have relied on Mehlert's argument, though, and as we'll see in Chapter 8, Woodmorappe is not the most reliable source for parsing content secondarily.

Another side player was Luther Sunderland (1929-1987), a grassroots creationist fond of attending evolutionary lectures to capture supposedly incriminating admissions for the quote mine. His activities did not extend to the factual level, though. Three quarters of a century after Robert Broom's dead-on prediction of probainognathids, in his final posthumous work Sunderland (1988, 91) proclaimed with utmost confidence: "But there is no convincing scenario that can even be conceived for getting the jaw bones across the jaw joint."

As we've seen from the fossils and the embryology, that was just plain wrong.

Sunderland (1988, 91) did a mini-Gish, first quoting from a book review by Eric Lombard (1979) taking to task the book's author for not doing a work useful for "constructing phylogenies of mammalian taxa." Not realizing this was a comment on the need for particular diagnostic characters for systematics, Sunderland jumped to the conclusion that this meant "there are absolutely no fossils showing the migration of the jaw bones of the reptile up into the ear of the mammal."

Sunderland (1988, 92) then authority quoted a *New Scientist* article by Tom Kemp (1982b, 583). It's revealing that Sunderland left off the end of Kemp's last sentence, which interestingly the equally creationist *Genesis Park* (2011d) thought to include in their quote-mined version (my **bold** on their addition): "Each species of mammal-like reptile that has been found appears suddenly in the fossil record and is not preceded by the species that is directly ancestral to it. It disappears some time later, equally abruptly, without leaving a directly descended species **although we usually find that it has been replaced by some new, related species**."

Either way, this was bumping into the Punctuated Equilibrium issue I noted earlier, where direct species-species transitions were understandably rare in the fossil record because vertebrates tend to speciate allopatrically. While *Genesis Park* glanced past it (even quoting from Eldredge and Gould on Punc-Eek), the earlier Sunderland couldn't get past the enticement of several dated "gap" quotes culled from the venerable Simpson (1944, 105, 107).

And all without Sunderland or *Genesis Park* discussing any actual fossils.

Which was interesting, because Sunderland (1988, 93) had to step over a whole page of them (pelyocosaurs and cynodonts) to get to the last paragraph of Roger Lewin (1981), an article in *Science* reporting on a Smithsonian conference of paleontologists and biologists, just the sort of note-comparing that would bear so much solid science work in the decades to follow (unlike the wheel-spinning stasis of antievolutionists).

I've highlighted the part Sunderland quoted in **bold**: "The transition to the first mammal, which probably happened in just one or, at most, two lineages, **is still an enigma**."

Note the implications of that "first mammal." That was the taxonomical issue of where you drew the line, and whether the monotremes and therians split after that "mammal" condition was achieved or before. The fossils and genetics in 1981 weren't in a position to settle that. Although Lewin's use of "enigma" was rather an over-statement, that toss-off was all Sunderland needed to nick it and run.

That this form of superficial analysis was not restricted to Christian creationists was shown by Harun Yahya at the *Signs of Allah* website. The pen name of oddball Turkish creationist Adnan Oktar, Yahya has dozens of websites and duplicative books sprinked around the Internet. One of his videos even turned up in the current lesson plan for a 10th grade biology class in Youngstown, Ohio, spotted by the diligent Zack Kopplin (2016). The *Youngstown City Schools* (2015, 3-4m 12-14) curriculum guide also dangled claims of "irreducible complexity" and a reference to Michael Denton proclaiming evolution to be a "theory in crisis." There'll be more to say about what is in crisis regarding Denton's work in chapters 5, 6 & 11, but I'll note that Ohio has been a battleground of Intelligent Design apologetics for many years, instances of which I examined in Downard (2015b, 35-47).

But back to the Lewin quote. The section on "The Origin of Mammals" in Harun Yahya (2004a) declared (Yahya's ellipsis): "Not surprisingly, not a single fossil to link reptiles and mammals is to be found. This is why evolutionist paleontologist Roger Lewin was forced to say that 'the transition to the first mammal ... is still an enigma.'"

It was the essence of secondary pseudoscholarship for Yahya to think a decades-old quote from a science writer (Lewin was not a paleontologist) could substitute for a discussion of the fossil and biological evidence that by 2004 had rendered any "enigma" description obsolete.

And speaking of secondary pseudoscholarship, Luther Sunderland's book hovered as an uncited source by a later creationist, Scott Huse (1997, 89), who at least had the virtue of brevity when he announced without even a stab at documentation, "There are no transitional forms between reptiles and mammals," and plowed on without further ado.

Even though Sunderland's book wasn't directly listed or even mentioned by Huse, the bibliography in Huse (1997, 209-213) testified otherwise, as dozens of often obscure sources Sunderland (1988, 184-188) had used inexplicably showed up in Huse's list—sources Huse never cited on anything. Quite a few of those related to mammal evolution, with one telltale entry being the Lombard book review, which Sunderland had incorrectly listed with the volume number wrong (93 instead of 92) and noting only the specific page for the quote, not the full pages for the article. Huse had exactly that Sunderland version in his bibliography.

Scott Huse was thus not only a vivid example of scholarly parasitism, but a visibly indiscriminate and untidy one.

Which brings us to Doug Bandow.

A Fellow of the *laissez-faire* conservative Cato Institute, and author of muckraking exposes like *The Politics of Plunder: Misgovernment in Washington*, Doug Bandow (1991) had boarded the antievolutionary express to enthuse about Phillip Johnson's pivotal work, *Darwin on Trial*: "Johnson lacks a technical background, but he makes up for that deficiency with his ability to deconstruct poor reasoning."

A skill Bandow obviously failed to apply to the "poor reasoning" Johnson showed while trying to deconstruct the RMT, as you'll see in Chapter 7.

Eight years after savoring *Darwin on Trial*, Doug Bandow (1999) recommend three books "demonstrating that religious faith does not mean checking one's mind at the church door."

The Genesis Question by astronomer Hugh Ross (1998) was essentially an unimproved retread of his prior Ross (1996), an Old Earth Creationist (OEC) apologetic claiming that modern physics discoveries somehow proclaimed the unique identity of the Christian God.

The Science of God by kabalistic physicist Gerald Schroeder (1997) flexed over on the Judaism end (where physics proof for Ross' explictly *Christian* God was evidently not so apparent after all) to correlate billions of years of geology and cosmology along a logarithmic scale keyed to the Genesis Creation Days.

That was no easy trick for either Ross or Schroeder, as both had to overlook the obvious snag of Genesis Day 4, which described the Sun, Moon and stars being made *after* the Earth and plant life. Compounding things, that cosmological goof wasn't even original to the Bible. Hermann Gunkel (1895) spotted how the celestial bodies appeared on the scene in that same late order in the much earlier Babylonian *Enuma elish* creation myth from the 2nd millennium BCE, elements of which apparently filtered into the developing Genesis story following the Israelites' traumatic Babylonian Captivity in the 6th century BCE, Frederick Greenspahn (1983) and Norman Cohn (1993, 45-51).

While more secular or liberal writers like Paul Tobin (2000) or B. Robinson (2011) readily accepted that Babylonian/Genesis connection, there's a much longer trail of religious authors downplaying them, from G. Michell (1932) and A. Millard (1967) through to major scholars like Alexander Heidel (1951) and Wilfrid Lambert (2013), and the practice continues with Pete Enns (2010) at the pro-evolution (but still definitely religious) *BioLogos Forum*. Even the extent of the impact of the Babylonian Captivity has remained an area of some contention, as Amy Marcus (2000, 154-178) discovered in her foray into modern biblical archaeology.

Creationists have bifurcated on this issue. YECers were happy to take the Genesis Days literally, and Henry Morris (1985, 238) flatly castigated nonliteral interpretations as "strained renderings," while OECers like Robert Newman (1999, 108), Walter Bradley (2001, 172-173) and Stephen Barr (2003, 45) kept up the straining side by claiming that it was an increasing transparency of the primordial atmosphere that made the already created celestial bodies visible to all those many people who didn't in fact exist yet on Earth to notice it.

Hugh Ross (1998, 42-45) took that same tendentious position, as did Schroeder (1997, 67), though with his own novel twists. While Ross shied away from tagging dates to his framework and only generally referred to plants, Schroeder's logarithmic calibration specifically pegged Day 4 to 1.75 bya down to 750 mya, and explicitly redefined "plants" to mean photosynthetic algae. While that got his calibration straight, it still remained that Schroeder offered no physical evidence that the Earth's atmosphere had remained *opaque* that late, unlike geophysicists William Hartmann & Ron

Miller (1991, 81) who depicted the atmospheric view from 3.9 bya (a full *2 billion years* before Schroeder's Day 4 cutoff) as clear enough for the much larger Moon to be seen.

Even had it all been relentlessly overcoast, though, such misplaced concreteness wouldn't have explained why *God* wouldn't have known the Sun, Moon and stars were there, even if the eyeless photosynthetic "algae" weren't taking note for posterity.

As for birds not appearing until the next Day 5, Ross (1998, 47-50) skipped past that very gingerly, while Schroeder (1997, 69, 94) again tripped over his all-too-specific chronology, which had ended the Fifth Day at 250 mya. No birds back then, but Schroeder resolved that by deciding the *oaf* of Genesis 1:20-21 actually referred to "winged (insect) life."

By the way, Michael Behe enthused on the dust jacket that "Schroeder vindicates the fruits of sophisticated biblical scholarship with the tools of modern science." Tools which in Schroeder's hands were used solely as a precision buffing instrument to grind the Old Testament down until what was left no longer looked scientifically silly to modern eyes.

If Ross and Schroeder were problematic recommendations, Doug Bandow's third witness for a scientific faith fell off the wagon completely: Scott Huse's *The Collapse of Evolution*.

In March 2001 I emailed Bandow asking whether he was familiar with the background of any of the books he cited. He replied that he "found them to be better argued and researched than the typical 'creation science' and young earth tracts, which aren't well-founded," plainly oblivious to the fact that Huse's book *was* a Creation Science one, and not even slightly well-founded.

While source tracking has not proven to be one of Bandow's enduring interests, the same could not be said of his strident social conservatism, a *Kulturkampf* (culture struggle) drum which Bandow (2013) continues to beat.

Well, enough of Bandow. For some comic relief, there's Richard Milton's *Shattering the Myths of Darwinism* from 1997, my poster child for scholarly incompetence.

Fairly peripheral on the antievolution scene, Milton is a strange bird, a nonreligious person who nonetheless is functionally a Young Earth creationist. Milton pulled that trick off by credulously accepting the arguments of actual YECers, all without any sign of fact checking, as I explored often in the text and notes of Downard (2004c).

Relying solely on British chemist and Flood Geologist A. J. Monty White, Richard Milton (1997, 199) eased past the reptile-mammal landmine by assuring us that "No fossil remains have been found" for mammal ancestors, even though "recognizing a transitional skeleton ought to be straightforward if, as Darwinists claim, mammals evolved from reptiles."

Knowing what you do about the fossil players, I think that one deserves a full ROFL.

As an aside, Ronald Numbers (1992, 327-328) noted White's infighting with other British creationists less enamored of his Biblical literalism.

Some creationists deflected the RMT as Milton did, by not getting too close to the technical literature. The Nuttings' *Think & Believe* (1990b) creationist newsletter dismissed a new ancestral fossil (*Adelobasileus* from 225 mya) as but "a mosaic" of mammal and reptile, based solely on a local newspaper mention of it. As the newsletter did not follow the paleontological literature directly, the eventual technical paper, Spencer Lucas & Zhe-Xi Luo (1993), escaped their attention. We'll look at that critter again in Chapter 10.

Perhaps the most oblique richochet off the reptile-mammal data during this period occurred when Bert Thompson (1995, 214) positioned the first "mammal-like reptiles" in the Carboniferous Period on a chart "based primarily" on Stephen Jay Gould (1993) *The Book of Life*, but did not otherwise discuss them. The chart turned out to be from the chapter by Michael Benton (1993a, 25). Evidently Thompson lacked the stamina to press ahead the seventy pages to a later chapter in the book, also by Benton (1993c, 96), where the evolution of the mammal jaw was addressed.

The Jehovah's Witness *Life—How did it get here?* book, *Watchtower* (1985, 69, 253n) bumped into the RMT via very short authority quotes nicked from two *Time-Life* general publications from 1963.

The first consisted of a clause nipped from Archie Carr (1963, 40-41). Misattributed by the *Watchtower* citation to page 37, the part they use is in **bold**, along with their bracketed insertion:

> **There is no missing link** between [that connects] **mammals and reptiles**, nor any single fossil type which, as *Archaeopteryx* does for birds, stands out clearly as half reptile, half mammal. If each mammalian feature could be traced to its point of origin, we might hope to put a finger on the first mammals. As it is, the case rests mostly on bones and teeth, and relying on such skeletal characters alone, we only see the ancestral forms slowly acquiring the skeleton and dentition that today we associate with mammals. We can only deduce the scheduling of the less solid attributes not susceptible to preservation in the rocks.

To get to that snippet, *Watchtower* had to step over what Carr had just written: "The record of the derivation of mammals from reptiles is both far longer and far more detailed than the history of birds—and also, unfortunately, it is far less clear. It begins with the pelycosaurs of the late Carboniferous—a group not far removed from the old stem reptiles. From these there radiated a great array of types known as mammal-like reptiles, and during the late Paleozoic and early Mesozoic these creatures were the dominant vertebrates on the land."

How that record could be called "far less clear" said more about the inadequacies of 1960s *Time-Life* science condensation than the fossil data, but *Watchtower* wasn't slowing down enough to notice as they culled another bit from Richard Carrington (1963, 36-37), this time at least getting the pagination right. The part they liked was: "Fossils, unfortunately, reveal very little about the creatures which we consider the first true mammals."

That was a debatable contention at best, no fair characterization of even

1960s paleontology, but it was a downright misrepresentation of the science known by the 1980s.

Nor did *Watchtower* quote (or think about) what Carrington had written in a previous paragraph: "The structure of the teeth, among other features, shows that pelycosaurs and therapsids, although not necessarily the direct ancestors of mammals, may certainly be regarded as their ancient uncles and aunts."

But then, *No Cousins* are ever allowed in the *Watchtower* creationist menagerie, are they?

Watchtower's liberal use of out of context and dated quotes came under fire by Malcolm Levin (1992), including *Watchtower* (1985, 69, 253n) mining from the older Ledyard Stebbins (1971, 146) on bird evolution (*Watchtower*'s extract in my **bold**): "**The transition from reptiles to birds is more poorly documented** than are the other transitions between classes of vertebrates." Levin reminded that Stebbins had gone into the evidence for those "other transitions," and further took *Watchtower* to task for not even thinking to use the more recent revised edition, by then available as Stebbins (1977, 217). But the 1971 version of Stebbins still offered an adequate measure of *Watchtower*'s thin scholarship regarding the RMT, since the therapsids' jawbone-ear connection showed up in Stebbins (1971, 148, 158).

Levin also mentioned the overtly apocalyptic tone of the Jehovah's Witness book, such as *Watchtower* (1985, 248) declaring "that the theory of evolution serves the purposes of Satan." Levin contrasted that with the more circumspect Creation Science publications of the period. Though, to be honest, Henry Morris' ICR believed pretty much the same thing about satanic evolution, reflected in Henry Morris (1963, 77-78), but also in a parade of creationists since then, including Glen Chapman (2004), Max Younce (2009), *San Antonio Bible-Based Science Association* (2010) and *Answers in Genesis* (2016b), still holding that "The lie Satan has especially spread in our age is the lie of evolution and millions of years, which implies that God's Word is wrong and that man's ideas are more reliable."

Such a medieval mindset doesn't obsess only on evolution, of course, not with so many demons bustling about in the End Times, represented by nervous works from Hal Lindsey & C. Carlson (1970; 1972), Nicky Cruz (1973) and the *Fundamentalist Journal* (1988), to the more current Lucifer-mongering of Henry Morris III (2013b).

Nor are those the only players. Harun Yahya (2002; 2004b) plays the Satan card as easily in his Islamic apologetics as Christians do in theirs, and biblical geocentrist Marshall Hall (2011) detects Satan's hand in the success of the heretical heliocentrism that has grown so popular in science over the last 400 years. I'll have some more to say about geocentrism down in Chapter 12.

And let's not forget the world of contemporary *Kulturkampf* politics. Because it won't be forgetting *you*.

Former Pennsylvania Senator Rick Santorum shares a lot of that apocalyptic creationist view, believing that Satan schemes behind the drapery to tempt America into decline and peril, prompting raised eyebrows from secular critics like Kyle Mantyla (2012), Eric Reitan (2012) and Brian Tashman (2016b). Tashman (2013a-b; 2016a,c) has further reported on how the resurgent gay rights and marriage equality issues have been seen by fundamentalists as part of the satanic plotting, and how during the 2016 presidential campaign, preacher Darrell Scott even assembled a prayer circle around Donald Trump to protect that shy wallflower from the "Concentrated Satanic Attack" Scott firmly believed was directed against the GOP nominee.

I wish I were making all that up. Or that Mr. Trump hadn't won the election (or at least the quixotic Electoral College).

A Word to the Wise

When theologian John Moreland wanted to marshal scholars favoring the developing Intelligent Design movement for his 1994 anthology on *The Creation Hypothesis*, they were in something of a pickle when it came to paleontology.

None were paleontologists.

So they brought Kurt Wise on board as their only viable alternative.

Wise is a Young Earth creationist with a paleontology degree (he studied under Stephen Jay Gould at Harvard, which must have been interesting for them both). Inside Wise's head, the fossils of dinosaurs and trilobites and the RMT are all just the happenstance debris of the Biblical Flood, which YEC doctrine typically pegs at around 4500 years ago.

Which means Kurt Wise might be really good at mentally stagemanaging a data set to whittle it down until it could be fitted into that much smaller Flood Geology geochronology box.

So, what did Kurt Wise (1994, 226-228) have to say on the RMT for the Intelligent Design set?

If anybody would have had home court advantage on paleontology, it would have been Wise, who could not be counted unaware of the fossil data or reluctant to show off his expertise in countering their evolutionary implications.

Except, he was a Young Earth Creationist, with his own nonnegotiable axes to grind, and doctrinal baggage to carry. Not to mention a deeply flawed methodology.

Wise began with quite a concession (my **bold**):

> Series of fossil species like the horse series, the elephant series, the camel series, the **mammal-like reptile** series, the **early birds** and **early whales** all seem to be strong evidence of evolution. Another class of fossil evidence comes in individual *stratomorphic intermediates*. These are fossils that stand intermediate between the group from which they are descendent and the one to which they are ancestra—both in stratigraphic position and in morphology. They have a structure that stands between the structure of their ancestors and that of their descendants. However, they are also found in the fossil record as younger than the oldest fossils of the ancestral group

and older than the oldest fossils of the descendent group.

Well, that would seem Game, Set & Match, wouldn't it? Evolution takes the trophy home.

Ah, now for the back pedal:

> Once again, the existence of stratomorphic intermediate groups and species seems to be good evidence for evolution. However, the stratomorphic intermediate evidences are not without difficulty for evolutionary theory. First, none of the stratomorphic intermediates have intermediate structure, it's the *combination* of structures that is intermediate, not the nature of the structures themselves. Each of these organisms appears to be fully functional organisms of fully functional structures. *Archaeopteryx*, for example, is thought to be intermediate between reptiles and birds because it has bird structures (e.g., feathers) and reptile structures (e.g., teeth, forelimb claws). Yet the teeth, the claws, the feathers and all other known structures of *Archaeopteryx* appear to be fully functional. The teeth seem fully functional as teeth, the claws as claws, and the feathers as any flight feathers of modern birds. It is merely the *combination* of structures that is intermediate, not the structures themselves. Stephen Jay Gould calls the resultant organisms "mosaic forms" or "chimeras." As such they are really no more intermediate than any other members of their group. In fact, there are *many* such "chimeras" that live today (e.g., the platypus, which lays eggs like a reptile and has hair and produces milk like a mammal). Yet these are not considered transitional forms by evolutionists because they are not found as intermediates in stratigraphic position.

We've heard this argument before. It was Duane Gish's line of attack, that *mosaic* argument. All Wise did was slip in his "stratomorphic intermediate" terminology, but the shell game was the same.

Parenthetically, off his own deep end, Scott Huse (1997, 148) took exactly the opposite position as Wise, averring (again with no documentation) that "Evolutionists insist that the duck-billed platypus is an evolutionary link between mammals and birds." Creationists like Huse have been muddling the platypus for some time, covered by Jim Foley (1997), playing on the relatively limited monotreme fossil record. More fossil platypuses are known today, such as reported by Timothy Rowe *et al.* (2008), though still largely represented by isolated teeth. The sequencing of the platypus genome by Wesley Warren *et al.* (2008) suggested monotremes were diverging sometime in the Jurassic, but Robert Carter (2008), *Creation Tips* (2008), *Truth in Science* (2008a) and Andrew Schlafly's *Conservapedia* (2016) reflect how more recent creationists initially pushed aside such scientific work, only to pay no further attention to it. We'll return to the monotremes in Chapter 10.

The big problem for Wise's case against the RMT is that we know just how directly intermediate and numerous those reptile-mammal characters were, from lengthening and coiling cochlea to the Meckel's cartilage retained in highly specific contexts and evolving into the later state along with the morphing synapsid jaw. And how substantial parts of this evidence were known in the paleontology of the 1990s, that field Wise had an official degree in.

It is inconceivable that a trained paleontologist could be completely oblivious to them, so Wise's circumlocutions on this are even more suspect.

And the consistently missing antievolutionist conceptual piece is still off the field: at no point did Wise offer any description of what even *Archaeopteryx* would have needed to look like to qualify as a legitimate intermediate, let alone wade through the much larger body of reptile-mammal critters.

As for the functionality of extinct life, were transitional features supposed to be *non-functional*? Does Wise expect that one to fly?

Sorry, that is exactly what evolutionary thinking would *not* expect or allow.

After again noting how "Mammal-like reptiles stand between reptiles and mammals, both in the position of their fossils and in the structure of their bones," Wise insisted that such transitional forms were rare, specifically listing them: "*Pikaia* among the chordates, *Archaeopteryx* among the birds, *Baragwanathia* among lycopods, *Icythyostega* among the amphibians, *Purgatorius* among the primates, *Pakicetus* among the whales and *Proconsul* among the hominoids," and that this "list is very nearly complete."

Wise offered no technical citation for this compendium, by the way.

A parenthetical note: for those unfamiliar with lycopods (club mosses), Wise's reference sequence was presumably not to suggest they were deemed morphologically intermediate between birds and amphibians (or that *Pakicetus* was somehow a transition between primates and hominids).

Wise's choice of the Late Silurian *Baragwanathia* was curiously dated for someone supposedly so up on the geological record, though, as by then further more informative fossil specimens were on hand, such as the Middle Silurian *Cooksonia*, Dianne Edwards *et al.* (1992). A sense of where the paleontology was by the mid-1990s could be seen by comparing Lambert & The Diagram Group (1985, 38-39) summarizing early vascular plant evolution, especially showing the change in leaf configuration, with Rich *et al.* (1996, 374-375) putting *Cooksonia* as representative of primitive plants.

But by now Wise's mind may have already been too intent on sprinting towards his necessary object, the reduction of the dataset to a manageably small list he could crumple and toss without straining.

You'll notice how in the space of only a few paragraphs, Wise's references to "early birds" and "early whales" (*plurals*) got trimmed to only single examples when he rolled out his list.

It reminds me of the scene from Charlie Chaplin's classic 1940 satire on Hitler and Fascism, *The Great Dictator*. Required by the rules of diplomacy to dance with the overweight wife of the visiting rival thug Benzino Napaloni (played with chin-jutting bravado by Jack Oakie), Chaplin's tyrant Adenoid Hynkel proceeded to goose-step her around the floor. That perfunctory obligation done, Hynkel then paid Mrs. Napaloni a snarkily deescalating compliment: "Madam, your dancing was superb ... excellent ... very good ... good."

That's pretty much what Kurt Wise was saying about that fossil record, reducing the formidable real examples down to as few as he could manage, and (sort of) actually talking about only one, *Archaeopteryx*.

And did you notice the absence of the *mammal-like reptiles*?

He mentioned them several times, remember, but somehow all those synapsids slipped his mind when he got around to compiling that "very nearly complete" list.

Think about what Wise was trying to do here.

With that grandiose omission of the synapsids, Wise was trying to minimize the significance of transitional forms being known among primitive plants (*Baragwanathia*), along with an anchor for the very earliest chordates (*Pikaia*), plus one for a major mammal habitat adaption (*Pakicetus*), and two more for our own human lineage (*Purgatorious and Proconsul*)—though we might well inquire, what about those later even more relevant Australopithecines, don't they count?

And don't forget those *birds*.

Put *Probainognathus* beside *Archaeopteyx* on the table, which Wise obviously didn't, and he was trying to marginalize having acknowledged intermediates turning up in 100% of the only vertebrate classes to have emerged in the last quarter of a billion years (mammals and birds).

It's hard to get more comprehensive than 100%.

And that's even accepting his disingenuous claim that those examples pretty much exhausted the field.

Having carved down the taxa, at least in his own YEC imagination, Wise dangled a more traditional creationist explanation for this arbitrarily truncated state of affairs: "The very low frequency of stratomorphic intermediates may be nothing more than the low percentage of mosaic forms that happen to fall in the correct stratigraphic position by chance—perhaps because of random introduction of species by a Creator or the somewhat randomized burial of organisms in a global deluge."

Yes, that inexplicably evolution-friendly Creator who kept on deliberately slipping fossils of the form expected by evolutionists onto the "random" scene, to be further "somewhat randomized" to just coincidentally align during the (geologically preposterous) Flood to end up in that "stratomorphic intermediate" state Wise never explored in too much detail.

Or maybe it was just the Fossil Genie, after all.

Kurt Wise (2006) eased around the RMT again when he sought to hijack for creationism a Middle Jurassic swimming docodont that had turned up. Described by Qiang Ji *et al.* (2006) with commentary by Martin Thomas (2006), *Castorocauda* was larger than the average Mesozoic mammal and had several aquatic adaptations, including a beaver-like tail and a coat of beautifully preserved hair (at 154 mya that put it some 30 million years prior to the former oldest, *Eomaia*, which we'll be bumping into again later). Claiming that such a form somehow posed a problem for evolution (not enough time for that, really?), Wise boldly claimed that creationists "expect a wide disparity of mammalian form to exist at the time of the dinosaurs (because God created that diversity in the Creation Week)."

Which was particularly disingenuous, because there was no "time of the dinosaurs" in creationist chronology, as everything was contemporaneous, especially fossils, which must all have occurred by being buried during the Flood. In the YEC scheme, that "time of dinosaurs" dated to exactly the same time as trilobites, and early tetrapods, and hominids, and the pyramids of Egypt. Unless rollers were slipped under all their chronology to glide it past Noah's time, as a few creationists have taken up as their labor, such as John Ashton & David Down (2006).

But more characteristically, Wise only mentioned the bits about the new fossil that could be spun within the typological frame (beaver-like tail or seal-like piscivorous teeth). Wise did *not* mention anything about *Castorocauda*'s characteristically intermediate docodont jaw layout.

While Wise's creationism has shown no great interest trying to work out why any of these docodonts (or 500-plus genera of dinosaurs) didn't manage to survive after their boat ride only 4500 years ago, evolutionary paleontology hasn't flinched in working to explain those extinct forms.

When new docodont fossils turned up, a burrowing model described by Zhe-Xi Luo & Qing-Jin Meng *et al.* (2015) and an arboreal sap-feeder covered by Meng *et al.* (2015), a lot more could be noted of them than just marveling at Creation Week one-offs. As the Luo paper explained, the reduction of the digit segments in the burrower would have involved the "fusion of the proximal and intermediate phalangeal precursors, a developmental process for which a gene and signaling network have been characterized in mouse and human. Docodonts show a positional shift of thoracolumbar ribs, a developmental variation that is controlled by *Hox9* and *Myf5* genes in extant mammals."

That's how science works. You seek out as much of the data as you can, and then forthrightly explain what you think happened. Other scientists study that work, adding effort of their own, and the science moves on to understand more and more.

And then there's creationism.

Kurt Wise circa 1994 (or 2006 too) was showing how even a paleontology degree couldn't entice his creationist imagination onto the field of data appreciation, let alone explanation. In all respects he was simply retooling the argument of Duane Gish, only with suprisingly *less* attention to detail.

Now a stroll down the daisy chain.

Theologians John Ankerberg & John Weldon (1998, 219) relied on Wise's "stratomorphic intermediate" argument to dismiss the RMT. Which was interesting, as they were aware of lawyer Phillip Johnson's much lengthier account (which we'll explore in Chapter 7). But they may have held fellow-creationist and accredited paleontologist Wise as more authoritative.

Two other Young Earth creationists, Paul Nelson & John Mark Reynolds (1999b, 97) similarly invoked Wise's expertise, noting Wise "has a stronger knowledge of the contemporary geological record than many of his old earth critics." Pity that so little of that purported knowledge had filtered down to his written arguments, especially in *The Creation Hypothesis* (nonetheless "a splendid book" according to Nelson & Reynolds).

Nelson has performed an equivocal role as a prominent *Discovery Institute* apologist for Intelligent Design, by the way, where his YEC proclivities are consistently downplayed. Whether that has been by mere random accident, or by some conscious Intelligent Design, one may ponder.

But it does remind us of the consistent circumstance that the *Discovery Institute* apologetic format has never substantively criticized any of the main YEC claims of Creation Science (and thereby bump into the fact that both camps actually employ the same flawed picky-choosy evidential approach). I surveyed this curious state of affairs (up through the 2005 Dover *Kitzmiller* case) in Downard (2015b), but there have been no counterexamples in the years since Dover.

For instance, David Klinghoffer (2016d) extolled Marvin Olasky (2016) for his survey "of recent literature challenging Darwinism and offering alternatives, including a variety of books from authors arguing for intelligent design."

On that, Olasky included Stephen Meyer (2009; 2013), David Berlinski (2010) and Michael Flannery (2011), as well as the venerable Michael Denton (1985; 2016a), Phillip Johnson (1991; 1997) and Michael Behe (1996; 2007). But Olasky comes from the Jerry Falwell branch of *Kulturkampf* apologetics, which fully embraced Young Earth Creationism, and Olasky (1987; 2007; 2011, 2012a-b; 2015) has slipped all over the antievolutionism resource field, drawing little distinction between full blown creationism and Intelligent Design.

So it was that Olasky's 2016 entry listed a range of YEC books, from a 2011 reprint of the venerable (and still pseudoscientific twaddle) of John Whitcomb & Henry Morris (1961), to more recent apologists redressing the YEC dogmas in fresh print, such as Terry Mortenson (2004), Henry Morris III (2013a), Henry Morris III *et al.* (2013), Steven Boyd & Andrew Snelling (2014) and Tim Clarey (2015). Olasky balanced that with several Old Earth works from about the only place he could on that shrinking OEC turf, Hugh Ross (2008; 2014; 2015a-b), Fazale Rana (2011) and Rana & Ross (2014).

But it was methodologically indefensible to treat *any* YEC work as viable, so Olasky's tossing in a few OECers to balance them out disingenuously begged that fundamental scholarly methods question.

At least the OECers were trying to address the problem, though, challenging the loony geochronology and cosmology (even if they had to arbitrarily adopt the very Methodological Naturalism that antievolutionists decry to do it). The ID works Olasky invoked have never got that far, and included Klinghoffer, who equally disingenuously proclaimed that "Olasky does a great job in presenting a comprehensive picture of the books that are out there. He also includes creationist works in his omnibus review, none of which I've read so I leave it to others to judge their merits."

That wriggle was especially ironic given that Olasky (and Klinghoffer secondarily) took Bill Nye to task for visiting and criticizing Ken Ham's new *Ark Encounter* theme park in Kentucky, instead of focussing on the "great intellectual ferment" going on among critics of evolution—a ferment that Olasky obviously thought included the many YEC authors he included. Authors who were, after all, providing the "scientific" underpinnings for the very *Ark Encounter* whose scientific idiocy Nye was taking issue with.

So what's our take home on how the resurgent late 20th century Creation Science approached the RMT?

They either didn't notice it, or tried to flick it aside as quickly as possible so they could go on not noticing it.

What then of the new rival Intelligent Design movement that was coming on the scene in the 1980s? Would their "ferment" that Marvin Olasky was so impressed with in 2016 do any better?

And maybe even more importantly, what had been going on in the study of the classification of life during that period, and would they pay due attention to that either?

5. Michael Denton: a Methodology in Crisis

The Creation Science movement hit its high water mark of potential cultural and educational impact in the mid-1980s, when several states tried to put the creationist version of biology, geology and cosmology into their science curricula. The Supreme Court finally said nay.

There are plenty of works covering that history, from various angles (legal, philosophical, sociological, and scientific), including Edward Larson (1985), Raymond Eve & Francis Harrold (1991), Ronald Numbers (1992), Christopher Toumey (1994), Robert Pennock (1999) and Niles Eldredge (2000). I'll plug my own #TIP online summary, Downard (2015c).

Underlying that Creation Science worldview was a deep-seated *Kulturkampf* dissatisfaction with an increasingly secular and materialistic society, one that seemed to fall away from traditional faith. That angst did not abate after the splashy failure of Creation Science in the courts, though, and found a new outlet among certain academics in what would become the Intelligent Design movement.

One of the first ripples on this pond came in 1984, when Charles Thaxton, Walter Bradley, and Roger Olsen's *The Mystery of Life's Origin* put new steam into a traditional antievolutionary argument: the "Origins or Bust" approach to dismiss evolution generally by yanking out its underpinnings at the start. If life couldn't originate naturally, so the argument goes, then there can be no "evolution" to worry about.

That was a faulty premise, of course, since nothing about the evolution of life required that its start had to be natural, even if arising naturally would be more consistently an extension of the processes seen to govern what happened to life after it appeared some three and a half billion years ago.

By focusing on *origins* rather than what happened to life afterward, Thaxton *et al.* (1984) operated far away from such issues as the RMT.

Not so the next player: Australian biochemist Michael Denton's extremely influential 1985 book, *Evolution: A Theory in Crisis*.

Although not everybody being so influenced got Denton's discipline right. In about as broad a spectrum as you could find, one time Supreme Court nominee Robert Bork (1996, 294) in *Slouching Toward Gomorrah*, and homeschooling Oregon mother Lynn Barton (2006) on "Why intelligent design will change everything" for the *Kulturkampf* website *WorldNetDaily*, decided Denton was a "microbiologist." Over at the *Discovery Institute*, fellow biochemist Michael Behe (2006, 40), *Evolution News & Views* (2015) and "Director of Development Operations" Kelley Unger (2015) dubbed Michael Denton a "geneticist."

Those fields are not the same, and the muddling of them suggested how fluid discipline identifications could be in Antievolution Land. Microbiology deals with the activities of microorganisms, and genetics with the sequencing and activity of functional units of DNA, while biochemistry focuses on the physical properties of the molecules used by living systems. All involve specialized skill sets, but not necessarily an awareness of how the biological pieces interact in the whole organism.

And all are a *long* way from paleontology, the fossil record of ancient and extinct life, which you'd have to look at to spot the RMT.

In any event, while Henry Morris and Duane Gish were wallpapering over their Biblical convictions to appear innocuous enough for secular consumption, and *The Mystery of Life's Origin* was digging fresh "Origins or Bust" trenches, *biochemist* Denton tried to reopen what had become settled ground by then in science: the recognition that the big sweep of life showed natural branching common descent.

Much as 1980s Creation Science Flood theory harkened back to outmoded 17th century geology, Denton sought to revive the pre-evolutionary notion of "typology" reminscient of Georges Cuvier (1769-1832) and Richard Owen (1804-1892), where living forms were thought to be fixed by some external parameters that supposedly could not be accounted for by speciation or natural selection.

A problem: typology had never been a going concern.

Carl Zimmer (1998, 20-21) remarked on the sad tale of the vertebrate "Archetype" Owen came up with, "a lampreylike thing" that "was the blueprint that God referred to as He guided the history of life." Owen struggled in vain to figure out how the turnover seen in the fossil record actually occurred, but never went beyond proposing vague "secondary laws he simply called 'creative acts.'" And that was in the 19th century, when there were way less fossils and biological data to incorporate.

Stephen Jay Gould (2002, 312-329, 1070-1076) contrasted the "European formalism" of Owen's Platonic "archetype" with the British "functionalist" tradition, represented, ironically, by both William Paley (1743-1805) of the "watch on the heath" analogy of Design, and Charles Darwin, whose close observation of living systems would generate the evolutionary concepts to supplant Design as the productive working science paradigm.

The biggest problem for typology was a data glut. There was an explosion of identified species to account for by the time Owen and Darwin came along, from the few hundred known in the 17th century of naturalist John Ray (1627-1705) to *hundreds of thousands* a century later, and the number was continuing to climb rapidly all through the 19th century, as noted by John Van Wyhe (2008, 8, 36).

Harriet Ritvo (1997) chronicled the fascinating philosophical and social background of the struggles going on in 19th century natural science as the Owens of the field tried to make sense of the overload of new taxonomical data before Darwin's evolutionary theory sorted everything out.

What Michael Denton's *Evolution: A Theory in Crisis* sought was a taxonomical Do Over.

His chapter on "The Typological Perception of Nature" tried to stuff life back into a nonevolutionary hierarchical structure, where finches are *birds* and cats are *mammals*, but both are simultaneously *amniotic vertebrates*. Blurred categories weren't allowed (think Musicals and Comedies in my film example). For Denton, the end categories could never transform one into the other, because everything was a fixed type (presumably *designed*, otherwise he wouldn't be so popular in the design movement).

When it came to the details, though, *Evolution: A Theory in Crisis* conceded to no rival creationist a knack for eluding both the paleontological data and their geological context.

Map of Time, meet the *Bermuda Triangle Defense* (Player #3).

Take this laundry list from Denton (1985, 136): "But surely no purely random process of extinction would have eliminated so effectively all ancestral and transitional forms, all evidence of the trunk and branches of the supposed tree, and left all remaining groups: mammals, cats, flowering plants, birds, tortoises, vertebrates, molluscs, hymenoptera, fleas and so on, so isolated and related only in a strict sisterly sense."

Who said they had been "eliminated"? Denton certainly didn't lob a barrage of technical papers at his readers to show it.

Fossil ancestors for mammals and birds were hardly imaginary, even in 1985. We'll look specifically at Denton's jabs at the RMT in Chapters 7 & 11, but in this and the next chapter we'll be delving into some non-mammal topics for what they reveal about Denton's analytical method, and how strikingly similar it is to the Gish-style creationist shell game.

I'll be focusing on the hymenoptera (ants and wasps) and birds, but at the risk of sounding petty, I do have to wonder where and under what circumstances Denton proposed science recover an adequate fossil representation for *fleas*. They do have "a questionable Cretaceous record in Australia" according to Rich *et al.* (1996, 235), but are otherwise no more likely to turn up in the available strata than any other soft-bodied animal of such miniscule size.

Not that the Fossil Genie lost all interest, though, as some comparatively large early transitional fleas have turned up from the Middle Jurassic, Diying Huang *et al.* (2012). They still retained non-jumping hindlegs from their stonefly ancestors.

Interestingly, the flea example did not show up in the revised Denton (2016a), sparing him the effort of trying to explain why their somewhat improved fossil record didn't count on the Darwin side of things.

Typology kicks an anthill

Not that quite specific intermediates were totally unknown to science in 1985, such as the "wasp-ant" *Sphecomyrma*, linking two of the Hymenoptera which Denton had so pointedly included on his still limited list of supposedly distinct typological exemplars.

Much like Robert Broom and the probainognathids, Edward Wilson, Frank Carpenter and William Brown had *predicted* the features a wasp-ant common ancestor should have, just prior to the Fossil Genie once again stepping in with a physical confirmation, an expeditious coincidence noted by Douglas Futuyma (1982, 55). For those itching for the technical paper, there's Edward Wilson *et al.* (1967).

That Futuyma noted the *Sphecomyrma* case is of importance in the avoiding inconvenent data department, as this typological mashup failed to crawl into the content of creationists Duane Gish (1993a, 67-69, 108, 207; 1995, 26, 345) and Henry Morris & John Morris (1996, 18, 118), or Phillip Johnson (1991, 17. 25-26, 31, 70-71, 76) over in ID Land, even as they had specifically cited Futuyma's book on other matters.

More peripherally, Tom Bethell (2002, 55-56) blithely dismissed the significance of *Sphecomyrma* (which he didn't mention by name) as not being of much importance to the credibility of Darwinian processes. Bethell did not reveal to his readers what would represent something of importance for evolution, if not such specifically described fossils *predicted in advance*. Then again, Tom Bethell's grasp of "science" is not that tight, as we'll see in Chapter 8 regarding his comparable skepticism about Einstein and relativity theory.

While Michael Denton's second book, *Nature's Destiny*, had seemingly dropped the typological hot potato, Denton never gave up on his beloved typology, and pressed his case once again in Denton (2013; 2016a-f).

The title of his revised book, *Evolution: Still A Theory in Crisis*, was oddly reminscent of Gish's 1995 revamp (*Evolution: The Fossils* STILL *Say No!*). But as we saw with Duane Gish, Denton should have quit while he was behind.

Without ever directly mentioning *Sphecomyrma*, Denton repeatedly stumbled around the critter in his 2016 revision, a most revealing tiptoe regarding both the limited utility of Denton's attempt to resuscitate typology, and the apologetically tactical character of his research scholarship. Keep *Sphecomyrma* in mind, though, it won't be off the field for long.

Denton (2016a, 51-52) trotted past ants with this brief section dangling a particularly focused series of their diagnostic features (I've included Denton's endnote sources in **bold**):

> Even less-inclusive taxa are often defined by very curious novelties that are not led up to via innumerable intermediate forms. For example, even the relatively low-level taxon of the ants (Family Formicidae), one of the families of the Hymenoptera, possess the following defining characteristics: "Male and fertile female (queen) nearly always winged... workers wingless. 1st or 1st and 2nd abdominal segments of gaster scale-like or nodiform and well-separated from the part behind (the gaster). Male gaster without an upturned terminal spine; worker with spur of fore tibia not much curved and not externally pectinate. Antennae elbowed and less clearly so in male." **[Owain Richard & Richard Davies (1977, 1234).]** As well, all ants possess a unique novelty, a metapleural gland on the underside of their thorax. **[Christian Rabeling *et al.* (2008).]** These defining features of the ants were in place at least sixty-five million years ago when the major adaptive radiation of ants occurred at the end of the Cretaceous era and have remained unchanged since then to the present day. **["Chapter 2" of Bert Hölldobler & Edward Wilson (1990), no specific pages cited.]**

Could Duane Gish have done any better in bringing up technical issues that he didn't discuss further, or parsing his own cited sources as tactically as Denton did here?

Let's start with those "abdominal segments of gaster scale-like or nodiform and well-separated from the part behind (the gaster)."

Honestly, would many of his intended readers have the foggiest idea what any of that meant? Didn't it just scream to be *explained* so that they could understand it? Isn't that what informative science writing is all about, especially if your object is to present a cogent case for a typological design model to supplant the supposedly clunky evolutionary one?

But Denton did *not* explain it. So was he waving this jargon in front of his readers merely to impress them: "Male gaster without an upturned terminal spine; worker with spur of fore tibia not much curved and not externally pectinate." Hoping readers would buy into the typological frame because such snippets sounded oh so technical?

Let's take a closer look at some of the details.

A "nodiform" is just a knot or swelling, but even his own older source noted this phenomenon occurred on the "1st or 1st and 2nd abdominal segments," a clue that there was variation even here, rather than a fixed stereotypical ant model. As for the "gaster" behind them, that's the segment of the ant's tail containing the heart and stomach, along with chemical defenses and stingers in many species. But while that involves abdominal segment III on *most* ants, some make a "postpetiole" out of segment III, in which case the gaster begins with abdominal segment IV.

All of which is suggesting switches in regulatory cascades modifying the developmental anatomy in purely natural ways, just as has been seen in the evolution of vertebrate limb and vertebrae formation. But did that prompt "geneticist" Denton's typological imagination to explore what of that was known in the scientific literature available in 2016, such as Hélène Niculita (2006) on variations in usage of the *abdominal-A* gene in ants?

No, it didn't.

Just one indicator of how much biological activity is going on even with that ant gaster (and how creepy nature can be in the wild) is suggested by David Hughes *et al.* (2008), describing how a group of parasitical nematode worms have hijacked the biology of the giant South American "turtle ant" *Cephalotes atratus*. Laying its brood in the ant's gaster, the nematode initiates a cascade of change, not only transforming it from a matte black bulb into a bright red that resembles ripe fruit, but altering the ant's behavior so that it elevates its behind to make the "fruit" as visible as possible to the foraging birds that are the next stage in the nematode's parasitical life cycle. The nematode even weakens the abdominal attachment of the ant's gaster so that it is more easily pulled off the body when some bird swoops down for lunch.

Was any of this the product of intentional design? Or was this yet another iteration of a natural living world where organisms interact in complex ways, exploiting variations and mutations in a manner no less mechanistic even as they are amazing and/or disconcerting?

Just in case you think the *Cephalotes* example was some isolated fluke, the book by Carl Zimmer (2000b) remains an eye-opener on the varied and usually nasty world of parasites. Then again, animals (especially insects) have reproductive strategies of such weird variety that it rather puts our specialized branch of hominids to shame, as recounted with droll bluntness by Olivia Judson (2002). Ain't Nature a hoot!

But back to ants. So many things going on biologically there, data tackled by the evolutionary literature that somehow didn't make it under Denton's typological microscope. Consider the issue of *flight* that Denton thought suitable to peg as a typological ant diagnostic.

Ants are distinctive in having grounded worker castes that lack the wings their queen possesses. And there's quite an evolutionary dynamic going on there, as explored by Roberto Keller *et al.* (2014). The flightless worker ants are not just wingless versions of their flying genetic parent; they have very distinctive thorax connections, especially a heavily muscled T1 segment devoted to assisting their powerful and active jawed mouths. The Keller paper identified an ancestral configuration that was expanded in the worker caste but modified in two different ways among the still-winged queens, depending on how their colonies are spread.

In species where the queens go through a worker-like stage, actively foraging outside the nest (hunting for and capable of carrying large prey back to feed their first generation of worker larvae), they've retained the "intermediate T1" configuration. In other species, the nestbound ("caustral") queens have dispensed with that foraging phase because they'd evolved a way to draw on the flight muscles as a nutrient source to feed the first larvae. This had apparently developed at least twice independently in ants, and involved a tradeoff in the segment balance, for "only queens with a highly reduced T1 have an expanded T2 that constitutes most of the thoracic dorsum," noting further that "queens with intermediate T1 are also intermediate for T2."

Obviously a lot more work remains to be done on that topic, to pin down the precise genetics and biology of it, of course, such as that nutrient transformation of flight muscles. Anyone want to place some bets on who's going to do that work, though, evolutionists or typologists?

The Keller paper certainly wasn't drawing on typology for inspiration. They noted how a "morphological trade-off between adjacent body segments can occur due to competition for metabolic resources during adult development," processes which had been explored experimentally by Frederik Nijhout & Douglas Emlen (1998) regarding butterfly wings and beetle horns. Subsequent work on beetle horns and treehopper appendages identified more of those developmental mechanisms and tradeoffs, Armin Moczek & Debra Rose (2009), Moczek (2010; 2011) and Benjamin Prud'homme *et al.* (2011).

As for the implications of the queen ant case, the Keller team suggested that "the functional cost of enlarging T2 (reserves for colony founding) at the expense of T1 (reduced neck strength and work performance), occurred when founding behavior gradually shifted to claustral, with a decreased need to forage outside the nest."

Their observations about the distinctive proportions of TI in ant queens as a correlate of nest behavior had a predictive side, too. "For example, we lack data on queens of two early lineages, the extinct *Sphecomyrma* and the enigmatic *Martialis*, but based on our reconstructions we can predict that they will have an 'intermediate T1' and behave non-claustrally." That prediction has been supported by apparently foraging sphecomyrmid queens found in amber, Phillip Barden & David Grimaldi (2016, 516).

Interestingly enough, Michael Denton had encountered something about both of those taxa, given that Rabeling *et al.* (2008, 17916) had particularly noted the diagnostic importance of that big-eyed surface predator from the Cretaceous, *Sphecomyrma*, and that the subject of the paper was that peculiar new species of blind burrowing ant from the Amazon that they had found, *Martialis heureka*.

Martialis was similar to another new ant subfamily, the Leptanillinnae, that had been identified as the sister group to the other living ants. The mix of ancient and derived features in *Martialis* had posed quite a challenge for ant systematics, as Brendon Boudinot (2015, 43-44) noted.

But not apparently for Denton's typology, which achieved its simple clarity by not noticing any of them. Repeatedly.

Sounding not unlike Duane Gish or Kurt Wise from the previous chapter, Denton (2016a, 109) sought to dismiss the idea of real transitional forms by relegating them to the category of mere "mosaics, possessing a mixture of new and old traits," and cited the secondary Jerry Coyne (2009, 52) for this: "Coyne gives a good example in his *Why Evolution is True* of an ant, which combines older defining features (synapomorphies) of the Hymenoptera and newer taxa-defining novelties of the ants." That happened to be an illustration of *Sphecomyrma*, which Coyne had specifically described as "bearing almost exactly the combination of antlike and wasplike features that entomologists had predicted." A passage Denton conspicuously had not quoted.

Here we're at the nub of how revived Denton's typology was going to elbow aside evidence. If taxa acquire traits incrementally (and why shouldn't they?), they will be dismissed as a "mosaic." But if they seem to acquire them all in a block (probably because the earlier forms didn't get preserved, that *Bermuda Triangle Defense* again), then that will be deemed evidence of a typological discontinuity. Heads typology wins, tails "Darwinism" loses.

There's another trait the Denton type shares with the Gish/Wise type: neither ever described what a precursor would have needed to look like to violate their typological lines. If you're noticing a pattern here, it's because there is one.

Catching the sense of the metaplural gland

Let's continue to follow Denton's citation trail, though, and think about that Rabeling paper Denton had brought onstage, not for its substantive content, but only for a point that had been noted in it quite peripherally, that metapleural gland. What about that?

Once more, Denton wasn't moved to discuss the topic further, either to critique an evolutionary accounting of it, or to offer a meaningful typological characterization as a superior working alternative.

As it happens, that gland is another story which revealed as much about Denton's lack of curiosity as it did about the laborious process of scientific investigation.

The self-described "short publication" (a mere 24 pages) of Bert Hölldobler & Hiltrud Engel-Siegel (1984) summarized what was known about the gland at that time. First spotted by entomologists in 1860, the next century of study involved only selected ant genera, but by the 1970s the fossil ant *Sphecomyrma* and the example of the primitive living *Nothomyrmecia nacrops* (an Austrailian form known by only that one species) put the gland's origin deep in ant history.

It evidently wasn't showing up in male ants (which would have implications for its evolution based on sexual variation during development), and in several genera it has been reduced or was absent altogether. But Hölldobler & Engel-Siegel cautioned that their 1984 survey was "far from complete." That's because there are so many ant species, with new ones being uncovered all the time, and even by the more recent assessment by Sze Yek & Ulrich Mueller (2011) there were still many uncharacterized species and considerable uncertainty about exactly what the gland's varied secretions were doing in those ants.

It would seem a necessary stage for science to characterize what was going on first, in as full a detail as possible, before venturing to the explanatory phase of working out whatever evolutionary processes may have been involved.

Or, you can be a typologist, mention them as an object of mystery, and move on.

Given the limited nature of the existing Mesozoic fossil record (*Bermuda Triangle Defense* alarms flashing all the while), working out the history of the metapleural gland would necessarily entail interpolating the past based on the detailed study of living species, trying to connect their developmental biology and genetics. It's fair to say that, at the moment, science is still in the data collection stage on metapleural glands.

For example, William Hughes *et al.* (2010) explored the role of the metaplural gland in leafcutting ants, where great variation in the gland's size and activity related to warding off parasites. Building up comparable datasets across the thousands of ants species, and identifying exactly which genes are involved in the production of their features, are essential steps in that process.

But will antievolutionists pay much attention to that either?

Consider the cavalier way Denton tossed off a mention of that "major adaptive radiation" 65 milion years ago, as though everything was so typologically boring from then on. Without taking note of any of the 50 to 70 million years *preceding* that phase, of course, or investigating how fossil examples (like that *Sphecomyrma* he did not discuss) fitted into his typological boxes.

This was an especially telling omission given what one of Denton's few cited sources had to say, Hölldobler & Wilson (1990, 23). I've highlighted a few especially interesting bits in **bold**:

> *Sphecomyrma freyi* proved to be the nearly perfect link between some of the modern ants and the nonsocial aculeate wasps. In particular, the Cretaceous ants had the following primitive wasp-like traits: mandibles very short and with only two teeth, **gaster unconstricted**, sting extrusible, and middle and hind legs furnished with double tibial spurs. The *Sphecomyrma* were in fact a mosaic, for they also possessed distinctively ant-like character states: thorax reduced in size and wingless, petiole or "waist" pinched down posteriorly at its juncture with the rest of the abdomen (**but still primitive in form in comparison with later ants**), and--most important--an apparent **metapleural gland**, the possession of which is the key diagnostic trait of modern ants. The Sphecomyrma were **intermediate** between most modern aculeate wasps and almost all modern ants in the form of the antennae, which combined a proportionally short first segment with a long, flexible funiculus.

Hölldobler & Wilson had continued on that same page to lay out the *Map of Time* outlines of what happened biogeographically after *Sphecomymra*:

> The Mesozoic fossils unearthed to date seem to present us with the following picture. During middle and late Cretaceous times representatives of a few species belonging to the very primitive subfamily Sphecomyrminae ranged widely across the northern hemisphere in what was then the supercontinent Laurasia. They were evidently scarce in comparison with later ants in Tertiary and modern times. Only 2 individuals (Sphecomyrma canadensis) have been found so far among thousands of insects in amber from Alberta (J. F. McAlpine, personal communication).

Scientists were just beginning to fill in the holes in that limited ant dataset, summarized by Henry Cooper (2001), but in the years between Hölldobler & Wilson's book and Denton retooling his typology, a lot of additional specimens of *Sphecomyrma* and related early genera had turned up, revealing a lot more of their ancient diversity over those millions of years (including specialized ones whose mouth parts suggested feeding modes unlike living ants): David Grimaldi & Donat Agosti (2000), with commentary by Ted Schultz (2000), Michael Engel & David Grimaldi (2005), Vincent Perrichot and Andre Nel *et al.* (2008), and Phillip Barden & Grimaldi (2012; 2013; 2014). Barden & Grimaldi (2016, 518) represent a current phylogeny of these newly found early Late Cretaceous ants.

There's an obvious *Bermuda Triangle Defense* issue here, concerning how so many examples of what insects and ants were living all those millions of years ago depended on which of them had accidently got stuck in tree sap, and thus occasionally preserved in amber. That's the case for *most* ant fossils, and especially so for the critical Cretaceous period, charted by John LaPolla *et al.* (2013, 611).

But even with that constraint, just how "unchanged" were those post-Cretaceous ants leading up to the 8000-plus living species? Denton's cited source of Hölldobler & Wilson (1990, 9-19) had extensively surveyed the fossil specimens available at the time, and summarized that Big Picture in a cladogram on page 26. So everything I'm about to report was in principle known to Denton, too.

From an as yet unpreserved "ancestral vespoid," Hölldobler & Wilson's chart put sociality (living together in hives) and that metapleural gland as early traits that had appeared by the time of the two fossil Sphecomyrminae genera, after which fully "ant-like mandibles & antennae" emerged at some point (the lack of fossils meant scientists couldn't resolve the timing there).

The Formicinae would eventually add their "specialized poison gland" and "extreme reduction of sting." But just when and how that occurred couldn't be resolved by the fossil record, either, given that there were only fossil genera preserved from much later in the Eocene and Miocene.

Another branch involved a much larger range of ant families that developed a "pygidial gland," whose anatomy and functions have been teased out over the years by many workers, from Charles Kugler (1978) to Bert Hölldobler *et al.* (2013).

That's not an unusual gland for insects, as it crops up independently in many groups, including in the perennial creationist poster insect, the Bombardier Beetle. I examined Duane Gish's version of that tale in Downard (2004c, 203-205), where the creationist tried to use the noxious beetle sprays as analogous justification to bill the herbivorous hadrosaur dinosaur *Parasaurolophus* as a fire-breathing dragon. I'm not making this up.

But back to ants and the diversity Michael Denton wasn't noticing.

Running along that pygidial node, the Nothomyrmecinae added a special sound-making "ventral stridulatory organ," while the appearance of "Pavan's gland" (which secretes pheromone trails) marked two further lineages on another branch, the Aneuretinae (known by a single living species) and the Dolichoderinae.

The Dolichoderinae were the more varied, having developed a "lobate poison gland" along with dispensing with cocoons and reducing the stinger. Hölldobler & Wilson noted one fossil Dolichoderinae genus was known from the Miocene, but another earlier genus has turned up from the Eocene, Gennady Dlussky *et al.* (2014). Some fossil genera are now known for the Aneuretinae too, Michael Engel & David Grimaldi (2005).

The next node after the pygidial gland involved those ants that had added "tubulation of abdominal segment IV."

One branch there concerned another new trait, two ant families with "distinctive mandibles, postpetiole." The Myrmeciinae were known by one fossil genus from the Early Tertiary (though that would change, as we'll see below, regarding an intermediate ancestral form), while no fossils were noted for the Pseudomyrmecinae (featuring "extreme arboreal adaptations").

Another set branched off from the Ponerinae. Three families (Dorylinae, Ecitoninae and Leptanillinae) had developed the Army ant "legionary behavior" and a "dichthadiiform queen" featuring a highly enlarged gaster. None had a fossil record.

Over on the Poinerinae side, only 2 genera out of 50 had left fossil traces (again late in the game, in the Miocene and Early Eocene). And there were their nodal cousins, the *Myrmicinae* (with 2 genera and an ant nest preserved from the Miocene). More examples of those have turned up from the Eocene, Gennady Dlussky & Alexander Radchenko (2006; 2009). One may notice how close that subfamily spelling is to the *Myrmeciinae*, meaning how careful you'd need to be to not confuse the two labels. The *Myrmicinae* showed "enlargement of frontal lobes" along with "strong postpetiolar construction" and a convergent "loss of cocoons" (as had the Dolichoderinae).

Yes, all these ants looked generally *ant-like,* in the same way humans and marmosets and gorillas look *primate-like,* but the ease with which Denton dismissed the millions of years of variation among those ants, covered by his own cited sources, told much more about the lack of curiosity in Denton's typology than it did about the supposed limits of evolutionary classification.

As it happened, I found still more pertinent information on the biological diversity of the Myrmicinae (and the evolutionary reasons for them) during the course of tracking down Denton's primary source citations for this work. Denton (2016a, 256, 336n) included H. Frederik Nijhout (1999) among a catalog of papers on self-organization in biology, as though they somehow represented something other than more discoveries on the many natural ways biological systems can come to be the way they are.

It was certaintly an interesting piece, although not necessarily in quite the way Denton wanted. True, Nijhout noted "that embryonic development is almost entirely self-organizing," in that each cell at that stage responds only to signals it receives from its immediate neighbors, but "The situation is quite different during postembryonic development," because such cell-to-cell signaling "becomes increasingly inefficient" and longer-range hormonal signaling takes over.

The Nijhout paper was actually a commentary piece, referring to a primary source work in that same issue of *Proceedings of the National Academy of Sciences* (PNAS) that Denton hadn't cited: Jay Evans & Diana Wheeler (1999) on differential gene expression in caste formation in honeybees (where the difference between making a queen instead of a worker was what food their larva was fed during a narrow window of development). Rather significantly, Nijhout (1999, 5349) noted how their work was "the first to specifically identify the nature of the proteins that are differentially expressed at the time of the developmental switch; in fact, it is the first specific information we have on the molecular events that accompany a developmental switch between alternative pathways in insects in general."

Think about that. If the idea was to make sense of the biological processes operating in insects, up until then highly critical pieces of the connective puzzle hadn't been identified. Evans & Wheeler's work was seminal, and set into motion a cascade of connected research that followed up on their technical findings, including quite a few papers working out the biological mechanisms underlying caste variations in ... drum roll ... the Myrmicinae ants.

Glennis Julian *et al.* (2002) identified genetic shuffling from hybridization contributing to caste differentiations among the harvester ant branch, *Pogonomyrmex barbatus*. William Hughes *et al.* (2003) and Sanne Nygaard *et al.* (2011) found more direct genetic roots for the adaptive move to fungus farming in *Acromyrmyx* leaf-cutting ants 50 million years ago (based on the degree of difference in the genes), and a subsequent transition to multiple queen mating 40 million years later. And Lino Ometto *et al.* (2011) uncovered a very specific mix of selection factors governing the evolution of caste genes in the *Solenopsis* fire ants.

Seen at this detailed level, the mechanisms all looked pretty natural and unguided, and thus, dare we say, "Darwinian."

Another thing. All of this technical work had only been one step away from Denton, as they were among the many papers listed as citing Evans & Wheeler at the end of the *PNAS* abstract. That's how I found out about them. So close at hand, and yet not close enough for Denton.

Another take away from the few works Denton had bothered to cite on ants was how limited the fossil record was for certain families, how that understandably complicated the task of making sense of them, and how you'd only know about that circumstance if you stopped to think about it.

On the bottom of their cladogram page, for example, Hölldobler & Wilson (1990, 26) had shown the segmentation layouts of three ants still retaining primitive characteristics: *Nothomyrmecia nacrops*, the *Myrmecia* and *Amblyopone* ("a worldwide genus of primitive ponerines thought to be derived from the sphecomyrmine-nothomyrmecine clade"). Among these three, the fourth segment had developed a notable derived crimp in the *Amblyopone*, but there was also the matter of how the petiole (segment II) attached to the remaining gaster segments III-VII. This was quite broad in *Amblyopone*, unlike the narrower attachment in the ancient sphecomyrmines.

Seeking to actually account for such variations, Hölldobler & Wilson (1990, 27) pressed forward with predictive diagnostics:

> Either the broad petiolar attachment of *Amblyopone* and related genera in the tribe of Amblyoponini represents an evolutionary reversion from the primary narrow petiole of the Ur-Formicidae, or the Amblyoponini (and possibly the remainder of the Ponerinae, Myrmicinae, and related higher subfamilies) are an independent clade of ants predating the Nothomyrmeciinae and Sphecomyrminae. We favor the first, more conservative hypothesis. The matter will, however, remain open until more

evidence is obtained. For example, the discovery of *Amblyopone*-like forms with fully-constricted petioles from post-Cretaceous deposits would favor the first hypothesis. If, on the other hand, *Amblyopone*-like forms are discovered in deposits at least as old as those containing *Sphecomyrma*, the second, more radical hypothesis will receive strong support.

So far the Fossil Genie hasn't delivered on that deep root, though close cousins of that peculiar *Nothomyrmecia* have turned up: *Prionomyrmex* trapped in amber all the way across the globe in the Baltic, Philip Ward & Seán Brady (2003), and *Archimyrmex* from the Americas, Gennady Dlussky & Ksenia Perfilieva (2003).

Such a spatial spread reminds us of a *Map of Time* truth: all the ant fossils *missing* from the biogeographical scope, representing the real estate in between. Limited distribution of amber-forming tree sap would be one obvious factor, while another would be the difficulty of stray ants leaving body imprint fossils in the rest of their environment.

More recent finds of Eocene *Archimyrmex* species from Germany showed that their diagnostic abdominal proportions, mandible lengths and wing venation patterns were *intermediate* between those of their modern Myrmeciinae counterparts and the related Pomeromorpha subfamily, Gennady Dlussky (2012, 189). Wing venetion can be quite revealing, by the way, as Sandra Schachat & George Gibbs (2016) worked out regarding the ancestral layouts in fossil moths.

In other words, the natural variation that evolution builds on, one microevolutionary ratchet at a time, still appeared to be going on in the examples that have turned up.

All of which data would need accounting for by any revived typology, wouldn't it?

Which brings to mind something else for typology to tackle, something once again buzzing under Denton's vertebrate eye: the *wasp* side of the Hymenoptera.

Specializing as active stinging predators, cladistic analyses put wasps as diverging from the presumed earlier Mesozoic antlike line, but the basal Chrysidoidea wasps were already well-established and diverse by the Late Cretaceous, showing more wing venation patterns than their modern counterparts, Denis Brothers (1999; 2011). Brian Johnson *et al.* (2013) offers a recent summary of the wasp-ant phylogeny.

Whether the Fossil Genie will smile on paleontologists with some earlier amber finds to clarify their prior evolutionary history, only time will tell. But there are already some revealing fossils to hint at the deeper root of things.

Not only do we have the Sphecomyrma "wasp ant" to play with (that are deemed "ants" chiefly because they possess that metapleural gland), but there are several groups of very *antlike wasps* (the "false ants" Falsiformicidae, and the Armaniidae) that have been hard to classify because they lacked that gland (or its presence was unclear due to the preservation state of the fossils). For example, one Cretaceous armaniid (known only by an imprint fossil, and not a body neatly glopped in amber) was pegged as an early ant by Gennady Dlussky (1999) due to that preservational ambiquity, while its group fell among the most basal "stem" ants in the analysis by Philip Ward (2007).

Thus the lines in the real world between these early wasps and ants are so blurry that LaPolla *et al.* (2013, 618) cautioned that "the only definitive synapomorphies that can be used to define true ants seem to be the presence of the metapleural gland, a distinct petiole weakly attached posteriorly, and the differentiation of females into queen and worker castes."

Would all this information impact Denton's typology?

Probably not, as his typology couldn't get beyond its rigid *ants are ants* view, "unchanged" only because the Denton *kind* would not assess any of the changes.

In this regard, I was struck by something Ross Crozier had noted in his commentary on a new ant phylogeny by Seán Brady *et al.* (2006), one that did pay attention to all that dataset in a way the typologist Denton didn't. Crozier (2006, 18029) pointed out how "Early thought on ant phylogeny was bedeviled by the belief that all or most of the genera with armored cuticles and strong stings belonged in a single subfamily," not realizing how many of those traits could arise independently.

Denton's typology was making much the same mistake, only on steroids, focusing on specialized details (those gasters or the metapleural gland) but not relating them to the full range of the existing forms that objectively varied far more widely than Denton was prone to notice.

Was this to be the hallmark of the New Typology, not including most of the data?

I have one last stop on the genetics and developmental biology that had been uncovered regarding the physical anatomy of ants, wasps and their kin.

By the time of Michael Whiting (2004, 352), sawflies and "wood wasps" were identified as basal forms reflecting the ancestral configuration preceding the bees, wasps and ants, and work was already underway exploring the connections there, such as Daisuke Yamamoto *et al.* (2004) identifying the *Ar dpp* variant in sawflies of a gene well known in insects as a big player in appendage and segmentation formation (and so presumably relevant to any closely related taxa). Higher up in the developmental process, more of the connective details were studied by Susanne Schulmeister (2003), tracing what incremental modifications were needed to transform the basal muscle connections in the genitalia and end segments of sawflies (down there in that "gaster" zone) into the various layouts seen in the later hymenoptera.

A *Map of Time* reminder: insects are so diverse, and have been around for such a long time, that even the branch of just one of the current North American sawfly genera apparently traced back over 20 million years, Catherine Linnen & Brian Farrell (2008). Fossils found recently from way farther back in the Permian suggested to Dmitry Shcherbakov (2013) that the Raphidioptera ("snake flies") were near the root of the Hymenoptera, reflecting a repeated macroevolutionary "miniaturization bottleneck" in tiny insects that constrained the later form of their descendants. A Cretaceous sawfly has also been found exhibiting the enigmatic *nygmata* structures found in assorted insect wings, Mei Wang *et al.* (2014). What these cellular clusters do (if anything) is still not known, though the suspicion is they may perform some glandular role.

Now I am not a professional entomologist. So if I (as a curious if annoying layman) could find such a range of scientific research after only a few days of Internet mouse-clicking from the seclusion of my writing den, exactly what was preventing Michael Denton from stumbling onto them for his own newer book?

Unless, of course, it never occurred to him to look.

Failing to see the forest or the Tree (or the birds nesting in them)

The ant story from Denton's 2016 book showed a continuation of the procedure he had followed in his earlier work.

Michael Denton (1985, 195) was able to bypass so much of the fossil record not only by not paying much attention to it, but because he argued that the acceptance of transitional forms was in the end just a matter of ideological predilection, not a conclusion compelled by objective characteristics. "As evidence for the existence of natural links between the great divisions of nature, they are only convincing to someone already convinced of the reality of organic evolution."

For Denton (1985, 132), "the order of nature betrays no hint of natural evolutionary sequential arrangements, revealing species to be related as sisters or cousins but *never* as ancestors and descendants as is required by evolution. The form of the tree makes explicit the pre-evolutionary view that it is discontinuity and the absence of sequence which is the most characteristic feature of the order of nature."

Now the "tree" Denton referred to was a very simple radial chart on the next page, where a "vertebrate archetype" branched into the "anamniotic" and "amniotic" archetypes. The anamniotic one in turn diverged into the "amphibian" and "fish" archetypes, while the amniotic side split into the "mammalian" and "avian" archetypes. These categories yielded terminal examples arrayed around a circle: "frog," "toad," and "newt" for the amphibians; "salmon," "lungfish," and "hagfish" for the fish; "dog," "man," and "whale" for the mammals, and "penguin," "duck," and "eagle" for the birds.

Denton gave no indication that any of these "archetypes" might once have been physically represented in the past by identifiable organisms, such as some prehistoric "amniotic" creature that was neither mammal nor avian, but ancestral to both. The "archetypes" functioned as conceptual templates only, the essential blueprints with which an Intelligent Designer might have contrived a particular typological example (and by means Denton studiously failed to describe).

All very prosaic and familiar. But where exactly do the known members of the *reptile* class fall in this tidy scheme, and how are the fossil diapsids and synapsids to be arranged on it? Just as with Kurt Wise's whittling down his examples last chapter, their glaring absence from Denton's 1985 chart served to guide the eye away from the obvious question: what ever was to be done with extinct intermediates like *Archaeopteryx* or the therapsids, which stand both chronologically and morphologically as links attaching two of Denton's immutable types with the antecedent reptiles.

Isn't that just the sort of paleontological sequence that he avers "*never*" happened?

The spread of the typological problem for mammals was available to the Morrises or Gish too via their common cited source of Futuyma (1982, 77):

> Most of the modern orders of mammals are represented by less specialized species as we go back in time, until when we reach the Paleocene, they become so unspecialized that it's harder and harder to distinguish one from another. The condylarths, for example, appear to be ancestral to various groups of hoofed animals; but the condylarths are similar to the creodonts, which appear to be primitive carnivores; and many of the creodonts could equally well be classified as insectivores.

None of which caught the attention of Gish or the Morrises. Or Denton's typology, as Denton (1985, 105) treated the living world in just as conventional a way as those Creation Scientists: "No one, for example, has any difficulty in recognizing a bird, whether it is an eagle, an ostrich or a penguin; or a cat, whether it is a domestic cat, a lynx or a tiger. Moreover, no one can name a bird or a cat which is in any sense not fully characteristic of its class. No bird is any less a bird than any other bird, nor is any cat any less a cat or any closer to a non-cat species than any other cat."

Exactly like a Creation Scientist insisting on immutable kinds, Denton left out any discussion of fossil carnivores in order to keep his "cat" classification from getting fuzzy. Whether his choice of examples also meant that "birds" and "cats" were themselves created "types," Denton did not volunteer.

But it was interesting that he mixed two such disparate taxonomical categories in the same breath. Birds (*Aves*) comprise a whole class involving some 9000 species, while cats (*Felidae*) represent only a family, suggesting how arbitrarily accordian-like Denton's choice of exemplars were.

Is a "cat" any less a quadrupedal carnivore than a dog, or less a mammal than an elephant? Why then aren't mammals a "type" in which the variation from cat to Michael Denton was merely a matter of microevolutionary degree, much as aerial hunters like eagles and flightless aquatic penguins can safely reside in the "bird" archetype? As we'll see down in Chapter 11, that sort of trajectory is exactly what Denton allows to be the case too.

Denton (1985, 175-176, 211-213) admitted that *Archaeopteryx* "hints of a reptilian ancestry" without explaining why a bird with reptilian teeth, tail and clawed wings didn't completely overflow the "avian archetype."

Instead, Denton quickly focused on the feathers and flight anatomy that permitted him to safely tag *Archaeopteryx* as "bird." But he didn't wonder why *Archaeopteryx*'s lacking the complete flow-through skeletal air sac lung system Denton deemed absolutely characteristic of birds didn't blur the "avian archetype." Nor did he relate any of the Mesozoic birds known at the time to the typological *Aves*, to see how they stacked up against modern ones regarding their lungs or other anatomy. That reluctance might be compared with the review of Mesozoic craniofacial air sac systems by Lawrence Witmer (1990)

The typological quicksand only got deeper in the 1990s as still more not-quite-yet birds were discovered along with a flock of feathered dinosaurs, which I took note of in Downard (2003b, 83-84). And the field hasn't got smaller in the years since.

On the feathered theropod side, there were new taxa like *Anchiornis*, *Caudipteryx*, *Protarchaeopteryx* and *Sinosauropteryx*, documented by Pei-ji Chen *et al.* (1998), Qiang Ji *et al.* (1998), Qiang Ji *et al.* (2001), Mark Norell *et al.* (2002), Martin Kundrat (2004), Gareth Dyke & Norell (2005), Norell and Xing Xu (2005), Dongyu Hu *et al.* (2009) and Xing Xu & Qi Zhao *et al.* (2009). And even without feathers, newer theropod fossils suggested *Archaeopteryx* was far from alone anatomically, representing just one variant on a quite diverse theme of small dinosaurs, usually with mobile enlongated front limbs, Xing Xu & Hailu You *et al.* (2011), but occasionally short limb models too, uncovered by Xiaoting Zheng *et al.* (2010) and Junchang Lü & Stephen Brusatte (2015).

Meanwhile, primitive early birds like *Apsaravis*, *Confuciusornis*, *Eoalulavis*, and *Liaoningornis* were picking up in the Cretaceous where *Archaeopteryx* left off in the Jurassic, gradually losing their ancestral features (their reptilian teeth, a bony balancing tail, and the distinctive gastralia bones that lined the belly walls of their dinosaurian predecessors), Lianhai Hou *et al.* (1995; 1996; 1999), José Sanz *et al.* (1996), Mark Norell & Julia Clarke (2001) and Zihui Zhang *et al.* (2009).

And something else to take note of: the major bird form during the Cretaceous were the now extinct *enantiornithines*. With some 80 species found so far, they differed from later birds in their shoulder articulation, having a convex knob articulating with a concave shoulder blade, the reverse of ornithurines where it's the shoulder blade that has the convex bulge, Cyril Walker (1981). Working out how all that happened would depend on some Fossil Genie luck, of course, along with a lot of working backwards on the genetic end, though Ruijin Huang *et al.* (2000) and Sanne Kuijper *et al.* (2005) have begun some of that for the scapula of birds and mammals. More enantiornithines have turned up since, Matthew Lamanna *et al.* (2006), Dongyu Hu *et al.* (2011), Jingmai O'Connor *et al.* (2011; 2013) and Xia Wang *et al.* (2011), with the enantornithine *Xiangornis* and the basal ornithuromorph birds *Zhongjianornis* and *Archaeorhynchus* blurring the lines between the two groups, Hu *et al.* (2012), Zhonghe Zhou *et al.* (2010) and Min Wang & Zhonghe Zhou (2016). The enantionithines also appear to have had a less developed respiratory system than modern birds, Zihui Zhang *et al.* (2014).

And there's the *four-winged* fish-eating dino glider *Microraptor* (feathered on its hind legs as well as arms) to further hint at feathered theropod diversity during the Cretaceous, Xing Xu *et al.* (2003), Gareth Dyke *et al.* (2013) and Lida Xing & Scott Persons *et al.* (2013).

An aside is warranted on *Microraptor*, a specimen of which figured in the "Archaeoraptor" hoax that caught *National Geographic* with its pants down in 1999. The discovery of world-class fossil sites in China, and the enthusiasm some of their municipalities showed for building museums, encouraged some crafty entrepreneurs to give nature a nudge now and then, and concoct particularly attractive slabs for sale.

The Archaeoraptor case was one where two perfectly legitimate fossils (a bird and what later was found to be a *Microraptor*) were spliced together. Stephen Czerkas (1951-2015) and *National Geographic* failed to look that gift dinobird in the beak, doing a front cover spread on it, Christopher Sloan (1999). But Storrs Olson (part of the anti-dino-bird cadre at the Smithsonian) was primed to spot such hyped fakery, which led to some rapid investigation and hangdog retraction, Rex Dalton (2000a-c) and Lewis Simons (2000).

Antievolutionists were happy to pounce on this embarrassing flap, of course, from Jonathan Wells (2000a, 123-126), who dubbed it the "Piltdown Bird," to creationists Brad Harrub & Bert Thompson (2001). As for Stephen Czerkas, he seemed oddly accident prone when it came to birds. Czerkas & Czerkas (1991, 83) had been one of the few who had accepted the problematic Triassic "Protoavis" fossil that several creationists (including Duane Gish and the parasitical Hank Hanegraaff) hoped would undercut the importance of *Archaeopteryx*, a tale I recounted in Downard (2003b, 85-88). And Czerkas would come to work with Alan Feduccia to try and stuff some of those feathered dinosaurs into a flightless bird coop.

But all that was green eyeshade stuff, far removed from the active business end of evolutionary paleontology and biology, where more of the fossil archosaur, dinosaur and avian dots have been connected up, from physiology (including the beginnings of those avian-style "intrathoracic air sacs") and growth rates to brooding behavior, David Varricchio *et al.* (1997; 2008), Patrick O'Connor & Leon Claessens (2005), Alan Turner & Diego Pol *et al.* (2007), Jonathan Codd *et al.* (2008), Paul Sereno *et al.* (2008), Gregory Erickson *et al.* (2009), Emma Schachner *et al.* (2009; 2013; 2014), Richard Butler *et al.* (2012), Stephen Brusatte *et al.* (2014) and Colleen Farmer (2015).

Of course there are outstanding paleontological issues (there are *always* going to be outstanding paleontological isssues), such as when birds developed the powerful sternum modern ones use as an anchorage for the muscles needed for a full powered take off. Xiaoting Zheng *et al.* (2014), Christian Foth (2014) and Jingmai O'Connor *et al.* (2014) have thrashed over the difficulty in identifying the relevant connecting cartilage in fossils. That too is work not involving antievolutionists.

But other features have started to give way to actual *experimental* investigation, as the biological toolkit has expanded to the point where paleogenomic retroengineering can directly reconstruct some of the stages involved in the transformation of the avian skull form, wrist and leg bones from their dinosaur ancestors, represented by Bhart-Anjan Bhullar *et al.* (2012; 2015) and João-Francisco Botelho *et al.* (2014; 2016). This may be compared with creationist Elizabeth Mitchell (2016a) summarily dismisssing the significance of the Botelho work; we'll get back to the Bhullar side of things down in Chapter 11.

And something else from another direction: Alan Brush (1996; 2000; 2006) and Richard Prum (1999; 2005) had been independently investigating feathers from the developmental biology end. A lot of knowledge pooling ensued, including Prum & Scott Williamson (2001), Prum & Brush (2002), and Prum & Jan Dyck (2003) integrating the new fossil finds with what had been uncovered on the developmental side. Prum was also involved in the work of Matthew Harris *et al.* (2002; 2005) on the genetics of feather development, such as the role of *sonic hedgehog* and *BMP* noted back in Chapter 2.

Much like Robert Broom pondering the evolution of the mammalian jaw, Brush & Prum reasoned that any evolutionary origin for feathers would have to be consistent with their developmental biology. Instead of worrying about which form of scale feathers might have evolved from, or dwelling on what adaptive function the intermediate stages might have been serving, they worked out a series of transitional stages that were consistent with the way feathers actually formed. By that approach it turned out that the simplest line of evolution didn't start out from any familiar scale at all, but with the *placode* from which scales and feathers differentially developed.

Most interestingly, the earliest proposed protofeathers were thought to be *unknown in any bird*, fossil or living.

Or at least they *were* unknown, until the Fossil Genie puffed in with *Sinornithosaurus* described by Xing Xu *et al.* (2001). Prum took part in that work too.

Now that paleontologists knew more what to look for, the Fossil Genie had fun, as some wonderfully preserved deposits were uncovered. Protofeathers of the form Prum & Brush identified on developmental grounds turned up in other dinosaur groups: dromaeosaurs near the bird root, and in their therizinosaur cousins, and on the tyrannosaurs still farther removed, and in *Ornithomimus* farther out on the theropod line, and even way over on the ornithischian side (the two main dinosaur groups, saurischian and ornithischian, are distinguished by their pelvic shape). It's quite a string of papers: Xing Xu & Zhi-lu Tang *et al.* (1999), Xu & Xiao-lin Wang *et al.* (1999), Xu *et al.* (2004; 2012), Xu & Gou Yu (2009), Xu & Xiaoting Zheng *et al.* (2009; 2010), Oliver Rauhut & Christian Foth *et al.* (2012), Darla Zelenitsky *et al.* (2012), Pascal Godefroit & Andrea Cau *et al.* (2013), Godefroit & Helena Demuynck *et al.* (2013), and Godefroit *et al.* (2014a-b).

All of which suggested the bird model of feather, that highly specialized flight-adapted form characterized by asymmetrical plumes, was just one of the many ways the underlying developmental processes were branching among dinosaurs.

The more traditional morphologically based ornithologists were (to say the least) slow to embrace the dinosaurian origin for birds, partly because, many of them had not been properly acquainted with paleontological evidence and methods, as noted by Prum (2002).

Alan Feduccia, the Fred Hoyle of bird evolution

Fred Hoyle (1915-2001) was a prominent physicist who to his dying day held fast in rejecting Big Bang cosmology (Hoyle initially coined the term as an insult). And Young Earth Creationists continue to invoke Hoyle as an authority to dismiss the Big Bang.

Over in bird Antievolution Land, the poster child counterpart is Alan Feduccia, who favored a much earlier origin for birds in some as yet unspecified lineage of archosaurs, even trying (without much success) to dragoon to their cause *Longisquama*, an odd Triassic archosaur with tall vanelike spikes along its back, contending they were "nonavian feathers," Terry Jones & John Ruben *et al.* (2000). Prum & Brush (2002, 282-283) summarized the case against that argument.

Feduccia (1999a-b; 2001; 2002; 2003; 2012; 2013), Theagarten Lingham-Soliar (1999; 2003; 2010; 2011; 2013; 2014), Terry Jones & Larry Martin *et al.* (2000), Feduccia *et al.* (2005), Lingham-Soliar *et al.* (2007) and Stephen Czerkas & Feduccia (2014) have held the line on objecting to the dinosaur paternity suit. While observers like Brian Switek (2007) were less than impressed with their arguments, because of his prominent opposition to the dinosaur end of things, Feduccia became the inevitable authority to be invoked by antievolutionists, such as creationist Brian Thomas (2014) waving Stephen Czerkas & Alan Feducccia (2014) over at *ICR*.

Though my favorite is Jonathan Wells (2000a, xiii) thanking strange bedfellows Alan Feduccia and creationist Ashby Camp (!) for their assistance on the technical details of *Archaeopteryx*! This was not unprecedented in the world of antievolutionary apologetics, where (like Kurt Wise showing up in the Moreland anthology back in Chapter 4) one has to backscratch allies as you find them.

It's also noteworthy that Kevin Padian & John Horner (2002) commented how Feduccia was wedded to a "typological" way of thinking, which put the ornithologist right up Michael Denton's alley.

The corners a typological approach could paint itself into was illustrated by how Feduccia handled the dromaeosaur dinosaurs. These agile predators included such exemplars as *Deinonychus*, which famously appeared in Steven Spielberg's *Jurassic Park* under the name of *Velociraptor*—a systematic quirk that turned on Michael Crighton having relied on Gregory Paul (1988), who advocated some nomenclatural renaming that hadn't actually taken on in dinosaur systematics.

Alan Feduccia (1999b, 132, 394-398) maintained "no feathered dinosaur has ever been found" and that those that had been found were really flightless birds ("Mesozoic kiwis"). For that reason, Feduccia (1999b: 382, 396) specifically insisted dromaeosaurs (lacking feathers in the fossils) were dinosaurs having only a coincidental resemblance to birds. That is, until the Fossil Genie got busy and so many feathered examples started turning up in China. Whereupon Feduccia (2002, 1196) summarily *reclassified* dromaeosaurs as flightless birds having only a coincidental resemblance to dinosaurs. Apart from the feathers, nothing about their overall anatomy had changed in the meantime, and the irony of this taxonomical flip-flop was not lost on Richard Prum (2003, 553-554).

With evidence for feathers turning up even in *Velociraptor*, cladistic studies put the dromaeosaurs deep within the paravian branch of dinosaurs, Alan Turner & Peter Makovicky *et al.* (2007; 2012). And Darren Naish (2016a-c) has done a useful survey of the manoraptoran theropods and how the glut of new data is being sorted out by the paleontologists regarding bird and feather origins.

Another of the temporary controversies Feduccia figured in was the bird digit issue, which gave a measure not only of the technical science but which side in the debate was getting the better of the experimental evidence. Birds have retained fingers 2-3-4 on their hand, while the reduced digits of theropod dinosaurs involved 1-2-3, Alan Feduccia (1999a) and Martin Kundrát *et al.* (2002). Such evidence notably contributed to the dinosaur-bird doubting, reflected in Richard Hinchcliffe (1990; 1997; 2008).

That digit arrangement didn't actually rule out a dinosaur origin for birds, though, as some theropod could have dropped the first and later fifth digits, but Günter Wagner & Jacques Gauthier (1999) suggested another explanation, that a rare (though not unprecedented) "frame shift" had occurred in bird development, where the digit formation had jumped one *anlagen* bud down. Subsequent research confirmed that idea, including direct experimentation tweaking the relevant genes, Frietson Galis (2001), Galis *et al.* (2003; 2010). Monique Welton *et al.* (2005), Alexander Vargas & John Fallon (2005), Wagner (2005), Monique Welten *et al.* (2005), Vargas & Wagner (2009), Vargas *et al.* (2009), Xing Xu & James Clark *et al.* (2009; 2011), Rebecca Young et al. (2009; 2011), Koji Tamura *et al.* (2011), Matthew Towers *et al.* (2011), Zhe Wang *et al.* (2011), Young & Wagner (2011), Xing Xu & Susan Mackem (2013).

So, a lot of scientific work had flowed under the bridge since 1985. Data that would have to be accounted for by any revived typology seeking to out-explain evolution. What did Denton do with all this in 2016?

Denton (2016a, 24) first bounded into Richard Prum & Alan Brush (2002, 289) for an extended quote, which I'll reproduce as he had it (Denton's ellipsis and *italics*):

> Recently, Wagner and colleagues... proposed that research on the origin of evolutionary novelties should be distinct from research on standard microevolutionary change, and should be restructured to ask fundamentally different questions that focus directly on the mechanisms of the origin of

> qualitative innovations. This view underscores why the traditional neo-Darwinian approaches to the origin of feathers, as exemplified by Bock (1965) and Feduccia (1985, 1993, 1999) have failed. *By emphasizing the reconstruction of a series of functionally and microevolutionarily plausible intermediate transitional states, neo-Darwinian approaches to the origin of feathers have failed to appropriately recognize the novel features of feather development and morphology and have thus failed to adequately explain their origins.* This failure reveals an inherent weakness of neo-Darwinian attempts to synthesize micro and macroevolution. In contrast, the developmental theory of the origin of feathers focuses directly on the explanation of the actual developmental novelties involved in the origin and diversification of feathers (Prum 1999). Restructuring the inquiry to focus directly on the explanation of the origin of the evolutionary novelties of feathers yields a conceptually more appropriate and predictive approach.

Clearly it was Prum & Brush's jab at "neo-Darwinism" that prompted Denton to quote them. But while they offered two examples of this flawed "neo-Darwinian" approach (Bock and Feduccia), Denton showed no curiosity about either of them.

Feduccia, remember, was exemplifying a *typological* way of approaching the problem, the very mode Denton was trying to push to the front of the queue. As for Walter Bock (1965), although it was an influential paper, it was also very dated, and Kevin Padian (2001, 598-599) noted how Bock's preference for the ornithologist-favored "arboreal" versus the paleontologist-supported ground runner "cursorial" theory of flight origin not only lacked animal examples, but presented a distracting false dichotomy.

Prum & Brush (2002, 265) similarly noted that "the fundamental problem with these combined scenarios is that they conflate the analysis of these complex issues and eliminate many plausible combinations."

Denton (2016a, 172, 315n) obviously could have known of these caveats, since he quoted from that very page of Prum & Brush in his main feather section, but was too busy perhaps eliminating "many plausible combinations" of his own as he once again authority quoted the authors only to highlight the aspects of feathers he wanted up front for his typological approach (again Denton's *italics*):

> Over the last half of the 20th century, neo-Darwinian approaches to the origin of feathers, exemplified by Bock (1965), have hypothesized a microevolutionary and functional continuum between feathers and a hypothesized antecedent structure (usually an elongate scale). Feathers, however, are heirarchically complex assemblages of numerous evolutionary novelties—the feather follicle, tubular feather germ, feather branched structure, interacting differentiated barbules—*that have no homolog in any antecedent structures*.

Now Prum and Brush (2002, 270-280) had meticulously laid out the developmental approach that uncovered the origination of those novelties and related that to the incoming fossil evidence. Unless he shut his eyes selectively while reading, Denton must have known about that congruent data, but we'll see how he dodged around all that shortly.

Kevin Padian (2001, 599) noted something relevant that bypassed the arboreal versus cursorial cliches: "The central problem of the origin of bird flight is the evolution of the flight stroke, which generates thrust." Cladistic analyses of the newer fossil data had homed in on the maniraptoran theropods as a highly relevant group, whose grasping arms had a range of movement that naturally supplied that motion. Add feathers as an airfoil, and flight became an obvious (if not unavoidable) adaptive option. And a fairly popular one, given later fossil finds like *Microraptor*.

It was especially revealing that Walter Bock (2000) was still not paying attention to the developmental work (no Brush or Prum) nor discussing any of the newer paleontological finds. "Neo-Darwinian" he may have been, but Bock's method of data avoidance was all too similar to what Denton was similarly failing to do in his neo-Typology.

Michael Denton's feather typology, *Still a Theory in Crisis*

Michael Denton split his revised (I cannot say *improved*) argument on feathers into two sections of the book. The first consisted of a single paragraph in Denton (2016a, 46), which stands as one of the most acrobatic evasions I've ever come across. I've noted the positioning of the endnotes Denton (2016a, 293n) had in **bold**.

> All modern birds, and some related groups of reptiles **[10]**, possess closed pennaceous contour feathers consisting of a central shaft or rachis. (See Figure 3-2.) Fused to the rachis are barbs, and attached to each barb are hooked distal barbules pointing towards the tip of the feather and interlocking grooved proximal barbs pointing to the base of the feather. (See Figure 3-3.) All organisms possessed of this defining features can be unambiguously assigned to a unique clade belonging to the more inclusive dinosaur clade Theropoda **[11]**.

The figures were straightforward illustrations of feather structure. But the issue by 2016 was not just how modern bird feathers *looked*, but how their internal genetics and developmental biology offered clues to their evolution. A topic which had been explored at length by Prum & Brush in the very paper Denton had earlier quote-mined.

Endnote 11 was just a link to the *Wikipedia* article on dinosaurs ("accessed on August 18, 2015," which at least gave a terminal date for when Denton was still actively researching).

As it happens, the *Wikipedia* post offered quite a thorough summary of current dinosaur data, including a section on the *evolution of feathers*, referencing such papers as Xu *et al.* (2004) on fossil finds, and even more interestingly, Lorenzo Alibardi *et al.* (2005) on the presence of feather keratins in the early stages of skin development in alligators (they are not expressed later once the actual scale develops).

You may recall my citing Alibardi's work back in Chapter 2 on the common developmental substrate for feathers and hair. Alibardi's 2005 paper was *experimental confirmation* of the evolutionary model Prum & Brush had worked out, but somehow that content didn't flag its way to being noticed by Denton, who did not cite any of that work in his 2016 book.

A curious omission, isn't it.

But even more interestingly, evidently a *Wikipedia* post was all we were going to get on 150 million years worth of dinosaur paleontology. And not a clue on what *types* there might have been among them.

Now in the paragraph above, it sounded like Denton was conceding that birds (with their feathers) were just members of that theropod dinosaur clade. If so, didn't that utterly shred the utility and history of Denton's own 1985 typology, which had not given even the smidgiest hint of birds being really just feathered dinosaurs?

Ah, but wait, Denton's endnote 10 had slipped a banana peel under this seeming admission: "There is controversy about whether there are 'reptiles' with pennaceous feathers. Some researchers have argued that they are actually secondarily flightless birds and NOT reptiles/dinosaurs. For references" Denton referred the reader to (another drum roll please) ... Casey Luskin (2008b) at *EN&V*.

This was the *Discovery Institute* pundit's take on a new dinosaur find, reported by Fucheng Zhang *et al.* (2008). *Epidexipteryx* was a pigeon-sized Chinese theropod dating to the Late Jurassic, and thus a contemporary of *Archaeopteryx* in Europe. It also had long "ribbon-like" tail feathers, and science reporting on the find naturally highlighted how this was one more avialan dinosaur at the origins of feathered birds.

Intelligent Design advocate Luskin understandably took issue with that interpretation (his **bold**): "Unreported in the media is the fact that **the paper contains language directly hinting that *Epidexipteryx hui* could also be 'interpreted as secondarily flightless.'** In other words, *Epidexipteryx* may not have been a 'feathered dinosaur' at all, but instead a bird that lost its ability to fly while retaining feathers."

Luskin neglected to specify what those features might have been. An understandable oversight, given that the paper hadn't mentioned any.

Here's what it *had* noted.

Epidexipteryx had the basic theropod body layout, including teeth but with a somewhat shortened tail. While it had feathers, its limbs lacked the quilled pennaceous types, unlike that of flyers (including flightless birds). The presence of the long plumes on the tail, though, suggested they were for sexual display, a feature not uncommon in birds. Finding that so early in the game, in an almost-bird, had implications that Zhang *et al.* (2008, 1107) concluded their paper with. It was that end sentence that Luskin tactically "quoted," but leaving out a most relevant first word (which I have marked in **bold**) that significantly changed its meaning:

"**Unless** *Epidexipteryx* is interpreted as secondarily flightless, the absence of pennaceous limb feathers in this taxon suggests that display feathers appeared before airfoil feathers and flight ability in basal avialan evolution."

As a particularly reliable secondary source, lawyer Casey Luskin was not setting high marks to justify Denton's confident citation of him.

Incidentally, the group *Epidexipteryx* belonged to is known by only a few partial examples, but going by the fossils described by (ironically enough) Stephen Czerkas & Chong-Xi Yuan (2002) and Fucheng Zhang *et al.* (2002), they're a funky lot of apparently arboreal animals, with unusual toothed skulls and especially enlongated arms. One of the most intriguing has turned up only recently: little *Yi qi*, with a batlike wing membrane along with the feathers, Xing Xu *et al.* (2015). Darren Naish (2015) noted paleontologist Andrea Cau had suspected that group might have featured skin flaps on their wings, and *Yi qi*'s membrane represented "yet another of those fossil animals predicted to exist prior to its discovery."

The Fossil Genie never rests.

What of the remainder of the references Denton wanted his readers to consult in Luskin's post, his *only* source on this important issue of whether there existed truly feathered dinosaurs (oh, sorry, "reptiles")?

Luskin had invoked two congenial (but dated) authorities: Alan Feduccia (1999, 396) and a paper coauthored by Feduccia, Terry Jones & Larry Martin *et al.* (2000), for the claim that the bipedal *Caudipteryx* had a center of gravity like that of birds, not theropods, and consequently that it (and the Oviraptosauria to which it belonged) had to have been flightless birds, which in their view meant not dinosaurs.

Given the overall dinosaurian anatomy of the Oviraptosauria, though, this was not a particularly persuasive argument, as Richard Prum (2003) noted. And Per Christiansen & Niels Bonde (2002) pointed out another problem: the Jones paper had got a lot of their data wrong, such that *Caudipteryx*'s center of balance wasn't where they'd claimed it was. The years since have not changed that evidential situation, as Naish (2009; 2015) has noted of Feduccia and the band of ornithologists digging in their heels on the dinosaur connection.

Whether Luskin stopped to study any of these criticisms of Feduccia's argument, he certainly didn't cite any, which put an ironic cast to Luskin's firm conclusion that "It seems that the 'feathered dino' interpretation may be driven by an attempt to fit these fossils into the standard evolutionary paradigm, not the data. Unfortunately, the view that these fossils are *not* feathered dinos but are rather secondarily flightless birds is a possibility that is not being communicated to the public in the media."

You'll notice how Luskin slipped in the arbitrary and questionable *typological* presumption that the two classifications (bird and dinosaur) were necessarily *mutually exclusive*. If you think in terms of natural branching common descent, though, a person is *simultaneously* the child of their parents, and *all* of their ancestors, and related thereby to *all* of their cousins. Trace that lineage back far enough, and humans are mammals *and* aminiotes *and* vertebrates *and* chordates *and* metazoans *and* eukaryotes (cellular life whose DNA is contained in a nucleus). It's a big family.

Pegging a group as a "flightless bird" depended then on what you meant by "bird." In the phylogenetic analysis by Teresa Maryańska *et al.* (2002) that Feduccia had alluded to, the Oviraptorsauria were being put squarely in the *Avialae* clade, but idiosyncratically farther down on the nodes than even *Archaeopteryx*. Maryańska had not incorporated later birds, and curiously enough, of the many anatonomical character states they evaluated, *feathers* were not among them. But Alan Feduccia clearly liked their conclusion, since it reinforced his bird *or* dinosaur distinction, and Casey Luskin was happy to authority quote Feduccia in his 2008 posting.

But that classification by Maryańska hadn't held up in more recent cladistic analyses, especially as more taxa became available, such as the *Epidexipteryx* specimen that was the topic of the 2008 Zhang paper. By then, the Oviraptosauria were placed among the derived *Maniraptora* (not way down in the *Avialae* clade as in the older Maryańska work had them). Naish (2016b) offered a comparable assessment.

By the time Denton was writing, it was clear that flight repeatedly evolved in feathered dinosaurs, and so flightless versions may have cropped up several times. And the *Avialae* clade had correspondingly grown to incorporate those feathered dinosaurs most closely related to the living birds. The ones that neither Denton nor his secondary sources of Luskin and Feduccia seemed keen to discuss.

In that one Endnote 10 of Denton's 2016 book, we are getting yet another snapshot of what is so wrong about antievolutionist apologetic "scholarship." Denton drew on Luskin's misplaced authority, without checking out the accuracy of his interpretation. And Luskin was already relying on Feduccia's misplaced authority, without checking out the accuracy of *his* interpretation.

Moreover, we know Denton could have been aware of the controversy swirling around Feduccia, because Denton (2016a, 174, 316n) had cited Thor Hanson's book on feathers, which had mentioned the feathered dinosaur brigade extensively. Hanson (2011, 57-58) had even mentioned the Feduccia flap, though it's possible Denton was in too big a hurry gleaning authority quotes to have spotted it. We know Denton got to the very page where Hanson mentioned Feduccia's tailspin reclassification of feathered dinosaurs, and it is interesting that Denton's endnote got one of the page numbers wrong at this very spot, listing the last of his Hanson quotes as occurring only on the previous page 56.

What got lost in this Feduccia-Luskin-Denton daisy chain of conclusion jumping was the curiosity and gumption to look at as much of the data as possible, and subject all claims about them to the test of rigorous analysis. I surmise that non-paleontologist Denton wanted to get away from all that glut of new fossil data as quickly as possible. But did Denton really think that single lame *EN&V* posting by non-paleontologist Casey Luskin could do the job for him?

Well, it didn't.

Feathers fall faster than facts in the typological vacuum

Michael Denton (2016a, 120) listed a variety of "taxa-defining novelties" that in his estimation posed intractable problems for their gradual evolution. These included the perennial origin of life and "higher mental faculties," along with rather specific features like "the wing of the bat" and "the life cycle of the eel," and ... "the feather."

"As I show in every case," Denton assured his readers, "it is very difficult to envisage how these novelties might have come about by a series of small adaptive steps."

Denton (2016a, 170-180) devoted over ten pages to delivering on this promise in the feather case, putting it as one of his star witnesses for the correctness of typology. Since he did not elevate the RMT to this status (which we'll get to in Chapters 7 & 11), it's relevant to see how Denton approached a subject (feather origin) that he pegged as a major illustration of his typological position. Even more than his brief flip at the paleontology noted earlier, though, those pages represent an amazing corkscrew of selective quoting and elbowing aside of relevant information.

Denton began with a lengthy quote from Alfred Wallace (1910, 287-288) on feathers, as though we really could benefit from knowing a scientific opinion written before dial telephones or DNA. After which Denton (2016a, 171-172) proclaimed (with my **bold** this time):

> The origin of the feather is as puzzling as the origin of the tetrapod limb or the enucleate red cell. To be sure, unlike the enucleate red cell or the 'primal pattern' of the tetrapod limb, **the feather is clearly an adaptive form**, useful for flight, for insulation, for sexual display, etc. Further, **no one doubts the utility of the stages that led from the simple follicle through the plumaceous feather to those closed pennaceous countour feathers of modern birds** (see Figure 9-2). Yet this taxon-defining novelty appears to be just as inexplicable in Darwinian terms. Cumulative selection cannot even begin to account for the origin of the series of novelties that make up the avian feather.

Blimey. What a pile of bold concessions and unqualified assertions.

Denton's Figure 9-2 was the simplest of stick-figure schematics showing four feather stages, from (1) a "single cylindrical filament" to (2) an unbranched plumaceous feather followed by (3) parallel vaned feathers and finally (4) feathers with branching barbule splits. No sources were given for this picture, but it was way more rudimentary than the detailed drawings in a source we know was available to Denton, Prum & Brush (2002, 271-274), that included explicit breakdowns of follicle cross sections illustrating how the developmental variations were leading to each new very explicable stage.

So was all that incremental evolutionary development the thing that "no one doubts"? Including Denton (and the *Discovery Institute* that published him)? And if so, why *can't* "cumulative selection even begin to account" for so adaptive a series in "Darwinian terms"?

That Denton wanted that to be true was plain enough, but what any of that *meant* to Denton was left out in the cold, for he promptly went off on a journey through more older writers on the obsolete "frayed scale" model of feather origins, including Philip Regal (1975) and the aforementioned Bock (1965), work that "a few insightful scholars" like the "landmark article" by Prum & Brush had helped to deconstruct. This was the context of Denton's second quote on Bock I noted earlier, where Denton was highlighting the novelty aspect of feathers without venturing any further.

But Denton was writing in 2015, after so much scientific water had flowed under the evidential bridge. Denton (2016a, 173) stepped only a bit closer to that biological substrate when he breezily acknowledged that Prum & Brush had shown how "many of the genes and developmental systems utilized in feather morphogenesis pre-existed the origin of feathers," including our old friends *Shh* and *Bmp2*. Denton (2016a, 315-316n) had three sources for this: Prum & Brush (2002; 2003) and Matthew Harris *et al.* (2005). The 2003 Prum & Brush was their general article for *Scientific American*. Earlier, Denton (2016a, 58, 296n) had also peripherally cited Prum (2005), but only among examples of Evo-Devo (evolutionary developmental biology) and evolutionary biology papers whose *titles* purportedly "betray the fact that genuine novelties are a primal fact of the biological universe."

So we have transitional stages building on known feather development, and the genetic toolkit for that already in existence (and so not requiring some newly designed DNA mechanisms other than mutations in their regulation and deployment), and fossil forms confirming the existence of those transitions uniquely in the group closely related to birds on general anatomical grounds (as Denton sort of conceded in his earlier section). So where exactly was the evolutionary problem?

Denton asserted (my **bold**), "But despite the co-option of pre-existing gene circuits, all the evidence points to feathers being genuine novelties, **not homologous to reptile scales or any known antecedent structure**," and cited of all people Prum & Brush (2002) as his sole justification for that claim.

Wrong. So wrong.

Consider Prum & Brush (2002, 285), their *italics* and my **bold**:

> How *are* feathers homologous with scales? The presence of a derived, morphologically distinct placode is only shared by feathers and avian scutellate scales (Maderson and Alibardi 2000). The patterns of gene expression specific to feather and avian scutate scale placodes are also shared with avian reticulate scales, however (Chuong et al. 1996; Widelitz et al. 1999). Thus, a molecularly defined placode is shared by all avian scales and feathers regardless of the presence of the morphological characteristics of a feather placode. These developmental features support **the homology of avian scales and feathers at the level of the placode**.

Having accepted Prum & Brush's insight that feathers are fundamentally tubular follicles, while having apparently missed those same author's identification of its core variation on the developmental processes of the archosaurian placode, Denton (2016a, 173-175, 316n) insisted: "This first novelty is without any antecedent structure in any reptile scale or any other vertebrate skin appendage," then stepped back from citing technical work to cull some general quotes about feather follicles and Prum's views on them from that secondary source noted earlier, Thor Hanson (2011, 37, 56-57), before concluding (his ellipsis): "As Prum and Brush point out: 'Avian reticulate scales... and all reptilian scales examined to date lack a morphologically definable placode.' Neither Prum nor any other author to my knowledge has provided a Darwinian scenario in which an adaptive continuum leads from the placode to the actualization of the feather follicle."

Ah, it's that "to my knowledge" part that may be the revealing point here, since Prum & Brush's segment on reticulate scales was extracted from a long paragraph immediately preceding the one Denton didn't quote, the one where Prum & Brush noted the *homology* of avian scales and feathers at the placode level.

The game Denton was playing at this point was something I call *Zeno Slicing*, named in honor of the famous motion paradox posed by the classical Greek philosopher Zeno back in the 4th century BCE.

Zeno pitted the fleet-footed Achilles against a Tortoise (ironically appropriate for my "Tortucan" model of conflict avoidance). Granting the slower turtle a head start, by the time Achilles reached the spot where the Tortoise had been when the human racer set out, the tortoise must have moved on some additional distance. And by the time Achilles reached *that* spot, the turtle had pressed on some still smaller distance, and so on, so that the Tortoise always seemed to be keeping ahead of Achilles. Despite how this flew in the face of common experience, Zeno thought he had proven by sheer reason that Achilles could *never* catch up, and such problems have long offered a useful guide to trace the limits of "logical" reasoning, Douglas Hofstadter (1979, 29-32, 610), Eli Maor (1987, 3-4, 17-18), William McLaughlin (1994) and John Barrow (2000, 55-57).

The big flaw in Zeno's argument turned on his failure to conceptualize the idea of *velocity* (where the faster Achilles was certain to overtake the slower Tortoise at some point). Underlying that was something that the mathematics of the time didn't allow Zeno to do: add up an *infinite series* of ever smaller increments that were piling up as the Tortoise and Achilles moved forward. What the 4th century BCE Greek was stumbling over was a problem in *differential calculus*, a tool which wasn't invented until the 17th century by Isaac Newton (1642-1727) and Gottfried Leibnitz (1646-1716). Their independent (and rather contentious) discovery was itself a case study in how much our prickly human nature plays out in even the most "abstract" of advanced mathematics work, reflected in Maor (1987, 12-13) and Brian Blank (2009).

No such complicated mathematics was involved in Denton's case, though. Instead, Denton was *Zeno Slicing* at a much simpler level, pulling apart each component to keep everything from connecting to anything else. But the data was out there, just outside the typological Tortoise's range.

For example, Roger Sawyer *et al.* (2005, 257-258) described the observational and experimental evidence building on the discovery that "the embryonic layers of the scutate scale epidermis are homologous with the periderm, sheath, and barb ridge cell populations that make up the epidermis of the embryonic feather filament," suggesting that "the ectodermal placodes of feathers, which direct the formation of unique dermal condensations, co-opted the archosaurian embryonic layers to form the feather epidermis." And remembering the deep split between diapsids and synapsids paleontologically, it's interesting that the placode stage becomes suppressed over on the mammal side of things, as another paper Denton didn't spot, Kentaro Suzuki *et al.* (2009), discovered regarding the role of Wnt/β-catenin signaling in hair follicle formation. The dermal placodes were especially hard to identity in reptiles, as they were expressed only briefly, but recent work by Nicolas Di-Poï & Michel Milinkovitch (2016) pinned that side of things down too.

Appearing after Denton's publication, it does reinforce how non-predictive his typology was turning out to be. But worse, Denton (2016f) stepped onto that minefield on his own, posting on the new Di-Poï paper—though only secondarily (via a *Washington Post* summary link), unlike creationist Elizabeth Mitchell (2016b) who lit into it by direct citation.

It's fun to compare Mitchell with Denton reacting to this cutting edge science finding.

Mitchell labored to pigeonhole the new work by conceding (my **bold**) that "Clearly the discovery of embryonic prescale placodes in reptiles is a great example of observational science. Careful observations confirm that something **no one knew existed** really is present after all." What Mitchell left out was that their work was being done precisely because they were following up on a substantial and ongoing *evolutionary* line of inquiry. What had been learned about the origins of feathers and hair implied reptiles ought to have retained that placode substrate, which inference Di-Poï & Milinkovitch had now *confirmed* in a particularly interesting case. Mitchell did not allude to any of the relevant work by Prum or Alibardi they had drawn on, but tried to wave the new paper aside as though it were just part of the "fraudulent 'proof' of" Ernst Haeckel's recapitulation theories (we'll return to that recurring antievolutionary shibboleth down in Chapters 9 & 10).

Denton was just as obtuse as Mitchell, though, only wafting to a different breeze as he now conceded their homology (though not noting his own opposite claim about that very point in his own new book, the ink barely dry there). Denton then greased the wheels under his typological goalpost to ease it down past the latest data, declaring "It is difficult to imagine that the placodes were gained one by one until they covered the body surface. What would be the selective advantage of one placode, and indeed what could be the selective advantage of cell condensations of this sort before the developmental trajectory leading to fur, feathers, and scales was in place?"

As the placodes occured in *all* those dermal cells, what made Denton think a mutation at the regulatory end would necessarily effect only *one* placode at a time? If there was a difficulty in imagining things, it clearly wasn't on the part of Di-Poï & Milinkovitch.

Denton's spin encapsulated the deadening trajectory of his main book, where Denton (2016a, 176) was by then totally hung up on the notion of what an evolutionary "novelty" meant, quoting from Prum & Brush (2002, 287-288, 316n) again on whether feathers qualified as one according to the definition of Gerd Müller & Günter Wagner (1991). At the end of that, Prum & Brush had explicitly directed their readers back to their caveats on feather homology that I've previously quoted. In other words Denton had to step past that placode homology issue *twice* in the same paper.

Being a "novelty" in the sense Müller/Wagner and Prum/Brush were discussing was not to claim that the feature involved ultimately unprecedented genetic or developmental mechanisms, and thereby falling into Denton's typological trench, but only that their application in the new form was unlike any other previous feature. It was a *novelty*, a new *combination* of antecedent processes, not a typological miracle. It was looking as though Denton was arguing that an evolutionary novelty couldn't really be an evolutionary novelty solely because it was a *novelty*, never mind that science had identified so many of the genetic and developmental components that accounted for the evolution of that "novelty".

Denton hit on allusions to "novelty" with exactly the same quote-mining tunnel-vision as any creationist had "stasis" or "abrupt appearance." Thus Denton (2016a, 58, 296n) selectively invoked Massimo Pigliucci (2008, 888) when he discussed the many features that one or another scientist had offered as an example of such novelty, from "new patterns of wing colors in butterflies" and "variations in jaw morphology of cichlid fish" to "the sexually dimorphic horns of horned beetles." Denton didn't allude to the examples Pigliucci had given, instead extracting only the last part of Pigliucci's next sentence (Denton's quoted part in **bold**): "It is difficult to see what all of these features have in common, partly because they **span all levels of biological organization, from morphological to behavioral to molecular traits**."

Pigliucci was in fact *criticizing* the vague spread of those purported novelties, not affirming their status as novelties or suggesting that genuine ones were intrinsic to any of those levels. In fact, farther down on that very page, Pigliucci noted of the cichlid example, "while the various jaws of African cichlids are clearly linked to this group's radiations, it is questionable in what sense variations of a preexisting structure (the jaw) can be considered a 'novelty.'"

It's worthwhile to quote what definition Pigliucci (2008, 890) had proposed, since it covered the subject quite nicely (my **bold**):

> Evolutionary novelties are new traits or behaviors, or novel

> combinations of previously existing traits or behaviors, arising during the evolution of a lineage, and that perform a new function within the ecology of that lineage. This definition does not imply any specific mechanism or long-term evolutionary effect, since the first one is precisely the objective of empirical research in this area and the second one is a matter to be settled based on the historical record. However, my definition (1) makes explicit the fact that **often novelties are not absolute discontinuities but can be built on previously existing parts**, (2) indicates that they are a phenomenon that affects the evolution of **certain lineages without implying that all derived characters are in fact novelties**, and (3) requires some kind of ecological function to eventually be coupled with the novelty, although it does **not imply a necessary link between novelties and adaptive radiations.**

Ah, but we know how hung up Denton has been on that *adaptive* issue too.

For example, early on in the book, Denton (2016a, 70, 298n) listed our monotonous seven mammal cervicals among "non-adaptive patterns, quite beyond the explanatory reach of any adaptationist or selectionist narrative," yet cited only the paper by Frietson Galis *et al.* (2006) in his endnote, a study reporting how common such variation in cervical vertebrae were in our own species, but noting how *selection* worked against that because of the many pleiotropic effects that led to illness and death.

By now I was also wondering what was up with Denton's repeated insistence on some explicitly "Darwinian scenario." Was he thinking of specific experiments identifying how selective pressures would preserve an adaptive innovation? And how would a writer need to phrase things to meet Denton's requirements here?

There's that recurring antievolutionist methods problem. Denton didn't stop to explain what he meant by a "Darwinian scenario," as opposed to say, a *non*-Darwinian scenario. Or, better still, a "typological scenario" of comparable explanatory detail, so the reader could *compare* the two. How about that?

At this stage in Denton's argument the "novelty" trench had been dug down to just about the same depth as the "transitional form" hole among antievolutionists generally, where lots of text was offered to say what is supposedly unsatisfactory, without ever getting around to conceptualizing what manner of evidence or argument would be acceptable.

Instead Denton did what generations of antievolutionists have done, (most infamously Duane Gish), stressing the *spaces* between the identified stages of feather evolution rather than the affirmed existence of the stages themselves (the initial protofeather notably showing up in those feathered theropods) that he sort of accepted on one page, only to dismiss it on another.

"The fact that all the above stages are quite distinct, and that there is an empirical gap between each stage—that no known feather structures bridge the stages in either living or fossil species—is highly suggestive," Denton (2016a, 176).

But Denton was being too clever by half here. Each of Prum & Brush's stages already involved the most incremental of variations, as the initial protofilament prong successively split and frayed in ways both developmentally justifiable and (with the new dinosaur examples) physically corroborated in the fossil record. As Alan Brush (2006, 123, 124) reminded, "It is not coincidental that the earliest fossil feather matches the simplest feather phenotype on extant birds," and that "The earliest stage of the follicle, with the complex folding and invasion of mesodermal tissue, is sufficient to produce a single filament. This matches the filaments on coelurosaurian dinosaurs (e.g., *Sinosauropteryx*), and is presumably the most primitive feather."

One could interpolate further degrees of variation, such as ones showing more barbule snagging, but that would raise the question of how likely would it be to find such evidence as *fossils*?

As it happens, all too many of even the best-preserved fossil feathers only show that they *are* feathers. If their shape was overtly asymmetrical (required to function as an airfoil), they were putatively flight feathers; if not, they were otherwise covering plumage. But beyond the blurry traces of vanes and putative barbule connections in fossil feathers, none show the detailed microstructure sufficient to resolve the question Denton was now demanding of them. A case in point would be *Archaeopteryx*, where the recent microanalysis by Ryan Carney *et al.* (2012) illustrated just those limitations in the fossil data.

So, given the odds, we'd be ready to cue the *Bermuda Triangle Defense*, except for the fact that the Fossil Genie wasn't prepared to give up filling in what antievolutionists insisted they shouldn't.

A cache of seven curious feathers ended up in some French amber back in the Early Cretaceous (about 100 mya), Vincent Perrichot & Loic Marion *et al.* (2008), showing "a flattened shaft composed by the still distinct and incompletely fused bases of the barbs forming two irregular vanes." It was impossible to tell whether these were from some early bird or non-avian dinosaur, but they did represent "the first fossil evidence of the intermediate stage between the very distinct stages II and IIIa defined by Prum (1999) in his theory of evolutionary diversification of feathers. Stage II is characterized by non-ramified barbs attached at their base to the calamus, without barbules. Stage IIIa corresponds to the appearance of a central shaft formed by the fusion of non-ramified barbs and the appearance of the planar form," Perrichot & Marion et al. (2008, 1199). Interestingly enough, this intermediate structure eliminated the need for a further IIIb stage proposed by Prum.

Cancel that *Bermuda Triangle Defense* this time.

While creationist Elizabeth Mitchell (2015) was still insisting at *AiG* that "the fossil record does not reveal an evolutionary progression in feather development," Michael Denton pulled a double *Garrett Hardin* by citing works that were pinning down the detailed genetics of feather evolution from the underlying placode, such as the pioneering work of Mingke Yu *et al.* (2002)

and Matthew Harris *et al.* (2005).

"Clearly a great many other, yet-to-be identified changes must also be in place to cause the helical displacement toward the anterior of the feather," Denton (2016a, 177, 316-317n) wrote of Harris' work, but relegated the details to an extended endnote quote that included this from their conclusion, Harris *et al.* (2005, 11739):

> The first branched feather barbs evolved through the establishment of an activator-inhibitor interaction between the plesiomorphic, or preexisting, interacting Shh and Bmp2 signaling systems in the basal epithelium (i.e., marginal plate) of the feather germ to produce meristic patterning and morphogenesis of the barbs. The subsequent plumulaceous-to-pennaceous morphological transition evolved through the derived integration of an additional short-range inhibitor and a D/V [dorsal/ventral] polarized signaling gradient. Thus, evolutionary novelty of pennaceous structure required the coupling of the plesiomorphic, previously independent Shh/Bmp2 module and D/V signal gradients. The signaling mechanisms that produce complex barb branching in pennaceous feathers were an inherent potential of the molecular mechanisms previously evolved with the origin of the simpler, plumulaceous feathers. Integrated signaling between modular developmental systems provide both stable mechanisms of morphogenesis and inherent capacities for the generation of morphological and evolutionary novelties.

All this sounded quite splendid, laying out specific genetic interactions involved in the evolution of feathers, connected directly to experimental work. But that was still not good enough for Denton's moving typological target.

Denton (2016a, 177) also quoted from the abstract of Yu *et al.* (2002, 307) on the role of *BMP* and *Shh* in feather formation. But Denton *didn't* quote this rather relevant part: "The fact that the barbs and the rachis can be converted experimentally in the laboratory favours the barb to rachis model. Our data suggest that a radially symmetric feather is more primitive than the bilaterally symmetric feather in terms of molecular and developmental mechanisms, and may have been the prototype of feathers," Yu *et al.* (2002, 311), who added that "Some fossilized primitive skin appendages on *Sinornithosaurus* also favour this model."

Subsequent work by Yu *et al.* (2004, 186)—a paper Denton did not cite, by the way—further illustrated the experimental isolation of which elements came from *Shh* and which from *Bmp* 2 and 4, along with the contributions from the *noggin* gene.

That would seem to represent a huge amount of work clarifying the evolutionary framework. But not from down in the typological trench. "Every aspect of feather origins bristles with challenges to Darwinian scenarios," Denton (2016a, 177-178) insisted, and continued with this (my **bold**):

> What came first, the cellular condensation, that created the barbs (which occurs first in the development of extant feathers), or the apoptosis (which occurs after the development of the barbs in extant feathers) that

> separated them into discrete filaments? Only if both developmental processes are in place can the adaptive end of a branched feather be actualized, which again raises the specter of evolution *per saltum*. Again, **severe problems** arise in **trying to imagine a Darwinian scenario** for the origin of the barbules and their subsequent differentiation into distal barbules with hooks and proximal barbules with grooves, which interlock together, binding the pennaceous feather into a closed vane. In attempting to reduce feather origins to Darwinian scenarios, we are led not only into "endless absurdity," but into **direct conflict** with what is known of the developmental processes underlying feather ontogeny for one-to-one, bit-by-bit Darwinian functionalist approaches.

Just as Duane Gish had with those jawbones, demanding they march into the evolving mammal ear one at a time, Michael Denton seemed unable to imagine that the barb and apoptosis processes might have been occuring *concurrently*.

Denton (2016a, 179) concluded his foray into feathers with another of his give with one hand but take back with the other declarations: "Although it is clear that the feather was actualized in stages, which can be followed in the fossil record, there are no known adaptive sequences leading gradually to each of the novelties or new homologies."

And he cited Prum & Brush (2002) once again on that.

I'm scratching my head.

6. Wrestling with Cladistics: The saga of Colin Patterson

That Michael Denton's "typology" was ultimately a meaningless concept was affirmed by how it couldn't be practically applied to anything, by anyone.

Take Phillip Johnson, who freely stepped in Denton's shadow, channeling both his arguments and the secondary sources Denton relied on, such as the British barrister and ornithologist Douglas Dewar (1875-1957) on the supposed impossibility of whale evolution, Denton (1985, 216-218) and Johnson (1991, 178-179), which I discussed at fuller length in Downard (2004c, 355-385).

Ronald Numbers (1992, 145-152) recounted Dewar's distinctly *Kulturkampf* background. Abandoning his half-hearted Darwinian beliefs because of the harm it was doing "to the morality of the white races," in 1935 Dewar helped form the "Evolution Protest Movement" with a group of like-minded British eccentrics keen to combat Darwinism's purported goals of moral degradation (promoted by psychoanalysis), human extinction (via birth control), and political revolution (through communism). Not all that dissimilar panic buttons from those still being pressed by the ID movement and Creation Science in the 21st century.

That the Intelligent Design argument on whales was grounded on the opinions of an oddball creationist writing half a century earlier didn't bother either Denton or Johnson in the least. Neither stopped to figure out how the cetaceans could be categorized *typologically*, and Denton's vagueness pervaded Johnson's approach to all aspects of evolution.

In the Epilogue added to the 1993 edition of *Darwin on Trial*, Johnson (1993a, 157-158) ingenuously stressed that microevolution was "change within the limits of a pre-existing type, and not necessarily the means by which the types came into existence in the first place. At a more general level, the pattern of relationships among plants and animals suggests that they may have been produced by some process of development from some common source. What is important is not whether we call this process 'evolution,' but how much we really know about it."

At no point did Johnson lay out anything at all about what these types were within the ID framework. And the situation hasn't changed notably in the decades since, with creationists at least blundering into this area now and then via their baraminology work (noted later in this chapter, and again in Chapter 10), but absolutely *nothing* substantively on the idea anywhere in the ID literature.

Kenneth Miller (1999, 99) offered what remains a valid *Map of Time* observation: "I have never read, nor do I ever expect to read, an explanation of any event in natural history in which the explanation of design is correlated with actual events."

So many questions arose in the typological framework, and yet Johnson did not stop to think about them. Whatever did he mean by "some process of development"? If a physical lineage was involved, was there direct manipulation at the genetic level? Or didn't animal "types" reproduce in the way known today?

There are real conceptual consequences attached to any process of physical descent (in or outside of a "type," whatever that may be) which antievolutionists have yet to think much about. Though Johnson (1993b, 39) did intimate in a *First Things* exchange with theistic evolutionist Howard Van Till (1993) that "it does not necessarily follow that we are referring to the ordinary process of reproduction that we observe in today's world, where ancestors give birth to descendants very much like themselves."

How instructive it would have been for Johnson to have explained what the heck he meant by any of that.

The same problem surfaced in Michael Denton's retooled 2016 typology, as he sought not only to have his cake while eating it, but subcontracting to evolution to do all the fact baking first, and then show no willingness to accept any of the cake afterwards.

Denton (2016a, 103) declared (his *italics*):

> It is a major fallacy to think that belief in distinct Types as immanent features of nature and the notion of descent with modification are incompatible. On the contrary, just so long as Types are defined by unique novelties (homologs or synapomorphies), which are *themselves not approached gradually via transitional forms*, then the Types themselves can be said to be absolutely distinct and can arise suddenly at that moment in phylogeny when members of any lineage acquire a novel homolog.

But what does that "suddenly" *mean*?

Take those metapleural glands in ants, a defining novelty that appeared at some point, and by its presence automatically yanks even the most wasplike arthropod over into the ant category by the necessity of classification (like a film can't be a Musical and a Comedy at the same time).

If you were to get in the old Time Machine, and go back to see what happened with that first bona fide "ant," what would Denton say you would have seen happening? Did an entire population of non-ants "suddenly" acquire the gland, or just one Ur-Ant which somehow bred its way to universality? Are either of those prospects genetically or biologically defendable, based on what we know about the functional dynamics of living organisms?

Denton hadn't thought that part through. Not even a little bit.

If there was no genetic system for the production of glands in play at that spot on the ant thorax, and no incipient cascade of "almost gland" in their immediate biological ancestors, not only how *did* it come about, we would have to ask how *could* it come about? Do glands ever appear out of nowhere, *poof*, in living organisms? Denton certainly didn't offer any evidence that they did, or look into what genetic underpinings may have been involved in the metapleural case (which, as we've seen, was no simple matter as so much of the fundamental developmental genetics of it have yet to be identified).

After a few more pages of not indicating what he thought happened with such issues as the RMT jaw-ear shift (which we'll explore in Chapter 11), Denton (2016a, 111) rolled out this "Caveat" (with a somewhat ungrammatical first sentence lacking an "of" that suggested some slack time in the pre-publication proofreading department):

> In this context, I have a clarification to make regarding the title my earlier book *Evolution: A Theory in Crisis*. As that book was a critique of Darwinian causation and not the theory of common descent, a more appropriate title might have been *Darwinism: A Theory in Crisis*. I think I used the words "evolution" and "Darwinism" too loosely, conveying the impression that they referred to the same thing, when of course these are two very different concepts. In many places, where I should have used the term "Darwinian evolution" rather than "evolution," I created the impression that I doubted the notion of common descent.

"Too loosely" was an understatement.

But this put Denton at odds with the bulk of the Intelligent Design Movement (let alone Creation Science) where doubting the reality of common descent is very much the norm. The only other major figure in the Design movement who ostensibly accepts common descent is Michael Behe (1996, 5): "I find the idea of common descent (that all organisms share a common ancestor) fairly convincing, and have no particular reason to doubt it."

Yes, Denton and Behe "accept" descent with modification, but not so much that they mean anything like what evolutionists do by it: that all living organisms are related by a process of natural branching speciation that does *not* involve the discontinuous appearance of features "suddenly."

Take Denton's "types." If all living things are related by natural branching common descent, then all *types* are too. Grumping about the detailed mechanisms of that branching, or the extent to which they are "adaptive," would be irrelevant to the overall reality of descent. To say there are "types" that have *not* emerged by natural means, that *cannot* have been bridged by intermediate forms, is to reject common descent fully and in principle.

The same caveat would apply to Michael Behe's Irreducible Complexity candidates.

So why did they think they were "accepting" common descent?

Both Behe and Denton may have fallen into this perfunctory "acceptance" of common ancestry because to suggest otherwise erased the difference between their imagined discontinuities (Behe's Irreducible Complexity and Denton's Types) and the *Kinds* favored by creationism.

But the rest of ID isn't buying it. After noting the two Michaels (Denton and Behe) who "affirm the idea of common descent" (and which effectively exhausted the list), John West (2016) noted that "we have other affiliated scholars who are strongly critical of universal common descent, the claim that all living things are descended from one original primordial organism. I think that our diversity on this issue is a good thing."

Even here West was being disingenuous, as I know of no IDer willing to concede that any living organism traces back to significantly different precursors (say, mammal to reptile, or bird to dinosaur, or human to primate), let alone all the way back to "primordial" ones. But the lure of "Origins or Bust" is strong.

The endnote in Denton (2016a, 304n), in which he thought to explain more of what he meant about his sundry 1985 misphrasings, clarified nothing at all, though it did show the tread on the latest set of rationalizing wheels he had installed on his movable evidential goalpost:

> Phrases such as "the evidence.... does not provide convincing grounds for believing that the phenomenon of life conforms to a continuous pattern" (page 194), "if gradual evolution is true" (page 228), and "Nature has not been reduced to the continuum that the Darwinian model demands" (page 357) do convey the impression of denying evolution and common descent. This was not my intention. I do, however, reject any sort of gradual bit-by-bit evolution, for reasons I am presenting here. I also reject the term "evolution" in the mainstream meaning of the word, i. e. "gradual adaptive change." The tree of life is not connected by innumerable adaptive pathways leading from the trunk to all the most peripheral twigs.

So which is it? Was Denton now accepting *some* bit-by-bit evolutionary transformations (just not "innumerable" ones) while rejecting their import because they fail somehow to represent "adaptive pathways"?

With the addition of that adjective "adaptive," Denton was putting one more hurdle between the fossil and biological runners and the dreaded "Darwinian causation" finish line he doesn't want any of the evidence ever to cross. Cue the *Zeno Slicer*.

On that issue of adaptation, Denton (2016a, 13) had explained (his *italics* but my **bold**):

> I use this term to refer to any feature or characteristic of an organism which does not appear to serve any conceivable *specific* adaptive end—in other words, any feature that makes no contribution to the fitness of the organism. Such features are invisible to natural selection because natural selection only sees traits which serve some adaptive end. Examples might be the **shape of the maple leaf (a non-adaptive feature restricted to an individual species of plant)** or the pentadactyl limb (an example shared by many thousands of different vertebrate species).

Right off the bat, Denton had misstepped theoretically, by being overly restrictive in his definition. Traits which are adaptive in a particular condition are certainly subject to getting preserved by natural selection, but only if they get preserved to be selected first. It is *reproductive success* by individuals within a population that is the playing field of natural selection. And any individual doesn't just reflect that adaptive trait, but how it plays out in the whole organism in its ecological context.

Not everything needs to be instantly adaptive, and a feature can eventually come to have its particular characteristics not because every DNA nucelotide was incrementally adaptive, but because happenstance channeled the dynamic landscape in which it developed. Think of that arbitrary skin buckling that later channeled hair and feather shapes many millions of years later.

Or take heterochrony (that differential growth rate factor noted back in Chapter 3). An acceleration of the ancestral rate in plants has channeled developmental patterns in the leaves of their descendants, Kathleen Pryer & David Hearn (2009).

I brings up leaves because the *Discovery Institute*'s David Klinghoffer (2016a) repeated Denton's fauly assumption concerning a new Intelligent Design video that drew on Denton's views, Klinghoffer characterizing the maple leaf as "A 'Nightmarish Scenario' for Darwinism—the Curse of Non-Adaptive Order."

Interestingly, though, like Duane Gish never quite getting around to the details of Punctuated Equilibrium, Denton's new book had never quite got around to discussing the details of this supposedly non-adaptive "Nightmarish Scenario" represented by the shape of maple leaves.

Not that there was a shortage of work in the field to notice. Over the last couple decades many scientists had been working out the genetic players in leaf development, from homeobox genes and microRNAs to WOX proteins: Ju-Jiun Chen *et al.* (1997), José Micol & Sarah Hake (2003), Javier Palatnik *et al.* (2003), Sharon Kessler & Neelima Sinha (2004), Tracy McLellan (2005), Hirokazu Tsukaya (2005), Angela Hay & Miltos Tsiantis (2006), Alexandru Tomescu (2009), Paolo Piazza *et al.* (2010), Adrienne Nicotra *et al.* (2011), Hao Lin *et al.* (2013), Daniel Chitwood *et al.* (2014), Maya Bar & Naomi Ori (2014), Yasunori Ichihashi *et al.* (2014) and Kathleen Ferris *et al.* (2015).

Along the way, it's proven possible to recreate leaf shapes *experimentally*, such as Daniela Vlad *et al.* (2014) regarding the model plant *Arabidopsis* (a relative of cabbages and mustard) and its evolutionary cousins. Needless to say, the plants most commonly used for such studies are small ones as easily cultivated as possible. One would wonder whether Michael Denton would expect scientists to grow stands of maples (for decade-long studies?) before his typology would permit the insights drawn from other plants to be recognized or applied.

Denton (2016a) cited none of this work.

With unintended irony, while extoling the "cavalcade of examples of non-adaptive forms in nature" purportedly shown in Denton's new book, *Evolution News & Views* (2016c) bumped into the latest installment of some of the work being done by the actual researchers in this field. Daniel Chitwood & Neelima Sinha (2016) surveyed the state of the science and covered the many possible biological and *adaptive* components underlying the diversity of leaves (not just the narrow maple case Denton mentioned), and laid out "a synthesis explaining both historical patterns in the paleorecord and conserved plastic responses in extant plants."

All of which the anonymous *EN&V* author glibly dismissed as "little more than suggestions and just-so stories."

Denton (2016a, 152) similarly dangled meaningless distinctions on the subject of angiosperm (flowering plant) origins in general: "I am not arguing that these curious patterns could not have arisen by descent with modification from earlier plant species, but I am insisting that they could not have arisen gradually via long sequences of tiny advantageous mutations."

And why ever not?

It would have been interesting, if not exciting, to read what Denton imagined could have been going on with that. We know what occurs in actual speciation, and it does not entail vast macromutations. So how could any organism have been physically descended from another without each link in the speciation descent chain doing only what we observe in all known species?

As for the adaptive reasons why the specific shapes we see emerged, how would modern scientists go about identifying the exact adaptive contexts for extinct plant life? Jump in the old Time Machine and go back to see what the ecological circumstances were in which the adaptation occurred? Sorry, not possible.

OK, what then would be the manner of experimental work that could be done now, other than identifying the genetic players that make an individual leaf look as it does, and where feasible, experimentally generating some of the observed variation? Just the sort of work that has been done to a great degree, even if Denton wasn't noticing.

Take some of the available examples Denton might have explored had he exercised his online mouse clicking.

Leaf shape is measurably open to natural selection, as the field work of Kerry Bright & Mark Rausher (2008) showed for morning glory leaves. As for the selective variables, Tsukaya (2005, 547) noted how rheophytes (plants adapted to life beside water habitats prone to flooding) "show two types of adaptations to two opposite types of environmental stress. Leaves of rheophytes are narrower than closely related species and are thus able to resist the strong flow of water. On the other hand, leaves of rheophytes are thicker than those of their ancestral species in order to tolerate dessication during exposure to high levels of sunshine in fair weather." Tsukaya (2005, 552-553) went on to note how light intensity and gravity affected the shape and growth of leaves.

When the specific genetics can be pinned down, the picture again looks more like a tradeoff of competing factors, including "Darwinian" natural selection, rather than a static appearance of typological novelty.

For example, Piazza *et al.* (2010) identified the "key information as to when and how the characteristic leaf form of *A. thaliana* evolved," involving a loss of KNOX protein expression due to a natural selection sweep. As for what adaptive value the evolved leaf shape had, Piazza *et al.* (2010, 2226) cited Dana Royer *et al.* (2005) concerning how "leaf margin complexity" was related to increasing temperature. "This raises the possibility that unlobed leaves may have been selected during a warm phase of the global climatic cycle that followed the divergence of *A. thaliana* from its lobed leaved relatives," which Mark Beilstein *et al.* (2010) had shown occurred during a warming phase in the Miocene.

Because so many factors can come into play in evolution, identifying the combined effects regarding even a single species is daunting, from thermoregulation issues to leaf resistance to predatory herbivores, Nicotra *et al.* (2011). Temperature gradients (either by latitude or elevation) appear to play a role, where overall leaf size tends to decrease with altitude, Rubén Milla & Peter Reich (2011), and temperate leaf forms are more toothed on the margins, Samuel Schmerler *et al.* (2012). Surveying 3549 plant species from six continents, Royer *et al.* (2012) found "toothed" leaf forms tended to be of the deciduous leaf dropper variety, "and to have thin leaves, a high leaf nitrogen concentration, a low leaf mass per area, and ring-porous wood."

As for maples specifically (a *genus*, not a *species*, as Denton miswrote above), there is no typologically uniform "maple leaf," as the shape of the warm climate red maple *Acer rubrum* already differs visibly from cold climate ones *in that same species*, but experimental replantings in the opposite climate changed their serration edges, with warmer climate versions being more rounded, Royer *et al.* (2009). Maples aren't unique in this plasticity, as the same phenomenon showed up in the *Viburnum* shrubs surveyed by Schmerler *et al.* (2012).

Just how experimentally modified does the maple leaf need to be to be seen as "adaptive" by Denton? And if every stage in any particular leaf's evolutionary history requires only micromutational variation, how does it rescue Denton's typology even if that sequence isn't overtly "adaptive" enough for Denton's never-specified criteria?

Denton really hadn't thought his own position through, not in 1985 or 2016, or the manner by which science would present evidence sufficient for Denton to change his mind. It is *his* conceptual problem, not one of the limits of micromutations or "Darwinian" natural adaptation. And it is the common mental block that pervades antievolutionist apologetics.

We'll see in Chapter 11 how Denton tried to adapt that dismissive "adaptive" idea to the RMT evidence, but first it's good to dig a little deeper into why Denton thought his views were reflecting cutting edge scientific thinking in the first place, and how that has tapped into some recurring quote mine tropes of modern antievolutionism.

Digging Cladistics

Some of the reason why Michael Denton thought that pre-evolutionary typology was on the brink of a revival was because the Austrialian biochemist was tripping over a revolution in taxonomy that he had completely misunderstood (or at least misrepresented), the new phylogenetic systematics, more succinctly known as *cladistics*.

The old Linnean classification system was showing its age by the mid-20th century. Based initially on looking at overall features, as the centuries wore on and more developmental data emerged, the old categories (like those "reptiles") were proving cumbersome.

Enter Willi Hennig (1913-1976), who in 1950 came up with a new technique of analysis that sought to include every scrap of data you could get your hands on, all turned into numbered character states that could be crunched mathematically.

For a sampling of takes on the history and application of cladistics, there's Richard Dawkins (1986, 275-284), Linda Gamlin & Gail Vines (1986, 32-33), Philip Whitfield (1993, 176-177), David Fastovsky & David Weishampel (1996, 51-54. 61-63, 70, 90), Simon Conway Morris (1998, 176-180), Henry Gee (1999), Niles Eldredge (2000, 202-203n) or Colin Tudge (2000, 33-62).

Those itching for the more technical side can dive into Michael Lee (1998), Paul Sereno (1999a) or Richard Hudson & Jerry Coyne (2002) discussing things like the difference between "crown" and "stem" groups, how "nodes" figure in cladistics, and the effect different definitions have on the interpretation of genetic loci data. Joel Hagen (2003) surveyed the often bumpy process of applying mathematical analysis to systematics, and James Valentine (2004, 12-31) laid out how the techniques were to be applied in evaluating the origin of phyla back in Deep Time.

Beyond such green eye shade technicalities, the basic idea of cladistics is that such systematic techniques could be applied to *anything*, even things that weren't genealogically related (or even alive), like rock formations or dining room chairs. Or whether movies were "musicals" or not depending on how they treated their singing section (my example from last chapter).

Anything that could be characterized by definable numerical properties could in principle be run through a cladistic mill to identify similarities and differences Where living things are concerned, though, evolutionists want to know as best they could what their actual lineage was, and that a true systematics would reflect the real ancestry of the organism. Ironically, cladistics proved a spectacularly useful tool in working out such evolutionary relationships by *not* depending on evolutionary assumptions, and by avoiding presupposing anything at all about specific ancestor-descendant relationships.

How could that be?

In part, it's because it's really good at measuring *closeness*.

Remember, all those character states are numerical, and cladistic analyses can easily pile up hundreds of them in a single study, such as the 275 features Zhe-Xi Luo *et al.* (2002, 48-78) used to evaluate early mammal relationships (requiring *thirty-one pages* of double-column fine print).

For a given character trait, the numbers are either alike, or nearly so, all the way down to not very similar at all. Put more and more features on the measuring scale, and evolutionists can step back and use that independent cladistics tool to test likely lines of descent and closeness of relationship based on the whole range of data. Character state *x* of taxon *a* is definably closer to, or farther away from, character state *x* of taxon *b*. Now multiply that by multiple taxa each represented by hundreds of character states.

Cladistics is a kick ass tool.

But you still had to pay attention to the *biology*.

A particular trait might be relatively easy to emerge developmentally (say, a tiny protrusion from an already existing bone). The presence of that feature might be due to its having been inherited from the first one to develop that mutation (a shared "plesiomorphy"), or it might have been some convergent "homoplasy" that originated independently (though hardly an evolution buster, given that it wasn't biologically difficult to produce).

Other traits, though (say, those interconnected ear bones in mammals), are much less likely to be popping up independently, because they're a *suite* of features, though individual independent variations in lineages would inevitably show up in the finer details of the configuration. All that would become clearer the more you looked at it from the standpoint of *what* had appeared *when*. Which means applying cladistics to fossil cases is hip deep in getting your *Map of Time* straight first.

Paying attention to the developmental biology and genetics of the features in living organisms would obviously be of great relevance to making sense of the parade of fossil life. Which we've already seen, antievolutionists don't seem very good at, that whole *what* and *when* thing.

Cladistics allowed numbers and degree of probability to be attached to all this, with the ideal being a "parsimonious" picture, where the minimum of distinctive features were arising, to be carried along in recognizably descendant lineages. So it was that cladograms (charts of acquisition of character states) were already being used by Tom Kemp (1982a, 297) to illustrate mammal evolution, George Gaylord Simpson (1983, 169-170) showed how cladistic "parsimony" quickly sorted out the evolution of the elephant family, and Leonard Radinsky (1987, 7-8) took note of how cladistic assessments of the fossil sequence were matching up.

Obviously the cladistic toolkit was being actively used even as Michael Denton wasn't one of those doing it in service of typology.

That notwithstanding, move on down to Michael Denton (2016a, 52-55, 293n, 295n) and he was continuing to argue that cladism was furtively affirming the "Reality of the Types," though only by propping up the husks of several dated "VHS vs Beta" era issues.

It's been a long time since cladists squabbled over the "defining characters" of systematic groups, as critic John Beatty (1982) had with advocate Norman Platnick (1982), and Denton's concession that things "may have cooled" down since was putting it mildly. Nor had Rupert Riedl (1925-2005) prodding science on the persistence of fundamental body plans in Riedl (1977) ushered in a typological revolution. Instead, it was the discovery of conserved homeobox genes that put Riedl's insights and Evo-Devo along the same track, as recounted by the Günter Wagner & Manfred Laubichler (2004) retrospective.

Nor has Denton been the only antievolutionist to nurse their own petrifying misconceptions on cladism.

Cornelius Hunter (2001, 40-41) was most extreme in missing the point when he claimed that independent cladistic analyses of molecules could not provide "strong support" for common descent because the technique could also be applied to objects, like cars, obviously not related in a genealogical sense.

Hunter contended hypothetically that "data from automobiles disguised as molecular data" would be fitted into an imaginary evolutionary framework. Or even "given random, uncorrelated data," the evolutionist would merely waffle "that the maximum parsimony model was a bad assumption because the molecular evolution was too fast."

His presumption that designed systems would mimic the hierarchical relationships of natural lineages remained for him to document. The steel, plastics and paints used by a particular car manufacturer would not possess differential mutations that tracked independently to common origins, as natural molecules tend to do. New components would be restricted to makers and particular model years, showing the same static character values no matter how the relationships were shuffled to fit an "evolutionary" framework.

Niles Eldredge (2000, 144-146) commented on exactly this taxonomical aspect of designed objects, that they are often hard to organize into the sort of tidy hierarchical nests so characteristic of nature (and recall my musical film examples). Eldredge happened to be "an expert in the history of design of the coronet," and that field manifested exactly the opposite characteristic of a Darwinian system. Because coronets are definitely the product of intelligent design, improvements could (and were) made without any deference to where the idea might have come from originally. The resulting murky taxonomy was exactly what doesn't happen in a Darwinian framework of "descent with modification," where changes can only be *inherited* rather than *copied*. Francis Arduini (1987) has made similar points.

As for purely random datasets, Hunter did not seem aware that cladistic studies already took such comparisons into account, such as Rudolf Meier *et al.* (1991) assessing how convergent features (homoplasies) were detectable in datasets. And cladists continued to do so, as Alexei Fedorov *et al.* (2003) checked against literally *tens of thousands* of random sets to measure the significance of intron placement (the noncoding segments of DNA that get snipped out prior to protein assembly) in relation to ancient gene boundaries.

Convergence can occur at any level of biology, of course, all the way down to viruses and mitochondria, James Bull *et al.* (1997) and David Mindell *et al.* (1998). In their commentary on papers by Nicolas Gompel & Sean Carroll (2003) and Elio Sucena *et al.* (2003), Michael Richardson & Paul Brakefield (2003) noted the convergent tendency of pigment patterns and hair distribution in fruit flies, and Régis Chirat *et al.* (2013) called attention to the underlying developmental and mechanical constraints on the convergent appearance of spines in seashells.

If frisky is more your thing, Raymond Huey *et al.* (2000) spotted the very rapid convergence of wing length in a species of fruit fly only introduced in the Americas a few decades ago. The shift was predicted as a correlation of latitude, by the way, but the segment of the wing doing the lengthening turned out to be different than its Old World counterparts.

The important distinction is that "convergence" is not something casually argued. Whitfield (1993, 176-177) explained "primitive" versus "derived" traits in classification, and anyone can follow how such concepts are used in works like Simpson (1983, 196-200), Dawkins (1986, 100-107), Gamlin & Vines (1986, 13, 78, 81, 95, 126), Neil Shubin (1998) regarding Jennifer Clack (1998), or Ernst Mayr (2001a, 222-226).

While some older taxonomists used to the morphologically based Linnean classification rules chafed at the tight strictures of cladistics, newer analysts found the rigor of it liberating, and took to the new methods with enthusiasm. This was especially true in paleontology, where so many bits and pieces are found rather than the fully assembled working packages seen in living organisms, so that a data-crunching tool like cladistics proved most useful.

Not surprisingly, Alan Feduccia wasn't inclined to adopt cladistics in his own evaluation of bird orgins, Feduccia (1999b; 2012) reacting to the practice mainly by criticizing specific examples of its use.

Being a biochemist, Michael Denton's discipline was even farther removed from the fray than Feduccia's ornithology, and Denton showed little appreciation of what the systematics convulsion was about, apart from the fact that the technique itself didn't depend on evolutionary assumptions.

But it was in mistaking cladistics as somehow *abandoning* or *rejecting* evolution that Denton most seriously ran off the rails, along with other antievolutionists of like confusion.

Rather like bears to honey, antievolutionists had long been drawn to evolutionary tussles over phylogenetic classification (what was thought more closely related to what), going at least as far back as Henry Morris (1963, 87-90), though a generation later, Henry Morris & John Morris (1996, 35-37, 244) were more circumspect in alluding to cladism's arrival on the scene.

Creationist Gary Parker frequently invoked Denton's misplaced authority on matters ranging from homology to molecular clocks, Morris & Parker (1987, 48-50, 54-55, 60-61, 99-100, 145-146). As for phylogenetic systematics, Parker burbled in Morris & Parker (1987, 122) that "one of the most brilliantly and perceptively developed themes" in Denton's *Evolution: A Theory in Crisis* concerned "how leaders in the science of classification—after a century of trying vainly to accommodate evolution—are returning to, and fleshing out, the creationist typological concepts of the pre-Darwinian era."

No they weren't, but you could see how the secondary addicts among the antievolutionists managed to talk themselves into thinking that they had.

Duane Gish (1993a, 315) cited a 1980 piece by Colin Patterson which quoted his colleague Gareth Nelson on how cladistics was "rediscovering" or "fleshing out" *pre-evolutionary* systematics. Gary Parker had riffed off that 1980 example but changed the phrase to *creationist*. We'll have more to say on Patterson and his version of cladistics shortly.

Farther out on the apologetic trail, Wendell Bird (1989, Vol. 1, 144) blithely submerged cladistics in a discussion of supposedly conflicting "Approaches to Macroevolution." Indeed, Bird maintained that its "proponents refuse to use and in some cases reject macroevolution entirely."

This would have been news to the founder of cladistic analysis, Willi Hennig (1966, 225), who readily concurred with the principle that "macroevolution" not only occurred, but ultimately consisted solely of *microevolutionary* processes.

But what in the heck did Bird mean by "use" macroevolution?

Macroevolution is an observation about the range of accumulated disparity in life. It's something you see and appreciate and try to understand. It's not itself a process or tool that you *use* like cladistic analysis, or an Allen wrench.

Bird was in a muddle.

An explanatory footnote in Phillip Johnson (1991, 134) was only a tad less effusive than fellow lawyer Bird: "Some Darwinists of the old school think that cladism predisposes the mind to think of evolution as a process of sudden branching rather than Darwinist gradualism, and a few cladists have said that, as far as their work is concerned, the hypothesis of common ancestry might as well be abandoned."

Time for a second opinion.

In his final major opus, *The Structure of Evolutionary Theory*, Stephen Jay Gould trenchantly offered one of the reasons why cladistics had stirred up trouble, and it had nothing to do with "macroevolution" or rejecting common descent. It *did* involve how species were tagged. Gould (2002, 605):

> Many evolutionary biologists have failed to recognize that the so-called cladistic revolution in systematics rests largely upon this insistence that species (and all taxa) be defined as discrete historical individuals by branching (leading to the rule of strict monophyly)—and not as classes with "essential" properties by appearance (leading to the acceptance of paraphyletic groups). Many biologists reject (and regard as nonsense) the cladistic principle that no species name can survive the branching off of a descendant--and that both branches must receive new names after such an event, even if the ancestral line remains phenotypically unchanged. But this counterintuitive rule makes sense within cladistic logic—for cladists define new entities only as products of branching (the word *clade* derives from a Greek term for *branch*). A transforming species that does not branch cannot receive a new name even if the final form bears no phenotypic resemblance or functional similarity to the original ancestor. Thus if such extensive transformation occurs in unbranched lineages, a cladist, by failing to designate a truly different anatomy with a distinctive name, retains the technical individuality of species at the price of a severe assault against legitimate intuition.

Leigh Van Valen (1978) offerered similarly wary comments about the limits of cladism when applied to paleontology, and Peter Dodson (2000, 506-508) did likewise apropos working out bird origins.

This may all sound very arcane, but such things were more than quibbles. Unless species splits or significant changes in them were designated separately, the evaluation of how their character states related over time couldn't be conducted as unequivocally when some species were kept as lumped blocks because they hadn't visibly branched. Olivier Rieppel (2011) noted the impact of this issue, which has taken on greater significance when studying life at the prokaryotic bacteria level—wee little things that don't reproduce sexually, but are capable of rampantly exchanging genetic information and thereby obtain new functions that can make their taxonomy especially difficult to work out.

No wonder that outsiders, unaware both of the taxa involved and how the specific technical classification rules were applied in the stricter forms of cladistics, licked their chops and decreed confusion reined in Evolution Land.

Colin Patterson (1933-1998)

Probably the most notorious and persistent example of authority quoting in the cladistic systematics venue concerns the remarks of the late Colin Patterson, who stuck his foot in his mouth on several occasions and has been the citational darling of antievolutionists ever since.

When all the pieces are put on the table, it's quite a tale of obtuse scholarship and talking at cross purposes.

As a combination cladist and evolutionary gradualist, Patterson candidly reflected the transatlantic consternation British neo-Darwinist paleontologists experienced in the 1970s and 1980s under the persistent siege of cocky American paleontologists introducing Punctuated Equilibrium—topics that subsequently dominated Niles Eldredge (1995), *Reinventing Darwin*.

Patterson also favored a particularly restrictive version of cladism, "transformed" (or pattern) cladism, which specifically left common ancestry out of the picture as a guiding assumption.

Huzzah, now we are on the trail of Phillip Johnson's sources, and Gary Parker's "fleshing out" notions. Because transformed cladism sounded so congenial to the antievolutionary objections of creationism, Duane Gish (1993a, 317) was downright giddy to report "evolutionary biologists despise transformed cladism!"

Well, less despised than *ignored*. For there have been very few practioners of unalloyed pattern cladism. Patterson's collegue Gareth Nelson was one, and Andrew Brower another. But even Brower (2000, 143) wasn't rejecting evolution as a well supported scientific concept, only contending that by not using evolutionary assumptions in the method, their technique "allows inference of a scientific theory of evolution."

The Patterson affair started in 1979 when the paleontologist replied to a letter by quote-mongering Luther Sunderland quizzing Patterson on a variety of evolution topics, including the supposed absence of transitional forms in Patterson (1978).

Ah, what could go wrong?

To what extent Sunderland revealed his creationist proclivities up front to Patterson is hard to say, but Sunderland (1988, 101-102) was obviously delighted with the paleontologist's response, since he quoted three paragraphs of it. Patterson had begun with:

> I fully agree with your comments on the lack of direct illustration of evolutionary transitions in my book. If I knew of any, fossil or living, I certainly would have included them. You suggest than an artist should be used to visualize such transformations, but where would he get the information from? I could not, honestly, provide it, and if I were to leave it to artistic license, would that not mislead the reader?

This was an amazing thing for Patterson to have written, and not only for his surface agreement that transitional fossils didn't exist. Was Patterson suggesting that the practice of fleshing out extinct fossils by informed illustration was impossible, inevitably misleading, or merely that he had no skill or inclination to provide the necessary input himself?

It's hard to decide based on the letter, but in either case Patterson was skating on thin ice. Forensic reconstruction of bones builds on the real way in which skin and muscles attach, and this can be rigorously applied to extinct life. The most that could be argued was that artists of 1979 weren't up to that task, or Patterson didn't know much about the issue to contribute himself--neither boding well for Patterson's expertise wading into his interaction with a tendentious American creationist.

The field of fossil reconstruction has of course moved on since 1979, as Ian Tattersall (1995, 16-17) illustrated concerning hominids. Current paleoartists like Elisabeth Daynes or Gregory Paul draw on every scientific clue to arrive at their painstakingly researched reconstructions, covered by Helen Thompson (2014) and *Smithsonian* (2015).

Meanwhile, underinformed creationists like Harun Yahya (2009) are stuck in a Sunderland mindset, griping about "fantastic drawings" without examining any of the underlying science. We'll bump into that attitude again with Kenneth Poppe down in Chapter 9.

The next longer paragraph in Patterson's letter was a jawdropper, and Sunderland must have been rubbing his hands in gotcha delight to have read this:

> I wrote the text of my book four years ago. If I were to write it now, I think the book would be rather different. Gradualism is a concept I believe in, not just because of Darwin's authority, but because my understanding of genetics seems to demand it. Yet Gould and the American Museum people are hard to contradict when they say there are no transitional fossils. As a paleontologist myself, I am much occupied with the philosophical problems

> of identifying ancestral forms in the fossil record. You say that I could at least "show a photo of the fossil from which each type of organism was derived." I will lay it on the line—there is not one such fossil for which one could make a watertight argument. The reason is that statements about ancestry and descent are not applicable in the fossil record. Is *Archaeopteryx* the ancestor of all birds? Perhaps yes, perhaps no: there is no way of answering the question. It is easy enough to make up stories of how one form gave rise to another, and to find reasons why the stages should be favored by natural selection. But such stories are not part of science, for there is no way of putting them to the test.

But look at how the book in question, Patterson (1978, 131-133), had taken note of *Archaeopteryx* and other transitional forms:

> In several animal and plant groups, enough fossils are known to bridge the wide gaps between existing types. In mammals, for example, the gap between horses, asses and zebras (genus *Equus*) and their closest living relatives, the rhinoceroses and tapirs, is filled by an extensive series of fossils extending back sixty-million years to a small animal, *Hyracotherium*, which can only be distinguished from the rhinoceros-tapir group by one or two horse-like details of the skull. There are many other examples of fossil "missing links", such as *Archaeopteryx*, the Jurassic bird which links birds with dinosaurs (Fig. 45), and *Ichthyostega*, the late Devonian amphibian which links land vertebrates and the extinct choanate (having internal nostrils) fishes."

That kinda sounded like Patterson accepted transitional forms in the fossil record, didn't it? And he had even included a picture of *Archaeopteryx* in the book, so why had Sunderland written of their absence, or Patterson not noted that he had done just that with that bird "missing link"? Instead, he'd responded to Sunderland solely in terms of *testing*. That was Patterson grumbling under his transformed cladism hat, wasn't it? No way to put them to the test for *Patterson*, at least. But was it really untestable, in principle?

Let's get specific. Would an absolutely perfect bird ancestor have looked notably *different* from *Archaeopteryx*? If so, in what ways? Both would have had some identifiably distinctive anatomy, wouldn't they? If they did not differ substantially, in what sense then would *Archaeopteryx* not represent a perfectly valid surrogate for whatever actual ancestor never got preserved? Or would transformed cladist Patterson not allow it to be identified as such in any case?

The same conceptual problem would apply to the RMT players, of course (which Patterson's 1978 *Evolution* book so notably did not mention in his skimpy listing of transitional examples), an issue only compounded because there were so many more of them.

Some of what Patterson wrote to Sunderland clearly turned on the dictates of pattern cladism, which precluded identifying transitional forms *in principle*. But in his allusion to Gould and Eldredge, Patterson was showing that he might not have fully understood that problem either. Gould and Eldredge had not claimed that no intermediate forms existed at all, only that they were comparatively rare insofar as *species* transitions were concerned.

They weren't even opposed to Patterson's favored mode of gradualism. As Gould & Eldredge (1977, 121) put it, "The model of punctuated equilibria does not maintain that nothing occurs gradually at any level of evolution. It is a theory about speciation and its deployment in the fossil record." As noted last chapter, I covered more on the Punctuated Equilibrium issue (and how all antievolutionists have tripped over it, and I do mean *all*) in Downard (2016). Maybe I needed to have included Colin Patterson, too.

To be blunt, in discussing *Archaeopteryx* and the fossil record, Patterson sounded a lot like antievolutionists. And by this I mean, given how much information was available, he was being oddly *vague*.

I ran into exactly the same "can he ever be pinned down on anything" problem on this very point (*Archaeopteryx* and identifying bird transitionals) with Intelligent Design antievolutionist Jonathan Wells in 2001, which I recounted in Downard (2003b, 90-92). Like Sunderland-Patterson, mine was a letter exchange, where I tried to get Wells to explain what a perfect reptile-bird intermediate would need to look like to satisfy him. Wells wouldn't say, and hopped back onto the same rock to insist *Archaeopteryx* was inadequate, without ever saying what *would* be adequate.

Sunderland's quoting of the Patterson letter ended with: "So, much as I should like to oblige you by jumping to the defense of gradualism, and fleshing out the transitions between the major types of animals and plants, I find myself a bit short of the intellectual justification necessary for the job."

I would be inclined to agree, but not for the most charitable of reasons.

Patterson's 1979 letter was juicy enough for creationist purposes, with Percival Davis & Dean Kenyon (1993, 106-107), Paul Taylor (1995, 108), Gish (1995, 349), James Perloff (1999, 11), Ken Ham *et al.* (2000, 28-29) and Vance Ferrell (2001, 459-460) readily invoking the Sunderland letter for the claim that Patterson didn't believe there were any transitional fossils.

But things got even curioser as Patterson was interviewed directly by Sunderland a few months later. Sunderland (1988, 102-103) catalogued his questions and Patterson's disconcertingly terse replies.

"You stated in your letter that there are no transitions. Do you know any good ones?" Patterson replied, "No, I don't, not that I would care to support, no."

"Throughout the interview he denied having transitional fossil candidates for each specific gap between the major different groups. He said that there are kinds of change in forms taken in isolation but there are none of these sequences that people like to build up," but Sunderland did not elaborate on which of these "sequences" they discussed. If the RMT was among them, was the sequence missed because they were being taken "in isolation"? Plug in the *Zeno Slicer* again?

Sunderland then quoted Patterson: "If you ask, 'What is the evidence for continuity?' you would have to say, 'There isn't any in the fossils of animals and man. The connection between them is in the mind." Yipes. Knowing what we do of the fossil record (including primates and hominids), it's difficult to think Patterson was much of a paleontologist or evolutionist at all, to have given such a vacuous answer.

What Sunderland quoted next only served to reinforce this. Did Patterson "know of any documented evolution going on today in the macro sense where we're looking for a new structure that previously did not exist—like an arm forming?"

Patterson answered, "No, not of an arm forming, not in the macro sense."

"Then you know of no structure that you could classify as developing and not fully functional?" Patterson replied, "No."

We've heard this song before. This is the tune of Duane Gish and Kurt Wise last chapter, caroling the idea that evolutionary transitions would require *non-functional* halfway intermediates (the opposite of the evolutionary mandate).

If all Patterson said to these leading rhetorical questions were flat "nos", once again we have to wonder just how up on things Patterson was. An evolutionist should have challenged Sunderland's initial flawed assumption, not tacitly reinforce it.

That Sunderland hadn't got the memo that neither cladism nor Punctuated Equilibrium weren't actually a new version of creationism was evident as Sunderland was soon invoking Patterson and Eldredge as *supporting* creationism, promoting letters of disavowal from them both, reported by Stanley Weinberg (1980, 4, 7). Years on, Eldredge (2000, 17-18, 129-134, 187n, 208-204n) took note of the persistent Mr. Sunderland with similar disaproval.

While Patterson never got interviewed by Sunderland again, he stepped into trouble on his own in 1981 when he spoke at an informal gathering of systematists at the American Museum of Natural History in New York. Unbeknownst to the participants, another creationist was on hand for this, tape recorder at the ready, and shunted the resulting barely audible tape on to Luther Sunderland, where a transcript ultimately percolated through a receptive antievolutionary community, as covered by Arthur Strahler (1987, 354-355).

The creationist reaction was predictable. Gary Parker decided Patterson believed evolution "has been *falsified*," Morris & Parker (1987, 58). Wendell Bird (1989, Vol. 1, 152) highlighted the unflattering parallels Patterson found between creationism and evolution, such as when Patterson contended that evolution "not only conveys no knowledge but it seems somehow to convey *anti-knowledge*."

Phillip Johnson (1991, 9) also drew on this "remarkable lecture" (whose "bootleg transcript" was circulated by "somebody") for the conclusion that both creationism and evolution were "scientifically vacuous concepts which are held primarily on the basis of faith. Many of the specific points in the lecture are technical, but two are of particular importance for this introductory chapter." The derivative Ankerberg & Weldon (1998, 139) siphoned Johnson's account of the Patterson AMNH talk secondarily.

The first of Johnson's points was Patterson's provocative claim that evolutionists weren't able to identify anything about "evolution" that was "true," and the second concerned the supposedly shady character of the Darwinian natural selection mechanism.

Concerning the context of the debate, the Research Notes in Johnson (1991, 157) said "I discussed evolution with Patterson for several hours in London in 1988. He did not retract any of the specific skeptical statements he has made, but he did say that he continues to accept 'evolution' as the only conceivable explanation for certain features of the natural world."

This certainly fitted in with Johnson's philosophical view that evolution wasn't based on a solid grid of interlocking evidence, but never considered that maybe Patterson was not the most disinterested witness on this. I have to wonder whether Johnson was playing yet another Sunderland role, trying to discuss issues he did not understand with someone ill-disposed to clarifying them.

The key to Patterson's discomfiture was briefly (if not inadvertently) touched on by Johnson (1991, 10) himself, when he mentioned that "now, according to Patterson, Darwin's theory of natural selection is under fire and scientists are no longer sure of its general validity."

That wasn't really true. What *was* getting shaken up was the gene-centered traditional Darwinian model Patterson was steeped in, and the convulsion of the period was in part driven by the very cladistics Patterson was on the radical fringe of.

None of which filtered down to conservative boosters Tom Bethell (1985; 1999b, 20; 2001a, 2005, 218-219) and Doug Bandow (1991) who invoked Patterson's seeming apostasy for their more political audiences, or YECers Paul Taylor (1995, 115), Hank Hanegraaff (1998, 33, 170n, 177-178n) and Bert Thompson & Brad Harrub (2002, 6-7) banging the Patterson drum for their creationist readers. Farther down the apologetic daisy chain, conservative foghorn Ann Coulter (2006, 2001) got her version of Patterson secondarily from Bethell (2001a).

Australian creationist Carl Wieland continued to field the Sunderland edition of Patterson's opinions in the 1990s, prompting critic of creationism Lionel Theunissen (1997) to track down the particulars. Contacting him in 1993, Patterson explained the context of his views expressed in the 1979 letter (which was indeed turning on Patterson's transformed cladism strictures on identifying transitionals within the method). Patterson wrily acknowledged to Theunissen that "I seem fated continually to make a fool of myself with creationists."

No argument from me there.

Patterson then explained that "The famous 'keynote address' at the American Museum of Natural History in 1981 was nothing of the sort. It was a talk to the 'Systematics Discussion Group' in the Museum, an (extremely) informal group. I had been asked to talk to them on 'Evolutionism and creationism'; fired up by a paper by Ernst Mayr published in *Science* just the week before. I gave a fairly rumbustious talk, arguing that the theory of evolution had done more harm than good to biological systematics (classification). Unknown to me, there was a creationist in the audience with a hidden tape recorder. So much the worse for me. But my talk was addressed to professional systematists, and concerned systematics, nothing else."

Likely prodded by Theunisson's inquiry, Colin Patterson (1994, 7) incorporated a short note on the kerfuffle in a subsequent contribution of a systematics book, which Paul Nelson (1998) quoted when he dived onto the Patterson affair. I have highlighted some noteworthy additions to the 1981 version in **bold** (the bracketed comment was Nelson's inclusion):

> In November 1981, after an invitation from Donn Rosen [a fish systematist at the American Museum, now deceased], I gave a talk to the Systematics Discussion Group in the American Museum of Natural History. Donn asked me to talk on "Evolutionism and Creationism", and it happened that just one week before my talk Ernst Mayr published a paper on systematics in *Science* (Mayr 1981). Mayr pointed out the deficiencies (in his view) of cladistics and phenetics, and noted that the "connection with evolutionary principles is exceedingly tenuous in many recent cladistic writings." For Mayr, **classifications should incorporate such things as 'inferences on selection pressures, shifts of adaptive zones, evolutionary rates, and rates of evolutionary divergence.' Fired up by Mayr's paper, I gave a fairly radical talk in New York, comparing the effect of evolutionary theory on systematics with Gillespie's (1979, p. 8) characterization of pre-Darwinian creationism: "not a research governing theory (since its power to explain was only verbal) but an antitheory, a void that had the function of knowledge but, as naturalists increasingly came to feel, conveyed none."** Unfortunately, and unknown to me, there was a creationist in my audience with a hidden tape recorder. A transcript of my talk was produced and circulated among creationists, and the talk has since been widely, and often inaccurately, quoted in creationist literature.

Nelson showed no curiosity about Ernst Mayr (1981), who had suggested the best aspects of several techniques (including cladistics) should be used in concert, or wade into the more provocative unspoken presumption in Nelson's own creationism, that there existed (or could ever exist) a classification system (kinds or baramins as you will) that could explain in detail evidence like the RMT. M. Aaron (2014) is the only creationist so far to take a stab at it, and we'll look at his effort in Chapter 10.

It is also interesting that, despite the experience Patterson had with creationists already, that he thought to underscore his own ill-chosen rhetoric by reminding his readers of his allusion to Neal Gillespie (1979, 8). If Patterson seriously thought current evolutionary theory's explanatory power was on a par with "pre-Darwinian creationism" (let alone post-Darwinian modern creationism) his conceptual perspective was seriously askew.

Which brings to mind: what exactly do modern creationists say about systematics?

Being Kind to Baraminology

The subject of antievolutionist non-taxonomy is wide enough for another book, but a few examples from the recent creationist camp will suffice to illustrate the scale of the YEC quandary.

By the 1990s, creationists like Kurt Wise were trying to work out how many initially created *kinds* there were (bearing in mind how many of them are still alive today, only 4500 years after the Flood). To this end they have revived the 1940s work of creationist Frank Lewis Marsh (1899-1922), who coined the term "baraminology" (*baramin* being a Hebrew neologism for "created kind"). Ronald Numbers (1992, 124-133) chronicled Marsh's role in the creationism of the period.

Because Kurt Wise was trained in paleontology, he helped retool the revived form (dare we call it "transformed baraminology"?) to include subcategories that reflected the nomenclature of standard systematics.

Thus there are "apobaramin" and "polybaramin" groupings of unrelated organisms that would include one or more "holobaramins" (the preserved representatives of the actual created "baramin" type), riffing off standard terminology like the paleontological "holotype" that refers to the benchmark fossil representative of a genus.

Within a baramin there could be "monobaramins" (naturally related lineages) as subsets. This is the crucial subcategory, for if a monobaramin spreads past the species level, this means *accepting natural evolution within a created kind*.

Though rather as stuck on the same common categories (dog, cat, turtle, sunflower) as Henry Morris or Duane Gish in their traditional YEC apologetics, or Michael Denton in his propped up typology, baraminologists Todd Wood and his collaborator David Cavanaugh have tried to apply topological algorithms to morphological data sets (cribbed from the standard cladistic fieldwork, one might add), under the presumption that holobaramins would stand out by the discontinuities in their spatial plotting.

But so far their work hasn't isolated notably impressive holobaramins—in Chapter 10 we'll look into the curious exceptions of Todd Wood (2010) on hominids and M. Aaron (2014) on synapsids (both turning on fossil examples and operating up at the family level). Most baraminology studies have focused on living animals, though, and kept much lower in the taxonic ranking. Down at the generic level, things look like a series of monobaramins. In other words, *evolution*.

The sunflower case was typical, illustrating just how much of the available dataset the creationists had to step around in their own analyses.

Wood & Cavanaugh (2001) and Cavanaugh & Wood (2002) determined that sunflowers were apparently related (as a *monobaramin*) but couldn't say how far up the phylogenetic ladder their *holobaramin* went. This posed a further theoretical problem, as even these plants showed a range of photosynthetic systems (C3, C3-C4 intermediate, and C4) that would have to be submerged in an initially perfect created type. Consequently Cavanaugh and Wood decided the genes for all the systems had to have been designed in at the Creation, providentially anticipating the Flood whereby the sunflowers would be prompted to evolve (within their holobaramin, whatever that may be) the variant photosynthetic pathways we observe today.

All that variation zipping along in less than 4500 years? Is that genetically plausible? Wouldn't this require super-fast mutation rates beyond anything seen in living biology, dwarfing the microevolutionary processes evolutionary theory builds on for its far slower dynamics of change—a process that creationists like Wood and Cavanaugh insist cannot happen?

Yep, Todd Wood (2002a-b; 2003) is very serious about this contradictory idea, as he would have to be, to square the observed range of diversity with the compact *kinds* required by the theologically mandated Flood event.

How much data the baraminological view needed to have accounted for in just this one sunflower case sprawls across biogeography (where things live), endosymbiotic inheritance (biological mergers creating novel biology), gene duplications (which mutate to generate new genetic information), positive Darwinian selection (the differential survival of improvements) and other issues. A technical sampler: Kåre Bremer (1987; 2000), Ki-Joong Kim & Robert Jansen (1995), Yrjö Helariutta *et al.* (1996), Maurice Ku *et al.* (1996), J. Marshall *et al.* (1996), Kåre Bremer & Mats Gustafsson (1997), Michael Clegg *et al.* (1997), María Drincovich *et al.* (1998; 2001), Hyi-Gyung Kim *et al.* (1998), Jose Panero *et al.* (1999), Lien Lai *et al.* (2002), David Remington & Michael Purugganan (2002), Lorraine Tausta *et al.* (2002), Ji Yang *et al.* (2002; 2004) and Loren Rieseberg *et al.* (2003).

What is a tangled problem for creationists trying to infer plant biology in Noah's day becomes a shipwreck when it comes to vertebrate *kinds*, for they were to be explicitly preserved aboard the Ark, and there's only so much room. The need to whittle down the number to keep within the putative stall plan of the Ark has been the proverbial elephant in the room (or on the gangplank in this case).

John Woodmorappe (1996) gave it the good old Young Earth Creationist college try in *Noah's Ark: A Feasibility Study*, but even there what we were missing was a plain listing of how many *kinds* there were supposed to be (let alone how their number had been determined). There'll be much more to say on John Woodmorappe's apologetics apropos our RMT subject in Chapter 8.

Alan Gishlick (2006) noted the many technical problems inherent in baraminology, but the most interesting has turned out how often baraminology has tended to confirm evolutionary relationships, to the point where Phil Senter (2010; 2011) marshalled the baraminology work in support of evolution, with not unexpected demures from Tood Wood (2010b; 2011a-c). A particularly juicy example of this was Wood & Cavanaugh (2003, 4-5) and Cavanaugh *et al.* (2003) effectively throwing in the towel on the famed horse evolution sequence by pegging its *monobaraminic* status.

And where to stop?

The early dog-sized horse ancestor *Hyracotherium* was very similar to several contemporaneous critters, such as *Homogalax* at the base of the tapiroid superfamily, Leonard Radinsky (1969). As George Gaylord Simpson (1960, 123) put it, *Homogalax* and *Hyracotherium* represent "not only the common parent of horse and tapir but also the common ancestor of two major divisions (suborders) of mammals."

If baraminology can't keep *Hyracotherium* out of the horse monobaramin stable based on the range of identified character states, how could the so-similar *Homogalax* be excluded, and by extension, all of its descendants down to the living tapirs?

Interestingly, Ken Ham (2012) was still preaching the invalidity of the horse evolution sequence a decade after Wood and Cavanaugh's gobsmacking concession. Is it possible that the *Answers in Genesis* creationist didn't bother to keep up with his own side's technical literature? As I said, *interesting*.

Or take the iconic poster child for evolution, Darwin's Finches that live on the Galápagos Islands in the Pacific. Where did they come from?

Evolutionary scientists had not been idle, applying their taxonomical and genetic analysis skills to track them back to several candidate genera within a group of seed-eating birds in the Emberizidae family, the tanagers and grassquits, such as the West Indian black finch *Melanospiza richardsonii* and the more common grassquit *Volatinia jacarina* of Central and South America, noted by Jonathan Weiner (1994, 221)—and Colin Patterson (1999, 92) too, showing he *could* venture an opinion on evolutionary relations now and then.

Later genetic analyses further pressed the Darwin Finch origins through to the yellow-faced grassquit genus *Tiaris* as closest living relatives, Joanna Freeland & Peter Boag (1999), Akie Sato *et al.* (1999; 2001) and Kevin Burns *et al.* (2002).

Now Michael Denton (2016a, 33-36, 290n) fully accepted the natural evolution of the finches, recognizing a lot of the technical work establishing how their beak variations arose strictly by Darwinian natural selection from such familiar genes as *Bmp*: Arhat Abzhanov *et al.* (2004; 2006; 2007), Ping Wu & Ting-Xin Jiang *et al.* (2004; 2006), Bailey McKay & Robert Zink (2015). Denton (2016a, 35):

> As far as the evolution of finch beaks is concerned, there is no need either at the morphological or genetic level to call for any causal agency other than cumulative selection. Here I concur with classic Darwinism. The beaks are clearly adaptations and their evolution is entirely explicable within a classic functionalist framework. As the different beak forms are clearly contingent adaptations which evolved to meet the unique environmental demands on a group of volcanic islands that only emerged from the Pacific a few million years ago, their evolution is beyond any structuralist or "laws of form" type of account.

Ah, but what Denton giveth with one paragraph, he immediately taketh away with another, insisting that such "adaptive sequences, either empirically known or hypothetical, *are* lacking in the vast majority of cases of macroevolution," including from "the scale of a reptile to the feather of a bird," Denton (2016a, 35-36).

Now you'll notice Denton was uncharacteristically doing a *Map of Time* thing with the finches, acknowledging their duration and the geological impact of their habitat. But that's only because we have so much data in play: a dozen species of living finches (more or less, as their number depends on how their status is defined), enabling scientists to *directly* examine their genetics and even tinker experimentally with them to show precisely how their fluctuating variations arose naturally.

What Denton has not shown an inclination to do (either in his 1985 book or its 2016 reincarnation) was apply such inferences to subjects in the far more distant past where we have only isolated *fossil* representations of the players, not living ones, and often millions of years separating the examples that have turned up. The idea that ancient life couldn't have avoided doing just what we can see the living finches doing today, though, naturally adapting in places where the climate and competitors were likewise changing over time, perpetually eluded Denton.

And *all* antievolutionists.

Just like Denton, they can easily recognize isolated instances of contemporary "microevolution," but appear incapable of using that mental yardstick to consider what must have been going on in the past. Every fossil animal would have represented a suite of genetics and ecological interactions as rich as anything physically observable in the Galápagos Islands today. Yet the antievolutionist invariably fails to work out in their own heads what manner of *Map of Time* they think was going on. They end up perceiving every data scrap only as isolated Zeno-sliced blips, never relating them to the underlying processes than can, fortunately, still be seen operating in their living descendants.

It's this *compartmentalization* of thought, hunkered down under an impenetrable Tortucan mindshell, that puts a Michael Denton circling the same limited cognitive terrain as a Jonathan Wells, giving a softball interview for Todd Butterfield, linked by *Evolution News & Views* (2016c). Ever ready to act as the ID fortress gatekeeper, Wells easily dismissed the extraordinary range of evidence showing the precise adaptive genetics of finch beaks as merely microevolutionary fluff, with no implications for the dreaded macroevolution.

And where do Young Earth creationists stand on this finch paternity suit?

They have a rather different problem, accounting for all the existing diversity while simultaneously keeping an eye on how few stalls were available aboard Noah's Ark. It makes for a quirky taxonomy.

Jean Lightner (2010; 2013) at *Answers in Genesis*' technical venue, the *Answers Research Journal*, lumped and clumped the 9000-odd species of birds into 196 bird "kinds." That included the "Sparrow/Finch Kind" that embraced over *twelve hundred species*! With a flick of the creationist wrist, Lightner had just accepted that *all* the finches on earth were in fact genealogically related, exactly as evolutionists had long been insisting.

And yet in the creationist mind, even that extensive speciation doesn't bring them pause regarding whether a "kind" means much of anything, if it can flow so readily along the speciation escalator. Nor did Lightner include any *fossil* birds in her survey, and didn't get too technical on whether those 196 "kinds" could survive strict baraminological scrutiny.

Map of Time again.

So where does all this put Paul Nelson and his playing of the Patterson card? Had he wanted to wade in on the issue of creationist explanations for taxonomy, the water may have been pretty deep (with or without a Flood spiggot overflow). Nelson had different interests in mind anyway, though, further quote mining Patterson (1994, 175, 188-189) on his equivocal feelings about the meaning of the patterns in molecular data that had been such an object of his prior interest. By this means Nelson sought to cast doubt on the ability of determining evolutionary relationships at that level.

Nelson added in a note this passage from Patterson (1988, 615):

> Convergence between molecular sequences is too improbable to occur, just as similarity between sequences is improbable to be explained except by common ancestry. Some might view this argument as viciously or vacuously circular, but the same argument is routinely advocated in morphology....This is the argument from complexity: if two structures are complex enough and similar in detail, probability dictates that they must be homologous rather than convergent.

Although Patterson's 1988 paper was about how using several independent methods could help distinguish homology from convergence (sounding rather like Ernst Mayr), Nelson did not go into that or any of the specific examples Patterson cited. Need we dwell on how molecular genetics had progressed in the decade since Patterson's paper, or belabor how it had not stopped as a discipline in the many years since then?

What we're left with in the Patterson affair was a technically minded scientist whose parsings of complex issues often generated quote-mine friendly text strings, coupled with only the dimmest sensibility that there were so many antievolutionists circling at the ready like vultures.

Patterson's transfomed cladist associate Gareth Nelson (1998) eulogized his colleague's legacy for *Nature*, but I'll conclude with some remarks on the 2nd edition of Patterson's *Evolution* book, which appeared posthumously in 1999. The work was a nice enough summary of the basics of genetic inheritance and microevolutionary variation, but not really a vast improvement over Patterson's 1978 edition, and oddly quiet on several relevant issues.

In a brief comment on the resurgent creationism of the 1980s, Patterson (1999, 121-122) made no reference to Sunderland's letter, his prickly AMNH speech, or awareness that antievolutionism had already pressed on to the

Intelligent Design path of apologists like Phillip Johnson or Paul Nelson. Even more interestingly, regarding transitional forms, Patterson (1999, 108-109) simply reprised his old 1978 views, adding only a picture of a new Devonian tetrapod, *Acanthostega*. Was that the best he could do in 1998?

Not only was there still no mention of the RMT characters, Patterson didn't even include such amazing newer finds as the transitional whales that were hot news by the 1990s, covered popularly by Philip Gingerich (1994) or Carl Zimmer (1998) in *At the Water's Edge*. These new fossils were filling in the picture on the transformation from legged ancestors to flippers, their nostrils migrating up the ceteacean skull as blowholes, and showing adaptation to seawater.

For the technically minded, I submit Philip Gingerich *et al.* (1983; 1990), Edward Mitchell (1989), J. Thewissen & S. Hussain (1993), Gingerich & Mahmood Raza *et al.* (1994), Thewissen *et al.* (1994), Thewissen & Sandra Madar *et al.* (1996; 1998), Thewissen & Roe *et al.* (1996).

Further whale evolution fossils turned up after Patterson's non-discussion of them, summarized by Thewissen *et al.* (2009). That represents quite a swim fest for the technical papers: Thewissen & Madar (1999), Thewissen & Hussain (2000), Gingerich *et al.* (2001; 2009), Thewissen & Sunil Bajpai (2001), Thewissen *et al.* (2001; 2006; 2007), Madar *et al.* (2002), F. Spoor *et al.* (2002), Thewissen & E. Williams (2002), Sirpa Nummela *et al.* (2004; 2007), Emily Buchholtz (2007), Lisa Cooper *et al.* (2007), Noel-Marie Gray *et al.* (2007), Madar (2007), Bajpai *et al.* (2009), Zhe Wang *et al.* (2009), Felix Marx & Mark Uhen (2010), James Dines *et al.* (2014) and Maya Yamato & Nicholas Pyenson (2015).

And that's just the fossil data and their implications. Don't forget the genetic clues (recall those baleen tooth enamel genes noted back in Chapter 2).

So, the legacy of the Patterson affair is not quite the trope antievolutionists would like. It was not the case of a premiere figure in the field revealing the rattling skeletons in the evolutionary closet, but rather a scientist who couldn't quite temper his language and didn't always think about things deeply enough.

Patterson's final 1999 edition of *Evolution* showed this in another even more telling way. It not only failed to mention Punctuated Equilibrium, it didn't even discuss *cladism* as a subject (transformed or otherwise). That was particularly odd, given how cladistic analysis had become ubiquitous in paleontology by then, so why not mention it? Or was that omission because the cladism that everybody was doing wasn't the version he so steadfastly championed? So leave it out?

There are a lot of peculiarities to the Patterson Case, but not really ones that call into question fundamental findings of evolutionary thinking. So some advice to antievolutionists (which I am sure they will not take): it's time to retire those Patterson quotes from the mine.

7. An ID Prosecutor Tries to Grill Therapsids

In giving antievolutionism a fresh coat of non-creationist paint, Michael Denton's *Evolution: A Theory in Crisis* quickly became a staple of antievolutionary citation, from creationist lawyer Wendell Bird (1989, Vol. 1, 220-221) to the core of the Intelligent Design movement, especially Phillip Johnson.

Being another in the long line of non-paleontologists thinking to elbow fossils aside without discussing much about them, Michael Denton (1985, 180-182) could freely (and briefly) acknowledge the specifics of fossil sequences (including the RMT) because he promptly upped the ante by requiring the preservation of internal organs before conclusive inferences could be drawn.

Denton (1985, 180) even suggested that, "The possibility that the mammal-like reptiles were completely reptilian in terms of their anatomy and physiology cannot be excluded. The only evidence we have regarding their soft biology is their cranial endocasts and these suggest that, as far as their central nervous systems were concerned, they were entirely reptilian." For this he drew on an irrelevant 1968 comparison of the overall brain size of the mammal-like reptiles with the earliest mammal known at that time, the Late Jurassic *Triconodon*.

But *Triconodon* lived about *50 million years* after the first Triassic mammals (already on the scope by the time Denton was writing in the mid-1980s), including *Morganucodon*. Brain studies were hampered by the need to destroy the skull to make endocasts (the brain itself not being preserved, but its form reflected by the impressions on the skull around it). Zofia Kielan-Jaworowska & Terry Lancaster (2004) represented that mode of analysis, regarding the Cretaceous multituberculate *Kryptobaatar*. Tom Kemp (2007c; 2009a) did considerable work on the Triassic *Chiniquodon*, identifying its limited low frequency hearing range and a brain that suggested mammal ones evolved in two main stages. Also called *Probelesodon*, this critter may be kept in mind when John Woodmorappe bumps into it next chapter, and Charles Creager takes a swipe at it in Chapter 10.

That endocast issue changed in the decades since, as non-invasive scanning techniques allowed more information to be gleaned, as Glenn Northcutt (2011) noted of the restudy of *Morganucodon* and *Hadrocodium* by Timothy Rowe *et al.* (2011) that continued to suggest there were several main stages in the evolution of the early mammalian brain, turning around expansion of olfactory sensing.

While such technological assists weren't available in Denton's 1980s, that didn't mean there weren't a suite of anatomical features on hand from which reasonable inferences could be drawn about metabolism or brain activity.

Specifically, Denton failed to take notice of their upright stance that implied a more than reptilian activity level (a similar inference applies to the more agile bipedal dinosaurs), or the suggestive presence of hair and whiskers on late members in the field that distinguished the later synapsids from their now very distant cousins, the anapsid and diapsid reptiles. Just like muscle attachments, larger sensory hairs leave telltale marks on the bones (foramina openings for the whiskers' neurological connections), and which in turn implied things about the brain where such sensory information would have been processed. These anatomical points were noted by Stahl (1985, 397-399), so Denton writing at the same time could have known about the primary source work Stahl drew on.

That is, if he'd stopped to pay attention to his own sources (this is getting to be rather a *pattern*, isn't it).

Ironically, Denton (1985, 112-115) had touched on a variety of biological clues suggesting how developmental switches might have flipped to generate the mammal transition, but saw these again only as difficulties. Denton (1985, 113) for example: "The major vessel leaving the left ventricle in a reptile, which is the major vessel carrying aereated blood from the heart, is formed from the fourth right aortic arch, while in a mammal it is derived from the left aortic arch."

Denton did not speculate about what he thought an intermediate form would look like here (the vessel forming from some *middle* aortic arch?) which might have clarified what if any role he was willing to grant the developmental process in macroevolutionary change.

As the genetics of axial asymmetry were still only poorly known in the pre-*Hox* 1980s, Denton was operating not unlike the way Duane Gish had with the organ of Corti and diaphragms. That long antievolutionary tradition of goalpost moving: never lay down criteria relating to what you do have, but always wade upstream as far as you're comfortable to something the science doesn't (yet) know.

Some research in that area was proceeding anyway, and for a most humane reason: whole body human organ reversals (*situs inversus*) occur in one of every ten thousand people, Enrico Coen (1999, 270-271), and those can have serious health repercussions in Rieger's syndrome. If not a product of natural evolution, might that be regarded as a "design defect" warranting a morally affronted lawsuit against the Designer?

Oh well.

In any case, the genetics of Rieger's syndrome were being uncovered in the years since 1985, Mei-Fang Lu *et al.* (1999) and Joseph Marszalek *et al.* (1999), as were the various axial symmetry genes noted back in Chapter 2. Far downstream from their involvement in forming paired appendages in vertebrates, the T-box transcription factors (*Tbx*) play out more directly in heart development, with work like Zheng-Zheng Bao *et al.* (1999), Kazuko Koshiba-Takeuchi *et al.* (2009) and Bjarke Jensen *et al.* (2013) investigating the developmental genetics underlying the evolution of vertebrate hearts.

But more of a theoretical problem was Denton's implicit presumption that *modern* reptiles represent the *ancestral* layout from which mammals would have had to have developed, and he had to literallly step over and edit his own source to do it.

Alfred Romer (1970, 408-414) had explicitly indicated the contrary, but Denton drew on that work only for a Figure 317 (from page 411) illustrating heart layouts. While Denton retained most of the long caption for the edited figure he removed some of Romer's text that had noted this:

> The mammalian condition has apparently arisen directly from the primitive type preserved in the Amphibia, for in modern reptiles the conus arteriosus shows a division into three vessels, rather than two; one, returning venus blood back to the body, leads only to the left fourth arch. In crocodilians the ventricular septum is nearly complete, and the elimination of the left fourth arch would give the avian condition.

Denton did not discuss the crocodilian example, either, and deleted its heart layout from the illustration. To our point, Romer (1970, 413) had gone on to explain, "Mammal ancestry diverged from that of the reptiles at an early date, and there is no reason to believe that the system of three heart orifices seen in modern reptiles was ever present in the mammalian line."

Much as Duane Gish had, Michael Denton (1985, 178) waved the convergence issue to dismiss the RMT fossil data that he paid even less attention to than Gish had. Denton's sole example was the similarity between the skulls of placental and marsupial dogs. "Anyone who had been privileged to handle, as I have, both a marsupial and placental dog skull will attest to the almost eerie degree of convergence between the thylacine and placental dog." Indeed, "in gross appearance and in skeletal structure, teeth, skull, etc," Denton declared they were "so similar in fact that only a skilled zoologist could distinguish them."

Oh, really?

In his classic criticism of creationism, *Science on Trial: The Case for Evolution*, Douglas Futuyma (1982, 46, 48) pointed out that the Tasmanian thylacine "wolf" has the marsupial dental layout of 3 premolars and 4 molars, while placental dogs have 4 premolars and only 2 molars in the upper jaw (3 in the lower though). Different *number* and *arrangement*. Did Denton's privileged observation not extend to counting the gross number of teeth? See Maureen O'Leary *et al.* (2013a, 666-667) for a tidy summary of the evolutionary development of the differing marsupial and placental dental layouts.

Frank Sonleitner found Denton's argument especially glib, forwarding to me a contemporaneous publication from Denton's own Australian backyard, Michael Archer & Georgina Clayton (1984, 588, 643-647), which noted the many diagnostic features unique to marsupials that separated the two taxa. These ran from the specialized tarsal bone in the foot to a host of distinctive features in their skulls. Besides the obvious dental differences, another item was especially apparent even to yours truly (a certified non-zoologist): the telltale holes in the palate found in all the Australian marsupials but in no placental mammal. We'll get back to Denton's notions about mammal dentition down in Chapter 11.

Of Pandas and People's Davis & Kenyon (1993, 117) and Richard Milton (1997, 192-193) similarly noted the correspondences between the skulls of North American wolves and the marsupial thylacines. And like Denton, neither mentioned the diagnostic traits that otherwise distinguished them.

Paleontologist Michael Benton (1990, 250-251) rightly related convergences to lifestyle: "even though a kangaroo looks very different from a deer or antelope, it lives in roughly the same way!" But while lifestyle can dictate a lot about the adaptive features an animal has, the internal structure still betrays its specific evolutionary lineage.

We don't have to *imagine* that a few "mammalian" traits could develop now and then through convergence, since they objectively did with the notosuchian crocodiles in the Early Cretaceous of central Africa, which had similar dentition and side-mounted eyes as the therapsids, noted by Czerkas & Czerkas (1991, 181). Small animals, the notosuchians were adaptively filling the sort of ecological niches in the southern hemisphere that mammals were occupying in the north, noted by Patrick O'Connor *et al.* (2010).

Secondary palates even developed in crocodilians, enabling them to breathe through their nose while the mouth remained open underwater. Interestingly, Roger Jankowski (2013, 63) noted that the "progressive anteroposterior closure of the embryonic mammalian palate actually resembles the various stages of crocodilian palatal evolution from protosuchians through mesosuchians to living eusuchians," suggesting many shared developmental pathways. But the resulting crocodilian features are not identical to their mammal analogs, and more importantly, the rest of their anatomy was still diapsid crocodile.

These details were what John Woodmorappe (2001) notably did *not* explain when he cited fellow creationist Mehlert (1993, 132) on these notosuchians, describing them as "an extinct Mesozoic crocodilian from Malawi" whose "dentition shows clear resemblaces to mammalian cheek teeth, and these crocodilians also contain another mammalian trait—a secondary palate."

Convergence simply doesn't ripple through the whole animal, clear down to the last tooth and bone, which is what Denton or Woodmorappe needed to establish to dismiss the RMT critters as merely convergent.

A similar distance from applied taxonomy dogged Cornelius Hunter (2001, 29-31; 2003, 46-48, 123-124) when he claimed fossil convergences somehow violated the idea that evolution was unguided, and thus were better explained by special creation. Since Hunter (2001, 48, 180n; 2003, 95, 160n) had specifically cited Futuyma's pages 46 & 48 (for quotes on the implausibility of God having designed living systems with the quirky patterns observed), Hunter's omission of the diagnostic reality of his own example of convergence may be chalked up either to obtuseness or intelligently designed evasion.

If you want another measure of just how flimsy the antievolution position is on this: just compare the superficial generalizations of Denton, Davis & Kenyon, Milton, Woodmorappe or Hunter to the stringent level of fossil and biological detail in a regular science paper, such as Rubidge & Sidor (2001) on convergent elements in therapsid evolution.

Darwin on Trial

Although Michael Denton's *Evolution: A Theory in Crisis* was inadequate even at the time, and his dismissal of the RMT evidence an examplar of how not to rigorously discuss a subject, his book engendered a broader impact because downstream redactor Phillip Johnson was at the nexus of the Intelligent Design movement, encouraging and promoting a group of *Kulturkampf* activists out to slay the Darwinian dragon once and for all.

These included "Irreducible Complexity" biochemist Michael Behe (someone else greatly impressed by Denton's 1985 book), mathematician William Dembski, and philosopher Stephen Meyer (who would help expand George Gilder's *Discovery Institute*), all covered in the enthusiastic *festshrift* hagiography by Dembski (2006).

A note on George Gilder's strident *Kulturkampf* opinions is warranted regarding the philosophical stance of the *Discovery Institute* on evolution. Gilder (2004) fumed that "Darwinian materialism is an embarrassing cartoon of modern science," and was contributing to a decline in educational excellence in biology classes that "espouse anti-industrial propaganda about global warming." The *Discovery Institute* posted a later opinion piece by Gilder (2006) for William F. Buckley's *National Review* that sported an equally sweeping subheading: "The Darwinian theory has become an all-purpose obstacle to thought rather than an enabler of scientific advance."

As we've seen already, though, the RMT was one scientific advance that was not a popular subject for antievolution discussion, apart from detail fidder Duane Gish, and you've seen how well that went. So how would Phillip Johnson, the patron saint of the Intelligent Design movement, do when Johnson (1991, 75-78) took on this "crown jewel of the fossil evidence for Darwinism"?

Johnson began with a (pardon the pun) jaw-dropping concession, turning royal much as Gish had on *Morganucodon*: "We may concede Gould's narrow point" concerning the existence of the transitional double-jaw-jointed therapsids.

But this was no "narrow point" for Johnson to cavalierly concede. To get from a reptile to a mammal *required* exactly that novel intermediate to have existed, and paleontologists had found the fossils showing exactly that. This was no trivial happenstance to be flipped aside.

Did Johnson think if only he pitched it fast enough, we wouldn't notice this foul ball?

The end of his concession paragraph sounded as much like Duane Gish as Michael Denton in goalpost moving: "there are many important features by which mammals differ from reptiles besides the jaw and ear bones, including the all-important reproductive systems. As we saw in other examples, convergence in skeletal features between two groups does not necessarily signal an evolutionary transition."

Only Johnson was trying to race the field a little too fast this time, for he had not actually discussed *any* convergent skeletal examples in his book, let alone ones justifying what he was trying to do in dismissing "Gould's narrow point."

There were several spots in Johnson's book where the issue of anatomical convergence might have come up, but didn't. Homology was mentioned on page 65 ("superficially similar body parts"), but no specific examples were given. There was a quote (page 70) by Gould again on the Australian marsupials, but that was about *biogeography* ("Why should all the large native mammals of Australia be marsupials, unless they descended from a common ancestor on this island continent?"), and didn't allude to their convergent members either.

Nor did any of the usual instances of fossil convergence turn up in that guise. Johnson's coverage of the *Coelacanth* or the amphibian *Seymouria* (pages 74-75) did not frame the argument in convergent terms. Finally, "convergent evolution" was not among Johnson's index topics.

Ah, but Michael Denton (1985, 181) had remarked that "many quite separate groups of mammal-like reptiles exhibited skeletal mammalian characteristics, yet only one group can have been the hypothetical ancestor of the mammals. Again, as with the rhipidistian fishes, the similarities must have been in most cases merely convergence."

Here may lie the source for Phillip Johnson's spectral "other examples" the lawyer only thought he had mentioned. Had Johnson so enthusiastically absorbed Denton's argument secondarily that he had lost track of where *Evolution: A Theory in Crisis* left off and *Darwin on Trial* began?

Not that this claim would have rescued Johnson's prosecution case anyway, since Denton had used "convergence" in the same invalid manner Duane Gish had, applying it to collectively derived therapsid characteristics rather than to ones of significantly independent origin.

Similarly to Johnson, Fred Hoyle & Chandra Wickramasinghe (1993, 158-159) dismissed the synapsid ancestry of mammals solely on the even more general grounds that "the genetic material of a mammal is grossly different from that of a reptile." They did not offer documentation for any this (a tough sled given what was by then coming to be known about the *commonality* of the metazoan genetic toolkit), but that had been a practice of long standing for that idiosyncratic pair.

Fred Hoyle was a physicist and Wickramasinghe a mathematician. Neither showed particular expertise when opining on evolutionary subjects. Frank Sonleitner (personal communication) called my attention to their earlier 1980s theory that both birds and mammals had somehow originated from infection by space bacteria attending the K-T extinction event 65 mya—a melodramatic notion fatally compromised by the fact that both classes had appeared over a *hundred million years earlier*.

Failure to have a properly working *Map of Time* seems to be a recurring problem among antievolutionists, doesn't it?

Anyway, both Phillip Johnson and Hoyle & Wickramasinghe were glossing over the actual range of mammalian variation, such as the urogenital system that differs significantly among the egg-laying monotremes, the marsupials and placentals, noted by Lambert & The Diagram Group (1985, 156) or Colbert & Morales (1991, 241). One can toss in Robin Dunbar & Louise Barrett (2000, 74) on primate specialties.

So which is the "mammalian" form that is supposed to differ so from the "reptilian" one? And how are scientists to make such distinctions when such data don't get preserved in the fossils? Or present an argument of detail that won't persuade the antievolutionist anyway, because they won't be stopping to pay attention to any of those details?

Even extant reptile metabolism is far from uniform, noted by Lambert & The Diagram Group (1990, 178). Crocodile hearts are intermediate between basal reptiles and birds, for example, and their sprinting and aquatic Mesozoic forms suggest modern crocodiles are the less-adventurous remnant of a once more physiologically diverse lineage, Colleen Farmer (1999, 585-586) and Roger Seymour *et al.* (2004).

While antievolutionists have remained stuck in their typological rut, though, the scientific imagination has kept going, and one can track the idea of metabolic evolution along sources as diverse as Richard Ellis (2001, 159-166), Mary Schweitzer & Cynthia Marshall (2001, 322, 326) and Carl Zimmer (2001). Warren Burggren (2000) described the diversity of vertebrate hearts, and Zimmer (2000a) summarized recent thinking on their underlying evolutionary development. David Norman (1994, 183-185) is also of interest, contrasting the hearing, lungs, and reproductive systems of birds and mammals, showing the various advantages and drawbacks each inherited from their divergent paths along the diapsid/synapsid divide.

Parenthetically, Mary Schweitzer *et al.* (1997; 2005; 2013; 2014), Schweitzer & Zhiyong Suo *et al.* (2007) and Schweitzer & Jennifer Wittmeyer *et al.* (2007) figure in one of the ongoing tropes of recent creationism, the preservation of soft tissue in dinosaurs and whether that supposedly disproves the standard geological paradigm. Despite the fact that the amounts involved are literally microscopic, and found only in sealed bone cavities, a parade of creationists have nonetheless marvelled over the fragments as though they represented a full body mummy of a *T. rex*, Walter Brown (2008), Vera Everett (2009), Sean Pitman (2010), Marcus Ross (2010), Brian Thomas (2011a-b; 2013a-b) and Elizabeth Mitchell (2013b), joined by the peevish *Kulturkampf* video blogger Bob Enyart (2013).

Similar tissue residues found in a triceratops horn by Mark Armitage & Kevin Anderson (2013) raised a kerfuffle at their university. With creationist publications like Armitage & Luke Mullisen (2003) and Armitage & Snelling (2008) under his belt, his playing of the triceratops matter was one step too far, leading to his dismissal and subsequent elevation to a plinth in the pantheon of persecuted unconventional scientists, reported by a sympathetic Jennifer Kabbany (2014).

Certainly subjects for another book, on the forensics of tissue preservation and the literally microscopic vision of creationists anxious to see only what they want to see. But we have so much more to investigate on the RMT.

The "No Cousins" mantra lives on

Underlying the issue of natural branching common descent is a conceptual point that should be rather obvious. No one is "directly descended" from their aunts or uncles. Not one. Likewise no one is "directly descended" from any of their distant cousins. Not a one.

And yet would it be correct to insist that no one is *related* to those aunts and uncles and cousins? Of course not. That would be really silly. They are related in that secondary way, as indisputably your relatives as any of your direct parents or grandparents.

What seems so simple a relationship to recognize for cousins within a species breaks down in the antievolutionary imagination the moment they try to tackle closeless of relationship higher up the scale, in genera, families, orders or classes. Ultimately it's a failure to conceptualize what it means for speciation to happen, but also a genuine inability on the part of the antievolutionist to work through *in their own heads* what intermediate forms would have needed to look like, and actively compare those theoretical templates with what has been found, to check for similarities or differences.

How close would a real fossil need to get to recognize its similarity to a theoretically expected intermediate? Or does it have to be absolutely exact? And what if the antievolutionist never even progresses to imagining what an absolutely exact match would be either? In other words, *nothing* is ever going to cut it for them? *Something* will be seized upon to allow them to push away what has been found, no matter what the similarities or cumulative patterns.

That's the "No Cousins" rule. And it pervades antievolutionist apologetics. It's the underlying structure of Duane Gish's creationist hairsplitting. And it's what Phillip Johnson was doing all the way through *Darwin on Trial*.

Having thought to dismiss the RMT features as insufficient biologically and inconclusive because they could be convergent (but without properly investigating either issue), Johnson dove into phylogenetics by contending that "Darwinism" required a specific line of descent to be valid, but evolutionists could only offer an "artificial" one. Or if that argument didn't work, Johnson dangled the idea that ancestry from the therapsids came at the unacceptable price of mammalian polyphyleticism.

This was the issue of whether certain groups were composed of members that were actually not that closely related (and thus another version of the convergence argument). This can be a hot topic, especially when you are trying work out relationships where a lot of the fossil blips haven't been found yet.

That was the case with the "Carnosauria" (big bipedal predatory dinosaurs like *Allosaurus* and *Tyrannosaurus*) that were still thought to be fairly closely related in the 1980s, reflected in David Norman (1985, 64-65, 70-71). But newer fossil finds and the application of cladistic analysis soon recognized three separate groups, ceratosaurs, allosaurs and coeleurosaurs (of which the giant tyrannosaurs were a jumbo version), Ralph Molnar et al. (1990, 187), Halszka Osmólska (1990), Fastovsky & Weishampel (1996, 263-279) and John Hutchinson & Kevin Padian (1997).

Yet all those dinosaurs were still *monophyletic* if seen higher up on the hierachy, as all were variant theropods. Likewise for the mammal players. Track back far enough and *everything* is "monophlyetic." Or a *monobaramin*, while we're at it. That's what evolution by natural common descent means.

The monophyly/polyphyly thing would have been an evolution-breaker only if there were separately created kinds or types involved, genuine discontinuities in the lineages. But Johnson showed no inclination to lay out that side of the equation (in no small measure due to the fact that nobody on the antievolution side had ever got to the point of doing that either, leaving him no substantive argument to copy).

Worse for Johnson's argument on the RMT, by the time Johnson was writing, that polypheticism matter was a dead letter. As Stephen Jay Gould (1992, 121) acerbically pointed out in his critical review of *Darwin on Trial*, the monophyly of the three main mammal groups (monotremes, marsupials and placentals) was a settled issue, accusing Johnson of tilting at the "rotted windmills" of obsolete subjects. "He attacks Simpson's data from the 1950s on mammalian polyphyly (while we have all accepted the data of mammalian monophyly for at least 15 years). He quotes Ernst Mayr from 1963, denying neutrality of genes in principle. But much has changed in 30 years, and Mayr is as active as ever at age 87. Why not ask him what he thinks now?"

The entire rejoinder by Johnson (1993a, 209) consisted of: "These quotations (pp. 77, 89) are placed in historical context to show how prestigious Darwinists dealt with or anticipated issues at the time." Actually, Johnson hadn't *quoted* Simpson at all, only abstracted his position; nor were there any appropriate references even for that in Johnson's Research Notes.

Are readers supposed to be impressed by such "scholarship"?

To add insult to injury, the ninth chapter of Barbara Stahl's *Vertebrate History*, on which Johnson had purportedly based his fossil treatment, had discussed that very point at length. Which means Johnson had to step over this in Stahl (1985, 410-411):

> By the late 1960s, Crompton and his colleagues F. A. Jenkins, Jr., and J. Hopson had formed a new opinion concerning the origin of the Mammalia. They think now that the triconodonts, docodonts, symmetrodonts, pantotheres, and perhaps even the multituberculates can be traced to a single line which emanated from one specific family among the Therapsida

> and thus that class Mammalia is monophyletic in a much narrower sense than that understood by Simpson. Their definition of a mammal continues to depend on the presence of the dentary-squamosal jaw joint rather than the absence of the articular from the mandible. However, by requiring that a mammal possess teeth that are not repeatedly replaced, that are (or primarily were) differentiated posteriorly into premolars and molars, and that are (or were at some time in their history) characterized by a primary cusp set between accessory ones in front and behind, they have excluded from the class Mammalia the tritylodonts, the cynodonts with crowned molars, and *Diarthrognathus* and its immediate relatives.

As a reminder, the triconodonts and docodonts were orders in the Eotheria subclass; symmetrodonts, pantotheres, and multituberculates were primitive marsupial orders. New fossil finds have continued to put more blips on the scope to remind us of their former diversity and range, such as the multituberculates and symmetrodonts coming to light in recent Spanish digs, Ainara Badiola *et al.* (2012) and Gloria Cuenca-Bescós *et al.* (2014). Kurt Wise's spin on the recent docodont fossils were noted back in Chapter 4, and they'll be coming up again in Chapter 10 regarding the evolution of mammalian dental replacement, and in Chapter 11 when we take a Grand Tour of the RMT with creationist Charles Creager.

One may chalk off this recurring cast of characters in Stahl's summary (human and fossil), the recognition of how relative minutia were being evaluated in refining the relationships in an increasingly cladistic context, and the complete inability of Phillip Johnson to notice any of it.

As a *Map of Time* note, Stahl (1985, 411) had also commented on how new fossil discoveries had played their part in improving the resolution:

> The conviction on the part of Crompton, Jenkins, and Hopson that the mammals did constitute a monophyletic group stemmed from studies of newly discovered Upper Triassic fossils and reexamination of others of approximately equivalent age. The new material, which came from red beds in Lesotho in southern Africa, consisted not just of teeth but of skulls and postcranial bones belonging to animals eventually named *Erythrotherium* and *Megazostrodon*. From the structure of their teeth these animals proved to be mammals, rather than reptiles like the majority of the forms at the site.

That situation hasn't notably changed since, such as this bit from Colbert & Morales (1991, 228) that I quoted a snippet from back in Chapter 3: "In recent years many students of this problem have tended to favor the monophyletic origin of the mammals, with the cynodont genus, *Probainognathus*, selected as representative of what the ultimate mammalian ancestor may have been like. This concept is based upon the evidence of numerous fossils collected in recent years, and as Crompton and Jenkins have shown, logically replaces the polyphyletic theory for mammalian origins which was based upon limited fossil materials."

Newer fossil finds have continued to fill in more blips on the early mammal scope, including Yaoming Hu *et al.* (1997) and Luo *et al.* (2002). Erik

Stokstad (2002) and Anne Weil (2002) highlighted an important new fossil, *Eomaia*, described by Qiang Ji *et al.* (2002) and illustrated by *Prehistoric Life* (2009, 357), clarifying aspects of the early eutherian radiation (including fur impressions). Another basal example found in Cretaceous China, *Acristatherium*, suggested a lot of early diversification had occured by 125 mya, Yaoming Hu *et al.* (2010). There will be more to say about *Eomaia* in further chapters.

The diagnostics of mammals' hardy teeth have played an important role in this study (many mammals being known primary by the teeth that get fossilized more easily than the rest of their remains). The section in Stahl (1985, 412-419) on this topic could also have been known to Phillip Johnson. Representative coverage since then could include John Hunter & Jukka Jernvall (1995), Abigail Tucker *et al.* (1998), Jernvall (2000), David Polly (2000) on Jernvall *et al.* (2000), Isaac Salazar-Ciudad & Jernvall (2002), Aapo Kangas *et al.* (2004), Tucker & Paul Sharpe (2004) and Sharpe (2007). Alexander Averianov & Alexey Lopatin (2014, 428) illustrated the four main tooth cusp formats found in fossil and living mammals, which show combinations of triangular and linear layouts.

By this time, quite specific genetics were being added to the toolkit to begin to account for how and why particular forms arose, such as the role of *Bmp* and *Fgf8* in the evolution of the specialized dentition found in toothed cetaceans, Brooke Armfield *et al.* (2013). Incidently Michael Denton (2016a, 50, 294n) cited the Armfield paper only to mention the cetacean and armadillo cases as exceptions to the mammalian norm of living placental mammals (three incisors, a canine, three premolars and four molars), but did not delve into the paper's genetic content and how that *accounted for the specific cetacean variation*. Everything is static in Denton's typological framework, you see, except for the things that aren't.

Outside the typological box, though, the science was moving on. Karthryn Kavanagh *et al.* (2007) had pinned down more of the mechanics of mammalian tooth formation, and armed with such information, investigators have moved ahead to experimentally *retroengineer* ancient tooth forms by mutating the *Fgf3* and *Eda* genes involved in tooth patterning, starting with Cyril Charles *et al.* (2009) and continuing with Enni Harjunmaa *et al.* (2012; 2014). Their 2014 paper was accompanied by an informative commentary by Zhe-Xi Luo (2014).

Interestingly, *Evolution News & Views* (2012a) richocheted off the early stages of that work, deciding that the findings of Harjunmaa's 2012 paper represented "an Argument for Intelligent Design in All but Name." They claimed that because Harjunmaa's team had found that evolving especially complex tooth cusps required increasingly unlikely genetic combinations, and so were less likely to occur and more easily lost in subsequent evolution. Which *EN&V* promptly translated into Michael Behe's ID position that "coordinating multiple independent mutations is too unlikely, hence never observed."

We'll be seeing how Behe employed such "logic" to tackle the RMT in Chapter 9 (Spoiler Alert: it doesn't involve his assessing any vertebrate paleontology), but the glaring snag for *EN&V* was that the scientists were not only *observing* those variations in the physical biology of living animals, they were *recreating* the fossil versions of them. And despite such ID dervish spin, those dynamics of tooth variation (where increasing complexity could be at odds with adaptive fitness) were completely in line with the latest evolutionary modeling, as David Polly (2013) noted of Issac Salazar-Ciudad & Miquel Marin-Riera (2013).

The evolutionary biology of tooth formation or the utility of genetic modeling were the last things on Phillip Johnson's mind back in *Darwin on Trial*, though, disinclined as he was to think through even his own position. Johnson (1991, 76) hit a particularly tall conceptual brick wall with this:

> Douglas Futuyma makes a confident statement about the therapsids that actually reveals how ambiguous the therapsid fossils really are. He writes that "The gradual transition from therapsid reptiles to mammals is so abundantly documented by scores of species in every stage of transition that it is impossible to tell which therapsid species were the actual ancestors of modern mammals." But large numbers of eligible candidates are a plus only to the extent that they can be placed in a single line of descent that could conceivably lead from a particular reptile species to a particular early mammal descendant. The presence of similarities in many different species that are outside of any possible ancestral line only draws attention to the fact that skeletal similarities do not necessarily imply ancestry.

Rather than not enough ancestors available, now there were *too many*? What exactly would the correct number be to satisfy Johnson's mobile sensibilities? And where exactly would paleontologists go to look for them? At no point did Johnson think to investigate the temporal and biogeographical distribution of the available fossil data. Or the fossils themselves, for that matter.

This is the *Bermuda Triangle Defense* (Player #4) again, on stilts.

Perhaps the most striking feature about Johnson's treatment of the therapsids is that it was *not* a trimmed version of a technical argument, a "Gish Lite." Instead, it resembled more the concise hit-and-run of a Scott Huse, pumping up a rhetorical dismissal without actually making good on the claims.

Toss in the *No Cousins* rule (where nothing could ever be "proven" to be on an *exact* line of descent) and antievolutionists like Phillip Johnson had the perfect apologetic vise to squeeze out all trace of evolutionary implication from any dataset. By never working through what a *perfect* evolutionary sequence would look like, Johnson functionally excused himself from the task of explaining in what way that would differ from what was already found. All fossils could be rejected in principle, however closely they might resemble the ideal intermediates that were never allowed to occupy the antievolutionary conceptual box in the first place.

Take the gorgonopsids, a side player on the mammal evolution track. As Rubidge & Sidor (2001, 465) put it, "The unsatisfactory state of gorgonopsian taxonomy has been the single largest impediment to a broader understanding of this group's evolution. Indeed, gorgonopsians possess such a stereotyped cranial morphology that ontogenetic changes appear to have been used to identify species and even genera." Robert Sloan (1983, 270) used a gorgonopsid to illustrate the skull changes seen in mammals, though, as a perfectly valid illustration of the increasingly mammalian snout of synapsid reptiles at an early stage in the process, when the jaw elements were still essentially reptilian.

Gorgonopsids objectively existed, at a particular time and place. But would a more closely related cousin on the required *direct* line of evolutionary descent have looked notably different from them? No wonder antievolutionists who are so fond of this *No Cousins* mantra liked to invoke Colin Patterson. They want their *Get Out of Fossil Evidence Free* card, saving them all that exhausting labor of specifying what the fossils would need to look like to satisfy them.

It also explains why antievolutionists (even ones with paleontology degrees like Kurt Wise) have contributed *nothing* notable to the field of paleontology over the last hundred years.

Johnson (1991, 76-77) wound all his ill-founded arguments up into one tight ball with this (my **bold**):

> It seems that the mammal-like qualities of the therapsids were distributed widely through the order, in many different subgroups which are **mutually exclusive** as candidates for mammal ancestors. An artificial line of descent can be constructed, but only by arbitrarily mixing specimens from **different subgroups**, and by arranging them **out of their actual chronological sequence**. If our hypothesis is that mammals evolved from therapsids only once (a point to which I shall return), then **most of the therapsids with mammal-like characteristics were not part of a macroevolutionary transition**. If most were not then perhaps all were not.

If this sounds oddly like the reasoning of Duane Gish regarding the Allin paper back in Chapter 4, a *convergent* resemblance of Intelligent Design and Creation Science, it's because it does. And we may likewise ask whether the claims Johnson was making were supported by any more of the actual evidence than Gish's version.

Ah, there we have a snag. Johnson did not offer any evidence. None.

Here's where things get really fun, where (just as I did with Duane Gish) we may delve into Johnsonian *source scholarship*.

Setting a low bar on "Research Notes"

The sources used by *Darwin on Trail* were collected in Research Notes at the end of the book. The Reptile-Mammal segment in Johnson (1991, 173-174) revealed he had exactly five:

1. The general book by Pierre Grassé (1977) for quotes.
2. Some quotes critical of creationism by Stephen Jay Gould (1981) on "Evolution as Fact and Theory" in *Discover* magazine.
3. The book by Douglas Futuyma (1982) for his assorted quotes critical of creationism.
4. The criticism of creationist claims on the RMT by James Hopson (1987).
5. And the book by Barbara Stahl (1985) on *Vertebrate History: Problems in Evolution*.

Only Grassé, Hopson and Stahl were particularly relevant to the specific paleontology and systematic claims Johnson was fielding, though none of them were primary source technical works. So, where among them were we to find evidence on those "mutually exclusive" candidates Johnson had alluded to? What qualified these fossils as such, and had subsequent work altered that perspective? What "different subgroups" had been put forward, by whom, and how varied were they really? And finally, what would their "actual chronological sequence" be, and would Phillip Johnson show any inclination to change his mind even were that to be done?

We have not a clue on any of these topics because Johnson offered *no evidence for any of them*. Instead, we got a single authority quote from Grassé (1977, 35), as evasive as any employed by creationist quote miners. The ellipsis was Johnson's:

> All paleontologists note ... that the acquisition of mammalian characteristics has not been the privilege of one particular order, but of *all the orders of theriodonts*, although to varying degrees. This progressive evolution toward mammals has been most clearly noted in three groups of carnivorous therapsids: the Therocephalia, Bauriamorpha and Cynodontia, each of which at one time or another has been considered ancestral to some or all mammals.

So, these three groups had been thought "at one time or another" to have been "ancestral to some or all mammals." That's a lot of spread. What were the reasons for this, and how did things fare in 1991 when Johnson was writing?

It is interesting that Johnson (1991, 178) had just begged off using "more specific technical terms" (such as cynodonts or theriodonts) that "would distract the general reader unnecessarily," only to immediately quote Grassé using just such jargon. Might some explaining have been in order then to clarify what Grassé meant? Or had Johnson avoided belaboring the terminology not to spare his readers angst, but to let himself off the hook?

In either case, since he brought the subject up via Grassé, what was going on here?

The fact of paleontology is you start out with no data at all (there was a time when no dinosaurs or therapsids were known). Gradually you dig up more, learn more, with the object of connecting more together to understand it. Or at least that's the idea in science.

In a work obviously known to Johnson, Stahl (1985. 303) grouped the bariamorphs with cynodonts as "animals which paralleled one another in evolving traits approaching those of mammals. The bariamorphs, apparently descended from therocephalians, but fade from the fossil record by the middle of the Triassic period. The cynodonts survived somewhat longer, preying upon the few remaining herbivorous dicynodonts and competing with the diapsid reptiles which were becoming predominant."

In other words, all three of the names Grassé mentioned but which Johnson elected not to explain or investigate were closely connected in time and phylogenetic rank, after the basal *Dimetrodon* bunch but before the more advanced cynodonts (you can compare the temporal sequence with the listing laid out regarding Gish in Chapter 4 when he tried a similar shell game). And this information was in one of Johnson's primary cited sources.

Another resource known to Johnson was Hopson (1987, 20-22) who had specifically included a primitive therocephalian as a transitional example showing the enlargement of the coronoid process and zygomatic arch. It would have been illuminating for Johnson to explain in what respect a perfect identified ancestor for mammals could have looked noticably different from it at that stage.

Had Johnson wanted to turn over a few of the rocks and investigate further, he could have learned more about what these little ratlike predators were suggesting about what biological processes were going on back in the Triassic. Information which was not hot news bulletin class in 1991.

Consider David Watson (1931), regarding one of the early bauriamorphs found. Watson (1931, 1163) had seen the implications the fossil had for the unseen genetics, based on the repeated appearance of similar elements in this group: "Such features as the reduction of the hinder part of the lower jaw and of the quadrate, the attainment of a secondary palate, the reduction of the phalangeal formula to 2. 3. 3. 3. 3., the appearance of an obturator foramen, which are all changes in a mammalian direction."

Without benefit of the genetic knowledge a Phllip Johnson might have had at his disposal sixty years later, Watson presciently reminded, "The independent appearance of such identical, or at least similar, changes at once recalls the familiar fact of the widespread appearance of identical mutations in closely allied animals, and even in animals which are not extremely near to one another in a systematic sense."

To get such features, genetic systems were trending repeatedly and adaptively in a direction that in one lineage down the line would eventually be labeled fully mammalian. Is all that to mean nothing because the mutations and variations were appearing too regularly, too consistently or too commonly? Once again, what is the *correct* amount of morphological change and rate to hit Johnson's mobile target?

A parenthetical note, a lecture on adaptation by David Watson (1929,

231-233) has been a recurrent subject of quote mining by creationists, from Henry Morris (1975) to *CreationWiki* (2015). Watson had written that evolution was "universally accepted, not because it can be proved by logically coherent evidence to be true, but because the only alternative, special creation, is clearly incredible."

That was in the context of something Watson quite rightly noted at length in his lecture: how much of the nuts and bolts details of biology were still unknown at the time Herbert Hoover was president. This was a generation before DNA was identified and long before homeobox genes were uncovered in the 1990s. But the quote miners *never* included his explicit caution that evolution nonetheless powerfully explained "all the facts of Taxonomy, of Palaeontology, and of Geographical Distribution" known at the time, including the RMT characters that had been uncovered by 1929 and were, after all, the objects of his own direct effort.

But back to the taxonomical polyphyleticism trail.

By the time Kemp (1986) described a new example of the Therocephalians, cladistics was coming on the seen, and the group was being placed as the "sister group" of the more advanced Cyndontia. Work in the years after *Darwin on Trial* have supported that, such as Adam Huttenlocker (2009).

It's also relevant to recall that the Therocephalians and their kin are known from fossils sprinkled across a *twenty million year* timeframe, with quite a few coming from relatively few notable deposits like the Karoo Basin, Fernando Abdala *et al.* (2008, 1017-2020). All to be filed in that *Bermuda Triangle Defense* folder.

While we're on this *Map of Time* issue, though, what of Johnson's claim (so like that of Gish on Allin) about the fossils not being arranged *chronologically*? Immediately after the Grassé quote noted above, the Research Notes in Johnson (1991, 174) thought to make good on that declaration (my **bold**):

> James A. Hopson of the University of Chicago is a leading expert on the mammal-like reptiles, and he argues the case for their status as mammal ancestors in his article "The Mammal-like Reptiles: A Study of Transitional Fossils," in *The American Biology Teacher*, vol. 49, no. 1, p. 16 (1987). Hopson is not testing the ancestry hypothesis in the sense that I do so in this chapter, but attempting to show the superiority of the "evolution model" to the creation-science model of Duane Gish. To that end he demonstrates that therapsids can be arranged in a progressive sequence leading from reptilian to mammalian forms, with the increasingly mammal-like forms appearing later in the geological record. So far so good, but Hopson does *not* present a genuine ancestral line. Instead he mixes examples from different orders and subgroups, and ends the line in a mammal (*Morganucodon*) which is **substantially older than the therapsid that precedes it**. The proof may be good enough to make Hopson's specific point, which is that for this example some form of evolutionary model is preferable to the creation-science model of Gish, but his argument does not qualify, or purport to qualify, as a genuine

testing of the common ancestry hypothesis itself.

The creationist textbook *Of Pandas and People* pressed this argument even farther, Davis & Kenyon (1993, 100): "in actuality, the first three of Hopson's Therapsids are contemporaries from two separate orders, and some are not thought to be mammalian ancestors. Rather than older, the fourth is more recent than the fifth, and the final therapsid is more recent than the mammal (Morganucodon) presented as its descendent!" Not incidentally perhaps, *Darwin on Trial* was listed as recommended reading by Davis & Kenyon (1993, 89).

All this was a hoot. Not only because it was the closest Johnson (or any Intelligent Design advocate) got to taking note of the creationists and their arguments (Davis & Kenyon didn't allude to the Gish factor), but because Johnson and the *Pandas* authors were doing exactly what Gish had with the Allin paper: making bald accusations on Hopson's specific chronological sequence that *they* (not Hopson) had got *wrong*.

Hopson (1987, 19, 21, 24) had graphed the appearance of the fossil groups illustrating the fourteen taxa noted on his two cladograms of mammal evolution. I posted an image of the chart page in question (with my yellow highlighting of the most relevant groups) in Downard (2014, 7).

In no case was the next group on the cladogram appearing before its predecessor. In fact, with considerable understatement, Hopson (1987, 24) had even remarked that "given the known imperfections of the fossil record, the correlation between degree of advancement toward mammals and time of appearance is surprisingly high."

Most specifically, in no sense whatsoever could *Morganucodon* be considered to support Johnson's claim that it was "substantially older than the therapsid that precedes it," or Davis & Kenyon's version that the Ictidosauria (the sister group next up on the chart) were first appearing later than *Morganucodon*.

Though it was nice at least to see the creationists calling it a mammal, which was more than Gish had been willing to do. John Woodmorappe (2001) would likewise identify them as mammals, in his Table games discussed in the next chapter.

So move over Duane Gish mangling Allin, Phillip Johnson and the *Pandas* authors were cramping your turf in the *Garrett Hardin* department of egregiously misrepresenting their own cited source. But Davis & Kenyon (1993, 121-122) couldn't even stop with trampling the Hopson article. They fumbled the jaw-ear issue itself (my **bold**):

> Notably, the skulls and mandibles (lower jaws) of the Therapsids are said to have bones homologous to those of the first mammals. The upper and lower jaws of reptiles articulate (fit together) with two bones (one each located at the back of each jaw) not found in mammals, According to Darwinian theory, these two bones have become relocated in the middle ear of the mammals through evolutionary descent (see Figure 5-7). Yet there is **no fossil record of such an amazing process**. Consider that to make this

> change, one of these bones had to **cross the hinge from the lower jaw** into the middle ear region of the skull, where Darwinian mechanisms reshaped and refined them into highly specialized, delicate instruments of sound transmission. Such an occurrence would be extraordinary enough by itself, but some Darwinists propose that this happened more than once!

Oh, really? Ironically, they were already looking at parts of it, but had *got their labels wrong.*

Figure 5-7 showed a reptile skull "sectioned through the back to show the small articular and the quadrate bones," with those pointed out by arrow, and a mammal skull "shows their relocation according to Darwinian theory" with lines pointing out their "relocation as incus and stapes." All these drawings deriving from work by Alfred Romer.

Only the "reptile" skull they showed was actually that of an *early therapsid* (representing one of the very transitional forms they were insisting didn't exist), with the quadrate and articular bones already relocated next to the sound-conducting stapes. And they got the fate of the articular bone wrong. As you may recall, that bone is held to have become the *malleus* in the middle ear, not the stapes (a hearing bone of long lineage in tetrapods).

An interesting Big Picture aspect of the evolution of the jaw and ear involves how the stapes (which started out directly connecting to the skull wall in fish) became associated with the spiracle. Remember those back from Chapter 2?

The spiracle is an opening in the fish skull connected to the throat by a passage along the first gill arch. By the time you get to early amphibians, the spiracle was now covered by a tympanic membrane, and the stapes had pulled up to be associated with that old fish gill arch passage, leaving the quadrate-articular jaw joint on its own farther down on the skull.

It was in the synapsid line (and *only* in the synapsid line) that this stapes-filled tunnel from the middle ear to the spiracle opening shifted down until it was beside the quadrate-articular in their highly-morphed jaw layout. In the early therapsid that *Of Pandas and People* were inadvertently using as their "reptile" example, the tympanic membrane was already attached to that articular/malleus bone, which along with the quadrate/incus next to it were nestled snugly at the end of the stapes.

The failure of Davis & Kenyon to closely inspect what it was they were supposedly looking at presumably inspired their silly assertion that "one of the bones had to cross the hinge from the lower jaw into the middle ear of the skull." Because of their own mistaken identification of the bones, Davis & Kenyon were apparently imagining that the pair had to *jump* inside the ear, when what was actually happening was a shortening of the stapes that *pulled* the attached quadrate/incus and articular/malleus along with it, as they became increasingly specialized for hearing rather than jaw chomping.

Davis & Kenyon might have noticed their mistake if they had bothered to identify more of the bones in the first place, as Arthur Strahler (1987, 416) or Douglas Futuyma (1998, 150) did when they illustrated the same classic Romer drawings. It was just a matter of taking note of what the little labels (s) and (sp) and (a) and (q) and (tm) signified on Romer's drawings. For those interested, I included the Strahler and *Pandas* images in Downard (2014, 5).

But Davis & Kenyon were in too big a hurry to disqualify the therapsids to take careful note of the *four* skull examples Romer had so meticulously laid out for the attention of anyone of curious temper. And there's also that mammalian jaw embryonic development mentioned back in Chapter 3. You'd think that would have been pertinent to note, wouldn't you? But *no one* in the antievolutionist game had shown the slightest awareness of it. That would include those who could have known about it because it was *in the sources they cited*.

Christopher McGowan (1984, 139) highlighted it, but Duane Gish (1993a, 163) mentioned McGowan only on thermodynamics, not any of his extensive criticism of creationist pseudo-biology. And there was James Hopson (1987, 18) lurking under the very noses of Phillip Johnson and the *Pandas* authors. But as we've all too plainly seen, paying close attention to the content of what they were supposedly reading hasn't exactly been their best suit, has it?

Now it should be noted that *Of Pandas and People* had been written by Percival Davis and Dean Kenyon for John Buell's Texas-based *Foundation for Thought and Ethics*, specifically to serve as a *school textbook*. Given that book's incompetence on so many of the factual details of the RMT, we might ask where was the *Icons of Evolution* umbrage of Jonathan Wells (2000a, 249-258), who excoriated regular science textbooks for their purported repetition of evolutionary errors?

Something else: *Of Pandas and People* was slated to be the supplemental Intelligent Design text for students in the Dover school district in Pennsylvania, until the Kitzmiller court case nixed that in 2005. Matthew Chapman (2007) and Lauri Lebo (2008) offered firsthand coverage of that dismal affair (entertaining or disturbing depending on how happy you are to see creationist views almost injected into a public school science curriculum). I also surveyed the issues and players in Downard (2015b, 91-109).

Jonathan Wells (2006, 154-155) grumped about how the Kitzmiller decision didn't favor the Intelligent Design view of things, but showed no inclination to evaluate whether the factual problems with *Of Pandas and People* fully justified its deliberate rejection.

As we'll see in Chapter 9, it's even worse that that, as Wells went on to coauthor with Bill Dembski *The Design of Life* retread of *Pandas*. Stay tuned for a replay of the same antievolutionary mistakes.

Phillip Johnson's mastery of the meaningless concession

As mentioned earlier, Stephen Jay Gould (1992) had written a scalding review of *Darwin on Trial* for *Scientific American*. Lighting into Johnson on both philosophical and technical grounds, towards the end Gould got around to what I considered the hot button issue, Johnson's numbing concession on the therapsids:

> On page 76, he admits my own claim for intermediacy in the defining anatomical transition between reptiles and mammals: passage of the reptilian jaw-joint bones into the mammalian middle ear. Trying to turn clear defeat into advantage, he writes: "We may concede Gould's narrow point." Narrow indeed; what more does he want? Then we find out: "On the other hand, there are many important features by which mammals differ from reptiles besides the jaw and ear bones, including the all-important reproductive systems." Now how am I supposed to uncover fossil evidence of hair, lactation and live birth? A profession finds the very best evidence it could, in exactly the predicted form and time, and a lawyer still tries to impeach us by rhetorical trickery. No wonder lawyer jokes are so popular in our culture.

Now that the battle was joined, how would Johnson respond to it?

When *Scientific American* bluntly refused his rebuttal, Johnson presented his commentary in an Epilogue to the 2nd edition of *Darwin on Trial* in 1993. In the grand tradition of genius unjustly denied (from catastrophist crackpot Immanuel Velikovsky to Ancient Astronaut groupie Erich von Däniken), Johnson enthusiastically excoriated his critic. As for the RMT, though, all Johnson (1993a, 167) thought to do was to double down:

> Pending an unbiased review of the evidence that I hope to encourage, I accept the therapsid example for now as a rare exception to the consistent pattern of fossil disconfirmation of Darwinian expectations. My point was that any single example of this sort cannot be conclusive, and even this "crown jewel" of the Darwinian fossil evidence illustrates points on a putative "bush"; rather than a specific ancestral line leading to an identified first mammal. That an army of researchers dedicated to finding confirmation for a paradigm has found some apparently confirming evidence here and there is not surprising. To evaluate the paradigm itself we have to consider also the mountains of negative evidence—like the absence of any pre-Cambrian fossil ancestors for the animal phyla. We also have to consider whether the accepted description of the therapsid sequence has been influenced by Darwinian preconceptions.

At this point Johnson had jumped his own apologetic mark by citing the absence of Cambrian ancestors as though their nonexistence (or existence, for that matter, for those that have turned up) somehow invalidated the significance of the RMT a *quarter of a billion years later*.

You see, it was one of Johnson's recurring complaints about naturalistic evolution that its proponents conflated agreement with modest microevolutionary change into proof of the larger atheistic metaphysical worldview. By jumping back to the Cambrian—an area where *Darwin on Trial* was considerably less conclusive than promoted, as I explored in Downard (2003b, 16-63; 2004b)—to avoid dealing with the therapsids, Johnson was practicing the very sin he accused Darwinists of.

On the science side, it's no small problem trying to work out what was going on in a series of Precambrian and Cambrian ecosystems dominated by small soft-bodied creatures guaranteed not to leave much of a fossil record. For that reason, most of the science on the origin of phyla has had to work backwards from the features of what living descendants have remained. Much of the thinking on this couldn't have even begun making notable headway in the pre-*Hox* era, which means only works after the mid-1990s can be considerably especially relevant.

The upshot is that the burst of animal phyla during the Cambrian reflected a changing ecosystem and adaptive landscape following the collapse of the Ediacaran biota. These new forms (especially a clog of buggy arthropods) were not apparently representing some radical appearance of new genes, but instead were drawing on a changing regulatory deployment of metazoan patterning and metabolic genes that had long been in existence.

The origin of phyla is too big a topic to condense even into a digressionary list, but some of the trajectory of scientific thinking can be followed in James Valentine (1989; 1994, 1995; 1997; 2001; 2002; 2003; 2004; 2007), Derek Briggs et al. (1992), Gregory Wray *et al.* (1996), Francesco Ayala *et al.* (1998), Lindell Bromham *et al.* (1998), Michael Lynch (1999), Graham Budd & Soren Jensen (2000), Chris Cameron *et al.* (2000), Kevin Peterson *et al.* (2000; 2004; 2008; 2009), Kevin Peterson & Eric Davidson (2000), Lindell Bromham (2003), Graham Budd (2003; 2008), Jaime Blair & Blair Hedges (2005a-b), Kevin Peterson & Nicholas Butterfield (2005), John Welch *et al.* (2005), Charles Marshall (2006), Joseph Ryan *et al.* (2006; 2007; 2010), Joseph Ryan & Andreas Baxevanis (2007), Juan Couso (2009), Xavier Fernández-Busquets *et al.* (2009), Charles Marshall & James Valentine (2010), Michael Lee *et al.* (2013) and Derek Briggs (2015).

But back to Phillip Johnson, and how he was dancing around what data he did bump into.

In *Darwin on Trial* Johnson (1991, 79) had written that *Archaeopteryx* was the "lonely exception to a consistent pattern of fossil disconfirmation." That was take on the origin of class Aves. In the space of a few years, and with *no new information* under his belt, solely for the purposes of his rebuttal to Gould, the therapsids were suddenly accepted ("for now") as another "rare exception to the consistent pattern of fossil disconfirmation of Darwinian expectations." That's the origin of class Mammalia.

The original text of Johnson's *Scientific American* rejoinder had fired both barrels at once: "The therapsid reptiles and Archaeopteryx are rare exceptions to the general absence of plausible transitional intermediates between major groups, which is why it is important to understand that even these Darwinist trophies are inconclusive as evidence of macroevolution."

Recalling just how few new vertebrate classes had appeared over the last *quarter of a billion years*, just as Kurt Wise had back in Chapter 4, Johnson apparently didn't realize how he had just run out of "rare exceptions" to concede.

Nor did Johnson improve his game (or paleontological acumen) in his later works, as he just kept repeating the congealing mantra of dismissal at arm's length.

In *Reason in the Balance*, Johnson (1995, 107) wrote briefly:

> Because of this way of thinking, even the notorious discrepancies between the facts of the fossil record and Darwinian expectations do not matter so long as there is *some* evidence (*Archaeopteryx*, Lucy, the "mammal-like reptiles") that can be interpreted to fit the paradigm—and the critics are unable to propose a credible mechanism for evolution by big jumps. If the contest is between Darwinism and "we don't know," Darwinism wins.

Two years later, in *Defeating Darwinism*, Johnson (1997, 59-60) had to add to his list of the "small number" of fossils:

> I've long been fascinated by the conflicting messages Darwinists provide concerning the fossil evidence. On the one hand, they proudly point to a small number of fossil finds that supposedly confirm the theory. These include the venerable bird/reptile *Archaeopteryx*, the "whale with feet" called *Ambulocetus*, the therapsids that supposedly link reptiles to mammals, and especially the hominids or ape-men, like the famous Lucy. These examples, all from vertebrate animals, are pressed very insistently on me in debates as proof of the "fact" of evolution and even of the Darwinian mechanism."

That "all from vertebrate animals" point was another application of the *Bermuda Triangle Defense*, since Johnson showed no interest in delving into the fossil record of invertebrates and the degree to which they tend not to get fossilized at all. Were scientists not supposed to pay attention to the data at hand, just because they're vertebrates (like Johnson and yours truly)? And slough off the "small number" of fossils documenting the RMT in the process, involving *400 genera* and thousands of specimens, which comprised the dominant land animals of the Permian period, Robert Sloan (1983, 264).

Johnson (1998, 30) continued to flick aside all the RMT data he never discussed, as just some isolated phenomenon in the macroevolutionary history of life, his attitude a consistent testament to the common absence of a functioning curiosity-fueled *Map of Time* among antievolutionists, regardless of intellect or level of education. Though in his case there may have been a trace of simple and sustained ignorance, as Johnson (1991, 71) claimed that "Most of the evidence relied upon by today's Darwinists was known to Darwin's great contemporary, the Swiss-born Harvard scientist Louis Agassiz."

Really? Agassiz was born in 1807 and died in *1873*. Almost all the strong evolutionary cases in the fossil record turned up since then, from dinosaur taxa to the mammal-like reptiles at the turn of the 20th century, to the Cretaceous birds and intermediate whales in the 1990s that Johnson peripherally alluded to. Johnson might have taken note of something his own source of Stahl (1985, 399) spotted, about Agassiz's attitude toward finds from the pioneering geologist and paleontologist, the Reverend William Buckland (1784-1856), dismissed by Henri de Blainville (1777-1850):

> When Buckland proposed that two lower jaws retrieved from the Middle Jurassic Stonesfield Slate in England might have belonged to mammals, the possibility was dismissed by the French biologist Blainville on the supposition that the warm-blooded tetrapods did not arise until the heyday of the reptiles was over. Agassiz, when he was consulted, sidestepped the question by saying that the material was too fragmentary to decide so crucial an issue, but Cuvier agreed with Buckland that the bones and teeth might be the remains of some primitive mammalian form.

Oh, and modern evolutionists certainly seem to pay a lot of attention to that DNA stuff, something utterly unknown in Agassiz's day.

In one email exchange I had with Johnson in the late 1990s, I specifically invited him to say what a genuine transitional sequence would have looked like in the RMT case (carefully wording my inquiry to not position myself as either pro or contra evolution, so as not to trigger a "circle the wagons" defensive response). Johnson not only declined to answer my question, but warned how such an approach was falling into an "evolutionary trap", playing into Darwinist hands by focusing on individual examples instead of looking at the "evidence as a whole."

Just how you're supposed to see the "evidence as a whole" without including any of the individual pieces (or allow any of those, or collection of them, to ever mean that evolution might be going on) was a persistent *gap* in Johnson's reasoning far wider than anything seen in the RMT.

At the 1998 "Creation Week" seminar Stephen Meyer organized at Whitworth University (where he was at the time an Associate Professor of Philosophy), I listened while Johnson pilloried Gould once more for criticizing his book only on niggling details, during which he did not volunteer the therapsid matter.

In the question period that followed I explained how Gould had brought up that "narrow point." Then I tried to get Johnson to explain in what respect the RMT and Gould's allusion to it failed to qualify as a relevant countreexample to the claims of *Darwin on Trial* about the lack of genuine macroevolutionary evidence. I grew quite peeved as he stuck to his philosophical guns, and Johnson finally cautioned the audience that my evident annoyance was due to his treading on my supposed "evolutionary religion," and moved on to the next raised hand.

While no one in the audience seemed aware that Gould's review had not been as gauzy as Johnson intimated, it was interesting that Stephen Meyer made no move to correct Johnson about leaving out the therapsid issue. Had I not raised it in my own question, the assembly would have been left with Johnson's version uncontradicted.

Afterward Meyer and I had a lengthy chat (with a small audience of holdovers from the main lecture). I asked Meyer whether it was *philosophically* legitimate for Johnson to proceed as though he had successfully disposed of the therapsids without actually mentioning any of them, either there or in his books.

I never did get an answer.

We'll see in Chapter 9 how Meyer thought to address the RMT when he finally got around to it a decade later, in the other *Pandas* retread, *Explore Evolution*.

But the situation was even worse than the fiddly bits of therapsids, since Johnson was in effect second-guessing why I found the RMT so persuasive. To contend that I was impressed only because my evolutionary presumption dictated it was the opposite of the truth. I had become *convinced* of evolution largely because of such evidence. I tried to explain that distinction to Johnson in my e-mail exchanges, and also brought the point up at Whitworth, all to no avail.

For Johnson, it was simply inconceivable that anyone could arrive at an evolutionary sentiment because they were motivated by the quality of the supporting data. So my reflection on my own reasoning could not be valid.

Johnson was going way beyond rejecting my views as being merely wrong. He was denying the very reality and integrity of my own thought processes in reaching them. Putting words in people's mouths is unsanitary enough without Johnson's "theistic realism" coming along, trying to inject them directly into the brain.

The final stop in this sorry journey concerns the "unbiased review" of the evidence Johnson proposed to encourage—not undertake himself, mind you. His thinly veiled accusation that all the solid work ("apparently confirming evidence here and there") of professional paleontologists up until then was tainted by their paradigmatic bias was already insulting. But beyond that lay the more fanciful hope that somehow the jaw articulations of therapsids might relocate if only one stared at them long enough with a sufficiently unbiased eye.

We'll find out next chapter just what sort of critical review Johnson was willing to consider "unbiased."

Dangling loose ends as the Millennium turned

Paleontologists Kevin Padian & Kenneth Angielczyk (2007, 216) summed up their view of Michael Denton and Phillip Johnson's hairsplitting when it came to spotting fossils relating to identifying evolutionary lineages. "The difficulties in defining and diagnosing Mammalia do not reflect a lack of evolutionary understanding," but the understandable challenge of making "a sharp distinction in a continuum of organisms, because the division reflects the arbitrary placement of a name, not the understanding of genealogy."

That lack of understanding has been shared by *everybody* in the antievolutionary pool, not just Denton and Johnson, though the cluelessness was there for ready export, as when creationist bumpkin Paul Taylor (1995, 43, 284) relied on Johnson (1991, 77-78) for his dismissal of the Repile-Mammal Transition,

Another was creationist Ashby Camp (1998)—that chap Jonathan Wells extolled in his *Icons* book, noted back in Chapter 5. Camp included Johnson

(1991) in his short references, but didn't cite him directly as he employed the traditional barrage of authority quoting in his own relentless version of the *No Cousins* argument. Regarding the RMT, we've already encountered a lot of the gang Camp cited, including Robert Carroll (1988), Tom Kemp (1982a), Roger Lewin (1981) and even that old chestnut Romer & Price (1940). A few of Camp's examples will illustrate his all too familiar quote-mining method.

Following the "fully formed" mantra, Camp quoted Carroll (1988, 193) with surgical precision (the Camp extract in **bold**): "In sharp contrast with the fossil record of amphibians, modern amniotes are linked to their Paleozoic ancestors by a relatively complete sequence of intermediate forms. **The earliest known amniotes are immediately recognizable because of similarities of their skeleton to those of primitive living lizards.**" And Carroll (1988, 397), again my **bold** for what Camp used: "**The transition between pelycosaurs and therapsids has not been documented.** It may have involved an environmental shift as well as changes in morphology and physiology. The therapsids are already quite diverse when they first appear in the Upper Permian of Russia."

As Camp did not discuss the available deposits and their relation to the continental distribution of these forms, this was yet another play of the *Bermuda Triangle Defense* (Player #5).

Like Duane Gish, Camp insisted that Morganucodontids "have a fully-functional reptilian jaw joint (quadrate-articular) which distinguishes them from all living mammals," but offered no sources for the claim (not even Gish). That was the closest Camp got to addressing the ear matter, name dropping lots of mammal taxa via the quote mine, but not those relevant probainognathids or any of the new fossil finds that his dated source base couldn't possibly have known about to have discussed.

Ashby Camp joined the long procession of antievolutionists not mentioning the embryological data on the jaw-ear transition, of course, most likely because he never investigated deeply enough beyond the quote mine to have encountered it. And again, like all antievolutionists (including Gish and Johnson), even though he was dancing through a thicket of specific taxa, Camp never explained what any of the fossils would have needed to looked like to satisfy him.

A big part of the antievoluionary scholarly pattern is this overreliance on secondary citation. And Cornelius Hunter (2001, 77, 182n) presented yet another illustration of that with his very condensed version of the sort of argument Phillip Johnson padded out to several pages. Hunter averred that "with evolution we must believe that across the reptile-mammal transition organisms evolved so rapidly that they appear fully formed and diverse in the fossil record, that there are large gaps between the reptiles and the mammals, and that convergent evolution must have occurred many times."

To that end Hunter used the same Futuyma quote Johnson had, along with the accusation that tracing lines of descent was compromised by an abundance of candidates. Hunter went beyond Johnson, though, by sprinkling in a variety of paleontologists, such as Romer (1966, 184-185) and Carroll (1988, 377, 397-398), but seldom quoted any of them, and definitely did not discuss any specific fossil examples or features. The possibility arises that he had culled much of this from one or another creationist quote mine, but was too exhausted to copy more of the text.

In that way Hunter artfully bypassed the detailed taxonomical characters Romer (1966, 184) offered to support the general point that "In the varied therapsid types, we span nearly the entire evolutionary gap between a primitive reptile and a mammal."

Romer had mentioned a late Triassic "evolutionary 'no-man's-land,' a time when the mammals were occurring. Unfortunately, our knowledge of this transition is still poor," and Hunter pounced on that. As we know a lot of that situation in 1966 (or 2001 for that matter) involved a shortage of available Triassic deposits to look in, we can once again file under the *Bermuda Triangle Defense* (Player #6).

Incidentally there was another dearth of fossils spanning some fifteen million years in the Early Carboniferous that was called "Romer's Gap" in his (dubuous) honor. I say "was," though, because the Fossil Genie is ever busy, and fossils have turned up since to plug that "gap" too, filling in more of what was going on before *Tiktaalik* in the Devonian, Jennifer Clack (2002), Timothy Smithson *et al.* (2012), and Jason Anderson *et al.* (2015). Regarding the bigger picture, Peter Ward *et al.* (2006) identified "a low oxygen interval" during the Carboniferous that had cramped early arthropod and vertebrate colonization of the land.

But back to hunting down Cornelius Hunter's source base.

Although ostensibly an Intelligent Design advocate, Hunter (2003, 8, 41) exemplified the narrow incestuousness of lower echelon antievolutionism by listing creationist Ashby Camp among those to whom "I am indebted," while reprising his earlier *Darwin's God* reptile-mammal conclusions without elaboration or additional documentation.

That Camp was likely among Hunter's quote mining resources is borne out by the nature of Hunter's references. Like Scott Huse and Sunderland noted back in Chapter 4, source parallels between Camp's 1998 piece and Hunter ranged from the various Carroll citations that Camp had actually quoted but Hunter hadn't, to an obscure secondary attribution of George Gaylord Simpson from a 1972 Time-Life book *Life Before Man* that appeared in Hunter's notes, but with no sign of the Simpson quote Camp had used.

Why then list them at all? Unless, of course, just like Scott Huse, Hunter was angling to pad his even slimmer references without bothering about the content or relevance. And just to complete this clubby daisy chain with some of the Intelligent Design gang, Michael Behe, William Dembski and Stephen Meyer filled the back cover of Hunter's *Darwin's God* with great praise for his analytical prowess.

Not that Behe (1996) or Dembski (1998b; 1999; 2002) were showing any hurry to discuss paleontology on their end, let alone the RMT. Behe's

Darwin's Black Box was keen on defending "irreducible complexity" at the molecular level, while Dembski's work was (and would remain) in the realm of abstract mathematical justification for Intelligent Design. The closest Dembski (2002, 246, 303n) got in *No Free Lunch* was a glancing blow, insisting as a reality "the scarcity of transitional forms in the fossil record." And who did Dembski rely on for this claim in his endnote? The questionable authorities of just Michael Denton (1985) and Davis & Kenyon (1993).

Similarly, the only stab at paleontology in Dembski & Kushiner (2001, 65-69) was a reprint of a superficial *Touchstone* magazine article by Robert DeHaan & John Wiester (1999) retelling the Intelligent Design spin on the Cambrian Explosion (that the origins of phyla were somehow beyond the naturalistic Darwinian pale). Thus was the subsequent half billion years of life (including the RMT) again left outside the ID gated community.

A survey of other antievolutionists writing in the late 20th century showed how little they were willing to engage the RMT, either to refute their evolutionary implications or (better still) actually account for them from within their non-evolutionary framework.

Ralph Muncaster (1997) was just a slim twenty-eight-page pamphlet, so can be excused for not hitting on it, and a very "simplified presentation of macro-evolution" charted by the zany German catastrophist Hans Zillmer (1998, 177) zipped by mammal evolution in his haste to find evidence that dinosaurs and people rubbed shoulders before the Flood.

But the much bulkier tome by Walter ReMine (1993, 417) was quite another matter, oh so briefly tossing off that "evolutionists often use geological age of appearance to identity ancestors from descendants. This remains an important criteria for many groups including those at issue here, the mammal-like reptiles." ReMine cited Kemp (1982a, 12), but offered nothing further about the RMT, biological or geological. Left dangling was ReMine's creationist implication that paleontologists *shouldn't* be paying attention to the order in which fossils were appearing?

Gerald Schroeder didn't even get to the level of copying somebody else's mistakes. As he decided the Bible classified birds as reptiles, Schroeder was able to accept *Archaeopteryx* as the only intermediate form. Otherwise, Schroeder (1997, 95) didn't field any documentation at all to support his certainty that "In the entire fossil record, with its millions of specimens, no midway transitional fossil has been found at the basic levels of phylum or class."

Sorry Gerald, class Mammalia.

Still other antievolutionists who have grumped about the fossil record at times still haven't shown any curiosity about the RMT. Old Earth creationist (a dying breed) Hugh Ross (1994; 1998; 2009) never mentioned it, and as of this writing, Ross' *Reasons to Believe* website has absolutely nothing on the subject. Fence-straddler (Young Earth but Old Universe) Hank Hanegraaff (1998) is another revealing omission, given that he presumably knew of the issue, since Hanegraaff credulously cribbed several of Gish's 1995 book claims.

Apart from Kurt Wise's deflection noted back in Chapter 4, none of the antievolutionary contributors to Moreland (1994) or Moreland & Reynolds (1999) mentioned the RMT. The only one in that latter anthology who did was theology professor John Jefferson Davis (1999a, 81; 1999b, 139), citing Tom Kemp (1982a), Robert Carroll (1988) and Colbert & Morales (1991) to castigate the Young creationist offerings of Paul Nelson & John Mark Reynolds (1999a) and the Old Earth creationism of Robert Newman (1999) for so plainly failing to deal with the clear evidence of macroevolution exhibited in such fossils.

James Perloff's *Tornado in a Junkyard* was another monument to secondary citation and credulity. Perloff (1999, 14) quoted the anachronistically dated Denton (1985, 180) passage I noted earlier in this chapter on the supposedly reptilian metabolism of therapsids. Perloff (1999, 15, 17, 31, 282-283n) had elsewhere cited the older Stahl (1974, 349-350) volume on feather evolution, yet another area where antievolutionists juggled the facts until they dropped all the balls, as I covered in Downard (2003b). Had Perloff only stopped to read the more recent edition of Stahl, of course, he could have caught up to where Michael Denton should have been on the RMT, but wasn't.

But Perloff (1999, 54) was clearly channeling Duane Gish (who he cited elsewhere) even though he didn't explicitly credit him (or anybody in fact) for this: "Evolutionists claim that, over time, bones from the reptilian jaw must have migrated to the ear and become mammalian ear bones. Yet there are no fossils of numerical intermediates (i.e. animals with two ear bones, or two or three bones in the lower jaw). And how did reptiles hear and chew during the transition? This is what I mean when I call natural selection plastic. Nevertheless, such theories tend to become 'fact' once enunciated."

Perloff's own scholarly attention span had to be especially plastic by this time, since Perloff (1999, 14, 16, 19 , 282-283n) quoted or cited Romer (1966, 260-261), Kemp (1982a, 319), Carroll (1988, 463) and the aforementiond Barbara Stahl without noticing any of their content relating to the inaccurate position he was so obviously repeating unchecked from Gish.

With writers like Hunter and Perloff, antievolutionism's ecumenical "any old port in a storm" ethos had arrived, indiscriminately cribbing both ID and YEC authors.

Gish has remained a favored authority over in YEC-land, of course. Duane Gish (1995) was the sole authority for Ken Ham, Jonathan Sarfati and Carl Wieland's *The Revised & Expanded Answers Book*, and David Catchpoole, Don Batten, Sarfati and Wieland's *The Creation Answers Book*, when both works identically opined: "Claimed evidence of fossils linking different kinds of organisms does not stand scrutiny," Ken Ham *et al.* (2000, 130) and Catchpoole *et al.* (2007, 121). *The Creation Answers Book* was compiled to defend the scriptural and scientific veracity of the Flood, though, so dinosaurs and hominid fossils were higher on their target list there than the RMT, which never got a mention in either work.

Although the *Answers Book* has had many iterations since, the revising part has yet to progress to the *dealing with most of the science data* stage, repeating the YEC dogma and their conservative religious apologetics even as they fall farther behind the data curve. Chunks of the 2010 edition are available online, with entries by Bodie Hodge (2010), Jason Lisle (2010), Elizabeth Mitchell (2010) and Andrew Snelling (2010) on Flood Geology and related YEC themes.

The 2013 volume is available in even larger swaths, with more of the newer *AiG* apologetic writers joining the Old Guard, like Steven Austin (2013), David DeWitt (2013), Don DeYoung & Jason Lisle (2013), Danny Faulkner (2013), Joseph Francis (2013), Ken Ham & Roger Patterson (2013), Bodie Hodge & Tim Lovett (2013), Hodge & Georgia Purdom (2013), Hodge & Paul Taylor (2013), Jason Lisle (2013), Lisle & Mike Riddle (2013), David Menton (2013a-b), John Morris (2013), Terry Mortenson & Roger Patterson (2013), Michael Oard (2013), Purdom (2013a-b), Riddle (2013), Andrew Snelling (2013a-b), Snelling & Hodge (2013), Snelling & Tom Vail (2013), and John Woodmorappe (2013).

Lots of chapters for the reading there, repeating familiar YEC talking points. And yet still nothing on the RMT. Though Roger Patterson's involvement in the *Answers Book* parade served to reinforce the incestuous nature of creationist "scholarship." The mentions of the RMT and therapsids were both sporadic and unsourced in his critique of secular textbooks, Patterson (2009, 111, 194, 197, 216), but as Gish (1995) was the only source among the "Tools for Digging Deeper" on the origin of vertebrates that mentioned them, Gish's mirage of paleontology was presumably off somewhere in the distance when Patterson insisted "Some unknown ancient reptile gave rise to what we know today as reptiles, birds, and mammals."

British creationist Geoff Chapman (2002) at least *quoted* the still wrong Duane Gish (1995, 171): "Not a single intermediate between an animal with a powerful, fully functional reptilian jaw-joint and a powerful fully-functional mammalian jaw-joint has been found." Chapman waved the Organ of Corti, too.

Which brings us to Don McLeroy. Busy dentist by profession, confirmed creationist by conviction, and Texas Board of Education member from 1998 to 2011 (Governor Rick Perry appointed him chairman in 2007), McLeroy's dismal tenure navigating the board ever Rightward has been chronicled by *Texas Freedom Network* (2009a-b; 2013a-b).

On the subject of the RMT, McLeroy (2003) relied on a short quote from the 1989 edition of Davis & Kenyon (1993, 101) as his sole lever to flip it aside. But he then went on to ask (no source this time) "what about the origin of the mammalian organ of corti?"

Where did he get that from? Besides *Of Pandas and People*, McLeroy cited a line of distinctly Intelligent Design authors who had not brought it up—Thaxton *et al.* (1984), Denton (1985), Johnson (1993a), Behe (1996) and Dembski (1999). So dropping that in without a source for it suggested McLeroy may have been dipping in the same Gish-fed creationist data stream as Chapman, but didn't want that connection to show in his printed text aimed at the Texas Board of Education.

If Perloff, the *Answer Book*, Chapman's wee pamphlet or McLeroy's school board apologetics represented condensed authority quoting, Vance Ferrell's *The Evolution Cruncher* was a nine hundred page steroidal big print quote mine orgy. And yet for all that, the RMT slipped under his creationist radar as "mammal-like reptiles" appeared inconspicuously on a chronology of life chart Ferrell (2001, 419) copied. That he apparently didn't notice that, or at least was not about to be prodded by the details, was clinched when Ferrell (2001, 476) later asserted (his **bold**): "**Here is a sampling of what you will find in the complete strata of the 'geologic column'--but remember that this 'complete' strata is to be found in its entirety nowhere in the world.**"

For the Permian, Ferrell listed "beetles, dragonflies." That's it, no vertebrates, like *Dimetrodon* and the gang. And for the Triassic, just "pines, palms." Nothing on those therapsid furballs, or the early dinosaurs and crocodiles. Ferrell only got to "crocodiles, turtles" for the Jurassic—still no dinosaurs, or mammals! And there were but "ducks, pelicans" in the Cretaceous. The dinosaurs needed to sue (a *T. rex* Sue perhaps?). Only with the Paleocene epoch did Ferrell finally mention "rats, hedgehogs."

Not surprisingly, Ferell's creationist imagination seemed unable to get beyond the commonest of familiar modern life. No chance of him populating any *Map of Time* in his head with anything close to all that we know were living over those long ages.

The recurring absence of the RMT in antievolutionist apologetics stood out for Donald Prothero (2007, 280), who noted how Jonathan Wells' *Icons of Evolution* book and several of Jonathan Sarfati's works in the period failed to mention "this extraordinary transitional series."

We'll see how Wells thought to remedy that omission in Chapter 9, but Sarfati is coming up to bat now.

It wasn't quite true that Sarfati hadn't mentioned the RMT at all, though. Sarfati (1999, 54-55) slipped a very short section on it into his *Refuting Evolution*, quoting "a specialist on these creatures," the same Kemp (1982b) that Sunderland quoted back in Chapter 4. Sarfati then cited (but did not quote) Colin Patterson (1982) to nudge the reader even farther (my **bold**): "Evolutionists believe that the earbones of mammals evolved from some jawbones of reptiles. But Patterson recognized that there was no clear-cut connection between the jawbones of 'mammal-like reptiles' and the earbones of mammals. In fact, **evolutionists have argued about which bones relate to which**."

No they haven't. And Sarfati offered zero documentation for anyone who had. Which would have been a tough slog anyway, given that the bone homology had been worked out *one hundred and seventy years earlier*.

As the 2008 2nd edition of *Refuting Evolution* left the RMT out entirely, what a miserable measure of just how superficial Sarfati's creationism was that his "scholarship" couldn't have progressed beyond trying to use a quarter-century-old propped up Patterson as a sock-puppet, instead of honestly diving into the by then vast pile of actual science research.

You know, the stuff I called attention to in Chapters 2 and 3. If I could find all that, what exactly was putting the brakes on Sarfati's mouse clicking? This is the same lethargy seen in Michael Denton on ants. Were antievolutionists (whether YEC creationists or ID advocates) literally incapable of honestly engaging the evidence?

It's beginning to look that way, isn't it?

8. *AiG* Takes a Swing: Less Mighty Woodmorappe Strikes Out

John Woodmorappe is an interesting character. Like "Harun Yahya," Woodmorappe is the pseudonym Jan Peczkis uses to assail evolution.

Which is peculiar, as Peczkis (1993; 1994) seems to accept the standard evolutionary and geological framework, published in regular science venues. But once his creationist hat goes on, the Young Earth Dr. Jekyl side of Woodmorappe shows.

In one of his apologetic pieces, Woodmorappe (1999), endeavored to show the absence of teleological purpose in evolution. But look who he quoted on it: "Illinois high school science teacher Jan Peczkis writes: 'The misconception that evolution works towards a pre-determined goal is held by many high school and college students. This is understandable because evolution is an abstract and generally non-observable phenomenon, and living things do seem well-designed for their environments.'"

Gee, it's almost as if Jan Peczkis was carefully tailoring his writing to be of maximum utility for diligent quote miners like ... John Woodmorappe (AKA himself).

The anticreationust *AiGbusted* (2008) website noticed this disingenuous manuever, evidently prompting *CreationWiki* (2011) to revise their entry on Woodmorappe to remove their mention of the Peczkis doppelganger.

I told you Woodmorappe was interesting.

But pseudonymous sock puppetry is not the reason for Woodmorappe getting his own chapter here. Following the tradition of Duane Gish (and what a success that was), Woodmorappe (2001) decided to dispute the RMT at the primary source technical evidence level (once again, my **bold**):

> Evolutionists repeatedly claim that their assembled chain of mammal-like reptiles shows a step-by-step morphological progression to mammals. Despite this, a close and simultaneous examination of hundreds of anatomical character traits shows no such thing, even if one takes basic evolutionary suppositions as a given. **Very many, if not most**, of the pelycosaur and therapsid traits used in recent evolutionistic studies to construct cladograms actually show a **contradictory pattern of progression towards, followed by reversion away from**, the presumed eventual mammalian condition. Furthermore, **gaps are systematic** throughout the pelycosaur-therapsid-mammalian "sequence", and these **gaps are actually larger** than the existing segments of the "chain". These sobering facts demonstrate that, however the supposed evolutionary "lineage" of mammal-like reptiles towards mammals is interpreted, it is **divorced from reality**.

Strong claims, requiring equally strong documentation.

Woodmorappe began with an end-run to slip some YEC doubts on the table:

> The highly-touted, alleged succession of mammal-like reptiles towards increasing "mammalness" is not found at any one location on Earth. It can only be inferred through the correlation of fossiliferous beds from different continents. Judgments are made as to which stratum on one continent is older than another stratum on another continent. Moreover, intercontinental correlations are made even when the fossil genera do not correspond with each other. Instead, the correlations are based on the general similarity of specimens, as well as their assumed degree of evolutionary advancement.

Woodmorappe cited only Bruce Rubidge (1995, 4-5) on this point (which was a discussion of the links between Russian and African therapsids), and did not offer any specific examples to support his creationist inference that such correlation work was in any sense haphazard or invalid. But in his citation note Woodmorappe did slip in this gratuitously lame observation: "There are no therapsid genera common to the Upper Permian deposits of Russia and South Africa."

And why should there be? Siberia was up by the pole, even then clear across the planet from South Africa. It would not be an evolutionary requirement of a genus to infest a continent.

Why focus all of a sudden on *genera* (a collection of closely related species) anyway, rather than some higher taxonomical level? Was Woodmorappe trying to plant the idea of unrelatedness by picking two locations far enough apart and hunkering down at almost the base taxonomical level? We've got an abortive attempt at a *Bermuda Triangle Defense* (Player #7) here.

Let's investigate what Woodmorappe wouldn't.

The genera of the two regions were distinct, but only if taken at that grade. Step up just one notch to their Burnetiidae *family* and things started looking different.

For decades this basal therapsid family was known by only two genera, *Burnetia mirabilis* in South Africa, and *Proburnetia viatkensis* from Russia. At around the time Woodmorappe was not applying creationist typology to a subject he had brought up, paleontologists had uncovered three more from South Africa and another Russian genus.

Thanks to our recurringly obliging Fossil Genie, though (who either didn't get Woodmorappe's memo on the limits of Permian biogeography or preferred to ignore it), a team including Bruce Rubidge (!) later unearthed in South Africa a new species of one of those Russian genera, *Paraburnetia sneeubergensis*, "that is remarkably similar to *Proburnetia*," Roger Smith *et al.* (2006).

That spatial relationship of life (fossil or living) is a subject antievolutionists fail miserably at, probably because they're too used to pulling datasets apart to keep the evolutionary picture too fragmented to see, rather than putting the pieces together to make a clearer *Map of Time* image.

Take the *Dicynodon* genus, which we've already encountered. They're known from the Late Permian of Eastern Europe, but with a limited sampling it wasn't clear whether that was a valid genus, or more a collection of very similar cousins. As more examples were assessed, the cladistic analyses by Kenneth Angielczyk & Andrey Kurkin (2003a-b) suggested they were likely not a monophyletic bunch, though still closely related to a particular genus, *Kannemeyena* from Russia, filling in the biogeographical relationships of how the species of that time were spreading and evolving in adjacent regions.

More recently, Christian Kammerer *et al.* (2011) and Marcello Ruta *et al.* (2013) have sorted out a lot of the "dicynodont" relationships and their varied history through the Permian mass extinction and into the Mesozoic.

Since Woodmorappe thought to cast doubt on the bigger stratigraphy issue in the way he did, as a seed of doubt, turnabout fairplay, and we may ask what has gone on over in the creationist "technical literature" on this subject of the dicynodonts?

As of this writing, *nothing*. No accounting for how many "kinds" were involved or whether they were just another of those elastic monobaramins. And M. Aaron (2014) did not include them in his analysis (the only creationist stab at it so far). More on Aaron's piece in Chapter 10.

The fossil data Woodmorappe skipped around were just temporary stage props, invoked but not studied (nor apparently, genuinely understood). But the really big scenery was about to be rolled in, as Woodmorappe promised to undertake the "comprehensive approach" that was needed. "*All* the anatomical features must be considered, not just a few."

Ah, but that is one thing Woodmorappe was not disposed to do, as he took every opportunity to step around those anatomical features.

Oversplitting apomorphies (Three Card Monte Edition)

Take this trimmed Woodmorappe quotation from Zhe Luo (1994, 101). The creationist may have been in such a hurry this time that his note only gave the pages for Luo's entire chapter, not the specific page. Here's what Luo wrote, with the sentences quoted by Woodmorappe in **bold**):

> **By oversplitting apomorphies in its favor, one hypothesis can dominate over its rival without gaining any biological insight. One way to guard against this fallacy is to show how the apomorphies in support of a given hypothesis are biologically associated.** If apomorphies can be grouped by their anatomical association, that should show whether or not the number of characters truly reflects the complexity of the anatomy. That will expose the idiosyncratic oversplitting of apomorphies. In this study, I categoize apomorphies supporting each alternative hypothesis by the anatomic areas to which they belong (Table 6.2).

Woodmorrape explained in a note that "An apomorphy is a trait that appears for the first time at a given position in the cladogram, as reconstructed by evolutionists."

Not quite.

Apomorphies refer to inherited derived features that are consistently found *only in a particular lineage,* not just the first time they show up in the cladogram. They also specifically exclude ones that have been lost secondarily among any of the descendants. So already Woodmorappe was wandering off the road at the basic terminology level. And Woodmorappe was leaving that data far behind by illustrating the cladistic concepts with abstract shapes along a simple line, far from the level of detail Luo laid out in that Table 6.2 Woodmorappe left out of the quote.

Table 6.2 was a listing of apomorphies relating to features in two mammallike groups, the *tritheledontids* (AKA ictidosaurs) and the *tritylodontids*, and whether they were uniquely shared by mammals. Because they're going to be cropping up again in further antievolutionist apologetics, to simplify things I'll tag the *tritheledontids* as "T1" and the *tritylodontids* as "T2".

The T1 Ictidosauria/tritheledontids were small insectivorous predators with a long run from at least the Late Triassic into the Jurassic, and included double-jawed *Diarthrognathus* (from back in Chapter 4) and *Pachygenelus*. The range of the latter was fairly broad, showing up in North America as well as across Gondwana to the south, Sankar Chatterjee (1983) and Neil Shubin *et al.* (1991). Further finds have helped sort out ictidosaur relationships, Chris Gow (2001), Agustín Martinelli *et al.* (2005), Christian Sidor & John Hancox (2006) and Martinelli & Guillermo Rougier (2007), with Fernando Abdala & Ana Ribeiro (2010) surveying cynodont diversity and distribution across Gondwana during the Triassic.

You may recall the largely herbivorous T2 tritylodontids from last chapter, where Barbara Stahl noted how their teeth structure put them and other groups off the main line of mammal evolution, even as so much of their anatomy was the way it was because they shared common descent in that synapsid line.

Sorting out just which group was more *closely* related to mammals was the purpose of Luo's paper. T1 and T2 had many traits in common with full mammals, but Luo left those off the list because those wouldn't be as informative in deciding that closeness of relationship. Six main skull areas were involved, keyed to the illustrations on page 103 of two cynodonts (*Probainognathus* and *Kayentatherium*) and the early mammal *Sinocondon*. Let's take a stroll through Luo's list, to see what pieces were showing up on the paleontology scope, at the top of the very page Woodmorappe had quoted from, so we can compare it to his creationist response. Some of the bone names should be familiar already, but the stuff in quotes are obviously getting more technical.

(1) In the orbitotemporal region (the eye area), only one feature showed up for T1 (a "slender zygomatic arch"), while four where shared by T2, involving a specialized enlarged palatine bone that contributed parts of the eye opening, along with an "ascending process" flange, and separate openings for the "greater and lesser palatine nerves."

(2) Down in the temporomandibular joint and lower jaw, while T2 had no

apomorphies here, T1 had four notable ones: the "lateral ridge" on the dentary bone was enlarged at the end; the indentation in the squamosal bone where the dentary attached and the motion line on the lower jaw faced the same way; and the cartilaginous joint "symphysis" connections were similarly "mobile."

(3) Regarding the palate, T2 again had no relevant novelties, while T1 showed two: the end of the palate was larger and the "pterygopalatine ridges" there extended to an adjoining basisphenoid bone (an embryonic component of the skull base).

(4) The petrosal bone was a dead heat, as both had two different apomorphy sets. These bones house the structures of the inner ear, so any similarities or differences among them could be quite diagnostic about phylogeny and lifestyle. Unfortunately, they're also small bones and not readily preserved, though meticulous care and advances in scanning technology have culled much knowledge from them, such as Eric Ekdale *et al.* (2004) and Elijah Hughes *et al.* (2015). Back in 1994, though, among the T1, the "round window" and "jugular foramen" openings could be seen to be farther apart and the jugular bone was starting to separate from the "hypoglossal foramina" opening. Over in T2, there were suggestions of the "hyoid ('stapedial') fossa" and a split in the "paroccipital processes." In other words, subtle variations but not vast structural differences.

(5) Down where the quadrate bone attached to the jaw assembly, once again there were specialized variations. In T1, the edge of the dorsal plate was rounded, and the bone's orientation was "most similiar to that of *Morganucodon*." For T2, there was a "stapedial process" bump on that bone, and the quadrate appears to have contacted the "paroccipital process" on the front of the petrosal bone.

(6) Finally, there were the teeth. T1 showed particular wear marks on the upper and lower teeth, while the features in T2 involved how the roots were shaped and implanted in the jaw. Recallling Stahl's point above about how T2's teeth differed from mammals, these apomorphies would be indicating their relative common ancestry with mammals, which was of course the taxonomical issue Luo was addressing.

None of these diagnostic features involved grand new structures, just the quite subtle morphing of their common anatomy. Trying to scope out exactly which of these extinct groups represented a closer lineage to the later surviving mammals depended ultimately on the development underpinnings of those variations and how they had come about genetically. And remember, nobody has the DNA of a T1 or T2 to look at to open up that biological hood.

Taken all together, T1 and T2 represented therapsid lines that had great overall resemblance to one another (and nothing at all like their now highly diverged distant diapsid cousins, like the dinosaurs or emerging reptiles). Telling which one of these similar groups was more closely positioned on the mammal lineage depended entirely on how many fossil examples were available and how well preserved they were. The data available to Luo in 1994 suggested those T2 tritylodontids were a bit more likely to be the sister group to mammals than the T1 tritheledontid cousins, though by the time of Kielan-Jaworowska *et al.* (2004, 157) the accumulating data had trended more in favor of T1. Further finds have tended to nest the mammals and T2 next to the earlier branching T1, as reflected in Jun Liu & Paul Olsen (2010) and Marcello Ruta *et al.* (2013).

But Woodmorappe was trying to dismiss any of them being related at all, near or far, and for that we have to look more closely at the features Luo's chapter had covered in great detail.

What did Woodmorappe have to say about that? Just three sentences: "Of course, the phrase 'biologically associated' smacks of evolutionistic just-so stories. However, in this study, I do not attempt to make any anatomical judgments, but rely on datasets provided by evolutionists. In this way, the negative conclusions regarding evolution become all the more compelling."

In other words, Woodmorappe was not going to pay any attention at all to what the features were, or how they related to the larger anatomy. He was going to abstract the numerical side of the datasets until it would look "all the more compelling."

So this is what he meant by a "comprehensive approach"!

Nothing about Luo's list involved "evolutionistic just-so stories." They were *observations* about the facts of the anatomy, the data that would need to be accounted for as much by a creationist explanation as any evolutionary one. Luo was doing that *paying attention to all the facts* thing that Woodmorappe professed to be aiming for, until he pointedly excused himself from the duty.

Here, quite literally, is where you can see Woodmorappe putting down the shells for his creationist version of Three Card Monte. Only we've just spotted him pulling out the factual peas.

Now for Woodmorappe's game. And we get to watch. In slow-mo, to catch the moves.

Overall skeletal characters from Sidor & Hopson (1998)

Table 1 listed Woodmorappe's calculations of "mammalness" for 19 of the 21 fossil examples from Christian Sidor & James Hopson (1998, 259), a paper examining how quickly varied mammal traits were being acquired in the known lineages, and whether any of that suggested other than a gradual process of acquisition. On that point, Sidor & Hopson (1998, 254) had concluded: "These correlations are consistent with the hypothesis that rapid accumulation of derived features occurred relatively infrequently within the synapsid lineage leading towards mammals and that gradual character evolution predominated."

Gradual evolution, was it? High five one for Darwin!

Not being an evolutionist, though, Woodmorappe understandably could care less about any of that content, but drew on their extensive dataset (not any work he'd undertaken himself, remember) for the raw numbers laid out in

Sidor & Hopson (1998, 269-270).

Of the 181 character traits they'd examined, Woodmorappe excluded 77 "reversing characters" as well as another 16 for reasons he never got around to specifying. As for the remaining 88, all Woodmorappe did was generate a bulk number based on the *sum* of the values assigned to them for the cladistic analysis, with an adjustment for characters that weren't known for that particular specimen.

We'll be looking more closely at those "reversing characters" in a bit, as well as checking some of Woodmorappe's totals.

Up front, though, you need to know that most of the individual entries in the three works Woodmorappe was using were either 0 or 1, but could range higher (all the way up to 6 for some traits in the Sidor & Hopson survey). A zero usually denoted the absence of a trait, which could represent an ancestral condition. But not necessarily. As we'll see below, an individual trait disappearing could be related to the evolution of a new feature that dispensed with it. In that case the value for that trait would *drop* to 0 in a descendant.

But it's worse than that. Adding up the numbers only works if they are *numbers*. But they aren't, they're symbols used to represent a character state, which can certainly be evaluated cladistically in relation to other instances of that same character trait, but which ceases to have any meaning at all if you treat them as if they *were numerical.* For example, the "progressive" item 144 in Sidor & Hopson (1998, 272) is a postcranial character: "Humeral head shape: broad and strap-like (0), elongate oval (1), subspherical (2)."

Character state 2 (subspherical) is not one more than being oval. These could just have easily been labeled A, B & C. How then would you add letters, let alone "normalize" them?

All of which challenged the very idea that Woodmorappe could be telling much of anything useful just be adding up the raw numbers without paying attention to what the numbers represented.

And something else.

Wouldn't it have been the most obvious thing in the world to do, when claiming the supposed inadequacy of the evolutionary accounting of the RMT data, to show at the same time how effectively the *creationist* version explained those exact details?

While Woodmorappe included brief descriptions for the examples he did use, he only cautioned that these (shown on my version in parentheses below) "reflect evolutionary notions of the mammalian condition. The descriptions are not intended to endorse these evolutionary notions." But they certainly didn't give any clue as to what *kind* Woodmorappe thought any of them were, either. Unless, of course, working that out was something else on his "not ever going to think about" list.

Not that Woodmorappe was averse to quibbling over evolutionary systematics, as when he contended that "genera of mammal-like reptiles are inflated by taxonomic oversplitting, a fact that is substantiated by more recent studies," and citing Gillian King & Bruce Rubidge (1993) and Barry Cox (1998), papers trying to sort out which descriptive names were valid based on a fuller assessment of the evidence.

Such oversplitting (or undersplitting) of the taxa has yet to trouble the creationist's ill-defined kinds, though. Nor would any form of splitting make the fossils disappear, such as *Eodicynodon* that both papers had covered, and which Cox specifically mentioned as being just what an *ancestral form* would look like for four other valid taxa that his analysis had identified.

So no "kinds" are going to be forthcoming from Woodmorappe. OK.

But what about *chronology*? While Sidor & Hopson had noted the time frame for each of the samples, Woodmorappe left that out too. I've put that factor back in, along with noting the occasional typo (possibly due to the conversion of the article to its online version), and noted by *asteriscs several taxa on the Sidor & Hopson list that Woodmorappe decided not to include.

Woodmorappe's "mammalness" numbers are on the left.

Woodmorappe Table 1 data

5	Ophiacodontidae ("primitive pelycosaurs") Mid-Pennsylvanian Period (310 mya), mispelled as "Ophiacondontidae"
	Ten million years later
0	Edaphosauridae ("advanced pelycosaurs") Late-Pennsylvanian (300 mya)
1	*Haptodus* ("primitive sphenacodont"), lopped off as "*Haptodu*" genus
3	Sphenacodontidae ("overall sphenacodont")
	Thirty million years later
29	*Biarmosuchia* ("primitive therapsids") Late Permian Period (270 mya)
*	*Eotitanosuchus*
32	Anteosauridae ("primitive therapsids")
32	Estemmenosuchidae ("primitive therapsids")
33	Anomodontia ("varied therapsids"), truncated as "Anomodonti" family
	Four million years later
43	Gorgonopsidae ("primitive therapsids") Late Permian Period (266 mya)
52	Therocephalia ("advanced therapsids")
	Eleven million years later
80	*Dvinia* ("primitive cynodont") Late Permian (255 mya)
81	*Procynosuchus* ("primitive cynodont")
85	Galesauridae ("medial cynodont")
	Five million years later
87	*Thrinaxodon* ("varied cynodont") Early Triassic Period (250 mya)
	Ten million years later
82	Cynognathia ("advanced cynodonts") Middle Triassic (240 mya)
	Six million years later
101	*Probelesodon* ("advanced cynodont") Middle Triassic (234 mya)

102	*Probainognathus* ("advanced cynodont")
	Twenty million years later
109	Tritheledontidae ("sister-group candidates") Late Tiassic (214 mya)
120	Morganucodontidae ("mammals")
	Fourteen million years later
*	*Sinoconodon* Early Jurassic Period (208 mya)

By Woodmorappe's own table, apart from the high "5" given for the Ophiacodontidae family at the start, and the relatively slight drop to "82" for the Cynognathia in the Triassic, his numbers traced a fairly steady progression of "mammalness." Given the features of the taxa (including that *Probelesodon/Chiniquodon* noted last chapter) none of this would seem a terribly bad showing for evolution, especially given that Sidor & Hopson weren't trying to map a single beeline to "mammalness" in the first place, but had surveyed the rate of trait acquisition in varied proto-mammal lineages, not just the groups thought to fall on the direct line to modern mammals.

Would no internal variation be expected in so diverse a group sampled over a hundred million years? Of the sort we can see physically in, say, those Galápagos finches from back in Chapter 6?

Ah, but Woodmorappe doesn't think there were all those millions of years. But neither was he keen to actually apply that conviction to this data set, such as explaining what these critters were all doing only 4500 years ago when they were supposedly getting sloshed around in the Flood, and why none of them ended up surviving from the Ark, where members of their "kinds" should have been preserved.

Instead he never addressed the subject (even with an external reference), doing much as Phillip Johnson had in the last chapter when he declined to bog his readers down with technical terminology when it was likely Johnson who didn't want to be troubled with it.

At the very least though, Woodmorappe's own table had rammed those outlier Ophiacodontidae and Cynognathia under our nose. What was up with them? What did he have to say about what contributed to the 5 score of the first, or the slump in the second?

Nothing. Woodmorappe never mentioned them again.

Had he wanted to try being informative (for a change), Woodmorappe could have noted from Sidor & Hopson (1998, 271) that a big chunk of that 5 consisted of a rating of 2 for character state 65, which measured the orientation of the "paroccipital process," a mirrored pair of projections at the rear base of the vertebrate skull where jaw muscles are attached. That little protrusion rated a 0 in Sidor & Hopson's cladistic analysis if they were "strongly posteroventral and lateral" (meaning they stuck out to the front and to the side); a 1 if they were "moderately posteroventral and lateral"; or 2 if they were shorter and more flattened in a "transverse" layout.

That transverse state is what was found in the Ophiacodontidae. And that would turn out to be the most common mammal form, present in *all* the later thirteen animals listed by Sidor & Hopson. The process was a rather rounded nub by the time of *Morganucodon*, *Sinoconodon* or tiny *Hadrocodium*, shown in Luo & Crompton *et al.* (2001, 1537). It was only in some of the intervening examples *millions of years* earlier that their variations had earned them an atypical 0 or 1.

But wait, wouldn't that mean trait 65 was one of those "reversing characters" that Woodmorappe assured us he had excluded from the Table 1 summary?

Oops.

In fact, if you looked closely at the character listings, Woodmorappe had ended up including seven more of the "reversals" among the 88 "progressive" characters (15, 25-26, 39, 113, 129 and 169).

As for the *Thrinaxodon*/Cynognathia slump from 87 to 82, I'm not sure how he came up with those numbers at all. Based on the 88 character traits Woodmorappe said comprised his main set, *Thrinaxodon* came to only 77 while the later Cynognathia totaled 81. None of the "?" cases where the trait wasn't preserved in the Cynognathia fossils fell on the listings Woodmorappe said he was using, either, so his adjustment math shouldn't have significantly altered those numbers.

Now those 77 and 81 sums would have looked even more reversing had he put that in, except that a lower or higher individual state number didn't pose a problem for evolution in the first place.

Take character trait 61: "Pterygoid quadrate ramus contact the quadrate." The ancestral state was a 1 for its presence, and it progressed to 0 for its absence. Thus the number went *down* for the Cynognathia because that part of the bone was disconnecting from the quadrate (which was relocating as an earbone, remember). As for what these bone shifts in the Ophiacodontidae and Cynoganthia branches were saying about the underlying biology or genetics, was Woodmorappe decreeing that such modest morphings of those tiny protrusions and contact points somehow exceeded the load limit of natural variation?

We don't know what Woodmorappe thought about any of that. Which was interesting, since the paper Woodmorappe drew on next for his second table, Luo & Crompton (1994, 343-350), had discussed the placement of the pterygoid quadrate ramus at some length, illustrating its varied disassociation from the quadrate in *Procynosuchus, Thrinaxodon*, *Probainognathus* and *Massetognathus*.

Do you think Woodmorappe might have been in too much of a hurry constructing his tables to notice any of the content of the work he was skimming?

Quadrate skeletal characters from Luo & Crompton (1994)

Woodmorappe's next list involved 8 of the 26 animals from his Table 1, plus two new examples drawn from Luo & Crompton, which I've again marked with *asteriscs. You'll recall those T2 Tritylodontidae, of course, but also the critters just mentioned on that pterygoid quadrate ramus matter. Incidentally, the Luo & Crompton paper appeared in the *Journal of Vertebrate Paleontology* the very issue before Jan Peczkis (1994) was published, wearing his paleontology "evolution" hat.

Luo & Crompton had examined only 14 character traits this time, relating specifically to the evolution of the quadrate bone (details which Woodmorappe again did not identify or otherwise discuss). With so few listings, Woodmorappe devoted one column to the total values, and another for the five that underwent no "reversals" (NR). I put those numbers to the right of the main ones, and once more include the chronological frame Woodmorappe excluded.

Woodmorappe Table 2 data

	NR	
2	0	Anomodontia
		Four million years later
6	0	Gorgonopsidae
3	0	Therocephalia
		Eleven million years later
1	0	*Procynosuchus*
		Five million years later
5	3	*Thrinaxodon*
		Sixteen million years later
15	5	*Probainognathus*
13	9	**Massetognathus* ("advanced cynodont")
		Twenty million years later
20	9	*Tritylodontidae ("sister-group candidates")
21	12	Tritheledontidae
25	13	Morganucodontidae

The left numbers on this table were a little more jumpy. But did they mean what Woodmorappe wanted them to, the prevalence of "reversals"? What were the features that were rising and falling there?

All Woodmorappe had to say on this table's content was: "The small number of characters also necessitates a different approach, from that used in Table 1, in computing the Mammalness Index. Because there are only 14 traits, if one were to, as before, compute the relevant quotient and then multiply it by 100, it would cause serious distortion of the data. For this reason, the Mammalness Index in Table 2 is simply the sum of character polarities for each taxon."

"Serious distortion" of what data? The details Woodmorappe seemed oh so reluctant to mention?

Woodmorappe had a footnote to this passage that stayed equally abstract:

> This bias is essentially a small-numbers effect. For example, imagine a situation where one genus has a score of 8 relative to 10 characters, and the other has a score of 11 relative to the 10 characters. Now, in another situation, one genus has a score of 40 relative to 50 characters and the other has a score of 55 relative to 50 characters. In both cases, the Mammalness Index is technically the same: 80 and 110, respectively. But this has little practical meaning in the first case, as only 3 points separate the first and second genus.

Only 3 of what? The lack of "practical meaning" was all on Woodmorappe's shoulders at this point.

Since we're talking only ten taxa and fourteen traits, let's take a look at some of the original dataset in Luo & Crompton (1994, 360-364) and what they signified.

Take item 9 on Luo & Crompton's list, which Woodmorappe excluded from his catalogue of progressive traits: "00011111??".

That referred to the articulation of the quadratojugal bone, which in those first three cases abutted the edge of the quadrate's dorsal plate, which was listed as 0. In later cynodonts it had separated, earning a 1. Luo & Crompton (1994, 362) had noted that "The quadratojugal has never been found in any known specimens of *Pachygenelus* and *Morganucodon* and is coded '?'."

But that was 1994. Remember, paleontology has this habit of not standing still, especially with a busy Fossil Genie running around, depositing things to make evolutionists happy, and new finds have come on the scene since, such as further examples of the *Brasilodon* family, José Bonaparte *et al.* (2005) and Irina Ruf *et al.* (2014), and cousins like *Microconodon*, Hans-Dieter Sues (2001), "a transitional and possibly mammaliaform taxon, and its lower jaw appears to have a well-developed lateral ridge that approaches the condition of brasilodontids," Zhe-Xi Luo (2011, 360).

Where do things stand today?

"In most cynodonts, the quadrate (incus) is ancestrally associated with the quadratojugal; the quadratojugal and quadrate together reinforce the jaw hinge. A trade-off is that the complex quadrate and quadratojugal association can limit the mobility of the incus, thereby reducing auditory sensitivity," Zhe-Xi Luo (2011, 361). That trade-off was resolved in most of the later protomammal lineages by losing the little bone altogether, though it was retained in that sideline that included *Brasilodon*, where it had slipped down as a projecting spur, illustrated in Luo (2011, 359).

So that trait was not posing some obstacle to the natural evolution of the mammalian quadrate setup at all.

But what about those eight "reversing" traits?

Two of those (one quarter) were doing this *only* in the Gorgonopsia, a group living early in the sequence, and not held to be on the direct line of descent to mammals, even if pretty close cousins, as charted by Jun Liu *et al.* (2009, 396). Item 2 involved "Curvature of the contact facet of the posterior side of the dorsal plate: flat or nearly flat (0); convex (1); concave (2)." And trait 7 concerned "Dorsal margin of the dorsal plate: with a pointed dorsal process ('dorsal angle') (0); rounded (1)."

What did that jargon mean?

Without his discussing any of the details, by his exclusion of them in his analysis, Woodmorappe was trying to dismiss the gentle evolution in curvature of one bone and the rounding of another solely because the gorgonopsids had developed this same feature in their shared anatomy.

The micro scale of what was going on was even more evident in another of the "reversing" traits, number 10 on Luo & Crompton's list. It related to "Articulation of the pterygoid to the medial margin of the quadrate: the quadrate ramus of the pterygoid contacting the anterior face of the medial margin of the quadrate (0); the posterior end of the quadrate ramus of the pterygoid is laterally overlapped by the medial side of the quadrate (1); no articulation (2)," Luo & Crompton (1994, 373).

The raw numbers were "1210022222". That second number 2 in the list was from our old pal the Gorgonopsids, prone to convergent innovation.

As illustrated in Luo & Crompton (1994, 346, 348, 350-351, 366-367), concerning *Thrinaxodon*, *Probainognathus* and *Massetognathus*, this feature involved how close the end of the pterygoid bone was getting to the adjacent quadrate. All of the taxa shared the same general anatomy (as well they ought, since they were so closely related in an evolutionary sense). And because the animals concerned were very small (mouse-sized), the actual difference between "contacting" and "overlapped" and "no articulation" at all involved the tiniest of anatomical shifts, a bump of less than a *millimeter*.

Woodmorappe explained *nothing* about any of this to his readers.

He did take issue with Luo & Crompton's cladograms, though, insisting (my **bold**) that "the proliferation of reversing traits makes it difficult for evolutionists to decide which mammal-like reptiles, and inferred early mammals, are, evolutionarily speaking, closest to each other. This confusion is reflected in the construction of **widely-contradictory** cladograms," and cited "Luo and Crompton, Ref. 19, p. 340. Four different versions of cladograms are presented, with each one supported by one set of evolutionists. My descriptions involve two of these: (A) and (C)."

Another typo had cropped up here as the cladograms occurred on page 342 (not 340), but that was pocket change compared to the hyperbole of Woodmorappe deciding the nearly identical cladograms A & C (which differed *only* on whether those quirky basal Gorgonopsia had diverged before the Anomodontia or whether they ranked next in line after them, closer to the Therocephalia) were somehow "widely-contradictory."

Even if you looked at *all four* cladograms in Luo & Crompton, they couldn't be characterized as that. All agreed on the placement of the Permian cynodont *Procynosuchus* and the later Triassic *Thrinaxodon* in the middle of the parade. For context, by that time dinosaurs were just gearing up.

The trees differed slightly after than on the ordering of the so similar T1

Tritheledontidae and T2 Tritylodontidae, leading into the early mammal *Morganucodon* at the end of the parade. While the "Strict Consensus Tree" A and the "Preferred Tree" B put T2 before T1, C switched them. Cladogram D, meanwhile, (which Woodmorappe did not discuss) was much like the B cladogram, except it ranked *Probainognathus* farther along, between T2 and T1. All so shocking, so "contradictory"?

There was no controversy at all about which taxa needed to be included in these studies. For example, no one was showing genuine confusion by dragging in contemporaneous diapsid reptiles like archosaurs or dinosaurs. Any cladogram worth its nodes would put any of those as outliers. What "conflict" there was had arisen only because the taxa involved look so much alike, all nudging so close anatomically that the finest of discernment was needed to sort them out (helped along by a few Treelength values) to provisionally assign a taxonomical pecking order assembled from their scattered remains that survived the travails of geology hundreds of millions of years after they had scampered around alive.

Without further clarification or example, Woodmorappe moved on to his third table.

Dental and cranial characters from Luo (1994)

There were twelve animals this time, with several new ones (again with my *asteriscs) including *Sinocondon* oddly enough, that Woodmorappe had not included on Table 1 even though it was a subject of Sidor & Hopson. Once more, Woodmorappe had two data columns, both rather clumsily labeled, the first as "81 of 82" characters, and the second to the right as "53 of 81 of 82" progressive characters (Prog). This time, I've left out the chronological element to better see the order Woodmorappe put them in.

Woodmorappe Table 3 data

	Prog	
0	0	*Thrinaxodon*
18	7	*Probainognathus*
19	7	*Diademodontidae ("advanced cynodont")
35	7	*Traversodontidae ("advanced cynodont")
78	34	Tritylodontidae
58	54	Tritheledontidae
100	104	*Sinoconodon*
131	120	**Haldanodon* ("mammal")
139	131	*Triconodontidae ("mammal")
134	126	**Dinnetherium* ("mammal")
132	128	*Morganucodon*
117	122	**Megazostrodon* ("mammal")

There are more seeming reversals this time (four in the first number series, and only those last three for the progressive values), highlighted in gray. But just as interesting is the chronology, as Woodmorappe put two of the taxa in an odd order: placing *Morganucodon* (which lived around 214 mya) *later* than *Sinoconodon* (which lived around 208 mya). That six million year span is longer than it took the human species to evolve from the australopithecines, so is hardly a trivial oversight (unless you're a Young Earth Creationist, for whom all fossil life represented the same Flood-era time slice). Or, had Woodmorappe simply forgotten for a moment that in the regular chronological system, the bigger the number in a "years ago" thing, the older it was?

This was an even more ironic slip, given how finicky Duane Gish and Phillip Johnson had been on the therapsid sequencing issue. Anyway, here's what things look like when we correct for the time sequence:

Woodmorappe Table 3 data

	Prog	
0	0	*Thrinaxodon*
19	7	*Diademodontidae Early Triassic 250 mya
		Sixteen million years later
18	7	*Probainognathus*
35	7	*Traversodontidae Middle Triassic 234 mya
		Twenty million years later
78	34	Tritylodontidae
58	54	Tritheledontidae
132	128	*Morganucodon*
		Six million years later
100	104	*Sinoconodon*
		Eight million years later
134	126	**Dinnetherium* ("mammal") Early Jurassic 200 mya
117	122	**Megazostrodon* ("mammal")
		Thirty-three million years later
139	131	*Triconodontidae ("mammal") Middle Jurassic 167 mya
		Twenty-two million years later
132	120	**Haldanodon* ("mammal") Late Jurassic 145 mya

If anything, the numbers seem bumpier this time around (again noted in gray). Which only highlighted our curiosity to find out what (if anything) these lumped "mammalness" totals meant this time.

Woodmorappe certainly thought they meant a lot (my **bold**):

> Probably the most informative analysis of mammal-like reptiles as (alleged) transitional forms is the one which focuses, **in detail**, on the presumed changes from advanced cynodonts to the earliest mammals (Table 3). The sister-group cynodonts (Tritylodontidae and Trithelodontidae) rival each other for the status of the closest non-mammalian relatives to mammals. Yet, when all of the characters are considered, one is struck by the **chasm** between these sister-group advanced cynodonts (58 and 78) and the earliest presumed mammals (100–139). However, the "bottom falls out"

> when only the progressive characters are considered in Table 3. Here, a **giant evolutionary leap** is required to make the presumed change from fairly advanced cynodonts (7) to the advanced sister-group cynodonts (34 and 54). From there, another **great gulf** must be spanned in order to link the sister-group cynodonts (at 34 and 54) with the earliest mammals (104–131).

But when has Woodmorappe focused on any of that "in detail" stuff?

Let's have Woodmorappe feed out a little more rope, before we start tugging again, this time with *Woodmorappe* supplying the **bold**:

> One of the most striking findings uncovered by this analysis is that the **majority** of anatomical traits (the ones actually used by evolutionists in the construction of their cladograms) **do not** show a unidirectional progression towards the mammalian condition! Of the 181 anatomical characters considered by Sidor and Hopson, 165 were deemed to be sufficiently complete, in terms of data, for further consideration in the present study (Table 1). Of these 165, 88 were found to be progressive. In stark contrast, no fewer than 77 of the 165 showed reversals of character.

Woodmorappe had a footnote for this: "Some characters used by Sidor and Hopson (Ref. 15), notably those which they numbered 3, 10, 28, 41, 44, 52, 102, etc., underwent more than two reversals of progress that each had previously made towards the eventual mammalian condition."

He then went on: "This is not an isolated instance. As noted earlier, 9 of the 14 quadrate characters used by Luo and Crompton were likewise reversing (Table 2). Finally, in the analysis of 82 mostly dental and cranial characters, by Luo (Table 3), no fewer than 53 characters were found to be reversing." Woodmorappe had footnoted that too: "One additional character (No. 39 in Luo, Ref. 10) was rejected because its character-polarity was unknown for too many taxons. This left a total of 28 progressive characters and 53 reversing ones available for the present study."

As we've seen already, the details that Woodmorappe never thought to mention cast into doubt that these "reversals" were anything of the kind.

But we'll look again one more time just to be sure.

Let's take the character trait "3" that Woodmorappe listed as exhibting "more than two reversals of progress."

Sidor & Hopson (1998, 270-271) defined that as "Premaxilla vomerine process: absent (0), present (1)", and the numbers for that feature were "1110??001101111111111" (with the first being the Ophiacodontidae and the last belonging to *Sinoconodon* 100 million years later). You'll notice that whatever that meant, the "reversals" had stopped halfway through, as *all* the second half of the list showed it, a condition that continued for the next *sixty-six million years*, without exception.

What then was getting "reversed"?

Woodmorappe again never stopped to explain (hide that card).

The vomerine process (VP) is a small spur of bone that can extend from the premaxilla (which forms at the front of the dentary) onto the vomer behind it along the roof of the palate. The premaxilla goes way back in vertebrate history, visible in fish then and now, along with their descendants on both sides of the diapsid/synapsid line. The vomerine process varies a lot, though, growing or shrinking in various lineages over time, depending on the developmental and adaptive circumstances.

Snakes, for example, show way more range than the therapsids. In a more recent study, John Scanlon & Michael Lee (2011, 79) required more numbers to cover the VP there, where 0 represented "extensive overlapping contact with vomer," 1 was "non-overlapping, point contact with vomer," and a 2 for "not in contact with vomer. Inapplicable in some agamids (where process contacts anteromedial flange of maxilla), but others exhibit state 1."

The order of the numbers in Scanlon & Lee gave another clue to the developmental process, where the trend as the numbers grew higher suggested a tendency for the VP to form less contact with the adjoining vomer.

Back in Sidor & Hopson, describing taxa that lived a couple of hundred million years ago, it was the 0 state that denoted the failure of a VP to develop in some of the therapsid lineages.

The paleontology continued to move on also, of course.

Christian Sidor (2003a, 979-980) described the VP as a "tapering wedge" in the therapsid example he illustrated, a new fossil from the Biarmosuchia. That find meant the Fossil Genie had filled in the first of the "?" marks above, turning it into another 1 were the chart to be made today, meaning only 5 of the 21 examples in Sidor & Hopson (1998) ultimately failed to exhibit that little VP sliver, and all of those 0s were earlier in the show.

Just how developmentally shocking was it for a few of the therapsid cousins over a span of a hundred million years to ratchet back the slim VP wedge in their palate now and then?

The genes governing jaw formation were just being explored back in 2001. Woodmorappe never paused to think about that, but in the years since, more work has been done (again by the regular scientists applying their evolutionary model as fruitful guide, not by any of the antievolutionist bystanders quote mining from arm's length).

Kevin Parsons *et al.* (2014) have identified the *Wnt* signalling pathway as playing an important role in generating cranial variability, including the premaxilla and vomerine process, which in the modern cichlid fish "supports the upper jaws as they open and close." The Parsons team were *experimentally generating* such variations at the microevolutionary level, seeing shifts in the shaping and degree of mineralizing ossification that resulted from their tiny tweeks. Just how implausible was it that comparable natural variations were occuring in that ancient life, all those millions of years ago?

One thing seems clear, antievolutionists like Woodmorappe weren't in much of a hurry to work any of that out.

Like Duane Gish, Woodmorappe brought his *Answers in Genesis* argument to a close with some appeals to uncertainty, which only underscored the failure of the creationist imagination in second guessing what the Fossil Genie was going to be up to in the 21st century.

"Several creationist scholars have pointed out the lack of evidence for gradational change in the mandibular-auditory mechanism of the 'advanced' mammal-like reptiles towards that of the presumed early mammals. Interestingly, a few evolutionists have actually acknowledged this fact in print," Woodmorappe claimed, and pulled a *Bermuda Triangle Defense* in quoting Edgar Allin & James Hopson (1992, 608): "Intermediate stages in the transference of postdentary elements to the cranium are poorly documented. Indeed, the only fossil evidence on this critical interval is the presence of persistent attachment sites for the anterior end of the postdentary unit in the primitive therians *Amphitherium* and *Peramus*."

We'll bump into those critters again later, in Chapter 10 regarding creationist *kinds*.

While the fossils showing that final phase of the "transference" (the retention and evential discarding of cartilage holding the auditory bones to the jaw) hadn't turned up as of 1992, we *know they did later* (covered back in Chapter 3), filling yet another of the strenously dug creationist potholes.

Finally, Woodmorappe decided to dismiss the relevance of stratigraphic correlation, by which fossils fall into a temporal sequence that relates to their geological appearance. We have to remember that Woodmorappe is a Young Earth creationist, for whom such sequencing is anathema, and it is by such means that doubts are sown without actually defending the YEC Flood framework that he actually believes in.

So it was that Woodmorappe withdrew one sentence from a paragraph in Peter Wagner & Christian Sidor (2000, 473) referring to *simulations* they had run to evaluate the limits of cladistic analyses of datasets. I've put the part Woodmorappe used in **bold**:

> **Stratigraphic correlations, like phylogenetic relationships, must be inferred from data and are not actually observations themselves.** Thus, the simulations assign first appearances more accurately than sometimes is possible for real taxa. The extent to which correlation error and uncertainty affect real data varies enormously among taxonomic groups and preservational realms. However we did not explore those effects but concentrated on the effects of evolutionary and taxonomic parameters.

This statement was not about what Woodmorappe wanted it to be, doubts about geological correlation. By leaving out that they were discussing their abstract simulations (a tool to model possible scenarios in Deep Time), Woodmorappe showed once again how addictive quote mining apologetics remains in antievolutionism.

No simulation is needed to see that whenever Woodmorappe had an opportunity to explain fairly and honestly the information his own cited sources were discussing, he failed consistently to do it.

This is a feature that extends to anyone who relied on Woodmorappe to dispose of the RMT for them. We'll meet those on the creationist side in Chapter 10, but there were some from the Intelligent Design camp who bought into Woodmorappe's Three Card Monte too.

Woodmorappe's ID Fan Club: Phillip Johnson & David Berlinski

This is a particularly juicy scholarly trail, which revealed not only the sort of creationist reading matter that has filtered onto the Intelligent Design apologetic scope, but underscores the abysmally trivial nature of that ID lens.

Mathematician David Berlinski has been a dedicated (albeit quirky) contrarian who has questioned not only the underpinnings of Darwinism but has put in his multiple cents' worth of skepticism about Big Bang cosmology in a series of articles for *Commentrary* magazine, Berlinksi (1996a-b; 1997; 1998; 2001). A distinct thread of "mystery" ran through these pithy essays, objecting to the tendency of recent science to endeavor to explain too much, or at least whatever appeared to Berlinksi as stepping on the toes of the "ineffable" (whether God or not). Berlinksi does not appear to be a religious believer, though, putting him and agnostic Michael Denton considerably outside the vast majority of antievolutionists, who are anything but nonreligious.

In 1997 he was one of the participants in William F. Buckley's PBS *Firing Line* debate on evolution. The antievolution side was clearly a *Discovery Institute* affair, since Berlinski, Phillip Johnson and Michael Behe were all *DI* Fellows. The pro-evolution opposition consisted of biologist Ken Miller, *United Church of Christ* minister Barry Lynn of *Americans United for the Separation of Church and State*, philosopher Michael Ruse and anthropologist Eugenie Scott of the *NCSE*.

On that occasion Berlinski freely acknowledged that the fossil evidence for the RMT was strong, yet treated this as though it were some inconsequential microevolutionary fossil blip rather than the strongly supported macroevolutionary appearance of a whole new vertebrate class.

Interestingly, in comments on the *Firing Line* evolution debate, John Woodmorappe (1997a) "particularly enjoyed Dr. Berlinski's feisty style of debating. Reminds me of Dr. Duane Gish. Berlinski did not let the evolutionists get away with any of their baloney. A better debate, in my opinion, would perhaps have consisted of Berlinski and Dr. Duane T. Gish against any two evolutionists. I also enjoyed Phil Johnson's deft response to the evolutionists' provocation of showing him a book obviously intended for children, and dealing with a topic not relevant to this debate (dinosaurs and men)."

That allusion to a creationist book for children having been shown during the debate was not a point Woodmorappe should have brought up.

The unlikelihood of Berlinski agreeing to be on the same dais with any overt YECer, let alone that consummate detail-fiddler Gish, was exceeded only by the disingenuity of Woodmorappe's chortling over the evasive dodge of the YEC bomb by Johnson at the PBS debate, given that the idea that dinosaurs coexisted with humans remains a key dogma of the very creationism that Woodmorappe resolutely believes in. Whether drawn from one of Gish's works for juveniles, or from one of the more "technical" efforts by YECers, the question remains whether either Gish or Woodmorappe would have been so glib in flicking the issue aside had either been participants.

Woodmorappe's gushy ode to Berlinski and Gish may be contrasted with his distinctly less charitable "refutation" of criticism by ex-creationist Glenn Morton (1996) concerning Woodmorappe's 1996 book, *Noah's Ark: A Feasibility Study*. Woodmorappe (1997b) went way beyond the factual dispute about the forensic implausibility of the Flood or Ark, to fume that "Morton is attacking the very Word of God" and invoked Martin Luther ranting about similar "criminal monsters" who had attacked Scripture. Such eye-twitching fury may have accounted for Woodmorappe's having consistently left the second "n" off Morton's first name.

But back to Berlinski.

Berlinski (2002) buffed his antievolution cue sheet in the December 2002 issue of *Commentary* magazine, taking issue with the eye evolution modeling of Dan-E. Nilsson & Susanne Pelger (1994), accusing them (erroneously) of having cobbled together their data points, but also suggesting the Intelligent Design claims of Michael Behe and Bill Dembski might not be quite as conclusive as their side had supposed. Berlinski then touched on the RMT, offering in his typically rococo way this seeming concession: "The margin between the reptiles and the mammals appears far more friable today than it did a century ago."

I know, I had to look up *friable* too, which means "easily crumbled or pulverized."

But like Lucy Van Pelt yanking back the halpless Charlie Brown's football at the last minute, Berlinski immediately deflected the macroevolutionary impact with this obtuse splurt of florid prose: "When the dead are interrogated at length, then, oxygen levels do drop in certain chambers of conflict, but curiously enough, both Darwinian and design theorists seem to be turning blue as a result."

Berlinski's article occasioned some letters of criticism, prompting a rejoinder of his own in the March 2003 issue of *Commentary*, where along the way (page 24) Berlinski decided to retrench on that reptile-mammal thing (Berlinski's ellipsis but my **bold**):

> On the other hand, the reptile-mammal sequence, the jewel in the crown of Darwinian paleontology, is not without critics of its own in the intelligent-design camp. The **indefatigable Phillip Johnson** has drawn my attention to a paper by John Woodmorappe in TJ 15(1), 2001, pp. 44-52. (TJ is self-described as a "creation journal," a fact **of no relevance to an assessment of Woodmorappe's arguments**.) **Using cladistic analysis**, Woodmorappe investigated a discrete group of morphological characteristics that paleontologists have offered as evidence for the evolutionary nature of the reptile-mammal sequence. At issue is the claim that mammal-like reptiles, when arranged in succession from the pelycosaurs on up, "show an

> essentially unbroken chain of progressively more mammal-like fossils." With respect to 165 of the 181 anatomical characteristics C. A. Sidor and J. A. Hopson's "Ghost Lineages and 'Mammalness': Assessing the Temporal Pattern of Character Acquisition in the Synapsida" (Paleontology, 24 (2), 1998, Appendix 2, pp. 269-270), Woodmorappe argues that "the majority ... do *not* show a unidirectional progression toward the mammalian condition" (emphasis added).

Remember, Phillip Johnson was at the time the *only* critic of the RMT evidence in the "intelligent design camp," not some random sampling of a broad field of criticism. And as we explored last chapter, Johnson's critique was so much argumentative meringue.

But now we learned not only that Johnson included AiG's *Technical Journal* on his reading list (unlike the actual paleontology literature, from which he has remained resolutely aloof). We discover that Johnson felt confident enough about the gravitas of Woodmorappe's paper to *recommend* that work with a straight face to Berlinski (in a manner he has apparently not done with any actual paleontology or systematics work).

Thus armed, Berlinski went on: "I have not studied Woodmorappe's paper thoroughly, but I am quite sure that both his conclusions and the methodology upon which they rest will be widely disputed, if they are ever widely noted. I return to my own starting point: neither Darwinians nor design theorists can look to the fossil record with perfect equanimity."

What an astonishing statement.

First, Woodmorappe's piece was *not* a "cladistic analysis." It was true that Woodmorappe *criticized* several cladistic studies, but not by either using that technique or showing he even adequately understood the principles of it. Which makes one wonder whether Berlinski had studied Woodmorappe's contribution at all, let alone "thoroughly."

And what was there to say about Berlinski's *Get Out of Scholarship Free* card, that the article having appeared in a "creation journal" was "a fact of no relevance to an assessment of Woodmorappe's arguments"?

Answers in Genesis requires ideological and doctrinal purity from all their members and contributors. Woodmorappe is a Young Earth creationist who thinks he's cracked the stall-booking problem on Noah's Ark, and so might fairly be suspected of a rather sizable committment to skew any data covered to the preconceived mandates of that YEC dogma, which includes the inadequacy of fossil evidence for evolution.

Those facts about the author and the venue should have raised at least a few warning bells, putting the onus on anyone relying on it to have vetted it "thoroughly" before even thinking of calling attention to it as a worthy science argument. And yet, while being hyper-suspicious of everything about Nilsson & Pelger's *Royal Society* piece on eye evolution, Berlinski was all *a priori* open-mindedness when it came to Woodmorappe's shredding of the taxonomy papers for the AiG *Technical Journal*.

In Downard (2003a) I noted the irony of Berlinksi offering up

Woodmorappe even as he trashed Nilsson & Pelger, and after I alerted Nilsson to Berlinski's accusations, Nilsson (2003) responded to Berlinski himself. Berlinski (2006) responded to my critique in "The Vampire's Heart" (another of Berlinski's over the top titles)—but only repeating his assertions about the supposed mathematical indequacies of Nilsson & Pelger's eye modeling; nothing reflecting on his having cited Woodmorappe or whether he'd tempered his views on the RMT.

Casey Luskin (2006) in turn reprised Berlinski's thread on Nilsson's work, yet with no mention of the Woodmorappe issue. Meanwhile, Nilsson (2004; 2009) appeared undeterred by Berlinski's roadblocks, continuing his work to understand all that could be had about the evolution of visual systems.

For my part, the casual ease with which Phillip Johnson and David Berlinski could be impressed by the dithering of a John Woodmorappe spoke volumes about the character of their fundamental scholarship and curiosity, and still more to see how readily the secondary Casey Luskin could step past it all.

Was it even possible for Intelligent Design to get its analytical act together, on the RMT or anything else?

9. Intelligent Design Gets Serious (?)

While John Woodmorappe was parsing slices of the RMT technical literature, several authors were absorbing the claims of Intelligent Design and diligently recycling them into parasitical books.

Trinity College of Florida professor Thomas Woodward (1991) had been among the early promoters of Phillip Johnson's *Darwin on Trial* for *Christianity Today*, and reprised the various arguments of Johnson, Behe and Wells for Woodward (2003). While the ID version of the Cambrian Explosion got great play as a theoretical obstacle for evolution, Woodward showed no curiosity about discussing the actual fossils. And that extended to not mentioning the much later RMT, either, even though his own source Phillip Johnson had so prominently brought it up.

In his next book, Woodward (2006, 104-105) finally bumped into the edges of the issue by criticizing a *Natural History* article by Donald Prothero (2005) on the recent paleontology finds that had filled in so many of the fossil gaps of old, including the new whale intermediates.

"He also touts the alleged transitional fossils between dinosaurs and birds, between reptiles and mammals, and between primitive hominids and our own species, *Homo sapiens*. As I expected, among the gaps being filled in there is absolutely no mention at all of the Cambrian chasms," Woodward (2006, 105) declared.

Woodward neglected to specify what about any of those examples warranted that "alleged" tag, and cited no further works in response. But what Woodward "expected" apparently overshot what he had just read, since Prothero (2005, 56) had explicitly brought up the "frustrating" vertebrate fossil record regarding their origins, noting how the primitive living lancelets and sea squirts (tunicates)—forms considered to be relevant to vertebrate developmental origins—were not represented in the Cambrian as evolutionists expected.

Prothero continued: "But recent discoveries in China from the Middle Cambrian epoch, between 510 million and 500 million years ago, have included not only the earliest relatives of the lancelets, but also some soft-bodied specimens that appear to be the earliest vertebrates. Thus backboned animals can now be traced all the way back to the Cambrian, when most of the modern branches of animals originated." Prothero's examples will be popping up again next chapter regarding the *Truth in Science* case.

Even more peculiar, Woodward (2006, 101) had already written about those basal critters: "In the Cambrian, we find just a few very primitive (almost unrecognizable) chordates, which at best seem to be proto-fish." But instead of seeing them for what they were, the presence of "expected" primitive forms at the base of Cambrian chordates, Woodward went on to dismiss such apparent implications as merely *"fossil succession"* (his *italics*) and opined (again his *italics*): "What is *disputed* by most scientists on the ID side is whether a given new form arose from earlier animals—or was ancestral to later groups—*and whether natural processes drove the changes.*"

Since none of the "scientists on the ID side" ever specified just what they'd accept as evidence that anything had ever arisen from earlier animals, and offered no serious reflection on the detailed natural processes that might have been involved, Woodward would have had a hard time documenting any of that with direct citation. But we don't know what he might have considered appropriate on either point, because he offered nothing at all.

Something else not unexpected, critic of ID Jason Rosenhouse (2008) found Woodward's 2006 book superficial and inaccurate, noting its reliance on peripheral secondary sources rather than discussing the meatier arguments that had been leveled against Intelligent Design over the years.

A somewhat different brand of parasitical scholarship occurred with conservative ideologue Tom Bethell (1996; 1999a-b; 2000; 2001a-b; 2002), who had been grumping over relativity theory along with naturalistic evolution for some time at *The American Spectator*. The peculiar animus some conservatives have for poor Einstein has been surveyed by Brad DeLong (1997), while John Farrell (2000) and Chris Hillman (2001) delved into some of the physics issues and players in the anti-Einstein brigade.

Bethell's enthusiasm for fringe physicists included Tom Van Flandern (who is also convinced that the "face on Mars" was a bona fide clue to ancient civilizations on that planet). All of which puts into some perspective Bethell's ability to rely on certain people with tenacious credulity.

While he left his Einstein (but not his global warming) doubts out of *The Politically Incorrect Guide to Science*, Bethell (2005,199-235) showed his ability to copy without any critical examination most of the talking points of modern Intelligent Design fielded by Behe (1996), Dembski (1999), Johnson (1997; 1999) and Wells (2000a-b). With such condensed secondary parasitism, though, the absence of even Johnson's *Darwin on Trial* in his references meant the vast RMT evidence (indeed, virtually all fossil and biological data that wasn't mentioned by his very limited source base) escaped his attention as easily as it had Woodward's.

Another more generic ID groupie was medical doctor Geoffrey Simmons, whose several antievolution books dispensed with any internal references. The bibliography of Simmons (2004, 313-318) was laden with the familiar congenial sources: Behe (1996), Johnson (1997), Wells (2000a) and several from Dembski (1999; 2002) who supplied a lauditory foreward. But there were also the dated Francis Hitching (1982) and still more problematic Richard Milton (1997). The bibliography in Simmons (2007, 279-283) was even slimmer, but added an older 1989 edition of Davis & Kenyon's *Of Panda's and People*, along with creationist Jobe Martin (2002). Good grief.

Without direct citation in the text it wasn't clear just exactly where Simmons extracted his particular views, such as the obsolete opinion of Simmons (2004, 37) that "the whale, the largest animal on the planet, lacks signifcant fossil evidence." It's notable that while Carl Zimmer (2000b) on parasites was in his sources, the more obviously relevant Zimmer (1998) on whale evolution wasn't. Simmons (2007, 30-31) was still missing the whale fossils.

More specific to our topic, while Simmons (2004, 119-124) marveled in the details of human hearing, he showed no awareness that there was a distinct evolutionary history on that, documented at length in that RMT evidence he never addressed. Nor did Simmons (2007, 194) mention the changing synapsid or therapod metabolisms (noted back in Chapter 7) when he flatly declared, "The transition from cold-blooded species to warm-blooded species is enormous and there is no clear evidence of any lukewarm-blooded transition species ever existed."

Although Simmons never defended Young Earth creationist geology, perhaps a little of the YEC spirit of Jobe Martin leaked into this peculiar *Bermuda Triangle Defense* (Player #8) in Simmons (2007, 268): "The Cambrian period fossils are found thousands of feet higher than younger fossils in the Grand Canyon. Fossilized dinosaur bones are found on the surface in the Dakotas, while their possible predecessors are at the bottom of the sea. Sometimes fossils are found in the 'right' order; sometimes they are not. Sometimes a logical transition actually skips to an entirely different continent, yet that part is left out of the textbook."

I haven't a clue how Simmons managed to think there were Cambrian fossils found so high in the Grand Canyon's stratigraphy, or what dinosaur "predecessors" he thought had been found "at the bottom of the sea." But again, Simmons didn't bother with any references for us to follow to see where he had glitched up.

Chaplain Christopher Carlisle and journalist Thomas Smith strained at evenhandedness in their *The Complete Idiot's Guide® to Understanding Intelligent Design*. Carlisle & Smith (2006, 114-116) briefly tackled the fossil record, acknowledging the reality of the evolutionary fossil progression, but not going into any detail, and not specifically mentioning the RMT. But they tried to blunt the impact of that overall evolutionary fossil progression by dangling the "significant conflict" over Punctuated Equilibrium (which relates to the pacing of speciation issues, remember, not macroevolution) as though it referred to the supposed absence of "fossils that embody 'intermediate forms' between extreme forms." That Carlisle & Smith (2006, 116) further referred to a fully capitalized "Tyrannosaurus Rex" was a clue that the details of taxonomy and biology were not their gig.

And then there's Ann Coulter, who had steered clear of the evolution issue through most of her ferocious political punditry. That is until 2006, when she couldn't resist assailing the smug "Darwiniacs" and the presumed atheism and political lapses of their "cult" in *Godless: The Church of Liberalism*. Skeptics like Peter Olofsson (2007) marveled at Coulter's brave attempt to slay the Darwinian dragon by rhetorical excess, offering her the out that Coulter might have written it as a satire. But the two chapters on evolution in Coulter (2006, 199-245) caught my attention not only because of the ease with which she repeated and exaggerated the claims of the design movement, but because some of her improvisations on the way represented a truly ditzy ignorance.

Coulter (2006, 205) too readily nicked a quote secondarily (she offered no source for it) from Thomas Cavalier-Smith seemingly conceding Michael Behe's claim about the absence of "a comprehensive and detailed explanation of the probable steps in the evolution of the observed complexity." What she evidently didn't know (presuming she'd never read the original text) was that Cavalier-Smith (1997) had offered a detailed *criticism* of Behe's evasions about what had been uncovered nonetheless about the evolution of the immune system (Cavalier-Smith's discipline), including numerous misrepresentations of the scientist's own work by Behe (1996, 68, 179, 279n, 285n). Incidentally Behe has never responded to any of those issues since, even though Behe (2000) was aware of Cavalier-Smith's piece.

Things were no better for Coulter's understanding of paleontology.

"We don't have fossils 'connecting' the extinct to the extant," flatly asserted Coulter (2006, 217), and went on to insist that "Neither the creation nor the extinction of dinosaurs was accomplished by a gradual process of any sort."

This was a perilous area for Coulter to invade if her evidence rested solely on what she'd read in the generic Intelligent Design literature, because dinosaurs are a *non-subject* in the ID canon (apart from stray taxa lambasted on account of their connecting birds with dinosaurs).

The K-T check-out was certainly catastrophic and fairly abrupt, largely on account of the destabilizing Deccan Traps volcanic eruptions in India, aggravated at the end by the Chicxulub asteroid impact on the Yucatan peninsula (which apparently contributed a nice oil reserve to boot), David Carlisle (1995), José Nishimura *et al.* (2000), Greg Ravizza & Bernhard Peucker-Ehrenbrink (2003), Jan Smit (2008), Gerta Keller *et al.* (2009), Paul Renne *et al.* (2013) and Blair Schoene *et al.* (2015).

The appearance of dinosaurs back in the Triassic, however, was anything but abupt, involving a bunch of quite generalized bipedal archosaurs (think scaled-down versions of the "velociraptors" rampaging in *Jurassic Park*), as Coulter might have noticed had she bothered to read something on it first, such as Paul Sereno (1999b). Paleontology since then has filled in still more of the blips, courtesy of that obliging Fossil Genie, including the many *almost* dinosaurs that have been given their own tag, the Dinosauromorpha, surveyed by Max Langer *et al.* (2013) and Langer (2014). We'll be tumbling onto those dinosauromorphs again later in this chapter.

But an interesting scholarly bulletin first. Shortly after Coulter's book came out, William Dembski (2006b) enthused, "I'm happy to report I was in constant correspondence with Ann regarding her chapters on Darwinism--indeed, I take all responsibility for any errors in those chapters."

Taking him at his word, in April 2006 I inquired by email as to just what topics might have arisen in that correspondence that could have resulted in what ended up in her book. On that account I was accused of "sheer smarminess" by Dembski (2006c), with Dembski (2006d) sticking to his guns in the factual responsibility department. That constituted quite a litany, as I tracked in Downard (2006a-b,d), but no further accounting has been forthcoming from Dembski regarding any of Coulter's chapter content.

I warned Dembski in Downard (2006c) that Coulter's discussion of the RMT was among her numerous errors, but I hadn't covered that in the 2006 postings. Well, the time of reckoning has come on that.

Coulter (2006, 229) started off with this unsourced claim:

> Darwiniacs love to cite, for example, the progress from the reptile's multiboned jaw to the jaw of mammal-like reptiles with fewer bones, leading inexorably to the single-boned mammal jawbone with two bones moving to the ear. The jawbone metamorphosis didn't prove evolution, but here at least at last was one small part of the fossil record that was not wildly inconsistent with the theory of evolution—in contradistinction to the Cambrian period and the absence of transitional species, for example.

Unsourced it might be, but remember, Phillip Johnson's *Darwin on Trial* (which interestingly enough Coulter never directly cited or mentioned anywhere in her book) was nonetheless the *only* work Coulter could have vacuumed in the Design literature that discussed the RMT. Certainly not *Dembski*, who had not addressed that subject at that time. As her line exactly condensed Johnson's argument, though, complete down to that faulty conclusion jump on the relevance of the Cambrian (noted back in Chapter 7), it is not implausible that Coulter had drawn on *Darwin on Trial* even without having cited it directly.

Now you may recall Johnson's provisional concession of Gould's "narrow point" on the jaw-ear transition. Another clue that Coulter had been reading *Darwin on Trial* even without explicitly noting it comes from what she argued next. Being a fellow lawyer by training, it was very possible Ann Coulter didn't like the idea of letting that concession lay out there, seeming to give even an inch to those dreadful "Darwiniacs." So she decided to take it back for him (my **bold**):

> In fact and to the contrary, the much-celebrated migration of the reptile jawbone raises more questions for the theory of evolution than it answers. How did it happen? How, that is, did those bones **figure out just where to go**? One would think that **if they had perfect independence in migrating anywhere**, the bones would have landed all over the place, but no, we have no evidence, over the course of the reptile-to-mammal transition, that those **wandering bones** had any other destination in mind than the one they ultimately found.

This was incredibly stupid. Bones do not have "perfect independence in migrating anywhere." There are genetic mechanisms that guarantee exactly the opposite. And no one writing in 2006 who had claim to scientific literacy could have thought otherwise.

Take Timothy Rowe (2004, 402):

> In ontogeny the auditory chain differentiates and begins growth attached to the mandible. But the connective tissues joining them are torn as the brain grows, and the entire auditory chain (stapes, incus, malleus, ectotympanic) is carried backward during the next few weeks to its adult position behind the jaw. Transposition of the auditory chain is a consequence of its differential growth with respect to the brain. The tiny ear bones quickly reach adult size, whereas the brain continues to grow for many weeks thereafter. As the developing brain balloons, it loads and remodels the rear part of the skull, detaching the ear ossicles from the developing mandible.

As seen so relentlessly in the fossils that Coulter showed no curiosity about, none of the bones involved were ever "wandering" anywhere. They were following a deep evolutionary trajectory, which Coulter might have discovered if only she'd thought to catch up with the developmental biology that was, after all, known to science since the era of Andrew Jackson.

But as no one in the antievolution literature on Coulter's very limited reading list had ever bothered to do that, either, it is reasonable to surmise that the developmental biology of the mammalian middle ear was never among the subjects on which Coulter consulted the sage William Dembski.

Maybe things might be better for the anti-"Darwiniac" cause if some of the other *Discovery Institute* guns weighed in.

Stephen Meyer and Company Explore (?) Evolution

The failure of *Of Pandas and People* to get much traction as a supplemental school text, and its attendant notoriety in the Dover case, didn't discourage others in the Design movement from trying to produce an improved version that aimed to do better.

The first off the mark was *Explore Evolution* in 2007, a glossy paperback written by Stephen Meyer, microbiologist Scott Minnich, "freelance technical writer" Jonathan Moneymaker, microbiologist Ralph Seelke, and Paul Nelson (whose Young Earth proclivities were distinctly not alluded to).

Apropos our RMT theme, though, we may note there were no *paleontologists* among the book's authors (understandable enough, since none of them did that).

As there was a standard geological chronology chart in Meyer *et al.* (2007, 18), though, one must wonder what manner of conversations there were among the coauthors as to whether all those hundreds of millions of years actually existed. That was one prickly topic *Explore Evolution* was definitely not up to exploring.

There should be no wonder how Young Earth creationist Paul Nelson could be a coauthor on *Explore Evolution*, however. Meyer *et al.* (2003) had already coauthored a piece on the Cambrian Explosion with Nelson and another YEC believer, Marcus Ross (who did have a paleontology degree, after all), in which a standard chronology was used in the text that those two authors did not in fact accept. But the YEC squashed 4500-yr old Flood sequence has never stripped gears with Steve Meyer's multi-hundred-million-year model in either work because there was no functional *Map of Time* between then, a common cognitive landscape where they would all be marking things down to be understood and explained—and still less, *used* and *tested* to successfully anticipate subsequent paleontological finds in the way a Robert Broom or Neil Shubin had.

Instead, all the authors in *Explore Evolution* were keen on contriving a not unexpected amalgam, at once confident and superficial. Part Michael Denton's *Evolution: A Theory in Crisis* (inspirational even when not being directly cited). Part Michael Behe's *Darwin's Black Box*, where of course the flagellum got a big plug, with the added cachet that coauthor Scott Minnich was an actual flagellum researcher for a change, having contributed to the technical papers by Glenn Young *et al.* (1999), Bert Ely *et al.* (2000) and Steven Monday *et al.* (2004).

Interestingly, none of that Minnich work was cited by Meyer *et al.* (2007, 116-123) in their coverage of the rotating flagellum (suggesting how such technical efforts didn't directly bear all that much to undermine the evolutionary origins issue). Indeed, Milton Saier (2004) was *Explore Evolution*'s only technical citation on the flagellum (relating to another matter, how the evolutionary core of the supposedly "irreducibly complex" wiggler was a secretion pump, now found in the Type III protein secretion system). Although work like Mark Pallen & Nicholas Matzke (2006) had begun laying out the parameters of flagellar evolution, that side of things slipped past Meyer *et al.*'s attention, and subsequent work has continued to explore the evolutionary connections between the Type III secretion system and its flagellar cousin, such as Katsumi Imada *et al.* (2016).

Several hot button topics in *Explore Evolution* were direct cribs from Jonathan Wells' *Icons of Evolution*, though: those controversial embryo drawings by Ernst Haeckel (1834-1919) and the industrial melanism "evolution in action" case of the peppered moth.

Meyer *et al.* (2007, 66-70) followed Wells in reminding readers about Haeckel and one of his old 19th century drawings that had taken too many liberties with the general similarity of vertebrate embryos. Haeckel's fudging was done trying to support one of his pet theories, the idea that vertebrate embryos specifically and inevitably recapitulated the *adult* form of their ancestors. That had not panned out, as even the textbooks that had imprudently used his dicey drawing tended to note, and more recent editions of those textbooks easily bypassed the kerfuffle by replacing the drawing with modern microphotographs of the relevant embryos.

The technical side of Haeckel's embryological theories have been covered thoroughly and without the ID fireworks by Michael Richardson *et al.* (1997; 1998) and Richardson & Gerhard Keuck (2001; 2002). Only the Richardson *et al.* (1997) paper got briefly cited by Meyer *et al.* (2007, 69, 71n).

Regarding the peppered moth, Meyer *et al.* (2007, 88-89, 93-96) fanned the doubts that had been raised regarding whether the experimental protocols used by Bernard Kettlewell (1907-1979) back in the 1950s to show industrial melanism were up to snuff. No one denied that the population of moths had grown darker along with the soot-smudged trees, and tracked lighter again after pollution abatement actions were instituted. What was up for grabs was whether Kettlewell's experiments on differential bird predation as the agent of selection had taken proper account of all the variables (such as how many moths were prone to resting naturally on tree trunks during the day, and in a way where they might be seen or not by hungry birds).

That was a technical problem with a limited shelf life, as it turned out, as the scientist who highlighted the issue in the first place, Michael Majerus (1954-2009), set about replicating Kettlewell's experiments with improved methods, fully confirming the original findings shortly before his untimely death, Majerus (2009) and Laurence Cook *et al.* (2012). More recently still, Arjen Van't Hof *et al.* (2011; 2016) identified the transposable element responsible for the specific melanism shift that underlay the natural selection.

Which put the peppered moth firmly back on the evolution evidence list, all methods holes plugged. Though how many secondary citation addicts in the antievolution community will continue rehashing the point because they can't get past Wells' *Icons of Evolution* version, we'll see in future.

It *is* worth noting that all of this hard science work (the critical spotting of difficulties in the peppered moth case and the remedying of it by fresh and more rigorous experiment) was yet again done by *evolutionists*, not evolution's sundry critics carping on the sidelines.

There have been occasional criticisms of *Explore Evolution*, like Wesley Elsberry (2007), *NCSE* (2008) and Brian Metscher (2009), focusing chiefly on the book's tendentious quote-mining, accompanied by a torrent of defensive ID responses from Casey Luskin (2009a-f; 2010a-b), Jonathan Wells (2009a-e) and defending coauthor Paul Nelson (2009a-i). But probably because the book never actively caught on as a viable school supplement, it has faded into general obscurity even as the more antique *Of Pandas and People* continues to act as a critical lightning rod.

Some of *Explore Evolution*'s problems may have turned on how obsolete it was to begin with.

For a book written in the early 21st century, *Explore Evolution* was surprisingly mum on a variety of by then highly relevant biological subjects, from the pivotal role several billion years ago of endosymbiotic bacterial fusions that formed eukaryotic cellular organelles (mitochondria in cells generally, and cyanobacteria specifically as plant cell chloroplasts), to gene duplications and their subsequent diversification as a formation pool for novel genetic mixes.

There was certainly no shortage of technical literature available to them by 2007.

On endosymbiosis, a sampling could start with Robert Schwartz & Margaret Dayhoff (1978) for an early overview. Then a deep dive more specifically into plant and protist chloroplasts and their photosynthetic contribution: Stephen Giovannoni *et al.* (1988), Geoffrey McFadden (1990; 1999; 2001a-b), Kristin Bergsland & Robert Haselkorn (1991), G. R. Wolfe *et al.* (1994), Susan Douglas (1999), Akiko Tomitani *et al.* (1999), Claude Lemieux *et al.* (2000), Jan Andersson & Andrew Roger (2002), Hwan Su Yoon & Gabriele Pinto *et al.* (2002), John Raven & John Allen (2003), Debashish Bhattacharya *et al.* (2004), Ka Hou Chu *et al.* (2004), Charles Delwiche *et al.* (2004, 122-127) and Tomitani (2006).

One could follow with a stroll though mitochondrial origins by Sabrina Dyall & Patricia Johnson (2000), Toni Gabaldón & Martijn Huynen (2003), Dyall *et al.* (2004) and Thomas Richards & Mark van der Giezen (2006). Then explore subsequent endosymbiotic events in algae and amoeba with Hwan Su Yoon *et al.* (2002), Yoon *et al.* (2005) and John Archibald (2006). And, in case you were thinking endosymbiotic events were strictly a thing of the past, there's Birger Marin *et al.* (2005) studying "a plastid in the making."

All that data would still need to be accounted for as thoroughly and at a comparable depth of detail by any proposed Intelligent Design alternative. If the professed object of the book really were to ***Explore*** *Evolution*, honestly and imaginatively, that such a notable subject as endosymbiosis wasn't even coming up at all was not an auspicious sign.

And such work has not stopped since *Explore Evolution* wasn't taking notice of it. There's been a lot more work on plastids and chloroplasts, such as Debashish Bhattacharya *et al.* (2007), Christopher Howe *et al.* (2008), Sujith Puthiyaveetil *et al.* (2008), Monique Turmel *et al.* (2008), Katharina Händeler *et al.* (2009), Patrick Keeling (2010; 2014), Steven Ball *et al.* (2011), Richard Dorrell & Allison Smith (2011), Takuro Nakayama *et al.* (2011), Dorrell & Christopher Howe (2012a-b; 2015), Nakayama & John Archibald (2012), Peter Civáň *et al.* (2014), Geoffrey McFadden (2014), Archibald (2015), Raoul Hennig *et al.* (2015) and Raphaël Méheust *et al.* (2016).

As for mitochondria, Vladimir Hampl *et al.* (2008), Karin Hjort *et al.* (2010), Finlay Maguire & Thomas Richards (2014), Michael Gray (2014; 2015) and Ramya Purkanti & Mukund Thattai (2015) have continued to investigate more of their varied symbiotic activity. Gabaldón & Huynen (2007) and Davide Sassera *et al.* (2011) have worked on reconstructing the proto-mitochondrial metabolism and features, while Judit Prihoda *et al.* (2012) have looked into the metabolic "cross-talk" in diatoms between their mitochondria and chloroplasts. Axel Kowald & Tom Kirkwood (2011) and Cheryl Bender *et al.* (2012) are among those exploring the role of mitochondria in aging and *apoptosis* (the necessary cell deaths that underly much of multicellular life). And Elizabeth Pennisi (2016) has noted the argument of Justin Havird *et al.* (2015) that mitochondria may even have played a role in the origin of sexual reproduction in eukaryotes.

Incidentally, although Meyer *et al.* left endosymbiosis out of their exploration, as Meyer (2009; 2013) continued to do solo, other antievolutionists haven't been so coy. Endosymbiosis is too obviously a macroevolutionary dynamo to be left unchallenged on the apologetic circuit. So Young Earth creationists Georgia Purdom (2006), Daniel Criswell (2009) and Jeffrey Tomkins (2015b) have taken a swing at disputing the reality of endosymbiosis, as have Mike Gene (2005), Albert De Roos (2007), *Uncommon Descent* (2011a)—an ID website initially launched by William Dembski—and Jonathan McLatchie (2012a) over on the Intelligent Design side. Readers with time on their hands may compare at their leisure those efforts with the full body of technical literature to see for themselves which side has been getting the better end of the evidential argument.

The absence of gene duplication as an *Explore Evolution* topic was, if anything, even more astonishing, as that natural process plays a more pervasive role in the development of life, from bacteria to plants and animals, than the rarer endosymbiotic fusions that have ratcheted up biological complexity in their own ways. Duplicated genes can be more easily studied experimentally, for example, as Michael Neuberger & Brian Hartley (1981) showed fairly early on with bacteria.

Just a glance at the literature would glean examples like Sarah Teichmann *et al.* (1998), John Jelesko *et al.* (1999), Robert Friedman & Austin Hughes (2001; 2003), Michael Lynch (2002), François Mazet & Sebastian Shimeld (2002), Lukasz Huminiecki & Kenneth Wolfe (2004), Peter Stadler *et al.* (2004), Michele Morgante *et al.* (2005) and Jörn Petersen *et al.* (2006). The field is so large that many researchers specialize in particular features, such as Dietmar Lang *et al.* (2000) and Markus Hartmann *et al.* (2003) on the β-α *barrel roll,* whose diverse functions include mitochondrial pore openings.

Antievolutionists don't sit well with gene duplications, either, since it creates multiple genetic slates on which mutation and selection can write fresh "information" (which "Darwinism" isn't supposed to be able to do). Jerry Bergman (2006), Georgia Purdom (2008) and Royal Truman & Peter Borger (2008a-b) have poked at the gene duplication issue for YEC, while Michael Behe & David Snoke (2004; 2005) and Behe (2007; 2009) have tussled with Michael Lynch (2005) and Richard Durrett & Deena Schmidt (2007; 2008; 2009) on aspects of duplication and mutation dynamics.

None of those arguments have noticeably slowed the scientific juggernaut, which insists on discovering the connections despite antievolutionist certainty that they aren't supposed to be able to uncover anything productive.

Another quick sampling of subsequent gene duplication research: Roderick Finn & Borge Kristoffersen (2007), Chris Hittinger & Sean Carroll (2007), Xiang Gao & Michael Lynch (2009), Margrethe Serres *et al.* (2009), Baocheng Guo *et al.* (2012), Elisabeth Kaltenegger *et al.* (2013), Adam Hargreaves *et al.* (2014), Wenfeng Qian & Jianzhi Zhang (2014), Dan Andersson *et al.* (2015) and Ferdinand Marlétaz *et al.* (2015). While Carolina Minguillón *et al.* (2009) and Ajna Rivera *et al.* (2010) have worked on the role of gene duplication in the formation of vertebrate appendages and animal eyes, paleogenomics has pressed ahead to reconstruct some of the ancient enzymes generated along the gene duplication trail, such as Karin Voordeckers *et al.* (2012). And Loren Haarsma *et al.* (2016) have progressed to defining the generalized dynamics of diverging duplicated genes.

While Stephen Meyer (2009) had still left gene duplications out of his *Signature in the Cell* book, in the meantime the summary of recent findings by Manyuan Long & Kevin Thornton *et al.* (2003) had not only been cited by critics of Intelligent Design, their paper had figured in the Dover case as some of the evidence that had persuaded Judge John Jones to deliver the unfavorable verdict on the scientific credibility of ID in 2005. With all those years of steeping, Meyer (2013, 211-223) decided to knock Long & Thornton down a peg in *Darwin's Doubt*, in a variant of the "Origins or Bust" argument.

Meyer (2013, 212) insisted (his *italics*) that "none of these papers demonstrate *how* mutations and natural selection could find truly novel genes or proteins in sequence space in the first place; nor do they show that it is reasonably probable (or plausible) that these mechanisms would do so in the time available."

And how much time *was* available for all this, Steve? At no point did Meyer ever bring up a chronology against which we were to decide any of that.

In fact, Meyer hadn't offered much description at all on what was supposed to be going on in any of "these papers," particularly with the example Long & Thornton had discussed in the most detail, one they'd been investigating for a decade. Meyer (2013, 227) cited the original Manyuan Long & Charles Langley (1993) paper only to play grammar police and castigate them for writing that exons were "recruited" in evolution, but skipped their other related work, such as Wen Wang *et al.* (2000), Long & Michael Deutsch et al. (2003), Jianming Zhang *et al.* (2004; 2005) and Sidi Chen et al. (2010; 2013).

The pace and timing of a gene duplication certainly played a role here, for it helped the study to catch one that had occurred fairly recently, before further genetic modification blurred the splicing lines (as the unused exons could degenerate after millions of years once the old readthrough signals were bypassed by those of the insertions). Fortunately, the fruitfly's *Jingwei* gene had originated only some 2-3 million years ago in the common ancestor of two African *Drosophila* species, when the processed RNA of an alcohol dehydrogenase gene (ADH) was reverse-transcribed into a duplicated copy of the fly's *yellow emperor* gene.

In the *Jingwei* case, the chimeric result took advantage of some of the old ADH amino acid connections to interact with other hormones and pheromones, altering the ADH's substrate specificity in the fly's testes. Was over two million years not long enough for so common a transposition to further mutate some modified substrate specificity? Meyer didn't say.

The ADH side of the chimera hadn't popped on the scene just to undergo a gene fusion, either. ADHs had a very long metazoan pedigree in the diverse MDR superfamily, splitting off in the Cambrian from a formaldehyde dehydrogenase gene about *half a billion years ago*, Olle Danielsson & Hans Jörnvall (1992) and Bengt Persson et al. (2008).

Incidentally, on the paleogenomic front, Michael Thomson *et al.* (2005) had even retroengineered the ancient ADH precursor to a gene duplication in yeast (of obvious interest today to winemakers). Current ones have two ADHs with *opposite* specializations: Adh1 making ethanol, Adh2 consuming it. The resurrected ancestral form was similar to the Adh1 model, suggesting an adaptive role when placed in paleoecological context, as "Silent nucleotide dating suggests that the Adh1-Adh2 duplication occurred near the time of duplication of several other proteins involved in the accumulation of ethanol, possibly in the Cretaceous age when fleshy fruits arose."

More processes involving millions of years of windup time, it would seem, and involving natural processes whose fingerprints we can observe today in living organisms.

Take those retrotransposons. For those that carry an explicit "copy me" codon sequence, things can get pretty frisky, which is why about 10% of our own human genome consists of over a *million* copies of the primate *Alu* retrotransposon (and growing by about one every 200 births), Andrew Bennett *et al.* (2008), Nurit Gal-Mark *et al.* (2008) and Leslie Pray (2008). It originated over 80 million years ago when the *7SL RNA* gene was duplicated in the common ancestor of primates and rodents; the mouse version is known as *B1* (also with over a million copies, filling currently 7% of their genome), Aristotelis Tsirigos & Isidore Rigoutsos (2009). Herzel Ben-Shlomo *et al.* (1999) and Jan-Ole Kriegs *et al.* (2007) offer some perspective on what *7SL RNA* has been up to.

The *Alus* don't carry a "read me" code, fortunately, so mostly they just get copied. Their very presence changes the topology of the DNA sequence, of course, but if a mutation should flip the leading end of their nucleotides into a "start" codon, turning one on is more liable to gum up whatever protein or regulatory field it happened to be occupying, which is why *Alu*-driven diseases are more common than the occasional benign form, such as its role in variable substrate specificity in our brains, André Gerber *et al.* (1997) and Mark Batzer & Prescott Deininger (2002). Alu's capacity to glitch up cellular processes may be why they show up less frequently in cancerous cells, where rampant replication is the problem, Paula Moolhuijzen *et al.* (2010).

That ying-yang disease side of *Alus* doesn't play well in the world of Intelligent Design, where it has become a central dogma that essentially all DNA is the result of purposeful (and presumably benificent) *design*. No one has railed more in the ID camp on *The Myth of Junk DNA* than Jonathan Wells, in the book by that very name. But knowing how promiscuous retrotransposons are, the argument that even a majority of those million-plus *Alus* must have been intentionally designed into our and the other primates' DNA would seem quite a stretch.

So it was interesting to read how Jonathan Wells (2011a, 60-63) dealt with *Alu* and its mouse cousin *B1*. It involved alluding to functions only, as though they somehow automatically vindicated the design presumption, but not explaining enough about them to reveal how little of the data was being accounted for by that design inference.

For example, Wells (2011a, 60-61) cited Tsirigos & Rogoutsos (2009) not for their insights on the Cretaceous-era gene duplication that had generated B1, but only for the observation that there were biases in how Alu and B1 were distributed, Wells seeing this as "strengthening the inference that they serve important biological functions."

What Tsirigos & Rogoutsos had found was that *Alu* and *B1* were preferentially preserved in the *upstream* regions (the stretch that gets read first during transcription) of certain genes, but also in the *intronic downstream* parts of those genes (the DNA that gets edited out of the finished proteins assembled from the active genetic exons). The hotspot genes were ones understandably related to DNA processing (such as repair and recombination), exactly the areas where a "copy me" retrotransposon has a field day. The authors suspected these preferred *Alus* and *B1s* afforded "more binding sites to help increase the complexity of regulatory networks, or more transcript splice variants," on which selection could act. Which was what had happened with *Jingwei* in fruitflies, remember, where the novel binding sites obtained in the fusion expanded the regulatory framework.

Wells (2011a, 62) noted that "99% of the coding sequence of one human gene expressed in brain cells consists of *Alu* sequences," citing Roy Britten (2004), but again without further details. The gene in question was the "neuronal thread protein" *AD7C*, which apparently consisted of a string of five *Alus* that had also undergone some mutation in the process, though given the ubiquity of *Alus* in the genome, Jan-Ole Kriegs *et al.* (2005) initially warned that the suggested structure might have been an artifact of the detection techniques. In any case, AD7C represented the *only* identified gene composed of *Alus*, and with a nasty kick to it besides, since its overexpression was showing up in Alzheimer's disease, Suzanne de la Monte *et al.* (1997). Would that qualify as a "design defect"?

Or take Wells (2011a, 62) announcing how, "In 1986, Russian scientists reported that B2 elements help to regulate the transcription of rat ribosomal RNA," citing L. Yavachev *et al.* (1986). *B2*? That would be different from *B1* how? Wells didn't explain that either, but the B2 superfamily was yet another of those natural evolutionary variants of the 7SL RNA lineage, this one restricted to three rodent families, as worked out by Irina Serdobova & Dmitri Kramerov (1998). While that B2 bunch had a story to tell about the natural relationship of the animals that uniquely shared them, what did it mean in the truncated world of Intelligent Design? Nothing, apparently, which can be compared to the contemporary Kramerov & Nikita Vassetzky (2011) sorting out how the many known retroelement clusters have spread and modified throughout the placental mammals.

Now all of this tapdancing has involved biology. Our main object is paleontology, though, with that RMT football out by the goalpost.

Explore Evolution trailing behind the paleontology data curve

Overall, paleontology was a slim subject in *Explore Evolution*. Oh yes, there was the obligatory Cambrian Explosion, as inevitably expected in the antievolution gig by then as Noam Chomsky bringing up oppressed Palestinians and American imperialism in a lecture for the collegiate Left.

But the CE occurred half a billion years ago. The vast bulk of fossil life coming downstream after the origin of distinct animal phyla slid by Meyer *et al.* without detailed examination, including the completely absent dinosaurs, mentioned only on the fly while dismissing their potential relationship to birds, Meyer *et al.* (2007, 21, 128, 138). The RMT we'll see in a bit.

But even the substantial paleontological finds of early tetrapods (including *Tiktaalik* discussed back in Chapter 2) were very much in the news by the time Meyer *et al.* (2007, 28) breezed past them.

They started off with a *Bermuda Triangle Defense* (Player #9) argument worthy of Duane Gish by declaring how "the earliest fossil evidence of amphibians comes from sites in Greenland, South America, Russia, and Australia. During the Devonian period, these sites were 'separated from one another by thousands of kilometers of open ocean and land'," citing Malcom Gordon & Everett Olson (1995, 128-133, 262-264) and John Long & Gordon (2004).

A bit of paleogeography was in order. Taking a look at the map of the Devonian world in *Prehistoric Life* (2009, 110-111), for example, the globe was midway between supercontinents 390 mya when *Tiktaalik* and its cousins were paddling about. What would eventually become Russia was in middle northern latitudes back then, Greenland and Australia were along the equator, and South America was parked down by the South Pole.

Gordon & Olson were arguing in 1995 that tetrapods might have originated on several occasions in those separate locations, but that depended on whether the land legs were originating in the process of leaving the water. The fossil work by 2006 was suggesting something else: that the physical anatomy of tetrapods was evolving in an aquatic environment, long preceding the terrestrial habitat change. In that case, the amount of water separating the land on which the swimming tetrapods had yet to penetrate would be comparatively less of an obstacle for ancestral animals *that still lived in the water*.

Citing Long & Gordon so generally was rather a scholarly mistake for Meyer *et al.*, given that Long & Gordon (2004, 700) were already incorporating this change in perspective on tetrapod origins (their bracketed "[mya]" by the way):

> It is now widely accepted that the first tetrapods arose from advanced tetrapodomorph stock (the elpistotegalids) in the Late Devonian, probably in Euramerica. However, truly terrestrial forms did not emerge until much later, in geographically far-flung regions, in the Lower Carboniferous. The complete transition occurred over about 25 million years; definitive

emergences onto land took place during the most recent 5 million years. The sequence of character acquisition during the transition can be seen as a five-step process involving: (1) higher osteichthyan (tetrapodomorph) diversification in the Middle Devonian (beginning aboout 380 million years ago [mya]), (2) the emergence of "prototetrapods" (e.g. *Elginerpeton*) in the Frasnian stage (about 372 mya), (3) the appearance of aquatic tetrapods (e.g., *Acanthostega*) sometime in the early to mid-Famennian (about 360 mya), (4) the appearance of "eutetrapods" (e.g. *Tulerpeton*) at the very end of the Devonian period (about 358 mya), and (5) the first truly terrestrial tetrapods (e.g. *Pederpes*) in the Lower Carboniferous (about 340 mya).

Long & Gordon (2004, 700) highlighted how the fossil picture had changed: "Over the past 25 years, new discoveries have increased the number of known Devonian fossil tetrapod taxa from two (1932-1977) to 10 (1977-2003), plus three elpistotegalid fishes," and referred the reader to a chart of the chronological appearance of those animals on the next page. It's relevant that the later fossil finds like *Tiktaalik* were pushing the origins side farther back into time, something not at all unexpected given how sparse the examples were previously (only *ten* taxa known in 2003 to cover 20 million years). Since they did cite the paper, though, in principle all the many fossils discussed by Long & Gordon were in principle known to the authors of *Explore Evolution*. They just didn't think any of them important enough to share with their readers.

With 80% of the Devonian tetrapods then known coming from only partial remains, Long & Gordon (2004, 716) stressed "there is presently no unambiguous way to determine whether there was a single 'main line' of tetrapod evolution (as the cladistic analyses assume) or whether any of the problematic older fragmentary fossil forms (e.g., *Metaxygnathus*, from the late Frasnian of Australia) imply the possibility of a diphyletic origin of tetrapods."

Incidentally, *Metaxygnathus* was newly discovered at the time, a Late Devonian animal (355 mya) from China, and based initially on a partial jaw, Min Zhu *et al.* (2002). Later finds (surprise!) revealed yet more diversity to the group, Jennifer Clack *et al.* (2012), suggesting "their apparent scarcity in northern Gondwana and China is a sampling artifact that will be remedied by future discoveries."

I'll bet on the Fossil Genie coming through before antievolutionists make any sense of the data they already have.

Apart from the circumstance that the known Devonian tetrapods apparently lived in freshwater ecosystems, it's unclear based on such limited information what role variations in *salt tolerance* may have played in any adaptive radiation from one habitat to another, and whether once or multiple times. There is reason to suspect, though, that early stegocephalian tetrapods (those with toes rather than fins) had a fairly broad salt tolerance, Michel Laurin & Rodrigo Soler-Gijón (2010), and newer evidence for the stem tetrapod *Acanthostega* suggested they had a long fully aquatic juvenile stage, Nadia Fröbisch (2016) regarding Sophie Sanchez *et al.* (2016).

The discovery of tetrapod tracks 18 million years before *Tiktaalik* in a tidal zone supported the idea that early tetrapod evolution was occurring in a "marine intertidal and/or lagoonal zone," Grzegorz Niedźwiedzki *et al.* (2010). Not surprisingly, Michael Denton (2016a, 159, 313n) saw those same Devonian tracks as only another problem for evolution (since they implied many millions of years of "ghost lineages" that had yet to be filled in by the Fossil Genie), not as an opportunity to integrate more of that data into an ID framework that didn't exist.

The failure of *Explore Evolution* to stop and ponder the implications of their own sources, let alone make better sense of the data from their Intelligent Design perspective, reflected that recurring pattern in antievolutionist source analysis, and only got worse as they went on to cite Jennifer Clack (2004) regarding new fossils "that seemed to show a connection between fish and tetrapods—in particular, in the structure of the front fins of some bony fish and in the forelimbs of an early tetrapod."

But Clack's commentary was not the technical work itself. That was Neil Shubin *et al.* (2004), a paper Meyer *et al.* did not cite, let alone the technical shoptalk between Per Ahlberg (2004) and Michael Coates *et al.* (2004) on whether a particular bone from *Elginerpeton* qualified it as a tetrapod. Since they discussed none of its substantive content, *Explore Evolution* operated far from giving any superior accounting in their Intelligent Design solliloquy, ever rehearsing offstage.

And there's a deep methodological reason for such omissions. Antievolutionists don't really care all that much about past life. Apart from its peripheral utility as harbinger of mysterious Design, ancient life can come and go without snagging much in the way of their active curiosity. That's why the Cambrian is so hot on their topic list, and not the much busier Mesozoic. They are content to point to the places where the data is sparse, doing the *Bermuda Triangle Defense* thing, rather than trying to make sense of the times when the data are much richer.

Explore Evolution illustrated their deep lack of functional paleontological curiosity in a potshot taken on the occasional mismatch between the order in which ancient life first appears in the fossil record, compared to when they were expected to have first originated based on cladistic phylogenetics. From a biological point of view, any cladograms drawing on genetic data would be guaranteed to place divergence nodes long before the morphologically distinct forms show up (think how that evagination/invagination variation predated anything with visibly preserved pimples/dimples in the diapsid/synapsid split). Add to that how the odds are stacked in favor of the first fossil to show up in the record *not* being the very first one that ever existed in that group, and *Explore Evolution* was winding up for yet another *Bermuda Triangle Defense* pitch.

The primary source Meyer *et al.* (2007, 27, 38) were picking on was Mark Norell & Michael Novacek (1992a). Both of the notes in the quote below (indicated with my **bold**) were just to that single source. But as there were no further explanations in the notes nor specific page numbers cited, having two general notes to the same source looked more like padding than detailed documentation.

Here's how *Explore Evolution* played it:

> These predictions often do not match the actual appearance of animal fossils in the fossil record (though they are remarkably consistent for plants). Many "older" groups of animals (as depicted in cladograms) appear above, not below, the supposedly "younger" ones in the fossil record. The primate fossil record "poorly reflects" the predicted evolutionary sequence, say Norrell [*sic*] and Novacek. "Groups thought to have branched off early in primate history appear late in the record or have no fossil record." **[19]** The problem is not as serious with the mammal-like reptiles. However, five "intermediate" forms that cladograms predict should have arrived neatly in sequence over a long period of time actually appear suddenly at the same time in the fossil record. **[20]**

Let's stop a moment, before we look more closely at the original source.

So, *all of plant life* and the rather notable *mammal-like reptiles* adequately comport to the evolutionary cue sheet, apart from a handful that inconsiderately refused to get fossilized at their earliest appearance. The Fossil Genie in spite mode? Or just more *Bermuda Triangle Defense*.

But does this mean that *Explore Evolution* was all at the ready to *accept* that natural evolution, if only the fossil record of those unspecified examples nudged closer to their inferred sequence? That all would be OK for Meyer *et al.* if only they were sprinked out more, and not arriving "at the same time"? We'll see shortly that *Explore Evolution* meant no such concession when it came to the RMT.

Now the only named example of the "many" animal groups that supposedly bucked that sequential issue were the primates (our family group). But already Meyer *et al.* were running off the scholarly rails, for their direct quote wasn't exactly a direct quote, having left out a word. That wasn't so serious a problem for their cribbing as were the taxa they left out. Here's the passage where "Norrell" & Novacek mentioned those primates. I've omitted their references and put the parts *Explore Evolution* used in **bold**:

> The documented fossil record of primates is generally regarded as one of comparatively high quality based on the diversity and widespread geographic and geochronologic distribution of primate fossils and the amount of attention the group has received. Yet the primate fossil record **poorly reflects** higher level cladistic branching patterns. This is because some taxa (tarsiers and cheirogalines, for instance) **thought to have branched off** very **early in primate history appear late in the record or have no fossil record**.

So Norell & Novacek were mentioning specific *higher level* groups. But that was the view in 1992, fifteen years before *Explore Evolution* was writing. Had there been no newer research in the meantime? Or information that should have been taken note of to fairly assess whether this was quite the evolutionary obstacle Meyer *et al.* wanted it to be?

First, some background.

Lambert & the Diagram Group (1985, 158, 161) and Stahl (1985, 473) both reflect where the paleontology and taxonomy were prior to Norell & Novacek's paper. The prosimians were appearing around 60 mya, from which the lemurs and tarsiers would branch off by around 30 mya. But working out the precise divergence times would depend not only on the luck of the fossil draw, but which pieces of fossil data you had to work with.

The cheirogalines (dwarf and mouse lemurs) are a subgroup of the Strepsirhini (lemurs and lorises). It turns out those cheirogalines have quite a range of dental morphology, as Frank Cuozzo *et al.* (2013) later discovered in the first systematic survey of their living representatives. That level of variation would obviously affect trying to pin down their systematics if you only had mainly teeth to work with.

Which was certainly the case when Meyer *et al.* were writing. Of the thirteen strepsirhini fossil taxa available to Tab Rasmussen & Kimberley Nekaris (1998), only *one* had a skull. The rest were represented by stray jaws and teeth. The same held true for the lemurs, such as the teeth that turned up in Oligocene Pakistan to put another blip on the scope, Laurent Marivaux *et al.* (2001). A similar situation applied to the tarsiers, James Rossie *et al.* (2006) noting that "until now, the fossil record of tarsiers has been limited to a single jaw and several isolated teeth" from the Miocene.

The newly found fragments of facial bones from the Chinese fossil Rossie *et al.* had found did push the tarsier record back many millions of years, into the Eocene 45 mya, but couldn't resolve that "higher level" branching issue (that *Explore Evolution* notably left out of their parsed quote) because *none* of the fragmentary fossils preserved the base of the skull, the auditory region, or areas around the eyes that were the diagnostic features the scientists would *have to have available* to resolve whether the modern big-eyed nocturnal tarsiers were more closely related to the extinct big-eyed omomyids or the surviving anthropoids.

On the general systematic side, it is perhaps most revealing that *Explore Evolution* didn't even follow up on what was, after all, the topic of the Norell & Novacek paper they had taken such pains to bring up: how closely the fossil record tracked the systematics (whether based on molecular branching evidence or cladistic analysis).

Explore Evolution certainly didn't cite anyone else, but that's not because there wasn't work to note. Norell & Novacek (1992b) for example, had expanded their analysis from 24 to 33 cladistic studies, reiterating that overall there was good correlation of the cladogram dating of origination with the actual fossil record. They again stressed how, "Other groups, such as primates, show very poor correlations because certain major clades have either unreasonably short fossil durations or no fossil record at all."

Michael Benton & Rebecca Hitchin (1997) found a similar pattern in their review of a random sample of cladograms, including Norell & Novacek (1992a-b). Of course there were some disagreements. *Duh*. The various studies were often measuring *different* data sets or character features. No surprise, though, that lack of correlation tended to involve animals with sketchy fossil records (such as being soft bodied, small and fragile, or having a habitat like the sea), all of which would inevitably hamper getting a full grip on where and when they lived. Diego Pol *et al.* (2004) and Matthew Wills (2007) have made the same points, and later studies have refined the analytical metrics to reflect those limitations, such as Clint Boyd *et al.* (2011) and Anne O'Connor & Matthew Wills (2016).

But even with those occasional sore spots, overall the cladograms have consistenly matched up with the main outlines of stratigraphic appearance. Benton & Hitchin (1997, 889) concluded that "it would be hard to explain why the independent evidence of the stratigraphic occurrence of fossils and the patterns of cladograms should show such striking levels of congruence if the fossil record and the cladistic method were hopelessly misleading."

An expanded analysis by Benton *et al.* (2000) returned the same results. "Experience shows that major changes in the dating of fossils do not occur at the level of geological systems or stages, but at the finer divisions of substages and zones. Likewise, orders and families are often relatively stable, while new discoveries constantly alter the definitions of genera and species of fossils," Benton *et al.* (2000, 536).

And what about our vexing primates?

Well, the Fossil Genie hadn't closed up shop entirely. New finds have continued to push back the fossil ranges and fill in some of the old "gaps," such as Yaowalak Chaimanee *et al.* (2003; 2004; 2011; 2012), Stefan Merker *et al.* (2009), Rodolphe Tabuce *et al.* (2009), Blythe Williams *et al.* (2010), Sid Perkins (2013a) re Xijun Ni *et al.* (2013), James Rossie *et al.* (2013), Laurent Marivaux *et al.* (2014) and Mariano Bond *et al.* (2015). Relating the biology of living tarsiers to the adaptive parameters of their fossil predecessors, Amanda Melin *et al.* (2013) have even been able to work out the eye opsins of ancestral tarsiers. Such is the practical utility of interdisciplinary *evolutionary* study.

But the forensic fact remains that the earliest primates were tiny agile animals that lived in forests, not floodplains. Such habitats are not marveously conducive to fossil preservation, especially animals the size of a pygmy mouse lemur, and the whole of the anthropoid fossil record (ranging from the Eocene on) charted in Williams *et al.* (2010, 4798) fell into only a *dozen* localities sprinkled across Africa and southern Asia.

At no point has any antievolutionist bogged down their vaulting imagination with an appeciation of that fundamental fossil reality. And while the scientists have diligently attempted to make the best sense of that mixed physical record, *Explore Evolution* had brought the primate subject up only as a dated authority quote to cast a shadow, not light. Casey Luskin (2010a-b) has continued that apologetic tapdance on the whole issue of biogeographic dispersal, especially of primates. And with all that, we still have no idea from their Design perspective why any of those animals existed where and when they did.

At this point we may offer a clue on the scholarly detective story of how the "Norrell" & Novacek quote (their names were correctly printed in the notes at least) got into *Explore Evolution* in the first place. The book's coauthor Paul Nelson may have been the source, since Nelson (2010) trotted it out again, though this time at greater length (his **bold**):

> "The relationships examined here also reveal that the quality of the fossil record judged from other perspectives does not necessarily reveal its match with independently derived phylogenetic evidence.... **the primate fossil record poorly reflects higher level cladistic branching patterns**. This is because some taxa (tarsiers and cheirogalines, for instance) thought to have branched off very early in primate history appear late in the record or have no fossil record." (1992, p. 1692; emphasis added)

But as surveyed above regarding the subsequent technical literature, by this time Nelson was even farther behind the data curve, both on the systematics and the paleontology. And yet it didn't occur to him to "Explore Evolution" based on the evidence of *2010*, not 1992.

Explore Evolution and Ray Comfort, separated at methodological birth?

What the authors of *Explore Evolution* were up to in the lack of gumption department was exactly what goes on all the time in the antievolution literature.

For example, in 2013 evangelist Ray Comfort did an apologetic antievolution video, *Evolution vs. God*, which consisted of a string of gotcha interview questions aimed mainly at bemused college students waylaid on campus. It's available on *YouTube* for those anxious to see it. Being a video, though, it was notably shy on technical sources, but fortunately Comfort did a companion guide to supply some substance to the argument. Or so he thought.

Dinosaurs hadn't been mentioned in the video, but Ray Comfort (2013, 10) evidently felt he was being awfully detailed by showing a phylogenetic tree "from *The Dinosaur Data Book*" where the stretches known by fossil evidence were tinted gray. Comfort did not identity the author or publication date of that *Dinosaur Data Book*, or the page numbers of the chart. And that's because he was relying secondarily on fellow creationist Carl Kerby (2011), who hadn't given those details either.

Comfort (2013, 11) opined (his *italics*): "But, as Carl Kerby points out, notice the fine print at the bottom of the chart: 'Tinted areas indicate solid fossil evidence.' Another way of saying it is 'only the dark areas are proven fact.' The white areas have no proven evidence, so if they're not fact, they're merely speculation (fiction!). Keep in mind that these white areas are where evolution had to have taken place for Darwin's theory to be true—and those are the areas with *no evidence*."

Comfort's protege Kirk Cameron was also impressed with Kerby's dino chart argument, posting a version of it as Kerby (2013) on his own website. A former teen heartthrob who got religion after a reckless adolescence, Cameron gained a certain infamy in creationist circles by trying to field the preposterous "Crocoduck" (a photoshopped duck with a crocodile's snout) as somehow disproving bird evolution. Sorry, as seen previously in Chapter 5, paleontologists know a lot about what it takes to evolve from a reptile to a bird, and no crocoduck-like form is proposed anywhere on the chain. In fact, a chimera like the crocoduck is exactly what paleontologists would *not* expect such an intermediate to look like.

But couldn't an omnipotent Designer have made one? Presumably so, meaning that the *absence* of crocoducks in the fossil record is far more of a problem for the creationist position than for evolution.

Just as with *Explore Evolution*'s use of the dated 1992 Norell & Novacek article, though, the publication date of that *Dinosaur Data Book* was especially relevant to the similar evasive argument Kerby, Comfort & Cameron were trying with Mesozoic taxa. Unlike certain brands of religious apologetics, science work isn't frozen in time. New findings are made, the science moves on.

As it happened, even without the author being noted, I recognized the book by its *title*. I'm a dinosaur buff, you see, and had the volume in my library. All I had to do to check on this was trot downstairs to my basement dinosaur shelves, and pin down the chart specifically to Lambert & The Diagram Group (1990, 26-27).

That 1990 publication date is really important.

The *Dinosaur Data Book* was an excellent compendium of information available *at the time*, which meant mainly up to the late 1980s. It couldn't possibly be aware of fossils that only turned up later. But Carl Kerby and Ray Comfort and Kirk Cameron were not in 1990, were they? They were writing in 2011 and 2013. So what had happened in paleontology over that near quarter of a century?

While there were indeed a lot of blank spots for the dinosaurs and crocodiles and archosaurs on the *Dinosaur Data Book*'s 1990 chart, that was no longer the case for the paleontology of the 21st century.

Far more examples of basal crododiles, dinosaurs and archosaurs have turned up, as covered by Michael Parrish (1991), Paul Sereno & Andrea Arcucci (1993; 1994), James Clark *et al.* (2000), Nicholas Fraser *et al.* (2002), Ricardo Martinez *et al.* (2011) and Sterling Nesbitt (2011). In fact, so many new fossils had been discovered that the dividing line between those groups and their closest cousins had grown as blurry as between the first mammals and their relatives, meaning there were *crocodylomorph* and *dinosauriform* categories to cover them, just as there was that *mammaliaformes*.

Such distinctions are reflected in a range of work, including Alick Walker

(1990), Ursula Göhlich *et al.* (2005) and Randall Irmis *et al.* (2007). And even as Kerby/Comfort/Cameron weren't noticing by 2013, the paleontologists continued to ply their trade on all those fronts. Sterling Nesbitt & Richard Butler *et al.* (2013) on archosauriforms. Diego Pol *et al.* (2013) and Randall Irmis *et al.* (2013) on crocodyliforms. Max Langer & Jorge Ferigolo (2013), Langer *et al.* (2013) and Nesbitt & Paul Barrett *et al.* (2013) on dinosauromorphs.

Faced with an ongoing technical scientific issue they had taken such pains to bring up, what the academic Paul Nelson and Stephen Meyer *et al.* shared methodologically with Kerby/Comfort/Cameron down in the popular creationist trenches was an abiding lack of curiosity and gumption to simply *go and look*.

Explore Evolution goes *Gee Whiz*

Explore Evolution dribbled mention of the RMT in separate sections of their book. The first was the briefest of mention by Meyer *et al.* (2007, 20) in a chapter on the common descent issue. Just as Duane Gish had, a *Dimetrodon* was shown (in this case a photo of a skeleton and not a drawing), accompanied by Figure 1:6 on the next page showing a dozen skulls representing links in that evolutionary chain, drawn from a cladogram in Tom Kemp (2005, 89). That was a most unwise thing for them to do, as we'll see.

Although there were now a dozen taxa under their own nose, *Explore Evolution* never discussed any of them, not in that section or anywhere else in the book. Nor did they make any mention of the jaw-ear transformation, or the known developmental biology of those features in the mammal embryo. Was it possible they didn't know about any of that evidence? Not credibly when it came to the fossil end of things, given their various selective citation of Kemp (1982a; 2005).

The closest *Explore Evolution* got to bumping into that jaw-ear matter was an endnote much later in the book. "Here are some other features that mammals have and reptiles do not: hair or fur, a three-boned inner ear, a single-boned jaw, and highly differentiated teeth," reported Meyer *et al.* (2007, 140n).

That source-free note had been offered to supposedly document a main text point by Meyer *et al.* (2007, 129) which dismissively alluded to the "superficial resemblance between skeletons" (details of which they did not specify), followed by their contrasting the egg laying of reptiles with the live birth of mammals (*Explore Evolution* having apparently mislaid in their haste those egg-laying monotremes).

Having shunted aside unmentioned *all* the fossil evidence we actually do have tracing the evolution of the jaw-ear system in the therapsids (and which must have been known to them courtesy of the Tom Kemp books), Meyer *et al.* (2007, 130-139) turned to areas especially difficult to trace paleontologically: the soft tissue heart and lungs. This was the same *Bermuda Triangle Defense* dodge Duane Gish (Chapter 4) and Michael Denton (Chapter

7) had taken in their own earlier apologetics, and one may compare the talking points common to them, from bird and mammal diaphragms to the supposed difficulty of getting from the three-chambered heart of reptiles to the four-chambered mammal model.

Unlike Denton, however, who had slipped past the crocodile example in Romer (along with Romer's cautionary notes on the fallacy of thinking the ancient amniotes ever had the three-orifice system of modern reptiles), Meyer *et al.* (2007, 132) plowed fresh disingenuous turf by bringing up crocodiles to sow further doubts about the evolutionary origin of the mammalian heart, wondering how their "heart could evolve step-by-step from a three-chambered heart while avoiding a non-functional (lethal) intermediate form."

Explore Evolution then turned Gee Whiz. A *Gee Whiz* is any argument that tosses in the complexity of a biological system simply for its wow factor, as though that somehow invalidated its evolutionary origin or made a Design explanation any more viable, without stopping to look more closely at either their sources or the larger scientific context. Methodologically, that's a kissing cousin to the *Bermuda Triangle Defense*.

In this case, Meyer *et al.* invoked Craig Franklin & Michael Axelsson (1994), Franklin *et al.* (2000) and Axelsson (2001) on how "the crocodilian heart employs a cogged-tooth valve that opens and closes in response to the adreneline in the crocodile's system," and ended with a congenial authority quote: "As Craig E. Franklin, an evolutionary geneticist from the University of Queensland has explained: 'What we've shown here is an evolutionary novelty. The crocodile heart is the most complex and most bizarre in terms of its plumbing.'"

That last item was a quote from a *Science News* article, Susan Milius (2003), which was discussing (and rather belatedly) the findings of the technical paper, Franklin & Axelsson (2000) on those "cogteeth" valves—a misleading choice of words on their part in the Franklin *et al. Science* letter, as they were not shaped like a watch cogwheel, but rather two lozenge-shaped growths that could contact or not to adjust the blood flow. But then *Explore Evolution* may never have checked on what they actually looked like, since they drew on Milius' very brief secondary coverage, not the primary technical paper.

Even their Franklin *et al.* (2000) *Science* letter that *Explore Evolution* had cited held a few flip sides. It was directed at Paul Fisher *et al.* (2000), a controversial paper claiming to have found a preserved fossil dinosaur heart, which Timothy Rowe *et al.* (2001) suspected was a mineralized accretion. Franklin *et al.* weren't disputing the claim that it was a fossil heart, though, just quibbling with Fisher *et al.*'s characterization of crocodile heart dynamics. They concluded their letter: "There is no doubt that the fossil that Fisher *et al.* describe is an exciting find, but the four-chambered heart seems to point to an ancestral archosaurian condition shared by crocodilians and dinosaurs, and, hence, birds."

They cited only one source for that, Paul Sereno (1999), which had not discussed hearts at all (archosaurian or otherwise). Which does raise some warning flags on how quickly Meyer *et al.* should have jumped on their technical bandwagon.

Incidentally, the crocodilian ability to shut down oxygenated blood selectively (which was one of the features Franklin *et al.* were highlighting in their *Science* letter) may have developed adaptively to bring more CO_2 into play to facilitate digestion, Colleen Farmer *et al.* (2008).

What *Explore Evolution* was trundling around was a complex issue, to be sure, but not the one they were flagging. How can scientists work out the evolutionary history of organs, when those organs stand almost no chance of ever being preserved?

Working out the fine point details of that would snag on that *Bermuda Triangle Defense* issue, of course, meaning the hard science work would have to default to pinning down the genetics of the living versions and then doing a lot of heavy duty paleogenomic retroengineering somewhere down the line. Work on that side has been progressing, though. *Irx4* was identified as a player in heart chamber formation by Zheng-Zheng Bao *et al.* (1999), but other elements have only been spotted recently, such as the role of the *Tbx5* protein by Koshiba-Takeuchi *et al.* (2009). But it's a measure of antievolutionist curiosity that they haven't exactly been falling over themselves to discuss the implications of such findings for the origin of vertebrate hearts.

What we've seen so far is how *Explore Evolution* took every opportunity to dilute the RMT data in other topics they didn't discuss in detail either.

Meyer *et al.* (2007, 20) took another opportunity to try and slide past the RMT to something else when noted that pro-evolution scientists "point to fossils called 'mammal-like reptiles,' which appear in Permian and Triassic strata (200--300 million years ago). Mammal-like reptiles are extinct groups that appear to have mostly reptilian traits, mixed in with some mammalian features. Or consider another example. Recently, some scientists think they have discovered a transitional fossil sequence connecting land-dwelling mammals to whales."

There was one footnote to this, but it had nothing to do with the RMT, but only referred to the whale matter, and even that was suprisingly limited: two general magazine articles by Philip Gingerich (1994) and Carl Zimmer (1995), and two older technical papers, Philip Gingerich *et al.* (1990) and Gingerich & Mahmood Raza *et al.* (1994).

If you check back to my survey of whale evolution back in Chapter 6, there were over a dozen other technical works adding much more fossil evidence by the time *Explore Evolution* was hoping to elbow them all aside. So even on this narrow point, Meyer *et al.* were both superficial and dated. Then again, by the time they got to their "Reply" section, Meyer *et al.* (2007, 27) were breezily dismissive of "a possible whale-to-mammal transitional sequence."

Uhh, whales *are* "mammals," as were their ancestors tracking back 150 million years to the origin of mammals. Thus reminding us of the somewhat inattentive proofreading in *Explore Evolution* (that "Norrell" & Novacek article) along with the sluggish technical research.

As for those synapsids, *Explore Evolution*'s shell game had commenced with that claim that those extinct groups had "mostly reptilian traits, mixed in with some mammalian features." All of which was presumably to suggest some inconsequential random muddle rather than a coherent and objective progression of traits.

But if the authors did not know that was misstating the fossil facts, it could have only been because they didn't even bother to consult their own cited source, as Tom Kemp (2005, 1-2) had noted right at the start: "The earliest, most primitive ones have very few mammalian characters, just a small temporal fenestra in the skull and an enlarged canine tooth in the jaw. In contrast, there are later forms that possess almost all the modern mammalian skeletal characters, lacking only a few of the postcranial skeleton ones like a scapula spine, and fine details of the ankle structure."

And Kemp (2005, 2-3) continued:

> Around the end of the Triassic Period, about 205 Ma, a number of fossils are found of very small animals that have the great majority of the skeletal characters of modern mammals. The brain is enlarged and the postcranial skeleton differs from that of a modern mammal only in a few minor details. Their novel feature is the jaw mechanism. The dentition is fully differentiated with transversely occluding molar teeth that functioned in the unique manner of basic living mammals, and there is a new jaw hinge between the dentary bone of the lower jaw and the squamosal of the skull that permitted a much stronger bite.

Kemp had specifically noted *Sinoconodon* (which we met last chapter) as an example of that advanced almost-mammal condition. *Sinoconodon* will be scampering onto the field again next chapter, too.

But *Explore Evolution* showed no interest in exploring the fossils and their features. Instead, once again citing only generally Kemp (2005), Meyer *et al.* (2007, 29) embarked on the broadest of *Bermuda Triangle Defense* (Biogeography Edition): "Textbooks also frequently fail to mention that the different skeletons shown in transitional sequences (including the mammal-like reptiles) were not found close together geologically. In fact, some supposed ancestors and decendents [*sic*, another typo, should have been *descendents*] were found in widely separated layers of sedimentary rock, representing tens of millions of years of geologic time."

Are they kidding?

Was it now a mandate of Intelligent Design that a hundred-million-year transition could not be real only because it *wasn't* taking place *all in one spot and at one time*?

Remember, on the *previous page* Meyer *et al.* (2007, 27) were complaining because some of the transitions were supposedly *not* arriving "neatly in sequence over a long period of time" but appeared "suddenly at the same time."

It looks like no matter what the sequence the fossils arrive in, it won't pass muster by their ever-flipflopping Intelligent Design filter.

The labored parsing Meyer *et al.* had to do extended even to the authority quote they trotted out next (my **bold**). "As zoologist Henry Gee writes, **referring to fossil vertebrates in general**, 'The intervals of time that separate the fossils are so huge that we cannot say anything definite about their possible connection through ancestry and descent.'"

Only Gee (1999, 23) had *not* been writing about "fossil vertebrates in general," or the means by which their branching phylogenies were determined. Quite to the contrary, he had a very particular case in mind, and one which ironically undermined the logic of the argument they were trying to make. We'll bring that witness onstage in a moment.

But there's more to be gleaned from the content of the Kemp cladogram that *Explore Evolution* included in their book.

Kemp's cladogram had a dozen critters marching up to mammals, almost all of which we've encountered in previous chapters. Kemp began with the Early Permian *Eothyris*, known only by its skull, but morphologically at the base of the game, as noted by Robert Reisz *et al.* (2009). If its body proportions were typical of early synapsids, it would have been about palm-sized, as illustrated in *Prehistoric Life* (2009, 191). *Haptodus* followed on Kemp's chart, and another of those sphenacodonts (the group that also included *Dimetrodon*). Then the dog-sized predator *Biarmosuchus*, at the base of yet another of the diverging synapsid lines, which the Fossil Genie has helped fill in since by the time of Ashley Kruger *et al.* (2015). The gorgonopsids and therocephalians followed on Kemp's cladogram, then *Procynosuchus*, *Thrinaxodon*, the probainognathids, our T2 tritylodontids and T1 tritheledontids, and ending with the morganucodonts.

But it's important to recall that *all* of these animals involved in the RMT had still *existed*, in time and space, and presented biogeographical distribution patterns that needed to be addressed by Intelligent Design just as much as by any evolutionary model.

Take the dicynodonts falling earlier in the chronology, covered by Kenneth Angielczyk & Andrey Kurkin (2003a-b). They initially appeared in the southern hemisphere of the Pangea supercontinent, and began to spread and diversify afterward, clear into Asia. Why?

Was the Designer stuck in a dicynodont rut, churning out new genera that just happened to fit what would seem unavoidable in an evolutionary context? Species capable of arising naturally and evolving as their expanding habitat varied, and whose biogeographical distribution fell into logical place when correlated with continental plate drift movement, as Richard Aulie (1975, 27-28) pointed out long ago when exploring the significance of the therapsids in working out mammal origins.

But *Explore Evolution* wasn't trying to *explain* that physical dataset. All they had in mind was dismissal, and an entirely generic and useless one at that.

Only the dataset wasn't going to go away. Nor has it got any smaller.

Just take *Morganucodon*, late in the chronology showing the arrival of the first primitive mammals. Its range ran almost pole to pole, throughout Eurasia, North America, Africa and India (at that time still parked down by Africa and Antarctica), as charted by Thomas Rich (2008, 235). The proximate ancestral populations for them therefore had many continents to play around on, and a lot of time to do it, and every detail laid out in subsequent works like Jessica Whiteside *et al.* (2011) or Andrey Kurkin (2011) on the ebb and flow of the earlier taxa over space and time constitute a persistently accumulating pile of scientific data that would still need to be accounted for by Intelligent Design, not ignored.

Sliding off the scale

What *Explore Evolution* did next was especially interesting. They *repeated* the Kemp chart eight pages later, as Figure 1:8 which they captioned: "Sequence of Mammal-like reptiles, shown to scale (compare to Figure 1:6 on page 21)."

We'll get to the showing "to scale" part in a moment, but it's important to realize that this reprint chart was ironically right beside a rudimentary Figure 1:7 that showed a bunch of separate vertical lines labeled "Reptiles" and more separate lines dubbed "Mammals" with a great big "?" straddling the dotted line drawn between them, and captioned: "If the mammal-like reptiles are not transitional intermediates, then reptiles and mammals may be separate groups with independent starting points."

None of the Figure 1:7 "reptile" or "mammal" lines were identified as any specific taxa (were they intended to represent species, genera, families, orders?) but *Explore Evolution*'s readers needed only to have eased their eyes *slightly to the right*, to the reprint of Kemp's chart, to see some of the very animals that the "?" of 1.7 were insisting weren't there.

With this second chart, *Explore Evolution* was all too plainly pulling a *Garrett Hardin*.

Why would the authors be so ostentatiously obtuse to put the disconformation of their argument up *again*? Because they didn't think it was a disconfirmation, and that's because they had modified Kemp's chart in a particular way that they evidently thought was oh so momentous. By that adjustment "to scale" the gorgonopsian skull was now very big, for example, while the probainognathan was not much larger than the tiny *Eothryis*.

And so what? Meyer *et al.* (2007, 27, 29):

> Some textbooks alter the scale of pictures showing the order of appearance of groups such as the mammal-like reptiles. This makes the features appear closer in size than they really are, and creates the impression of a close genealogical relationship, and an easy transition between different types of animals. Presentations of the reptile-to-mammal sequence, in particular, often enlarge some skulls and shrink others to make them appear more similar in size than they actually are.

Wrong.

Such charts scale the drawings to make it *easier to see* the examples. The spotting of similarities would require only *looking*. Nothing about the skulls' features were altered or excluded, allowing the notable shift in forms over those millions of years to be better appreciated without eyestrain.

Now there was a footnote to the *Explore Evolution* passage, Meyer *et al.* (2007, 38n), but once again that was not to any technical literature:

> Some authors do include the scaling ratios they use, leaving it up to the audience's mathematical skills to calculate actual comparative scale. Other authors use a scale legend line, and it's up to the reader to notice that the same length line that represented 2 cm in Picture A represents 10 cm in Picture B. Still other authors simply put "Skulls not to scale," somewhere in the caption. Unless students read the fine print and do the calculations, they are often left with a misleading impression of the similarity of the animals in these alleged sequences.

Sorry, but the ones doing the "misleading" here were the authors of *Explore Evolution*.

Missing yet again a chance to actually discuss some of the fossils in the "alleged" chain and explain in what sense they were not progressively visibly similar in their jaw, dental and auditory configuration, Meyer *et al.* never offered any technical literature to support their implicit and necessary assumption that an *allometric* scale change in any of those lineages lay beyond the bounds of natural systems.

To claim that it would required that they ignore the fact that many individual species have objectively notable sexually dimorphic size ratios, from the extinct moas of New Zealand to the early hominids leading to our own *Homo* genus, along with living forms like weasels and lizards, Michael Bunce *et al.* (2003; 2009), Emiliano Bruner *et al.* (2003), Satoshi Suzuki *et al.* (2011) and Bartosz Borczyk *et al.* (2014).

In fact, allometry has been a topic of scientific interest since the 1930s, reviewed by Jean Gayon (2000). Work in this area includes Stephen Jay Gould (1966) and Geoffrey West *et al.* (1997; 1999; 2002), so how the club at *Explore Evolution* managed to miss it is anybody's guess.

The synapsids aren't exempt from such study, either, such as Marcelo Sánchez-Villagra (2010) and Sandra Jasinoski *et al.* (2015) writing since *Explore Evolution* came out.

Something about *Explore Evolution*'s scaling argument should sound awfully familiar, though, shouldn't it? It's what Duane Gish was up to back in Chapter 4, waving the *scale* issue in front of his readers to imply that somehow animals couldn't be naturally related if their size had shifted (even if that involved millions of years and multiple lineage chains to do it).

But there was a further kick to *Explore Evolution*'s "Recalculated" chart, beyond that convergent resemblance to Gish's Creation Science. Whoever had done the measuring had got some of the scaling *wrong*, most glaringly so in the case of the tiny morganucodonts, that were shown on the *Explore Evolution* rescale as way bigger than they were in reality, seemingly dwarfing the T1 tritheledontid that preceded it.

If you check Czerkas & Czerkas (1991, 118) where the *Morganucodon* skull was shown to scale, you could see it was barely an inch long, which would have made it notably *smaller* than most of the others on the cladogram, including *Thrinaxodon*, whose skull ranged from 1.5 to 3 inches (30-96mm) according to the extensive survey performed more recently by Fernando Abdala *et al.* (2013).

So whoever was doing the calculations for *Explore Evolution*, they were not doing a very good job of intelligently designing the product.

To add insult to injury, somebody else had apparently spotted the goofs too, and thought to correct them. The creationists at *Watchtower* (2010, 24) were apparently inspired by the *Explore Evolution* argument to pull the same stunt with the Kemp cladogram, complete with the same secondary authority quote cribbed from Henry Gee (1999, 23). But someone on their graphics staff must have taken the trouble to study *something* about the actual animals, presumably realizing that the *Explore Evolution* redraw was wrong, and so rescaling the skulls on their own. For those wanting a peek, I included the Meyer *et al.* and Jehovah's Witness *Watchtower* pages in Downard (2014, 10-11, 13).

What was all this chart-fiddling telling about Intelligent Design pretentions to serious method when the creationists were striving to get their figures closer to the mark than the ID effort? And what was it revealing about antievolutionist "logic" in general that neither one of them recognized that this was a flimsy argument to begin with? Especially given what authority quotee Henry Gee had been saying about a certain fossil species.

On a fossil hunt for human evolutionary remains in Kenya with Maeve Leakey, Gee (1999, 14) had found a skull that resembled that of

> the modern African civet cat (*Civettictus civetta*), a relative of the mongoose, except much larger. Tantalizingly, the skull could have belonged to the giant civit (*Pseudocivetta ingens*), an extinct relative of the African civet. Bones and teeth from this rare creature have been found at Olduvai Gorge in Tanzania and at Koobi Fora, on the eastern shore of Lake Turkana. If the skull really did belong to a giant civet, it would be the first example ever found on the west side of Lake Turkana and, at 3.3 million years old, the oldest example from anywhere by a million years: I would have held in my hands the only evidence of hundreds of thousands of generations of giant civet that must have existed between this example and the fossil closest to it in time. When my giant civet was alive, the next animal in its species to leave any trace of itself to posterity to date lived a million years in the future.

Gee (1999, 22-23) returned to the topic of ancestry as it related to our human part of the picture. If you found a fragment of an ancient hominid, Gee wondered, could you literally be holding your specific ancestor?

> It is possible, of course, that the fossil really did belong to my lineal ancestor. Everybody has an ancestry, after all. Given what the Leakeys and others have found in East Africa, there is good reason to suspect that hominids lived in the Rift before they lived anywhere else in the world, so all modern humans must derive their ancestry, ultimately, from this spot, or somewhere near it. It is therefore reasonable to suppose that we should all be able to trace our ancestors, in a general way, to creatures that lived in the Rift between roughly 5 and 3 million years ago. So much is true, but it us impossible to know, for certain, that the fossil I hold in my hand is my lineal ancestor. Even if it really was my ancestor, I could never know this unless every generation between the fossil and me had preserved some record of its existence and its pedigree. The fossil itself is not accompanied by a helpful label. The truth is that my own particular ancestry—or yours—may never be recovered from the fossil record.

So Gee was not talking about the ways scientists identified species or how phylogenetic relationships were deduced, but solely the impossibility of knowing if any one particular fossil was on your own personal direct line of descent (or "decent" if we adopt *Explore Evolution*'s misspelling convention). This is a *Genealogy Roadshow* matter, not a taxonomy hurdle.

Whatever your ancestor might have been, it would have looked enough like the found bone that it could easily have been the very one you were holding. That was the context of Gee's ancestry tale. Gee then reminded his readers of the giant civet fossils. It was this section *Explore Evolution* and *Watchtower* so briefly nicked (once more my **bold** to remind you of their quoted part at the end):

> The obstacle to this certain knowledge about lineal ancestry lies in the extreme sparseness of the fossil record. As noted above, if my mystery skull belonged to an extinct giant civet, *Pseudocivetta ingens*, it would be the oldest known record of this species by a million years. This means that no fossils have been found that record the existence of this species for that entire time, and yet the giant civets must have been there all along. Depending on how old giant civets had to be before they could breed (something else we can never establish, because giant civets no longer exist so that we can watch their behaviour), perhaps a hundred thousand generations lived and died between the fossil found by me at LO5 and the next oldest specimen. In addition, we cannot know if the fossil found at LO5 was the lineal ancestor of the specimens found at Olduvai Gorge or Koobi Fora. It might have been, but we can never know this for certain. **The intervals of time that separate the fossils is so huge that we cannot say anything definite about their possible connection through ancestry and descent.**

While Gee was rightly noting how you couldn't be sure a later fossil of the same prehistoric civet species was the direct descendant of the older fossil, *Explore Evolution* and the still more secondary *Watchtower* were trying to use that last sentence as a prybar to dislodge the entire RMT fossils that neither of them ever discussed.

Add to that how Henry Gee (1999, 147-156) accepted the strictures of cladistic taxonomy, where ancestor-descendant connections were not an object of their technical interest in principle, only sister relationships. Identified of course based on quite rigorous evaluation of all the taxonomical characters, which as we've seen, antievolutionists have not shown themselves particularly keen to do, prompted or not.

So Gee had every justification for the qualifying fence he put around those ancient bones. It's the unavoidable limitation of an incomplete data set. But, remember, Gee's example of *P. ingens* was a *single species*. No macroevolution was involved, not even a genus jump. Just a single fossil species, of which Gee had two examples separated by a million years.

Being a species, though, meant *of course* they were all *related*. That's what it means to be a *species*, an actively breeding population of organisms. That's the most fundamental attribute of being a species. And that would even be true in the Intelligent Design version of species (where we humans comprise one of them, after all, *Homo sapiens*). So Gee's caveats about the difficulty of identifying bone B as being lineally descended from a particular prior A in the same species would *not* be a justification for deeming the two specimens as unrelated. They couldn't be *unrelated*, since they were in the same species.

This ought to have been pretty obvious to *Explore Evolution* or the *Watchtower*. That is, if they had stopped long enough in their dip at the quote mine to have thought any of this through. But they didn't. Instead both invoked Gee as though he were discussing what they wanted him to (large scale phylogenetic systematics). The result was *Explore Evolution* and the *Watchtower* were inadvertently but unavoidably implying that members of a single species (*P. ingens*) were not related at all and that science could make no reasonable inference that they *were* related!

As Gee had made such a big deal about the fossil giant civets being merely enlarged versions of the current African civets, this issue also ran smack dab into *Explore Evolution* and *Watchtower*'s canard about the scale issue. They literally had an example under their collective noses that suggested how a particular body shape could enlarge or shrink allometrically, but none of them apparently thought to investigate anything further about Gee's star example, *P. ingens*.

Taxonomists had recognized the obvious connections for some time, such as Quintin Hendey (1973, 102) noting the difficulty in pegging the generic placement of a large fossil civet as to whether it was some variant of the extant smaller civets or fell among the larger extinct giant forms—the differences being a matter of allometric scale, not divergent anatomy. The fossil record has improved a smidge in the years since, Jorge Morales & Martin Pickford (2005; 280-283; 2011) suggesting the pseudocivettas are an offshoot of the Paradoxurinae, a group with distinctive "bunodont" teeth (ones with rounded crowns or cusps) that split off from the viverrids (the family covering the civets and their kin) in Africa around 18 mya, but spread widely there and even into Asia.

Yet more data for antievolutionists not to pay attention to.

Sorry about being so monotonous on this, but antievolutionist behavior on this is rather monotonous, isn't it? It's important that you notice that. It's the recurring pattern in the antievolutionist canon: of research trails not followed, of curiosity not sparked, of data simply not thought through and genuine explanations never offered, *even for the subjects the antivolutionists had gone out of their way to bring up.*

Far from honestly *exploring* evolution, Meyer *et al.* were retreading ground trod all the way back to Duane Gish, and doing so with *less* detail and even less imagination. Even their pretty graphics were a flop.

Meanwhile, the British YEC group *Truth in Science* (2009) offered to distribute free copies of the UK edition of *Explore Evolution* to every school library with a science section, in order to "Awaken Interest and Encourage Critical Thinking." And Krista Bontrager (2011) likewise recommended this "fair minded look at the most popular evidence for Darwinism presented in mainstream science texts" for the OECers at Hugh Ross' *Reasons to Believe*.

How much usage in any actual classrooms *Explore Evolution* has got is difficult to assess. It does appear that a science teacher at a charter school in Minnesota has been using it, noted by Josh Rosenau (2016b). The book had been adopted back in 2008 on the recommendation of a different teacher, to replace *Of Pandas and People*. Hardly an improvement.

One has to wonder, though, how much fact checking of the book was undertaken by *Truth in Science*, Bontrager or the Minnesota charter school teachers. Or would that have to left up to some especially diligent students? Oh, let's teach *that* controversy!

William Dembski & Jonathan Wells: 2 heads prove not better than 1

While the coauthors of *Explore Evolution* fielded one Big Gun (Steve Meyer) and a cast of lesser known characters in the Design movement, in 2008 *The Design of Life* was putting two of the main players together to kick the *Pandas* ball around the field one more time on their own.

Like *Explore Evolution*, many biological topics were notable for their absence, including endosymbiosis again. Other subjects were glibly dismissed without even a stab at substantive discussion, like *Archaeopteryx* which Dembski & Wells (2008, 62) flicked aside along with the egg-laying platypus as mere "oddities"—exactly as creationists Duane Gish and Kurt Wise had back in Chapter 4.

Rather interestingly, Dembski & Wells didn't even bother to cite coauthor Jonathan Wells (2000a, 111-135) on *Archaeopteryx*! Though that would only have pressed dangerously close to the vast topic of dinosaurs and their evolution, another notable paleontological subject missing from the *Design of Life* periscope specifically (and the Intelligent Design literature generally). One lone picture of a stegosaur did inexplicably pop up as Figure 3.14 on page 19 of the General Notes CD-Rom attachment. *Why* was unclear, as the figure was never alluded to anywhere in the text, nor were stegosaurs or dinosaurs discussed. Was the reader supposed to be impressed by such wandering lack of organization?

The Design of Life did press a baby-step closer to that topic of gene duplication that *Explore Evolution* had ignored completely, when Dembski & Wells (2008, 39) briefly mentioned the concept, again without technical documentation, followed by the gutsy claim that "it quickly falls apart when probed." And the evidence for this conclusive probing? *Only one source*, the congenial Michael Behe & David Snoke (2004), which had not conducted anything like a comprehensive refutation of the gene duplication literature. It may also be noted that Dembski & Wells did not mention any of the critical reaction to Behe & Snoke's claims (such as Michael Lynch's 2005 response noted earlier).

That *The Design of Life* was not really going to be about seriously investigating even the subjects they brought up can be seen in an interview coauthor Dembski gave to Devon Williams (2007) of James Dobson's *Focus on the Family*.

When asked by Williams, "How will your research affect the world of science?" Dembski confidently proclaimed the "overwhelming" argument of his new book with Jonathan Wells was "going to change the national conversation" because of its firm rejection of Darwinian materialism."

It hasn't. And there many reasons why.

First off, as noted back in Chapter 7, coauthor Dembski had shown no interest at all in paleontology and has yet to even try to apply any of his "Complex Specified Information" ideas to the glut of whole genome sequences piling up in the science literature. Which meant big swaths of the data set were off his scope, by both experience and inclination. And we saw earlier just how informative Dembski's "science" pep talks were with Ann Coulter.

You don't spark scientific revolutions by ignoring as much data as the Intelligent Design movement has.

The change in the "national conversation" Dembski had in mind was less technical than evangelical. *The Design of Life* was a sparsely footnoted lever to dislodge materialism (and its putative atheist agenda) from the science game, with only one goalpost on the approved end of the field. "The Designer of intelligent design is, ultimately, the Christian God," Dembski told Williams. So much for everybody else's religions.

And what about believers in that god or others, who stubbornly found the evidence for natural branching speciation (AKA evolution) fully persuasive? Just on the Christian end of things, for example, there were biologist Kenneth Miller (1999) and paleontologist Robert Asher (2012).

Dembski has no common cause with them, despite their faith. Naturalistic "Darwinism" is the deal-breaker for him, in a way creationist flushing of millions of years of geology isn't. And we know this because of how Dembski (2005) reacted to criticism of ID as so much craven Bible-denying apostasy by none other than Henry Morris ("a great man" according to Dembski). From Morris' perspective, ID failed because it refused to accept the primacy of God's revealed Word on the age of the world and the Flood, while embracing the materialistic (and hence "atheistic") geochronology system that allowed evolution through its very wide barn door.

"Despite my disagreements with Morris and young earth creationism," disagreements which Dembski took no pains to enumerate, "I regard those disagreements as far less serious than my disagreements with the Darwinian materialists. If you will, young earth creationism is at worst off by a few orders of magnitude in misestimating the age of the earth. On the other hand, Darwinism, in ascribing powers of intelligence to blind material forces, is off by infinite orders of magnitude."

You'd think that ought to have settled things, but when interviewed a decade later by Sean McDowell (2016b), it was clear that Young Earth creationists had not adopted Dembski's infinite measuring stick, cooling on ID in the meantime precisely because they were "not going to serve as a stalking horse for their literalistic interpretation of Genesis." Though regarding Ken Ham's new *Ark Encounter* theme park as "an embarrassment and waste of money," the fact remained that Dembski had never clarified on what grounds he had spurned the voluminous *scientific* pretentions of YEC.

McDowell rowed his similar boat in a circle: "Frankly, I don't care what you believe about the age of the earth. But if you use the young-earth position as a club to beat down the views of others (and this has happened to me repeatedly, with my job on the line), then I do have a problem with you."

Sean McDowell's disinterest aside, the problem with YEC was not merely faith or belief, but *method*, and Dembski's diaphanous "disagreements" involved more than just "misestimating" the age of things. Both Dembski and McDowell were avoiding the whole methodological underpinings of how YEC apologists like the "great" Henry Morris went about doing that. Young Earth creationism depended on the *a priori* banishment of natural explanation whenever it conflicted with Scripture, coupled with the tendentious reliance on secondary authority quoting and data parsing that removed *all* of standard geology and astrophysics and non-Biblical human history from the allowed True Facts book.

By never explaining *why* he was so willing to rely on tried-and-true methodological naturalism when embracing all those extra zeroes in the age of things (issues not directly falling into Dembski's area of expertise, mathematics), while chucking it overboard the moment the SS *Intelligent Design* hit the evolution ocean (yet another area not in Dembski's area of expertise), he wanted to calmly steer his separate craft along the same non-materialist heading toward the Hereafter as Creation Science.

Saddled with the weight of the "infinite," though, evolution (theistic or otherwise) was never going to be allowed to win on Dembski's watch.

Nor was *The Design of Life*'s coauthor Jonathan Wells looking to assess all the evidence with dispassionate reserve.

By the 1970s, Wells had converted to the distinctly cultic Unification Church, founded by the authoritarian control freak Sun Myung Moon (1920-2012). This religious conversion filled Wells with an abiding resolve. As Wells (1994) put it in a message posted on the church's *True Parents* website, "Father's words, my studies, and my prayers convinced me that I should devote my life to destroying Darwinism, just as many of my fellow Unificationists had already devoted their lives to destroying Marxism."

Just how far Wells' thinking on evolution comported to Unification Church doctrines is hard to tell, and not only because Wells has been shy on the point. One of their leaders, Sang Hun Lee (1914-1998), penned as vague a piece of antievolution apologetics as you could ask, Sang Lee (1991), within which parameters just about any antievolution claim could fit.

In briefly commenting on his 1994 *True Parents* "destroying Darwinism" post, though, Wells (2011b) acknowledged his religious motivation while downplaying its significance. Francis Crick (1916-2004) "freely admitted" his atheistic beliefs, Wells noted, but that didn't mean Crick was ruled out of the science game on that account. Though Wells notably didn't offer any indication that Crick was driven by some comparably obsessive motive towards "destroying" non-Darwinism.

"What ultimately mattered in Crick's case was not his motivation, but whether his biological claims were consistent with the evidence," Wells contended, and ended his apologia by doing a backhanded flip off the famous "Nothing in biology makes sense except in the light of evolution" view of Theodosius Dobzhansky (1973) while laying down a challenge: "I encourage readers not simply to take my word for anything, but to go to the scientific literature and check for themselves. After all, nothing in biology makes sense except in the light of evidence."

Oh, Jonathan, I'm more than happy to take you up on that.

Although both are stars in the Intelligent Design firmament, Jonathan Wells and Bill Dembski differed on what subjects they approached.and how they did it. Dembski favored a traditional scholarly format, footnoting claims, while Wells initially adopted the Phillip Johnson method of documenting sources in a Research Notes section at the end of the book.

Not that Wells always played that game so well. For example, if you tried to find out which sources Wells (2000a, 39, 271-272) relied on for his *Icons of Evolution* Cambrian Explosion chart showing the arrival times of the "major" animal phyla (a third of which Wells had actually got *wrong*), looking back in his Research Notes didn't help clarify where he'd gone awry, since it turned out he *hadn't given any sources* on that data. I went into the grisly details of this ineptitude in Downard (2003b, 25-27).

To Research Notes or not to research notes, that is the question

For *The Design of Life*, Dembski & Wells couldn't quite get their acts together, with what amounted to Wells' Research Notes being relegated to that "general notes" CD-ROM whose sixty-two pages could just as well have been printed directly into the book. Though if they had, it would have only highlighted how repetitious and superficial the overall argument was.

Take a sidebar in Dembski & Wells (2008, 53) that dangled "evidence that genetic programs do not control development." That consisted of five contentions:

1. Placing foreign DNA into an egg does not change the species of the egg or embryo. (The rare exceptions to this rule involve animals that could normally mate to produce hybrids.)
2. DNA mutations can interfere with development, but they never alter its endpoint.
3. Different cell types arise in the same animal even though all of them contain the same DNA.
4. Similar developmental genes are found in animals as different as worms, flies, and mammals.
5. Eggs contain several structures (such as microtube arrays and membrane patterns) that are known to influence development independently of the DNA. (See general notes.)

That's right, no Dembski-style footnotes up front. And no sources for them in the general notes, either, as it turned out, making it rather hard to check up on them, wasn't it?

Was it up to the reader to fill in the documentation like a blank crossword puzzle? Some Ralph Brinster *et al.* (1985) *down* on that "foreign DNA into an egg" thing, with maybe some John Gurdon & Ian Wilmut (2011) *across* as follow-up? Then have to figure out how Wells thought anything in current evolutionary biology was expecting that ramming DNA into an already established egg ought to make it into a new *species*.

Compounding the pile-up, in the section where the source documentation ought to have been, those general notes (pages 15-16) trotted out yet more *unreferenced* claims, this time about microtubules, centrosomes and membranes. Left to his own playing field, Wells had apparently forgot to bring the balls again.

On centrosomes, for instance, the general notes claimed that "Centrosomes play a central role in development: a frog egg can be induced to develop into a frog merely by injecting a sperm centrosome—no sperm DNA is needed."

Evidently coauthor Wells had slipped a cog somewhere regarding the content of that PhD in biology he had under his belt, forgetting so basic a fact that the frog egg itself *contains its own DNA*. It wasn't necessary for the centrosome to contribute any DNA to do what it does in development. Its structure alone was capable of triggering *parthenogenesis* (the asexual reproduction of an organism). That was a fairly established subject in the literature by then, such as Frédéric Tournier *et al.* (1989), Catherine Klotz *et al.* (1990) and Tournier & Michel Bornens (2001).

To imply that such development was somehow occurring independently of the "genetic programs" was twaddle.

Parthenogensis wasn't even mentioned as a topic in *The Design of Life*, incidentally, though at some stage Wells did bump into that Klotz paper, as he mentioned it (with a parthenogensis tag this time) in one of his subsequent ID lectures, along with a brief authority quote nicked from Frederick Nijhout (1990). Biologist (and caustic critic of creationism and Intelligent Design) P. Z. Myers (2009) took considerable pains to explain how Wells had misrepresented the substance of both papers (and you may recall Michael Denton's secondary misusage of Nijhout's commentary piece back in Chapter 5).

Wells' 4th point on the similarity of developmental genes in animals particularly caught my eye, as that was a backhanded allusion to the homeotic genes, which Dembski & Wells (2008, 51) had danced around a few pages earlier with this brief bit: "The universality of homeotic genes in development is supposed to be due to their presence in a common ancestor, but the preponderance of evidence suggests that the common ancestor lacked the features which those homeotic genes now supposedly control. From a Darwinian perspective, this is a serious problem."

Really? The idea that ancient genes started out doing something else, and then over time acquired new diverging functions in various lineages, was a "serious problem" for a "Darwinian perspective" that expected exactly that to have happened?

Once again, they offered *no sources* for any of this in the main text, and none of the few direct technical papers noted in the General Notes (pages 13-14) were offered to that end, either, representing only the most preliminary work in the field: Walter Gehring (1987), Rebecca Quiring *et al.* (1994), Sean Carroll (1995), Eric Davidson *et al.* (1995) and Georg Halder *et al.* (1995). Remember it was only in those mid-1990s that homeobox genes were first getting recognized. Which meant, exactly as *Explore Evolution* had on primate taxonomy (or Kerby/Comfort/Cameron on dinosaurs), a dozen years of developmental biology research had gone on without Dembski & Wells taking any note of it.

Just a sampling of relevant papers could include Sean Carroll (1994; 2000) and Carroll *et al.* (1995) on arthropod gene regulation (such as how insect wings developed). Jordi García-Fernàndez & Peter Holland (1994), Felix Beck *et al.* (1999), Panagiotis Prinos *et al.* (2001) and García-Fernàndez *et al.* (2007) had investigated vertebrate homeotic genes (including the developmentally diverse *Cdx* ParaHox pathways).

Mark Martindale *et al.* (2002; 2004) and Martindale (2005) worked on the deeper context of the bilaterian body plan and metazoan axial evolution (front/back, top/bottom stuff), Claire Larroux *et al.* (2006) looked into how the transcription factor complexes were used in the even earlier diverging sponges, and what that suggested about the origin of multicellularity, while Elena Simionato *et al.* (2007) drew in data on the several parallel duplication events involving the "basic helix-loop-helix" *bHLH* gene family (you may recall their involvement in our ear's sensory hair development, noted back in

Chapter 3).

At the core of all that was the origin of the homeotic genetic system itself, no easy act of biological archaeology to ferret out, given the hundreds of millions of years of gene duplications and specialization seen in their descendants (the only versions that can be directly studied today). But such hurdles have not deterred the scientifically curious. Sophie Pollard & Peter Holland (2000), Bernd Schierwater & Rob Desalle (2001), García-Fernàndez (2005a-b) and Daniel Chourrout *et al.* (2006) had begun teasing apart what the "ProtoHox" roots may have been of the Hox cluster, while Carolina Minguillón & García-Fernàndez (2003) had looked into how the deep-diverging *Evx* and *Mox* genes were related to that early split.

Early in the science work that *The Design of Life* wasn't noticing, Schierwater & Desalle (2001, 173) identified benchmarks that needed to be pinned down about how the *paraHox* and *Hox* genes functioned in the basal diploblastic phyla (the cnidarians and ctenophores). It's revealing how unknown their developmental genetics were when Gamlin & Vines (1986, 68-73) were writing, but much more was known by the time of James Valentine (2004, 55-56, 473-475). That included clues on the metazoan-like gene structure of the still more basal choanoflagellates, such as Nicole King & Sean Carroll (2001) and King *et al.* (2003), revealing something of what unicellular protistan organisms were doing with those genes that would later contribute to multicellularity in the hox-happy metazoans. Not surprisingly, choanoflagellates were yet another subject that did not arise in *The Design of Life*.

At the time Sean Carroll *et al.* (2001) summarized the burgeoning field, the science was still in the getting the factual ducks in a row stage, cataloguing the suite of proliferating *Hox* and *ParaHox* genes and what that distribution suggested about the order in which those functions were acquired in evolution. By the time of his more general book, *Endless Forms Most Beautiful*, Carroll (2006, 143-145) suggested that the ancestral Urbilaterian already possessed "at least six or seven *Hox* genes, *Pax-6*, *Distal-less*, *tinman*, and a few hundred more body-building genes." This suite of genes not only built the body layout, but contributed to visual sensing and contractile cells that lay the groundwork for a circulatory system with a beating heart. It was a cluster that has remained in play all the way down to the present.

Now genes that expanded the motility and sensory range of an organism would seem fertile ground for adaptive selection, but Dembski & Wells (2008, 52) restricted their discussion to a declaration that "Neo-Darwinism maintains that such genes evolved by encoding primitive adaptations that remain to be discovered, but this is ad hoc speculation." There were no sources for that, either, there or in the General Notes.

As for what remained "to be discovered," there was no danger of any of that coming from the Intelligent Design movement, given that they weren't doing any of the work.

Meanwhile, over in the gang actually doing the science work, Michaël

Manuel (2009) continued to clarify the order in which the front/back aspect of the *Hox* patterning system was evolving, and Mansi Srivastava & Daniel Lu *et al.* (2010) ferreted out more on the LIM homeobox family, whose Lhx transcription factors pervade metazoans, from arthropods and vertebrates to sponges and cnidarians (jellyfish).

Those LIM domains revealed yet another piece of the "Design of Life" puzzle. As Srivastava & Lu *et al.* (2010) noted in their abstract, "A tandem cluster of three *Lhx* genes of different subfamilies and a gene containing two LIM domains in the genome of *T. adhaerens* (an animal without any neurons) indicates that Lhx subfamilies were generated by tandem duplication. This tandem cluster in *Trichoplax* is likely a remnant of the original chromosomal context in which Lhx subfamilies first appeared."

Trichoplax is a placozoan, flat little invertebrate blobs thought to have diverged sometime after the sponges appeared in the Precambian, but before the big cnidarian/bilaterian split. Plain they may appear, but their rudimentary neurosecretory gene kit has proven as interesting as the still earlier choanoflagellates, Thorsten Hadrys *et al.* (2005) and Caroline Smith *et al.* (2014).

Beck & Emma Stringer (2010) and Ferdinand Marlétaz *et al.* (2015) have continued to work on *Cdx* ParaHox gene evolution too, and there's been plenty of further study of those interesting choanoflagellates, by Martin Carr *et al.* (2008), Nicole King *et al.* (2008). Alexandre Alié & Michael Manuël (2010), Pawel Burkhardt *et al.* (2011), Mark Dayel *et al.* (2011) and Stephen Fairclough *et al.* (2013), including the part their genetic toolkit would later play in the evolution of *neurons*, Burkhardt (2015).

Claire Larroux *et al.* (2008) and Srivastava & Oleg Simakov *et al.* (2010) investigated still more of the transcription factors in sponges common with later bilaterians, suggesting to Olivia Mendivil-Ramos *et al.* (2012) that current sponges had lost some of the ancient homeotic components along the way. More recent work on additional sponge genomes by Sofia Fortunato *et al.* (2014) confirmed that prediction when they established that one lineage did indeed still possess a *ParaHox* gene, along with a pile of transcription factors well used in the homeobox system by bilaterians, highlighting "the need to analyse the genomes of multiple sponge lineages to obtain a complete picture of the ancestral composition of the first animal genome."

The study by Morgane Thomas-Chollier *et al.* (2010) indicated that it was still unclear from existing descendants whether the units defining frontal positoning in the Urbilaterian were initially in one or two modules, and whether those governing the middle and rear of animal bodies originated as a single block or were initially separate units too. But whatever the sequence, Peter Holland (2015) noted how the Precambrian expansion of the Hox kit generated "bodies capable of high-energy directed locomotion, including active burrowing" that would eventually expand the biological options seen in that much later Cambrian Explosion that antievolutionists insist on seeing only as an inexplicable designer burst.

The goal of "Experimental Evo-Devo" is to actively retroengineer more of that ancient biology, and García-Fernàndez *et al.* (2009) applied what had been learned about the Hox and other genes to sketch out protocols on that regarding early chordate evolution.

But it's still no easy task. Science has to work with what is dooable, at the moment. Since so many interconnections have come downstream in the last few hundred million years, teasing apart even the most basic of relationships can take years of hard and careful effort. Take Georg Halder's area. Just to work out how the patterns in butterfly wings arose took over a decade, from Scott Weatherbee *et al.* (1999) through to Robert Reed *et al.* (2011) and Arnaud Martin *et al.* (2012).

Were Dembski & Wells seriously expecting scientists to retroengineer half a billion years of life in one spurt, resurrecting all the adaptive factors in play? And would they not rationalize all that away, anyway, were that accomplished?

We can see how Wells worked through one of these issues of forensic biology concerning those centrosomes that their book had brought up so lightly. Just as *The Design of Life* came out, Mark Alliegro & Mary Alliegro (2008) presented evidence that the microtubule centrosomes involved in cell division may have originated as an endosymbiotic system. "Centrioles, like mitochondria and chloroplasts, duplicate independently of the nucleus," their paper reminded, though in the centrosome component it involved RNA, not DNA, with genes "skewed towards nucleic acid metabolism."

The centriole signaling in turn resembled that found in "the nucleolinus, a poorly understood organelle proposed to play a role in spindle formation." First spotted in animal cells over 150 years ago, its role in cell division spindle formation was vaguely known but was difficult to study because it required special chemical staining to even see it. The nucleolinus in the surf clam turned out to be more easily visible without special coloring, which is where the Alliegros' work came in.

Centrosomes don't replicate in the manner of organelles like mitochondria or chloroplasts that hold their own DNA, though, as Laura Ross & Benjamin Normark (2015) have noted, but how they are assembled on the fly in cells remains an ongoing object of study. Mary Alliegro *et al.* (2010) and Mark Alliegro *et al.* (2012) have continued investigating the RNA and proteins found in the nucleolinus, and whether any of their work will shed light on centrosome dynamics (or support any deep endosymbiotic origin), only time will tell.

Now it happened that Jonathan Wells (2008) did not like the Alliegros' initial *PNAS* paper at all, fuming even that "the National Institutes of Health supported this medically useless speculation with our tax dollars, and the *Proceedings of the National Academy of Sciences* USA elevated it to the status of 'peer-reviewed" science—part of the 'overwhelming evidence' for evolution."

Oh, my.

Given that malfunctions in centrosomes play a role in many diseases—from brain disorders like microcephaly to too many centrosome bundles producing cancerous miotic spindles, reviewed more recently by Monica Bettencourt-Dias *et al.* (2011) and Pavithra Chavali *et al.* (2014)—one would think working out how the various components of the system functioned and originated would seem an eminently appropriate subject for NIH investigation. But apparently Wells' Intelligent Design wishes not only to circumscribe their own imaginative curiosity, but saddle govenment funding policy with it as well.

Wells rested his dismissal of the Alliegros on the absence of centrosomal DNA, citing only Wallace Marshall & Joel Rosenbaum (2000) on that point. But Alliegro & Alliegro (2008, 6996) had made a point of noting that paper as among the first to spot that *RNA presence*, in which Marshall & Rosenbaum had only speculated at the time on what it might have been doing there. Incidently, in the matter of follow-up on Wells' topic of "medically useless speculation with our tax dollars", one may note Konstantin Chichinadze *et al.* (2013), drawing specifically on the Alliegros' work on centrosomal RNA to probe more of its involvement in the processes of cell aging.

Along with disparaging the quality and import of the Alliegros' work, Jonathan Wells took that opportunity to dismiss the endosymbiotic origin of mitochondria and chloroplasts (my **bold**):

> Since these tiny organelles contain DNA and **look a bit like bacteria**, the hypothesis of symbiogenesis asserts that they were once free-living prokaryotes that were engulfed by other prokaryotes to form eukaryotic cells. The hypothesis has **lots of evidentiary problems**, but whatever the plausibility it seems to have with regard to mitochondria and chloroplasts completely evaporates in the case of centrosomes.

To say that mitochondria and chloroplasts "look a bit like bacteria" was to trample the data (which, it may be noted, neither *Explore Evolution* nor *The Design of Life* even bothered to mention), and handing the ball off at that point to his own Wells (2008) hardly helped, as he hadn't offered any documentation in that piece for any "evidentiary problems" with their endosymbiotic roots, let alone "lots" of them.

Like prokaryotes, mitochondria and chloroplasts hold their DNA in a circular ring, and replicate by fissioning using similar ribosomal transcription and translation systems, while cyanobacteria and chloroplasts further share their characteristic photosynthetic mechanisms, whose acquisitions and modifications in various lineages have been examined by Christopher Howe *et al.* (2008), while Raoul Hennig *et al.* (2015) have investigated more of the shared cyanobacteria/chloroplast membrane fusion lipids.

Both of these endosymbiotically derived organelles have a double membrane, too: the inner one with enzymes and transport systems like those of their ancestral prokaryotes, the outer one formed during the failed process of ingestion by the hosting prokaryote, a quite dynamic process surveyed since by Felicity Alcock *et al.* (2010). A flock of ubiquitin ligases have also been drawn into mitochondrial and chloroplast organelle regulation, studied by Qihua Ling & Paul Jarvis (2013).

As for the centriole and centrosome down on the cell division floor, the "highly conserved ninefold radial symmetry of the centriole/basal body" turned out to involve polymerized copies of a single coiled protein, but the evolutionary addition of over 100 proteins in the centrosome has engendered a lot of functional variants, reviewed by Michel Bornens (2012). Centrioles, cilia and flagella are part of a related ancestral amoebal system, Zita Carvalho-Santos *et al.* (2011) and Ralph Gräf *et al.* (2015), with the centrosomes a later development (especially among animals), Bornens & Azimzadeh (2007) Matthew Hodges *et al.* (2010) and Azimzadeh (2014). Not all animals have kept using centrosomes, though, such as the planaria (flatworms) whose diverging cell cleavage mode did not depend on polarity guidance, Juliette Azimzadeh *et al.* (2012).

Ah, so much science work to do. And being done. Just not by antievolutionists.

Dembski & Wells cut and paste *Morganucodon*

Jonathan Wells' two prior antievolution books, Wells (2000a; 2006), had managed not to mention the RMT at all, matching yawn for yawn coauthor Dembski's utter lack of interest in fossils generally. So how were these live wires to address that Star Witness for the macroevolutionary paleontological defense?

Like Phillip Johnson, Dembski & Wells (2008, 81) started off with a quote from Douglas Futuyma on the import of the RMT evidence, though from the textbook version in Futuyma (1998, 146) rather than from the more general Futuyma (1982, 189-190) as Johnson had. *The Design of Life* then contended:

> According to Darwinists, mammal-like reptiles called therapsids played a key role in the class-level transition from reptiles to mammals. Among the several therapsid lineages were the dominant land-dwelling vertebrates from the middle of the Permian period to the middle Triassic (see figure 3.10). Evolutionary biologist James Hopson describes a series of eight therapsid skulls that made up a sequence of intermediate types, supposedly leading to a ninth, an early mammal named *Morganucodon*.

They cited the venerable Hopson (1987) as their only source. Remember what we've seen of its contents back in Chapter 7, when Phillip Johnson and *Pandas* were contorting it.

Now *The Design of Life*'s "figure 3.10" on the next page was tagged "Reconstruction of Mammal-like reptiles" but consisted of only a painting of an unidentified skull (resembling that of the early synapsid *Dimetrodon*) and a photo of an unidentified garden sculpture of what appeared to be *Moschops*, another of those early synapsids noted back in Chapter 2.

Like Duane Gish before them, then, and *Explore Evolution*

contemporaneously, *The Design of Life* selectively focused their visual depictions only on the earliest (and thus most reptilian) of the synapsids. No illustrations of the later more relevant critters. Interesting.

As for Hopson's 1987 article, in order of appearance (with each new one in **bold** to keep track of how many were shown altogether), the advanced cynodont (1) ***Thrinaxodon*** had been shown on Hopson's page 18. Five skulls on page 19 showed a (2) "**primitive synapsid**," (3) a **sphenacodontid**, (4) a **biarmosuchian**, (5) an **eotitanosuchian** with its lower jaw reconstructed, and (6) a "**primitive theriodont.**" Dorsal (top) views of three skulls appeared on page 20: (7) a **gorgonopsian,** (8) a **therocephalian** and (9) ***Procynosuchus***. *Procynosuchus* showed up again in side view on the next page 21, along with another shot at *Thrinaxodon*, two other skulls, (10) ***Probainognathus*** and (11) ***Pachygenelus***, leading up to *Morganucodon*. Ventral (bottom) views on page 22 of (12) a **eutheriodont** were added to those of the aforementioned *Procynosuchus, Thrinaxodon, Probainognathus, Morganucodon* and a modern mammal (dog).

That's a total of *twelve* skulls shown, not just eight. Since Dembski & Wells did not list their eight, we have no direct measure of which ones they were missing, but even if the dorsal and ventral examples were removed, that would still make *nine* (as *Procynosuchus* was shown in side view). And would the exclusion of dorsal/ventral views mean that only *side views* were allowed for Intelligent Design evaluation? That any information gleaned from looking at the skulls from some other angle may be tossed aside?

That *The Design of Life* wasn't managing even that end of things all that well is indicated by where Dembski & Wells (2008, 82) went next, offering a further gloss (my **bold**) as they dashed for a metabolic defense in the manner of Gish or Denton:

> Hopson detailed several characters in the series that appear to progress together toward the mammalian body plan. These include: (1) **change in the way the limbs are connected**; (2) **increased mobility of the head;** (3) fusing of the palate; (4) improved musculature of the jaw; and (5) migration of the articular and the quadrate bones from the back of the reptile's jaw towards the middle ear (where in the mammal they would be transformed into auditory ossicles). The **simultaneous** transformation of those several traits is supposed to demonstrate that the therapsids constitute a continuous lineage to the mammal. (Of course, the fossils don't record the potentially vast differences in systems such as the reproductive and circulatory systems, or the organs, glands, and other soft tissues they entail.)

It's unclear what article Dembski & Wells were reading at this point, since Hopson's focus on skulls meant that (1) therapsid postcranial limb layout *wasn't discussed at all.* And what they thought (2) "increased mobility of the head" meant is a further puzzle, neck posture being another anatomical topic not addressed by Hopson. The sequential evolution of (3) a secondary palate in the more advanced cynodonts was noted by Hopson (1987, 23), along with the highly distinctive evolution of mammalian dentition, which feature did not make it onto Dembski & Wells' 5-point scope. Point (4) referred to the expansion of that uniquely synapsid skull opening and the muscles attached to it, Hopson (1987, 22) specifically noting it regarding *Eotitanosuchus* (which was before the further expansion of the coronoid process in later therapsids). And that last point (5) was the whole ballgame, after all, the telltale ear coopting.

I highlighted "simultaneous" in the quote above because if there's one thing that wasn't happening in the RMT, it's the arrival of the mammalian components in one dump. Indeed, Hopson (1987, 24) had stressed: "The series presented here show no great discontinuities anywhere—all morphological steps are relatively small (as, for example, in the transition of angular bone to tympanic ring)."

But then tight sequencing of the data was not something Dembski & Wells were showing great skill at by this time, for they went on about "several inconsistencies" in the Hopson fossil sequence that was part *No Cousins*, part *Bermuda Triangle Defense* (Player #10):

> The first three of Hopson's therapsids are contemporaries from two separate orders. What's more, evolutionary biologists dispute whether some of the therapsids Hopson lists are in fact mammalian ancestors. Also, there's the problem of getting structural similarities to match up with temporal ordering. Indeed, lining up fossils structurally so that transitions are as smooth as possible tends to throw off temporal ordering. For instance, the fourth of Hopson's therapsids is more recent than the fifth, and the final therapsid is more recent than the mammal (Morganucodon) that is supposed to be its descendant. One way around this is to assume that structural predecessors are in fact temporal predecessors, but that the fossil record (because of incompleteness or insufficient search) fails to indicate when the organisms in question first emerged. Such an assumption is, of course, entirely ad hoc.

And Dembski & Wells had quite a lock on that *ad hoc*, didn't they? All of this extraordinary paragraph was simply a very puffed up version of the argument originally fielded by Phillip Johnson (1991, 174) and Davis & Kenyon (1993, 100) quoted back in Chapter 7. Just who was cribbing from whom at this stage is hard to say, but all of them were still wrong.

Announcing that unspecified "evolutionary biologists" disputed "some of the therapsids" were not "mammalian ancestors" was especially disingenous, given that fifteen years worth of paleontological water had flowed along since to render the old Davis & Kenyon evasions even more *ad hoc*. The taxonomical issue would be whether these were all closely-related mammalian *relatives*, and Demsbki & Wells would have been hard-pressed to have cited any paleontologists disputing that relationship. Though as they offered none at all, it's again rather difficult to tell what (if any) candidates they had in mind.

And puffery is what we continued to get when Dembski & Wells (2008, 83-84) hit the critical jar/ear issue (again my **bold**):

> Let's examine the proposed evolution of the mammalian ear more closely. **The skull and mandibles (lower jaws) of the therapsids are said to have bones** similar **(homologous) to those of the first mammals. The upper and lower jaws of reptiles articulate (fit together) with two bones (one each located at the back of the upper and lower jaws) not found in mammals. According to Darwinian theory, these two bones relocated in the middle ear of the mammals** in the course of descent with modification (see figure 3.11). Darwinists describe the reptilian jaw bones as 'migrating' to their new locations in the mammalian ear. Nevertheless **there is no fossil record of such an amazing process.** Nor is it clear how the neo-Darwinian mechanism of natural selection acting on random genetic changes can cause bones to move and relocate. **Consider that to make this change, one of** the **bones had to cross the hinge from the lower jaw into the middle ear region of the skull.** Thereafter the neo-Darwinian mechanism would have had to **reshape and refine** these bones into a **highly specialized, delicate instrument of sound transmission. Such an occurrence would be extraordinary enough by itself, but Darwinists propose that this happened more than once** and without the need for any intelligent guidance.

This too was only an inflated version of the still very wrong original in Davis & Kenyon (1993, 121-122), copying it to the edge of plagiarism down to the phrasing (the **bold** above reflected their identical stretches in *Pandas*). All Dembski & Wells had done was pad it out to over twice the length, but with no better science study, repeating even that absurd "cross the hinge" claim that *Pandas* had made only because they had *misidentified the bones*.

And guess what Dembski & Wells' figure 3.11 turned out to be? Yep, a fresh version of the same *Pandas* picture, complete with a fresh new version of a "sectioned reptilian skull" that still showed the *same* intermediate *therapsid* bone layout they were continuing to offer as the reptile model. Though *The Design of Life* had at least corrected *Pandas'* mislabeling (noting the malleus and incus this time), they were still snagging on the biological reality that no hinge-crossing was ever involved and that there were indeed transitional forms that *The Design of Life* had all too literally been tripping over in that Hopson article they alleged to have read.

Hopson (1987, 25) couldn't have been clearer (criticizing Duane Gish, remember) that "the post-dentary bones of the lower jaw were already connected to the stapes via the quadrate even in the earliest mammal-like reptiles, and Allin (1975) has made a persuasive argument that these bones were already functioning in hearing while still part of the jaw. So the only step that had to be taken once the new mammalian jaw joint was established was to free the rod of post-dentary elements from contact with the dentary. This happens during the ontogeny of every living mammal by atrophy of the middle part of Meckel's cartilage," a reminder of that developmental information that Dembski & Wells were paying no more attention to than Phillip Johnson or *Of Pandas and People* before them (or anybody else in the antievolution parade to date, as we've been seeing).

Couldn't any of these authors have stopped to actually *read* Hopson once in awhile, just a bit?

As the *The Design of Life* neared publication, Casey Luskin (2007) enthused that Dembski & Wells (2008) presented "a potent critique of the alleged transition from reptiles to mammals." That could only be so if "potent" consisted of lazily copying arguments that were wrong to begin with and readers further enabled that laziness by not fact checking any of it. Though given the history of antievolutionary apologetics, all that was a pretty safe bet.

Incidentally, attending a 2013 lecture by Luskin at a Seattle church on the splendid content of Intelligent Design, I made known to Luskin afterward about the problems in both *Explore Evolution* and *The Design of Life* on the RMT evidence, which I recounted in Downard (2015b). At that time Luskin told me that he had not studied any of that RMT stuff, which didn't really surprise me, but invited me to send information on that to him. Which I did, from my own *Three Macroevolutionary Episodes* work, Downard (2003b), to a listing of some of the more recent paleontological work (part of the stuff covered back in Chapter 3). I never heard back from him.

Eight years after *The Design of Life*'s publication, the back-patting cluelessness at the *Discovery Institute* had only further petrified. With none of Dembski's revolution in "the national conversation" impending, Dembski & Wells' associates felt obliged to give the book some renewed hype. *Evolution News & Views* (2016f) quoted Michael Behe: "When future intellectual historians list the books that toppled Darwin's theory, *The Design of Life* will be at the top." And David Klinghoffer (2016c) burbled that *The Design of Life* "offers an accessible introduction to and survey of the evidence for ID. And it offers this service precisely because that evidence is so diverse, wide-ranging, and yes, often quite challenging for the non-PhD (to tell you the candid truth)."

As a non-PhD myself, I must candidly say that the main tool I found handy to assess how far-ranging the evidence was and whether *The Design of Life* got within a mile of it, was a honed curiosity to read the sources being cited, as well as the far larger range of science work Dembski & Wells missed completely.

Several snippets of *The Design* of Life were also put out online as Dembski & Wells (2016a-b). But as that 2016b happened to represent the summary section of Dembski & Wells (2008, 86-90) where they reprised their complaints against common ancestry, it's of relevance that not only had the book made such faulty claims in 2008, but that *EN&V* thought them worthy of repeating in 2016.

Besides riffing off Henry Gee (1999, 23, 32, 116-117) on systematics much as *Explore Evolution* had, Dembski & Wells 2008/2016b couldn't resist reprising their claim on the mammalian ear (my **bold**): "How exactly did those two bones from the reptilian jaw 'migrate' to the mammalian ear? The word 'migrate' in this context is **empty of scientific content**. What **genetic changes and selection pressures** were in fact operating, and how, specifically, did they bring about the evolutionary pathway in question? **No such details are known.**"

Is this the closest we're going to get to figuring out where the Dembski-coached Ann Coulter might have got her notions about migrating ear bones? Or was it equally possible Dembski & Wells were by 2008 resonantly channeling the "scientific" opinions of Coulter from 2006?

Given that *The Design of Life* couldn't even get up to speed on the developmental biology of the 1830s, let alone the genetics and paleontology of the 21st century, it's anybody's guess who by then was copying whom.

That same cloud of confusion infused "career biology instructor" Kenneth Poppe (2008) in *Exposing Darwinism's Weakest Link*, a book whose back cover contained Bill Dembski's praise for one of Poppe's earlier works, aimed at "Reclaiming Science from Darwinism." How much of "science" Poppe was up to reclaim was questionable, as this time around, the "Abbreviated Geological Timeline" in Poppe (2008, 281-282) alluded to the "earliest birds and mammals" in the Mesozoic, but no mention of those Permian therapsids in the previous Paleozoic.

But Poppe (2008, 134) had taken a shot at the RMT earlier with this juvenile assault on a secondary piece (my **bold**):

> You must also view with a critical eye even the most professional sources if they have an obvious axe to grind against any opposing theories, like Intelligent Design. Using one final *National Geographic* example, check out their April 2003 issue. The cover story—once again a Darwinistic message right up front—is titled 'The Rise of Mammals.' This highly imaginative piece, in the finest *NG* tradition, offers only **one true fossil** and three **reconstructions from fragments**. But it does include 22 different drawings, impeccably citing the names of the contributing artists in the **notable absence of contributing paleontologists**."

Like Thomas Woodward earlier in this chapter regarding Donald Prothero's article, Poppe seemed to have a reading problem, as Rick Gore (2003) mentioned plenty of paleontologists (both in the main text and in the illustration captions), including Michael Benton, Richard Cifelli (who noted the jaw/ear shift on page 16), Farish Jenkins and Zhe-Xi Luo.

So maybe it's not all that difficult to imagine how Poppe managed to so mischaracterize the fossils in Gore's article. Gore led off with a picture of the original rock slab containing the *full body fossil* of little *Eomaia*, the early placental mammal described by Qiang Ji *et al.* (2002). The legged manatee ancestor, *Pezosiren*, represented yet another nearly complete fossil, missing only its feet and the back end of the tail, Daryl Domning (2001). The extinct giant sloth *Megatherium*, was also hardly hypothetical, though the old British Museum had gathered together casts from various specimens to assemble their display back in 1850, as the current *Natural History Museum* (2013) recounted. The only fossil shown in the article that involved postcranial reconstruction was the model of tiny *Hadrocodium*, but the evolutionarily important part (its jaw and ear-bearing skull) was entirely real, as covered in

Zhe-Xi Luo & Alfred Crompton *et al.* (2001). *Hadrocodium* will crawl on the scene again next chapter concerning the claims of David Coppedge.

Was Poppe thrown off because Gore hadn't waved the technical papers under his nose? Or was Poppe unable to do a bit of googling on his own to fill in the forensics himself? And would he have reflected on any of their content had he done so?

Even though most of Gore's article was not about our particular slice of the mammal evolution story, it had usefully summarized the broad sweep of evidence from the earliest mammals through to so many branching offshoots. But Kenneth Poppe reflected the parochial obsession of most antievolutionists: it's got to be all about *us*. And so, without even a nod at any of the completely real evidence presented by Gore, Poppe projected his own methodological failings onto *National Geographic*: "Why does such an otherwise fine publication make mistake after mistake? Once again, the fault is in assuming from the get-go that fossils support any type of natural transitional advancement from apes to humans, when they do just the opposite."

As for rescuing science from Darwinism, if this is what Poppe and Dembski have in mind, don't be surprised at how unenthusiastic the paleontological community have been in not running to hop on the Intelligent Design bandwagon slowly circling a long-dry watering hole.

Michael Behe teeters on *The Edge of Evolution*

Arguably the gutsiest attempt to preclude the RMT without actually discussing any of it came in a chapter scribing out "The Mathematical Limits of Darwinism" in Michael Behe (2007, 46-63). Behe's argument in a nutshell: as it (a) took two mutations for the malaria parasite *Plasmodium falciparum* to acquire resistance to the chloroquine drug and (b) that this was incredibly unlikely to occur by Darwinian natural evolution, Behe concluded (c) that all of big scale macroevolution was rendered impossible because that would have required lots and lots of combined mutations far beyond that punitively improbable chloroquine pair.

Ta-dah! Trick, done. Wait for applause.

This has been Behe's primary stage show.

In his testimony for the Dover *Kitzmiller* Intelligent Design trial, and reprised in *The Edge of Evolution*, Behe (2005a-b; 2007, 236-237) tried to dig much the same "here and no farther" trench with another drug resistance case, that of imipenem, bouncing that time off only one science paper, Barry Hall (2004). I explored the background details of that in Downard (2015b, 113-115, 123-125), illustrating both how fascinating the actual biology was (involving just as many interacting pieces as we'll see in the chloroquine case), but also how uncurious Behe appeared to be about delving into any of that from his supposedly so inquisitive and informative Intelligent Design desk.

Michael Behe (2007, 77-82) even sought to minimize the implications of the evolution of antifreeze proteins in fish, discovered by Liangbiao Chen *et al.*

(1997a-b) and Chi-Hing Cheng & Liangbao Chen (1999). While accepting their natural origination via a point mutation from a gene duplication event, Behe nonetheless thought this mere "blood additive" of no proper evolutionary import because it was "coded by multiple genes of different lengths, all of which produce amino acid chains that get chopped into smaller fragments of different lengths—very much like the junk in my gutter," Behe (2007, 81). His fellow ID star Stephen Meyer (2013, 220, 438n) tangentially bounced past the topic as well.

Just how selectively Behe navigated his source field, though, was seen when Behe (2007, 82) parsed Peter Davies *et al.* (2002, 928) on the atypicality of AFP. I've highlighted the section Behe quoted in **bold**, but the *italics* were his:

> The remarkable diversity of AFP types in fishes shows that **a number of dissimilar proteins have adapted to the task of binding ice. *This is atypical of protein evolution*.** Most proteins that serve the same function in different organisms do so as a result of direct descent from an ancestral form. For example, citrate synthase, an enzyme in the tricarboxylic acid cycle, is essentially the same protein in all aerobes, having been required throughout their evolution.

That there was more to the evolutionary "slings and arrows" of AFP than were to be seen in Behe's philosophy was heralded by the commentary on the original Chen papers by John Logsdon & Ford Doolittle (1997), and chorused by the sizable body of technical work following up on that, including Cheng Deng *et al.* (2010), Joanna Kelley *et al.* (2010), Thomas Near *et al.* (2012) and Tianjun Sun *et al.* (2014) on the fish cases, and John Duman (2015) noting comparable processes among arthropods. Atypical AFP may have been, but purely natural it all still was.

While Casey Luskin (2011) dismissed that follow-up Deng *et al.* (2010) in much the same perfunctory way Behe had the original Chen work, it's interesting to consider how many regular scientists have failed to heed the ID stop signal, and plunged ahead to relate the "neofunctionalization" of AFP described in Deng's paper to their own investigations, such as Ruiqi Huang *et al.* (2012), Elisabeth Kaltenegger *et al.* (2013) and Hitashi Ito & Ayumi Tanaka (2014) on plant enzymes. Or Adam Hargeaves *et al.* (2014) on how snake venom toxins originated. On a still broader scale, Baocheng Guo *et al.* (2012) proposed how genetic indels (insertions and deletions) served as "important sources of genetic variation," Federico Abascal *et al.* (2013) explored how and why there were differences in vertebrate histone chaperones, Wenfeng Qian & Jianzhi Zhang (2014) experimentally verified fitness value predictions of duplicated genes and the part neofunctionalization played in that, and Dan Andersson *et al.* (2015) laid out more about what's come to be known about how new functions arose from preexisting genes.

A lot of detailed work there, as one side builds on information to steadily advance their understanding ... while the other trips over it and stays put.

But back to our chloroquine as RMT-slayer trail.

Michael Behe (2007, 46-51) was aware that the malaria parasite must degrade the (to it) toxic heme component of hemoglobin, that the chloroquine drug somehow inhibited the parasite's ability to do that, and that something about this resistance may pose problems for the parasite (and thus account for why the mutant strain tended to disappear when chloroquine wasn't used). It was known that resistance to chloroquine involved the parasite's *PfCRT* membrane protein, and Behe (2007, 49) cited Patrick Bray *et al.* (2005) that the mutation might allow chloroquine to leak out more easily. The downside of an adaption that interfered with internal resource pumping might also account for why the suite of mutations tended to reverse when chloroquine wasn't being thrown at it.

That Bray *et al.* were so tentative in their conclusions about what was going on at the biological floor should have raised a warning flag that a lot of the dynamics were not then known. But Behe has never been one to slow for signals, as I delved into in Downard (2004d, 335-354) regarding his Irreducible Complexity claims deployed in Behe (1996).

Behe (2007, 50) also drew on Bray *et al.* (2005, 324) for a simplified graphic of the *PfCRT* protein (424 peptides in length), showing the ten connected transmembrane domains, along with eleven protein spools extending out into what the original drawing had indicated was the digestive vacuole on one side (where the hemoglobin gets broken down), and the parasite's cytoplasm on the other. But Behe highlighted only the two key chloroquine mutations falling in the first and sixth domains, which he noted by arrows. He did not include *thirteen* other mutations that related to resistance to various antimalarial drugs, not just chloroquine, involving six more of the membrane nodes, as well as three of the digestive vacuole spools.

All of those would be clues as to where adaptive hotspots might be, relevant to the varying interactive effects of particular mutations or their combination, and how the membrane system might evolve in response to the dozen assorted drugs thrown its way, each with distinctive molecular structure and hence shape, as illustrated by Makoah Aminake & Gabriele Pradel (2013, 272). Complicating things still further: substantial "genetic cross talk" can arise between the "seemingly independent" respiration and biosynthesis pathways in the parasite, as drug cocktails target those internal systems in a tangle of possible outcomes, recently catalogued by Jennifer Guler *et al.* (2015).

The windup for his big pitch was in Behe (2007, 57) when he compared the chloroquine case with another antimalarial drug, atovaquone, about which "a clinical study showed that about one in a trillion cells had spontaneous resistance," citing Nicholas White (1999b).

Behe continued:

> In another experiment it was shown that a single amino acid mutation, causing a change at position number 268 in a single protein, was enough to make *P. falciparum* resistant to the drug. So we can deduce that the odds of getting that single mutation are roughly one in a trillion. On the other hand,

> resistance to chloroquine has appeared fewer than ten times in the whole world in the past half century. Nicholas White of Mahidol University in Thailand points out that if you multiply the humber of parasites in a person who is very ill with malaria times the number of people who get malaria per year times the number of years since the introduction of chloroquine, then you can estimate that the odds of a parasite developing resistance to chloroquine is roughly one in a hundred billion billion. [Behe citing generally White (2004).]

That mutation issue came up specifically in White (2004, 1085) and was not quite as Behe had presented it (my **bold**): "Resistance to chloroquine in *P. falciparum* has arisen spontaneously less than ten times in the past fifty years (14)," citing Xin-zhuan Su *et al.* (1997). "This suggests that the per-parasite probability of developing resistance de novo is on the order of 1 in 10^{20} **parasite multiplications**. The single point mutations in the gene encoding cytochrome *b* (*cytB*), which confers atovaquone resistance, or in the gene encoding dihydrofolate reductase (*dbfr*), which confer pyrimethamine resistance, have a per-parasite probability of arising de novo of approximately 1 in 10^{12} **parasite multiplications** (5)," citing White (1999a). "To put this in context, an adult with approximately 2% parasitemia has 10^{12} parasites in his or her body. But in the laboratory, much higher mutation rates thane [*sic*] 1 in every 10^{12} are recorded (12)," citing Pradipsinh Rathod *et al.* (1997).

Incidentally, Behe (2007, 48, 51, 280n-281n) had peripherally cited both the Su and Rathod papers, but not White (1999a).

In any event, Behe's "one in a hundred billion billion" odds of getting a dual chloroquine resistance was *not* a value per parasite (or for the whole population of them in all hosts), but per *replication event* of any parasite found. Nor was it a measurement per se of the incidence of the mutations involved, but only of the success of achieving functional resistance with whatever mutations came along in their biological context.

A surging sea of mutations might be going on down at the parasite's molecular level, mutations shifting back and forth at one or another site, but that would get reflected upstream epidemiologically only insofar as whether the infected patient was getting better or not. That distinction lies at the core of what Behe was about to do with chloroquine resistance, and how disinterested his argument had to be with the functional biology of the organisms. Behe already was failing to consider the paleontology of the RMT. By diving into this biological sidebar instead as a surrogate, his failure to evaluate more of the relevant factors put this play methodologically as a biological version of the *Bermuda Triangle Defense* (Player #11).

P. falciparum has a multistage reproduction cycle, a factor which needed to be considered as well, since the bodily environment where the parasite has to run the immune system gauntlet varies from host to host. Less than ten parasites were typically injected by mosquito bite, but each one soon produced tens of thousands of merozoites that then reproduced asexually in the bloodstream, Qin Cheng *et al.* (1997) and Nicholas White (2004, 1087).

That's how you get a trillion parasites in an individual infected person. Those merozoites that persist eventually produce gametocytes that provide the sexual stage to infect the next mosquitos that may bite the victim, a process offering further medical challenges, recently reviewed by Teun Bousema & Chris Drakeley (2011).

How many people are infected fluctuates. White mentioned 300 million malaria infected worldwide, while another of Behe's sources, Jacques Le Bras & Remy Durand (2003, 147), wrote that 500 million cases were known in Africa alone. Behe (2007, 59) upped his number to a round billion people.

It's all still a large number, though, and over the fast few decades nature has been playing around with thousands of billions of billions of not identical *P. falciparum*, all of whose genetic details play their individual parts in the natural selection game. White (2004, 1091) had summarized how modern medical practices could impact these numbers (again my **bold**):

> If two drugs are used with different modes of action, and therefore different resistance mechanisms, then the per-parasite probability of developing resistance to both drugs is the product of their individual per-parasite probabilities. This is particularly powerful in malaria, because there are only about 10^{17} malaria parasites in the entire world. For example, if the per-parasite probabilities of **developing resistance** to drug A and drug B are both 1 in 10^{12}, then a simultaneously resistant mutant will arise spontaneously every 1 in 10^{24} parasites. As there is a cumulative total of less than 10^{20} malaria parasites in existence in one year, such a simultaneously resistant parasite would arise spontaneously roughly once every 10,000 years—**provided the drugs always confronted the parasites in combination**. Thus the lower the de novo per-parasite probability of **developing resistance**, the greater the delay in the emergence of resistance.

That 10,000 years number represented a ceiling based on an idealized multiple usage of drugs. The rate for multiply resistant parasites to emerge without that protocol would be considerably less than 10,000 years, of course, and even Behe (2007, 60) acknowledged that the double mutations in chloroquine had developed in only a "few years."

But I highlighted that "developing resistance" aspect because that is what White was talking about. His numbers were *not* a measurement of the range of point mutations occurring in any replicating parasite, but a reflection of how often those translated as a *package* into a functional resistance you could measure in a test.

That distinction was utterly lost on Behe, who by then was off on a high hurdle sprint. With maybe only "a trillion creatures" in our human lineage over the last ten million years, Behe (2007, 60-61) yanked his mathematical tripwire:

> If all of these huge numbers make your head spin, think of it this way. The likelihood that *Homo sapiens* achieved any single mutation of the kind required for malaria to become resistant to chloroquine—not the easiest

> mutation, to be sure, but still only a shift of two amino acids—the likelihood that such a mutation could arise just once in the *entire* course of the human lineage in the past ten million years, is miniscule—of the same order as, say, the likelihood of you personally winning the Powerball lottery by buying a single ticket.

Well, I guess lotteries must never be paying out, then. And by that bungled analogy, Behe had thought he'd disposed of more than just human evolution. Acronyming a "chloroquine complexity cluster" as CCC, Behe (2007, 61) declared:

> Let that sink in for a minute. Mammals are thought to have arisen from reptiles and then diversified into a spectacular array of creatures, including bats, whales, kangaroos, and elephants. Yet that entire process would—if it occurred through Darwinian mechanisms—be expected to occur without benefit of a single mutation of the complexity of a CCC. Strict Darwinism requires a person to believe that mammalian evolution could occur without any mutation of the complexity of this one.

That is the closest Behe's Intelligent Design-level of curiosity ever got to exploring that hundred million years of life represented by the RMT. And he offered no sources for it, either.

Instead, Behe (2007, 62-63) upped the ante even higher by doing his multiplication trick for a "double CCC," which shoots up to values exceeding the number of all the bacteria that ever existed in the history of the Earth. "So if we do find features of life that would have required a double CCC or more, then we can infer that they likely did not arise by a Darwinian process."

By such escalating justification, a frontispiece chart in Behe (2007, v) restricted "Contingency in Biology" to microevolutionary elements within or below Species, while the origin of Genera, Families and Orders were plopped in "The Tentative Edge of Random Evolution." That left everything else beyond the Darwinian pale, from Classes (that would include Mammalia and Aves) up to Phyla and back through to the origin of life, nestled below a cosmic orgy of anthropic expectation, including the providential "Location of solar system in galaxy."

Nothing about this "Intelligent Design" spectrum would have worried Creation Scientist Duane Gish, by the way. The conclusion-jumping methods were common to both.

But all that depended on whether Behe's characterization of atovaquone and chloroquine resistance was on the mark. If not, his carefully lined up set of dominos would not fall over to make its big Design picture.

There was the obvious counterargument that focusing on the improbability of just two specific mutations missed how there might be multiple pathways to beneficial outcomes. Behe (2007, 61-62) firmly put his foot down on that one, and thereby crashed through his own theoretical gangplank (my **bold**):

"No. Many, many other mutations in addition to the ones we discussed have popped up by chance in the vast worldwide malarial pool over the course of a few years. **In fact mutations in *all* of the amino acid positions of *all* of the proteins of malaria—taken both one and two at a time—can be expected to occur by chance during the same stretch of time.** And other types of mutations besides just changes in amino acids would also occur (such as insertions, deletions, inversions, gene duplications, mobile DNA transpositions, changes in regulatory regions, and others, perhaps even including whole genome duplication—some of these types of mutations are discussed in the next chapter). Although some other mutations in some other proteins are thought to contribute to chloroquine resistance," citing La Bras & Durand (2003), "none are nearly **as effective** as that in PfCRT. That means that of all the possible mutations in all of the different proteins of malaria, only a miniscule number have the ability to **help at all** against chloroquine, and only one, PfCRT, is **really effective**. Natural selection gets to choose from **a staggering number of variation, yet at best only a handful help**. So a CCC isn't just the odds of a particular protein getting the right mutations, it's the probability of an effective cluster if mutations arising in an entire organism."

Behe was seeing the nature of the problem and yet not *seeing* it.

What Behe had failed to consider was the possibility that some of those other mutations might contribute little steps along an adaptive pathway that would eventually end up favoring the version of *PfCRT* that did the best job (the honing thing that natural selection is unavoidably prone to do). So was Behe paying all that much attention to what of that was going on in the "entire organism"?

Atovaquone mutation billiards, aiming for the quinol pocket

Consider that atovaquone resistance, its biology and selective environment. Atovaquone was developed in the 1980s to selectively target the parasite's own respiratory chain through its cellular mitochondria without bothering those in the host cells (but remember that mitochondria puts us knee-deep in endosymbiotic territory). The mitochondria of *P. falciparum* was rather unusual: at only 5966 nucleotide bases in length, it has the smallest known mitochondrial genome, and coded for only cytochrome *b* and the paired *c* units, Akil Vaidya & Michael Mather (2000, 284).

When atovaquone was used alone, resistance to it developed alarmingly quickly, and Michael Korsinczky *et al.* (2000) bred some test strains of *P. falciparum* to find out more about what was going on with its cytochrome *b*. Three (not just one) mutation sites significantly reduced atovaquone binding and therefore could boost resistance: M133I, V284K, and Y268S/Y268C, where either of two new amino acids would do there (serine or cysteine). Asparagine turned out to work at 268 too, noted in the later Mateusz Plucinski *et al.* (2014) study, which also identified a *fourth* mutation site (I258M, where an isoleucine was replaced with methionine) that brought about resistance to a mixed atovaquone-proguanil (AP) treatment. Gilles Cottrell *et al.* (2014) reviewed the mounting medical concerns over the failing AP combination, as the parasite was marshalling immunity to that too.

These mutation hotspots were located on different binding zones of the cytochrome *b* molecule, an obvious topological factor to consider if the idea was to fully understand the biological dynamics of drug resistance, along with the fact that the pool of available mutations was *deeper than first thought*.

Relevant single *and double* mutations were showing up in the experimental strains because the observed point mutation rate of around one per 500 million parasites per generation seen in Korsincky's sampling was *several hundred times higher* than that estimated by just the resistance work. Antoine Berry *et al.* (2006) encountered that high mutation rate in their own subsequent study of cytochrome *b* in 135 malarial isolates (work which also preceded the publication date of Behe's 2007 book). A factor contributing to that is how *P. falciparum* turned out to have a higher genetic recombination rate than had been previously thought, explored more recently by Hongying Jiang *et al.* (2011).

So elements at the base of Behe's uncompromising *Edge of Evolution* "double CCC" calculation were fraying on the *edges*.

Now it happened that of the three mutation sites that could bind atovaquone naturally, it was specifically the Y268C form (which exchanged the tyrosine amino acid with serine) that was showing up as the favorite one figuring in the drug resistance. That turned out to be related to its position on the quinol oxidation pocket, as Nicholas Fisher *et al.* (2012) later discovered with the benefit of advancing 3-D molecular modeling (more of the subsequent hard work conspicuously not being done by many antievolutionists).

That quinol pocket is part of the mitochondria's normal biological function, but the version in the malaria parasite already had *a prior natural mutation* (a loss of four amino acids compared to its otherwise nearly identical mammalian counterpart) that slightly altered its shape. Throw atovaquone at that mutant pore (whose ends bind to the amino acids at the adjacent spots 181 and 272) and the blocked pocket shuts the mitochondria down. And because those endosymbiotic organelles are critical to the cell's function, the parasite is toast too.

Now the differing serine (or cysteine or asparagine) at 268 dangles a modfied bond poking at the middle of the molecule, and so may interfere with the atovaquone docking. But wouldn't that also screw up molecules normally using the pocket? Yep. And that's why any drug resistance runs the risk of having a downside. The Fisher paper found that the drug-resistant strain had slower growth, but that the expression of *other genes* already in the cytochome complex (there are two models of *c* along with the slightly mutant *b*, remember) had ramped up to compensate for the metabolic drag of the impaired pocket.

New adaptive responses from a suite of existing systems interconnecting in a far from inert biology that reacted as it did because of one perfectly natural point mutation perturbing the see-saw of a previously mutated quinol pocket landscape. And don't forget (don't *ever* forget) that real people were (and are) getting sick, or even dying, should they get stuck riding that parasitical see-saw.

That's how biological systems function in the real world, for our particular good or ill, an often perilous carnival of trade-offs and adaptive twists, because lots of things are going on at the same time, and all those interacting factors needed to be explored in order to understand the significance of atovaquone resistance. Looking at just the gross number of parasites and the incidence of the Y268C mutation in isolation, as Behe was so primed to do while sprinting for his magic "double CCC" calculation, missed all the other connecting mutations that contributed to the quite natural (nay one say "Darwinian") evolution of that new response.

Concerning the chloroquine side, by the time of Mauro Chinappi *et al.* (2010) it was clear that the reason why mutations in the *PfCRT* membrane were so important is that changes in certain amino acids there determined the peripheral electric charges that allowed or restricted chloroquine circulating where it was needed, in or out of the digestive vacuole (DV) where it could slow the degradation of the toxic heme. The issue was no longer just the two mutations Behe was interested in (the ones most prominent in the measured chloroquine-resistant mutants) but the internal hurdles that would alter the adaptive landscape leading from one mutant form to the next.

When seen in 3-D modeling form (rather than the flat schematic Bray *et al.* had used in 2005) the two most relevant mutations were K76 (the one that could allow chloroquine to flow through the membrane and thus out of the DV) and S163 that blocked that passage (and thus would restore CQ sensitivity in the parasite). These turned out to be *adjacent* on the channel, acting as a very specific gate. Other factors in the cell could and did effect whether chloroquine actually got pumped out of the DV, which is why changes to the K76 spot were necessary but not sufficient for CQ resistance.

From what I've been citing of the more recent technical literature, the science work obviously hadn't stopped, and in April of 2014 a paper appeared in the *Proceedings of the National Academy of Sciences* on the *PfCRT* chloroquine transporter and its part in CQ resistance that brought Behe and his ID defenders out of the den, in force.

Robert Summers *et al.* (2014) investigated a range of existing plasmodium populations and worked out the many paths whereby the full chloroquine resistant version could have evolved. Interestingly, it was still not quite pinned down how chloroquine actually did its thing, Summers *et al.* (2014, E1759) writing that "is thought to exert its antimalarial effect by preventing the detoxification of the heme released from digested host hemoglobin." Jill Combrinck *et al.* (2013) represents pioneering work on that.

However the drug worked, though, the critical element in a selection process would be whether more of the chloroquine was getting pumped out of the digestive vacuole. *Any* improvement in that regard would increase the survival of that parasite, and hence the odds of it passing on those traits to descendants of the many merozoites/gametocytes eventually sexually reproduced in the mosquito host.

There were two main plasmodium lineages involved, though found in five regions, the CQ resistance having "arisen independently in Columbia, Peru, Papua New Guinea, the Philippines, and Southeast Asia," (page E1759). One of the main lineages showed much more range in how many of the alleles manifested substantial CQ resistance, though both could develop equally high resistance in some.

Would that mean those supposedly astonishingly difficult concatenated mutations were showing up on *five different occasions*? Wouldn't such frequency require them to be *separate* design events, by the strict mandates of Behe's CCC? If not, then wasn't that showing how unrepresentative the original probability estimates were when it came to what might be happening down on the genetic mutation floor?

Assuming design events were actually involved in any of them, though, that opens a still more unsettling can of worms, suggesting a designer positively dedicated to making sure medical science didn't make too much headway against *P. falciparum*.

Think about that. Kind of creepy.

It would make any putative Designer (and only just the *one*?) into something like Ian Holm's drunken self-pitying Napoleon in *Time Bandits*, anxious to see more of the violent Punch and Judy show because "that's what I *like*, leetle things hitting each other." If any of that recently emergent resistance to antimaliaral drugs were really acts of CCC-transcending intentional design, what would that say about the ethics or character of the Designer?

But back to the biology as explored in the new Summers paper.

"These haplotypes all contain a mutation at position 76—the replacement of the positively-charged lysine (K) with an uncharged residue, usually threonine (T)—but are otherwise diverse, containing at least 4 and up to 10 mutations," (E1759). That's relevant, because there were lots of known strains showing varying degrees of chloroquine resistance, and the Summers *PNAS* paper diligently pinned down lots more about which mutations those were and where they fell on the DNA. Testing on sixty experimentally generated plasmodium variations (only one of which, D17, wasn't already known from existing *Plasmodium* populations) suggested that the K76T and adjacent N75E spots (remember their critical location on the membrane pore) "play pivotal roles in enabling PfCRT to mediate CQ transport." Neither alone had that effect, though, but together they "produced modest but significant CQ transport activity," (E1760).

But what any of these mutations *did* depended on the whole biological kit.

If that K76T switch was thrown, replacing an asparagine at N326D with an acidic residue caused a 20% *drop* in CQ transport, (E1761), not good for the parasite. And if N75E, N76T and N326D occured *together*, those also decreased CQ transport. But a different amino acid at N326S didn't do that, even though that same N326S abolished the reaction completely if other mutations had occurred. Clearly something very interesting was going on here, suggesting how complicated the gain or loss of charges and residue configuration along the transport channel could be, affecting the overall response of the *package* to chloroquine.

Putting all those variables into play, Figure 4 on page E1762 laid out multiple pathways to achieve higher CQ transport (and hence resistance) in the two known *PfCRT* families. Temporary reversals in CQ transport level could be tolerated along the way, of course, provided there was still a net positive, and "it is important to note that mutations in PfCRT are likely to have been accompanied at various points by changes in other genes, some of which may have served to maintain or even increase the parasite's resistance to CQ when a new mutation in PfCRT decreased its CQ transport activity," (E1764).

But there were even paths available on the Summers' chart that showed *no* reversals at any stage, allowing an incremental ratcheting of chloroquine resistance. And as they involved examples known from plasmodium haplotypes seen in the wild, there was nothing hypothetical about investigating how those steps could link together mutation by mutation to result in a high level of CQ resistance. This was a biological analog of the RMT, except this time we had living examples of *all* the intermediate stages. The specific point mutations were listed conveniently enough on the previous page, in Summers' Figure 2, referring to eight relevant gene amino acid residue positions: 74, 75, 76, 220, 271, 326, 356 & 371.

As one of the two *PfCRT* families had a tougher go getting to the final highly resistant strains, let's follow one of the available trails on that hard case side to see what was involved (the cumulative acquisition of mutations underlined, with the new site mutations and transport ratings in **bold**, and marking the increasing resistance by the intensity of the box border):

Starting with D38, the one with that initial N76T mutation, the sequence in the variation zone was MNTAQNIR. D38 showed *no* CQ transport, so that can be our baseline. Any of the billions of parasites on earth might have had that trait for quite awhile, but unless they were exposed to chloroquine, they'd be clearing away the fatal heme just fine, so there'd be no obvious way to tell how long that N76T variant had been kicking around in any ancestral parasite's DNA.

One mutation in the adjacent 75 spot brings you the D32 strain, M**E**TAQNIR, which has a measurable CQ transport rating of **20**, meaning some of the chloroquine is getting shunted down the membrane and so not available to interfere with the heme degration going on back in the DV. Unless there were actually chloroquine used on it, though, it's again unknown when and how often D38s were in play, but selection would now have an edge to work on if chloroquine was used. Incidentally, in the game of genetic shuffleboard, D39 (which also has no initial CQ effect) would have done equally well for a starting point, as M**E**KAQNIR can also flip to the D32 allele in the sequence with only one mutation.

With another mutation at 220, METSQNIR, you have strain D26, with a CQ rating that surges to **55**. You think selection wouldn't notice that? Each of the next two changes remain at that selectable 55: an alteration at 371 to D20's METSQNII, and one further mutation at 271 resulting in D10's METSENII.

A mutation at 74 produces GB4, IETSENII, with a CQ transport value of **100**. Selection's going to be busy then, and parasite merozoites of that strain are definitely going to stand a better chance of making it to the gametocyte springboard stage. Which is confirmed by the fact that GB4 is one of the commoner alleles in the existing *Plasmodium* population.

Finally, the addition of a change at 326 brings us to the K1 strain, IETSESII, fielding an eminently selectable **120** CQ, at the top of the currently known plasmodium haplotypes.

Now remember, every single one of the intermediates involved exists in nature. No hypotheticals. Unless Behe's CCC meter wanted to spiggot off all natural mutations, none of them could be banned from having arisen from a single mutation from an earlier known strain, or be kept from mutating any further into yet another known strain.

So what was the reaction of Michael Behe (2014b-e) and the gleeclub of Ann Gauger (2014) and Casey Luskin (2014) to the Summers *PNAS* paper? All three of them repeated Behe's mantra that only two specific mutations were allowed, that their combined odds were vanishingly small, and that the new *PNAS* paper was an astounding validation of Behe's prescient analysis in *The Edge of Evolution*. Luskin even demanded that Behe's critics apologize for having doubted his sagacity.

The irony meter of Laurence Moran (2014) shot off the dial trying to gauge the collective obtuseness of Behe/Gauger/Luskin on the Summers study.

Incidentally, as Michael Behe (2014a) had, Gauger also lobbed Behe's CCC bomb at two other papers, claiming Sean Carroll *et al.* (2011) and Michael Harms & Joseph Thornton (2014) had encountered a similarly insurmountable mutational "canyon" with their work on glucocorticoids (secreted in the adrenal cortex of mammals along with mineralcorticoids).

Although Behe (2014a) had at least acknowledged that "Thornton's lab does terrific work," running "one of the most careful, thorough, meticulous experimental evolution programs in operation anywhere today," the Intelligent Design advocate hadn't gone into much detail on what they had accomplished. Which was having rather impressively *reconstructed* the form of the receptor after its gene duplication split from a common ancestral layout some 450 mya.

The more secondary Gauger took no note of that content at all. Clearly successful paleogenomic experimentation was not going to get much favorable press at *Evolution News & Views*.

Identifying the various mutations that contributed to and altered glucocorticoid affinity over those hundreds of millions of years so that current forms reacted only to higher threshhold concentrations, Thornton's team had drawn theoretical lessons on how rare and contingent (though still completely natural and obviously reconstructable) such pathway shifts were in the history of life. For Behe and Gauger, though, there were only "canyons" looming for "Darwinism."

For the record, the Thornton lab folk were not all that thrilled at Behe's muddling of their work over the years, and had previously expressed their considerable dissatisfaction to Carl Zimmer (2009).

But that still left all those known *Plasmodium* alleles in Summers' 2014 paper to account for. Where did they come from? Neither Behe nor Gauger nor the still more secondary Luskin seemed in much of a hurry to pin down any of that from their own Intelligent Design canyon.

Specifically, what about the multiple mutation pathways through them that could produce the final outcome, so graphically laid out in Summers' Figure 4?

Not one of them mentioned that. It was the primary subject of the *PNAS* paper, and yet none of them mentioned it.

Michael Behe (2015) went even further in avoiding the issue, responding to the criticism by Kenneth Miller (2014), who had reacted to Behe having seriously overplayed his rhetorical hand. Behe (2014e) had insisted that "it's utterly parsimonious and consistent with all the data—especially including the extreme rarity of the origin of chloroquine resistance—to think that a first, required mutation to PfCRT is strongly deleterious [Behe's post hyperlinking to Viswanathan Lakshmanan *et al.* (2005)] while the second may partially rescue the normal, required function of the protein, plus confer low chloroquine transport activity."

Miller (2014) had replied that Behe was wrong to claim that the K76T mutation was on its own deleterious, let alone strongly so, noting also that a single K76T had no effect on CQ transport either, and reminding how that was a specific data point (D38, as noted above) from the Summers' paper. Miller also reminded Behe of the core flaw in his mathematical argument: the failure to recognize how cumulative selection and inheritance rendered any multiplicative CCC number inapplicable. And Miller noted that these objections to Behe's argument were broadly shared, including by the equally crtitical Sean Carroll (2007).

As it happens, Miller did not mention Behe's citation of the Lakshmanan paper, but the only time K76T's stand-alone effect was alluded to there was in Lakshmanan *et al.* (2005, 2302), which speculated on how "other mutations may have been selected to compensate for loss/alteration of endogenous function associated with acquisition of K76T" and that their inability to clone one strain with only that mutation suggested "reduced parasite viability resulting from K76T in the absence of other pfcrt mutations." The lack of subsequent work to confirm that isolated incidence in the Lakshmanan study made it a problematic source for Behe redrawing his line in the sand in 2014.

Behe (2015) pressed on, though, skipping past what had been his own questionable point about the "deleterious" K76T to contort Miller's criticism concerning the effect of that as an isolated allele by bringing up the effect of *multiple* mutations that *included* K76T (my **bold**):

> It's nothing short of incomprehensible to claim that the K76T mutation has "no effect on transport activity." **Figure 2 of Summers et al.**—the very paper to which Miller is referring—shows that one variant of PfCRT (dubbed "**D39**") with a particular mutation (N75E) has no chloroquine transport activity. When the K76T mutation is added to it to make a double mutant (variant **D32**), it gains such activity. So it had no effect? Variant E1 has three mutations (none are K76T) but no activity. When K76T is added to those mutations to make the Ecu variant, activity appears. The only difference between the non-transporting and transporting variants is the addition of K76T, but Miller maintains it has no effect on transport activity?

Incomprehensible surely to Behe, who maybe shouldn't have so conspicuously called attention to that allele D32 and its adjacent N75E mutation, which was not among the more common resistant alleles (that was GB4 and K1) but rather one of the lower level adaptive players, a selectable second stage in the mutation trail tracked in the Summers paper that Behe otherwise showed no discernable curiosity about.

"Although I strongly disagree with nearly everything he wrote," Behe truly affirmed, Miller's "essay gives the public a chance to see directly how one informed Darwinist reacts to a basic empirical challenge to the theory."

I couldn't agree more.

Since Behe had explicitly cited Summers *et al.* (2014) and directly assailed Miller's 2014 posting that had reproduced Summers' Figure 4 mutation acquisition flow chart, it was Michael Behe who was in full *Garrett Hardin* mode. Any or all of the many synapsids that Behe has never stopped to ponder, may at this point let their evolving jaws drop.

The science work has continued, of course.

The dynamic of drug resistance can depend on the degree of parasitic inbreeding, such as in South America, leading to a fixation of the K76T resistant allele even after chloroquine usage was curtailed (the opposite of the African case, where the wild type K76 reasserted), Stéphane Pelleau *et al.* (2015). But despite the dominance of the K76T CQ resistant haplotype, the drug resistance has not recurred because of *other mutations* over in the transmembrane domain 9 (specifically Q352K or Q352R), whose added positive charge blocked binding with chloroquine, and so prevented its export from the DV.

And science even knows *when* that occurred. "A retrospective molecular survey on 580 isolates collected from 1997 to 2012 identified all C350R mutant parasites as being CQS. This mutation emerged in 2002 and rapidly spread throughout the *P. falciparum* population," Pelleau *et al.* (2015, 11672).

Nature, it would seem, continued not to get Behe's memo on what limits CCC was putting on how rapidly mutations arose or were compounded in natural populations.

The latest blip on the PfCRT evolution scope is Stanislaw Gabryszewski *et al.* (2016), who investigated the Ecuador Ecu1110 strain of *Plasmodium*. The difference between that and the initial wild type involved just four familiar mutations (at 76, 220, 326 & 356) and all sixteen of the possible intermediate stages were generated experimentally. Testing established that all but one of the 16 showed an increase in CQ resistance (and that outlier wasn't one of the K76T variants).

The resistance response, growth rate and other factors allowed simulation studies to be run to see how the variants would fare in thousands of different adaptive landscapes and at various levels of drug exposure. Many of these excursions hit "adaptive valleys" (one can imagine Behe or Gauger popping the champagne corks on those, if they ever get around to noticing the paper) but there were several paths that were most favorable even if the mutations were spaced out in the traditional way. Were several mutations showing up in fairly quick succession, though, that fixation time could kick in thousands of times faster.

In any case, Gabryszewski *et al.* suggested that I356L (which enhanced parasite growth on its own) or N326D were likely the first mutations in that parasite lineage, joined later by 220 and the ubiquitous K76T that made the CQ export feasible. You'll note that the 356 spot was *not* involved in any of the mutation chain links leading to the K1 lineage in the example I laid out above, and the 326 mutation was possibly the last link in the KI line, but both were implicated as the *first* ones in the Ecuador case.

More than one way to achieve an adaptive peak, then, at least in nature if not Intelligent Design.

Michael Behe's version of modern biology represents an Intelligent Design incarnation of the spirit of the mythic contrarians nearly a century ago who complained that bumblebee wings were too small to allow them to fly. That is, if all you were going on was standard aerodynamics, applying the formulas you would use for airflow over a fixed wing.

But bumblebees do fly, so the naturalistic assumption would be that the buzzing insects were not supernatural, and that some relevant factors were being left out of the calculations. That turned out to involve the fact that bumblebee wings were not fixed surfaces. They pivot constantly, generating low-pressure vortices that give them added lift. Not that this was all that easy a thing to figure out, taking extensive work like Richard Bomphrey *et al.* (2009) and Andrew Mountcastle & Stacey Combes (2013) to fill in the gaps.

It's that "filling in the gaps" thing that Behe and the Intelligent Design advocates required somebody else to do in the chloroquine case, and then dismissed the significance of the work when they did.

The Intelligent Design practice of repeating that which they had not actually shown, and ignoring big parts that didn't fit the ID picture, continued when Sarah Chaffee (2016) repeated Behe's *Edge of Evolution* chloroquine claim that "two simultaneous mutations is the most evolution can accomplish." Chaffee was waving Behe's authority to push back two pieces by anthropologist Barbara King (2016a-b) discussing a subject the ID movement would prefer go away: Ken Ham's new creationist *Ark Encounter* theme park. For some context, King (1994; 2007; 2013) reflects the range of her work on primate consciousness and culture, along with her views on the deep evolutionary roots of religion.

King did not specifically take issue with the claims of Intelligent Design, and that is what drew Chaffee's ire. Exactly as we saw with Marvin Olasky and David Klinghoffer back in Chapter 4, the ID camp doesn't take well to being upstaged by the flashier Young Earth creationists, but won't take on the less enviable task of explaining in what respect that YEC view is *wrong* (and risk splitting their *Kulturkampf* anti-Darwinian coalition in the process).

So Chaffee criticized King for her "narrow critique" that focused on a YEC movement that arguably has many more millions of active adherents than will ever copy an argument from *Evolution News & Views*, "rather than grappling with peer-reviewed research that questions neo-Darwinism" that none of her ID compatriots have shown any great aptitude to fact check first, up to and including the propositions of Michael Behe that Chaffee so freely invoked.

It would reveal a lot about the practical scholarly methodology of Chaffee or Olasky or Klinghoffer to see any of them explain in detail why a creationist argument such as the one by geologist Steven Austin (1994) on the reality of the Noachian Flood was scientifically flawed, and see if they could do so without resorting in any way to the dreaded Methodological Naturalism that ID has wanted off the table to allow their design inferences free entry.

I don't think that juggling act is possible, which is why we've never seen any in the ID movement try it. But if it ever happens, I'll bring popcorn.

10. Creationism Keeps Swinging ... And Missing the Ball

Once firmly into the 21st century, the Creation Science rollercoaster of RMT evasion gets bumpier by the year.

Bert Thompson & Brad Harrub (2001, 2; 2003, 5) didn't even bother with documenting their 2001 claim that "the fossil record does not satisfy for any such transitional forms" because all mammals were "equally mammalian," or realize how their slim 2003 *Gee Whiz* peroration on the "designed" ear was missing the fossils and embryology by a mile. A further measure of the shallow depth of their research skills and/or lack of curiosity: a 2013 text search of their *Apologetics Press* website turned up no hits for "therapsid" or "mammal-like reptiles," and this situation was unchanged in 2016 when I added "synapsid" and "Reichert" to the list. The only mammal-related hit was for Aaron Morrison (2007) offering a *Gee Whiz* piece on "The Intricate and Masterful Design of the Human Ear." No paleontology information was mentioned in it.

Other creationists have opted for folksy apologetic tropes, such as Bill Morgan (2005) recalling "When I was 14, my Biology teacher told me an amphibian turned into a mammal over millions of years. Frog + kiss = prince is a fairy tale, but amphibian + time = mammal is science, at least according to the Board of Education!"

The *San Antonio Bible-Based Science Association* (2007) reported on their president Scott Lane's inspiring lecture on "Biology and Intelligent Design." Lane was evidently running at about the same level of cluelessness as Ann Coulter over in ID Land, though, as the *SABBSA* "were amused by the ludicrous story evolution tells of how two of the three bones in the reptilian mandible were transported from the mouth to the inner ear and magically reshaped and resized, to form the extra two bones to make the mammalian inner ear bones. We saw how comparative anatomy, when examined on the genetic level, does not support evolution."

With Karl Reichert rolling over in his grave, it would have been informative to have learned what genetic or developmental resources Lane had in mind to document that one (presumably not Bill Dembski this time).

Meanwhile, Old Earth creationist Max Younce (2009) was even more superficial, concerned with combating the YEC views of Ken Ham but otherwise sounding like he was writing in 1920, not the 21st century (the book was subtitled "Don't let Satan Make a Monkey Out of You!"). Apart from some human evolution fossils (including the inevitable Piltdown hoax), paleontology was not on Younce's scope, meaning no RMT.

For those interested in a follow-up, the 1912 Piltdown affair has been the subject of many a *post mortem*, including Ronald Millar (1972), Frank Spencer (1990) and John Walsh (1996). I summarized the main points of the case (and how assorted antievolutionists have tried to spin the history) in Downard (2004e, 474-477), as have Glenn Branch & Eugenie Scott (2013) more recently. Isabelle de Groote *et al.* (2016) have pretty much clinched the case that the fossil's ambitious discoverer Charles Dawson (1864-1916) was the likely sole hoaxer.

The shrinking OEC apologetics of the Younces of the world were obviously not having an impact on young evangelists like Darek Isaacs (2009; 2010), who showed his ability to credulously repeat all the YEC mythology (including dinosaurs coexisting with people), but not spot anything like that RMT evidence that his creationist sources were ignoring for him.

A dictionary might have helped the loose spelling of Matt Slick (2011) when he presented a chart of the "Cretaceous Period" for his *Christian Apologetics & Research Ministry* (CARM), which noted "First reptiles" in the Pennsylvanian stage of the Carboniferous, and "First mammals" in the Triassic, but failed to note any of the RMT in the intervening Permian. The dictionary would have come in handy regarding the Triassic "ichtyosaur" he noted along with the "Suaropods" of the Jurassic.

Not that the heavy guns at the ICR's *Acts & Facts* were doing much better, or earning any points for fair or diligent scholarship.

Randy Guliuzza (2010, 11) wanted to prepare the Christian should they be confronted by someone offering up the mammal jaw-ear transition case as evidence of evolution. Citing (but not quoting) only the textbook by Scott Gilbert (2006, 17, 742), Guliuzza (whose background is in physics) suggested "Conversations can be readily started, for example, by saying, 'I read something about how the three tiny bones in your ear confer great capabilities for hearing.' Be sure to include enough details describing precise *fit* and *purpose* to highlight design and to keep the topic interesting."

Not interesting enough, perhaps, for those aware of what information Guliuzza was leaving out. And *intentionally* so, it would seem, given that Gilbert's textbook had explicitly mentioned not only Reichart's 1830s embryological findings but also the more recent paleontological work of Yuanqing Wang *et al.* (2001) regarding *Repenomamus* and its preserved Meckel's cartilage.

Guliuzza's fellow *Acts & Facts* contributor Frank Sherwin (2010) could have benefited from consulting Gilbert (or any science source, for that matter) when he tossed off this piece of systematic gibberish (my **bold**):

> Doesn't man's supposed journey from a single-celled ancestor clearly imply an upward progression from one kind to another? Not necessarily—now **some scientists** have surmised that evolution goes backward. Diapsids are considered "primitive reptiles," from which crocodiles and dinosaurs supposedly evolved. They are also seen as the **ancestors of synapsids**, some of which are believed to have **evolved backward** to become birds and dinosaurs. But others are thought to have maintained their synapsid anatomical features and evolved into people. The ruler must look like a Mobius strip if it can accommodate these possibilities.

It would have been genuinely exciting to have read who Sherwin had in mind for those "some scientists," so we could have seen just how it was that he had come to think any had claimed either (a) birds and dinosaurs in any sense had evolved from *synapsids*, or (b) that either of their lineages involved any evolving "backward." This would have been a nonsense position in 1910, but fielding it in *2010* was a clear measure of just how far behind the data curve Sherwin and the *ICR* had slipped.

David Coppedge ribs *Yanocodonon*

One character on the creationism front who waded in hip deep on one of the technical papers was information technology specialist David Coppedge (2007a) on the "Chinese scientists" Zhe-Xi Luo & Chen *et al.* (2007) who found *Yanoconodon*. If intended as a jingoistic jibe, one must caution that their lead author, Luo, works from the Carnegie Museum in Pittsburgh.

Coppedge became a *cause célèbre* in ID circles when the Jet Propulsion Lab let him go, the pot stirred vigorously by *Discovery Institute* (2010; 2011; 2012), Casey Luskin (2010c), Anika Smith (2010), *Evolution News & Views* (2012b-f) and David Klinghoffer (2012; 2016b), all of whom invariably depicted him as purely a persecuted ID advocate and never as a Young Earth creationist who just as tendentiously cast doubt on conventional geology and Big Bang cosmology, such as Coppedge (2005; 2006; 2007b-c; 2008a-c; 2010).

I can sympathize with how bored or exasperated his JPL colleagues might have grown to Coppedge's often pompous *Kulturkampf* hectoring, especially if he'd ventured anything like what he did at his website, such as Coppedge (2008b) grumping over the "mythical supercontinents Pangaea, Rodinia and Gondwana (which sound like characters in an earth religion)." Coppedge ultimately lost his wrongful dismissal lawsuit against the JPL, reported by Matt Young (2013).

As for the paleontology Coppedge was aiming at in his 2007 website post, recall from Chapter 3 that *Yanoconodon*'s separating middle ear bones fell on a distinct evolutionary continuum. But in Coppedge's antievolutionary universe, genuine transitional forms cannot exist (*No Cousins*!), and he shall find a way to dismiss everything about them. That proved a stiff climb this go-round.

Coppedge began by acknowledging the facts:

> First and most notable was the structure of the middle ear. They claimed it represents a clear transitional form between the attached bones of mammaliaformes ("mammal-like reptiles") and detached middle ear bones of mammals. Second, they noted the extra vertebra (26 instead of 19 for most mammals, and 22 for the nearest relative) and the presence of lumbar ribs, unusual for mammals but present in some widely-separated groups. This they explained by convergent evolutionary manipulation of *Hox* genes that govern the divisions of the vertebral column (sacrum, ilium, lumbar) and the presence of or absence of lumbar ribs. Experiments on lab rats show that these traits can be manipulated by knockout or

overexpression of these master-switch regulatory genes.

Well that would seem pretty solid, knowing the actual genetics in living mammals that allowed exactly the variations in vertebral number seen in the fossil, coupled with the transitional character of its middle ear. But Coppedge was ready to snatch all that back:

> There is no clear evolutionary transition in the vertebral characteristics. The authors noted that *Yanocondon*'s nearest alleged relative, *Jeholodens*, lacks lumbar ribs. Moreover, they found another pair of relatives on a different branch, one that has lumbar ribs and one that doesn't. For an animal to have 26 vertebra is "exceptional," they said; the only other one is *Repenomamus*, a dog-sized Cretaceous mammal that preyed on dinosaurs.

That a full skeleton was available for *Jeholodens* was good for seeing what a triconodont mammal looked like, Michael Benton (2005, 304-305) illustrating it in his summary of the many proliferating Mesozoic mammal clades. And you may recall Yuanqing Wang *et al.* (2001) from Chapter 3 on *Repenomamus*' rare preserved Meckel's cartilage. Coppedge didn't bring that up.

But what "clear evolutionary transition" was Coppedge demanding on the lumbar ribs? That he never said, but as Luo & Chen *et al.* (2007, 292) had explicitly shown, the ribs that form naturally along the thoracic vertebrae don't for the lumbar vertebrae of modern mammals, and that's because they express *Hox10* at that point. It's that simple. Turn that expression off, as had been done experimentally in mice, and the lumbar vertebrae appeared. All at once, *all along the lumbar vertebrae*, not just here or there. Other genes presumably fine-turned their length, which diminished from front to rear.

Since there were those ancient fossil mammals with exactly those lumbar vertebrae, though, it wouldn't take a rocket scientist (or a JPL information technician) to realize that *Hox10* wasn't getting expressed in their lumbar section. That's how they would have got those lumbar ribs. *Duh*. Furthermore, given the specificity of the trigger (the expression or not of an otherwise common homeobox gene), lineages could *independently* lose the ribs by gaining expression of *Hox10* at that point, which is what still happens with current mammals. And the converse: *any* group that had lost the ribs could easily have isolated members that regained them, just by losing *Hox10* expression there.

This was not speculation. This was the observed biology and fossil data, verified by direct experimentation, concerning all those fossil examples that no creationist had ever dug up or were ever illuminated by dint of their own study or effort.

That Coppedge was primed for rejecting such work of skill and evident application, with exactly the same hairsplitting enthusiasm as Duane Gish or Phillip Johnson before him, was made even plainer by how he tackled *Yanoconodon*'s ears.

Once again, Coppedge was not unaware of the data, summarizing it accurately enough, including that the inner ear bones were still connected to the jaw by Meckel's cartilage, and concluding with: "The arrangement in Yanocondon, they said, may be pedomorphic—a case of "arrested development" in which the embryonic attachment was maintained into adulthood."

On that hook Coppedge intended to hang the case, but not by exploring the issue of paedomorphism, such as the notable example of the Mexican salamander axolotl, which retains its external gills into adulthood, Randal Voss & Bradley Shaffer (1997). The axolotl had been an object of informative study for some time, as Bryan Hall (2009, 203) reminded regarding the 1893 discovery by the pioneering embryologist Julia Platt (1857-1935) "that the cartilages of the craniofacial and pharyngeal-arch skeletons of the mudpuppy" salamander arose from the neural crest cells, an observation generally accepted only half a century later when it was confirmed by scientists studying the axolotl.

Instead, Coppedge jumped to a *very* secondary source, and in the process revealed ever so much more than he intended about how his own Tortucan mind worked. The underlined section was a weblink, and the [bracketed] inclusions and **bold** were Coppedge's.

> So that's the story. How good a transition is it? The well-known skeptic and pseudoscience fighter James Randi thought this was "very cool" as a demonstration of evolution. He gave it a big write-up at his James Randi Educational Foundation where he reproduced the figures from the original paper. The figure he left out, however, is the cladogram (phylogenetic tree), which, suprisingly, shows "homoplasies [convergent evolution] of DMME [definite mammalian middle ear] in basal mammals" in six places on the tree. One of them is *Hadrocodium*, a lower Jurassic mammal lacking some of the typical features of mammals but having a complex hearing system (see Reference.com). Does this mean that an even less derived species had mammal-like middle ear bones, separated from the jaw? If so, *Yanoconodon* is too late to be considered a transitional form. Randi displayed some ignorance of modern evolutionary theory by using the pedomorphy hypothesis to conjure up the ghost of Haeckel: "It's one module in development that **flaunts a lovely example of embryonic recapitulation** of evolutionary history," his article boasts. (Recapitulation is dismissed by most Darwinists these days.)

Coppedge was making the same common mistake with these fossils' transitional features as have generations of antievolutionists, failing to distinguish relationship from chronology. The Figure 2 chart was *not* a chronology. True, individual fossils belonged to particular points in time (though YECer Coppedge might have challenged that, but didn't). Knowing that *Hadrocodium* (with its presumably detached adult ear bones) nested at the edge of the monotremes cladistically said nothing about how long either of those lineages had existed, only that the taxonomical characteristics were sufficiently similar.

In order for the mammalian middle ear to evolve, the bones had to be coopted from the jaw, and eventually their linkage to the jaw severed. The intermediate nature of *Yanoconodon* was in its preservation of that Meckel's cartilage as an *adult*, as no living mammal does, but which an early evolutionary transition would. That there were cousins living even earlier in the sequence that had seemingly already moved beyond that to complete detachment, as *Hadrocodium* was thought to, didn't diminish the significance of either fossil to illustrate the presence of transitional features in lineages that most certainly existed.

We know what was going on with the basal mammaliaformes shown on Luo's Figure 2, running from the early *Thrinaxodon* and *Probainognathus* to *Morganucodon* on the late end, and that the "definite mammalian middle ear" (which Luo abbreviated as DMME) was first marked with *Hadrocodium*. In all of these cases, though, the future ear bones were shrinking along the jaw, until finally only the Meckel's cartilage connected them, with disconnection evidently occuring at varying pace in the many near-mammalian lineages.

It's what happened next that was offering clues to the deeper developmental sequence. *Hadrocodium* was nestled at the edge of the node leading to the monotremes. But because the monotremes took their own time to originate their own DMME (only two of the later ten taxa listed were marked as having them), and only four of the dozen "Crown Mammals" next in the chart did so (that included *Yanoconodon*), the issue remained whether their respective lineages possessed an inherited DMME at the presumed *Hadrocodium* stage (with some lineages regaining the adult cartilage attachment on their own), or whether *Hadrocodium* represented what happened several times, a common detachment of the ear bones to form a complete DMME.

Neither of the scenarios would have required any astonishing new genetics, of course, since the living marsupials still do part of the trick by retaining the cartilage until later in juvenile development. The commonality of paedomorphosis in developmental biology couldn't rule out either of the two evolutionary options.

As noted back in Chapter 3, though, more recent CT scanning of *Hadrocodium* reduced the significance of that part of the puzzle by suggesting the postdentary trough may not have actually disappeared, and so rendering it unclear whether the Meckel's cartilage had become fully detached at that time in *Hadrocodium*. As for what developmental factors may have been driving the cartilage detachment in mammals generally, further work on monotreme and marsupial middle ear development by Héctor Ramírez-Chaves *et al.* (2016) identified the timing of molar eruptions as a likely major factor, rather than primarily brain expansion as had been formerly thought.

There were lots of biological factors to consider then in this seeming mystery. Yet rather than taking a stand on whether it was physically impossible for fossil forms of life to do what we know living mammals manage to do (disconnecting the ear bones from the jaw and doing so at differing rates in marsupials and placentals), Coppedge chose to filet the secondary Randi posting instead. Why?

It's likely it was that Haeckel sideswipe. Coppedge wanted to drag Luo's evidence down via guilt by association. The bad old Haeckel advocated "recapitulation" and now the silly old Randi was trying to do the same, using Luo's fossil example. Though technically we don't know that it was James Randi himself who posted on the website. But no matter.

Now there *was* some "ignorance" here, but the Randi web post was not the one showing it. It was Coppedge's inability to tell the difference between Haeckel's very specific (and wrong) version of recapitulation made over a century ago (that animal embryology tracked the *adult* forms of their ancestors) and the modern view of "recapitulation" that recognizes how living animals inevitably built on the long trail of developmental systems acquired along their given lineages (increasingly shown by experimental work, such as the Luo paper Coppedge was tripping over).

Though the term "recapitulation" is less used today, the concepts of heterochrony and heterotropy (variable temporal or spatial development) as well as paedomorphosis fall within that theoretical boundary, reflected variously by Ernst Mayr (1994), Michael Richardson *et al.* (2001), Richardson & Gerhard Keuck (2002) and Brian Hall (2003).

That this confusion on Haeckel is a common trope in antievolutionary writing, from Dave Nutting (2007) and Elizabeth Mitchell (2014b; 2016b) on the YEC side, to *Explore Evolution* (noted last chapter) and Jonathan McLatchie (2010) in ID Land, didn't make Coppedge's repetition of it here any less misleading, especially as those ideas figured in what Coppedge did next, a long quote from Luo & Chen *et al.* (2007, 291) referring to the two cladistically parsimonious evolutionary courses that could account for the relationships shown in that circular Figure 2 Coppedge threw at James Randi.

Coppedge's parlor trick depended on not recognizing that Luo's DMME wasn't a claim that the ancestral form hadn't already *coopted the jaw bones* for hearing. That process was already far along in the mammaliaformes, as Luo's work had so diligently established over the years, and which Luo's 2007 paper had usefully illustrated in Figure 3 on the *same page* as the quote used by Coppedge. Like *Explore Evolution* with Kemp's chart in the previous chapter, the reader needed only to ease the eye leftward this time to see what in the jaws needed further transforming in the adjoining middle ear to complete that official DMME. The issue now was only exactly when (and how many times) the adult detachment occurred.

That Coppedge wanted evolution to be horrified at the prospect of this happening several times instead of once, was clear enough. That he had to step over the known biological processes that made either alternative entirely plausible, was equally plain. What wasn't at all clear at any time was what *Coppedge* thought was going on here.

Coppedge being a Young Earth creationist, *Yanoconodon* and *Hadrocodium* (and dinosaurs and ancient whale ancestors and hominids, just to name a few) were not only alive and trundling about 4500 years ago, and their "kind" shephered gently onto the Ark by Noah and his family as the rain commenced, but that after the Flood *every single one of them went extinct*. As a preservation mechanism for terrestrial fauna, the Ark was a monumental washout.

But instead of presenting his audience with a fair reflection of whether Creation Science (or Intelligent Design) concepts could say anything informative about why those animals with those characteristics existed where and when they did (that *when* part being really important), Coppedge could only hammer away on what was in fact his own odd view of the scientific process.

Coppedge even challenged the paleontologists' skill at assessing the growth rates and age of *Yanoconodon*: "How do they know the developmental stage of this fossil? Maybe it was a juvenile and its ossified Meckel's cartilage had not yet dissolved. How much can you tell about the morphology, lifestyle and complexity of an animal from its bones?'"

Oh, David, do try and read the paper you've just cited, as Luo & Chen *et al.* (2007, 288) had addressed that very point: "the exposed interior of the lower and upper jaws shows no replacement at any tooth loci and its functional teeth (including the last tooth) are all permanent. The holotype is either a late-stage subadult or an adult with a delayed eruption of the last molar." Luo & Chen *et al.* (2007, 289) also noted some relevant skull bone mergers: "Many modern therians have relatively rapid early skeletal growth, but terminate this growth upon the fusion of the epiphyses to the diaphysis. By contrast, eutriconodonts lacked a similar mechanism for terminating skeletal growth that would presumably slow down (although not stop completely) in the adult, because of the absence of the epiphyseal growth plates."

Given such gauzy disinterest in the details of the science work he'd made such a point of citing, Coppedge couldn't stop spiraling into a smug anti-intellectual rant: "The paper is filled with jargon (plesiomorphy, apomorphy, pedomorphy, heterochrony, homoplasy), but no amount of hand-waving justifies their interpretation. Jargon is, at best, a convenience for those playing the game; at worst, it is bluffing and obfuscation. Most important, their story is based on the flimsiest piece of ossified cartilage and says nothing about how a complex system of hearing, employing finely-tuned ossicles, evolved."

Coppedge was now all *Gee Whiz*. Though on that theme of *jargon*, if Luo *et al.* were to be disallowed from employing technical terminology in what was, after all, a technical paper published in a technical journal aimed at their technically-savvy expert audience, would Coppedge be as critical of Michael Denton's sprinkling of unexplained jargon on ants back in Chapter 5, when his own target audience was objectively not going to be jargon-savvy?

A final observation is warranted on Coppedge, the underlined links were to more of Coppedge's opining, namely the entries on "Bluffing" and "Equivocation" in the "Baloney Detector" glossary of Coppedge (2011), which reaffirmed the tropes that circulate in Coppedge antievolutionary universe, many of which we've been bumping into already.

The "Bluffing" segment (defined as "Appearing to know more than you do") included "Using big words to sound smarter: *Ontogeny recapitulates phylogeny.*" Haeckel again!

Coppedge also had "Bold assertions made without evidence, e.g., American Museum of Natural History: 'Birds are dinosaurs.'" Readers may flash back to the evidence presented in prior chapters to see how far that antievolutionary trope has failed to fly.

"*It's been documented!* (Where? By whom? So what?)." Do you really want to go there, David, and address how firmly *you* (or your fellow antievolutionists) have properly documented your own assertions?

"In debates, some evolutionists have claimed the fossil record is filled with transitional forms (but Colin Patterson of the British Museum could not think of a single case for which one could make a watertight argument)." Pardon me, Mr. Coppedge, but your credulity slip is showing. Now you may see why I made such a point of investigating the Patterson case in detail back in Chapter 6.

The "Equivocation" segment (defined as "Confusing the issue by using vague terms or shifting the definitions of words") included "It depends on what the meaning of *is* is." And thus had Bill Clinton's impeachment-avoidance dissembling been enshrined on Coppedge's *Kulturkampf* version of Mount Rushmore.

This one really stood out for me, though: "Cults that redefine Christian terms, e.g. Mormon missionary *Why, yes, we believe in salvation by grace through faith* (which, being translated, means working your way up to godhood)." A cautionary heads-up flag for any Mormons employed at JPL, at least while Coppedge was around.

And then: "*Evolution is a fact, but we don't yet understand the mechanism*." Actually, we understood lots of the mechanisms, David, just not *all* of them (which is why evolutionary science is so interesting and rewarding to explore). So would *that* be an example of "Confusing the issue by ... shifting the definitions of words"?

And finally, "Shifting definitions of Natural Selection depending on which has better rhetorical effect in the context of the discussion (excellent treatment of this tactic in *The Biotic Message* by Walter ReMine)." Here we see Coppedge parasitically citing yet another secondary redactor (recall ReMine's flimsy jab at systematics noted back in Chapter 7). Check any of *his* sources, David?

All this inspires me to recommend two more platitudes might be added to Coppedge's list: *Look before you leap* ... and *Practice what you preach*."

Truth in Science displays neither

By the time the British creationist website *Truth in Science* (2008b) weighed in on the RMT, they were in the same boat as David Coppedge. The evidence directly available online had grown so extensive that it was getting difficult not to go all *Garrett Hardin* the moment you tried to link to anything specific in the scientific literature.

That's what happened at every step in their argument, most glaringly so when they stumbled onto the jaw embryology issue (that topic so otherwise completely ignored in the antievolution canon). It's important to remember that this *Truth in Science* piece represents the *only* antievolutionist source I've encountered so far that has alluded to the embryology issue even at this superficial level. Wow.

Truth in Science didn't even trip over it directly, but bounced into it secondarily, having read a general article for *New Scientist* by Donald Prothero (2008), on ten notable transitonal sequences known in the fossil record: the velvet worms (onychophorans) and lancelets (like *Amphioxus*) relating to the origin of Cambrian phyla, "Fishibians" like *Tiktaalik* for the tetrapods, our stars the synapsids, ceratopsians among dinosaurs, rhinos and giraffes in more recent mammals, and a trio of marine vertebrates (ichthyosaurs, pinnipeds and manatees).

Prothero's piece hadn't offered technical sources for that fossil evidence, though his 2007 book had, but even the brief paper mentioned enough fossil names and the scientists involved in that work to allow anyone to track down primary citations if they had an inclination to. Accent on that *if*.

Onychophoran relationships have been covered by many over the years, such as William Ballard *et al.* (1992), and Gregory Edgecombe (2009) nicely summarized their significant evolutionary implications. Lancelets represent a huge field of study on their own, further fueled by the full *Amphioxus* genome just then being sequenced by Linda Holland *et al.* (2008). Pressing on to the vertebrates, I refer the reader back to Chapter 2 for the *Tiktaalik* trail, and Chapter 4 on ceratopsian origins. The range of basal rhinos was reflected in Patricia Holroyd *et al.* (2006), though the revealing intermediate *Samotherium* giraffe found by Nikos Solounias' team didn't get a full systematic description until Melinda Danowitz & Rebecca Domalski *et al.* (2015) and Danowitz & Aleksandr Vasilyev *et al.* (2015).

As with *Tiktaalik*, finding marine reptiles was inevitably constrained by the availability of relevant ocean deposits of the right age, and work has often spread out over many years, such as the notable paleontology on ichthyosaurs by Ryosuke Motani *et al.* (1998; 2014; 2015). The Fossil Genie has lent a helpful hand on seals, Ulfur Arnason *et al.* (2006) and Natalia Rybczynski *et al.* (2009), and the many manatee finds by Philip Gingerich & Daryl Domning *et al.* (1994), Domning (2001) and Iyad Zalmout *et al.* (2003) traced their evolution from terrestrial quadrupeds.

After nicely summarizing all that, and the gist of the RMT, Donald Prothero (2008, 38) noted how "during embryonic development, the middle ear bones start in the lower jaw, and then eventually migrate to the ear."

Truth in Science pounced on that, harrumphing that this "remarkable statement" was one "the general reader has little choice but to take at face value," but couldn't resist pressing one dangerous step too far: "In mammalian embryos, however, the middle ear ossicles (bones) are derived from the 1st and 2nd arch mesenchyme (i.e. embryonic connective tissue of mesodermal origin)," and linked directly to the "recent article" (already a decade old) by Moisés Mallo (1998).

Using that "however" was their first great blunder, and citing the Mallo paper was their second one, going full *Garrett Hardin* here, because that quite fine work had explictly noted the homology of the middle ear bones with their reptilian counterparts, the articular and quadrate.

True enough, Mallo had noted that the mammalian malleus and incus form in the first pharyngeal arch (the stapes derives from the second), and that was specified because the paper was about the *mammal* side of things. But those bones develop from that arch only because that's where their *homologs* originate. Mallo hadn't belabored that point, presumably because his technical readership would have known that context, just as saying mammal ear bones were made of molecules and atoms couldn't be taken as evidence that articular and quadrate bones weren't also made of molecules and atoms. As for the maleus, it too comes from the first arch, though subsequent work by Stephen O'Gorman (2005) showed that a short attached bit from the second arch has come to be attached to it in mammals, reminding us of how complex the mix-and-match of developmental evolution can get, especially after a couple of hundred million years.

Nothing in the Mallo paper, or in his other ongoing work on the embryology of the mammal jaw, such as Mallo (2001), justified *Truth in Science*'s dissembling use of that *however*. Just to see how little "truth" or "science" was cropping up here, compare their evasion with any contemporary technical survey, such as Brian Hall (2009, 203-246) on the formation of vertebrate cartilage and facial bones from the first two pharyngeal arches, including those articular and quadrate bones. Or Creuzet *et al.* (2002) tracing in the diapsid birds how the *absence* of *Hox* gene expression regarding the first pharyngeal arch elements opened up opportunities for evolution of the facial skeleton.

Clearly, following in Duane Gish's carefully placed footsteps was not going to be easy for so lumbering a gang as *Truth in Science*. And it only got worse when they declared that "not every palaeontologist would agree with Professor Prothero's evolutionary assessment of the synapsid cranial morphology," and cited, of all people, Christian Sidor (2003b) as their sole example.

"In this article, Professor Sidor presents detailed findings based on the morphological analysis of an extensive number of fossil synapsid jaws. These included 19 pelycosaurs, six basal therapsids, 13 dinocephalians, 25 anomodonts, ten gorgonopsians, ten therocephalians and 25 cynodonts."

All the conclusions you're about to read were drawn from Sidor (2003b, 620), and *not one of them* contradicted Prothero's summary of the cranial facts.

None of the conclusions *Truth in Science* trotted out were put in quotation marks, but were functionally direct quotes, though with revealing tactical omissions. Starting with the very first one, when *Truth in Science* neglected to include the highly relevant last sentence.

Here's the original text, with the parts they left out in **bold**:

> 1. The lack of a well-supported phylogeny has exaggerated previous estimates of morphological convergence or parallelism in the synapsid fossil record. The hypothesis of multiple therapsid groups arising independently from pelycosaur-grade ancestors **(e.g., Olson 1962; Boonstra 1972)** necessitated rampant homoplasy and are now considered untenable **(Rubidge and Sidor 2001). Certain lower jaw characteristics and proportions are better viewed as broadly distributed synapomorphies indicative of common ancestry.**

Leaving out the *date* of the sources being discussed was not all that surprising for antievolutionists (those "Norrell" and dinosaur cases from the previous chapter), but in this instance that *Map of Time* omission was more important even than the identity of the papers themselves. Put into plain English, studies from *decades before* that had seen lots of seeming convergence were now superceded by improved cladistic analysis (that also dealt with *more fossil data* that had come along since), and so recognized those jaw features as just *ancestral* ones being inherited along those lines.

You may recall how that convergence issue arose back in Chapter 7 regarding the gorgonopsid cousins, and what Rubidge & Sidor (2001) had to say about them. And don't forget how John Woodmorappe contorted the content of that paper in Chapter 8.

Sidor's second conclusion (which *Truth in Science* copied exactly) was another non-starter if the idea was to undermine Prothero's summary.

> 2. Despite the striking differences between the lower jaws of basal synapsids (i.e., "pelycosaur") and mammals, mandibular evolution within synapsids was predominantly conservative. Except for dicynodont anomodonts, most therapsid subclades do not acquire substantial morphological novelty in their lower jaw structure.

Incidentally, *Truth in Science* inserted a bracketed "[emphasis added]" after the "predominantly conservative" phrase. Which was odd, since they hadn't put any *emphasis* in. But then missing their own point was par for the course by now.

The issue here concerned how much *directionality* there had been in the evolution of synapsid jaw proportions, as "only the ancestral lineage leading to mammals (i.e., along the backbone of the cladogram) shows a consistent dentary enlargement," Sidor (2003b, 611), "and that clades budded off from this line retained their ancestral proportions but did not systematically continue the trend."

That was all. The same bones existed in the animals, expanding very gradually lineage to lineage. There was no general trend for bigger jaws (remember how small most mammals were). The most notable feature in the long-term was the prominence of the coronoid process, and of course the coopting of the angular and quadrate into the middle ear, anatomical features Sidor at no point challenged. None of that required any radical change in jaw proportions relative to the skull, or a trend for greater size, which were the details Sidor was measuring.

Interestingly enough, one of those amonodont dicynodonts (*Placerias*) sauntered up the gangway at Ken Ham's *Ark Encounter* themepark, a picture of which Tim Chaffey (2016) waved at Josh Rosenau (2016a) for disparaging the "stuffed animals" on display there. The creationist version follows the conventional paleontologist depiction, such as *Prehistoric Life* (2009, 220-221), meaning at least for the purposes of booking the Ark's stalls, creationists were willing to accept the forensic reconstruction of the extinct animal. Which should remind us of that *Map of Time* problem.

The herbivorous tusked *Placerias* lived during the Triassic about 220 mya, though based on some fragmentary remains found in Australia, some stragglers in that lineage may have persisted down into the Cretaceous a hundred million years later on what back then was an archipelago of Gondwana, Tony Thulborn & Susan Turner (2003). Chaffey did not venture an opinion as to why there were none of the *Placerias* or other putative therapsid "kinds" surviving *today* though, it being the intended purpose of the Ark to have preserved them. Later in this chapter we'll see how little thinking baraminologists have been doing on therapsids generally, failing to work out much about how many of their "kind" there may have been, created and/or surviving.

Truth in Science freely adapted Sidor's next point, though leaving out how the jaws scaled "either isometrically or with slight positive allometry" that might have helped clarify what the paleontologist was talking about. Probably because of its reference to "smaller-than predicted" values, though, *Truth in Science* kept Sidor's last sentence intact. Here's their creationist version:

> 3. When comparing the dentary and postdentary bones with overall jaw length, the trends in body-size are not sufficient to explain the reduction of the postdentary bones in synapsid evolution. Importantly, when compared with other synapsid subgroups, cynodonts are characterized by smaller-than predicted postdentary areas.

All this was referring to some tables and charts occurring back on Sidor (2003b, 616-617), showing how the critters' dentary bones scaled just by size (isometrically), while most of the little postdentary bones showed much the same isometric scaling, with a few exhibting just a little allometric variation (where the bones would have been reducing comparatively faster than the dentary). Once again, given how tiny the actual bones were (think of *Probainognathus* or *Morganucodon*), all Sidor was identifying was the lack of significant deviation from the overall proportions along the line from early synapsids to late therapsids—a reduction of the postdentary bones that *included* their incorporation into the evolving mammalian middle ear.

Sidor's final point related to an exception to the overall conservatism of synapsid evolutionary trends (my **bold** to show the sources *Truth in Science* again left out):

> 4. Selection acting to decrease the size of the postdentary bones, and thereby improving high-frequency hearing, is still the most tenable mechanism for the evolution of the mammalian lower jaw **(Allin 1975; Allin and Hopson 1992)**. However, this mechanism by itself has difficulty explaining the converse pattern in anomodont therapsids (i.e., decreasing the size of the dentary and increasing the size of the postdentary bones).

Truth in Science did not stop to explain anything at all about those anomalous anomodonts, either so that their readers might have some clue what they were talking about, or more importantly to account for them from their own creationist perspective (fat chance there).

There were clues enough in Sidor (2003b, 605, 607, 610-618) as he touched on the anomodonts as a "side branch" not leading to mammals whose "substantial morphological novelty" included an expanded opening in the lower jaw for muscle attachments (which the other synapsids tended to close) and a more prominent angled bulge along their more substantial postdentary bones. Prothero (2007, 274) showed some of the anomodont skulls and their position as one of the sidebars in later Permian synapsid taxonomy.

Of relevance is something I noted back in Chapter 4, how the anomodont *Dicynodon* (which Woodmorappe had also tripped past back in Chapter 8) had dental plates that performed more of a plant chomping role than their less prominent dentary teeth. That specialized layout served them well enough when the Permian mass extinction obliterated so much of their competition, and the piglike anomodont *Lystrosaurus* proliferated (at least for awhile), illustrated by *Prehistoric Life* (2009, 220).

Trying to scope out to what degree selection for specifically higher-frequency hearing played among synapsids generally, the resurgence of more prominent postdentary bones *only* among the anomodonts suggested to Sidor (2003b, 618) "either that high-frequency hearing was not important to anomodonts or that selection for this feature was not exclusively molding mandibular evolution in this group." None of that anomodont side-issue refuted the observed developmental relationship between our middle ear bones and their ancestral jaw bones.

Sidor's final paragraph hadn't been numbered, but *Truth in Science* slapped one on anyway, rephrasing "epipodal" as "limb" but retaining most of Sidor's text in their version:

> 5. These conclusions, in combination with those of recent studies on long-term patterns of limb and cranial evolution, suggest that morphological trends within synapsids should be re-investigated within a quantitative and phylogenetic framework.

In other words, do the hard work and pay attention to all the data, to learn more about what actually happened. Something *Truth in Science* would have done well to have tried now and then, just for the novelty of it.

Truth in Science should have quit while they were behind, but they couldn't resist nosediving into the *Garrett Hardin* hole they had been deepening point by Sidor point (my **bold**): "In other words, the most detailed recent study suggests that there is **little supporting evidence** for the transition sequence proposed by Professor Prothero."

That was false. Cranial blockage of the rectum level false.

This degree of evasion was beyond some trivial typographical error, unlike the inadvertent misspelling of "anapsid" reptile as "anaspid" (an ancient group of fish) by Prothero (2007, 271) that *Truth in Science* ostentatiously nitpicked as evidence of Prothero's supposed ignorance of reptilian anatomy.

Truth in Science continued to dig their hole, though. "But this is just the beginning of the problem when one considers the emergence of mammalian characteristics in the fossil record," indulging in the same Gish-style hairsplitting of passages, only from the more recent Tom Kemp (2005, 1-2, 120-122, 126, 133) on what defined a mammal, and how the brain, dentition and thermoregulation developed.

The example of Gish and *Explore Evolution* in previous chapters showed how dangerous a path Kemp-culling could be, but *Truth in Science* went full steam ahead to argue that the diagnostic features of the synapsids were somehow arbitrary or cherry-picked, and in the process stepped right around the same details *Explore Evolution* had left out, along with the more extensive analysis of therapsid endothermy done by Kemp (2006b) himself.

Truth in Science tried to make a big deal out of two things: first, the fact that mammals have "determinate" growth (where there's an initial juvenile spurt), as opposed to animals like reptiles that grow all through life in an "indeterminate" way, and second, the way mammalian "diphyodont" two-stage tooth replacement (milk teeth followed by permanent ones) differed from the continuous "polyphyodont" tooth replacement found in reptiles and other amniotes.

They quoted (their ellipses, my **bold**) Kemp (2005, 121):

> It is not until the basal mammal *Morganucodon* that the **combination** of determinate growth **and** diphyodonty is known to have evolved ... The incisors, canines, and posterior postcanines are added sequentially at the back, not replaced, and therefore can be properly referred to as molar teeth. Given its correlation with growth pattern, it is assumed by this stage that lactation has evolved.

"In other words," opined *Truth in Science*, "there is no fossil evidence for the emergence of determinate growth and diphyodonty until the first true mammal appears in the fossil record. This fact is omitted in Professor Prothero's article."

It was interesting to see *Truth in Science* complain Prothero's article hadn't covered every nuance of synapsid evolution in only two-thirds of a page (eight paragraphs in two columns). Since the *Truth in Science* article ran for nine pages in my printout, lack of space won't be an available excuse for them. As with their specialized use of "however", the omission *Truth in Science* was accusing Prothero of turned on their tightly morphing a conjunctive "and" into the implication that *nothing* was known about either the determinate growth part or tooth replacement patterns occuring in the previous synapsids, or that the *Morganucodon* example told *all the story*.

And we can judge that simply by reading on through the rest of the paragraph *Truth in Science* so wanted us to peruse, as Kemp had continued (sorry, my **bold**, I can't resist):

> **However**, the story is complicated by the situation in *Sinoconodon*, which is basal to *Morganucodon*. It still had indeterminate growth, for specimens are found that range in skull length from 2.2 to 6.2 cm, corresponding to an estimated body mass range of 13-517 g (Kielan-Jaworowska 2004). The tooth replacement pattern is also more primitive in *Sinoconodon*, as indeed it had to be in order to allow for the very considerable growth in size of what must have been sub-adults not dependent on lactation for their growth. The postcanine teeth are not replaced, and there is loss of anterior and addition of new posterior postcanine tooth row of only three or four teeth (Fig. 4.12(b)). **This condition in *Sinoconodon* is therefore intermediate between the primitive tritheledontid and the fully mammalian conditions.** It may be speculated that the state of evolution of lactation was also intermediate, with maternal provision of milk limited to an early neonate stage only, after which the juvenile was weaned and relied on its own foraging, or perhaps on a more limited conventional food supply provided by the mother.

Kemp's Fig. 4.12 appeared on that same page, just *above* the extract *Truth in Science* quoted, and right beside the text you just read, all of which rather pulled the rug from under the creationists' argument. Or to paraphrase a bit: "This fact is omitted in" the *Truth in Science* article.

Had *Truth in Science* been especially serious about either truth or science, they might have investigated what their own quoted source of Kemp was talking about. Just as with the laying of eggs (or not), mammals and their cynodont precursors were more varied than *Truth in Science* made out. While *Sinoconodon*'s molars were replaced once, and the postcanines erupted sequentially, "other dental replacement characteristics are very primitive for mammals. Incisors in *Sinoconodon* show an alternate replacement pattern, and the canines were replaced at least three times (Crompton and Luo, 1993; Zhang et al., 1998), as seen in many nonmammalian cynodonts," Kielan-Jaworowska *et al.* (2004, 150).

Not that all current mammals have that diphyodont replacement pattern, by the way, notably some very small insectivores, along with toothless anteaters and baleen whales, and Zhe-Xi Luo *et al.* (2004, 160) added how that "can vary among genera of the same family." Nor were all cynodonts a monotony of indeterminate growth, significant exceptions being the diademodontids (who replaced their postcanine teeth front to back) and the T2 tritylodontids (who we've been recurrently encountering) that never replaced their postcanines at all, just letting the rearward ones drop off from wear, Zhe-Xi Luo *et al.* (2004, 163).

More specimens of the T2 tritylodontids have been found since, Yaoming Hu *et al.* (2009), Tim Fedak *et al.* (2015) and Rachel O'Meara & Robert Asher (2016). Sometimes these were just fragments, sometimes not, but the more data that became available, the more chance there was for making useful observations from that collective mix.

Take the Late Triassic *Oligokyphus*, which retained the ancestral joint between the quadrate and squamosal, along with "intermediate growth patterns, with more extended adult growth patterns than *Morganucodon* and slightly slower juvenile growth. This suggests a gradual evolution of mammalian growth patterns across the cynodont to mammaliaform transition, possibly with the origin of rapid juvenile growth preceding that of truncated, determinate adult growth. In turn, acquisition of both these aspects of mammalian growth was likely necessary for the evolution of diphyodont tooth replacement in the mammaliaform lineage," O'Meara & Asher (2016, 439).

The diademodonts are a rather more restricted case, as they involve only a single genus, but every additional example of them testifies to how much or little they varied. These six-foot-long nocturnal omnivores were a successful bunch, Jennifer Botha *et al.* (2005) tracing the paleoecology of the Karoo branch, and Agustín Martinelli *et al.* (2009) reporting on Argentinian examples (continental drift neighbors during the Triassic, remember).

"*Diademodon* is one of the first non-mammalian cynodonts to exhibit a combination of a mammal-like jaw adductor musculature and precise postcanine occlusion," along with a slower tooth replacement rate than the early nonmammalian cynodonts, Botha *et al.* (2005, 304, 307). Considering how *Diademodon* did its version of tooth replacement, Fai-Kui Zhang *et al.* (1998, 209) suggested that the full mammalian diphyodonty mode came in two stages, "the single replacement of postcanines" followed by "the single replacement of incisors and canines."

You'll notice both the Botha and Zhang papers predated *Truth in Science*'s 2008 pontification, meaning they could have in principle tumbled onto it in the course of their research. But we've already seen what level of research they did take a stab at.

Add in another player from Chapter 8, the T1 tritheledontids, information drawn from another paper predating 2008. Kielan-Jaworowska *et al.* (2004, 157) reported, "If tritheledontids are the sister taxon to mammals, as is preferred by the majority of workers, then the origin of mammals can be characterized by a shift from the primitive multiple alternating replacement of all postcanines in most cynodonts to a derived pattern of sequential single replacement of postcanines in mammals."

All of that diversity would be because of variations in their developmental genetics. That's what biological systems do. It's not magic.

Here's where some *Map of Time* perspective might have been of help too, had the YEC *Truth in Science* been up for it. Recall the players from back in Chapter 8 and when they lived. *Morganucodon* scampered around some 214 million years ago, and *Sinoconodon* (retaining its more cynodontian intermediate growth rate and intermediate tooth replacement pattern) was running long some six million years *later*. Then it's another 64 million years before the docodont *Haldanondon* teeters even more closely on the brink of the mammalian norm, noted by Kielan-Jaworowska *et al.* (2004, 153).

That's a 70 million year spread (the time separating us from the dinosaurs). During all that time could anything have prevented such natural variation from occurring? No radically new jaw bones were appearing, only the gentle morphing of ones already in play. And the mechanics of making those teeth were systems of very long standing, so we'd be seeing just an incremental shift of the existing components, and all as slow as molasses. It's hard to imagine a more gradual transition sequence than the RMT.

Come to think about it, would that mean a putative Designer (and still just the one?) couldn't quite make up their mind about what to do fiddling with tooth replacements over tens of millions of years?

You may already be ahead of me in supposing that the science work hasn't stopped, either, as dedicated researchers put more data on the field and draw further reasonable conclusions. We'll get back to that a bit later, when we see how Elizabeth Mitchell tried to build on *Truth in Science*'s faulty foundation, and what new information she had to overlook.

I'll end this section with the full *Bermuda Triangle Defense* (Player #12) *Truth in Science* pulled when they brought Kemp (2005, 126) on stage for the issue of therapsid hair (my **bold** indicating the partial sentence they quoted):

> The most unambiguous evidence would be the presence of of insulation because there is no other conceivable primary function for a furry covering other than maintenance of an elevated body temperature produced by internally generated heat. Unfortunately, because it is a protein, hair is rarely preserved, and **as yet no mammal-like reptile has been shown by direct fossil evidence to have possessed a pelt**. Convoluted and not overly convincing arguments for its presence have been offered, such as that the presence of foramina in the snout of cynodonts indicates the presence of vibrissae, which in turn indicates that hair had evolved, and therefore may have covered the body.

In another Kemp text *Truth in Science* managed not to quote, Kemp (2005, 182) referred to "Preservation of the impression of the pelt of the Early Cretaceous placental *Eomaia*, and of hair of a Late Palaeocene multituberculate"—all reminders of just how rare mammal pelts were in the fossil record.

Think about it. Quite independent of whether evolution was going on ("Darwinian" or otherwise), tens of millions of years of mammals were objectively living and dying, teeth and some bones showing up now and then, but for all that, only a scarce few with fur directly preserved for paleontologists to study. Following Qiang Ji *et al.* (2002) on *Eomaia*, Ji *et al.* (2006) had pushed the earliest example of hair tens of millions of years further back, to the Middle Jurassic, added to by Chang-Fu Zhou *et al.* (2013), all of which still reiterated the forensic fact that soft tissues like fur were rare for perfectly natural reasons.

Though there are other ways to capture them, as subsequent work illustrated, showing both how dedicated and curious paleontologists are, and how deadly useless antievolutionary apologetics are.

Just as stray feathers can float down and get included in a deposit, hairs can land as impressions or get trapped in amber, Vincent Perrichot & Loic Marion *et al.* (2008) and Romain Vullo *et al.* (2010). Feathers and hair have shown up that way from the Cretaceous. But a furball can also be something else's *dinner*, with the undigested bits ending up along with bone fragments in the carnivore's feces. Though not the most glamorous of pursuits, in the last few decades a whole field of paleontology has developed to study fossilized dung, known in the trade as *coprolites*. The size of a coprolite says a lot about the bulk of the animal pooping it, of course, but it's the relative presence of seeds, fish scales or bones that indicate whether the Number Two dropped from a carnivore or herbivore.

Occasional hairs have shown up in Cenozoic coprolites, noted by Vullo *et al.* (2010, 685), by which time full mammals were on the scene, of course. But what about earlier, much earlier? Remember, it was the *synapsids* that were the dominant vertebrates of the Permian, so anything big enough to leave a good coprolite from that period had their paws all over it, and anything smaller but still furry that ended up being eaten could only be another of those synapsids. In the years since *Truth in Science* was invoking Tom Kemp, Late Permian synapsid coprolites have been found containing filamentous traces "too large and straight to represent bacteria or fungal hyphae," but consistent with small mammal hairs, Roger Smith & Jennifer Botha-Brink (2011, 51). Piotr Bajdek *et al.* (2016) have since turned up similar hair traces from other synapsid coprolites.

And recall also from Chapter 2, the work of Julien Benoit *et al.* (2016) on the role of the *Msx2* gene relating to the inference of hair among therapsids. That work put a specific genetic base under the "presence of vibrissae" Kemp was skeptical of a decade earlier.

As it so happened, the Intelligent Design bastion *Uncommon Descent* (2016) flipped past the Benoit study via a secondary online summary at *EurekAlert!* (2016). The anonymous *Uncommon Descent* poster neither disputed the contents nor offered any ID explanation for those data, other than to grump how the idea of "evolutionary experiments" supposedly personified "evolution" as some "no-design Designer."

That they weren't paying too close attention to even that commentary, though, could be seen by how they labeled a CT scan of *Thrinaxodon*'s skull they copied from the *EurekAlert!* website. The skull was shown on the left to illustrate the sensory pits on its snout (the vibrissae suggestive of sensory whiskers), and the *same* image (representing the *same* fossil) was repeated to its right with the maxillary canal sections highlighted in green and pink so the reader could more easily see their anatomical connection.

Evidently unable to read even a simple caption correctly, or to ponder the implications of the two images being otherwise identical, *Uncommon Descent* clumsily tagged the image as "maxillary canal, mammal left, therapsid right."

When it comes to keeping mammals and therapsids straight, it would seem the confused spirit of *Pandas* and Dembski's *Design of Life* does live on.

Circling the Woodmorappe bandwagon

As we've moved into the 21st century, when online apologetics and googling have begun to supercede old school print resources, the relative usage of Duane Gish's venerable 1995 *opus magnum* has begun to fade away in comparison to the more obviously accessible online contribution of John Woodmorappe's 2001 *AiG* analysis.

But remember, Woodmorappe's study was completely wrong, which means *anybody* citing it on its substantive argument (as Phillip Johnson and David Berlinski had) was committing as great a sin of secondary credulity as the many traditionalists still relying on Gish's equally flawed *Evolution: The Fossils STILL Say No!*

Some are just Hit-and-Runners, such as the "Five concise responses to atheistic arguments" by Robert Carter (2011). In between dumping on the Big Bang, Carter declared "There is nothing preventing God from having created a bewildering assortment of species (e.g. the 'mammal-like reptile')" and passed the hot potato on to Woodmorappe (2001) with a weblink.

Over at *AiG*'s *Journal of Creation*, the much more extensive Shawn Doyle (2011) still drew on Woodmorappe (2001) as his only source for the claim that "the reptile-to-mammal 'transition' as a whole has many problems." One of the minor problems in Doyle's piece, by the way, was that the whole paragraph in which the Woodmorappe citation occurred was lopped off from the official pdf print version of the paper (though it was still in the online text edition). A methods note to *AiG*: "cut and paste" assembly of a copy requires you not skip that *paste* part. Just saying.

Doyle's paper was aimed at the newly discovered *Liaoconodon* fossil, and was titled "'Transitional form' in mammal ear evolution—more cacophony." You may recall *Liaoconodon* from back in Chapter 3, and how the anchoring of the eardrum allowed the middle ear to separate from the jaw and become devoted exclusively to hearing.

What did Doyle have to say about all that?

Doyle waved several red herrings, including an expected sideswipe at the irrelevant Haeckel solely because Jin Meng *et al.* (2011, 183) had correctly reminded how "the embryonic pattern of modern mammals recapitulates the phylogenetic changes." And Jonathan Sarfati (2000a) was cited as Doyle dangled some guilt by association: "since there is one well known fraud to have come from there (the *Archaeoraptor* hoax), one wonders if that is the only fossil fraud to have been perpetrated on the scientific establishment. I'm not suggesting that this or any particular find from the Jehol group is fraudulent, but merely pointing out that there is reason for *a priori* scepticism about fossils from there."

Would that be more or less "*a priori* scepticism" than Doyle showed by his acceptance of Woodmorappe's incompetent 2001 paper? In any event, no one (not even Doyle's fellow creationists) had offered a shred of evidence that any of the relevant fossils in the RMT were in any sense fraudulent. Nor did Doyle dispute any of the physical layout of the bones in *Liaoconodon* or any of the other fossil examples. He even reprinted an illustration from Meng *et al.* (2011, 183) showing the similarity of the bone layouts for *Morganucodon* with its mandibular middle ear (MME), followed by the transitional middle ear (TMME) of *Liaoconodon*, and finally the generalized therian definitive model (DMME). It didn't take a magnifying glass to see how no new bones were involved, just a shifting of their relative orientation, along with the downsizing of the middle ear bones.

How then could any transition from the MME to the DMME *avoid* looking like what was on display with that TMME? As usual for an antievolutionist, Doyle did not explain what the data would have needed to look like to satisfy him. But all that information under his nose did not stop Doyle (2011, 44) from insisting (his *italics*) that "The morphological disparity between the morganucodonts and *Liaoconodon* remains huge, and they've now added another large gap in morphology between some 'early' mammals and extant mammals! Note that we are talking about two *large* morphological gaps."

That was a bold and absurd exaggeration to offer in 2011, given what was by then known of the many fossils that had been found, the comparable range of developmental variety in living mammals (from monotremes through marsupials to placentals), along with the accumulating genetic information on how those changes came about. Doyle was similarly oblivious to the developmental and genetic ramifications of another paper he peripherally cited, Luo & Chen *et al.* (2007) on the role of homeobox genes in the heterochronic evolution of mammal vertebrae, showing Doyle could miss the point just as thoroughly as David Coppedge earlier in this chapter.

Doyle ended up with the very dregs of creationist apologetics when he rolled out yet another tattered creationist chestnut, Ashby Camp (1998), to source the claim that "The crucial morphological changes that are required to evolve a mammalian middle ear (whether a DMME or a TMME) from a jaw joint are still *conspicuously absent* from the fossil record."

Which was utter piffle, as Camp's parade of mined quotes had shown no such thing. Recall also from Chapter 7 how Camp had even repeated the Gishian mythology that the morganucodontids had "a fully-functional reptilian jaw joint." Nor had Camp showed the slightest indication that he was even aware of the embryonic development data, let alone make proper sense of it.

As none of Shaun Doyle's sources ever alluded to the subject, it can be presumed the secondary source-swallower Doyle knew nothing about it either.

Doyle did tiptoe perilously close to almost offering a suggestion of what a creationist explanation for the RMT fossil data might have involved when he slipped this into his argument: "There is widespread testimonial evidence for the biblical Flood in the universal spread of Flood legends, of which the Genesis account is the most realistic and reliable. This provides a solid conceptual basis for understanding fossil distribution, which evolution lacks."

His sources for this consisted of a creationist trio: Nozomi Osanai (2005) and Jonathan Sarfati (2006), purveyors of second-hand mythology that were conspicuously unaware of the fact that ancient Egypt in particular had no Noah-style flood legend at all, and the completely irrelevant John Reed (2008), offering a superficial rejection of Cuvier's approach to fossil evidence in the course of Reed's purely philosophical and religious justification for a biblical catastrophism that certainly ventured no explanation whatsoever for any actual fossil distribution (with or without evolution).

To put my two cents in on the Flood issue, drawing on works like Dorothy Vitaliano (1973, 150-177), Norman Cohn (1996) and William Ryan & Walter Pitman (1998), I surveyed the Flood myth terrain in Downard (2004c, 232-239), including the recurring Mesoptamian river flooding and earlier Black Sea postglacial infilling that provided the mythic tinder for the Israelite version long after the waters had drained, when they encountered the Flood and creation stories during the Babylonian Captivity.

The creationist citational daisy chain involving Woodmorappe and Camp got even longer with David Pratt. Pratt (2000; 2006) was already convinced that plate tectonics was on the ropes, and manifested comparably ill-grounded skepticism about evolution in Pratt (2014), where he freely mixed ID citations to Michael Denton (1985) with creationist sources from Duane Gish (1995) to Jonathan Sarfati's *Refuting Evolution*. Pratt disingenuously relegated the celebrated horse evolution sequence to "a relatively minor morphological transformation," which view he based only on Gish (1995, 189-197), not any of the more recent baraminological work (noted back in Chapter 6) that had pulled the typological rug from that case by recognizing it as a monobaramin.

Pratt's 2014 offering galloped into the RMT with this:

> The transition from mammal-like reptiles to mammals in the Triassic is usually claimed to be best documented case of macroevolution in the fossil record. However the "sequence" from pelycosaurs to therapsids to mammals contains many gaps, which neo-Darwinists assume will one day be filled by new fossil discoveries. Mammalian traits sometimes appear, disappear, and then reappear, they sometimes appear out of sequence, and they often emerge independently—and supposedly by chance—in different lineages.

His sources were to four weblinked works (my **bold**): "**Ashby L. Camp**, 'Reappraising the 'crown jewel', Creation Research Society, 1998, trueorigin.org; **J. Woodmorappe**, 'Mammal-like reptiles: major trait reversals and discontinuities', Journal of Creation, v. 15, no. 1, 2001, pp. 44-52, creation.com; **James Downard**, 'A tale of two citations' **(critique of Woodmorappe)**, June 2003, talkreason.org; The Fossil Record, pp. 154-60; **Shaun Doyle**, "Transitional form' in mammal ear evolution—more cacophony', Journal of Creation, v. 25, no. 3, 2011, pp. 42-5, creation.com."

James Downard?! Was he kidding? To slip my analysis in as merely a "critique of Woodmorappe" was certainly the height of moxey, but makes me wonder whether Pratt could have paid the slightest attention to its content, since I called into question the very idea of relying on Woodmorappe as a legitimate source to begin with. Adding the equally incompetent Camp and Doyle only compounded Pratt's scholarly problem.

Answers in Genesis (2016a) buffed their Woodmorappe parasitism by double-dipping two of his pieces, Woodmorappe (2001; 2002). That latter work attempted to use his prior criticism of the RMT as a justification for similar doubts on the taxonomy of whale evolution. But Woodmorappe (2002, 118n) confidently citing Woodmorappe (2001) as though he had in fact "shown that relatively few anatomical traits of mammal-like reptiles display a unidirectional trend towards mammalness" still wasn't any less evasive or misleading, no matter how many times it was repeated by Woodmorappe or *AiG*.

Not that *AiG*'s 2016 copiest was paying that close of attention to the topic by then, anyway, for they inexplicably included a picture of the "Fossil of *Tiktaalik*, a type of fish that evolutionists mistakenly supposed was a transitional form between reptiles and mammals," and cited their own David Menton (2007). Now while Menton had indeed disputed (unjustifiably) *Tiktaalik*'s status as a transitional form, it was regarding *tetrapods*, not mammals. Did the *AiG* redactor have no clue that there was a difference?

AiG did uphold the tradition of trotting out dated authority quotes, though: "As with all alleged evolutionary transitions, evolutionists themselves are unclear about what exactly evolved into what. They realize that this group of animals does not show an obvious transition from reptile to mammal," and cited their only technical reference, Zofia Kielan-Jaworowska (1992, 195), for this authority quote: "It is not known which cynodont family was ancestral to mammals, or whether all the mammals originated from the same group (family) of cynodonts. In the vast literature concerning mammalian origins, it is easier to find suggestions that one or the other therapsid or cynodont family cannot be ancestral to the Mammalia, rather than to find a positive answer."

They did not quote anything further, but in the very next paragraph Kielan-Jaworowska had called attention to an "exception" to that negative story, the work by James Hopson and Arthur Crompton recognizing the implications of the transitional *Probainognathus*. All you had to do was read into that section and it was clear that the issue of the moment was the placement of those T1 tritheledontids and T2 tritylodontids.

But 1992 was almost a quarter of a century before. Why then not at least investigate subsequent work, especially by that same author on that very issue, from her coauthoring of Luo *et al.* (2002) to Kielan-Jaworowska *et al.* (2004)? Unless, of course, understanding what was being discussed was no more on the *AiG* apologetic agenda than it was on that of *Explore Evolution* or Paul Nelson over at the *Discovery Institute*.

Elizabeth Mitchell and *Your Inner Reptile*

We've seen how much scientific information on the RMT there was on the scene as the second decade of the 21st century dawned, though you wouldn't know it in the creationist spin room, as they weren't keen to address much of it. Elizabeth Mitchell (2013a) was a rare outlier at *Answers in Genesis*, taking on Chong-Xi Yuan *et al.* (2013) regarding the evolutionary significance of the multituberulate fossil *Rugosodon*.

Up until the finding of this new virtually complete fossil, understanding the evolution of the multituberculates was hampered by the fact that so many of them were known only by their teeth. Later ones were known to have had highly flexible ankles, but without more full skeletal examples to trace the lineage, it couldn't be known whether that condition was a derived one or more ancestral. *Rugosodon* showed that ankle matter was indeed ancestral, and its teeth showed something else. While the earliest multituberculates were insectivorous, later ones were the first major mammal group to specialize with an herbivorous diet. As an omnivore, *Rugosodon* supplied another bridge from one mode to the other. Zhe-Xi Luo *et al.* (2016) continued the analysis of how *Rugosodon*'s basal foot layout allowed moving with feet planted flatly, setting the stage anatomically for later evolution of forms that walked on flexed toes.

Mitchell (2013a) pulled all that story apart like so much taffy, then stuck clumps of it back together to fit their creationist candy recipe, starting with her declaration that "the 'evolving' features ensuring multituberculate success were actually present from its earliest appearance in the rock record."

That's a flat No. The evolved bone shapes in *Rugosodon*'s foot were *not* just what was to be found in later forms. What had happened was that gradual shifts in them *permitted* the evolution seen later, joint modifications (such as the astragalus in the upper ankle) tracked in considerable detail by Luo *et al.* (2016, 157-163).

Mitchell hopped to another rock with the claim that "The diversity of these animals illustrates not evolution over millions of years but variation within created kinds. And their occurrence in many parts of the fossil record is not evidence of their evolutionary appearance on earth but instead is consistent with their presence in many of the habitats catastrophically buried during the global Flood, less than two thousand years after God created them."

A pretty neat trick for an alleged Flood (for which Mitchell cited not a shred of evidence), especially since the Tiaojishan Formation in which *Rugosodon* was found is riddled with volcanic deposits, Junling Pei *et al.*

(2011). While there were the odd deposition of volcanic ash washed by water (the area did have streams), as the main volcanic flows were not showing the sort of pillow lavas that form under water, was the catastrophic Flood spiggot somehow flipping on and off like some spasmodic marine lawn sprinkler?

And even had Mitchell offered some physical evidence for that tiny slice of Deep Time being due to the Flood, the rest of China's tectonically active geology all through the Mesozoic offered more of the same to account for, Shaofeng Liu *et al.* (2003), Timothy Kusky *et al.* (2007) and Yong-Qing Liu *et al.* (2015). Sprawling across the land and representing many millions of years of natural tectonic change (including interbedded pyroclastic flows and airborne volcanic ash), the known geology of China is by no means easy to stuff into the shallow catch-all Flood bucket.

Mitchell (2014) returned to the RMT zone with a critical review of the "Your Inner Reptile" segment of the PBS series based on Neil Shubin's book *Your Inner Fish* (mentioned back in Chapter 3). As she had with *Rugosodon*, Mitchell parsed her description for *AiG*: "He shows us their teeth and points out that this subclass of reptiles, unlike modern reptiles, had some features that would have allowed them to chew their food. Their teeth were not all identical little pegs but were differentiated into incisors, canines, and molars. Because modern reptiles do not typically have this arrangement, evolutionists consider this a transition toward mammal-ness."

But it was simply wrong for Mitchell to suggest that reptiles "do not typically have this arrangement." Reptiles do not have that distinctive layout *at all*. Only in the synapsids are we finding that specific mix of "incisors, canines, and molars," and they clearly did not arrive on the scene all in one fully-formed designed gush.

Shubin's video also called attention to our vestigial retention of the amniotic yolk sac. Mitchell didn't like that either:

> He tells an evolutionary tale to explain their presence and then tries to support it with genetic claims that are common but do not bear close scrutiny. Shubin's story is that fish ancestors evolving into land reptiles figured out how to keep their eggs alive in a shell but had to supply their embryos with protein-rich egg yolk. Then, once those reptiles evolved into live-bearing mammals, they didn't need the yolk sac for food so it just stayed around as a vestigial remnant.

Being a video aimed at a general audience, it was hardly the place for a parade of technical literature, but that didn't mean the scientific work wasn't there to notice. Vertebrate egg yolk genes had been modfied in mammals concurrent with the evolution of lactation and placentation that built in turn on a network of other genes (including caseins derived from *tooth genes*, interestingly enough), explored in papers that would include David Brawand *et al.* (2008), Anthony Capuco & Michael Akers (2009), Kazuhiko Kawasaki *et al.* (2011), Olav Oftedal (2012; 2013) and Oftedal & Danielle Dhouailly (2013). Tadasu Urashima *et al.* (2015) represents the ongoing scientific study, this time on more clues the biology of the platypus offered about the deep evolutionary roots of mammalian lactose usage.

Much work had also been done on the varied duplication and loss history of the related vitellogenin (Vtg) genes (and their assorted lipoprotein homologues) from fish to honeybees, including Vtgs retained in the platypus genome, Paul Richardson *et al.* (2005), Roderick Finn & Borge Kristoffersen (2007), Patrick Babin (2008), Finn *et al.* (2009) and Heli Havukainen *et al.* (2011). The prospect of isolated pseudogene remnants of those vitellogenins in *humans* garnered firm dismissals from Jeffrey Tomkins (2015a,c) on the YEC side, and similar ID foot-stamping by Anne Gauger (2016a-b) and *Evolution News & Views* (2016e) contesting theistic evolutionist Douglas Venema (2016a-e) at *The BioLogos Forum*, further indication that when push comes to shove, it's hard to tell creationist and Intelligent Design advocates apart based purely on their technical apologetics.

In the grand creationist tradition of changing the subject when confronted by inconsiderate data, instead of seeking out any of that ample and available documentation, Mitchell dodged via Guojun Sheng & Ann Foley (2012) and Shicui Zhang *et al.* (2011) to complain: "What Shubin fails to mention is that the human embryo's yolk sac performs many essential functions, being in place before any embryonic organs develop," and that the "VTG genes do not exist just for the purpose of making yolk protein."

And who said they should? Shubin failed "to mention" such details because none of them were relevant to his point, nor would they have made any of the physical evidence he was highlighting go away. Nothing in the notion of evolutionary descent with modification required ancestral biological components to summarily *turn off* in a new derived lineage. Quite to the contrary, such divergence within conservation is the norm in natural evolution, and nearly a dozen varied functions of Vtgs alone were identified by Chen Sun & Shicui Zhang (2015, 8820).

It was Mitchell's creationism that required *Zeno Slicing* the biology of nature into sufficiently small providential blocks whose isolated functions could be *acknowledged*, only to be nonnegotiably attributed to "our wise Designer" without explaining *why* those specific versions of molecules were doing what they were in such diverse lineages.

And we can see this with Mitchelll's having cited Sheng & Foley (2012). Cue *Garrett Hardin*. Mitchell cited that paper as the reference for her summary:

> Additionally, the yolk sac forms blood vessels while the embryonic heart is forming; when fused with the heart they form the embryo's cardiovascular system. Furthermore, in the earliest weeks of development the human's yolk sac is involved in transferring nutrients to the embryo, absorbing uterine secretions and keeping the embryo "fed" until the placental circulation can be fully established.

All of which was perfectly true, but somehow Mitchell managed to clip off all the *context* the authors had provided for that information. Sheng &

Foley (2012, 97) had reminded that "Ancestral mammals are believed to be more reptile like, and the monotremes (the basal extant mammals) share many developmental features with reptiles." Specifically on the yolk sac's nutrient role, Sheng & Foley (2012, 101) had noted how "Structurally, the mammalian yolk sac endoderm is very similar to the avian yolk sac endoderm and is indicative of active uptake from its apical surface and high metabolic processing within its cytoplasm," hardly a biological shocker given that, "Like its avian counterpart, the mammalian yolk sac endoderm actively produces major serum proteins."

Knowing how things were related in an evolutionary sense opened windows of experimental opportunity effectively closed to the shuttered creationist bandwagon. Regarding how nutrients are transfered into the mammalian embryo (the yolk sac being nutrient poor compared to its avian counterpart), Sheng & Foley (2012, 102) had reminded

> that this transport route evolved early in the amniote lineage, and that neither the chorionic ectoderm as the entry point for nutrient absorption, nor the allantoic vasculature as the nutrient transport venue, nor the association of these two as the main nutrient absorptive organ, is unique to the eutherian mammals. Comparison of the yolk sac-mediated nutrient uptake in birds and the chorioallantoic placenta-mediated nutrient uptake in mammals is relevant, therefore, to the cell biological, biochemical, and physiological understanding of the latter.

Since Mitchell cited Sheng & Foley (2012) we know those observations were there for her to read. So either she was dashing through the text too fast to notice, or she was aware of them but didn't want her *AiG* readers to know about them too. As with Duane Gish and so many others in the antievolutionary parade who thought to use technical science work as a blugeon to beat back Darwin, the moment Mitchell cited that paper, those were the only two *Garrett Hardin* alternatives: obtuse inattentiveness to, or deliberate evasive suppression of, the relevant evidence.

Scholarship is a contact sport.

But all that was just a windup for Mitchell's main evasions, concerning the RMT jaw/ear issue and the embryology and the many fossils showing it. As Gish and *Pandas* and *Design of Life* had before her, Mitchell's review tiptoed past the forensics by limiting what was shown and discussed. So it was that Mitchell drew on a *Wikipedia* illustration of the basic reptile and mammal jaw configurations, but did *not* include one for the intermediate therapsid form.

As for the embryology, it was monumentally disingenuous for Mitchell to leave out the Reichert evidence when she wrote only that "In reptile embryos Meckel's cartilage does not disappear; it ossifies to form part of the jaw. These are observable embryologic developments that, along with countless other characteristics, distinguish mammals from reptiles."

Mitchell offered no sources for any of this section, and fell into what had become the Pavlovian reaction in antievolutionary apologetics, slamming into

Shubin for supposedly resurrecting "the discredited embryonic recapitulation theory." Sorry, but the science has moved beyond Haeckel in the 21st century. Live with it.

As for the fossils, Mitchell next dismissed Luo & Crompton *et al.* (2001) on *Hadrocodium* with the bold and wrong assertion that "nothing about this animal's skull shows that it is a transitional form, only that it is a species of, as well as can be determined from just a tiny skull, an extinct mammal that appears deeper in the fossil record than evolutionists expected to find it." Not that Mitchell or anyone else in the antievolution biz had ever laid out criteria for what a transitional form *would* look like, in this or in any other fossil example (as we've seen all through the antievolution roadshow).

With perhaps the biggest splat of unjustified secondary citation in as short a space as possible, Mitchell also insisted that "the fossil record has failed to reveal the clear-cut trends in reptilian jaw bone changes that evolutionists often claim is present," and offered this as her reference footnote: "See C. Sidor, 'Evolutionary trends and the origin of the mammalian lower jaw,' *Paleobiology* 29, no. 4:605–640 and also "Synapsids and the Evolution of Mammals" at **Truth in Science** to learn more."

Sorry, Dr. Mitchell, but *Truth in Science* had totally misrepresented Sidor's paper, and you were a gullible fool for accepting their falsehoods.

Mitchell (2014c) further compounded her RMT daisychaining when she dismissed a new mammal fossil expanding the understanding of Cretaceous mammal distribution, described by David Krause *et al.* (2014), and decreed in a footnote that "'Mammal-like reptiles' is a term suggesting they were transitional forms, but they were not." She then referred her readers to her own flawed Mitchell (2014a), once more to the messed up *Truth in Science* (2008b) and the monumentally unreliable John Woodmorappe (2001).

Indeed the reader would "learn more" by consulting those sources, not only discovering in Sidor (2003b) more of the solid evolution evidence Mitchell chose to ignore, but directly measuring the shallow depth of her creationist secondary addiction by the very fact that she could rely on as superficial and mispresentative a foundation as *Truth in Science* or Woodmorappe in the first place.

And maybe bump into further research as well, something Mitchell might have tried in 2014. For example, there's that tooth formation issue her *Truth in Science* source had skirted six years before. Had no work been done on that since 2008?

Oh, yes.

Among those double-jawed transitional probainognathids, Agustín Martinelli & José Bonaparte (2011) had gathered together what was known of tooth replacement patterns in *Brasilodon* (the critter I noted back in Chapter 4) and how that correlated with its paleoecology. Their postcanine tooth replacement (which was slower than some of their cousins) ran back to front, unlike the early mammals that settled on front to back. This appeared to be related to skull growth and the impact of diet, where the chomping teeth of the probainognathids still had room to emerge back to front on their comparatively rapidly growing jaws. Concerning when the mammalian double-footed teeth evolved, Marlena Świło *et al.* (2014, 818-819) noted "*Brasilodon* and *Brasilitherium* have teeth that show root with a groove, which could possibly be an early stage of a double-rooted tooth formation."

Meanwhile, the genetic side of tooth production in vertebrates had progressed as well, surveyed by John Whitlock & Joy Richman (2013). Many of the specific triggers have yet to be identified, but among those found so far are our old pals sonic Hedgehog (*Shh*), *Bmp* and *Wnt*, just as busy in tooth production as they were in hair and feather formation. At the deepest level, mammals and reptiles generate teeth from the same dental lamina, "the ingrowth of oreal epithelium into the facial mesenchyme."

"The appearance of this primary mammalian dental lamina is virtually indistinguishable from the snakes and geckos that we have reconstructed," Whitlock & Richman (2013, 67) reported. "However, unlike mammals, the continuous dental lamina surrounding the jaw in reptiles persists into post-hatching animals." This is true of us humans too, though interestingly the rodent lineage shows more derived variation on this than we do, breaking down their lamina into discontinuous placodes. As a result, "It may be surprising to learn that the primary dental lamina of humans most closely resembles that of the reptile and not that of the mouse."

As we've seen, although the fossil data suggested there was no fixed form as to how tooth replacements appeared (alternating or sequential), there was some connection with jaw growth. And newer experiments with chameleons (that produce teeth sequentially but don't replace them) have supported that relationship by revealing "that areas of tooth initiation are not orderly in the anterior-posterior direction but instead seem to correlate with regions of increased jaw growth. In other words, when space is available, additional teeth form," Whitlock & Richman (2013, 68) citing the primary work by Marcela Buchtová *et al.* (2013).

Bmp and *Wnt* both play antagonistic roles in generating a "zone of inhibition" that can curb tooth replacement, ending up in a highly tissue-specific "dialogue between epithelium and mesenchyme which regulates tooth number," Whitlock & Richman (2013, 69). So it's looking like the transition from polyphyodonty to diphyodonty in mammals several hundred million years ago involved the changed fate of the dental lamina, and working out just what happened there on the genetic side is where the scientific train was headed at the time the many science researchers at *ICR* and *AiG* (and web pundits like *Truth in Science*) were not doing any of the work or paying attention to any of those who were.

Charles Creager depends too highly on the *kind*ness of *Talk.Origins*

It's a prime presumption of creationism that there are immutable categories of life. The "kind" is their baby, part of their own theoretical framework. So it is a fair measure of their scientific dedication to see how much trouble they've taken to assemble an authoritative listing of just how many created "kinds" there are. You'd think they'd have every reason to do so, if only for their own education and understanding.

If creationists were in need of models to emulate, evolutionary science offers a wide sampling. Lynn Margulis & Karlene Schwartz (1988; 1998) listed all the 35 known animal phyla in their *Five Kingdoms* work, along with many genera illustrative of each. In *Evolution of the Vertebrates*, Edwin Colbert & Michael Morales (1991, 428-437) laid out all the 181 vertebrate orders (fossil and living) falling under our own phylum *Chordata*. And from my favored niche down dinosaur way, in that *Dinosaur Data Book* Ray Comfort and his friends had skipped past so lightly back in Chapter 9, David Lambert & The Diagram Group (1990, 107-171) described 51 dinosaur families and what subsidiary genera were then known to fall within them. Even when a rival scientific taxonomy was being offered, like the reclassification of predatory dinosaurs proposed by Gregory Paul (1988), you got *more* terminological detail, not less.

So it is quite revealing that there exists no thorough analysis of the kinds within the creationist literature. We've already encountered the baraminology stabs at the field back in Chapter 6, and how the "monobaramin tag" disguised just how often their own work only backhandedly confirmed the evolutionary relationships their dogma insisted couldn't be real. That's because all of life is one gigantic *monobaramin*, connected through and through by natural branching common descent.

Moving onto the RMT field, it's still more revealing that the first (and so far *only*) effort to address anything like the full sweep of the data occurred in 2003 (with an equally disinformative string of revisions in 2008) by only a minor player in the creationism field, Charles Creager. And it was all decidedly half-baked.

Armed with a BS in physics earned from the creationist Bob Jones University in 1989, Creager went on to start his *Genesis Mission* website in 2000. He's also functioned as an administrator and contributor to *CreationWiki*, which should give some idea both of his *Kulturkampf* ideology and his practical dedication to scholarly method.

Creager (2003) crossed my RMT scope when he took on the animals listed as RMT transitional forms by Kathleen Hunt (1997), part of an extensive series on intermediate fossils at the venerable *Talk.Origins Archive*. So Creager was not organizing a list of the relevant characters *himself*, but simply riffing off the substantive research of somebody else. We've seen such opportunism before, especially with Duane Gish expanding his apologetic kit bag based on the technical papers thrown at him by his many critics. And we saw how well that went with his misrepresentations of the 1975 Allin paper.

Things got even more interesting, though, when we realize that Creager's original posting was not retained in later iterations of *Genesis Mission*. Instead, a revised version of it eventually appeared without attribution at *CreationWiki* in 2008, but its content (including exact phrasing and references to "I") made it likely this was his own revamp, and I am listing it as Creager (2008) as a result. What you'll be reading here is based on the hard copy of the 2003 version that I printed out while it was still in existence in cyberspace, with references to the Creager (2008) revision as needed.

Antievolutionism is a very *reactive* enterprise, more prone to respond to things than doing original research. And although Hunt's listing of forty taxa was up to date for the time and was still fairly current when Creager compiled his criticism, it was also the case that back in 2003 finding primary technical sources online was far more challenging than today. So my analysis is less of a "how come you didn't think to look" critique (though that will crop up on occasion anyway), than taking note of what Creager was proposing by way of *creationist* explanation, spotting what information he decided not to take note of that was certainly known to him via Hunt's secondary coverage, and checking whether his stray functional predictions about what "kinds" were involved held up based on what data came on the scene later, or even by the time of his own 2008 reboot.

Because there were so many taxa to cover, though, what Creager attempted to do (focusing on all the "gaps" and "differences" he encountered or imagined) ended up a particularly vivid illustration of the perils of *Zeno Slicing*, since we get to see all his equivocations step by step.

Hunt (1997) unequivocally laid down the RMT gauntlet (her *italics*):

> This is the best-documented transition between vertebrate classes. So far this series is known only as a series of genera or families; the transitions from species to species are *not* known. But the family sequence is quite complete. Each group is clearly related to both the group that came before, and the group that came after, and yet the sequence is so long that the fossils at the end are astoundingly different from those at the beginning.

Creager naturally hit on that absence of species links (hardly a surprise given the scattershot nature of small vertebrate preservation stretching over a hundred and fifty million years), though that lack of fossils for particular species-species transitions turned out not to be entirely the case, as we'll see. But Hunt was still fully on target concerning the overall arc of the dataset. She cited mainly general reviews, including "Kemp's more detailed but older book (1982)," but only occasionally offered direct primary source technical citations. Which meant that to investigate the background context of her examples would have required doing some of their own legwork. Creager was not up to that effort.

Hunt began with critters found in a Carboniferous deposit in Nova Scotia. **(1)** *Paleothyris* was "An early captorhinomorph reptile, with no temporal fenestrae at all," while the "earliest known synapsid" **(2)** *Protoclepsydrops haplous* showed just the beginnings of those temporal fenestra (our mammalian cheekbones) while still featuring "amphibian-type vertebrae with tiny neural processes."

So we have two very "reptilian" forms from the same time frame and location, but one showing a first trace (the skull opening) of the many diagnostic changes required to lead to mammals. The challenges of trying to make sense of so limited a fossil distribution can be seen comparing what little Romer & Price (1940) could say of then—that would be the paper Duane Gish bounced off of back in Chapter 4—with the careful progress of R. Paton *et al.* (1999) and Timothy Smithson *et al.* (2012). *Paleothyris* is known from a fairly complete skeleton, illustrated in its slab by *Prehistoric Life* (2009, 168), but the fragmentary nature of *Protoclepsydrops* ("one incomplete partial skeleton and skull") was still regarded as limiting its taxonomical utility by Michael Benton & Philip Donoghue (2007, 39).

Creager mentioned *none* of the anatomical details Hunt had highlighted, though, when he flatly declared "There are clear differences between Protocelpsydrops and Paleothyris. It would be quite a jump between the two, without any intermediates, but there are none, not even a gap." So right off the bat we got a *Bermuda Triangle Defense* and *No Cousins* double play, but with the absence of the factual bat in his hand when he swung, Creager struck out.

Hunt's next pitch was another Carboniferous Nova Scotia synapsid, **(3)** *Clepsydrops*, noting only that it was the first of the most primitive pelycosaurs, the "ophiacodonts." That particular genus was known just by an isolated femur (and some 19th century finds initially given that name turned out to be species of our old friend *Dimetrodon*), but there are fuller skeletons available if you included the rest of its family. One of the largest of the early synapsids was *Ophiacodon* (about 3 meters in length), living from the Late Carboniferous into the Early Permian, resembling a stout crocodile in the *Prehistoric Life* (2009, 168-169) illustration. Further members of the group have continued to turn up, such as the one described by David Berman *et al.* (2013), and more recent forensic analysis of the bone structure of *Clepsydrops*' femur by Michel Laurin & Vivian de Buffrénil (2016) helped trace the synapsids' adaptive move to a terrestrial habitat.

Back in 2003, though, Creager "could only find a web site that had a reference to a fossil Femur but no picture. The page is now gone but one had to wonder if they have any more of it." Unwilling to venture beyond the Web, though, Creager then replayed his *No Cousins* move: "It should also be noted that Clepsydrops would be contemporary with Protoclepsydrops and Paleothyris on the evolutionary time scale, so how could any of them be ancestor to the other."

(4) *Archaeothyris* was a fourth Nova Scotia beast on Hunt's list:

> A slightly later ophiacodont. Small temporal fenestra, now with some reduced bones (supratemporal). Braincase still just loosely attached to skull. Slight hint of different tooth types. Still has some extremely primitive, amphibian/captorhinid features in the jaw, foot, and skull. Limbs, posture, etc. typically reptilian, though the ilium (major hip bone) was slightly enlarged.

In other words, about as gently incremental a variation on the "reptile" cousins known from that time as one might want, which is why *Archaeothyris* was held to have more diagnostic utility by Benton & Donoghue (2007, 39).

Unless they were literally the only examples of their "kind," each one of them presumably had ancestral lineages *within that kind* stretching farther back into the past. But remember, being a Young Earth creationist, there was likely not much of a "back into the past" in any *Map of Time* conception Creager might have had. Though we could only know of any of that for sure if Creager stopped to explain something about it, which he didn't.

Instead of recognizing the important anatomical variety caught in those ancient Carboniferous fossil snapshots, and taking a swing at those facts from his creationist perspective, Creager stamped his foot at the plate:

> Not only would Archaeothyris be contemporary with Clepsydrops, Protoclepsydrops and Paleothyris on the evolutionary time scale, while out lasting the others. There is not evidence of a trend in the few areas the [*sic*] select for comparison. The fact that all four of the kinds are found in the same rocks, and the lack of any general trend, places a high degree of doubt on any claim of a transition among them.

Here it sounded like Creager thought those four animals represented separate *kinds*. And yet all were very similar reptilian tetrapods, except for the slight but specific skull and anatomical changes in some of them that Hunt had noted but which Creager failed to mention (*Archaeothyris* is regarded as the first clear synapsid, supplanting the more fragmentary *Protoclepsydrops*). So, on what *baraminological* basis was he distinguishing them? And what exactly would a "general trend" have looked like at this stage in synapsid evolution, so that concept could be compared to what fossils were found, allowing us to decide whether any of them qualified? Creager didn't stop to say.

Once again, Creager was taking broad swings without having that factual bat in his hands. And it only got worse as he continued through Hunt's examples, which began to refer to more of the specialized synapsid branches that were coming on the scene.

(5) *Varanops* dated from the early Permian, and Hunt noted (my **bold**):

> Temporal fenestra **further enlarged**. Braincase floor shows first mammalian tendencies & **first signs** of stronger attachment to rest of skull (occiput **more strongly** attached). Lower jaw shows **first changes** in jaw musculature (**slight** coronoid eminence). Body narrower, deeper: vertebral column more strongly constructed. Ilium further enlarged, lower-limb musculature **starts to change** (prominent fourth trochanter on femur). This animal was more mobile and active. Too late to be a true ancestor, and must be a "cousin".

You can imagine what Creager focused in on. "Note the large differences with *Archaeothyris*. At least they are admitting that they don't have the actual ancestor, but its existence is assumed."

But what were these "large differences"? Every feature Hunt pointed out referred only to slight changes, not radical novelties, and Creager made no attempt to illustrate his hallucination by example. Moreover, *Varanops* had a temporal range known to span 25 million years at the time of Sean Modesto *et al.* (2001, 257), but a decade later the *Fossil Genie* had continued to be active, extending the varanopsid lineage in both directions by another 15 million years, Modesto *et al.* (2011, 1031), meaning there were members of that group living back into the later Carboniferous. Further Permian examples have been uncovered also, such as by Hillary Maddin *et al.* (2006), Jennifer Botha-Brink & Modesto (2007) and Nicolás Campione & Robert Reisz (2010). *Varanops* was typical of the group in resembling a small monitor lizard, as illustrated in *Prehistoric Life* (2009, 186-187). Interestingly, a specimen of *Varanops* has been found that was scavenged by a temnospondyl amphibian, specifically proven as one of its teeth broke off in the process and was still stuck in the bone, Robert Reisz & Linda Tsuji (2006). Rainer Schoch (2010; 2013; 2014) has reviewed temnospondyl relationships, including the role of heterochrony in their development and their shifts from aquatic to terrestrial habitats.

Recalling back from Chapter 3 how YECers are committed to exclusively herbivorous critters prior to the Flood (or Fall), the stream of new paleontological data on even this narrow slice of fossil life will never make much sense to those whose patchwork *Map of Time* was already so well padded by the *No Cousins* rule.

With the next critter, **(6)** *Haptodus* (which we encounted back in Chapters 8 & 9), Hunt shunted back to the Carboniferous to note its relation to a particular branch of the synapsid line (my **bold**):

> One of the first known sphenacodonts, showing the initiation of sphenacodont features while retaining many primitive features of the ophiacodonts. Occiput still more strongly attached to the braincase. Teeth become size-differentiated, with biggest teeth in canine region and fewer teeth overall. Stronger jaw muscles. Vertebrae parts & joints more mammalian. **Neural spines on vertebrae longer.** Hip strengthened by fusing to three sacral vertebrae instead of just two. Limbs very well developed.

One of the things that stands out about the sphenacodonts was the fairly conventional reptilian upward swooping curve of their lower jaw (a shallow version of the vertical coronoid process found in much later mammals), visible in various examples surveyed by Michel Laurin (1993, 201), Jörg Fröbisch *et al.* (2011, 116) or Frederik Spindler *et al.* (2015, 25), and quite unlike the comparatively flattened jaw form found among the varanoptids, which is among the suite of characteristics leading to their placement as close cousins but not quite in the same group.

That's of interest because Creager (who again declined to discuss any of their specific anatomy) could not get over the *Map of Time* glitch in his head, and stumbled on with the claim that "Haptodus and Varanops are similar enough that they are probably the same kind. Varanops however is evolutionarily dated as younger than Haptodus and so it can not be an ancestor. The simple fact is they don't have any evidence of such an ancestor."

So without batting an eyelash, Creager skipped past visible differences between the two groups (ones which in regular systematics put them on separate branching lineages), to clump them together as the same *kind*, even while not applying the same measuring stick to the less disparate *Archaeothyris* and *Haptodus*.

That Creager's typology was going to be both arbitary and no more useful than Michael Denton's ID version showed as he plowed through the next sphenacodonts, **(7)** *Dimetrodon* and **(8)** *Sphenacodon*. As Duane Gish had back in Chapter 4, Creager didn't want the finback *Dimetrodon* to share the stage with their close sphenacodont relatives: "Dimetrodon and Sphenacodon seem to be distinct kinds, with no relationship. Sphenacodon seems to be the same kind as Haptodus or Varanops. Also Dimetrodon and Sphenacodon would be contemporaries with both Haptodus or Varanops and as such they can not be ancestor or descendent."

As those four taxa sprawled across millions of years from the late Carboniferous into the early Permian, it "seems" Creager's idea of "contemporaries" was especially pliable, along with the snag that by clumping *Sphenacodon* into the same *kind* as *Haptodus* and *Varanops*, from a baraminological view it was impossible to preclude their natural evolutionary interrelationship as potential *monobaramins*. Could it be Creager had no idea what it meant to be a "kind" in the first place?

Hunt was putting up the various fossils to illustrate the incremental acquisition of mammalian traits as you moved forward in time, and with **(9)** the biarmosuchia (mentioned back in Chapters 8 & 9) we were into the earliest therocephalian therapsids. The biarmosuchids definitely got around (known from South Africa and Russia), though represented by less than a dozen notable genera from those two locations by the time of Roger Smith *et al.* (2006, 340). Hunt's accounting reminded how the biarmosuchia retained the preceding sphenacodontid features ("jaw muscles inside the skull, platelike occiput, palatal teeth") while continuing to enlarge what would become the full mammalian zygomatic arch.

Their bone morphing included (Hunt's *italics* but my **bold**) how the "Occipital plate slanted slightly backwards rather than forwards as in pelycosaurs, and attached still more strongly to the braincase. Upper jaw bone (maxillary) expanded to separate lacrymal from nasal bones, **intermediate between early reptiles and later mammals.** Still no secondary palate, *but* the vomer bones of the palate developed a backward extension below the palatine bones. **This is the first step toward a secondary palate, and with exactly the same pattern seen in cynodonts.**"

So much information to assess, and Creager so unwilling to do so, as his entire comment on the biarmosuchia was a *Bermuda Triangle Defense* move: "Where is the mid Permian? It look [*sic*] like they have an unacknowledged gap, and the differences between Biarmosuchia gives no evidence of a relationship with Dimetrodon, or Sphenacodon. They need to fill in the gap before make [*sic*] such a claim."

Whether Creager was being coy and knew paleontologists had dubbed the dearth of fossils in the Middle Permian "Olson's Gap," the transition from sphenacodont to therapsid still fell within it. Some of the issue turned on how rock formations were correlated, globally and locally, and that could get pretty involved, as reflected in Spencer Lucas (2004; 2005; 2013) contending with Michael Benton (2012). New deposits have broadened the Permian field as well, such as finds in Zambia, Megan Whitney & Christian Sidor (2016) and Adam Huttenlocker & Sidor (2016).

Specifically regarding the sphenacodont/therapsid transition, though, unbeknowst to Hunt's 1997 *Talk.Origins* post (and therefore Creager's clumsy secondary redaction of it in 2003, or his still lamer repetition of it in 2008), Michel Laurin & Robert Reisz (1990; 1996) had already plugged a part of "Olson's Gap" by their reevaluation of the skull of the Early Permian *Tetraceratops*, which had been found almost a century earlier in Texas, but had not been properly studied because it was so difficult to extract from its surrounding rock matrix.

It was worth the wait. Laurin & Reisz (1990, 250) explained, "As the most primitive known therapsid, *Tetraceratops* bridges the large morphological gap between primitive and advanced mammal-like reptiles. This makes the fossil record of synapsids the most extensive of any group of terrestrial vertebrates, extending without any significant break through the last 320 million years of Earth's history." And Laurin & Reisz (1996, 102) concluded their later more detailed accounting: "The presence of a therapsid in Lower Permian sediments of North America eliminates unequivocally the temporal and geographic gaps that have separated the two large synapsid assemblages that dominated the terrestrial scene during Paleozoic."

The intermediate character of *Tetraceratops* has not been undermined by subsequent work, where debate has turned on which side of the synapsid/therapsid line it belonged, not that its transitional features weren't straddling it, such as the treatments by J. Conrad & Christian Sidor (2001) or Eli Amson & Laurin (2011), with Sean Modesto *et al.* (2011, 1031) nesting *Tetraceratops* outside the pelycosaurs and sphenacodonts, and beside the rest of the therapsids. On top of that, by then the Fossil Genie had stepped in with *Raranimus dashankouensis* from the Middle Permian of China, filling in another part of "Olson's Gap" in that region with its "unique combination of therapsid and sphenacodontid features," Jun Liu *et al.* (2009).

It's interesting to compare Creager's creationist dodge ball (where he failed to specify exactly what would qualify in his eyes as "evidence of a relationship" with anything) with the detailed analysis of the evolution of the biarmosuchid nose and palate by Christian Sidor (2003a). Drawing on additional fossils that preserved the relevant areas in greater detail than in other examples (including the *vomerine process* we encountered back in

Chapter 8 regarding some of Woodmorappe's supposedly reversing character traits), such evidence supported the idea that secondary palates evolved independently in the anomodonts, therocephalians and cynodonts.

Now think about what that means. Whether or not they were designed or evolved, there would still be developmental processes and underlying genetics involved—that's how living things roll. To get the form we see in that extinct life, there must have been a biology and genetics that generated those exact layouts. The power of an evolutionary perspective is that the processes observable in living forms can offer direct clues to what was going in those extinct but still related ancestors. The weakness of a creationist perspective is that it doesn't occur to them to look.

A lot of the genetics of palate formation have only been identified recently, such as by Jeffrey Bush & Rulang Jiang (2012) and Jeong-Oh Shin *et al.* (2012). A great deal of plasticity in the secondary palate has been discovered experimentally, though, by Junko Okano *et al.* (2006) and Rachel Menegaz *et al.* (2009), meaning the biological components *can vary as independent units*. Does it then require some vast leap to realize similar variety couldn't have been avoided over the millions of years of synapsid evolution? Or to be all that surprised that variant forms of secondary palates managed to originate several times?

But the process wasn't willy-nilly. The interacting anatomy couldn't be prevented from playing a part. For example, the fact that mammals had developed a larger more movable tongue than their distant archosaur cousins had an impact on how the related components evolved among the synapsids. "Looking back on how the palatal bones originated from dermal bones in amphibians and early tetrapods, it seems actually logical that the palatal shelves grow horizontally in crocodilians in an oronasal cavity where the nonmuscular tongue is small. It seems that the development of a big muscular tongue in mammals became an obstacle to palatal shelve growth, forcing them to grow downward," Roger Jankowski (2013, 64).

Beyond the "gap" with the cynodonts

Pressing on with Hunt's *Talk.Origins* list, **(10)** *Procynosuchus* represented the first of the cynodonts. A measure of how paleontology has changed concerns how one of the early skulls found had been ground to bits back in the 1960s to provide the skull sections, but advances in 3D imaging allowed Julien Benoit & Sandra Jasinoski (2016) to reconstruct it from the original detailed section drawings. The genus is known by a fairly complete skeleton, fortunately, described at length by Tom Kemp (1979; 1980), and illustrated in *Prehistoric Life* (2009, 191). The skull and body of *Procynosuchus* offered a wealth of data that Hunt reported in her summary, showing just how far from those basal amniotes the therapsids had come:

> Probably arose from the therocephalians, judging from the distinctive secondary palate and numerous other skull characters. Enormous temporal

> fossae for very strong jaw muscles, formed by just one of the reptilian jaw muscles, which has now become the mammalian masseter. The large fossae is now bounded only by the thin zygomatic arch (cheekbone to you & me). Secondary palate now composed mainly of palatine bones (mammalian), rather than vomers and maxilla as in older forms; it's still only a partial bony palate (completed in life with soft tissue). Lower incisor teeth was reduced to four (per side), instead of the previous six (early mammals had three). Dentary now is 3/4 of lower jaw; the other bones are now a small complex near the jaw hinge. Jaw hinge still reptilian. Vertebral column starts to look mammalian: first two vertebrae modified for head movements, and lumbar vertebrae start to lose ribs, the first sign of functional division into thoracic and lumbar regions. Scapula beginning to change shape. Further enlargement of the ilium and reduction of the pubis in the hip. A diaphragm may have been present.

On that masseter muscle matter, by the way, recall from Chapter 4 the still earlier basal *Charassognathus* found by Botha *et al.* (2007) that not only nudged the cynodonts back another *million years*, filling in more of the "ghost range" the evolutionary model predicted for them, but showed the first signs of the distinctive masseter muscle layout. The Fossil Genie seemed particularly puckish this time, continuing to favor Darwin over creationist Design yet again.

Meanwhile, what were Creager's thoughts on the information Hunt had reported in 1997? In its entirety: "Procynosuchus show significant differences from Biarmosuchia. In fact the legs would be an evolutionary reversal. In no way does this suggest a relationship."

Say what?

I haven't a clue where Creager got that "legs would be an evolutionary reversal" idea. It certainly wasn't from reading Hunt (who didn't mention legs), nor from any relevant technical paper I could find. Though their skull diagnostics justifiably distinguished them as close cousins, overall *Biarmosuchus* and *Procynosuchus* were still physically very similar, long-tailed flat-footed animals capable of a wide splayed stance, with the latter group having more robust front legs and possibly a more agile swimmer, per the extensive analysis by Kemp (1980). In the paw department, there were minor variations in phalangeal number, *Biarmosuchus* having a 2-3-4-5-3 layout, while the procynosuchids lost one in their 2-3-4-4-3 configuration, surveyed by James Hopson (1995). Rubidge & Sidor (2001, 460) identified two particular procynosuchid novelties: their cervical vertebrae were longer, and the fourth and fifth tarsal bones were fused in their feet. All interesting features hinting at both genetic and selective/adaptive factors, just as the gradual reduction by fusion of ancestral cranial bones in synapsids surveyed by Christian Sidor (2001) implied a parallel alteration of their developmental gene regulation over those many millions of years.

Creager (2008) dropped the stupid claim about leg reversal, getting much longer but no wiser (my **bold**):

> The description seems to indicate that *Procynosuchus* has so many mammalian traits that it was probably a mammal. **Apparently some paleontologists think it was a mammal.** The reptilian features seem to be in the lower jaw so **most likely *Procynosuchus* was a mammal**. Furthermore, *Procynosuchus* is based on a sufficiently incomplete skeleton to raise some questions about the details of its reconstruction. It leaves plenty of room for evolutionary assumptions in its interpretation. And there is a sudden appearance of these mammalian traits; **it goes from reptiles with a few mammalian traits to mammals with a few reptilian traits, in one step**. This shows a real gap in time but in traits between mammals and reptiles.

Creager's ability to not notice the content of the previous string of animals (that showed the features were hardly arriving "in one step"), or identify who those paleontologists were who thought it "was a mammal," was far exceeded by his 2008 zeal to toss the still transitional *Procynosuchus* into that mammal basket, a taxonomical conclusion that had completely eluded him five years before!

Hunt's next example was another early cynodont **(11)** *Dvinia* from the late Permian that showed the "First signs of teeth that are more than simple stabbing points—cheek teeth develop a tiny cusp. The temporal fenestra increased still further. Various changes in the floor of the braincase; enlarged brain," along with the reduction of the rear jaw bones which we know would eventually be critical to the evolution of the mammalian middle ear.

Like *Charassognathus* mentioned above, *Dvinia* is known only from its skull, which is still the case per M. Ivakhnenko (2013), and Creager seized on that circumstance in 2003 to complain about a website illustration he'd found that showed it as a small furry critter. "Nice trick," Creager sneered, "It's amazing what evolutionists can discovered [*sic*] from a skull. This is like looking at someone's teeth to see if they are flat footed."

But unless the animal's skull floated around like the disembodied Cheshire Cat from *Alice in Wonderland*, there shall have been a full carcass attached to it when it was alive. And that would be just as true for any wonderfully designed *kinds*. So what would his creationist typology predict it ought to have looked like?

Here is where Creager really dropped the creationist conceptual ball (or forgot to bring it to the field in the first place). Even a non-evolutionist like Baron Cuvier back in the age of Napoleon could reasonably infer things about an animal's overall anatomy from isolated parts (even a tooth!) because in nature things don't happen in a muddle. Having a whole skull was a big advantage because it put the animal into relationship with similar ones for which the rest of the anatomy was known (like *Procynosuchus*), and it was perfectly reasonable for a scientific illustrator to draw on that full dataset to show *Dvinia* as it would have looked as a complete animal, attached to a body typical of its cynodont group.

It wouldn't have had a dinosaurian body, now would it, or an arthropodal insect one? No one was putting flippers on it, or conjecturing a finback sail. Was Creager thinking it shouldn't have had a generally cynodontian body, appropriate for its "kind"? Or had Creager no conception at all of what being a *kind* meant, even at the Crocoduck photoshop level?

With five years to think on it, Creager (2008) dropped the sarcasm on paleontological reconstruction, but summarily decided *Dvinia* (like *Procynosuchus*) was a mammal too. Which, if he'd stopped to ponder that, meant Creager had just attached a body to the floating skull that his 2003 self would have fumed over had any forensic artist decided to depict it at that time.

We've bumped into Hunt's next example **(12)** *Thrinaxodon* already, and Hunt had a long section on that early Triassic fossil, which included its expanding dentary bone with more mammalian tooth form, and the beginnings of the mammalian jaw hinge, involving "a ligamentous connection between the lower jaw and the squamosal bone of the skull." It should be noted that the move from Permian to Triassic meant getting past the Permian mass extinction, which certainly shook up life in general, and obviously had an impact on what was getting preserved and what new adaptive niches were opened up because so many of the former competitors in the food chain weren't there any more. For all that, dozens of *Thrinaxodon* specimens are available, and improved analytical techniques have come into play in the years since, such as Fernando Abdala *et al.* (2013) CT-scanning to see more of their tooth replacement patterns, and Sandra Jasinoski *et al.* (2015) identifying growth stages in their cranial shape.

One part of Hunt's *Thrinaxodon* section connected the data points particularly well (Hunt's *italics*):

> Nearly complete skeletons of these animals have been found curled up—a possible reaction to conserve heat, indicating possible endothermy? Adults and juveniles have been found together, possibly a sign of parental care. The specialization of the lumbar area (e.g. reduction of ribs) is indicative of the presence of a diaphragm, needed for higher O2 intake and homeothermy. NOTE on hearing: The eardrum had developed in the only place available for it—the *lower* jaw, right near the jaw hinge, supported by a wide prong (reflected lamina) of the angular bone.

By the way, *Thrinaxodon* is presented as having an active near-mammalian lifestyle in *Prehistoric Life* (2009, 221).

What did Creager have to say about what information Hunt had brought to the field at *Talk.Origins*? Just this: "While they have complete skeletons of this type, there are large differences with *Dvinia*, with no evidence of a trend." Still nothing from Creager about what those "large differences" might have been, though, but there was *definitely* a trend in Creager (2008), who decided *Thrinaxodon* had to also be a mammal.

Hunt's next examples were the Early Triassic **(13)** *Cynognathus* and **(14)** *Diademodon*. Both of these advanced cynodonts were initially known only from the same African deposits, so were obviously contemporary both in time and geographical range, though later they've been found to have spread into the then-adjoining South America, Fernando Abdala *et al.* (2005), Abdala &

Roger Smith (2009) and Agustín Martinelli *et al.* (2009).

As recounted by Hunt in 1997, *Cynognathus*' cusped cheek teeth "met in true occlusion for slicing up food," and were rooted as mammal teeth were "though with single roots." The double jaw joint was in place, though with the postdentary bones "reduced to a compound rod lying in a trough in the dentary, close to the middle ear. Ribs more mammalian. Scapula halfway to the mammalian condition. Limbs were held under body. There is possible evidence for fur in fossil pawprints." For those interested in how cynodont tooth occlusion worked, Michael Benton (2005, 293) offered a concise explanation, and *Prehistoric Life* (2009, 221) noted that "*Cynognathus* was probably an extreme carnivore and ate nothing but meat." So much for the YEC designed herbivory dogma.

Creager's 2003 response was a repeat of his all-purpose avoidance mantra: "The differences between Cynognathus and Thrinaxodon are significant with no evidence of a general trend. The incompleteness of Cynognathus' skeleton makes it even less supportive of evolution." Creager (2008) similarly contended "the differences between Thrinaxodon and Cynognathus show a significant difference in general body shape making any relationship doubtful without intermediates."

Yet again, what were these "differences" and what would true intermediates have to look like to be acceptable? Creager continued his own "general trend" by not saying.

Comparing the two animals should have taken note of their different size (think allometry again), as *Thrinaxodon* varied from 30-50 cm (under a couple of feet head to tail) while *Cynognathus* was roughly three times as large, meaning some of the latter's body and skull robustusness owed to its being a more dynamic predator capable of munching bigger dinners. But apart from being mammals of indeterminate kind, Creager's curiosity-averse creationism offered no insights about them. Meanwhile, regular evolutionary paleontology recognized *Thrinaxodon* and *Cynognathus* as adjoining branches on the cynodont line, charted by Jennifer Botha & Anusuya Chinsamy (2005, 391) and Botha *et al.* (2007, 486).

Hunt had noted yet more transitional features for *Diademodon*, which had developed a full mammalian secondary palate, but as yet "didn't extend quite as far back," and had the mammalian toe bone arrangement of 2-3-3-3-3, though "with closely related species still showing variable numbers." The mention of that variable toe layout (once again the sort of thing to be expected in natural variation at the species level) meant more than just skulls were known too, which made Creager's hastily spelled response particularly strained:

> I can't find any real evidence of any post cranial bones other than figments [*sic*], so as far as I can tell Diademodon is known only from it's skull and some fragmented bones from the rest of the body. So a proper comparison can not be made with Cynognathus or any other animal. Further more Diademodon and Cynognathus would be contemporaries on the

evolutionary time scale and so make any assigning of a ancestor / decedent [*sic*] relationship arbitrary at best.

Creager (2008) decided *Diademodon* was yet another mammal.

It's interesting to compare Creager's 2003 hand-waving and 2008 mammal-tagging with what the scientists down on the work floor were able to glean from the evidence that always remained blurry under the unfocused creationist lens. Analysis of the bones of *Diademodon* and *Cynognathus* by Botha & Chinsamy (2000) affirmed that their postcranial skeletons were "indistinguishable" except for "slight differences" in their "neural spines and centra of the vertebral column," but "that *Diademodon* had a cyclical growth strategy whereas *Cynognathus* grew continually at a constant rapid rate throughout the year," justifying their categorization as separate genera. The cranial variations in *Diademodon* related to its being an omnivore, unlike the carnivorous *Cynognathus*, and Botha *et al.* (2005) went on to study the isotope balances in their fossils. Based on what comparable isotopic preferences reflected in living animals, *Diademodon* apparently "fed in shadier, damper areas, was nocturnal and/or depended more directly on environmental water" than its close contemporary cousin *Cynognathus*. "This suggests that marked seasonality prevailed in the Karoo Basin during the Middle Triassic, and that *Diademodon* was sensitive to these differences."

Botha & Chinsamy were showing by their work exactly why evolutionary thinking has been so productive, and will not go away no matter how often antievolutionists proclaim it a "theory in crisis." And Charles Creager was showing by his lack of effort the true measure of creationist method. Yawn.

Never forget that every fossil represents a once living animal, which existed in an environment as rich and real as the one you the reader are in now. And that, step by careful inferential step, the science of paleontology has recovered ever more detailed snapshots of those ancient worlds, even as the Flood Geology album of life we've been getting with Creager was nothing but a string of blank pages, missing in particular that perennial issue: where were these synapsid "kinds" *now*, post-Flood?

Meanwhile, the fossil parade was still passing by.

With Hunt's **(15)** *Probelesodon* (also known as *Chiniquodon*, remember), the secondary palate was "longer, but still not complete. Teeth double-rooted, as in mammals. Nares separated. Second jaw joint stronger. Lumbar ribs totally lost; thoracic ribs more mammalian, vertebral connections very mammalian. Hip & femur more mammalian."

Fossils in that group have been studied more recently at length by Tom Kemp (2007c; 2009a). Their transitional hearing capacity couldn't develop into the full mammal mode until the secondary jaw articulation finally relieved stress on the old reptilian bones being drawn into hearing, and what could be gleaned from the skull about its evolving brain continued to suggest a two-stage process that by then was starting to make use of skull space being opened up as the jaw musculature continued to shrink among these small near mammals.

With these fossils, Creager took another painfully tentative stab at using a term that he should have been employing substantively all along: "Based on available evidence Probelesodon and Diademodon would seem to be the same kind of animal. More information on Diademodon would be needed to be certain." But as Creager couldn't even get a grip on the information Hunt had presented, it seems a lack of "more information" was not Creager's primary stumbling block in sporadically pigeonholing fossilized things as distinct "kinds."

And now Hunt arrived at **(16)** *Probainognathus*. Hunt may not have been aware of Robert Broom's literal prediction of it three quarters of a century before, and she noted only a perfunctory "Still two jaw joints" about the fossil. But she went on with "Still had cervical ribs & lumbar ribs, but they were very short. Reptilian 'costal plates' on thoracic ribs mostly lost. Mammalian #toe bones," and all of that should have been a clue to Creager that much more was known about *Probainognathus* than just its skull.

Yet all Creager could do was repeat his content-free mantra (my **bold**):

> The apparent lack of post cranial bones makes a real comparison with Probelesodon **impossible**, but the differences in the skulls suggests that they are **different kinds**. The difference in the skulls do not provide any evidence of a relationship. Also both Probelesodon and Probainognathus are classified as mid Triassic making them contemporary on the evolutionary time scale and as such they can not be objectively classified as ancestor and descedent.

Hunt's example **(17)** *Exaeretodon* represented a siding in the mammal evolution story, the traversodonts, and Creager (2003) shunted past the anatomy to quote how she ended her description: "More mammalian hip related to having limbs under the body. Possibly the first steps toward coupling of locomotion & breathing. This is probably a 'cousin' fossil not directly ancestral, as it has several new but non-mammalian teeth traits."

"Note: they know they don't have an actual ancestor," Creager crowed in full *No Cousins* mode, then doubled down on his misunderstanding about *Probainognathus*' postcranial fossils for another foray on *kinds* (again my **bold**):

> Exaeretodon is clearly different kinds of animals from both Probelesodon and Probainognathus. The **lack of post cranial bones** for Probainognathus makes a real comparison **impossible**. Also Exaeretodon, Probelesodon and Probainognathus are classified as mid Triassic making them contemporary on the evolutionary time scale and as such they can not be objectively classified as ancestor and descendent; though the fact that Exaeretodon is also classified as late Triassic does support putting it last.

What a perfunctory nod at a chronological reality Creager showed no curiosity to explore. By the trend of his 2008 argument, Creager should have tagged *Probelesodon*, *Probainognathus* and *Exaeretodon* as "mammals," but perhaps that stretched a mite too far for what he had tagged as "different kinds" five years earlier, given that they still showed those transitional jaw layouts he just couldn't get around to taking note of, and Creager (2008) tersely decided the three animals were "contemporaries" that showed "no objective evidence of descent here."

Some of Creager's problem may have stemmed from his taking the occasional dates Hunt put on the field (such as "239Ma" for *Exaeretodon*) rather too literally, as though they were single stickpins spotted on a very small time chart. But even subsets of a geological age still meant a *lot* of time, and the offshoot traversodonts were known to have had a broad range. Found initially in South America, they turned up eastward in India and Madagascar, and northward into North America and Europe, Hans-Dieter Sues *et al.* (1999), James Hopson & Hans-Dieter Sues (2006), Jun Liu (2007), Christian Kammerer *et al.* (2012) and Jun Liu & Fernando Abdala (2014). In just the North American branch, Jessica Whiteside *et al.* (2011, 8973) graphed their presence across some *15 million years* (that's three times as long as it took for basal hominids to diversify into specialized bipeds like Donald Trump).

So it was interesting to see how Creager had pegged his stickpins in 2003, arbitrarily separating *Cynognathus* and *Diademodon* (which he concluded were just mammals in 2008) while clumping *Diademodon* with *Probelesodon/Chiniquodon* (which he didn't claim as mammals in 2008) yet excluding *Probainognathus*, and then leaving *Exaeretodon* out in the typological cold as yet another separate "kind." And all this without discussing anything substantive about them, in 2003 or 2008.

Now compare that with working paleontologists who did indulge in that "paying attention to the data" thing, such as Téo Oliveira *et al.* (2009, 114). Their "simplified phylogeny of the Cynodontia" covered 17 taxa, and by their cladistic analysis *Diademodon* and *Exaeretodon* were not apart, but nested together as the more derived relatives of *Cynognathus* within the Cynognathia node, while *Chiniquodon* was not placed within *Diademodon*'s group at all, but fell over on the adjoining Probainognathia node, as a more basal member just ahead of *Probainognathus*.

And when Agustín Martinelli *et al.* (2016) published work on more recently discovered probainognathids, Ken Ham showed his dedication to curiosity by sarcastically Tweeting on it several times—though not on the open access paper itself, but only on a secondary October 5th *LiveScience* weblink to it. Titled "Meet Granddad: Weird, Ancient Reptile Gave Rise to Mammals," *LiveScience*'s bouncy headline provoked the creationist's repeated sarcasm the following day: "Call a 'Weird' reptile 'Granddad' and that's considered science. Say God created us and that's considered ridiculous." And then a few minutes later: "Calling a reptile 'Granddad' illustrates evolution is religion. It's worship of the Creature instead of the Creator."

This was literally the only time in all these decades that I've spotted Ham coming that close to a technical work on the RMT (recall his collaborative reliance on Duane Gish's secondary authority in *The Revised & Expanded Answers Book*, noted back in Chapter 7), and that's about as deep as he was going to get on it. Which was no deeper, after all, than what Creager was up to, just mercifully shorter.

Once more into the "gap" with Charles Creager: in search of T1 & T2

Although we're still a long way from the mammal finish line, Hunt highlighted that there was a "GAP of about 30 my in the late Triassic, from about 239-208 Ma" during which "Only one early mammal fossil is known," after which our old pals the T1 trithelodontids and T2 tritylodontids showed up (and therefore bumping into the issue of how these last *almosts* were related to the stem mammals). "Bear in mind that both these groups were almost fully mammalian in every feature," Hunt cautioned, "lacking only the final changes in the jaw joint and middle ear."

But it was that "gap" remark that was all the red meat Creager wanted in 2003, who proclaimed "It is interesting that this gap is right were [*sic*] a creationist would expect to find it. Right between reptiles and mammals." Off on his later mammal-tagging jag, though, Creager (2008) restated his view of the "gap" as (my **bold**) "This shows that rather than being an **actual transition**, these are just **mammals with a few reptilian traits.**"

Creager was mired in that core failure of all antievolutionary apologetics. How exactly could "an actual transition" (which he refused to define) be one without having members toward the end that were almost mammalian while retaining the last ancestral reptilian traits? Or fail to touch base earlier exactly where *Probainognathus* had 234 million years ago, smack in the middle of that "gap," an intermediate form no creationist ever anticipated.

What we were hearing in the five years between Creager's 2003 argument and his 2008 version was the laborious grunting of necessary goalpost moving, a deep *gap* of their own that creationists were never going to get ahead of. And something else. While Hunt had not mentioned that there was a mass extinction at the end of the Triassic, it was still the case that there was one, and that shake up needed to be taken note of too.

Though there were a few impact splats during the Triassic, they appear not to have played much of a role in the extinction event, covered by Gordon Walkden *et al.* (2002), Martin Schmieder *et al.* (2010) and Tetsuji Onoue *et al.* (2012). The bigger culprit appears to have been the volcanism in the *Central Atlantic Magmatic Province* (CAMP) that was beginning to split up the Pangea supercontinent (destablizing the ecosystem much as the Siberian Traps had during the Permian, and the Deccan Traps would in the Cretaceous), explored by Andrea Marzoli *et al.* (1999), Blair Schoene *et al.* (2010; 2015), Jessica Whiteside *et al.* (2010) and Terrence Blackburn *et al.* (2013).

Whatever the many causes, there is a biogeographical pattern to mass extinctions that antievolutionists (either YECists or more general IDers) are conspicuously slow to appreciate. The tumultuous removal of large swaths of life generates ecological holes that offer opportunities for new adaptive relationships to form, but because evolution is neither directed nor planned, it can take many millions of years before things rebound, as adaptive selection depends entirely on what natural mutations happen to come up in whatever lineages managed to slip through the extinction sieve. Works reflecting this issue would include Christine Janis (1993, 170-185, 194), David Raup (1994, 6761), Douglas Erwin (1998; 2001), David Jablonski (1998; 2001; 2002), James Kirchner (2002), Josef Uyeda *et al.* (2011) and Randall Irmis & Jessica Whiteside (2012).

One might then ask whether an Intelligent Designer (especially an omnipotent one, with no budget overhead to fret over) should be so oddly reluctant to refill the emptied ecospace. Or whether it was even more peculiar for five of the carefully designed animal theme parks to fall apart in the first place. Kenneth Miller, (1999, 102) offered a succint analogy, envisaging a zoo's "lion and the lamb" exhibit requiring a continual supply of replacement lambs. But whoever was running the real zoological gardens of the past seemed slow on replacing either lions or lambs. Creager's failure to pay attention to such factors regarding fossil preservation and turnover in Deep Time when bringing up the "gap" (along with avoiding explicitly defending any Flood Geology hypothesis had he wanted to float that leaky boat) represented yet another grand play of the *Bermuda Triangle Defense*.

And then there's the Fossil Genie, for whom *gaps* so often seem more an invitation than a barrier.

You may recall back in Chapter 3 the transitional mammaliaform *Haramiyavia* reviewed by Zhe-Xi Luo & Stephen Gatesy *et al.* (2015). Dating back in the Triassic to 210 mya, that pushed the base of the known mammals back some 20 million years, charted by Shandong Bi *et al.* (2014, 582), and mammals of some form left their prints in China during that time too, Lida Xing *et al.* (2013, 191-192). Fragmentary teeth and jaws of the early mammal *Morganucodon* (which was known only from the Jurassic when Hunt was writing in 1997) have also been found in Triassic deposits in Germany and France, Spencer Lucas *et al.* (2001) and Maxime Debuysschere *et al.* (2015). And a morganucodont tooth showing "transitional morphology between advanced cynodonts and mammaliaforms" has turned up in Triassic beds in Greenland, Marlena Świło *et al.* (2014).

Looking at the photographs accompanying the coverage of the Greenland expedition by Tomasz Sulej *et al.* (2014), one must also give a nod of appreciation at the stamina and dedication required to dig in so bleak a landscape (something more antievolutionists might take a pick at now and then, to earn their field cred by direct effort rather than quote mining from their den).

Hopping past the "gap" then, Hunt's *Talk.Origins* post continued to explore the last phases of the RMT, which Creager danced around as vigorously as he had the first part.

Hunt's **(18)** *Oligokyphus* and **(19)** *Kayentatherium* were T2 tritylodontids, and she accurately reported their many features that put them so close to the main mammal track, and yet with specializations (especially their teeth) that suggested they were more likely offshoots (and which further paleontological study tended to support). Creager (2003) had not quoted any of Hunt's commentary on the two fossils, but Creager (2008) decided to extract just a few sentences from it, which again I've put in **bold** (the *italics* are Hunt's):

> Face more mammalian, with changes around eyesocket and cheekbone. Full bony secondary palate. Alternate tooth replacement with double-rooted cheek teeth, but *without* mammalian-style tooth occlusion (which some earlier cynodonts already had). Skeleton strikingly like egg-laying mammals (monotremes). Double jaw joint. More flexible neck, with mammalian atlas & axis and double occipital condyle. Tail vertebrae simpler, like mammals. Scapula is now substantially mammalian, and the forelimb is carried directly under the body. Various changes in the pelvis bones and hind limb muscles; this animal's limb musculature and locomotion were virtually fully mammalian. **Probably cousin fossils (?), with *Oligokyphus* being more primitive than *Kayentatherium*. Thought to have diverged from the trithelodontids during that gap in the late Triassic. There is disagreement about whether the tritylodontids were ancestral to mammals (presumably during the late Triassic gap) or whether they are a specialized offshoot group not directly ancestral to mammals.**

Creager's 2003 mantra had droned how "The gap between the reptiles to mammals shows that there is no clear evidence for a relationship, nor is there any real evidence for a relationship between Oligokyphus and Kayentatherium." Creager (2008) was just as brief in his new mammal trend: "Note that even Talk.Origins doesn't think that Oligokyphus and Kayentatherium are mammal ancestors. These are clearly 100 percent mammals with no evidence of reptilian traits."

Besides missing that intermediate *double jaw joint* Hunt noted, Creager's *No Cousins* play was stumbling past the same issue of transitional growth rate in *Oligokyphus* in the T2s (and *Sinoconodon* too) that *Truth in Science* tripped over earlier. As for *Kayentatherium*, Hans-Dieter Sues & Farish Jenkins (2006) continued the science process of actually trying to understand the animal, revealing how its robust anatomy was more geared for strong swimming than burrowing.

Hunt's next samples, **(20)** *Pachygenelus* and **(21)** *Diarthrognathus* (Broom's predicted form again), were Early Jurassic representatives of the T1 trithelodontids. She noted that the new fossil found by Neil Shubin *et al.* (1991) showed "that these animals are very close to the ancestry of mammals." There were more signs of the mammalian double-rooted teeth, but their ancestral reptilian jaw joint was "still present but functioning almost entirely in hearing," with the "postdentary bones further reduced to tiny rod of bones in jaw near middle ear."

Creager (2003) evaded all that by declaring how he could "not find any evidence of post cranial bones, so while the skull may seem transitional the rest of the animal may disprove the claim." Creager was incapable of specifying what "transitional" should look like, of course, but the partial skeleton of *Pachygenelus* described by Chris Gow (2001) further supported "the sister group status of tritheledontids and mammals previously determined from cranial characters" studied earlier in work like Zhe-Xi Luo (1994).

Given his lack of effort to assess any of the technical literature relevant to the RMT, it was unlikely Creager (2008) was any more aware of that data, but by then he preferred to quote the last of Hunt's remarks on those fossils: "These are probably 'cousin' fossils, not directly ancestral (the true ancestor is thought to have occurred during that late Triassic gap). *Pachygenelus* is pretty close, though." *Close* is never allowed under the *No Cousins* rule, of course, though Creager had to again step right over the animals' vestigial double jaw joint (that Hunt had noted) to conclude "These are 100 percent mammals with no evidence of reptilian traits."

Hunt's *Talk.Origins* post then jumped backwards to the Late Triassic (225 mya) with **(22)** *Adelobasileus cromptoni*, "A recently discovered fossil proto-mammal from right in the middle of that late Triassic gap! Currently the oldest known 'mammal.'" Hunt quoted from the abstract of Spencer Lucas & Zhe-Xi- Luo (1993) that "Some cranial features of *Adelobasileus*, such as the incipient promontorium housing the cochlea, represent an intermediate stage of the character transformation from non-mammalian cynodonts to Liassic [the earliest epoch of the Jurassic] mammals."

None of that information made it onto Creager's 2003 scope, where he reflected the limits of web searching in those days when he confessed his inability to "find any images, fossils or drawings. It is referred to in several places, but very little real information on this type. However it can not be a desendent [*sic*] of Oligokyphus or Kayentatherium, since according to evoultionary [*sic*] dating methods Adelobasileus is considered to be older than Pachygenelus and Diarthrognathus."

As the years rolled on, though, the evidence and Internet improved, accumulating much new information on the fine point anatomical details that the T1 trithelodontids and T2 tritylodontids were tracing in these late Triassic almost mammals, such as the fluctuations in the interpterygoid openings covered by Agustín Martinelli & Guillermo Rougier (2007, 450-453). Although the T1s had developed a dentary/squamosal contact by that time, they retained that primitive interpterygoid opening "that later in mammaliaforms and basal mammals will be occupied by the sphenoid recess (sinus)." The more fossils that were found, the more things looked just like the kind of natural variation that would be expected in lineages evolving over those millions of years, nursed along now and then by that so often obliging Fossil Genie.

But all the later Creager (2008) wanted to highlight was one quote from Hunt's 1997 conclusion (Hunt's *italics*): "Also note that this fossil dates from slightly *before* the known tritylodonts and trithelodonts, though it has long been suspected that tritilodonts and trithelodonts were already around by then. *Adelobasileus* is thought to have split off from either a trityl. or a trithel., and is either identical to or closely related to the common ancestor of all mammals."

Creager then harrumphed (my **bold**), "*Adelobasileus cromptoni* is dated as nearly 20 MA older than *Pachygenelus*, *Diarthrognathusm* [*sic*], *Oligokyphus*, and *Kayentatherium*. So *Pachygenelus*, *Diarthrognathusm* [*sic*], *Oligokyphus*, and *Kayentatherium* cannot be the ancestors of *Adelobasieus cromptoni* and other mammals. This helps to show that the entire series **is a sham**. They claimed a gap, but it has [*sic*] a full mammal whose placement there simply does not fit the theory."

This also "helps to show" that while his fingers tended to slip to *m* while doing commas (the same thing happens to me), Creager wasn't quite observant enough in proofreading to correct the slip whilst cut-and-pasting the taxa strings into his text. Nor was Creager apparently aware that the paleontology had moved on since 2003, as subsequent fossil finds by Agustín Martinelli *et al.* (2005) and Christian Sidor & John Hancox (2006) had pushed the branch with the T1 Ictidosauria/tritheledontids back some twenty million years into the later Triassic, charted by Martinelli & Rougier (2007, 454), exactly as Hunt's *Talk.Origins* piece had anticipated back in 1997.

Though Hunt didn't allude to it, it was interesting that *Adelobasileus* fell *after* both the T1 & T2 nodes cladistically in Lucas & Luo (1993, 326), no matter how those two contentious groups were ranked. That was still how they were falling as of Michael Woodburne *et al.* (2003, 365-367), and the more recent analyses of Jun Liu & Paul Olsen (2010, 159), Marina Soares *et al.* (2011, 332) and Marcello Ruta *et al.* (2013) continued to root *Adelobasileus* and its close kin in a thicket of similar T1 and T2 taxa right at the base of the node leading off to *Sinoconodon* and *Morganucodon* in the Mammalia.

Hunt's next example **(23)** *Sinoconodon* is a familiar critter from the previous two chapters, but more significantly regarding *Truth in Science*'s tapdance on tooth formation previously discussed. Teetering on the mammal side of the taxonomy, its dentary/squamosal jaw joint was strong, but Hunt reminded (her *italics*): "Reptilian jaw joint *still* present, though tiny." As usual, Creager (2003) didn't comment on that, but did repeat his mantra concerning his inability to "find any evidence of post cranial bones, so while the skull may seem to be transitional the rest of the animal may disprove the claim."

Five years later, the Fossil Genie had not supplied any fresh postcranial skeletons for Creager to sweep under his typological rug (and the situation remains the same today), but Creager (2008) was on another tack anyway: "Talk.Origins claims that *Sinoconodon* had a tiny reptilian jaw joint but no independent information is available on this—not even a picture—so it is possible that this is an interpretation of a feature that has nothing to do with a jaw joint."

This was about as close as Creager got to addressing that fundamental issue that had been literally staring him in the face with all the trail of fossils he'd been wandering past: the old reptilian jaw joint had been joined by the new mammalian one in the Probainognathids, and had reached the state in early mammaliaforms like *Sinoconodon* where the new one was now primary, leaving the old reptilian connection barely in place even as it was increasingly involved in sound conduction. *Sinoconodon*'s bone layouts had certainly not changed since Alfred Crompton & Ai-Lin Sun (1985) first described them, and for those so inclined to look, there's a conveniently color-coded illustration of their configuration in Luo (2011, 358-359). Incidentally, *Prehistoric Life* (2009, 279) noted the palm sized *Sinoconodon* was "one of the largest primitive mammals found to date."

We met Hunt's next critter **(24)** *Kuehneotherium* back in Chapter 4. Hunt described it (my **bold**) as "A slightly later proto-mammal, sometimes considered the first known pantothere (primitive placental-type mammal). Teeth and skull like a placental mammal. The three major cusps on the upper & lower molars were rotated to form interlocking shearing triangles as in the more advanced placental mammals & marsupials. **Still has a double jaw joint, though**."

Creager (2003) ran his standard play (my **bold**): "The fact is that not only are there no known post cranial bones for Kuehneotherium, but not even a complete skull. This makes it's use as evidence for evolution **a joke**, since there is no telling what the rest of the animal looks like." By the time Creager (2008) returned to the topic, though, it may have sunk in that Hunt was focusing on the animal's *teeth* for a reason (they were showing the distinctive triple crowned cusps known in early mammals), and amended his argument thus: "The fact is that not only are there no known post-cranial bones for Kuehneotherium, but there is only a fragmentary jaw and some teeth. This makes its use for evolution a joke. Its placement is based on the fact that its teeth are similar to what evolutionists think mammal teeth evolved from."

As for the fossil record of *Kuehneotherium*, its worth noting that both versions of Creager offered only one source, an older set of Toronto University college course notes posted by Robert Reisz (1993) that had noted how it was only known from a deposit in Wales. The original 1954 fossil find was indeed from Wales, with a formal description by Doris Kermack *et al.* (1968), but so dated a summary obviously couldn't reflect findings that had come along after 1993. But just as *Explore Evolutio*n and Carl Kerby had with their topics, five more years of research slipped by while Creager (2008) just copied the old weblink—and anyone trying to follow that link currently would reach a dead end, as the university file address had changed in the years since (the domain in my bibliography reflects its current cyberspace location). I also checked with Reisz by email and confirmed the likely date for his old class summary was circa 1993.

Isolated kuehneotherid teeth had been found in other European locales, notably in France by Pascal Godefroit & Denise Sigogneau-Russell (1999), and Maxime Debuysschere (2016 in press) has done a comprehensive survey of the available teeth. But the fact remained that the original *Kuehneotherium* fossils still included enough of the jaw to deduce a lot more than the "joke" Creager was allowing under his creationist prohibitions. Armed with new CT scanning tools and the experience of how tooth wear related to diet, Pamela Gill *et al.* (2014) would determine that *Kuehneotherium* was more a lighter moth-munching insectivore than its relative *Morganucodon*, capable of crunching harder covered beetles. And unlike the fictional constraints of Flood Geology, the actual paleogeographical distribution of those animals suggested they were pretty resilient in their niches, both weathering the Triassic extinction event without much of a bump, Debuysschere *et al.* (2015) and Debuysschere (2016 in press).

With Hunt's next examples, all dated by her to the Early Jurassic at 205 mya, **(25)** *Eozostrodon*, **(26)** *Morganucodon*, and **(27)** *Haldanodon*, we had a "very mammalian" bunch, especially regarding their teeth featuring those "Triangular-cusped molars." Still Hunt reminded they possessed a "Tiny remnant of the reptilian jaw joint. Once thought to be ancestral to monotremes only, but now thought to be ancestral to all three groups of modern mammals—monotremes, marsupials, and placentals."

That was a specific genealogical relationship that ought to have been commented on and contrasted by Creager's creationist typology. But all Creager (2003) could say (my **bold**) was that they "seem to be the same kind of animal, could be a [*sic*] **varieties of mouse**. They are also dated by evolutionary methods as contemporary with Kuehneotherium. As such placing Kuehneotherium first is arbitrary and based entirely on evolutionary assumptions." Creager (2008) stayed with that rote opinion.

Once again, Creager was taking Hunt's 205 mya date peg a little too literally. The charts in Woodburne *et al.* (2003, 365-367) positioned *Kuehneotherium* in the Late Triassic about 210 mya, and so many millions of years before *Morganucodon* and *Haldanodon* living in the Early Jurassic.

Creager also tossed off that "varieties of mouse" claim as though he'd stopped to study any of that, either. Rodents are the largest block of mammals, and even niche lineages like guinea pigs and their kin can include dozens of genera covering many millions of years, as surveyed by Pierre-Olivier Antoine *et al.* (2001). Genetic studies by Dorothée Huchon *et al.* (2002) and Gennady Churakov *et al.* (2010) put rodents and rabbits as close relatives. But the genes of living forms could only go so far, leaving it to the paleontologists to fill in some of the fossil end, starting with basal forms that include *Tribosphenomys*, David Archibald *et al.* (2001), Jin Meng & André Wyss (2001), Leigh Van Valen (2002), Alexey Lopatin & Alexander Averianov (2004a-b), Robert Asher *et al.* (2005) and Jin Meng *et al.* (2007). It's of interest to note that the tooth forms seen in *Tribosphenomys* were among those *experimentally retroengineered* by Enni Harjunmaa *et al.* (2014), showing how far back on the evolutionary track paleogenomics has already begun to probe.

Now rodents of the "mouse" variety date from about 150 million years after the taxa Creager was trying to stuff into that typological grabbag. Was it the truncated YEC time frame kicking in here, looking at *Eozostrodon*, *Morganucodon*, and *Haldanodon* and only able to see something close enough to "mouse" that he could drop them in the bag even as no regular taxonomist would justify that classification?

However functionless and arbitrary Creager's creationist measuring stick was, the science was not about to stop trying to understand the facts. Older works, such as Farish Jenkins & Francis Parrington (1976), were just coming to grips with *Eozostrodon*, *Megazostrodon* and *Erythrotherium* (recall those last two mentioned by Barbara Stahl back in Chapter 7). Like *Morganucodon*, *Megazostrodon* was another very small critter, not much larger than your thumb, illustrated by *Prehistoric Life* (2009, 279).

By the time of Qiang Ji *et al.* (1999) reported on the basal triconodont *Jeholodens* (which David Coppedge parsed above regarding its ribs), mammal phylogeny was becoming a lot clearer. And when Thomas Martin (2005) put his new species of *Haldanodon* on the systematic block, there were dozens of additional specific taxa to include to trace more of the many branches of those "contemporary" early mammals occupying the Late Jurassic landscape, where most were ground foragers, some were into tree climbing, and a few were turning to burrowing.

Subsequent work has ferreted out more about that "fossorial" hole-digging thing, as Irina Ruf *et al.* (2013) found *Haldanodon*'s "curved cochlear canal and pneumatized middle ear region support the hypothesis that *Haldanodon* had more effective low-frequency hearing as an adaptation to a fossorial mode of life." And the components of the evolutionary picture have continued to be studied at ever more depth, such as Guillermo Rougier & Sheth *et al.* (2015, 11-13) taking note of the small "angular process" protrusion on the back of the lower jaw in *Morganucodon*, *Haldanodon* and related taxa, whose homologous variation apparently reflected the shifting strength of their associated pterygoid muscle attachments (there will be much more to say about those mammalian muscle attachments in the next chapter).

Everywhere in the fossil record there were signs of microevolutionary variation contributing step by step over millions of years to what only seemed "macroevolutionary" when you pulled back enough to see how far things had come. But that required both the *looking close* and the *pulling back*, neither of which skills Charles Creager showed so long as he wore his brain-numbing creationist cap, capable only of *Zeno Slicing* everything he encountered.

Down Memory Lane with Nebraska Man

Hunt's next fossils, the Late Jurassic eupantotheres **(28)** *Peramus* (known only from teeth) and **(29)** *Amphitherium* (for which more of the skull was available) put us at the base of the marsupial and placental mammals, as reflected by Timothy Rowe (1993), Demberelyin Dashzeveg (1994), Percy Butler & William Clemens (2001) and Roger Close *et al.* (2016, 164-165). Despite Hunt's having explicitly included *Amphitherium* because its assorted jaws were better preserved and broadened the data field, Creager (2003) augered in on only *Peramus* (my **bold**): "Let me get this straight all they found are some teeth, but yet from its just teeth they know what jaw joint and are [*sic*] bones like? A little far fetched. Reminds me of **Nebraska Man**. Further more there is no reference to any finds from the **mid Jurassic**, an [*sic*] so there is no connection with other types. It is a gap that according to evolutionary dating methods is 50 million years." Creager (2008) tidied up his grammar but stuck with the play.

Instead of thinking about why the eupantotheres did not represent an informative transitional stage for the mammals that came *after*, regardless of what did or did not happen earlier, Creager pointed away from that to where the data seemingly wasn't. Pure Duane Gish. Only this time Creager might have checked beyond Hunt's account to discover that *Amphitherium* already dated from the *Middle Jurassic*, reflected in the charting by Woodburne *et al.* (2003, 366-367), as did many relevant teeth that had been found from eupantotheres and their close cousins from that period, such as by Eric Freeman (1976) and S. Metcalf et al. (1992), along with further examples continuing on into the Cretaceous, José Canudo & Gloria Cuenca-Bescós (1996).

Move on into the 21st century and Brian Davis (2012, 789) summarized the matter this way: "Of all the known groups of Mesozoic mammals, one taxon, *Peramus*, has a particular relevance to discussions of molar evolution for its generally accepted transitional morphology between well-known Jurassic groups (such as dryolestoids) and higher mammals." And in a literal "jaw-dropping" mode, the Fossil Genie continued to fill in the Middle Jurassic record with a new species from Scotland whose dental formula was "intermediate between the primitively larger postcanine count (p5:m6-7) of *Amphitherium* and the reduced number in peramurans and tribosphenidians (p5:m3)," Roger Close *et al.* (2016, 155).

As for the *Nebraska Man* that Creager felt obliged to bring up both times, that referred to a single worn pig's tooth that in 1922 prompted American paleontologist (and theistic evolutionist) Henry Fairfield Osborne (1857-1935) to mistake for one from a possible human ancestor, Osborne (1922). Given the convergent similarity in human and pig molars, though, that was not an obviously stupid mistake to have made, even as William Gregory (1927) firmly corrected Osborne on the error later.

Part of Osborne's gun-jumping turned on the fact that American scientists were itching to find a human ancestor on our side of the Atlantic, something splashy to compete with the European and Asiatic ones that had turned up (including the fraudulent British Piltdown noted earlier), John Wolf & James Mellett (1985), with shorter summaries of the affair in Roger Lewin (1987, 54-55), Ronald Ecker, (1990, 145-146), Stephen Jay Gould (1991, 432-447), Raymond Eve & Francis Harrold (1991, 75), and my own Downard (2004e, 476-477).

Antievolutionists reflexively pair Nebraska Man with Piltdown as representing the most blatant of evolutionary incompetence, but always without giving any of the background historical details that undermine that snap judgment. Examples include notorious creationist cartoon pamphleteer Jack Chick (1924-2016), whose slipshod "Big Daddy?" piece has a special place in pseudoscience infamy, Chick (1972). But more familiar names haven't been able to resist the Piltdown/Nebraska Man lure either: Duane Gish (1978, 119-

120; 1995, 327-328), Henry Morris & Gary Parker (1987, 155), Wendell Bird (1989, Vol. 1, 227), Phillip Johnson (1991, 5, 82), Paul Taylor (1995, 36), Scott Huse (1997, 134-135) and Hank Hanegraaff (1998, 49-50, 52-54).

Hunt's next example, **(30)** *Endotherium* from the end of the Jurassic, was another "advanced eupantothere" known from a single Chinese specimen, and she focused on what was most relevant about it, its distinctive teeth that were part of a chain of paleogeographical evidence, as "recent fossil finds in Asia suggest that the tribosphenic molar evolved there."

Merely talking teeth diagnostics drew only more bored yawns from Creager (2003; 2008), who asked rhetorically whether anything else had been found. Actually parts of a jaw and fragments of a shoulder (scapula) and front leg (humerus) were known from *Endotherium*, reported by Fai-Kui Zhang (1984, 2-3, 7-8), and many paleontologists over the years noted its similarity to the Late Cretaceous Mongolian fossil *Zalambdalestes*, a genus lurking around at the base of the rodents and rabbits, reviewed at length by John Wible *et al.* (2004) and summarized by Michael Benton (2005, 310-312). The rat-sized critter with its long narrow probing snout was shown in *Prehistoric Life* (2009, 357).

We were farther into the Early Cretaceous eupantotheres with Hunt's **(31)** *Kielantherium* from Asia and the European **(32)** *Aegialodon*. Hunt wrote that both were known only by their teeth, though a partial lower jaw had been found for *Kielantherium* by Delgermaa Dashzeveg & Zofia Kielan-Jaworowska (1984) that apparently Hunt missed in 1997, and Richard Cifelli (1999) subsequently compared that jaw with five other tribosphenic mammals of the period that had notably broadened their known paleogeographic distribution. Zhe-Xi Luo & Cifelli *et al.* (2001, 54) grouped *Kielantherium* with the near marsupial *Deltatheridium* (noted back in Chapter 4) among the "Boreosphenida," a placing followed by Michael Benton (2005, 300, 309).

None of that effected the point Hunt made about their teeth, though: "Both have the talonid on the lower molars. The wear on it indicates that a major new cusp, the protocone, had evolved on the upper molars. By the Middle Cretaceous, animals with the new tribosphenic molar had spread into North America too (North America was still connected to Europe.)"

Creager (2003) played his Cheshire Cat trick again (his **bold**): "They only have the teeth of both of these and yet above is a picture of Aegialodon. It is clear that evolutionists have learned **NOTHING** from Nebraska Man." Creager (2008) left out the reference to the picture but stuck to the Nebraska Man trope.

By that means Creager slipped by the point Hunt was making about those molars. A distinctive feature of these more advanced mammal teeth was how they *interlocked*, a shallow indentation (the *talonid*) on the lower jaw molar matching with the protroduing *protocone* cusp on the corresponding upper jaw tooth, improving bite and chewing capacity, summarized by Michael Benton (2005, 305-306). How many times that may have developed, and in which lineages, obviously related to the teeth, meaning preserved chompers was what you needed to trace how that feature was developing (even as there still would have been a body attached, designed or not). But in Creager's world, nothing really means much of anything, so the evidence of something interesting happening down among these munchers over a hundred million years ago (or 4500 years ago, care to defend that?) failed to excite Creager's moribund curiosity.

Aegialodon was the first mammal found that had firm signs of that talonid/protocone layout, Kenneth Kermack *et al.* (1965). Unfortunately, as Dashzeveg Demberelyn (1975) noted, that single molar was too worn to glean all that might be learned about even that feature. Fortunately the Fossil Genie was on the case, and Demberelyn was reprting on a second tooth that had been found from an Early Cretaceous mammal that was "almost identical to that of *Aegialodon*. Because it is exceptionally well preserved and practically unworn, it provides critical information on the structure and function of the very early tribosphenic molars." Later still, one of those upper molars finally turned up from *Kielantherium*, Alexey Lopatin & Alexander Averianov (2006; 2007). No surprise, there was that expected protocone, meaning "Morphologically, *Kielantherium* is truly intermediate between pretribosphenic *Peramus* and basal tribosphenic mammals like *Pappotherium*," Lopatin & Averianov (2006).

Pappotherium was another of those stray tooth finds that didn't make it on Hunt's list (and therefore never poked onto Creager's secondary one). Bob Slaughter (1965) had reported on it decades ago, but more recent discoveries in Asia and North America have put that fossil close to the *deltatheroids* that rest at the base of the marsupials, Guillermo Rougier *et al.* (1998), Gregory Wilson & Jeremy Riedel (2010), Brian Davis & Richard Cifelli (2011), Zhe-Xi Luo & Chong-Xi Yuan *et al.* (2011, 444), Shundong Bi *et al.* (2015) and Guillermo Rougier & Michael Novacek *et al.* (2015).

The monotremes entered the parade with Hunt's **(33)** *Steropodon galmani* from the Early Cretaceous (about 115 mya, though Hunt hadn't supplied a date on it). That was a comparatively recent find at the time, by Michael Archer *et al.* (1985), though a few fragmentary early monotremes were joining the show by then: *Kollikodon* and *Teinolophos*, the definite platypuses *Obdurodon* and *Monotrematum*, and a lone giant echidna *Megalibwilia*, studied by Michael Woodburne & Richard Tedford (1975), M. Griffiths *et al.* (1991), Archer *et al.* (1992), Rosendo Pascual *et al.* (1992a-b), Timothy Flannery *et al.* (1995), Anne Musser & Archer (1998), Thomas Rich *et al.* (2001; 2016), Pascual *et al.* (2002), Musser (2006), Timothy Rowe *et al.* (2008) and Rebecca Pian *et al.* (2013). Toss in the genomic data that suggested a Jurassic ballpark for monotreme origins, Musser (2003) and Wesley Warren *et al.* (2008).

What all the fossil monotremes had in common was that they still had *teeth* (which could be studied even if very little postcranial skeletal evidence accompanied them)—even if extremely small, such as the miniscule thumbnail-sized jaw fragment of *Teinolophos* illustrated by *Prehistoric Life*

(2009, 357). The three living monotreme species develop only a few temporary teeth (including an egg tooth) that are not retained into adulthood, illustrated by Michael Benton (2005, 308). That's why the fossil record for the later largely toothless monotremes is sparser, teeth being one of the main fossil markers for so many mammals. But the scientists have worked with what they have, of course (it's what they do), from George Gaylord Simpson (1929) and H. Green (1937) studying monotreme teeth in the microscope era, to Keith Lester & Alan Boyde (1986) and Lester *et al.* (1987) applying scanning microscopes to both living and fossil examples.

Though creationists like Creager clearly didn't think they ought to be able to do anything with just a tooth (flashcard *Nebraska Man*!), Lester *et al.* (1987, 401) had no problem finding how the specialized monotreme tooth enamel represented "a structural stage intermediate between that of known multituberculates and extant tribosphenid mammals." And Flannery *et al.* (1995, 420) were able to work out that *Kollikodon ritchiei* "was a platypus-size monotreme that fed on material that needed crushing but not shearing. Similar adaptations are evident in marine predators, such as sea otters, crabs and some fish that feed on hard-shelled animals." Baron Cuvier would have given all this tooth work a satisfied nod of approval.

Similar observations can be made about other bones, of course, insofar as they can offer clues to the microevolutionary developmental mechanisms varying over time in their lineages. Take the embryonic element noted by Musser & Archer (1998, 1066) concerning a pair of small nostril bones in the living platypus (*Ornithorhynchus anatinus*) compared to its bigger Miocene cousin (*Obdurodon dicksoni*) living 15 million years earlier:

> The right premaxilla and septomaxilla are complete showing that the shorter septomaxilla terminates lateral to the underlying premaxilla, which meets its opposite at the midline. This contrasts with the rostrum in adult *Or. anatinus* where the premaxilla and septomaxilla are fused together in the adult and the resulting prongs of bone do not meet at the midline. However, in foetal *Or. anatinus*, illustrated by Zeller (1989a), the premaxillae are fused ventrally and the septomaxillae, overlying the premaxillae and separated from them by the developing marginal cartilage, terminate lateral to the midline (in part to accommodate the os carunculae) as they do in *Ob. dicksoni*.

The Zeller reference was to his German monograph on the platypus.

The septomaxilla are positioned beside the premaxilla (previously mentioned back in Chapters 2 & 8) in the upper jaws of reptiles, and retained in modified form down into the monotremes. John Wible *et al.* (1990) and Ulrich Zeller *et al.* (1993) have reviewed the evolution of the septomaxilla in synapsids and mammals, including its developmental traces still to be found in the embryology of anteaters, sloths and armadillos.

No surprise, none of that anatomy engaged the interest of Creager (2003), who sloughed *Steropodon* off with "It seems to be the same kind as a Platypus, it is hard to tell for certain since they all seem to have are a few partial jaws. Even so id [*sic*] clearly does not lead to any thing other than a Platypus." Creager's 2008 revision left *Steropodon* off the table entirely, though given his superficial method this was more likely by plain inattentiveness rather than any intentional oversight.

It's interesting to see how a still more recent creationist related to this issue. *Conservapedia* (2016) just as blithely acknowledged that "Modern adult platypuses do not have teeth, but the discovery of platypus fossils in Australia had already identified that their ancestors did indeed have unique and distinctive teeth." This brazen acceptance of a natural descent lineage that crossed generic boundaries didn't stop them from insisting (all without sources) that "The anatomy of the platypus is a problem for evolutionists, because its bill clearly adheres to the same design as that of the duck, suggesting a common creator."

Oh really? Jim Foley (1997) had long ago reminded the *Talk.Origins* set that "scientists have always known that the bill has nothing in common with that of a duck except for the shape. The bill of a duck is a hard keratin structure, while that of the platypus is a soft flexible organ packed with electrical and touch censors."

Reflecting on the fossil and developmental evidence, Musser & Archer (1998, 1075) proposed that the platypus bill "may have derived from a rostrum such as that of *Oligokyphus* in which the incisive foramina fused to form a single opening through the premaxillae; enlargement of such an arrangement may have resulted in a bill form like that of *Ob. dicksoni* (and thus of monotremes) in which this ovoid space bounded anteriorly by the conjoined premaxillae became progressively enlarged." That would be the same (18) *Oligokyphus* Creager couldn't make sense of above. Evidently the working scientists hadn't got the creationist memo as to how much of a problem the platypus' bill was supposed to be posing for them.

More recent work by Masakazu Asahara *et al.* (2016) have pinned down more of what was going on in the evolution of the platypus bill sensory system, and how that came at the price of reduced dentition. While some sensory capacity was known in the fossil platypuses, the existence of that system went along with neural wiring that runs to the brain along an "infraorbital canal" in the skull. The one in the Miocene *Obdurodon* was much less developed than the enlarged one seen in the living *Ornithorhynchus*, which the authors related to an adaptation to foraging in more cloudy water. But that went along with a tradeoff: any increase in the size of that nerve channel cut down on the space available immediately below for tooth roots in the upper jaw, hence the loss of teeth in the specialized living platypuses, along with reduced eye opening size as the animal relied on electrosensitivity more than vision.

Nothing comparable to such analysis has surfaced in the Flood Geology science pond.

Hunt's next example was **(34)** *Vincelestes neuquenianus*, which she listed as "early Cretaceous, 135 Ma." Reflecting the uncertainty of the later 1990s concerning how to position the many new fossils that were broadening the known range of Cretaceous mammal diversity, Hunt described it tentatively as "A probably-placental mammal with some marsupial traits, known from some nice skulls. Placental-type braincase and coiled cochlea. Its intracranial arteries & veins ran in a composite monotreme/placental pattern derived from homologous extracranial vessels in the cynodonts," and directly cited the technical paper on that vascular issue, Guillermo Rougier *et al.* (1992). Hunt later also mentioned the Early Cretaceous *Protungulatum*, which has held its position at the base of the placentals, Robert Sloan & Leigh Van Valen (1965), David Archibald *et al.* (2011), Maureen O'Leary *et al.* (2013a, 665), Thomas Halliday *et al.* (2015) and Halliday & Anjali Goswami (2016).

That Hunt cited the Rougier paper is especially revealing, since Creager (2003) paid no attention to it, as he repeated his petrified mantra: "Vincelestes seems to only be known from its skull. While this is better than Steropodon, Kielantherium and Aegialodon since all four are classified as early Cretaceous and so would have been a [*sic*] contemporaries. This means that the order is based entirely on evolutionary assumptions." Creager (2008) repeated that with an interesting deletion: "*Vincelestes* is only known from its skull. While this is better than *Kielantherium* and *Aegialodon*, since all three are classified as early Cretaceous they would have been contemporaries. This means that the order is based entirely on evolutionary assumptions." No *Steropodon*.

Had Creager done a bit of research in the meantime, and realized that *Steropodon* dated some ten million years *after* the "contemporaries" *Vincelestes* and *Kielantherium*? We can't tell, because Creager was not one to supply sources for things. By that time, though, analysts like Woodburne *et al.* (2003, 365-367) had positioned *Steropodon* as a basal monotreme, while the earlier *Vincelestes* and *Kielantherium* were seen as cousins much closer to the therian Marsupial/Placental branch (which we know from *Protungulatum* stretched back quite a ways). José Bonaparte (2008) noted also how the "prototribosphenic" teeth of *Vincelestes* suggested that "the tribosphenic condition may have developed first amongst taxa on Pangea, before the separation of Laurasia and Gondwana." And later finds, like *Acristatherium* from the Early Cretaceous, described by Yaoming Hu *et al.* (2010), would continue to reveal more about that diversifying mammalian landscape.

"Compared to many Mesozoic mammals, *Vincelestes* is known from excellent fossils," reported *Prehistoric Life* (2009, 356), "Of nine specimens, six include skulls." And given the intermediate phylogenetic position of *Vincelestes*, Rougier *et al.*'s detailed study of the skull was able to trace the pathway of the major arteries and veins based on the characteristic openings known from the living monotremes and therians, and compare their transitional layout to that of several still earlier fossil skulls (including *Thrinaxodon*, *Probainognathus* and *Morganucodon*). The completeness of the available *Vincelestes* skulls also allowed Thomas Macrini *et al.* (2007) to assemble a digital endocast of the brain. Timothy Rowe *et al.* (2011) and Jon

Kaas (2013, 35-38) have reviewed what has been learned so far about the evolutionary brain innovations seen in the therapsid/mammal lineage.

Kathleen Hunt turned next to "The first definite marsupial" **(35)** *Pariadens kirklandi* from the Late Cretaceous ("about 95 Ma") of North America, first described by Jeffrey Eaton & Richard Cifelli (1988). A related contemporary genus (*Arcantiodelphys*) showed up later in France, Romain Vullo *et al.* (2009), though both taxa were known only by their telltale tribosphenic teeth. *Pariadens* and company were smaller members of a diverse bunch of basal marsupials, including *Didelphodon* from the stagodont family (larger animals known only by two genera), Richard Fox & Bruce Naylor (2006). Recently a complete *Didelphodon* skeleton has been found by the *Rocky Mountain Dinosaur Resource Center* (2010), though the fossil hadn't been been formally described as of this writing.

Rather than ponder what *Pariadens* might have looked like or what it might have been related to (even within a "kind" as a monobaramin), Creager (2003; 2008) launched into his *Bermuda Triangle Defense* noted back in Chapter 4. The 2003 version:

> Where is the mid Cretaceous? There are no reference to fossils from the mid Cretaceous. Pariadens is classified as "late Cretaceous" and Vincelestes is classified as "early Cretaceous." Why no reference to this gap? Could it be that relevant fossils that are classified as the mid Cretaceous didn't fit their evolutionary assumptions? How do they know it was a marsupial? Teeth have nothing to do with reproduction, so it would impossible to tell just from the teeth. The only answer is that the teeth resemble those of a known marsupial. If that is the case then why can't it be that marsupial? The only possible reason for not drawing such a conclusion is the assumption that evolution occurred.

"The only possible reason" for Creager writing these things is that he never paid the slightest attention to the geology or the strictures of tooth diagnostics. Leaving aside Creager's implication that the paleontologists were comparing *Pariadens*' teeth to just a single marsupial, rather than to the whole range of known marsupial dention (living and fossil), by the time Creager was writing, the Fossil Genie had already weighed in with the quite complete skeleton of *Sinodelphys*, whose furry frame was illustrated by *Prehistoric Life* (2009, 357). Dating to 125 mya, this even more basal marsupial fell right in that "Middle Cretaceous" (had there been such a designation). As Zhe-Xi Luo *et al.* (2003, 1934) noted in their analysis of the fossil, "*Sinodelphys szalayi* is more closely related to extant marsupials than to extant placentals and stem taxa of boreosphenidans in its many marsupial-like apomorphies in the skeleton and anterior dentition."

The next on Hunt's block were **(36)** *Kennalestes* and **(37)** *Asioryctes*, both generalized Late Cretaceous mammals from Mongolia, Hunt quoting Robert Carroll (1988) that they "Could have given rise to nearly all subsequent placentals." Creager (2003; 2008) could find only references to *Kennalestes*' teeth, while contending that *Asioryctes* was "known only from its skull and fragmentary post cranial bones, making any reconstruction of *Asioryctes* questionable at best." Creager (2008) finally decided, "The fact is that there is almost nothing to compare. They are classified as late Cretaceous as is its alleged immediate ancestor *Pariadens kirklandi*, which is itself also based only on teeth."

Hunt had given no mya dates for the two taxa, but Woodburne *et al.* (2003, 365-367) pegged *Asioryctes* at around 85 mya, which fell some *10 million years* after *Pariadens*, Vullo *et al.* (2009, 19913). The unavoidaable pitfalls of Creager once again taking the general dating terminology tossed off in a *Talk.Origins* post too literally.

As for the fossils, Creager never got far enough to specifying which bones were known, or proscribe what reasonable inferences about relationships might be drawn from them. Both fossils preserved at least some of the *atlas*, the first cervical vertebra that articulates with the skull (hardly a trivial detail), while sections of the vertebral column and the feet of the front and hind limbs were known for *Asioryctes*, all covered at length by Zofia Kielan-Jaworowska (1969), who found enough diagnostic bones to argue against an arboreal origin for therian mammals. David Archibald & Alexander Averianov (2006) have found additional teeth from more basal members of their group, and John Wible *et al.* (2004) highlighted their connections to the *Zalambdalestes* eutherians (back again to the base of the rodents and rabbits, remember).

Which brings us at last to the end of Hunt's *Talk.Origins* RMT parade, with a trio of "Primitive North American placentals with same basic tooth pattern": **(38)** *Cimolestes*, **(39)** *Procerberus* and **(40)** *Gypsonictops* (all from the "very late Cretaceous"), though with *Procerberus* Hunt was slipping into genera that extended on into the post K-T Tertiary Period.

None of which interested Creager (2003) who once again yawned: "I could not find any information on how much Cimolestes, Procerberus and Gypsonictops have been found in the fossil record. Given the fact how little evidence there seems to be for their alleged ancestors there nothing to compare them to any ways." Creager (2008) was only slightly less bored: "There is no clear information on how much of *Cimolestes*, *Procerberus* and *Gypsonictops* have been found in the fossil record, but it seems that they are based only on teeth. Given the fact how little evidence there seems to be for their alleged ancestors there would be nothing to compare them to even if complete skeletons are available."

Oh, that *comparing* thing. So much easier to declare if you don't actually compare anything.

Richard Fox (1977) had already explored the "Relationships between *Gypsonictops* and its presumed ancestor *Kennalestes gobiensis*," while Jaelyn Eberle (1999) had uncovered a new species, *Alveugena carbonensis*, that was "both morphologically and temporally intermediate between small, early cimolestids (such as *Procerberus* and *Cimolestes*) and the earliest documented conoryctid taeniodont *Onychodectes tisonensis*," thus representing "a transition between the suborders Didelphodonta and Taeniodonta." We've already encountered *Didelphodon*; the taeniodonts were an extinct group of early Paleocene herbivores, Michael Benton (2005, 329-331, 361, 401). Further fossil finds described by Deborah Rook *et al.* (2010) and Rook & John Hunter (2014) have reinforced that cimolestid/taeniodontid relationship, and Fox (2015) hasn't stopped making sense of each new scrap of evidence on the *Cimolestes* genus. It's what scientists do, they keep at it.

There are some twenty extinct orders of eutherians known already, keeping the Fossil Genie busy filling in more of the relationships. The anatomical variations put *Asioryctes* and another fossil *Maelestes* close to *Cimolestes* and *Gypsonictops* beside the aforementioned Zalambdalestid family, John Wible *et al.* (2007; 2009) and Anjali Goswami *et al.* (2011, 16334). And not to forget that each of these fossils lived in an actual world, such as the fluctuating prehistoric seaway Jeffrey Eaton *et al.* (1999) noted regarding the environment in which *Gypsonictops* resided in Utah back in the Cretaceous.

The fact that so many of these taxa are known by bits of their jaws and teeth may have been only a source of creationist shrugs from Charles Creager, but they've proven a bonanza for analysts trying to work out the shift from five to four molars in mammals, since some of these samples include *juveniles*, allowing the tooth formation sequence to be worked out in ever greater detail. It turns out this Cimolestes/Zalambdalestes crowd were smack in the middle of that transition, as worked out by Alexander Averianov & David Archibald (2015). The shift turned on how the decidicious *dp3* tooth was gradually left out of the adult tooth mix (this could happen independently in various lineages, of course). And they even found one of the stem placental mammals, *Kulbeckia kulbecke*, where the absorption or retention of the tooth varied *within that species*, exactly as one would expect from an evolutionary intermediate stage that built on microevolutionary variation.

A particularly interesting side branch of these early mammals was *Leptictidium* and its kin, dating from about 15 million years after the K-T bump, Michael Benton (2005, 337-338), Kenneth Rose (2006) and T. Meehan & Larry Martin (2010). These little critters were *bipedal*, which was quite a rarity among mammals (kangaroos and humans exhaust the living list), unlike their so distant (and by then extinct) cousins the dinosaurs, where bipedalism was a common layout. For those who think developmentally, that difference was saying something about the distribution of genetic variation deep in the diapsid/synapsid divide, relating to tail formation and leg bone stress, and the neurological balancing adaptations that had to accompany such a shift, all of which would have been playing out on their differing ecological landscapes. I'll be looking for future scientists to ponder more on that disparity, and maybe make some sense of it, even as I won't be holding my breath that any antievolutionist will give it even a passing typological thought.

Kathleen Hunt (1997) concluded her survey of the RMT fossils with something that was plain to anybody following it step by step:

> The decision as to which was the first mammal is somewhat subjective. We are placing an inflexible classification system on a gradational series.

> What happened was that an intermediate group evolved from the "true" reptiles, which gradually acquired mammalian characters until a point was reached where we have artificially drawn a line between reptiles and mammals. For instance, *Pachygenulus* and *Kayentatherium* are both far more mammal-like than reptile-like, but they are both called "reptiles".

Which may be contrasted with Creager (2008) finishing his *Zeno Slicing* treatment this way:

> When one can look at the entire animal there is no evidence of a transition here. Even though they can point to apparent trends in some features, there are many for which there is no evidence of a trend at all. There are several cases were [*sic*] the alleged trend from reptile to mammal does not follow the evolutionary dating scheme. In other cases the evidence is so fragmentary as to make the existence of the taxons questionable. Once again, when the available evidence is looked at without assuming evolution, no evidence is found that it took place. Yes, there are some reptiles with a few mammal-like traits and some mammals with a few reptile-like traits, but from *Biarmosuchia* to *Procynosuchus* there is a clear and sudden change in dominance from reptile to mammal traits. So rather than the smooth transition Talk.Origins wants you to think exists, there is a sudden shift from reptile to mammal as creationists would expect.

You've seen step by step how far away from the *Talk.Origins* dataset Creager kept himself, and in particular how that supposed "gap" between *Biarmosuchia* and *Procynosuchus* was just a creature of Creager's own expectations, already falling apart factually even as he claimed it. And while his 2003 draft presented no science references at all, Creager (2008) included weblinks to three familiar creationists: Duane Gish (1981), Ashby Clamp (1998) and John Woodmorappe (2001). As those predated his 2003 trial run, it's not unreasonable to suspect that was all the "research" Creager had under his belt from the start. Which meant Creager was trying to pile up his crumbling sand castle, not on any beach of solid evidence, but on top of somebody else's already crumbling sand castles.

Baraminology finally trips onto the field

We've seen how many lineages there are in the RMT, and how badly Charles Creager did trying to slog through (and around) them. But as all of that data *existed*, every bit of it would still need to be accounted for in a baraminological framework. So what has been done on that in the years since Creager's *CreationWiki* was stumbling over it?

As of this writing, I could find only one example of an attempt to apply the tools of Creation Science baraminology to the RMT, by an M. Aaron, who took as his subject none of the extensive examples Hunt touched on, from the probainognathids to the T1/T2 bunch, but instead hit on the *caseids*, a single family of herbivorous pelycosaurs so peripheral to the RMT story they weren't even on Hunt's long list of relevant taxa.

Aaron (2014, 19) started off with a litany of limits to the baraminological toolkit that may be compared to what little slack creationists have been willing to cut evolutionists over the years. "Hybridization is one of the primary ways that baraminological status has been determined for extant organisms," Aaron citing Tood Wood & Megan Murray (2003, 98), which was a light popular restatement of the early baraminological work that still made no mention of the RMT synapsids (or *Homogalax* for that matter).

> Unfortunately, the hybridization test is unavailable to researchers studying fossil groups. With fossils, only morphological and stratigraphic data are available for use in determining taxonomic relationships. In addition, fossils are often found incomplete, and there are still gaping holes in our knowledge of many taxa. Nevertheless, several researchers have approached the baraminology of fossil groups with some success, mainly with a focus on Cenozoic mammal groups including equids.

At this point Aaron cited David Cavanaugh *et al.* (2003), which was all he had to say about the paper that had functionally thrown in the towel on the historic horse evolution sequence (noted back in Chapter 6). To this Aaron added S. Mace & Wood (2005) and Wood (2010) for their claims about archaeocetes and hominids, subjects I'll have to leave to another occasion, thought it is fair to say their arguments had a restricted shelf life dependent on the Fossil Genie not keeping busy in the meantime. In the loose-end department, Aaron (2014, 22) also listed in his references Wood (2006) surveying the purported progress of the baraminology field, but didn't cite it anywhere in the text. And there were minor temporal oddities with citation dates, as he tagged Wood's paper as 2007, and similarly nudged the Tom Kemp (2011) book chapter forward into 2012.

Aaron concluded his introduction with considerable understatement: "The vast majority of fossil groups have not been studied. This may be due to a lack of creationist paleontologists, or a lack of easily obtainable datasets of ancient organisms. However, with the increased use of the internet by paleontologists and other researchers, many new datasets have become available."

Many new datasets indeed, as I hope you've been seeing in this present work. Just how little of the available "new datasets" Aaron was willing to include, though, was another matter.

To start, as so many other creationists have (from Gish to Mitchell) Aaron (2014, 19) focused on the seeming discontinuities more than the evidence, declaring in his briefest of summary of the diapsid/synapsid divide that "The two major clades of modern amniotes, Diapsida and Synapsida, appear suddenly in the fossil record simultaneously," and cited Kemp (2011, 3). Of course, whether what was appearing "suddenly" or "simultaneously" meant what YECers do by it (Flood event 4500 years ago) or was inexplicable by natural evolution depended on the details, which Aaron did not stop to include.

It was true that *Archaeothyris* (noted earlier in Hunt's list) fairly qualified as the oldest known synapsid, and dated from the same Late Carboniferous time frame (about 305 mya) as the earliest diapsids, the Araeoscelidia family that includes the better preserved *Araeoscelis* from the Permian some 30 million years later, Robert Reisz *et al.* (1984) and Reisz & Sean Modesto *et al.* (2011). But that classification turned on paying attention to all the fossil data, as *Araeoscelis* didn't possess both distinctive diapsid skull openings (the upper one was filled in with bone, possibly related to a diet of extra crunchy insects), unlike later forms like the Permian *Petrolacosaurus* where both were fully open, Reisz (1977). *Lanthanolania* from around 265 mya is the current oldest definitive diapsid, Sean Modesto & Reisz (2002), though its upper opening was also quite slim, as was the case with *Orovenator* from around 290 mya that turned up since to bridge the 38 million year morphological gap between the Araeoscledia and the more obviously diapsid Permian examples, Reisz & Modesto *et al.* (2011). Those dates and those so gradual variations cover an awfully wide spread for any workable "simultaneously" or "suddenly" to hold.

It's a toss up then whether *Araeoscelis* had closed a second opening already present in as yet unpreserved ancestors (cue *Bermuda Triangle Defense*), or represented an "almost" diapsid that needed only that new space not to fill in to qualify as one (wave the *No Cousins* banner). We're in the same situation as the intermediate wasp/ant (or would you prefer ant/wasp?) clambering around Michael Denton back in Chapter 5, aren't we? And that's the unspoken presumptions just in Aaron's preliminary staging, before we got to the main event.

Aaron (2014, 20) maintained "There is little morphological variation within the family Caseidae," citing Hillary Maddin *et al.* (2008) & Kemp (2011, 4-5). Well, not quite. Maddin et al. (2008, 160) put it this way: "Members of Caseidae are superficially conservative in morphology, differing from just under one meter to greater than three meters in total body length." Variation by *size* ... memo to *Explore Evolution* and *Watchtower*: baraminology Aaron-style allows substantial *allometric size variation* within a baramin.

The caseids consisted of only six well-supported genera, with the best known species being *Cotylorhynchus*, a pudgy beast with a remarkably small head, Willis Stovall *et al.* (1966), and examples of that genus have continued to be found, Ausonio Ronchi *et al.* (2011). Because their fossil record was so sparse over a 40 million-year run, Maddin *et al.* (2008, 161) confessed "Virtually nothing is known of the origin and early history of the group," which may have been why Aaron homed in on the caseids as a target for baraminological pigeonholing. The fewer examples to be catalogued, the better the chances of defining an artifical "kind." Especially if Aaron were up to the same parsing tricks as John Woodmorappe showed back in Chapter 8, which indeed he was.

The Maddin paper was the first phylogenetic analysis of the caseids, focusing in particular on how *Ennatosaurus tecton* figured in the group. To that end a cladistic analysis was done, comparing 106 characters (73 cranial and 33 postcranial) for the six main caseid genera. Subsequent work by

Robert Reisz & Hillary Maddin *et al.* (2011) reclassified one of the two species of the *Casea* genus as properly beloning to its own new genus, *Euromycter*, which Aaron noted in his study. But the Reisz paper had also identified a new genus, *Ruthenosaurus*, which had been previously missed because it was only known from its partial postcranial skeleton. Apparently a juvenile, *Ruthenosaurus* was still already a large critter, hinting at just how much allometric variation had occurred within the caseid family, and *Alierasaurus*, another new large caseid found by Marco Romano & Umberto Nicosia (2014) evidenced the foot bone size adaptations relating to its bulk, which was part of an overall trend in the caseids toward larger size, graphed by Frederik Spindler *et al.* (2016, 613) in their study of several new possible caseids. And then there was *Trichasaurus*, another fragmentary fossil which Roger Benson (2012) suggested may have been a caseid, though more recently Romano & Nicosia (2015) have marked their doubts.

Why not include some of that new data (at least what was known by 2014) in his caseid baraminology study? For that, Aaron would have had to have done more of his own research and data evaluation, not just copy the numerical work assembled by the professional paleontologists, as he had with the Maddin 2008 dataset (or as Woodmorappe had likewise done back in 2001, with the effect we've seen). Baraminology in play was still very much a secondary dilettante apologetic enterprise.

The Maddin paper had also covered several outgroups for comparison: three closely related pelycosaurs (including *Eothyris*), the Reptilia and two "diadectomorpha"—that latter an even more basal bunch of reptile-like herbivorous amphibians that lived over a fifty-million-year span from the Late Carboniferous to the Late Permian, charted by Jun Lie & Gabe Bever (2015). Those other animals were included because the full measure of any relationship could only be done by *comparing* them to other things. That's how science investigation works: you broaden the scale of view, not diminish it. Not so in Aaron's baraminology, though, where data reduction via a "taxic relevance cutoff" test was the order of the day, relying on Ashley Robinson & David Cavanaugh (1998) and Todd Wood (2005) for his inclusion criteria.

Because two of the caseids (*Angelosaurus* and *Oromycter*) were known only from fragmentary material, Aaron summarily removed them from his study, effectively erasing their existence for those diagnostic pieces of them that had been preserved. That would be pieces of the body and upper skull for *Angelosaurus* and the skull of *Oromycter*, Everett Olson (1968, 244, 247-266, 293-301) and Robert Reisz (2005). And as 30 of the 106 character states in the Maddin study had four or more instances of "?" whenever no value could be assigned for a particular taxon because of an incomplete fossil, Aaron further whitled down his field by excluding 23 of those anatomical characters altogether (even though there were still a lot data for the features from most of the rest of the taxa).

Some word on the paleontology is in order. *Eothyris* and its relatives have an even sparser fossil record than the caseids—in fact, only it and the similar *Oedaleops* are known, covered by Wann Langston (1965) and Robert Reisz *et al.* (2009). I mention those sources, though, because both of those papers were cited peripherally by Aaron, meaning their content was at some point under his eye, along with Maddin *et al.* (2008).

Like the caseids, *Eothyris* and *Oedaleops* had wide skulls, but were carnivores, not specialized herbivores like the caseids. Yet their skulls were showing elements that would be expanded in the caseids. And while *Eothyris* had a more conventional pelycosaur nasal layout, its relative *Oedaleops* already sported a wider nares that was "more aberrant than in any pelycosaurs except the caseids," Langston (1965, 20). The caseids would take this even farther, with an especially enlarged nasal opening and pronounced bony snout, well illustrated by Everett Olson (1968, 231, 237). Regarding the skull shape, Reisz *et al.* (2009, 41-42) remarked how "In cross-section, the gently convex skulls of *Eothyris* and *Oedaleops* are very low and broad by eupelycosaur standards (Berman et al., 1995). The midpoint in skull length lies just ahead of the posterior margin of the orbit. Caseids inherited this primitive cranial morphology and retained its general configuration while modestly evolving towards increasingly isometric proportions."

For such reasons, the evolutionary taxonomy in Maddin *et al.* (2008, 172) charted *Eothyris* as a basal member (both phylogentically and chronologically), as did Adam Huttenlocker *et al.* (2011, 574) and the still more recent Romano & Nicosia (2015), as well as two other papers Aaron cited: Robert Reisz & Michel Laurin (2001, 1230) and Sean Modesto *et al.* (2011, 1031). Reisz & Laurin (2001) was a rather distant source for Aaron to have used, since it was about a different non-RMT subject, the parareptile *Macroleter* from the Late Permian, a poorly known group that showed features possibly relating them to turtles.

Turtle evolution is a fun topic for another work, but the upshot is that additional fossil finds and a lot of developmental and genetic studies have come along to suggest turtles might be derived diapsids, leaving the turtle-like pareiasaur features a convergence. For those wanting a lengthy tortoise trail of evidence to follow on that ongoing debate (Achilles optional), there's Michael Lee (1993; 1994; 1996; 1997; 2013), Rafael Zardoya & Axel Meyer (1998; 2001), Olivier Rieppel (1999) re Blair Hedges & Laura Poling (1999), Rieppel & Robert Reisz (1999), Scott Gilbert *et al.* (2001), Rieppel (2001; 2009), Grace Loredo *et al.* (2001), Julien Claude *et al.* (2004), Judith Cebra-Thomas *et al.* (2005), Linda Tsuji (2006), Hiroshi Nagashima et al. (2007; 2009), Chun-Li Li *et al.* (2008), Tyler Lyson & Gilbert (2009), Walter Joyce *et al.* (2009; 2013), Lyson *et al.* (2010), Tsuji *et al.* (2010), Jérémy Anquetin (2011), Hedges (2012) re Ylenia Chiari *et al.* (2012), Nicholas Crawford *et al.* (2012), Lyson & Joyce (2012), Naomi Lubick (2013), Lyson & Torsten Scheyer *et al.* (2013), Lyson & Bhart-Anjan Bhullar *et al.* (2013), Zhuo Wang *et al.* (2013), Daniel Field *et al.* (2014), Lyson et al. (2014), Gabe Bever *et al.* (2015), Bruce Rothschild & Virginia Naples (2015), Rainer Schoch & Hans-Dieter Sues (2015) and Ritva Rice *et al.* (2016).

The Modesto 2011 paper was closer to the RMT field, ranking the various pelycosaur lines nested at the base of the split into the therapsid branch. Following that trail through Deep Time, their pelycosaur cousins *Mycterosaurus* and the later *Varanops* lived further on in the later Early Permian (about 275 mya), by which time the caseids *Oromycter*, *Casea* and *Cotylorhynchus* were on the scene (and contemporary with their basal therapsid cousin *Tetraceratops*). *Angelosaurus* branched off from *Cotylorhynchus* a few million years later, leaving *Ennatosaurus* as the final known late straggler living in the Middle Permian, even as the therapsids were differentiating.

All that modified anatomy meant things. Decades ago, Everett Olson (1968, 325) identified details of diet and ecology for one of the caseid taxa Aaron lopped off from his study: "It would appear that *Angelosaurus* had departed farther from the primitive dietary habits of caseids than had any other North American genus. Not only are the teeth patterns moderately distinctive, but the strong wear of both anterior and cheek teeth suggests a crushing or grinding action. Possibly hard seeds or cones of conifers may have become an important element in the diet."

Maddin *et al.* (2008, 176-177) had tracked the variations in caseid tooth structure, from the simple conical pegs of *Eothyris*, through the more serated tops of the caseids, of which *Angelosaurus* represented a more derived form. Given how few taxa were involved, the authors recognized that it couldn't be decided how many of the shifts were due to convergent innovation or to reversals of an inherited tooth form, and of course we don't have the genes of these long-extinct animals to study and retroengineer in the way Enni Harjunmaa *et al.* (2012; 2014) have been able to do with therapsids on the line leading to recent mammals. At least not for the moment (the Science Progress clock ticking all the while).

Ah, but Aaron represented the creationist baraminological perspective. So what did he have to say about the *when* and *where* of these taxa from his own YEC framework, their implications for Flood-time palaeoecology, and what happened to them in the post-Flood environment, so that we could compare the two perspectives to see which one had the better of the evidence? Nothing. Nothing at all.

As for his baraminological topic, Aaron had chopped off a third of his caseid examples, and then sliced away a fifth of the numerical character data pertaining to what remained—and without saying exactly which character features he was excluding. Meaning readers of Aaron's short paper couldn't assess whether he'd overlooked significant features or not.

Though if you looked at the original Maddin source, we do know some of what he needed to be paying attention to, since Aaron listed the "five synapomorphies" Maddin *et al.* (2008) identified as diagnostic of a monophyletic Caseidae, and each of those related to specific character traits, which Aaron stepped past by giving them new numbers. Aaron did not explain what any of the jargon meant, either, but we'll chalk that up to his writing for a technically literate readership. Right.

Aaron began with "1. Medial projection of the lacrimal contributes to narial emargination." This feature was (17) in the Maddin study, and involved bony ridges around the nose (remember how much bigger those were in the caseids) formed by extension of the lacrimal bone. Here's how the Maddin paper tracked that feature through the dozen taxa (the caseids in **bold**): Reptilia (0), *Diadectes* (0), *Limnoscelis* (0), *Mycterosaurus* (0), *Varanops* (0), *Eothyris* (0), ***Oromycter*** (1), ***Casea broilii*** (?), ***Casea rutena***, AKA *Euromycter* (1), ***Cotylorhynchus*** (1), ***Angelosaurus*** (?) and ***Ennatosaurus*** (1). "The degree of development of the emargination varies within Caseidae, including variation in the number of circumnarial elements contributing to this feature. All caseids possess at least a minor contribution from the lacrimal (1). This suggests the early development of this feature of caseids began with a modified lacrimal and was followed by the other elements later," Maddin *et al.* (2008, 173).

All rather microevolutionary, isn't it.

"2. Absence of an anterior tapering of the lower jaw" was the next item Aaron listed. This feature was (53) in the Maddin study, and related to how the front of their jaw wasn't tapered (think again of that previously mentioned synapsid jaw curve tendency). Here were the numbers (same taxa order as for 17, the caseids still in **bold**): 001000**1111?1**. You'll notice that although (1) was diagnostic of the caseid clade, the feature was not unknown in earlier forms, and Maddin *et al.* (2008, 173) offered a reason for it: "This morphology evolved convergently in the diadectomorph *Limnoscelis*. Similarly, edaphosaurids possess a relatively deep anterior portion of the lower jaw compared to most basal synapsids, suggesting that this may be an adaptation to an herbivorous diet."

"3. First premaxillary tooth is the largest in the tooth row." This was (68) in the Maddin study, and differed from the basic synapsid layout where the first tooth was not the largest. The Maddin numbers: 110111**0000?0**. Once again you'll notice the plant-muncher *Limnoscelis* developed that enlarged tooth convergently. Don't let Michael Denton know it, but all this variation around *herbivory* sounded rather like a Darwinian adaptive feature, didn't it?

"4. Shape of the lingual surface of the marginal dentition is spatulate." This feature (72) concerned the shape of tooth indentations on the side facing the tongue, another feature that evolved convergently in many lineages, though it was "highly diagnostic for caseids in the context of basal synapsids," Maddin *et al.* (2008, 173). The numbers: 010000**111111**. This time it was *Diadectes* that showed that modified dentition convergently.

"5. Marginal dentition with lingual shoulder and lingual curvature." This feature (73) involved the crowns of caseid teeth expanding with a bit of a curve toward the tongue, though once again "There is variation in the point of onset and the degree of the curvature among caseids," Maddin *et al.* (2008, 173). Finally a clean sweep for the caseids for this incremental variation: 000000**111111**.

So the five diagnostic features that characterized the caseids in particular all involved microevolutionary variations on existing anatomical features. But so too did the five additional characters (which Aaron did not bring up) that similarly nested the caseids in a larger clade that *included* good old *Eothyris*. Character state number (1) related to their protruding snout shape, (33) & (40) involved a widened squamosal bone with a related projection, and two particular adaptations concerned herbivory (ponder, if you will, the needs of chomping into resistant shrubbery): (50) a shortened preorbital opening, and (58) a coronoid prominence on the jaw shifted more toward the front, changing the bite dynamics.

Having trimmed the dataset down and discussing nothing of what I've just reported from his assorted technical sources, Aaron claimed to have isolated four utterly disconnected baraminic blocks: the four Caseids, the two Didectomorpha, and *Eothyris* standing alone from the other two Eupelycosauria. A neat trick, but the data already had *Eothyris* too close to the caseids anatomically to be boxed off so clearly except by a shredder that left so much of the facts out.

Aaron concluded his paper with some defensive trench digging:

> I conclude that the family Caseidae is a holobaramin. However, it must be remembered that this conclusion is based solely on current skeletal morphological data (a very limited subset of the data available from extant organisms), and it may be subject to change as more fossils are discovered. Currently, *Eothyris* is excluded from the caseid holobaramin; however, it is possible that future fossil discoveries and baraminological analysis may overturn this exclusion. As this is another example of a holobaramin at the family level, this gives further support to the idea that the created kind is approximately equivalent with the taxonomic rank of family in both extant and extinct organisms.

It remained to be seen why the fossil discoveries already made about *Eothyris* didn't already challenge that exclusion from the caseid holobaramin, but Aaron's use of that creationist classification opened still more interesting doors. A "holobaramin" is a much higher hurdle in the baraminology track meet than a mere *monobaramin*, that latter representing an evolutionary subset of a presumed created *kind*. Even with so much of the dataset left off Aaron's analytical scope, it was hard to see how he could have avoided grouping the animals as overlapping monobaramins, except by just treating them as numerical abstractions.

But more interesting was Aaron's following the baraminological trend by treating the *family* rank as the gold standard of a kind, given that our own *Homo sapiens* is but a member of the larger primate family, along with chimpanzees, gorillas and orangutans. Historically, creationists have never wanted to allow that family grouping to hold for us, but the graphic parsing of modern baraminology isn't nearly fine enough to keep the lines from blurring down at the generic level, let alone our particular species, and has the potential (or threat) to lead them down many a theologically problematic

Garden of Eden path.

For instance, while Kurt Wise (2005) skipped past the primate family issue, Todd Wood (2010)—which Aaron had cited—used the same techniques Aaron used to box up the caseids, with the result that *Homo habilis*, *Homo rudolfensis*, "and most surprisingly" *Australopithecus sediba* ended up with us in the human holobaramin. But Wood still lumped the rest of the australopithecines with gorillas and chimpanzees into their own holobaramin, even though anatomically there is far more disparity between the upright australopithecines and the living apes than what separates us from *Homo habilis* or any australopithecine. Frankly, to plop even *Homo habilis* in the human "kind" would have been anathema to creationists of the Morris/Gish era, for whom humans needed to be their own "kind" in Eden with no pre-human connections, showing how far (but not yet usefully) the spin capacity of baraminology has *evolved*.

So how much of the Aaron-style baraminological methodology has filtered down into the practices of general creationism? Well, not much. *CreationWiki* (2014) offered a source-free entry on "Synapsid" that started off with the expected finback *Dimetrodon* picture, which we know looked enough like reptiles to make that appelation plausible for their creationist audience. After that, though, *CreationWiki* dived off a particularly high board to land in an empty pool (my **bold**): "anatomical evidence suggests that the *theriodont* animals (AKA those within the suborders *Gorgonopsia*, *Therocephalia* and *Cynodontia*) were actually mammals, and thus should only tentatively be considered synapsids by Creationists (if at all). Features **unique** to the theriodonts which are either not found, or are **not fully present** in other synapsids are a larger, mammalian dentry [*sic*], ear bones, fully erect pillar-like rear legs and a mammalian temporal fenstra."

By that caveat of adding a "not fully present" into their list, *CreationWiki* managed not only to give "unique" a fresh creationist definition that allowed them to nudge the oh-so-almost mammalian later cynodonts into the allowed mammal typological box, but also inexplicably sucked the far less mammalian *gorgonopsids* along with them (animals that no creationist before them, from Gish to Woodmorappe to their own Creager had contended were mammals).

But by that point *CreationWiki* was off on a wild animal hunt. They suggested the Acambaro figurines of Mexico that surfaced in the 1940s depicted authentic Precolumbian synapsids like *Dimetrodon* along with sauropods and other dinosaurs, and not modern imaginative forgeries. Although they failed to pass close scrutiny years ago, discussed by Charles Di Peso (1953) and Gary Carriveau & Mark Han (1976), a few more rambunctuous spirits since then have been willing to accept them as authentically ancient, from Dennis Swift (1994; 1999) and Don Patton (2011) in the creationist subculture, to of all people mystery novelist Erle Stanley Gardner (1969a-b), showing that the elderly creator of Perry Mason could still miss the boat when it came to identifying genuine Precolumbian ceramics. But then, the author of the hyper-analytical Sherlock Holmes, Arthur Conan

Doyle (1922; 1926; 1930), similarly dented his reputation when he came to defend the reality of *fairies* along with the validity of spiritualism, falling for the Cottingley Fairies photographs concocted by two very industrious girls who snipped pictures from a children's book and posed them on garden bushes, the story nicely summarized by James Randi *et al.* (2012).

CreationWiki also offered up a trio of pictorial links to Roman era paintings and a mosaic that in their estimation depicted therapsids surviving into that time. The idea that the artists might simply have been depicting wild beasts they had only heard of but never seen (such as crocodiles) and did the best they could on hearsay didn't occur to the cryptozoology-friendly set at *CreationWiki*. Oddly enough, they didn't include another example of a supposed Mesozoic survivor that has been making the rounds lately: a "stegosaurus" among the decorations at Angkor in Cambodia, dating to the 12th century. Not suprisingly, the fringe creationists like Don Patton (2008) fielding this beast failed to notice the Khmer artistic convention that filled in most every square inch of open pictorial space with decorative motifs, leafy ones reasonably accounting for the supposed stegosaurian dorsal plates that were festooning what was a passable depiction of a domestic pig, Steven Novella (2008) and Glen Kuban (2011).

The imperative among Ark believers to have prehistoric beasts surviving into historic times is why creationism can't stay away from cryptozoology, from Phillip O'Donnell (2002; 2007) and Mace Baker (2003) to *Genesis Park* (2011a-c) and Dale Stuckwish (2012). Daniel Loxton & Donald Prothero (2013) have energetically surveyed the broad credulity field from Bigfoot to the Loch Ness monster.

Brad Harrub (2015) is another current purveyor of creationist dino survival, confidently insisting (without giving any sources) that "We have scientific proof of elephant prints and dinosaur prints in the same location. We have scientific evidence of dinosaur petroglyphs etched into stone walls by men living just a few hundred of years ago." Although the posting had a 2015 date, its possible Harrub had written it back in 2005, since he referred to the Dover ID trial furor as a current controversy, asking "if the ACLU has stopped its legal intimidation tactics long enough to actually look at the latest scientific data?"

What Harrub had in mind was the reevaluation of *Hadrocodium* by Luo & Crompton *et al.* (2001), and the discovery of the dinosaur-munching *Repenomamus*, citing Hu *et al.* (2005) and the accompanying commentary by Anne Weil (2005). While David Coppedge had been concerned only about *Repenomamus*' ribs, according to Harrub the very existence of *Hadrocodium* and that larger mammalian predator "challenged everything evolutionists have ever maintained regarding dinosaurs and mammals."

No it didn't. It had been known for over a century that mammals had coexisted with dinosaurs, as I have made note of throughout this book. But it was also true that mammals never proliferated notably until after the K-T extinction cleared away their dinosaur competition. Even more ironically in

Harrub's not connecting the fossil dots, *Repenomamus*' dinner was a juvenile *Psittacosaurus* (the group from which the later ceratopsids would evolve, noted back in Chapter 4 regarding Duane Gish's sidestepping them)—a distinctly carnivorous thing to do for an animal supposedly herbivorous before the Flood.

Harrub thought to clinch his case by appealing to the views found in two dated general textbooks, Colin Pittendrigh *et al.* (1957) which George Gaylord Simpson had coauthored, and Peter Raven & George Johnson (1989), but even there Harrub stumbled on his own sources (his **bold**):

> Consider the following evidence. Most evolutionary timelines have the mammals evolving from reptiles. Indeed, the textbook I used in my freshman general biology class noted: "During the Mesozoic Era the reptiles, which had evolved earlier from the amphibians, became dominant and in turn gave rise to the mammals and the birds" (Raven and Johnson, 1989, p. 432). George Gaylord Simpson contended that, according to standard evolutionary theory, no advanced mammals were present in the "age of the dinosaurs." The dinosaurs, he suggested, became extinct in the Cretaceous Period, and the only mammals that had evolved up to that point (even at the very end of the period) were "**small, mostly about mouse-sized, and rare**" (1957, p. 797, emp. added). This is a logical explanation if you are going to contend that mammals evolved from reptiles, because it would require that mammals appeared much later in the picture.

How often we've seen creationists lacking a functional *Map of Time* in their heads seize on such qualifiers as "mostly" to act as though that meant *never*, or throw up their hands in amazement when some new fossils come along to broaden our understanding of what went on outside that "mostly."

Theoretically Harrub would have had a *Map of Time*, a painfully rigid one, but his readers were spared that backstory. Instead he showed only his own muddle, for the issue wasn't the origin of mammals (*that* had happened 150 million years before the Cretaceous) but when the *modern* therian lineages emerged during the Cretaceous. Nothing in the half century separating the 1957 Pittendrigh/Simpson textbook, or the intervening 1989 Raven & Johnson a quarter century on, had changed any of those systematic benchmarks. Nor had the RMT facts gone away by the time the later edition of Raven & Johnson (2001, 975) touched on the issue.

We're well into the 21st century now, and with all that time to *practice*, still the Creationism team was playing no better on the field than Duane Gish had a generation before.

11. Intelligent Design Can't Even Dribble

We've seen how big and growing the RMT dataset is, and how a whole generation of antievolutionists (YEC and ID both) have managed to trip over and around it all without ever seriously coming to grips with even a fraction of the available facts. And the reason for that is because in a most fundamental way, their apologetics have never been about the facts, the details of fossils and the biology of living animals, but rather a set of tropes marshaled in service of the *Kulturkampf* "culture war" that's about social issues and theological worldviews.

At that lofty philosophical level it's relatively easy for YEC and ID to rub shoulders, as they share a common disdain for "materialism" and its biological science embodiment of "Darwinism." Fold in the common failure of any antievolutionist to work out a substantive *Map of Time* in their own heads, and you have a playing field where lots of grassroots antievolutionists busily conduct a strenuous game under their own rules but without an opposing evolutionary science team allowed on the field to compete.

Evolution Dismantled (2014) tackling the topic of transitional fossils is the sort of thing you get as a creationist website these days, freely drawing on a smorgasbord of sources, from Duane Gish (1995) on the YEC side to Casey Luskin (2008a) culled from ID Land, with a thin sprinking of technical works like Alan Feduccia *et al.* (2005) dropped along the way—primary science sources that the anonymous authors likely knew about only because they had read about them in their secondary mining of the antievolution literature.

There was no sign that the writers ever looked into the reliability of their specific secondary sources, from the extensive criticism of Duane Gish over the years, down to that Luskin (2008a) *Evolution News & Views* piece opining on how to identify homologous tetrapod digits. If *Evolution Dismantled* knew about Carl Zimmer (2008) or P. Z. Myers (2008) explaining exactly why Luskin's analysis was shamefully wrong, they did not share those contrary takes with their readers. But it's more likely that the *Evolution Dismantled* team never thought to look, as they followed the practice of long standing in antievolutionary apologetics, where you let one faction represent both sides on the field.

More revealingly, even though Gish (1995) had addressed the RMT (badly, as we've seen), we know the more recent Casey Luskin hadn't been interest in it, and that may be a good measure of how disinterested current antievolutionists are with merely explaining data like the synapsids. Lacking perhaps the grabby panache of media stars *Tiktaalik* or *Archaeopteryx*, the gang at *Evolution Dismantled* couldn't be bothered with them.

Evolution News & Views explores the mammalian middle & inner ear

Casey Luskin may have had other interests than a hundred million years of transitioning synapsids, but his fellow *Evolution News & Views* author Jonathan McLatchie (2012b) was drawn onto the field by someone's question (my **bold**):

> A correspondent recently asked me about the evolution of the mammalian middle ear in relation to the fossil record. **Based on data gathered from embryology**, it is widely thought that the bones of the mammalian middle ear (the region just inside the eardrum) evolved from bones of the reptilian lower jaw joint. **Besides the paleontological data**, this hypothesis is based on the fact that, in mammals, Meckel's cartilage plays a role in forming the middle ear bones and mandible before subsequently disappearing. In reptiles, it ossifies to become part of the jaw.

Well, well, well, so someone at *EN&V* was seemingly aware of that embryological data (though McLatchie didn't cite any sources on it) and the corroborating paleontological data, or at least two papers: Zhe-Xi Luo & Peiji Chen *et al.* (2007) on *Yanoconodon*, and Jin Meng *et al.* (2011) on *Liaoconodon*, both taxa which a few creationists had hit on previously, as we saw last chapter with David Coppedge and Shaun Doyle.

McLatchie sought to pull apart the fact that the Meng paper had discovered "the first unambiguous ectotympanic (angular), malleus (articular and prearticular) and incus (quadrate) of an Early Cretaceous eutriconodont mammal," still linked by its ossified cartilage whose existence was expected based on the way the ear had been evolving in those lineages. McLatchie's counterpunch began (my **bold**):

> There are a few points that are worth raising here. Firstly, even supposing that the hypothesis of common ancestry is valid, this **lends little traction to neo-Darwinism** (one has to distinguish between pattern and process) and it does **nothing** to undermine the hypothesis of design. ID, in its purest sense, has **nothing to say about common ancestry**. **ID does, however, open up the possibility that universal hereditary continuity may be false, perhaps radically so. Many of us Darwin critics, therefore, also happen to be skeptical of common ancestry. But it would not invalidate our position on ID if common ancestry turned out to be true.**

This was meaningless drivel, but was entirely consistent with the decades of antievolutionary arguments going back to the Duane Gish era, where common descent was swept aside not by discussing the data, but by tactically worded rhetorical hand waving.

At this point McLatchie hit on the issue of whether the mammalian middle ear evolved independently in monotremes and eutherians (marsupials and placentals). McLatchie sang a familiar refrain: "Multiple occurrences of difficult evolutionary trajectories is something that is not easy to square with the standard neo-Darwinian narrative."

Says who? You already know the antievolution field from the previous chapters. No one McLatchie could have cited had pulled that trick off. All he had was the repetition of it as something antievolutionists would *like* to be true. So it's understandable that he didn't cite what he couldn't, and pressed on with a Gishian detail fiddle: *Liaconodon*'s middle ear "differs from that of *Yanoconodon*."

And why shouldn't it, given that *Yanoconodon* lived some *2-5 million years* before *Liaoconodon*? None of that *Map of Time* aspect stopped McLatchie from then pulling a *Garrett Hardin* by quoting from Meng *et al.*'s supplemental information, which he ran as a set of continuous text, even though he left out a whole paragraph. Here's the original Meng text, with the part McLatchie quoted in **bold** and several assorted [typos] noted along the way (suggesting he manually typed in the text rather than copy/pasted):

> **In general, what have been interpreted as ear ossicles in *Yanoconodon* differ significantly from the middle ear elements of *Liaoconodon*. In *Yanoconodon* all ossicles are broken, fractured and displaced. These elements are embedded in the matrix and only their broken sections and impressions are** [McLatchie had a typo here, as "anre"] **visible** (Luo et al., 2007; Part C in the Supplementary Information).
>
> Comparing with the malleus of *Liaoconodon*, the isolated element interpreted as the main body of the malleus of *Yanoconodon* is proportionally quite large and robust in relation to the ectotympanic; it does not have the hook-like configuration, and the malleoincudal articular facet is not seen in the fossil, nor is it from the reconstruction. The fragments identified as the goniale ("prearticular") of *Yanoconodon* do not match the anterior process of *Liaoconodon*.
>
> **The element identified as the incus of *Yanoconodon* was considered similar to modern mammals in having a crus longum (stapedial process) and a crus breve (for basicranial articulation). This element is proportionally much smaller than that of *Liaoconodon* and shows a different morphology. The articular facet for the malleus is not clearly shown. Because only the impression of the element in *Yanoconodon* was illustrated, it is difficult to make any detail comparison with the incus of *Liaoconodon*.**
>
> **The ectotympanic and its impression are probably preserved more completely in *Yanoconodon*; the outline of the element is similar to that of *Liaoconodon*. The dorsal crus of the ectotympanic and the prearticular element of the malleus in *Yanoconodon* were** [McLatchie typo had this as "was"] **interpreted as being fused to each other and both are further connected (fused) anteriorly with the ossified** [McLatchie mispelled that as "fossified"] **Meckel's cartilage. This differs from the condition of *Liaoconodon* in which the ectotympanic, the malleus and the OMC are clearly not fused.**

And what did all of this *mean*? Again, in the Grand Tradition of detail-fiddlers from Gish to Denton, McLatchie explained nothing whatsoever about any of this.

The Meng paper correctly noted where fossil details were limited, starting with the crushed state of *Yanoconodon*'s ear bones, but that didn't mean there was nothing to be gleaned from them. McLatchie certainly had the overall bone layouts available to him, as Luo & Chen *et al.* (2007, 291) and

Meng *et al.* (2011, 183) illustrated those of *Yanoconodon* and *Liaoconodon* respectively. There you could see that the angular/prearticular/articular/quadrate units were arranged the same way in both animals, forming a C-shaped arc still attached to the jaw at the angular by a long sliver of cartilage. The two fossils differed in how much broader the C-shaped conformation was in the earlier *Yanoconodon* than it was in *Liaoconodon*, where the points of the C were curved together tip to tip (as in the living monotremes and eutherians).

As you'd expect in a gradually evolving system, where incremental variation couldn't be prevented in their inherited anatomy, the incus in the earlier *Yanoconodon* showed signs of the tiny bumps that related to cranial articulation in modern mammals, and which were notably enlarged in the later (and better preserved) *Liaoconodon*. Meng *et al.* also reminded in their supplementary information how that fossil provided "the first direct evidence in Mesozoic mammals that the malleo-incudal articulation is associated with the jaw suspension and corroborates the Reichert-Gaupp theory on homology of the mammalian middle ear ossicles, a subject that has been discussed in numerous studies"—that was the physical fossil *confirmation* of that bone shift observed in the embryological work McLatchie did not otherwise discuss.

Rather than plunge into any of those data, McLatchie must have thought the information he quoted from Meng somehow presented a real obstacle to their natural evolution, or at least got ID off the hook to account for any of it themselves. In that respect, he really was sounding like Gish and generations of creationists here. McLatchie went on to an equally familiar argument that presumed lots of what he never bothered to establish (my **bold** but McLatchie's *italics*):

> Finally, in the absence of a **viable materialistic mechanism** to account for the transition in question, the supposition that one can slap these different structures down on a table and draw arrows between them seems highly suspect. The methodology is circular—it *assumes* that these structures are connected by descent. When one's entire interpretative framework presupposes common ancestry at the outset, it is no wonder that any and every observation is taken as supportive of that paradigm. **When some of the most impressively documented transitions in the fossil record (<u>such as the evolution of the whale</u>) occur far too quickly to be feasibly attributed to a naturalistic mechanism, do we not have reason to question whether that interpretative framework is actually valid?**

Here's where ID's insular worldview rubs methodological shoulders with YEC without even realizing it. Disallowed *assumptions*? Just the one, common ancestry? Doesn't everybody get to play this game?

Read a paper on *geology* and there too the scientists will not be rearguing the entire history of their discipline. They will be *assuming* (horrors!) the great age of the Earth, because they are drawing on all that prior science work that had established that paradigm to any reasonable person's satisfaction. Unless, of course, you were someone like Andrew

Snelling or John Woodmorappe, whose YEC dogmas did not allow that to be true.

If McLatchie gets to play this game, though, why can't YECers? Or, if they are ruled off court, despite their invocation of science data with just as much enthusiasm as IDers do, how does that not catch McLatchie's argument in the same net?

McLatchie certainly carried a giant assumption of his own, of course, that someone had actually done that successful challenge of the RMT dataset. Readers of this book don't have to speculate on that failure—you've read their arguments, and seen how far all of them have been from dealing with both the fossil evidence and the details of that "materialistic mechanism."

And like so many antievolutionists before him, instead of doing any of that hard work himself, McLatchie relied on a secondary authority to remove the problem for him. That underlined section was a link not to any technical paper that could be source fact-checked for accuracy, but to a ten-minute video lecture ("Whale Evolution Vs. Population Genetics") done by ID advocate Richard Sternberg in 2010. The only science work Sternberg alluded to directly in the video was what must have been Richard Durrett & Deena Schmidt (2007; 2008; 2009), which you will recall from Chapter 9 was part of a critical dispute with Michael Behe & David Snoke (2004) and Behe (2007; 2009) over the dynamics of mutation fixation.

But a video is *not* a technical paper, any more than a lecture by Kent Hovind. Even Neil Shubin's *Your Inner Fish* video series was only a steppingstone to the data, not a substitute for it. Without a fuller exposition of the argument, how exactly had McLatchie fact checked Sternberg's calculations to be so certain they were correct? Indeed, *had* McLatchie fact-checked Sternberg's calculations? Or had he just accepted that authority, granting Sternberg all *his* assumptions in a way he denied evolutionists?

Half a decade on from Sternberg's video lecture, there's still no substantive paper to inspect. What, no space in the lone current Intelligent Design technical venue, *BIO-Complexity*? Whether Sternberg's claims on the mutational impossibility of whale evolution occurring in "only" 5-9 million years would fare any better than Michael Behe's chloroquine show we looked at back in Chapter 9 remains to be seen.

McLatchie's post went far beyond just a double standard. Whether he realized it or not, McLatchie wanted his side to carry scientific weight without having done the work, while not allowing their opponents the weight of their own work that they *had* done.

And then there's *Evolution News & Views* (2013), which hit on our "amazing" mammalian ear again to criticize a technical paper on cochlea by Dáibhid Ó Maoiléidigh & James Hudspeth (2013) that suggested how the mammals' high-frequency perception had evolved. *EN&V* quoted the concluding passage of the paper, highlighting in **bold** some wording they deemed particularly objectionable: "Active hair-bundle motility is ubiquitous among tetrapod vertebrates and likely occurs in more basal clades as well.

The present model **suggests an evolutionary progression** in which high-frequency hearing arose when a preexisting auditory amplifier supplemented by active hair-bundle motility was augmented with an additional source of mechanical energy stemming from somatic motility."

EN&V pounced on that:

> Our amazement now turns to the illogic of such explanations. They just said that "active hair-bundle motility is ubiquitous among tetrapod vertebrates and likely occurs in more basal clades as well." This implies that the system in basal clades are already fine-tuned for high sensitivity. Where is there room for an "evolutionary progression"? How did neo-Darwinism achieve that fine-tuning in the first place? Where is the evidence of unguided evolutionary progression?

The unidentified *EN&V* author had just slipped a really big cog, and most clumsily, as O Maoileidigh & A. Hudspeth had *not* claimed that basal clades had *high frequency* acuity, only that they possessed those active hair bundles. In fact they *couldn't have* heard high frequencies, since it was the origin of that exclusively mammalian perception in later animals that was the subject of the paper.

The authors had specified that "only low-frequency spontaneous oscillations have been observed in non-mammals," O Maoileidigh & Hudspeth (2013, 5474). They noted *somatic motility* as the ability of outer hair cells to alter their length when stimulated by a change in membrane potential, and that occurred because of the millions of *prestin* molecules in their base. It couldn't have been put more plainly: the hearing evolution of mammals involved the *addition* of that energetic factor acting on a *previously evolved* hair bundle system which *lacked* that high-frequency perception. Their article was not expressly about that evolutionary path, though, but rather was a technical paper targeted to clarifying the physical parameters of the high-frequency perception system, expanding on earlier work, such as Gerhard Frank *et al.* (1999).

The concluding paragraph in the O Maoileidigh & Hudspeth paper had cited Geoffrey Manley (2001) on the cochlear amplification system in mammals. Which meant *EN&V* could have followed up on that trail, had it been their purpose to fully understand the topic before launching a diatribe on it.

Let's start with the abstract of Manley's 2001 paper, where we encounter again an old friend, that *organ of Corti*, and the role it played in all this:

> The last two decades have produced a great deal of evidence that in the mammalian organ of Corti outer hair cells undergo active shape changes that are part of a "cochlear amplifier" mechanism that increases sensitivity and frequency selectivity of the hearing epithelium. However, many signs of active processes have also been found in nonmammals, raising the question as to the ancestry and commonality of these mechanisms. Active movements would be advantageous in all kinds of sensory hair cells because they help signal detection at levels near those of thermal noise and also help

> to overcome fluid viscosity. Such active mechanisms therefore presumably arose in the earliest kinds of hair cells that were part of the lateral line system of fish. These cells were embedded in a firm epithelium and responded to relative motion between the hair bundle and the hair cell, making it highly likely that the first active motor mechanism was localized in the hair-cell bundle. In terrestrial nonmammals, there are many auditory phenomena that are best explained by the presence of a cochlear amplifier, indicating that in this respect the mammalian ear is not unique. The latest evidence supports siting the active process in nonmammals in the hair-cell bundle and in intimate association with the transduction process.

It's important to recognize that the outer hair cells in mammals lacked a substantial nervous system "afferent innervation," where the *sensory* neuron input signal was sent, but mammals do have an "efferent innervation," where the *motor* neuron input signal stimulated their movement, and it was this specialization that underlay how the system had evolved. Some reptiles show responses to higher frequencies (despite lacking the extra middle ear bones the mammals coopted), and geckos have been studied the most, being a rare lizard family that uses vocal signals in estabilishing territoriality. Significantly, their specialized hair bundle pairs formed from their *basilar papilla* (remember those regarding the organ of Corti from back in Chapter 4), and these also lacked that afferent nerve innervation in one set, while the other showed the efferent connection, studied by Eugenia Chiappe *et al.* (2007) and Manley & Johanna Kraus (2010). The working idea is that a mechanical resonance gets going in their interaction, amplifying the signal in a way also seen in their very distant mammal cousins.

Incidentally, in a superficial stab at defining the various lizard "kinds," creationist Tim Hennigan (2015) peripherally cited Manley & Kraus (2010) regarding the "unique hearing apparati" in what Hennigan insisted was the gecko "kind" but discussed none of the physical details the authors had noted, showing yet again how *Gee Whiz* expositions on creationist taxonomy involved paying attention to noticably less *data* than their evolutionary counterparts.

The evolutionary process of what was going on down in the mammal ear when high frequency was involved had been worked out in considerable detail by 2013, but *EN&V* wasn't getting close to that so long as they were stuck on the cosy typological mantra that there was a stereotypical "mammal" ear that could hear high frequencies from the Designer get-go, or slow down to investigate whether any paleontology had been done on the subject.

As we've seen in prior cases, while the older fossil sectioning method could easily destroy the delicate features of the inner ear, newer micro-CT scanning technology allowed those to be studied non-invasively, reviewed by Manley (2012). Add to that, the voluminous data on the molecular evolution of the biology used in hearing, such as by Lavanya Rajagopalan *et al.* (2006), Xiaodong Tan *et al.* (2011; 2012) and Zhen Liu *et al.* (2012) prior to the EN&V post, and Dmitry Gorbunov *et al.* (2014) still more recently, and you have a lot of data on the table, which the scientists have not been shy about making sense of.

The critical prestin protein started out as one among the diverse anion transporter superfamily, proteins that preferentially moved molecules with negative charges. While it no longer transported anions across the cell membrane in mammals, experiments showed that it still retained that family function, and simple amino acid mutations in a particular section of the prestin molecule determined whether it functioned as an anion transporter or changed to a motility generator.

Farther down the mammal trail, Manley (2012, 740) noted prestin evolution figured in the development of echolocating skills in bats and whales, citing Ying Li *et al.* (2010) and Yang Liu *et al.* (2010), adding the Fossil Genie confirmation of Nancy Simmons *et al.* (2008), where the oldest bats from 60 mya lacked the large cochlea needed for echolocation, while their descendants ten million years later possessed more robust ones suggesting that faculty was developing.

Now for some of the Big Picture. Manley (2012, 737) pointed out that mammals went through the first half of their history with no ability to hear the high frequencies known in modern mammals, and the gradual increase in that capacity in therians was spread out over some *fifty million years*. To give you an idea of how microevolutionary those changes were, the cochlea in Late Jurassic fossil mammals were only about three millimeters longer than their ancestral forms *eighty million years* earlier. Gradual enough for you?

It was their coiling up into a 270° spiral that opened up broader acoustic perception, and Manley (2012, 738) noted that

> the earliest mammalian cochleae (as those also of modern nonmammals) were smooth-walled bony canals that had no firm contact to the soft tissues. In modern nonmammals, it is possible to insert a small hook into scala tympani and pull out the entire cochlear "tube," something inconceivable in therian mammals. Yet it would have been possible in the earliest mammals and would even today be possible in monotremes.

In researching for this book, again and again I have encountered such a wonderful array of science knowledge. But *none* of it was coming from the hand of antievolutionists, and least of all from the webpages of *Evolution News & Views*.

Denyse O'Leary plays mix & match with "Intelligent Design" sources

Canadian Intelligent Design popularizer Denyse O'Leary is far more like creationist David Coppedge than a detail-fiddler like Duane Gish or John Woodmorappe (or even Casey Luskin on the ID side), in that she criticizes evolution without really paying close critical attention to the content of the many primary technical papers she alludes to (often linked secondarily via general press coverage). Many of her *Evolution News & Views* & *Uncommon Descent* postings can go quite overboard with the scattershot, involving a barrage of links to dozens of targets, such as done in O'Leary (2013a-c; 2015a,d-f).

O'Leary (2015b) veered close to our RMT subject with this:

> Similarly, whether large bird and mammal brains arise from common descent or convergent evolution is actually uncertain. Two distantly related groups of reptiles are thought to have given rise to mammals and birds, both featuring a much higher brain to body weight ratio than in their ancestors. Paleontologist R. Glenn Northcutt writes that the matter is 'contentious and unresolved,' because brains rarely fossilize.

Her underlined link was to a secondary source, Glenn Northcutt (2011), which was a *Science* magazine commentary on the paper by Timothy Rowe *et al.* (2011) that had presented new research on the brain endocasts of *Morganucodon* and *Hadrocodium* that went a long way to resolve that very issue (mentioned previously back in Chapter 7). Indeed, it wasn't the mammal side of the brain evolution thing that posed the big conundrum, but that of *birds*. Northcutt concluded his commentary (my **bold**):

> The endocasts of *Morganucodon* and *Hadrocodium* provide the first solid evidence of the stages in mammalian brain evolution. Unfortunately, far less is known about the emergence of the brains of living birds. The endocast of the oldest known Late Jurassic bird, *Archaeopteryx*, reveals reduced olfactory bulbs, large cerebral hemispheres that are in contact with an expanded cerebellum, and lateroventrally displaced midbrain lobes (*11*). **All of these neural traits also characterize living birds, and, not surprisingly, the relative size of the brain of *Archaeopteryx* appears to have been intermediate between that of living reptiles and that of living birds** (*11*, *12*). [Citing Patricio Alonso *et al.* (2004) and James Hopson (1977).]

Drawing further on a new dinosaur endocast described by Martin Kondrat (2007), Northcutt continued (again my **bold**):

> All these derived neural traits, plus its skeletal characteristics, suggest that *Archaeopteryx* was capable of flight. The missing part of this story is the appearance of the brain in the coelurosaurian theropod dinosaurs that gave rise to birds. The endocasts of these dinosaurs have not been described, but the endocast of an oviraptorid theropod, *Conchoraptor*, revealed **reduced olfactory bulbs, enlarged cerebral hemispheres, displaced midbrain lobes, and an enlarged cerebellum (*13*), as in living birds.** Furthermore, the relative brain size in *Conchoraptor* was similar to that in *Archaeopteryx*. This suggests that the neural traits of *Conchoraptor* may have evolved parallel to those of avian theropods, but it is also possible that these neural traits were widely distributed among theropods. If so, many 'avian' neural traits may have already been present in coelurosaurian theropods and were co-opted for flight.

And what did O'Leary have to say about that information, all of which needed to be accounted for by any proposed Intelligent Design alternative? As usual, *nothing*. So not only was she not proceeding to the main Rowe paper's content, she didn't even grapple with the insights Northcutt was putting before her in that secondary summary. Or do follow-up, either, as

O'Leary was writing in 2015, years after Northcutt's original publication.

Others have been far less reluctant to connect up the information. It's those working scientists again.

Take Ken Yokoyama & David Pollock (2012), who included Northcutt's observations on the parallel brain expansion in birds and mammals as part of a persistent trend in those classes, where mutations and the selective factors acting on them arose repeatedly and independently over the several hundred million years spanned by their separate lineages. In particular, the SP1 *zinc finger* transcription factors they were studying had undergone quite a convergent binge in birds and mammals, where *800 regulatory regions* had evolved in parallel ways based on "a single causative amino acid replacement at the same SP1 position along both lineages."

Yet more signals of the microevolutionary mechanism at work, cascading like so many small mutational rocks plopped naturally again and again, contributing by their interacting ripples to the variety and diversity of life.

It's still good to remind everybody of the scale of the problem, though. Just think about how difficult a task it is, working out the changes in extinct organisms when the only developmental and genetic data you have comes from their highly modified living descendents. And only a sampling of brain endocasts are available to survey the broad configuration of those fossil forms, while telling nothing at all about the cellular transmission connections that existed within their once-living brains.

Northcutt (2011, 926) put one of those fine point issues on the table when he mentioned the neocortex, and how "comparative neurobiologists continue to vigorously debate whether these centers evolved from the same ancestral neural centers," citing Harvey Karten (1997) and Anton Reiner *et al.* (2005) for that view, "or from different ones," referencing his own Northcutt & Jon Kaas (1995) along with Georg Striedter (1997) and L. Bruce (2007). Just how "vigorously" things were debated could be seen in the critical response of Ann Butler (1996) and Anton Reiner (1996) to how their positions were not quite accurately represented in Northcutt & Kaas' 1995 paper, prompting a conciliatory reply and clarification by Northcutt & Kaas (1996). A lot of the fracas turned on identifying more about the detailed developmental biology and genetics of the known brain systems, such as when and how the "dorsal ventricular ridge" specialized in mammals and birds, issues reviewed in detail by Francisco Aboitiz & Daniver Morales *et al.* (2002) and Aboitiz & Javier Lopez *et al.* (2002). Still more recent work has identified the ancient cellular and genetic networks supporting the convergent similarities of mammals and birds, reviewed by Thomas Finger *et al.* (2013) and Karten (2015).

What, if anything, could Denyse O'Leary's Intelligent Design have to say about any of that work, other than to find some entry point of uncertainty to dismiss it all as yet another intractable obstacle for "Darwinism"? Could ID ever in principle deduce anything predictively about living or fossil forms in the way modern evolutionary science has done as a matter of course? So far ID has failed to even strike out at that, following the YEC precedent of seldom trespassing onto the playing field in the first place, and so avoiding the embarassment of taking clumsy swings at the many newer factual balls whizzing by, pitched by all those many scientists doing the work.

What O'Leary did think to swing at came a bit later in her 2015 *EN&V* post, when she wrote, "A number of recent marketing strategies promise sales through appealing to a customer's 'self-centered' reptilian brain. But that piece of business folk wisdom is based on a myth," quoting from the opening paragraph of the secondary source linked in the myth link, the blog of Janet Kwasniak (2014). I've highlighted what O'Leary quoted in **bold**:

> The reptilian brain is a myth that should not be taken seriously and yet is referred to by many writers and is even seen in educational sites for children. **It is the idea that we have three brains: a reptilian one, the paleomammalian one and the mammalian one. The story goes that these were acquired one after another during evolution. The details differ with the writer. But it is all a myth based on an idea from the '70s of Paul MacLean which he republished in 1990. Over the years in has been popularized by Sagan and Koestler among others.**

Here O'Leary was slamming into a topic that was simple enough to understand, but could easily lead to blatant misunderstanding if, like O'Leary, one blundered into an area that used terminology in a specialized way that you've never bothered to think through. Remember my warnings right in the first chapter about how modern cladistic systematics does not use certain terms in the way popular usage does. No "fish" as a monophyletic clade—and "reptile" refers only to a particular subset of aminiotes.

Unfortunately for O'Leary's credibility as an attentive reader of her own cited (secondary) sources, Kwasniak had been quite explicit on this very point, having explained all about it only a couple of paragraphs down from the passage O'Leary nicked (my **bold** for an especially relevant bit):

> One problem with the reptilian brain is that we are not evolved from reptiles. The last common link between mammals and reptiles is called amniotes. They were like amphibians but did not need to lay their eggs in water. In other words, they were the first truly land-dwelling vertebrates and all terrestrial vertebrates evolved from them. They did not have a neocortex but they had all the other anatomical parts of the brain. The amniotes evolved into two groups: the diapsids which further evolved into four lines—turtles, lizards/snake, crocodiles, birds; and the synapsids which evolved into mammals. Mammal evolution is separate from reptiles from the earliest terrestrial vertebrates. **What is more, the neocortex makes its appearance very early in the synapsids line.** The triune story of what animals had what sort of brain is simply not what evolutionary biology has found.

Later Kwasniak noted, "All the descendants of amniotes have essentially the same architecture of brain with the same functions. There are differences in proportions, sizes, connections, fine-scale anatomy but not a gross difference of kind in the brains of land vertebrates."

How had O'Leary managed to miss all that? Unless, of course, she wasn't reading the content for comprehension, but only for *ammunition*. Once she thought she'd found it, O'Leary raced on to the next weblinked paragraph in her argument.

A further measure of O'Leary's tendency for glib secondary redaction was seen in that same 2015 posting when she wrote "The popular science literature claims that a near identity between the human and chimpanzee genome is irrefutable evidence of common descent. Why then do we hear so little about any of these findings, which muddy the waters? Why are science writers not even curious?"

That "irrefutable evidence" underlined text was a weblink that proved to be quite a daisy chain—and can of worms. It was to her own O'Leary (2014) post at the antievolution website *Uncommon Descent*, titled "Picture that terrifies creationists?" and criticizing Chris Mooney (2014) writing at *Mother Jones* (hardly a technical source).

In that 2014 *Uncommon Descent* post she opined:

> Naturally, I thought they had discovered a man actually morphing into a fly. It turns out, Mooney provides only primate gene sequences showing similar chromosomes of humans, chimpanzees, gorillas, and orangutans. Which mainly shows what genetics *doesn't* do. That is, if someone wants us to know that a (frequently claimed) 98% similarity between the guy fixing a computer and the chimp throwing poop proves something, I'd say it sure does. It proves that genes are only a tiny part of the story of inheritance. Seems we got a long ways to go to understand that.

Thus had O'Leary quickly sidestepped the *content* of Mooney's piece, which involved recounting not merely the similarity of the chromosomes, but that one of our chromosomes was literally formed by a *fusion of two of the old primate sequences*, and how scientists knew this because ours still retained fragments of the old centromere and endcap telomeres left over from the ends and middle of the formerly separate genes.

But O'Leary added to the depth of her hole when she added a PS to the 2014 post: "In fairness, some people dispute the high figures. *See also:* Genomics scientist Jeffrey Tompkins takes issue with BioLogos' we are 98% chimpanzee claim." That was a link to an anonymous *Uncommon Descent* (2013) posting that had in turn cited Jeffrey Tomkins (2012c), *More Than a Monkey: The Human-Chimp DNA Similarity Myth*. Yes, it's *that* Jeffery Tomkins, the creationist mentioned in previous chapters.

Tomkins (2011a-c; 2013a,c,d) has repeatedly hammered the YEC spin on the new genetic data in AiG's *Answers Research Journal*, as have Jerry Bergmann & Tomkins (2011) and Tomkins & Bergman (2011). Tomkins has also authored more general treatments for the ICR's *Acts & Facts* magazine and postings on their website, Tomkins (2011d-e; 2012a-b; 2013b,e). That confident spin may be compared to some of the regular technical work on the human chromosome fusion topic, from J. IJdo *et al.* (1991) and Yuxin Fan *et al.* (2002), to Julie Horvath *et al.* (2005) and Mario Ventura *et al.* (2012). Suffice it to say that the subject of chromosome fusion and mutation, and the ways by which antievolutionists dervish around it all, is deserving of another book.

"In fairness," Tomkins' short book O'Leary cited hadn't gone out of its way to make clear the author's doctrinal committment to Young Earth Creationism, though O'Leary ought to have known about that, since Mooney's article had explicitly mentioned not only Tomkins' attempt to refute the human chromosome fusion evidence, but also highlighted biologist Kenneth Miller's video lecture criticisms of those creationist claims. So was O'Leary not paying that much attention to even her own sources? Possibly not, since both her *EN&V* and *Uncommon Descent* offerings consistently *misspelled* Tomkins' name as "Tompkins."

Is there any wonder why I'm not particularly impressed with antievolutionist "scholarship"?

Denyse O'Leary (2015c) richocheted past the RMT again in a long posting arguing that the fossil record was littered with examples of "stasis" (that catchphrase which in Design circles is taken as an absence of evolution). One instance of which was: "*160 million years ago:* A Jurassic placental mammal 'redefines mammal history,' showing mammals are much older than formerly thought." That underlined link was to Jonathan Amos (2011), a *BBC* piece reporting secondarily on the new fossil find of an early placental mammal, *Juramaia*, by Zhe-Xi Luo & Chong-Xi Yuan *et al.* (2011).

How *Juramaia* redefined things was explained plainly enough by Amos: "The Liaoning specimen is especially significant because it means the fossil record now sits more comfortably with what genetic studies have been suggesting about the timing of the emergence of the different mammalian lineages. These DNA investigations had indicated that eutherians should have been in existence much earlier than the previous oldest-known eutherian fossil—a creature called *Eomaia*, which lived about 125 million years ago." And had O'Leary ventured from Amos' piece to look at the original *Nature* paper, the abstract there would only have reinforced this point, stressing how the fossil was "reducing and resolving a discrepancy between the previous fossil record and the molecular estimate for the placental-marsupial divergence," and how its features "provides the ancestral condition for dental and other anatomical features of eutherians."

O'Leary was teetering on the edge of a *Garrett Hardin*, since the new fossil not only filled in the paleontological gap (*Juramaia* lived 35 million years before *Eomaia*), it fulfilled the predictions of the genetic analyses that suggested placentals had evolved earlier than the fossil record had shown up to that time. But then, the Fossil Genie has been fond of that sort of thing, such as the 125 mya early marsupial relative that in one jump similarly pushed their fossil record back *50 million years*, Richard Cifelli & Brian Davis (2003) commenting on Zhe-Xi Luo *et al.* (2003).

These fossils were bumping into the tribosphenic molar issue Creager tripped over in the last chapter, as those tooth forms were evolving independently in two ancient holotherian mammalian groups with different geographic distributions during the Jurassic and Early Cretaceous: an *australosphenidan* clade endemic to the Gondwanan landmasses in the south (survived by the isolated monotremes); and the *boreosphenidan* clade populating the northern Laurasian continents, including the living marsupials, placentals and their closer extinct relatives, laid out by Zhe-Xi Luo & Robert Cifelli *et al.* (2001), Zhe-Xi Luo & Crompton *et al.* (2001) with commentary by André Wyss (2001), and Zhe-Xi Luo & Qiang Ji *et al.* (2007).

Depending on which fossil or genetic data were seen as most relevant to the sequence and pacing of these differentiating critters, scientists like Robert Meredith & Jan Janecka *et al.* (2011), William Murphy *et al.* (2012) and Mark Springer *et al.* (2013) perceived quite a "Long Fuse" burning over tens of millions of years before the K-T event shook things up. Modern mammal lineages took their time to show up over tens of millions of years after the K-T, highlighted in the "delayed rise" view of Olaf Bininda-Emonds *et al.* (2007) and Bininda-Emonds & Andy Purvis (2012), while Maureen O'Leary *et al.* (2013a-b) focused on the timing of the basal differentiation of their larger related groups, which clustered shortly after the K-T in an "Explosive Model." None of these ways of looking at the data were mutually exclusive, just as historians can regard the 18th century as a garden of slow social change compared to the 20th century world, even as it was an era of great social convulsion if you took the American and French revolutions boiling at its end as the benchmarks.

The biogeographical and temporal patterns of the later RMT animals that the science work was clarifying, year on year, stood in stark contrast to the absolute vacuum represented by the Denyse O'Leary secondary brand of ID during those same years. None got even as far as Creager's creationist potshotting, and taken together, none managed to do any better than Duane Gish had in the detail fiddling department back when VCRs and fax machines were cool products.

I can't resist giving you one final example of O'Leary's "scholarship" from her 2015 fossil post. Sounding a lot like conventional creationists who made way too much of the "living fossil" trope, O'Leary (2015c) tossed off how "a lacewing insect fossil, found at the 120 million year stratum in China, prompted the comment that, 'Seems that at this rate of evolution, we'd need trillions of years, maybe lots more, to see any substantial change,' because of its similarity to current lacewings."

The underlined link was to a *ScienceDaily* (2011) posting on the technical paper by Yuanyuan Peng *et al.* (2011). But there she'd slipped a big cog, because the comment she quoted hadn't occurred in the *ScienceDaily* secondary piece at all, but was actually something *O'Leary* had evidently nipped from an *Uncommon Descent* (2011b) posting (possibly by herself?) on the same *ScienceDaily* source, where it was stated: "A friend writes to say, 'Seems that at this rate of evolution, we'd need trillions of years, maybe lots more, to see any substantial change,' offering this lacewing fossil from the present day." The underlined part linked to a photo of the insect in question.

Exactly what qualifications this "friend" had to assess what it would take to transform one ancestral lacewing into another is unknown, but the hyperbole of "trillions of years" suggested they were unlikely to be much of a scientist. Though maybe they had read Behe's *The Edge of Evolution*, or seen a certain Richard Sternberg video dismissing whale evolution.

Just how biologically static the silky lacewings have been over the last hundred million years or so is hard to tell just from the stray wings that have turned up, and Peng *et al.* (2011, 218) noted the difficulty in assigning fossil examples to the small Psychopsidae family (several dozen possible fossil genera known, but only 5 living ones with 27 species). Examples with unusual wing structures continue to turn up, such as the Eocene lacewing described by Vladimir Makarkin & Bruce Archibald (2014).

But why bother to study any of the niggling *Bermuda Triangle Defense* details of that formerly more abundant and varied lacewing bunch, when "a friend" can dismiss the data for Denyse O'Leary secondhand?

At every stage in the antievolution game, we're seeing their *method* fully on display, a parade of authority quoting and inuendo rather than a marshalling of anything like the full data set, so that all might see whether Design arguments could ever make honest sense of it, especially when compared to all the hard science effort of the evolutionary perspective they are so keen to dismiss. Was there *no one* in the ID camp who could do any better?

Michael Denton: Typology *Still* in Crisis

Here is where Michael Denton's revised *Theory in Crisis* supplied exactly the test measurement of the proposition that antievolutionism suffered from intrinsic methodological flaws that no mere stage managing could ever overcome. Of all antievolutionist authors over the last thirty years, the agnostic Denton had the most scientific credibility insofar as he was not defending some traditional theologically mandated antievolutionism. And writing in 2015, Denton had every incentive to show the deep explanatory power of his typological alternative to account for the available data (such as I have been trotting out in previous chapters) in a way manifestly superior to the "Darwinian" alternative.

We've already seen how Denton tackled one of his main topics, the origin of feathers, and the side issue of ant evolution. In the ant case he dangled jargon and a limited array of sources that in no sense did justice to the evidence. In the feather episode, Denton pulled apart the data set, allowing Casey Luskin of all people to act as his sole defense for minimizing the existence of feathered dinosaurs, while repeatedly stepping over and around relevant developmental and genetic data present in his own cited sources.

What was he to do, though, with the RMT, which has a lot more fossil data on the board (just recall my lengthy Creager walk-through last chapter), and far fewer Design-friendly sources to draw on to do the dismissing part for him (no Casey Luskin RMT post on hand this time)?

As with feathers, Denton (2016a, 110-111) did the same pull-apart thing, bringing up the jaw/ear bone element first. Citing only Masakai Takechi & Shigeru Kuratani (2010), which was a quite detailed summary of the developmental and fossil evidence underlying the RMT jaw/ear transition, Denton introduced the anatomical players accurately but carefully, not taking note of any fossil examples:

> Intriguingly, perhaps the most widely cited functional transformation alluded to as 'evidence for evolution' is the functional transformation of three skeletal components of the upper and lower jaw in primitive fishes into the three ear ossicles—the malleus, incus, and stapes, respectively—in mammals. These skeletal components are Meckel's cartilage, the palatoquadrate (formed from the ventral and dorsal parts of the first pharyngeal arch, the mandibular), and the hyomandibular bone from the dorsal part of the second pharyngeal arch (the hyoid), which played a role in bracing the two parts of the jaw together. In reptiles, these three skeletal components are called the quadrate (which articulates with the articular bone in the lower jaw); the articular bone in the lower jaw; and the stapes (which transmits sound from the outer to the inner ear in all tetrapods). So these elements of the first and second pharyngeal arches were utilized firstly to form the jaws and a bracing hinge in primitive fishes; they were utilized secondarily in reptiles to form the stapes in the middle ear and part of the lower jaw linking the lower jaw to the skull; and they were utilized thirdly in mammals to form the three bones in the middle ear.

With that, Denton withdrew to his typological redoubt (my **bold**): "But yet again, although these changes do support the notion of descent with modification, when these transformations are **considered in depth**, it is clear that a series of adaptive **masks** have been imposed on a **changeless** underlying *Bauplan*," and then quoted a *very* dated source, Rupert Riedl (1977, 354), who had rhetorically asked of that bone relationship retained over 400 million years, "what is this superimposed coherence which so steadfastly keeps them together?"

Riedl was writing in 1977, back when cars had vinyl roofs and no collision avoidance systems, long before homeobox genes and other biological findings provided the very "coherence" Riedl was wondering about, and which Takechi & Kuratani quite adequately surveyed in the paper whose contents Denton did not otherwise consider "in depth" while waving scientifically meaningless buzzwords like "masks" and "changeless" about.

Denton concluded his brief coverage of this topic with a repetition of what he so wanted to be true: "The co-option of the same three bones for very different purposes during vertebrate evolution [here he had a note 15] illustrates descent with modification, but, ironically, it also provides one of the most stunning examples of invariance of an underlying *Bauplan*." The degree of Denton's tunnel-vision at this point was indicated by what that note 15 consisted of, no technical source, but just a redundant rephrasing: "The co-opting of these three skeletal elements for different functions during vertebrate evolution from fish to mammal is shown in all major texts of vertebrate paleontology and evolution."

That Denton thought what little he had just written justified his "stunning" claim of typological "invariance" in the RMT case was confirmed by the fact that he then went on to his caveats on "common descent" previously noted back in Chapter 6, notions with more baggage Denton was never in any hurry to follow through on. Denton (2016a, 114) picked up the "invariant" theme a few pages later with a discussion of mammal teeth, another subject which involved a lot of detailed biology and paleontology, as we saw particularly in the previous chapter. Denton began with a broad preemptive concession:

> I also acknowledge that some apparently non-adaptive Type-defining traits may eventually be explained in terms of Darwinian functionalism. As mentioned back in Chapter 3, the dentition in all extant mammals (except for extant Cetacea and armadillos) is based on the same basic dental pattern: no more than forty-four teeth, and no more than three incisors; one canine; four pre-molars; three molars in each dental quadrant; and precise dental occlusion (the exact fitting of the projections of the cavities of the molars to allow efficient mastication of food). Gradual adaptation for more efficient mastication of food and the need for precise dental occlusion must have played a significant role in the evolution of at least some features of this pattern, and the differentiation of the teeth into different categories (heterodonty)." [Citing Zhe-Xi Luo (2007) and "chapter 7" of Peter Ungar (2010).]

That was quite an understatement, given that in a chapter later than 7, Ungar (2010, 223) reminded that "Some morphological solutions to adaptive problems may be easier to arrive at than others and turn up again and again in living and fossil mammals. In fact, recent advances in evolutionary developmental biology show just how little genetic change is needed to radically alter tooth form. A few drops of signaling protein may be all it takes to grow whole new cusps on teeth developed in petri dishes. This approach is revolutionizing our understanding of the evolution of mammalian dentitions," as would be seen in the years since Ungar's book, in the work of Enni Harjunmaa *et al.* (2012; 2015) reported in Chapters 7 & 10. But apparently not so much in the unrevolutionary world of Denton's Typology, stuck back in the conceptual days of Richard Owens' 19th century Archetype.

The persistent lack of the tidy boundaries mandated by Typology was something that was obvious when all the data was set on the table. As Ungar (2010, 224) put it:

> the evolution from the reptilian to the mammalian feeding system is one of the best-documented transitions in all of paleontology. The pelycosaurs show the first glimpses of synapsid heterodonty and jaw reorganiuzation, especially in more derived forms such as the sphenacodonts. Substantial changes came during therapsid evolution, with the reduction of the quadrate bone and the postdentaries, as well as the development of the bony secondary palate. The cynodonts in particular showed more marked

heterodonty and separation of the temporalis and masseter muscles for improved control over movements of the lower jaw. Some more derived cynodonts had contact between the squamosal and dentary bones, with their implications for changes in the jaw joint, successive addition of postcanine teeth, and even prismatic tooth enamel. Finally, the earliest mammals by definition had true temporomandibular joints, diphyodonty, and precision occlusion.

For Denton, there must be *something* deeply typological and non-adaptive lurking somewhere down in that relentless parade of evolving fossils, and for the RMT dental layout Denton (2016a, 114) decided it was:

> the numeric pattern itself—the three, one, four, three constraint—there must have been underlying non-adaptive developmental constraints involved in imposing this specific numeric pattern early in mammalian evolution. Why not two, one, four, three? Although the premolars and molars perform different functions (the premolars hold and grind, the molars grind) [Denton cited Ungar's introductory Chapter 1 here], it is hard to see on purely adaptive grounds why the basal pattern of four pre-molars and three molars was preferred and fixed.

Except Denton's "fixed" dental arrangement *wasn't* quite fixed. Back in Chapter 7 we saw how the marsupial thylacine "wolf" had 3 premolars and 4 molars, while the placental dogs have 4 premolars and developed only 2 molars in the upper jaw (3 in the lower). That so many modern mammals are placental doesn't make the marsupial exception go away. Moreover, Maureen O'Leary *et al.* (2013a, 666) noted how the eutherian ancestor varied from 4 to 5 premolars, so while there has come to be a general conservatism in the placental mammal lineage *now*, the synapsids certainly didn't start out that way, and the placentals only fell into that "fixed" mode (more or less) after *millions of years* of natural branching speciation. As Denton himself might have spotted just by counting the range of teeth showing up in some of the fossils depicted by several sources otherwise cited by him, Christian Sidor (2001, 1421; 2003b, 606), including *Ophiacodon*, *Dimetrodon*, *Biarmosuchus*, the Late Permian therocephalian *Ictidosuchoides*, *Thrinaxodon* and *Morganucodon*.

There is only so much space in a jaw, though, and there appears to be an inevitable tradeoff between the needs of chomping and hearing, along with the rates at which the components form developmentally, as Héctor Ramírez-Chaves *et al.* (2016) noted concerning the timing of molar eruptions. So how much of the natural variation available to them could *avoid* affecting the overall success of the individual or the population of its descendants? We never learn from Denton what the minimum criteria are for "adaptive" success, and he obviously never illustrated that limit by concrete example, fossil or living. But was he seriously trying to pin his typological hopes on so peripheral a feature as a "fixed" mammalian dental arrangement that hadn't started out that way to begin with?

All along, Denton has been assuming a dichotomy entirely of his own making: that if something wasn't explicitly *adaptive*, then it had to be a typological given. But the issue was really whether any of the biology involved anything other than purely natural (dare one say *unguided*) processes. A naturalistic non-adaption would be no more congenial to a Design framework than a naturalistic adaptative one. Dangling the typology trope was just an act of semantic gymnastics. It still didn't explain anything.

Convergent evolution continued to play a hobgoblin role in Denton's thinking, as Denton (2016a, 220-221, 327n) cited Jakob-Christensen-Dalsgaard & Catherine Carr (2008) on tympanic hearing having evolved separately in major vertebrate groups (a point covered back in Chapter 3), and proclaimed (my **bold**): "If the conventional view was correct that evolutionary novelties arise gradually through interminable series of intermediate forms, then surely at least one tympanic origin would indicate this. **But no! None of the five origins of this remarkable adaptive device are led up to via Darwin's 'interminable number of intermediate forms.'"**

This plowed straight into the *Bermuda Triangle Defense*, for where exactly were paleontologists supposed to find those most ephemeral of bones? Something could be learned of the otherwise unpreserved tympanic membrane by what they would have been attached to, of course, especially the otic arch, as John Bolt & Eric Lombard (1985, 90-92) discussed concerning the fossil amphibians they studied. Bolt & Lombard (1985, 92) further suggested that "the dorsal quadrate process in the fossils is in an early stage in the evolution of the tympanic annulus," and related that directly to what happens to those counterpart bones as they grew and shifted position from tadpole to adult frog. "This is the growth pattern one would expect if the tympanic annulus was derived evolutionarily from an enlarged dorsal quadrate process," Bolt & Lombard (1985, 93).

Elements of that amphibian system would have been carried along in all their vertebrate descendents, with as many modifications along the diverging lineages as millions of years might allow, but finding specific traces in any one of those would depend on the luck of the fossil draw. Just how rare it has been to find fossilized bones so small was affirmed by the case of *Liaoconodon* noted last chapter, regarding Jin Meng *et al.* (2011).

But while there may be few fossils to track the process, the *genetics* of the tympanic membrane is another matter, as Taro Kitazawa *et al.* (2015) laid out the trail for mammals and diapsids. And there's the newer digital technology, where Michael Laaß (2015; 2016) has been able to reconstruct some of the internal details of several Late Permian therapsids, revealing bone-conduction hearing in *Kawingasaurus* and a more mammal-like cochlear cavity in *Pristerodon*. Both were anomodonts, that diverse and quirky group of highly successful herbivores discussed back in Chapters 8 & 10. Because they're not directly on the main mammal line, but just close cousins, they reflected in their separate ways developmental processes quite reasonably likely to have occurred more broadly in their therapsid relations.

And what if the Fossil Genie went berserk and provided case after case of transitional forms for that tympanic detail, to match that for the jaw/ear transition? Well, wouldn't Denton be able to attribute it all to that catch-all "changeless underlying *Bauplan*"?

Or, if worse comes to worse, demand more adaptive scenarios. That's what Denton (2016a, 238-241) tried when he addressed the RMT one last time. He began with another of his rhetorical deck-stacking (my **bold**): "Another classic case of a long-term trend that is **hard to explain** in terms of adaptation, i.e., as being imposed by a **constant** adaptive constraint, is the reduction of the post-dentary bones in the jaws of successive clades of mammal-like reptiles (the ancestors of the mammals) over approximately 150 million years," Denton (2016a, 238-239, 330n), citing the aforementioned Sidor (2001).

Now, who was saying anything about there being a *constant* adaptive constraint? As opposed to what, an *intermittent* one? Sidor (2001) certainly hadn't laid down any such parameters, nor was it a presumptive claim in the four other works Denton (2016a, 330-331n) cited at that point: Alfred Crompton & Farish Jenkins (1979), Hans-Dieter Sues (1985), Blaire van Valkenburgh & Ian Jenkins (2002) and Christian Sidor (2003b).

After describing more of the reptile and mammal jaw articulations, and the shrinking articular and quadrate, Denton (2016a, 239, 331n) slipped very gingerly past the highly transitional probainognathids with a single sentence acknowledging that "As the reduction occurred, a second joint originated between the squamosal and the dentary in several synapsid lines," but citing only a non-technical secondary source, Douglas Theobald (2011) at *Talk.Origins Archive*. What typological imperative caused an entirely novel double-jointed jaw system to appear at just the right time to straddle the transition between the reptilian and mammalian jaw systems (and gratuitously burnish the reputation of Robert Broom in the fossil prediction department), Denton did not venture.

Instead, Denton went on to wave the convergence and adaptive flags together, citing Qiang Ji *et al.* (2009) on that Early Cretaceous *Maotherium* (with its adult ossified Meckel's cartilage), and Zhe-Xi Luo (2011) as (my **bold**) "recent studies suggest that the definitive mammalian middle ear was also acquired independently and in parallel in several different lineages of early mammals. There is **no doubt** that parallel evolutionary trends for which adaptive scenarios are **hard to envisage** are marked in the synapsids."

Really, "no doubt" at all? According to whom? Once again it appeared to be *Denton* who found things "hard to envisage," but only because he didn't slow down long enough to describe what that involved. All the bones required for any parallel evolution already existed in comparable configurations in these later mammaliaformes, or didn't Denton really stop to read Luo (2011) closely? By what typological decree would Denton proscribe what could or couldn't continue to mutate in any of those lineages? As for adaptive utility, was Denton under the impression that being able to hear better, especially in more specialized frequencies useful for small nocturnal predators, would have no selective advantage at all? Really?

Instead of supporting his contentions with concrete technical details, relating the genetics and fossil information to what he thought had happened in all those forms based on his typological inferences, Denton (2016a, 239-240) did what so many other antievolutionists had done in a similar position: he stepped *backwards*, rolling out an older authority quote from Hans-Dieter Sues (1985, 116), touching on many buzzword issues we've bumped into already regarding the antievolutionary viewpoint. Here's how Sues' 1985 view appeared in Denton's book:

> Parallelism is a common phenomenon in phyletic evolution. Eldredge goes even further, stating that "parallelism turns out to be a *far more common evolutionary phenomenon than even most if its more ardent aficionados had thought*." Alberch has pointed out in an elegant review that the production of morphological novelties is constrained by developmental programmes. Characters can be viewed as end products of developmental pathways; selection is responsible for "fine tuning" of adaptations within limits defined by intrinsic epigenetic properties. Such shared epigenetic programmes can result in parallelisms in particular character complexes among related lineages... Extensive parallel evolution in features of both the skull and the postcranial skeleton is evident among advanced synapsids, regardless of the preferred hypotheses of tritylodontid relationships.

Denton should have truncated that quote at the ellipsis, since that last sentence let slip the issue Sues was discussing, the relationship of our old T2 gang that John Woodmorappe, Shawn Doyle and Charles Creager had tripped over in previous chapters. But 1985 was a long time ago. Why then did Denton trot out that decades-old view, rather than investigate where the science stood in 2015? There was certainly a lot of it on the table by then. As it happens, Alfred Crompton & Zhe-Xi Luo (1993, 31) had taken note of that very section of the Sues paper regarding the status of those T2 tritylodontids, and cautioned regarding that and other work done during those years how "lack of knowledge of the structure of many critical taxa and the uneven emphasis the authors have accorded to various characters have contributed to these conflicting interpretations."

Uneven emphasis could certainly describe what Denton (2016a, 240-241) offered in his next paragraph, which concluded his RMT section. Denton continued to toss the same pitches he had all along (his *italics*, my **bold**):

> An adaptive scenario to account for the reduction of the post-dentary bones is **hard to envisage**. It is **hard to convince oneself** that the reduction was for enhanced hearing (to assist in transmitting sound vibration to the middle ear) as **some authors** propose. Even if this explanation might work for the final stages when the post-dentary bones were already very small, it cannot explain the earlier stages of the reduction when the bones were still substantial parts of the jaw. Whatever the reason, the reduction and ultimate loss could hardly have been an adaptation to *strengthen the lower jaw* **for some obscure purpose**. Species with powerful bites, such as extant

> crocodiles and the iconic *Tyrannosaurus rex* (which had the most powerful bite force of any extant or extinct terrestrial predator), need very strong lower jaws able to sustain the enormous stresses associated with such powerful bites, and have "**typical" reptile jaws** composed of four different bones. Again, it is **hard to see** how the reduction could have been driven by the evolution of mammalian mastication and dentition, as it is **hard to see how such tiny bones could have played any relevant adaptive role**. Moreover, the reduction appears to have started early in synapsid evolution, before the development of "mammal-like mastication."

Denton had put so much of this story *backwards*.

First, his presumption that the shrinkage of the postdentary bones had to be the primary adaptive feature completely overlooked what was right under his nose, literally so, as that's where our own *dentary* bone resides in us mammals. It obviously never occurred to Denton's typological imagination that it might have been the strengthening of the dentary for eating that was an adaptive driver initially, with the postdentary bones shrinking as an increasingly less dominant component that only eventually took on an auditory role as very tiny bones.

We've previously touched on the sequential adaptive nature of this, from Crompton & Parker (1978, 200) in Chapter 3 on how mammalian jaw evolution was constrained by the balancing act of chewing and hearing, to Allin (1975, 416) in Chapter 4 noting that "Only long after the postdentary unit had departed from the mandible in primitive mammals could the angular (typanic) element be stablized as a static (non-vibrating) drum-supporting entity; this never took place in monotremes." And from Chapter 9, Hopson (1987, 25) reminding Duane Gish that "the only step that had to be taken once the new mammalian jaw joint was established was to free the rod of post-dentary elements from contact with the dentary. This happens during the ontogeny of every living mammal by atrophy of the middle part of Meckel's cartilage."

As for "how such tiny bones could have played any relevant adaptive role" in mastication, no one was saying they had. They were playing a role in *hearing*, however, and that didn't start with them being quite so small. Moreover, Denton never documented that assumption. Just how small would those postdentary bones need to have been to have contributed something to auditory sensation via bone-conduction? A feature known to occur in animals, as we saw back in Chapter 4 regarding the living and fossil examples covered by Renaud Boistel *et al.* (2013) and Michael Laaß (2015).

But Denton was especially disingenuous when he waved that phrase *for some obscure purpose*, as though successfully eating things was an inscrutable mystery, and to do so only to let the much larger predatory *T. rex* and crocodiles on the field—a particularly cheeky thing to do, reminding us of all the RMT fossils he never thought to discuss, by bringing up what were in fact irrelevant examples.

Denton (2016a, 331n) cited only one technical source for this contention, Karl Bates & Peter Falkingham (2012) on the bite force of tyrannosaurs and alligators. A situation which turns out to be due in large measure to their *size*.

Bates & Falkingham noted that human jaws, if scaled up to jumbo predator level, had strengths comparable to alligators. Ramping an alligator up to *T. rex* size still put the dinosaur as a more powerful chomper, though without knowing the exact strength of the muscles involved, they couldn't be certain that even *T. rex* had an unusually powerful bite for a carnosaur, only that its size put it at the upper range.

Bite force in living animals is related to their environment, though, the ways in which they interact with whatever it is they aim to have for dinner, and Bates & Falkingham reminded (my **bold**) that "Living carnivores preying on large animals have relatively high bite forces, while carnivores preying on small prey have more moderate bite forces for their size, suggesting that bite force represents an important **adaptation** to differing feeding ecologies, at least throughout carnivoran evolution," citing Per Christiansen & Stephen Wroe (2007).

Now the largest of the Permian synapsids, like *Dimetrodon*, while about the same full-grown length as a modern alligator, were still only about half as massive by weight, and most of the players in the RMT were far smaller than that, so dragging in apex predators as Denton had didn't on their own bear on the dynamics of much smaller animals pursuing comparably smaller prey (like insects that might involve the need for a specialized jaw shape or adaptively selected crunching teeth for bug munching). Which brings up something Christiansen & Wroe (2007, 356) noted of the living mammals they studied (my **bold**), "In marked contrast to large-prey carnivores and specialized herbivores are the **insectivores**, which have significantly lower BFQs [bite force quotients] than all other groups."

You may recall how many of the RMT were *insectivores*.

And then there's the really *obvious*: both the giant *T. rex* and the less giant alligator were *diapsids*, not *synapsids*. Did their different skull conformations play no role at all in the comparative evolution of these forms over so many millions of years? Denton clearly wasn't stopping to look (not a single RMT fossil was even mentioned in his 2016 book, let alone examined in depth), but that didn't mean there wasn't a body of work on the subject, starting with the basics of muscle attachment shown in textbook accounts like Kenneth Kardong (2005, 275-277, 281-282) or Peter Ungar (2010, 91). Moving on down to animals more the size of the RMT relevant synapsids, Mehran Moazen *et al.* (2009, 8274) and Peter Johnston (2014, 576) showed the assorted external adductor muscles in snakes, lizards and that "living fossil" *Sphenodon*, which are configured behind the dentary bone, and so pull more vertically than the angled way seen in the dentary-dominated synapsids that still retained the postdentary bones.

So, what has been learned about that mammal masseter issue? We can start way back with R. DeMar & Herbert Barghusen (1972), laying out the mechanics and evolution of the *coronoid process* in the synapsid jaw. You remember that, it's the upward jut of our own jaw, a bulge that started out much more modestly as a slight angled bend early in the synapsid lineage.

That difference changed the way forces were transfered at the leveraging "moment arm" and the effect of the associated muscles tugging on those bones. As their abstract put it:

> Selection for a posteriorly directed line of muscle action was the primary cause for the initial development of this process, and such selection occurred in conjunction with the predatory habits of the advanced sphenacodontids. In sphenacodontids, the height of the coronoid process was raised past a threshold and eventually allowed selection for a higher moment arm by further rise in the process to become effective. Cynodont evolution is characterized by extensive development of the coronoid process which then migrated posteriorly. The model shows that these are understandable events. Other events in cynodonts include the development of the masseter muscle, first as an extension of the external adductor, and later with an added superficial component acting anterodorsally. Selection of the masseter muscle and for the rise of the jaw articulation in advanced cynodonts may have involved the use of the cheek teeth in rudimentary chewing.

Crompton & Parker (1978, 197) picked up on this force distribution thread:

> In early stem reptiles the vertical forces generated at the jaw joint were as large as or larger than those generated at the point of bite. The progressive decrease in the size of the postdentary bones was accompanied by the development of a coronoid eminence and, later, a coronoid process. This permitted a change in the direction of pull but not in the leverage of the temporalis muscle. The insertion of the pterygoideus muscle shifted progressively forward and eventually moved from the postdentary bones onto the angle of the dentary. As the intersection of the lines of force of these muscles migrated forward, the force generated at the jaw joint during chewing would have decreased in magnitute, permitting a reduction in the size of the postdentary bones and jaw joint.

Of course, for Denton to have considered the potential adaptive significance of shifting jaw positions that directly pertained to how it was that the postdentary bones could become decoupled from articulation, he'd have had to pay attention to some of them once in awhile. Which he didn't.

Holger Preuschoft & Ulrich Witzel (2005, 409) pointed out another pertinent factor, regarding the *shape* of the synapsid dental and jaw arrangement (my **bold**): "An enlongated and narrow dental arcade offers **obvious advantages for longer reach and higher speed** of the approaching anterior teeth, as well as a reduction of potentially dangerous torsional moments about the longitudinal axis of the jaws."

All of those features related to the dentary bone and the teeth set in them, and how they related to what it was the animals might be trying to grab and crunch into digestible bits. None of this depended initially on reducing the size of the postdentary bones, which to start were just tagging along for the ride.

More importantly, at no point along this adaptive path was it necessary for mammal ancestors the size of a mouse to develop a ferocious bite like a tyrannosaur thousands of times their weight in order to scamper after crunchy-shelled bugs. What *was* happening, though, was the change in bite dynamics as the masseter muscle worked increasingly on the *outside* of the jaw bone (in a way unlike the diapsids), so that "each side of the lower jaw is held in a sling of muscles," as Crompton & Parker (1978, 194) put it. This made "it possible for the jaw to move from side to side or from front to back," allowing a *completely new way* of chewing to evolve.

Was that new arrangement of absolutely no adaptive value to the animals possessing it? For the eating thing, which critters kind of depend on, for not starving to death? Michael Denton implicitly thought so, but not because he investigated any of it in detail first.

Let's look some more at what failed to exercise the curiosity of the avatar of modern typology, the specifics of how the masseter muscles came about in the synapsids.

Herbert Barghusen (1968) took special note of how muscles form and attach in living reptiles and mammals in considering what was going in *Dimetrodon* and other synapsids, where the adductor muscles started out comparable to those in living reptiles. Because of how the synapsid zygomatic arch had bowed outward past the jaw, muscles at that early stage had the adaptive option of descending from the arch to the *angular* bone (as may have happened with the gorgonopsids) or expanding onto the *coronoid* as they would in the later cynodonts. There were also several locations in the skull where muscles could come to link up to the extending dentary, but one in particular played along already existing connective tendons. Eventually the connections in the advanced cynodonts took over the whole zygomatic region while covering still more of the coronoid process.

Fernando Abdala & Ross Damiani (2004) studied more of that zygomatic expansion and its effects on the muscle changes, identifying how the downward migration and division of the masseter muscles had occurred in tandem by around 250 mya. Illustrating *Procynosuchus* as a synapsid starting point, where the masseter attachment on the lower jaw was forward of its arch attachment, Abdala & Damiani (2004, 25) tracked how the orientation had changed in contemporary forms like the Galesauridae, where that jaw attachment had moved *backward* relative to its mount on the skull above, and come to be focused on a new "angular process" projecting from the dentary. That focus of the masseter muscles on that dentary prong persisted over millions of years into later cynodonts like *Thrinaxodon* and *Probainognathus*, and on into mammals, by which time the angular process itself had flattened out on a dentary bone whose multiple masseter muscles operated quite nicely without it.

Incidently, the recent stress modeling of synapsid and mammal skulls by David Reed *et al.* (2016) showed that even the tradeoff of bite force and hearing needs in the dual-jawed *Probainognathus* did not compromise either function.

Now what about that "mammal-like mastication" matter Denton tried to use as a handkerchief to wave in front of his typological disappearing act?

The development of the masseter muscle attachments took place long before the appearance of specialized tooth occlusion, as even Denton sort of recognized (though avoiding as usual going into much detail), and it was that jaw dynamic that allowed the reduction in the postdentary bones. As for how those postdentary bones affected the system, you have to remember how they came to be one of two jaw hinges in forms like *Probainognathus* (that layout that had been explictly predicted by Robert Broom on evolutionary grounds, and so obligingly confirmed by that evolution-loving Fossil Genie). How might a naturally evolving biological system deal with the juggling act of relying on one jaw articulation over another, while still retaining both?

Abdala & Damiani (2004, 26) noted how "Many innovations associated with the feeding system in late non-mammaliaform cynodonts and early mammaliaformes are interpreted as safety devices that allow the postdentary bones and the quadrate to reduce in size while maintaining a viable, but weak, craniomandibular joint (CMJ)." This extended even to the details of musculature, as "The orientation and short fibres of the superficial masseter are also probably related to decreasing the stress on the weak CMJ by preventing exposure to anteroposteriorly directed tensile forces that could displace the joint." In other words, just about everything about the changing jaw and muscle configurations in the transitional synapsids over those millions of years appeared to be *adaptive* in character, all microevolutionary ways to keep the dynamic forces in check while that munchy-crunchy and evolving hearing things went on functionally.

Jaw muscles have evolved a lot in vertebrates over all those millions of years. Studies working to identify the homologies and developmental biology underlying the variety in living forms (from fish and amphibians to reptiles, birds and mammals) include John Friel & Peter Wainwright (1997), Casey Holliday & Lawrence Witmer (2007), Rui Diogo & Yaniv Hinits *et al.* (2008), Mehran Moazen *et al.* (2009), Nikolai Iordansky (2010), Juan Diego Daza *et al.* (2011), Robert Druzinsky *et al.* (2011), Peter Konstantinidis & Matthew Harris (2011), Aléssio Datavo & Richard Vari (2013) and Peter Johnston (2014). For contrast, limb musculature has been relatively conservative from fish to tetrapods like us mammals, investigated by Diogo & Virginia Abdala *et al.* (2008).

As for the genetics of how muscle attachments can come to relocate on the bones, the recent study by Masayoshi Tokita *et al.* (2013) was most interesting. A masseter-like attachment has originated independently in parrots, involving modified expression patterns in a particular set of genes, including the homeobox *Six2*, and our stalwarts *Bmp4* and *Tgfβ2*. It doesn't require a rocket scientist (or a typologist evidently) to suspect that similar processes went on in the developmental biology of synapsids all those millions of years ago.

So much wonderful information was available for Michael Denton to have thought about. So much wonderful information that Denton *should* have thought about, since all of it would still have to be accounted for in any typological framework in order for that perspective to be taken as seriously as the body of regular science work Denton was out to displace. But in so many ways, from the RMT to feather origins to the amazing world of ants, Denton's *Evolution: Still a Theory in Crisis* was a model of failure and evasion, not a benchmark of solid accomplishment.

Evolution: Still a Theory in Crisis as the latest secondary ball

It's an amazing measure of how little investigative gumption there's been in the antievolutionary parade that even after Michael Denton's new book, it was still correct to say that no antievolutionist has dived into more of the RMT data than Duane Gish had a third of a century ago. Much ink has been spilled, yes, but not more light shed.

Not that anyone in the ID camp was seeing things that way, where Denton's retread was quickly recommended by a chorus as the latest heavy gun in their arsenal to pulverize the supposedly crumbling Darwinian fortress. As the *Discovery Institute* was the publisher of Denton's book, the squibs by Denton (2016b-f) were inevitable, as well as side-references by John West (2016), and the cursory video interview with Denton linked by *Evolution News & Views* (2016b). About a minute in, Denton was on about adaptation in exactly the way he did in his book (my transcription and *italics* reflecting Denton's accenting the term):

> The problem is, that it shows you what you need, to apply Darwinian ideas to microevolution. You first of all got to show the forms were *adaptive*, and it may well be you got to show they were adaptive in ancestral forms in the distant past. Generally speaking it's extremely difficult to show that a lot of the deep patterns of nature are adaptive. And on top of that it's very difficult to imagine sequences getting to these forms as well.

I can easily imagine viewers of the short video who already accept ID conventions nodding their heads at Denton's authoritative vaguery. But how many would have pressed on to reading Denton's book for real, and done enough fact checking to discover that the *difficulty* Denton proposed was painfully self-descriptive?

David Klinghoffer (2016a,e,i) was a complete embodiment of that uncritical acceptance, not only repeatedly channeling Denton's view of things, but praising others for doing likewise. Klinghoffer (2016i) was thrilled to hear the literary critic A. N. Wilson accept its science claims wholesale, deeming it "A truly great book," *Spectator* (2016). And Klinghoffer (2016e) extolled the favorable review by Sean McDowell (2016a). "Dr. McDowell is a professor of Christian Apologetics at Biola University. He admits the book is 'challenging' to read, and says he disagrees with Denton on 'some points.' But he gives an admirably lucid summary of its argument and judiciously weighs its significance." Klinghoffer asked, "Now why can't a Darwinist do something like this? Review a book you don't entirely agree with, but state its thesis correctly and evaluate its contents fairly."

But McDowell hadn't evaluated Denton's contents at all, only *repeated* a few of them, noting that "Denton provides a number of examples in nature that lack Darwinian pathways, such as the cell, limbs, feathers, wings, flowering plants, language, and more. Let's briefly consider a few of his examples." That consisted of him quoting Denton a bit on the supposedly non-adaptive character of angiosperms, the origin of the tetrapod limb, and how bats got their wings. At no point did McDowell investigate whether the dataset justified Denton's assertions.

So "evaluate" in Klinghoffer's lexicon was a branch of *cut-and-paste*, and "fairly" meant not disagreeing with any of it substantively. Even the caveats Klinghoffer alluded to weren't much to sneeze at. McDowell had found Denton's book "a challenging book to read. It's not written for the novice! But if you have the interest and time to wrestle with an important scientific critique of the neo-Darwinian synthesis, it is well worth the investment. Even if you end up disagreeing with Denton (as I do on some points), his book is thoughtful and timely."

That was it. No mention of what points McDowell disagreed with (scientific or philosophical?), or even whether he considered himself one of those *novices* the book wasn't written for. While it certainly was laden with jargon and science references, we've seen that alone was no guarantee that Denton was actually on the mark on any of it. And nothing in McDowell's short review represented other than a Glee Club, accepting that "He's right. The Darwinian model faces significant hurdles, which seem to get increasingly higher. Either the naturalist needs to answer the challenges raised by Denton, pose another naturalistic model (as Denton does at the end of his book) or be open to special creation. There are only so many available options." If that was an allusion to the final two chapters of Denton (2016a, 247-282), that was just the typologist's affirmation of a nebulous non-gene-based sort of epigenetic self-organizational "structuralist" view of life, that when push came to shove explained nothing much about anything, least of all the RMT.

Of the several bits Klinghoffer quoted from McDowell's short review, this stood out for me: "Simply put, there are not the innumerable transitional links Darwin predicted, and in many cases, there is not even conceivable links to account for various 'structures' in nature. According to Denton, this is one of the major unsolved challenges for Darwinian evolution." As we've seen in considerable examples (including especially the RMT), that was incorrect when Denton asserted it, was still so when McDowell enthusiastically repeated it, and hadn't got any less wrong when Klinghoffer added his own stamp of approval.

Klinghoffer concluded his gloss of McDowell with a sop of impartiality: "There is indeed a limited selection of responses to Denton and the challenge he poses. You can choose any of them, but you must choose. We don't demand agreement, but merely accuracy and fairness. And why, tell me, is that within reach of a Christian apologist, but not a Darwin apologist?"

Oh, brother! *"We don't demand agreement"*? The Intelligent Design movement generally, and Klinghoffer's apologetics specifically, have never shown the slightest inclination to credit any disagreement with ID tenets as credible. Theirs has been one long digging in of heels on every point of contention, never willing to give an inch.

Incidentally, Sean McDowell represents a philosophical juggling act of his own as the son of venerable Christian apologist Josh McDowell. McDowell (1972; 1975; 1989a-b), McDowell & Don Stewart (1981), McDowell & David Stoop (1982), and McDowell & Bob Hostetler (1992) have covered a gamut of evangelical themes, including trying to convince skeptics that the evidence for Christianity was unassailable. Son Sean has apparently moved away from his father's embrace of tradtional YEC doctrines, though, things like Josh McDowell (1986) trying to fit dinosaurs into the Biblical framework, but given his willingness to accept Denton's criticisms at face value, the son's analytical methodology may not be all that different.

A parenthetical note on prophetic methodology. It's interesting to see how similar Josh McDowell's defenses of the supposed fulfillment of Biblical prophecy were compared to how occultists reported their own lore, such as the many Nostradamus groupies I compared McDowell to in Downard (2004f, 672-674). Charles Ward (1940), Rolfe Boswell (1941), André Lamont (1943), Henry Roberts (1949), Stewart Robb (1961), Jeffrey Goodman (1979), Erika Cheetham (1985) and John Hogue (1997; 2002) have all gymnastically teased history to retroactively fit the Seer's predictions, and McDowell (1972; 1975) played exactly the same picky-choosy tricks with the Gospels. Belief in Nostradmus shows no more sign of abating than organized religions have, by the way, as Benjamin Radford (2001) noted regarding bogus Nostradamus prophecies circulating after the 9/11 terrorist attack, and anyone with cable TV has probably tripped over some of the credulous shows on the fellow that keep popping up just when you think he's gone away. For those insisting on playing quatrain hunter, Edgar Leoni (1961) remains a useful impartial compendium of Nostradamus' texts and the assorted historical minutae pertaining to them.

David Klinghoffer (2016g) bumped into Sean McDowell (2016b) again, concerning his interview with Bill Dembski, Klinghoffer reporting how the ID star had "largely retired from intelligent design," though only after having "shown (as in demonstrated and not merely gestured at) that naturalistic evolution is a failed intellectual and scientific enterprise." Such is the view down on the *Discovery Institute*'s Mobius Strip offramp.

Another *EN&V* contributor who couldn't resist invoking Denton's new book was Donald McLaughlin (2016), excoriating the *National Center for Science Education* for objecting to a new "Academic Freedom Bill" in Oklahoma that offered the current antievolutionary trope about the need "to inform students about scientific evidence and to help students develop critical thinking skills" concerning "some subjects such as, but not limited to, biological evolution, the chemical origins of life, global warming and human cloning." By the way, Nicholas Matzke (2016) has done a tidy taxonomy tracing the "evolution" of these antievolution "Science Education Act" wordings over the last dozen years, as legislators copied and modified the terminology from precursors in Alabama and Oklahoma, to keep one step ahead of disapproving litigation.

This time around, the NCSE had quoted the *AAAS'* Alan Leshner that there was no scientific controversy over evolution, which prompted McLaughlin to suggest "someone should give Leshner and the NCSE a copy of Michael Denton's forthcoming book." That *forthcoming* meant McLaughlin was evidently presuming the accuracy of the work before even reading it, let alone fact checking it.

Or the Oklahoma bill, for that matter.

McLaughlin showed no great curiosity to take note of who it was proposing that legislation, or whether they had a track record to consider. Oklahoma House Bill 3045 was authored by Representative Sally Kern (2016), but that was only her latest foray into the field. There was the similiar HB 1551 in Kern (2011), and before that HB 2211 on religious freedom in the classroom by Kern (2008), which hadn't mentioned evolution or science at all, but still endeavored to protect the religious student from discrimination by effectively forbidding teachers to grade unfavorably test answers where the student might have given a response that represented their firm religious belief but disagreed with the accepted science—say, on the age of rocks or the universe. As HB 2211 died in committee, Oklahomans never had a chance to see how this might have played out in actual classrooms. But the state has a long tradition of antievolutionism, recounted by Stanley Rice (2014), and Oklahoma's public school science teachers have had a mixed record when it comes to biological knowledge, surveyed by Tony Yates & Edmund Marek (2013).

The upshot: legislators like Sally Kern do not operate in an ideological vacuum, and it was disingenous of McLaughlin or the *Discovery Institute* to pretend HB 3945's drafter did not. Kern belongs to an entrenched conservative *Kulturkampf* block in the Oklahoma legislature, one that had been challenging not only scientific fields like evolution and climate change, but also social issues (most notably homosexual equality), as covered by Matt Corley (2009) and *Climate Progress* (2011). Kern had previosuly cosponsored several similarly worded antievolution bills with fellow *Kulturkampfers* Gus Blackwell and Senator Josh Brecheen, Blackwell *et al.* (2014a-b), and just on his own, Breechen (2011; 2012; 2014; 2015; 2016) had been a busy beaver indeed over in the Senate trying to get evolution taken down a peg in Oklahoma. None of these efforts have been particularly successful, but that obviously hasn't stopped them from trying (or discouraging their conservative constituencies from reelecting them).

It's not always easy to identify on what basis antievolutionist legislators like Kern or Breechen have come to their convictions. Locals papers may not ask the right questions, and those being interviewed may not be inclined to be all that forthcoming. But in the case of Josh Breechen (2010a-b) it was clear enough that he had superficially channeled familiar Intelligent Design tropes about there being "Renowned scientists now asserting that evolution is laden with errors" into a more doctrinally familiar "creation vs. evolution" context that raised serious doubts about how finicky he would be about source checking the credibility of any scientist, renowned or not. The problem all their legislation posed for sound science education was in how the line would be drawn (or if one would be drawn at all) concerning which sources would be permittable for a science teacher in Oklahoma to make use of while assisting their students in "objectively" assessing evolution or climate science or whatever else happened to fall into their ideological crosshairs.

It's been one of the persistent features of the *Discovery Institute* brand of Intelligent Design apologetics to pretend that modern creationism is irrelevant to grassroots efforts to "critique and review in an objective manner the scientific strengths and weaknesses of existing scientific theories," as the Oklahoma act put it. At no point over years of lobbying has anyone at the *Discovery Institute* ever laid out exactly what the criteria are for being "objective", or how that would play out in a classroom should the teacher want to *objectively* trot out an argument culled from Duane Gish or John Woodmorappe rather than one cribbed from *Explore Evolution* or *The Design of Life*.

And would that actually make much of a difference, if the underlying "logic" of all those antivolutionary resources were equally flawed? As I contend we have seen most abundantly in the previous chapters, concerning both YEC and ID.

Here it may be informative to look at a *test case*, an instance of a specific piece of technical science work, and how ID looks at that evidence compared to more conventional creationists, to see what (if anything) distinguishes them methodologically.

A Madness to their Method: The Antievolutionary Bird Beak Playoffs

You may recall from back in Chapter 5 the papers by Bhart-Anjan Bhullar *et al.* (2012; 2015) investigating the evolution of modern bird beaks from their dinosaur ancestors. The 2012 paper had laid out the case for birds having *paedomorphic* skulls, where juvenile developmental characteristics were retained into maturity (a point touched on in the previous chapter regarding David Coppedge's point-missing). That process involves another topic we've bumped into in previous chapters: *heterochrony*, where alteration in the scale and pacing of developmental timing produce changes in morphology.

The Fossil Genie has continued to be of help, of course, as the Early Cretaceous ornithurine *Yixianornis* "spans the gap between" the skull shapes of maniraptoran theropods and the avialans proper, Bhullar *et al.* (2012; 224). See Julia Clarke *et al.* (2006) on *Yixianornis*, and Zhonghe Zhou & Fucheng

Zhang (2005), Zhonghe Zhou *et al.* (2009), Hai-Lu You *et al.* (2010) and Min Wang et al. (2015) on more of these early ornithurines that have turned up since.

Regarding those beaks, "Anatomically, the internal skeleton of the upper beak is composed of fused, elongated premaxillary bones; these are paired, small, and form the tip of the snout in ancestral reptiles," Bhullar *et al.* (2015; 1666) reported. By then a lot of work had also been done on the genetics underlying their formation and natural variation, such as Arhat Abzhanov *et al.* (2004; 2006; 2007), Otger Campàs *et al.* (2010), Ricardo Mallarino *et al.* (2011; 2012) and Nathan Young *et al.* (2014), identifying Fibroblast growth factor 8 (*Fgf8*) and the ubiquitous Sonic hedgehog (*Shh*) as dominant genetic players.

Given all that, Bhullar's team took a crack at retroengineering some of the ancestral dinosaur skull from which the bird version had evolved. For this task they naturally had to pay close attention to the forms seen in fossils while experimenting to pull the beak formation genes out of the picture to see if they could reproduce what was known to have occurred in those fossils. Here they found another protein in the game, the *Lef1* lymphoid enhancer binding factor knocking around in the WNT signaling pathway, and when Bhullar's team succeeded in inhibiting that and *Fgf8*, the result was a bird where the premaxillary bones didn't fuse nor were enlongated, revealing a skull much more like their living archosaur cousins and theropod ancestors, showing once again the practical success of a method that combined the latest in biological evidence with a full appreciation of the evolutionary nature of life.

Michael Denton (2016a, 85, 300-301n) bumped into the Bhullar paper along with Matthew Harris *et al.* (2006), citing them regarding this: "Further evidence of the conserved genetic and developmental commonalities is illustrated by atavisms (i.e., 'throwbacks' to an ancestral condition) such as the remarkable case of 'chicken with teeth' and the even more remarkable recently genetically engineered chicken with what the authors claimed was a 'dinosaur snout.'"

He did not venture whether he thought the presence of fragmented tooth genes cluttering up the bird genome was to be expected or not in a typological framework, but right after his rhapsody on the evolutionary novelty of tympanic hearing discussed earlier, Denton (2016a, 221, 327n) returned to the Bhullar work (my **bold**):

> The second report comes from a paper published recently in the journal *Evolution* (alluded to in Chapter 5) in which the authors created what they claimed as a "dinosaur snout" by genetic **tampering** with the expression of a set of genes involved in the developmental module which generates the beak in a bird. Inhibitors of the gene products involved in beak development caused the chicks in the study to develop a reptilian snout, an apparently genetically engineered atavism. Evolutionary biologists might reasonably see such an experiment as supporting the notion of descent with modification, but **it would be wrong, as always, to conflate evidence for descent with modification with evidence for Darwinism.** In fact, as the authors themselves comment on their work:

At which point Denton quoted from the paper. We'll get to that quote presently, but first let's unpack Denton's windup. I highlighted that use of "tampering" because of how creationists described the process, as we'll see shortly, but it's that "as always" lambaste at "Darwinism" that reminds us how Denton never explained what exactly *would* be "evidence for Darwinism" in this case. Darwinism *is* natural "descent with modification." It's only Denton who insists on "Darwinism" requiring all biological attributes be purely adaptive, but this time he didn't even bother to wave the adaptation flag.

Instead, Denton homed in on one passage that obviously appealed to his preference for typologically congenial terminology, Bhullar *et al.* (2015, 1673) having used "abrupt", "rapid", "comparatively saltational", and "discontinous" all in a go. I've put the section Denton chose to extract from the Bhullar paper in **bold**:

> We submit here that the phylogenetic distribution of gene expression suggests a role for FGF and WNT in the FNP [frontalnasal premaxilla] in producing the distinct facial morphology of birds. We acknowledge, however, that the transition to the avian rostrum was undoubtedly complex—as shown by multiple transitions to a premaxilla with some bird beak like features in the fossil record. The results reported here represent one part of a manifold transition. Additionally, the characterization of the beak as a key evolutionary innovation is made more complex because its components were assembled over a longer period of time than that represented by the proximal stem of Aves—a caveat that applies to many such transformations or putative innovations (Donoghue 2005). However, **the abrupt geometric gap between nonbeaked archosaurs and birds and stem birds with beaks may suggest a rapid, comparatively saltational transformation. The difference in ontogenetic trajectories of shape change between nonbeaked forms, in which the premaxilla becomes shorter and broader with time, and beaked forms, in which it becomes longer and narrower, also suggests a discontinuous distinctiveness to the beak.**

What all that meant, though, related to the various figures Bhullar *et al.* (2015, 1667, 1672-1673) used to geometrically graph the snout and beak dimensions (such as premaxilla versus palate shapes) of the fossil and living forms, from archosaurs to dinosaurs and birds. Part of that potential abruptness Bhullar's work was addressing involved the concept of *pleiotropy*, where changes in relatively few genes could trigger multiple effects (in this case "both the premaxilla and much of the palate") with "implications for morphological systematics and modeling of morphological character evolution, which generally assume character independence." At this point, Bhullar *et al.* (2015; 1674) ironically cited as an example of that older more restrictive view, non other than the arch-cladist Colin Patterson (1982)!

Denton (1985, 149) had bounced past the concept of pleiotropy before: "Not only are most genes in higher organisms pleiotropic in their influence on development but, as is clear from a wide variety of studies of mutational patterns in different species, the pleiotropic effects are invariably species specific." *Invariably* species specific? That should have suggested that pleiotropisms might actually playing a role in the speciation process Denton purportedly accepted, but that connection never occurred to him. Nor did it when Denton (1998, 333) briefly trotted past the issue again.

As it happens, the pleiotropic aspects of speciation had been addressed by many over the years, such as Montgomery Slatkin (1982), William Rice & Ellen Hostert (1993), reprinted in Mark Ridley (1997, 174-186), Menno Schilthuizen (2001, 139-141) and Masato Yamamichi & Akira Sasaki (2013). Pleiotropic factors pervade mating pheromones in insect speciation, Chris Jiggins *et al.* (2005), Kerry Shaw *et al.* (2011), and Nadia Singh & Kerry Shaw (2012) commenting on François Bousquet *et al.* (2012), Andreas Wagner (2000a), David Liberles *et al.* (2011), Nicolas Lonfat *et al.* (2014) and Shengzhan Luo & Bruce Baker (2015) have described similarly overlapping gene and protein functions, while Michael Travisano (1997, 477-478), Craig MacLean *et al.* (2004) and Wenfeng Qian et al. (2012) offered analogues in bacteria and yeast. Up on the vertebrate floor, Irma Varela-Lasheras *et al.* (2011) noted the process in mammals, and Frietson Galis & Johan Metz (2007) and Zhi Wang *et al.* (2010) upped the stakes to relate it to the generation of evolutionary novelties and complexity.

That Bhullar *et al.* (2015; 1672) were not thinking of saltational in a way that precluded gradually arriving at it was borne out when they proposed "that the experimental embryonic phenotypes actually predict the morphology of yet undiscovered early avialan palatines and that new fossils will eventually be found that will show a gradual morphological transition toward modern birds."

Time will tell whether the Fossil Genie will take note of that, but Denton clearly wasn't when he tried to marshal Bhullar's work for his always vague and useless typology. Denton (2016a, 221-222) concluded his section on Bhullar's paper by connecting this beak issue with something he thought he'd pulled off (his *italics*, but my **bold**):

> Only future work can determine just how discontinuous the evolution of the beak actually was, and what role selection might have played, but it seems clear that the *actualization* of the beak only occurred because the basic developmental system of the reptilian snout was **pre-figured or pre-adapted for the transformation**. Yet again, "internal causal factors" were paramount in the origin of the beak and in the **channelling of evolution towards the modern avian Type**. As a result of this work, there are now six well-characterized taxa-defining novelties in birds which would appear to have been **primarily the result of internal causal factors and not gradual cumulative selection** to serve a succession of environmental constraints: (1) the feather follicle; (2) the plumaceous feather with unbranched parallel barbs; (3) the branched pinnate feather; (4) the open pennaceous feather with barbs and barbules; (5) the closed pennaceous feather with interlocking barbules; and now also (6) the beak. In none of these cases is there an **adaptive continuum leading to the novelties as would be required by an externalist Darwinian account**. Again, "internalism rules."

In all four of the bits I highlighted, Denton was repeating a framework that he never got around to specifying. On the first point, Bhullar *et al.* (2015, 1669) had specifically noted that "Ornithomimosaurs and oviraptorosaurs are two dinosaurian lineages that independently acquired a toothless, probably rhamphotheca-covered rostrum. Despite overall similarity with bird beaks, premaxillary shapes of these taxa grouped with those of more conservative archosaurs." So what then did it mean for a transformation to be "pre-figured" if it was only one of the many ways the archosaurian skull could change adaptively? In describing something as "pre-figured," Denton was merely restating the obvious evolutionary lesson: *everything* is a variation of a previously existing system.

Bringing up the "channelling of evolution towards the modern avian Type," Denton once again slipped past the prep work. The origin of the ornithurine beak lay in the Cretaceous, leaving the current range of birds as just what was left over after all the competitors went extinct. Why then was there an "avian Type" at all, rather than all of that being just variations on the "archosaur Type"? Or "tetrapod Type." Or "Chordate Type." Or "Metazoan Type." Or "Bacterial Type." Just how far back on the common descent lineage Denton kinda sorta accepted were things "pre-figured," if all he meant by that was that it had built on previous biology?

Denton's juxtaposition of "primarily the result of internal causal factors and not gradual cumulative selection" begged the question of how you would identify or define either, and whether that distinction meant anything at all. What "internal causal factors" could be at play in the archosaur skull intending a bird expansion of the premaxilla when in almost all archosaur skulls none of that ever happened? And what exactly would "gradual cumulative selection" look like at the fossil evidence floor, something so clearly part of an "adaptive continuum" that Denton would accept it?

Here we're just waiting for that *Bermuda Triangle Defense* and the *No Cousins Rule* to come on the field, but Denton has never got that far in his own game yet.

Finally, it's interesting to compare Denton's gloss on the Bhullar paper with their own concluding paragraph, because of the connection they'd made with a source and context we've already encountered. Bhullar *et al.* (2015; 1674) noted (my **bold**):

> The set of transformations affecting the avian feeding apparatus, which apparently occurred at the same node or at very closely spaced nodes in the avialan tree, suggests an integrated morphofunctional complex like that long posited to have arisen in the form of **multiple related anatomical transformations at the base of Mammalia (Crompton and Parker 1978)**. Precision tip gripping combined with kinesis, a novel avian jaw function, could ultimately have permitted the complete loss of teeth and the final transformations toward the face of living birds.

Whether Denton thought about them or not, there were adaptive and ecological contexts involved in both the mammal and bird lineages. As living organisms, they couldn't have avoided it. Animals that lived in specialized niches, of course, from nocturnal synapsid insect hunters that needed to chomp hard shells while keeping an ear out for predators, to diapsids that could fly when their rivals couldn't, able to drop down and precisely nip at the smallest of seeds or bugs. None of those lineages were occurring in isolation—they were just the ones that ultimately made it past millions of years of competition, and a mass extinction or two. Culling out only their so-derived descendents and slapping a Type label on them did a disservice to all those ancestors that had come before, showing the path each had taken from their oh-so-different beginnings.

That glib Type label was all we were ever going to get from Denton on the Bhullar paper. And no other ID advocate has taken a swipe at it either, to my knowledge, but three creationists weighed in on it. Since none of them accepted the common ancestry of life as Michael Denton supposedly did, how would their arguments differ from Denton's ID typological approach?

First off the mark was Fazale Rana (2015) playing for the OEC *Reasons to Believe* team.

Rana accurately summarized the body of Bhullar's work, and acknowledged that the scientists "believe they have gained important insight into key changes that contributed to bird evolution and have demonstrated the power of reverse evolution as a strategy to understand life history from an evolutionary framework."

That ball thrown, Raza promptly took it back: "Though this work and the case for bird evolution seem compelling, the reverse engineering studies' results and the observations from the fossil record can be readily explained from a creation model perspective." Raza had a footnote at this point, but like Denton above, it wasn't to any technical source, but just another claim: "It's important to recognize that the feathered dinosaurs—interpreted as transitional intermediates—appear in the fossil record *after* the first bird appears. Paleontologists refer to this as the temporal paradox. The out-of sequence fossil record justifies skepticism about the evolutionary explanation of bird origins."

Well, it didn't take Rana long to go for the *Bermuda Triangle Defense* (Player #13), did it? The note had a link to Rana (2000) on the redating of the Chinese Yixian Formation where many feathered theropods had been found, Rana citing Pei-ji Chen *et al.* (1998), Qiang Ji *et al.* (1998), Carl Swisher et al. (1999) and Paul Barrett (2000). Work at the Yixian Formation hadn't stopped in 2000, though, and further exploration found Jurassic strata underlying the Cretaceous deposits. It was in those older rocks that the feathered theropod *Anchiornis* was found by Dongyu Hu *et al.* (2009), dating slightly *earlier* than *Archaeopteryx*. The Fossil Genie had kept busy.

Offering the older technical work as though the science had ground to a halt since the turn of the century was a distraction anyway, since Bhullar's work was not about *feathers*. It was about the origin of beaks, and none of that "temporal paradox" applied to the fossil illustrations Bhullar made use of in 2015, since both the archosaur skull and its more derived theropod model long predated the appearance of birds. Was Rana's dated and self-referential approach on this point notably different from Michael Denton's cosy appeal to Casey Luskin to dismiss feathered dinosaurs for him?

Much as Denton had, Rana went on to prop up the ghost of Richard Owen's Archetypes:

> Though Owen's ideas were sound, they were largely abandoned in favor of Darwin's theory because many biologists preferred a mechanistic explanation for life's history and the origin of biological systems. In fact, one could argue that Darwin's theory is an adaptation of Owen's archetype, replacing the canonical blueprint of the Creator's Mind with a hypothetical common ancestor.

One could argue that, but not successfully, because Owen's ideas *weren't* sound, and justifiably lost traction with succeeding scientists because they never actually explained anything, let alone press on to predict new things (like those probainognathids) that the evolutionary approach did as a matter of course.

In a sense, the chief failing of the typological view (or the specifically Creator version Rana favored) was that it could account for absolutely everything independent of its detail, but always without appreciating how the actual science was conducted. So it was that Rana smugly insisted "the researchers merely stumbled upon differences in the developmental program" and "unwittingly reverse engineered a snout from a beak based on design princples."

There was nothing *unwitting* about the Bhullar work, nor stumbling to it. But there was a notable difference in functional outcome between what Bhullar's team had done, and the spin being put on the data by Rana. Bhullar's retroengineering explicitly offered a *prediction* about what should have existed in the past, and Rana made no more mention of that than Denton had. We'll see in the years to come whether the Fossil Genie ever makes good on that, or whether Rana or anybody else in the antievolution camp would change their mind based on such a finding.

But until then, we have Rana's creationist perspective that demanded a very different picture:

> So, is it reasonable to think that unguided, historically contingent processes could carry out such transformations when such small changes can have such profound effects on an organism's anatomy? Because evolutionary mechanisms can only change gene expressions in a random, haphazard manner, the best such processes could achieve would be the generation of 'monsters' with little hope of survival.

The genetic changes Bhullar's work identified involved the shape of the premaxillary bone. Nothing about any intermediate form, in which the gradual input of *Fgf8* or *Lef1* couldn't have avoided altering the size or fusion of that feature, would have constituted a "monster" incapable of eating or reproducing, and so be slated for an early death. And Rana offered no evidence whatsoever that any of that should have been the case. The *monster* tag was purely a dictate of his own dogma, in which all that existed must have been due to "the Creator's handiwork" and so could never have shown any flaw or defect. That left the rival evolutionary framework to be saddled with the arbitrary need to produce legions of non-functional monstrosities before they could arrive at the perfect idealized end product.

And what if the Fossil Genie jumped in to supply exactly what the Bhullar team had specified as a natural evolutionary intermediate? We may suppose a future Rana piece would merely fold that specimen in as yet another variation on the created Archetype. In Rana's hands, the concept is the reverse of *Zeno Slicing*, a featureless bag into which all life no matter what their form or sequence of appearance can be dropped, never to be thought of again. It's another case of "Heads Creation wins, tails evolution loses," and never mind the data.

So we have ID Denton and OEC Rana's versions of the Bhullar paper. Who's next?

Elizabeth Mitchell (2015) came on the field from YEC Land, and played the *Bermuda Triangle Defense* (Player #14) as well: "Without a beak there can be no bird, and evolutionists generally believe that premaxillary bones in non-bird ancestors lengthened and fused to form the bird's upper beak early in the evolutionary history of birds. The fossil record does not contain any record of this transition." Like Denton and Rana, Mitchell did not mention that Bhullar's team were specifically predicting what that transition ought to look like. Instead, Mitchell hyperventilated in how she described what they had done (Mitchell's *italics*, but my **bold**):

"Bhullar and Abzhanov **sabotaged** beak development to see what would grow in its place. 'It shouldn't produce some kind of monster,' Bhullar said, but instead should in essence reverse evolution and reveal the face of the chicken's ancestor. Bhullar and Abzhanov did not tinker with the chicken genome. Instead they **poisoned** two gene products—transcription factors, which regulate expression of other genes—associated with beak development." Repeatedly Mitchell referred to the experimental chickens as "defective," as when she asserted (her *italics*, but again my **bold**), "It is worth noting, however, that they consider their method to have accurately rewound evolution because the **defective** chicken embryos resemble **various morphologies** in animals that they already *assume* to be birds' evolutionary cousins and ancestors. Their reasoning is circular at best."

May we call upon Jonathan McLatchie to opine on shared assumptions?

Mitchell's prejudicial language was as revealing of her rationalizing mindset as MIchael Behe's minimizing of antifreeze proteins in the previous chapter was for his. Even were this work merely acts of *sabotage* and *poisoning*, it still showed what it showed. And Mitchell undercut her own argument earlier when she acknowledged that "The beakless snouts on the defective birds resemble those of dinosaurs and alligators." They weren't closely resembling synapsids, were they? So were the scientists not supposed to remind anyone how their experimental work related to the very features that needed to have modified to generate bird beaks, and that they were able to replicate that prior archosaur/dinosaur configuration?

The rest of Mitchell's argument followed the same track Rana used over in OEC Land, starting with trying to dismiss feathered dinosaurs. Painfully behind the science data curve in 2015, Mitchell decreed without offering any technical sources (even creationist ones) that "Much of what is today assumed to be true of dinosaurs is based on evolutionary conjecture. For instance, though the fossil record does not reveal an evolutionary progression in feather development, evolutionary scientists assume that fibers found on some dinosaur fossils represent early stages in feather evolution and encourage artists to depict dinosaurs covered with feathers."

Mitchell didn't use the word *monster*, but her next point was similar to Rana's (her *italics*):

> And as the researchers freely admit, they only targeted one small part of the animal's anatomy, not the countless changes that would be required to engineer a dinosaur-like animal from a bird embryo. But note we use the word *engineer*, not *reverse engineer*! The ability to *do* this sort of procedure in no way demonstrates that random processes did or ever could ever transform an animal into some other kind of animal. It illustrates just the opposite!

And just how many changes would that be? Hardly *countless*, as there are only a finite number of genes involved. Indeed, were any new genes needed at all, or would it suffice for just mutations in regulatory genes that altered their expression? It only took the changed expression of two (*Fgf8* and *Lef1*) to make those bird beaks. Like the specifics of *kinds*, creationism has not been quick to draw lines in the sand, instead hoping that by keeping the obstacles "countless" they can get themselves off the hook of *counting* anything.

Mitchell ended on what remains the immovable YEC promontory: "Observational science supports not tales of evolutionary transitions but instead affirms God's eyewitness account of the origin of birds on the fifth day of creation about 6,000 years ago."

And for good measure on this Bhullar paper, there's *Do-While Jones* (2015), an anonymous creationist newsletter writer. Rathering interestingly, this minor player on the antivolutionist scene managed to dive deeper into the dataset than Denton, Rana or Mitchell. First, he made a big deal about the fact that the Bhullar experiments hadn't raised the chicks to maturity. Never mind that the Bhullar team wasn't trying to make a "dino-chicken," and that the embryos showed no sign that they would have been unviable had they chosen to do so, *Do-While Jones* assailed the media, "The point is, they did not actually create viable dino-chickens—they just claim they could have; and that's good enough for lazy journalists."

Do-While Jones sounded like a milder Mitchell when he claimed the scientists had only "discovered how to mess with the DNA to affect the shape of whatever grows on the front of the face to make it more like a snout or a beak." But then got more critical:

> Our complaint is that, instead of devoting all of their efforts to making practical use of this knowledge, many scientists are wasting their time trying to figure out how all these genetic pathways developed by accident. If these pathways didn't develop by accident, their search is doomed to failure. If they did develop by accident, what does it matter? Either way, it is a waste of valuable scientific resources.

At least he took note of the predictive aspect of their work, but only to dismiss it: "They can't be sure they have replicated the steps found in the fossil record because the steps can't be found in the fossil record!" No, only that one particular step in the fossil record still needed fossil confirmation. The other stages in the game were not only known, they were actively illustrated in their paper.

Then *Do-While Jones* quoted almost all of the Bhullar *et al.* (2015, 1673) saltational paragraph that Denton had nipped only the last bit. That aspect prompted the creationist to harrumph, "the Punctuated Equilibrium problem has reared its head again"—reminding us of how carefully scientists should be in chosing their words, given the number of antievolutionists there are just itching to quote-mine them to their own purpose.

Most interestingly, *Do-While Jones* hit on an aspect of the Bhullar paper that neither Denton, Rana nor Mitchell had: that it had been *criticized*. Elizabeth Pennisi (2015) reported on the scientists who were surprised that Bhullar's team had found so prominent a role for the WNT pathway, given that they had pinpointed *Shh* instead, Nathan Young *et al.* (2014). On this matter, *Do-While Jones* showed he could selectively quite-mine on his own, offering up just the first sentence of one of Pennisi's paragraphs, which I've put in **bold**:

> **Marcucio, a developmental biologist, also worries that the changes in facial structure observed by the Harvard team may stem from unintended cell death caused by the inhibitors they used.** "Adding the fossil record to this work is really an important step, but I think they are just looking at the wrong pathway," he says. Abzhanov and Bhullar counter that *Fgf8* and *SHH* are often coexpressed and may work together; also, they saw no excess of cell death.

Having stepped over the caveats to Marcucio's criticism, *Do-While Jones* concluded, "We aren't sure, but Marcucio might be saying that messing with the DNA might cause fatal unintended consequences." All of which fitted into the creationist mindset where the world consists only of delightfully designed static forms, that one meddled with to one's peril.

So, what can we make of this quartet? It's revealing that none of the antievolutionists were willing to explore the fossil side of things, to offer a positive explanation for why those extinct critters existed at all, or why modern birds should possess so specific a mix of developmental genes that fiddling with a microevolutionary bit of them should so readily replicate their ancient form.

Which is the only game antievolutionists have been willing to play here. Hide the ball.

12. A Taxonomy of Error: How People Believe Things That Are Not True

With a remarkable consistency, antievolutionists have been so far behind the data curve that they haven't realized there was a curve they were behind. Impaired by their own limited secondary reading, they've consistently imagined they were on a straightaway, uncluttered by problems (or much of the evidence).

How do they do that? I mean, it's an interesting problem, isn't it? Something distinctive must be going on in the heads of antievolutionists. And that's why antievolutionism is such an informative field to investigate, because there has been so much ink spilled, so many lectures given, over so many years, that it's possible to detect aspects of those internal mental processes based on the sheer volume of external behavioral output.

You've seen all the parts of it here, on the RMT and the other subjects I've brought up, and in my decades of #TIP research on the "Troubles in Paradise" project, I've narrowed the problem down to four fundamental methodological mistakes that turn out to afflict not only antievolutionists, but more broadly describe just about anybody who manages to believe things that are *really* not true, from trivial if zany misbeliefs like the Flat Earth to nastier and more consequential ones like Holocaust Denial.

Any one of the four mental snags poses a hurdle for anyone with ambitions to be taken seriously as a careful thinker, but when all four are running at once, at full strength, you have the makings for an intractable ideology of error. Antievolutionists hit all four running.

#TIP Methods Foul Up 1: Overreliance on secondary sources

Though I didn't realize it initially, one of the most influential and lasting lessons I ever learned in my intellectual life came from one of my professors in college. His discipline was historiography, which is the history of history, studying how the practice of history had changed over the years, and identiying the strengths and weaknesses of all the various historians. Yes, I choose that "strengths and weaknesses" term deliberately, because historians have to grapple with exactly the issues that Intelligent Design advocates press upon us today as part of their turf, even though in practice that's never been their game.

My historiography teacher stressed a seemingly dull and ponderous scholarly point, that you get into trouble the moment you confuse a primary with a secondary source.

So you read or hear or watch somebody who tells you something you want to be true. Good. But how do you know that claim is actually true? Well, do they offer evidence for it? If not, isn't that a problem? But even if they do, wouldn't you have to *check* those sources to see if they were as claimed? That's where primary sources come in, the nuts and bolts of reality, where evidence and observation and experiment are put forward by people in a field, to be further tested and measured by subsequent work. We've bumped into a lot of that sort of thing in this book.

Unless you try to ground an argument in primary sources, you're implicitly buying into the take that the secondary redactor was trying to make, substituting an echo chamber for a microscope. In the antievolution case, this problem is a pathology bordering on addiction, where quote mining is a cottage industry for millions of people who can believe themselves knowledgeable about a serious scientific subject they actually know nothing substantive about.

Fortunately I don't have to guess at this. I've been measuring it in the course of compiling my #TIP dataset. Of the over 2000 antievolutionist writers I've tallied so far (generating some 7500 citations in my current #TIP reference bibliography over at *www.tortucan.wordpress.com*), a whopping 94% of them don't cite primary sources at all. And as far as I can tell, none of them seem to think that is a problem.

Over and over again in the antievolution case, we've seen a degree of ideological certaintity that was in inverse proportion to how closely they thought to look into their own sources. This has been painfully true with lower echelon creationists like Scott Huse (copying wholsesale Luther Sunderland's bibliography) or Hank Hanegraaff (channeling Duane Gish without wincing), but shows up as well over on the Intelligent Design side with the likes of Jonathan Wells or Denyse O'Leary. These are people who adopt the certainties of others with abandon, based not upon a fact check, but because it reinforces what they want to be true. It's the psychological tarpit of Confirmation Bias.

With only 6% of antievolutionists bothering to cite primary sources in their paper trail, the apologetic bottleneck is a tight one. I've further discovered that by overall volume, 56% of the 7500 antievolution citations were generated by only 87 source authors (a smaller 4% subset than even the 120 people citing primary work at some point). This narrow source base would be a problem even if they were paying attention to all the relevant data. But they're not.

#TIP Methods Foul Up 2: Limited Dataset

As of this writing, those 120 source-citing antievolutionists have brought up around 2300 technical papers. That work spans several decades, by the way, and includes the cosmological and geological areas being challenged by Young Earth creationists, not just the biological evolution work that IDers and YECers focus on in tandem. You can judge for yourself how well antievolutionists have tackled the RMT and related subjects I've touched on in this book, from YECers Aaron, *AiG*, Camp, Coppedge, Creager, Davis, Doyle, **Gish**, Kenyon, Mehlert, **Mitchell**, **Nelson**, *Truth in Science* and **Woodmorappe**, to **Behe**, **Dembski**, **Denton**, Johnson, McLatchie, **Meyer**, Minnich,

Moneymaker, O'Leary, Seelke and **Wells** on the ID side. That's only a few dozen people over all those years, and now you've had a good look at them. If I've managed to miss anybody who's weighed in on the RMT, let me know. My #TIP website has comments tabs, open to all.

The nine I highlighted in **bold** belong in turn to another equally limited category that I've been identifying in my #TIP project: the three dozen antievolutionists who can be counted core fact claimants in the modern antievolution movement (the ones who field the novel claims that get copied by the larger body of followers). Which means *three quarters* of those core apologists never thought to weigh in on the RMT at all.

Now I don't suggest that you should simply take my word for everything I've said. You can and should do your own source checking. In fact, I highly recommend it—that's how you get to properly understand a subject. That's how I did it. It's not magic, just effort and care.

So, what about those 2300 science sources antievolutionists have pulled from their ammo pile? That may sound like a lot, but that's pocket change compared to the relevant dataset I've been amassing in #TIP, which as of this writing covers over 18,000 technical papers (and I've cited over 1300 of that set in just this book). Which means at best about 13% of the relevant science work is getting noted by antievolutionists. And since a chunk of that has been tossed up secondarily or merely as the mine for a quote, a rough heuristic take home from my #TIP project is that antievolutionists are missing roughly 90% of the available dataset.

You've had a chance to see a lot of that mismatch in the RMT case, where most antievolutionists don't mention the subject at all, and those who did operated miles away from the relevant science work. And sometimes literally stumbling past it, right before their own eyes, in the sources they went out of their way to cite but clearly had not investigated clearly or fairly.

Yet antievolutionists are certain that they're paying attention to the data, or at least the important stuff, and are able to better appreciate that because they're unfettered by the philosophical baggage of Darwinian naturalism.

They're not the only group that thinks that way, by the way. In *Troubles in Paradise* I called that rationalization the *Von Däniken Defense*, for the former Swiss hotel manager who in the 1960s popularized the idea that Ancient Astronauts had meddled in human history, at least insofar as oversized stone monuments were concerned, Downard (2004a). Just like antievolutionists (I used Phillip Johnson as the example in the old #TIP chapter), Erich Von Däniken insisted he paid attention to the same information that regular archaeologists did, only using fresh assumptions to arrive at novel insights. Only he hadn't. Not even close.

Which brings us to an even deeper methodological diagnostic. Paying attention to all the data means more than just taking potshots at it. It means giving an account of it, pulling all that information together to make sense of it. It means figuring out what you think happened.

#TIP Methods Foul Up 3: No *Map of Time*

At some point a rival view of things has to explain the data. To make a valid *Map of Time* in their own heads. This happened, then this happened, then this did.

We know what the modern evolutionary perspective has to say. I've been lobbing it at you throughout this book. It involves all of life through all of time, fossil and living, placed into a temporal and geological context, drawing inferences about adaptations to climate and ecolological competition, and grounded even in the very stuff of life, the genes and their natural variation which have laid the microevolutionary groundwork for even the most sweeping of macroevolutionary transformations. Like that RMT.

There's no doubt whatsoever that both Intelligent Design and Young Earth creationism also have very specific ideas about what they don't want to have had happen with life in the past. An inadequate number of "adaptive Darwinian scenarios" for a Michael Denton; a catastrophic global Flood for a John Woodmorappe. But that's pretty general. What did they think happened?

It's that *positive* explanation that has been so lacking in both the ID and YEC perspectives. This isn't an accusation, it's an observation. I've been lobbing that at you too, all through this book. At every turn, calling attention to opportunities where the antievolutionist advocate could have—and *should* have—put on display what they thought happened. And every time, a failure.

Knowing that antievolutionists are running off a very limited dataset only makes matters worse, of course. They could only be taking a fair stab at the Darwinian dragon if they were showing how superior their explanatory framework was with as broad a dataset as that "theory in crisis" they were out to replace. But knowing what short shrift the RMT and dinosaurs have been given across the antievolutionist spectrum, together missing much of all the complex life that has ever lived, the absence of a working *Map of Time* in the antievolution context is a fundamental and fatal conceptual defect.

The song and dance we saw with Charles Creager on Hunt's RMT parade in Chapter 10 was no aberration. Michael Denton atop the ID foodchain had better grammar and more science citations, but was no more prone to grappling with the data than Creager was muttering on his own inability to spot *general trends*.

I seriously doubt that any antievolutionist will ever put forward what they think happened in the *Map of Time* sense I am demarking here, and that's because of something still more fundamental in what plagues the antievolutionary mindset—and, by extension, anyone who comes to believe things that are not true. Something which isn't at all dependent on or derived from whatever higher level philosophical or doctrinal beliefs they may so fervently hold, even if it has come to be inextricably entwined with them.

It turns on the simplest of methodological questions: What would change your mind?

#TIP Methods Foul Up 4: Contrary evidence is *not* allowed

Antievolutionists never think about what would cause them to change their minds. And I mean *never*. That's especially so for transitional forms (fossil or biological), where I literally know of not a single counterexample in all the antievolutionist apologetics, from St. George Mivart (1871) in Darwin's day down to the present swirl of Intelligent Design claims by Behe, Dembski or Denton. Someone laying out in detail what the fossils or other evidence would need to look like to cause them to slap their head and exclaim, "Well, *that* sure looks like evolution, no doubt about it!"

All we've got over decades is nitpicking and quote parsing, rejecting and denying, always saying what they *won't* accept, not what they *would*. I'd call that a *pattern*.

But do antievolutionists not describe such things just because they're shy or coy?

It's fair to say that a Young Earth creationist might be reluctant to come clean about what they actually think, especially in an academic or work environment, and YECers like Paul Nelson and David Coppedge have operated under the radar for years. But that wouldn't explain the absence of clear Flood Geology analysis in their own in-house literature or group discussions. Creationists like Andrew Snelling or John Woodmorappe do write fairly detailed technical studies on how particular rock formations were due to the Flood, but even then lack a description of what a disconfirmation of their views would need to look like.

That's because for them there cannot be a disconfirmation. Period. God's Word overrides everything, no matter what the evidence is, as thoughtful creationists like Tood Wood (2009) have acknowledged. Or take the case of Hugh Ross over in OEC Land, who does not defend a literal Biblical Flood. Ross (1994, 118) complained:

> Talk radio host John Stewart asked John Morris (a geological engineer) in my presence if he or any of his associates had ever met or heard of a scientist who became convinced that the earth or universe is only thousands of years old based on scientific evidence, without any reference to a particular interpretation of the Bible. Morris answered honestly, "No." Stewart has since asked the same question of several other prominent young-universe proponents, and the answer has been consistent: *no*.

And yet in his 2000 debate with Kent Hovind, gleefully reported by Jonathan Sarfati (2000b), Ross confessed regarding the age of the universe: "If the Bible clearly taught that it was young, I would believe that in spite of my astronomy." Not much arguing with that attitude.

As for Intelligent Design, over the years I've made a point of investigating whether they were any more prone to thinking this *contrary evidence* thing through than their creationist cousins, asking three of the major ID figures about particular cases they had covered in their own writings to see whether they had some standard for acceptance. Whether by physical letter or email, I always framed the question in a benign inquiring way so that they couldn't tell whether I favored or opposed evolution, just inviting their own input. In that way I hoped they wouldn't feel a need to go all defensive on me in the face of a critic, and just give an honest answer.

So it was that I asked Phillip Johnson about the RMT evidence (noted back in Chapter 7), and Michael Behe on whale intermediates, recounted in Downard (2004d, 383-385), along with Jonathan Wells on *Archaeopteryx*, noted in Chapter 6 and covered in fuller detail in Downard (2003b, 90-92). In every case, they not only hadn't thought about what evidence would change their mind, they positively couldn't be prodded to do so.

While it wouldn't be that surprising to find individual writers not thinking about this issue, to discover it as a universal property across the entire antievolutionary spectrum was quite another, and it demands an explanation.

Matthew Harrison Brady Syndrome: Under the Tortucan mindshell

In my "Troubles in Paradise" work I had come to identify the four fundamental flaws outlined above, but that raised another question. What was going on in an antievolutionist's head that they could operate that way? Intellect didn't seem to be a relevant factor, since people like Phillip Johnson (who got into Harvard when he was 16) or Michael Denton were by no means stupid nor unaccomplished academically. Nor did religious convictions appear absolutely necessary, even if the list of secular antievolutionists was a short one (David Berlinski, Michael Denton and Richard Milton pretty much exhausting the field).

What united them all, though, was something that reminded me of an exchange between William Jennings Bryan and Clarence Darrow in 1925 during the Scopes "Monkey" Trial on evolution. Regular science evidence for evolution having been excluded from the proceedings by the judge's ruling, Darrow manuevered the devout Bryan onto the stand to testify as expert witness on the only work that remained, the Bible. The back-and-forth that went on between Bryan and Darrow inspired the famous play and film, *Inherit the Wind*, where Bryan was fictionalized as Matthew Harrison Brady (played by Frederic March), and Spencer Tracy played the Darrow surrogate Henry Drummond. Regardless of your opinions on evolution, Stanley Kramer's 1960 film version is still worth a watch just for the splendid acting.

At one point, Darrow was trying to pin Bryan down on just how long the Days of Creation were in Genesis. The play and film didn't need to change a word from the juicy 1925 original:

> **BRADY**: "The Bible says it was a *day*."
> **DRUMMOND**: "Well, was it a normal day, a literal day, a 24-hour day?"
> **BRADY**: "I don't know."
> **DRUMMOND**: "What do you think?"
> **BRADY** (following a long pause): "I do not think about things I do not think about."
> **DRUMMOND**: "Do you ever think about things that you *do* think about?"

Now in the original trial, Bryan quipped "Sometimes" in reply, which I'm sure got a laugh. But that still left his original sentiment. Was it possible that there really were things that William Jennings Bryan simply didn't think about? It would certainly explain the difficulty Darrow had in getting Bryan to deal with the prehistoric world known by 1925 to have predated the Biblical 6000-year timeframe. What if Bryan's declaration that "I do not think about things I do not think about" was completely honest--that there were things Bryan *literally* didn't think about?

The psychological concepts of how minds resolve Cognitive Dissonance or how they can fall into a Confirmation Bias mode have been the subject of much scientific study, such as Louisa Egan *et al.* (2007), Keise Izuma *et al.* (2010), Bradley Doll *et al.* (2011) and Johanna Jarcho *et al.* (2011). But either attitude might be all too easy for anyone capable of what Bryan showed on the stand in 1925. In honor of him, or rather his *Inherit the Wind* doppelgänger, I dubbed such a hypothetical cognitive property (the actual ability to not think about things you don't think about) *Matthew Harrison Brady Syndrome*, or MHBS for short. I do confess to having liked the serendipitous implication of the "BS" part too.

There appear to be separate brain centers devoted to belief, disbelief and uncertainty, Sam Harris *et al.* (2008), and I suspect that my hypothetical MHBS component would involve at least the *anterior cingulate cortex*, a primate-derived social problem-solving system which plays a significant role in human self-deception and conflict resolution, Cameron Carter *et al.* (1998; 2000), George Bush *et al.* (2002), Vincent Van Veen & Carter (2002), Kristin Laurens *et al.* (2003), Phan Luu & Michael Posner (2003), Nobuhito Abe et al. (2006), Peter Rudebeck *et al.* (2006), Posner *et al.* (2007), John Anderson et al. (2009), Benjamin Hayden *et al.* (2009), Thilo Womelsdorf et al. (2010) and Jill O'Reilly *et al.* (2013).

Whether or not something like MHBS actually exists as a neurobiology component (and only time and neuroscience could settle an issue like that), it was an interesting concept to play around with, and certainly made a lot of sense of what I was seeing in the many antievolutionists I have studied, where not thinking about things was on display for all to see, again and again.

I could imagine cases in which a little MHBS might be in force, and even a good thing. Holding on to a purpose even when everything seems to be against you, from a resilient Churchillian leader in wartime to the average Janes and Joes who keep plugging away at things one day at time. It may even be a cousin to optimism and hope—Scarlett O'Hara deciding "Tomorrow is another day" in *Gone With the Wind*.

We all know people who have that *thing* they can't abide, and will never allow evidence to change their mind on it. In a trivial context, those who swear by Chevy over Ford cars or trucks—and those who as vehemently defend the contrary. We were a Chrysler family, by the way. A similar dynamic possibly plays a part in intense identification with particular sports teams. And of course, there are religious and political convictions.

But what might happen if *most* of a person's cognitive landscape were dominated by MHBS? Their worldview consisting of things they wanted to be true, and a mind that framed their own perception to never perceive contrary evidence as *contrary*? Such a person would be supremely impervious to mind changing because their MHBS intervened, like some internal sedative. The most intractable doctrinal creationists (YECers of course, but still more for geocentrists and Flat Earth believers) offer ample illustration of such minds at work, though nothing in the ID field appears notably less intense when it comes to not seeing things they don't want to.

As there was no word for what I was describing, I gave a provisional name to that too: a *Tortucan*, from the Latin for "turtle." People with a comforting mindshell, and often a rather narrow view of the world, hunkered down under that protective carapace.

Besides the obvious issue of how easily a *Tortucan* mind could reject any and all evidence for evolution, such a feature would explain the equally distinctive absence of any substantive criticism of Young Earth Creationism by the Intelligent Design movement. Insofar as they shared a common *Kulturkampf* view of the world, YEC and ID can and do readily rub shoulders, even to coauthoring antievolution papers and books, as we've seen with Paul Nelson at the *Discovery Institute*. Outside of that frame, well, it's something they just don't think about.

For example, about as close as Michael Denton (2016a, 57) got to addressing the issue of the rival creationism was to quote-mine a bit from "a recent anti-creationist post," the nine-year-old commentary by Penny Higgins (2006), in which "the author has to concede: 'Importantly, groups are united based on shared 'derived' characteristics.' In other words, groups of organisms are indeed distinguished from each other on the basis of unique sets of traits." But that was no concession, it was a recognition of the nested evolutionary structure of life, which Higgins ironically illustrated by the diapsid/synapsid divide. And we've seen how much attention Denton paid to that.

Freed of the nuisance of changing their mind in the face of evidence, antievolutionism functions as a dynamic social network, hunting bigger game than how cervical vertebrae evolved in mammals. The current antievolution movement pictures themselves as the vanguard of a scientific revolution, growing in numbers as exemplied by the "Dissent from Darwin" list begun by Ray Bohlin over a decade ago. The *Discovery Institute* (2014) version now has over a thousand evolution-skeptical scientists. Though creationists play that numbers game too, as Bob Enyart (2014) exemplifies. And a creationist or two pops up on the "Dissent from Darwin" list as well, such as John Baumgardner who's been on it all the way back to the roots of the list, Henry Schaefer *et al.* (2002). Baumgardner's YEC resume runs over many years, ranging from Flood Geology to genetic mutation claims, Baumgardner (1994; 2003a-b; 2016), Baumgardner & Daniel Barnette (1994) and Baumgardner *et al.* (2008; 2013).

Now a thousand Darwin skeptics may sound like a lot, especially mashed into fine print on a long pdf. If you wanted to gather them all together, you'd need a good sized metropolitan convention room to hold them, but that doesn't look quite so formidable when you compare them to the body of working scientists. Remember that my own #TIP collection has over 18,000 technical papers, and those were written by over 48,000 individual scientists. For them, we'd need a stadium.

And if you cross-checked to see how many of the scientists active in the antievolution movement have a seat over in the working scientist stadium, well, you'd be hard-pressed to scrape together even a hundred. They're people like Michael Behe, Jonathan Wells or Michael Denton on the ID side, along with creationists like Jeffrey Tomkins or Mark Toleman, all of whom have published works in evolution-relevant fields. They're just not usually major players in their fields, and you'd still be lacking a notable paleontologist or a heavy gun in the genetics discipline.

The "Dissent from Darwin" warrants a full analysis of who's on it and on what basis they came to sign it, but that's not so easy a task given how many of the signatories show no clear scholarly trail apart from the fact that their name's on the list. How many of them came to their convictions based on what they'd read in the popular antievolution literature versus a religiously motivated certainty is very hard to tell for most of them, as *RationalWiki* (2015) found when they tried to track down their backgrounds and publications. For any of them channeling an *Icons of Evolution* or *Evolution: Still a Theory in Crisis* secondarily, though, the fact that they possess an advanced degree counts no more than a brick layer or a clerical assistant doing the same thing.

However small a ripple antievolutionist arguments have made among regular scientists, there is a modest artistic gleeclub support group. Novelist Tom Wolfe has recently jumped onto the ID bandwagon, noted by Andrea Denhoed (2015), which was seen as such a good thing by David Klinghoffer (2015; 2016f) and considerably less so for Jerry Coyne (2016).

Intelligent Design's gone directly *fictional*, too (though one might say it's been doing that all along). Bruce Buff's 2016 *The Soul of the Matter* got an enthusiastic plug by Klinghoffer (2016h), pleased how this new "Dan Brown-Style Thriller" (and is another *Da Vinci Code* potboiler really a good thing?) embodied Intelligent Design concepts of teleologically engineered genetic information as its scientific underpining. According to Klinghoffer, "Buff has done his homework," which apparently consisted of extracting arguments from such ID stalwarts as Stephen Meyer (2009). Sorry, but that was less *research* than *stenography*.

I've found no indication David Klinghoffer has done enough of his own homework in this area to accurately spot others doing likewise, but the fact remains that engaging in genuine research for even a novel in the 21st century ought to involve more than just cherrypicking bits you want to be true. One need only look at Dan Brown, whose implausible buried mystery puzzles were no better grounded historically than Nicholas Cage's kitschy (albeit still entertaining) *National Treasure* movies.

But even in the novelization department, ID was a latecomer.

Antievolutionism partied with UFO Ancient Astronaut lore in Walt Becker's 1998 *The Link*, about a paleoanthropologist finding fossil evidence connecting humans and apes far too early for hidebound science. Fortunately Becker (1998, 413-416) included a Bibliography to see what got dumped into his particular thresher: elements of paranormal lore proffered in the 1979 version of John Anthony West (1993), credulously mixed with the strained claims of ancient advanced astronomy fielded by Robert Temple (1976) and Robert Bauval & Adrian Gilbert (1994), seasoned with heavy doses of the very dated neo-catastrophism of Charles Hapgood & Joseph Campbell (1958), Hapgood (1966) and Richard Milton (1997), along with the lengthy but wildly inaccurate *Forbidden Archaeology* of Michael Cremo & Richard Thompson (1993).

For those curious about the ancient astronomy claim, the critical Bernard Ortiz de Montellano (1996) is informative, and Wade Tarzia (1994) and Tom Morrow (1999) have surveyed the impact of Cremo & Thompson's curious book. Oh yes, and Becker had one isolated evolution skeptic, none other than Michael Denton (1985). What a mix.

But the reference that really dropped my socks came at the end of Becker's list: "*The Mysterious Origins of Man*, produced by Bill Cote, Carol Cote, and John Cheshire. Broadcast by NBC Network TV in February 1996. New York: BC Video." Narrated with portentious zeal by Charlton Heston (before his stint as Gun-Toter in Chief for the NRA), explicit creationists like Carl Baugh (disingenuously billed as an "archaeologist" and "anthropologist") and John Morris rubbed talking head shoulders with the newer breed of silly, represented by David Hatcher Childress (described innocuously as an "author/researcher" known for his "numerous articles on the coexistence of humans and dinosaurs").

Frank Sonleitner (1996) commented on this preposterous show at the time, and I included quite a dissection of it in Downard (2004e, 536-540). The jejune Dissertation by Carl Baugh (1989) gave a clue to the creationist's limited apologetic writing skills, while Baugh (2009) has shown his avuncular video lecture style. The *Institute for Biblical & Scientific Studies* (2008) and *San Antonio Bible-Based Science Association* (2008a-b) are examples of how Baugh's ideas continue to percolate through the grassroots creationist grapevines. Nor has David Childress disappeared in the decades since, as his now graying mug often shows up as an expert pundit in the many ancient astronaut outings still appearing on the *History Channel*.

This persistent intersection of fringe creationism and pseudohistory surfaced in another fictional YEC thriller, Christopher Lane's klunky 1999 *Tonopah*, in which digging by a dedicated creationist Flood paleontologist (there are so many!) unknowingly trespassed onto a secret Nevada base, whereupon homicidal government flunkies out to prevent disclosure of a 1950s atomic test mishap filled most of the novel with gratuitous mayhem in a sort of "Green YEC meets Rambo in Area 51" plot.

I wonder just how many of the "Dissent from Darwin" signatories have a clue about what goes on in the noggins of millions of creationists who might draw on such novels? Or Ken Ham's *Answers in Genesis* rather than wading through the footnoted tomes of Stephen Meyer or Michael Denton (that they'd never fact check anyway), let alone cribbing the latest from the *Discovery Institute*?

A particularly influential creationist in the *Kulturkampf* subculture that the *Discovery Institute* pays no attention to is the prolific Kent Hovind. A generation of grassroots believers have been "educated" by the video lectures of Hovind (1996; 1998a-f; 2000; 2002a-b; 2003a-d; 2006; 2007a-b; 2010; 2012a-b), and his popularity hasn't been slowed in the least by a prison stint for tax evasion, diligently covered by Peter Reilly (2012; 2013a-e; 2014a-f; 2015a-j) of *Forbes*.

It's not that the Discovery Institute hasn't interacted with Hovind followers—they just don't bother to notice. In Downard (2015b, 25-35) I discussed Hovind's influence on Joe Baker, whom Phillip Johnson and the Intelligent Design movement disingenuously tried to coopt as one of their own. As for Hovind's science arguments, you can get the gist of his method by one of his sillier claims, the idea that the solar system couldn't be billions of years old because, at the rate the Sun burned up hydrogen, it would have been as large as the Earth's orbit back then. This argument required Hovind to have been completely incapable of basic long division, working out how trivial even billions of years of fusion would be on the solar diameter, as I went into in Downard (2004c, 253-255).

Hovind is such a loose cannon that even "mainstream" creationists can't take him seriously. Jonathan Sarfati & Carl Wieland (2002) and Wieland *et al.* (2002) slammed Hovind in the name of "Maintaining Creationist Integrity" at *Answers in Genesis*, while on his radio show, the *Bible Answer Man* Hank Hanegraaff (2002) felt obliged to defend conservative Christian orthodoxy by accusing Hovind of weak scriptural exegesis and persistent duplicity (such as advocating strained Bible Code predictions only to deny it later). For those unaware of that prophecy niche, Michael Drosnin (1998; 2003) contended there were hidden predictions encoded into the original Hebrew text of the Mosaic Bible, an extraction procedure that particularly relied on that old Nostradamus-style retrofitting skill, as skeptical critics like David Thomas (1997), Matt Young (1998) and Randall Ingermanson (2004) have duly noted. And for what it's worth, William Dembski (1998a) flirted with the Bible Code too, seeing that as yet more deeply embedded divine information.

It's fair to say that anyone who could take Kent Hovind seriously can't be taken seriously, and yet I bump into Hovindistas a lot, from local High School graduates to Hovind groupies online in assorted Twitter jousts. They tend to be minor players on the creationism scene, but do share Hovind's hyper-conservatism. Some are quite ephemoral, such as Teno Groppi (2002a-c) lauding Hovind ("the **BEST** Creation Scientist I've ever heard") and recycling his confident tropes about the fossil record ("how did sea life supposedly evolve up into mammals?") along with acting as Wisconsin's 6th District Representative for the reactionary *Constitution Party* (AKA "Taxpayers Party") fielding their recurrent candidate Howard Phillips ("a saved Baptist").

While Groppi's *God and Country Center* had persisted since 1997, it was no longer online by 2014, but don't for a moment think the views represented by the Groppis of the world have disppeared just because they lost their webserver. The "Religious Right" movement that Howard Phillips (1914-2013) helped found, covered in Julie Ingersoll (2013), is what I came to characterize more generally as the *Kulturkampf* subculture of antievolutionism. A current example would be Andrew Schlafly's *Conservapedia*, Schlafly being the son of conservative activist Phyllis Schlafly (1924-2016), who shared her son's enthusiasm for Young Earth Creationism. This included Phyllis Schlafly (2001a-b) seriously recommending Hovind's "science" lectures.

The *Kulturkampf* movement sustains quite a fringe, including the small but persistent band of *geocentrism* defenders (yes, they don't think the Earth going around the Sun is settled science). In fact, some 20% of Americans haven't got the memo on that heliocentrism thing, reported by Allan Mazur (2010) and *National Science Board* (2014). And if you believe at least that such people don't actually *affect* anything, one of the geocentric-friendly folk was Tom Willis (2000), the fellow who wrote the antievolutionary "science" standards for the state of Kansas in 1999, a disconcerting tale I recounted in Downard (2015b, 11-16).

That situation has only got more pervasive in the years since. While geocentrists are not a popular subset in modern *Kulturkampf* conservatism, more generally creationist sentiments are quite another matter. Evolution itself need never have come up for the bad method of the Tortucan brains beneath to rattle off the cliff. Climate policy, public vaccination efforts to counter disease outbreaks, issues of weapons proliferation and GMOs (Genetically Modified Organisms) are all areas where a grounding in science (both its *content* and its *method*) is essential for elected and appointed officials to steer carefully ahead.

What then does it say about the state of the modern Republican Party, that the members it deemed qualified to serve on the House Science Committee have included bumpkin Ralph Hall and "legitimate rape" Todd Akin, climate change skeptics Mo Brooks and Dana Rohrabacher, as well as the venerable Paul Broun, who ranted that evolution and the Big Bang were "lies straight from the pit of Hell," Alana Horowitz (2012). Astronomer Phil Plait (2010a-c; 2013; 2016) has often called attention to the scientifc illiteracy of politicians like Ralph Hall, and recently excoriated Louie Gohmert of Texas for endeavoring to protect America from gay space colonies.

The 2016 election did nothing to change that reality. The newly-elected Republican Vice President, Mike Pence, has a pro-creationist background, noted by Dave Bangert (2012) and Ari Rabin-Havt (2016). And then there's the President-elect, Donald Trump, a man whose conception of climate policy seems to be restricted to Tweets, and who will have a chorus of "alt-Right" and *Breitbart* ideologues to advocate for pseudoscience and ignorance throughout his term. What affect all that will have in the end, only time will tell. It will certainly be *interesting*.

Donald Trump may very well be the Kent Hovind of modern American politics, someone who cannot be taken "seriously", except by millions of people who cannot be taken seriously because they took him all too seriously. And still, over 59 million people voted for Donald Trump in 2016 (a hairs-breadth less than the popular vote winner Hillary Clinton, yet enough for the Electoral Vote), which may be as graphic a measurement if ever there was one of aspects of the *Kulturkampf Tortucan* demographic, as well as the equivalent lethargy and ineffectiveness of liberal and progressive Democratic opposition.

It's not something I take any pleasure in, but in all honesty I suspect there has never been a more *Tortucan* mind to occupy the White House than Mr. Trump. Never for a moment think that the *Tortucans* of the world do not matter, or shouldn't be taken note of.

Well, enough of politics, and the Tortucan footballs of that most strenuous of participant sports.

Parting thoughts

It's been quite a journey, following the synapsids and their trail over more than a quarter of a billion years, across shifting continents and cataclysmic mass extinctions, outlasting even the dinosaurs (or at least the non-flying ones). We've seen the evidence from paleontology and genetics, compiling a "family portrail" not just of extinct life, but of ourselves, a mirror to our own history as highly specialized endothermic vertebrates.

We're the first species on Earth capable of discovering and understanding our own history in that way. And you've seen how generations of scientists have pieced together that story, even as so many others in and out of the scientific world have not been at all happy to see what's turned up in the family selfie. It's a mismatched competition, the science game arriving at one outcome, with all the evidential players openly on show—while the evolution skeptics play out their game on another field, a more political one, with carefully propped up cardboard versions of just the bits of evidence they need for their competition, the end game scores all decided in advance, of course. Guess who wins.

I came only slowly to the realization that evolution was true. It took things like the dinosaurs and the synapsids to do the trick. But once I saw them, in all their magnificent variety, I realized that natural branching common descent was true in the same way the Earth revolving around the Sun is (Tom Willis notwithstanding), or that continents move, about as fast and far in a year as your fingernails grow. And even though we don't feel the land moving under us, except in the rare earthquake when it slips too hurriedly, still the science can measure it now, down to the millimeter, extending our "observation" of things beyond our familiar expectations.

Paleontology and biology have similarly extended our reach to measure the past and present in ways no less wonderful than the great particle accelerators that reveal the innermost turbulence of matter, or the instruments of astronomy that revealed a universe vaster than even the most extravagant visions of myth and legend. All of these human-made tools have opened up windows to long ago times and spaces both infinitesimal and cosmic. What could be better for a species like us, curious to know what actually is, and was, and might be, and be able to craft the tools and science to actually do it?

In this book you've met more of our family gallery, ones I hope you won't be soon forgetting.

Especially those *probainonogathids*.

Maybe before reading this work, you might not have given them a second thought if you'd seen them in a science text or documentary. Little mouse-sized things, inconsequential.

Not any more, I hope. They've been retrieved from Deep Time, despite the difficulties and circumstances that had obliterated so many other living things, to remind us of who we are, how we came to be, and just how the work of science had rescued them from oblivion. To thumb their little noses perhaps at all the antievolutionists who never imagined they could have existed and still can't give them the time of their ideological day.

Hats off to Robert Broom, too, someone else I hope you'll not soon forget. What a thing to have done, to have used mind and reasoning to recognize from the fragments of past lives how one of the great mysteries of macroevolution had to have come about, and live to see his prediction of it confirmed by the happenstance of discovery, against all the odds.

The Fossil Genie is no enemy of Charles Darwin, it would seem. Nor of any of his friends.

Maybe the jinn has a soft spot for people who do the work, admiring the effort and diligence of all those who approach the Book of Nature knowing whatever is written on its pages will always be what it is, not just what some other book or creed might have commanded for it. And it was always the duty of the honest scientist never to start ripping pages out, or try to rewrite it to suit one's expectations.

The physicist Wolfgang Pauli was said to have dismissed a particularly inept physics hypothesis by snarking that "It's not even *wrong*." That's proven equally true for antievolutionist claims about the nature and history of life. Antievolutionists quit the playing field of paleontology a long time ago, and even those creationists with degrees in the field, like Kurt Wise, behave like the curators of a museum with no exhibits.

Not that there aren't actual antievolution museums. Ken Ham runs a big one, and many a dedicated creationist has their local counterparts, though any displays within couldn't be described as Cabinets of Curiosities, could they? Genuine scientific curiosity being in such short supply on that side.

Should there be visitors, though, most would not be asking to see the probainognathids. Though if there were any, you'd know the little critter would have had to have been most carefully recreated based on the *evolutionary* efforts of the working paleontologists, even down to that predicted double jaw.

And if still confidently tagged as an object of Design, someone of a sarcastic temperament might yet object, "If God didn't want people to believe in evolution, He shouldn't have created therapsids. That was just plain dumb."

References

Aaron, M. 2014. "Baraminological Analysis of the Caseidae (Synapsida: Pelycosauria)." Creation Biology Society *Journal of Creation Theology and Science Series B: Life Sciences* **4**: 19-22.

Abascal, Federico, Armelle Corpet, Zachary A. Gurard-Levin, David Juan, Françoise Ochsenbein, Daniel Rico, Alfonso Valencia, & **Geneviève Almouzini.** 2013. "Subfunctionalization via Adaptive Evolution Influenced by Genomic Context: The Case of Histone Chaperones ASF1a and ASF1b." *Molecular Biology and Evolution* **30** (August): 1853-1866.

Abdala, Fernando, & **Ross Damiani.** 2004. "Early development of the mammalian superficial masseter muscle in cynodonts." *Palaeontologica Africana* **40** (December): 23-29.

Abdala, Fernando, P. John Hancox, & **Johann Neveling.** 2005. "Cynodonts from the uppermost Burgersdorp Formation, South Africa, and their bearing on the biostratigraphy and correlation of the Triassic *Cynognathus* Assemblage Zone." *Journal of Vertebrate Paleontology* **25** (March): 192-199.

Abdala, Fernando, Sandra C. Jasinoski, & **Vincent Fernandez.** 2013. "Ontogeny of the Early Triassic cynodont *Thrinaxodon liorhinus* (Therapsida): dental morphology and replacement." *Journal of Vertebrate Paleontology* **33** (November): 1408-1431.

Abdala, Fernando, & **Ana Maria Ribeiro.** 2010. "Distribution and diversity patterns of Triassic cynodonts (Therapsida, Cynodontia) in Gondwana." *Palaeogeography, Palaeoclimatology, Palaeoecology* **286** (15 February): 202-217.

Abdala, Fernando, Bruce S. Rubidge, & **Juri van den Heever.** 2008. "The Oldest Therocephalians (Therapsida, Eutheriodonta) and the Early Diversification of Therapsida." *Palaeontology* **51** (July): 1011-1024.

Abdala, Fernando, & **Roger M. H. Smith.** 2009. "A Middle Triassic cynodont fauna from Namibia and its implications for the biogeography of Gondwana." *Journal of Vertebrate Paleontology* **29** (September): 837-851.

Abe, Nobuhito, Maki Suzuki, Takashi Tsukiura, Etsuro Mori, Keiichiro Yamaguchi, Masatoshi Itoh, & **Toshikatsu Fujii.** 2006. "Dissociable Roles of Prefrontal and Anterior Cingulate Cortices in Deception." *Cerebral Cortex* **16** (February): 192-199.

Aboitiz, F., J. Montiel, & **J. López.** 2002. "Critical steps in the early evolution of the isocortex. Insights from developmental biology." *Brazilian Journal of Medical and Biological Research* **35** (December): 1445-1472.

Aboitiz, Francisco, Juan Montiel, Daniver Morales, & **Miguel Concha.** 2002. "Evolutionary divergence of the reptilian and the mammalian brains: considerations on connectivity and development." *Brain Research Reviews* **39** (September): 141-153.

Abzhanov, Arhat, Dwight D. Cordero, Jonaki Sen, Clifford J. Tabin, & **Jill A. Helms.** 2007. "Cross-regulatory interactions between *Fgf8* and *Shh* in the avian frontonasal prominence." *Congenital Anomalies* (Kyoto) **47** (December): 136-148.

Abzhanov, Arhat, Winston P. Kuo, Christine Hartmann, B. Rosemary Grant, Peter R. Grant, & **Clifford J. Tabin.** 2006. "The calmodulin pathway and evolution of elongated beak morphology in Darwin's finches." *Nature* **442** (3 August): 563-567.

Abzhanov, Arhat, Meredith Protas, B. Rosemary Grant, Peter R. Grant, & **Clifford J. Tabin.** 2004. "*Bmp4* and Morphological Variation of Beaks in Darwin's Finches." *Science* **305** (3 September): 1462-1465.

Adachi, Noritaka, Molly Robinson, Aden Goolsbee, & **Neil H. Shubin.** 2016. "Regulatory evolution of *Tbx5* and the origin of paired appendages." *Proceedings of the National Academy of Sciences* **113** (6 September): 10115-10120.

Ahlberg, Per Erik. 2004. "Comment on 'The Early Evolution of the Tetrapod Humerus." *Science* **305** (17 September): 1715.

Ahlberg, Per E., Jennifer A. Clack, & **Ervins Lukševics.** 1996. "Rapid braincase evolution between *Panderichthys* and the earliest tetrapods." *Nature* **381** (2 May): 61-64.

AiGbusted. 2008. "The Dishonesty of John Woodmorappe." *Answers in Genesis BUSTED!* 27 January posting (online text at *aigbusted.blogspot.com* accessed 4/17/2016).

Alcock, Felicity, Abigail Clements, Chaille Webb, & **Trevor Lithgow.** 2010. "Tinkering Inside the Organelle." *Science* **327** (5 February): 649-650.

Aldredge, Robert D., & **David M. Sever,** eds. 2011. *Reproductive Biology and Phylogeny of Snakes.*

Boca Raton, Florida: CRC Press.
Al-Hashimi, Nawfal, Anne-Gaelle Lafont, Sidney Delgado, Kazuhiko Kawasaki, & **Jean-Yves Sire.** 2010. "The Enamelin Genes in Lizard, Crocodile, and Frog and the Pseudogene in the Chicken Provide New Insights on Enamelin Evolution in Tetrapods." *Molecular Biology and Evolution* **27** (September): 2078-2094.
Alibardi, Lorenzo, Luisa Dalla Valle, Alessia Nardi, & **Mattia Toni.** 2009. "Evolution of hard proteins in the sauropsid integument in relation to the cornification of skin derivatives in amniotes." *Journal of Anatomy* **214** (April): 560-586.
Alibardi, L., L. W. Knapp, & **R. H. Sawyer.** 2006. "Beta-keratin localization in developing alligator scales and feathers in relation to the development and evolution of feathers." *Journal of Submicroscopic Cytology and Pathology* **38** (June-September): 175-192.
Alié, Alexandre, & **Michaël Manuel.** 2010. "The backbone of the post-synaptic density originated in a unicellular ancestor of choanoflagellates and metazoans." *BMC Evolutionary Biology* (online @ biomedcentral.com) **10** (3 February): 34.
Alliegro, Mark C., & **Mary Anne Alliegro.** 2008. "Centrosomal RNA correlates with intron-poor nuclear genes in *Spisula* oocytes." *Proceedings of the National Academy of Sciences* **105** (13 May): 6993-6997.
Alliegro, Mark C., Steven Hartson, & **Mary Anne Alliegro.** 2012. "Composition and Dynamics of the Nucleolinus, a Link between the Nucleolus and Cell Division Apparatus in Surf Clam (*Spisula*) Oocytes." *The Journal of Biological Chemistry* **287** (24 February): 6702-6713.
Alliegro, Mary Anne, Jonathan J. Henry, & **Mark C. Alliegro.** 2010. "Rediscovery of the nucleolinus, a dynamic RNA-rich organelle associated with the nucleolus, spindle, and centrosomes." *Proceedings of the National Academy of Sciences* **107** (3 August): 13718-13723.
Allin, Edgar F. 1975. "Evolution of the mammalian middle ear." *Journal of Morphology* **47** (December): 403-437.
Allin, Edgar F., & **James A. Hopson.** 1992. "Evolution of the Auditory System in Synapsida ('Mammal-Like Reptiles' and Primitive Mammals) as Seen in the Fossil Record," in Webster *et al.* (1992, 587-614).
Alonso, Patricio Dominguez, Angela C. Milner, Richard A. Ketcham, M. John Cookson, & **Timothy B. Rowe.** 2004. "The avian nature of the brain and inner ear of *Archaeopteryx*." *Nature* **430** (5 August): 666-669.
Aminake, Makoah N., & **Gabriele Pradel.** 2013. "Antimalarial drugs resistance in *Plasmodium falciparum* and the current strategies to overcome them," in Méndez-Vilas (2013, 269-282).
Amos, Jonathan. 2011. "Fossil redefines mammal history." *BBC News* posting 25 August: online text accessed 3/19/2016 (*www.bbc.com*).
Amson, Eli, & **Michel Laurin.** 2011. "On the affinities of *Tetraceratops insignis*, an Early Permian synapsid." *Acta Palaeontologica Polonica* **56** (2): 301-312.
Anan, Keiti, Nobuaki Yoshida, Yuki Kataoka, Mitsuhara Sato, Hirotake Ichise, Makoto Nasu, & **Shintaroh Ueda.** 2007. "Morphological Change Caused by Loss of the Taxon-Specific Polyalanine Tract in Hoxd-13." *Molecular Biology and Evolution* **24** (January): 281-287.
Anderson, Bernhard W., ed. 1984. *Creation in the Old Testament.* Philadelphia: Fortress Press.
Anderson, Jason S., Tim Smithson, Chris F. Mansky, Taran Meyer, & **Jennifer Clack.** 2015. "A Diverse Tetrapod Fauna at the Base of 'Romer's Gap'." *PLoS ONE* (online @ plosone.org) **10** (April): e0125446.
Anderson, John R., John F. Anderson, Jennifer L. Ferris, Jon M. Fincham, & **Kwan-Jin Jung.** 2009. "Lateral inferior prefrontal cortex and anterior cingulate cortex are engaged at different stages in the solution of insight problems." *Proceedings of the National Academy of Sciences* **106** (30 June): 10799-10804.
Andersson, Dan I., Jon Jerlström-Hultqvist, & **Joakim Näsvall.** 2015. "Evolution of New Functions De Novo and from Preexisting Genes." *Cold Spring Harbor Perspectives in Biology* **7** (June): a017996.
Andersson, Jan O., & **Andrew J. Roger.** 2002. "A Cyanobacterial Gene in Nonphotosynthetic Protists--An Early Chloroplast Acquisition in Eukaryotes?" *Current Biology* **12** (22 January): 115-119.
Andrey, Guillaume, Thomas Montavon, Bénédicte Mascrez, Federico Gonzalez, Daan Noordermeer, Marion Leleu, Didier Trono, François Spitz, & **Denis Duboule.** 2013. "A Switch Between Topological Domains Underlies *HoxD* Genes Collinearity in Mouse Limbs." *Science* **340** (7 June): 1234167 (online at *sciencemag.org*).

Angielczyk, Kenneth D. 2009. "*Dimetrodon* Is Not a Dinosaur: Using Tree Thinking to Understand the Ancient Relatives of Mammals and their Evolution." *Evolution: Education & Outreach* **2** (June): 257-271.

Angielczyk, Kenneth D., & **Andrey A. Kurkin.** 2003a. "Has the utility of *Dicynodon* for Late Permian terrestrial biostratigraphy been overstated?" *Geology* **31** (April): 363-366.

———. 2003b. "Phylogenetic analysis of Russian Permian dicyonodonts (Therapsida: Anomodontia): Implications for Permian biostratigraphy and Pangaean biogeography." *Zoological Journal of the Linnean Society* **140** (March): 383-401.

Angielczyk, K. D., & **L. Schmitz.** 2014. "Nocturnality in synapsids predates the origin of mammals by over 100 million years." *Proceedings of the Royal Society of London* **B** (Biological Sciences) **281** (22 October): 20141642.

Ankerberg, John, & **John Weldon.** 1998. *Darwin's Leap of Faith: Exposing the False Religion of Evolution.* Eugene, OR: Harvest House Publishers.

Anquetin, Jérémy. 2011. "Evolution and palaeoecology of early turtles: a review based on recent discoveries in the Middle Jurassic." *Bulletin de la Société Géologique de France* **182** (May): 231-240.

Answers in Genesis. 2011. "Odd Saber-Toothed Beast Discovered—Preyed on ... Plants?" Answers in Genesis *News to Note* posting 2 April (online text at *answersingensis.org* accessed 2/20/2014).

———. 2016a. "Mammal-Like Reptiles: Transitional Forms?" *Answers in Genesis* undated "Kids" posting (online text at *answersingensis.org* accessed 7/25/2016).

———. 2016b. "The Great Delusion." *Answers in Genesis* undated posting (online text at *answersingensis.org* accessed 9/27/2016).

Anthwal, Neal, Leena Joshi, & **Abigail S. Tucker.** 2013. "Evolution of the mammalian middle ear and jaw: adaptations and novel structures." *Journal of Anatomy* **222** (January): 147-160.

Antoine, Pierre-Olivier, Laurent Marivaux, Darin A. Croft, Guillaume Billet, Morgan Ganerød, Carlos Jaramillo, Thomas Martin, Maëva K. Orliac, Julia Tejada, Ali J. Altamirano, Francis Duranthon, Grégory Fanjat, Sonia Rousse, & **Rodolfo Salas Gismondi.** 2012. "Middle Eocene rodents from Peruvian Amazonia reveal the pattern and timing of caviomorph origins and biogeography." *Proceedings of the Royal Society of London* **B** (Biological Sciences) **279** (7 April): 1319-1326.

Archer, Michael, & **Georgina Clayton**, eds. 1984. *Vertebrate Zoogeography & Evolution in Australasia.* Marrickville, Australia: Hesperian Press.

Archer, Michael, Timothy F. Flannery, Alex Ritchie, & **R. E. Molnar.** 1985. "First Mesozoic mammal from Australia—an early Cretaceous monotreme." *Nature* **318** (28 November): 363-366.

Archer, Michael, Farish A. Jenkins Jr., Suzanne J. Hand, Peter Murray, & **Henk Godthelp.** 1992. "Description of the skull and non-vestigial dentition of a Miocene platypus (*Obdurodon dicksoni* n. sp.) from Riversleigh, Australia, and the problem of monotreme origins," in Augee (1992, 15-27).

Archibald, J. David, & **Alexander O. Averianov.** 2006. "Late Cretaceous asioryctitherian eutherian mammals from Uzbekistan and phylogenetic analysis of Asioryctitheria." *Acta Palaeontologica Polonica* **51** (2): 351-376.

Archibald, J. David, Alexander O. Averianov, & **Eric G. Ekdale.** 2001. "Late Cretaceous relatives of rabbits, rodents, and other extant eutherian mammals." *Nature* **414** (1 November): 62-65.

Archibald, J. David, Yue Zhang, Tony Harper, & **Richard L. Cifelli.** 2011. "*Protungulatum*, Confirmed Cretaceous Occurrence of an Otherwise Paleocene Eutherian (Placental?) Mammal." *Journal of Mammalian Evolution* **18** (September): 153-161.

Archibald, John M. 2006. "Endosymbiosis: Double-Take on Plastid Origins." *Current Biology* **16** (5 September): R690-R692.

———. 2015. "Endosymbiosis and Eukaryotic Cell Evolution." *Current Biology* **25** (5 October): R911-R921.

Arduini, Francis J. 1987. "Design, Created Kinds, and Engineering." *Creation/Evolution* **7** (Spring): 19-24.

Armfield, Brooke A., Zhengui Zheng, Sunil Bajpi, Christopher J. Vinyard, & **J. G. M. Thewissen.** 2013. "Development and evolution of the unique cetacean dentition." *PeerJ* (online @ peerj.com) **1** (19 February): 24.

Armitage, Mark Hollis, & **Kevin Lee Anderson.** 2013. "Soft sheets of fibrillar bone from a fossil of the supraorbital horn of the dinosaur *Triceratops horridus*." *Acta Histochemica* **115** (July): 603-608.

Armitage, Mark H., & **Luke Mullisen.** 2003. "Prelininary observations of the pygidial gland of the Bombardier Beetle, *Brachinus sp.*" *Journal of Creation* (AKA Answers in Genesis *Creation ex Nihilo*

Technical Journal) **17** (1): 95-102.

Armitage, Mark H., & **Andrew A. Snelling.** 2008. "Radiohalos and Diamonds: Are Diamonds Really for Ever?" in Snelling (2008, 323-334).

Arnason, Ulfur, Anette Gullberg, Axel Janke, Morgan Kullberg, Niles Lehman, Evgeny A. Petrov, & **Risto Väinölä.** 2006. "Pinniped phylogeny and a new hypothesis for their origin and dispersal." *Molecular Phylogenetics and Evolution* **41** (November): 345-354.

Asahara, Masakazu, Masahiro Koizumi, Thomas E. Macrini, Suzanne J. Hand, & **Michael Archer.** 2016. "Comparative cranial morphology in living and extinct platypuses: Feeding behavior, electroreception, and loss of teeth." *Science Advances* **2** (12 October): e1601329 (online at *sciencemag.org*).

Asher, Robert J. 2012. *Evolution and Belief: Confessions of a Religious Paleontologist.* Cambridge: Cambridge University Press.

Asher, R. J., K. H. Lin, N. Kardjilov, & **L. Hautier.** 2011. "Variability and constraint in the mammalian vertebral column." *Journal of Evolutionary Biology* **24** (May): 1080-1090.

Asher, Robert J., Jin Meng, John R. Wible, Malcolm C. McKenna, Guillermo W. Rougier, Demberlyn Dashzeveg, & **Michael J. Novacek.** 2005. "Stem Lagomorpha and the Antiquity of Glires." *Science* **307** (18 February): 1091-1094.

Ashton, John, & **David Down.** 2006. *Unwrapping the Pharaohs: How Egyptian Archaeology Confirms the Biblical Timeline.* Green Forest, AR: Master Books.

Augee, Michael L., ed. 1992. *Platypus and Echidnas.* Mosmon, AU: Royal Zoological Society of New South Wales.

Aulie, Richard P. 1974a. "The Origin of the Idea of the Mammal-like Reptile." Part 1 of 3. *The American Biology Teacher* **36** (November): 476-484, 511.

———. 1974b. "The Origin of the Idea of the Mammal-like Reptile." Part 2 of 3. *The American Biology Teacher* **36** (December): 545-553.

———. 1975. "The Origin of the Idea of the Mammal-like Reptile." Part 3 of 3. *The American Biology Teacher* **37** (January): 21-32.

Austin, Steven A., ed. 1994. *Grand Canyon: Monument to Catastrophe.* Santee, CA: Institute for Creation Research.

———. 2013. "Why Is Mount St. Helens Important to the Origins Controversy?" *Answers in Genesis* 18 July 2014 posting of *The New Answers Book 3* Chapter 26 (online text at *www.answersingenesis.org* accessed 2/17/2015).

Averianov, Alexander O., & **J. David Archibald.** 2015. "Evolutionary transition of dental formula in Late Cretaceous eutherian mammals." *Naturwissenschaften* **102** (October): 56.

Averianov, A. O., & **A. V. Lopatin.** 2014. "On the Phylogenetic Position of Monotremes (Mammalia, Monotremata)." *Paleontological Journal* **48** (4): 426-446.

Axelsson, Michael. 2001. "The crocodilian heart: more controlled than we thought?" *Experimental Physiology* **86** (November): 785-789.

Ayala, Francisco Jose, Andrey Rzhetsky, & **Francisco J. Ayala.** 1998. "Origin of the metazoan phyla: Molecular clocks confirm paleontological estimates." *Proceedings of the National Academy of Sciences* **95** (20 January): 606-611.

Azimzadeh, Juliette. 2014. "Exploring the evolutionary history of centrosomes." *Philosophical Transactions of the Royal Society of London* **B** (Biological Sciences) **369** (5 September): 20130453.

Azimzadeh, Juliette, Mei Lie Wong, Diane Miller Downhour, Alejandro Sánchez Alvarado, & **Wallace F. Marshall.** 2012. "Centrosome Loss in the Evolution of Planarians." *Science* **335** (27 January): 461-463.

Babin, Patrick J. 2008. "Conservation of a vitellogenin gene cluster in oviparous vertebrates and identification of its traces in the platypus genome." *Gene* **413** (30 April): 76-82.

Badiola, Ainara, José Ignacio Canudo, & **Gloria Cuenca-Bescós.** 2012. "New Early Cretaceous Multituberculate Mammals from the Iberian Peninsula," in Godefroit (2012, 409-434).

Bajdek, Piotr, Martin Qvarnström, Krzystof Owocki, Tomasz Sulej, Andrey G. Sennikov, Valeriy K. Golubev, & **Niedźwiedzki Grzegorz.** 2016. "Microbiota and food residues including possible evidence of pre-mammalian hair in Upper Permian coprolites from Russia." *Lethaia* **49** (October): 455-477.

Bajpai, Sunil, J. G. M. Thewissen, & **A. Sahni.** 2009. "The origin and early evolution of whales: macroevolution documented on the Indian Subcontinent." Indian Academy of Sciences *Journal of Biosciences* **34** (November): 673-686.

Baker, Mace. 2003. "No. 362—Sea Dragons." Institute for Creation Research *Impact* (August): i-iv.
Ball, Steven, Christophe Colleoni, Ugo Cenci, Jenifer Nirmal Raj, & **Catherine Tirtiaux.** 2011. "The evolution of glycogen and starch metabolism in eukaryotes gives molecular clues to understand the establishment of plastid endosymbiosis." *Journal of Experimental Botany* **62** (6 March): 1775-1801.
Ballard, J. William O., Gary J. Olsen, Daniel P. Faith, Wendy A. Odgers, David M. Powell, & **Peter W. Atkinson.** 1992. "Evidence from 12*S* Ribosomal RNA Sequences That Onychophorans Are Modified Arthropods." *Science* **258** (20 November): 1345-1348.
Bandow, Doug. 1991. "Fossils and Fallacies." *National Review* **43** (29 April): 47-48.
———. 1999. "Bedtime Christmas Readings." *Cato Institute* 23 December posting of 13 December Copley News Service article (online text at *www.cato.org* accessed 4/17/2016).
———. 2013. "As Persecution of Christians Intensifies, Freedom of Thought Is at Risk." *Cato Institute* 30 December posting (online text at *www.cato.org* accessed 1/20/2014).
Bangert, Dave. 2012. "The evolution of Gov. Pence starts here; another creation science bill looms." *Lafayette Journal and Courier* posting 12 November online text accessed 11/21/2012 (*www.jconline.com*).
Bao, Zheng-Zheng, Benoit G. Bruneau, J. G. Seidman, Christine E. Seidman, & **Constance L. Cepko.** 1999. "Regulation of Chamber-Specific Gene Expression in the Developing Heart by *Irx4*." *Science* **283** (19 February): 1161-1164.
Bar, Maya, & **Naomi Ori.** 2014. "Leaf development and morphogenesis." *Development* (AKA *Journal of Embryology and Experimental Morphology*) **141** (15 November): 4219-4230.
Barden, Phillip, & **David A. Grimaldi.** 2012. "Rediscovery of the bizarre Cretaceous ant *Haidomyrmex* Dlussky (Hymenoptera: Formicidae) with two new species." *American Museum Novitates* **3755** (September): 1-16.
———. 2013. "A New Genus of Highly Specialized Ants in Cretaceous Burmese amber (Hymenopetera: Formicidae)." *Zootaxa* **3681** (June): 405-412.
———. 2014. "A Diverse Ant Fauna from the Mid-Cretaceous of Myanmar (Hymenoptera: Formicidae)." *PLoS ONE* (online @ plosone.org) **9** (April): e93627.
———. 2016. "Adaptive Radiation in Socially Advanced Stem-Group Ants from the Cretaceous." *Current Biology* **26** (22 February): 515-521.
Barghusen, Herbert R. 1968. "The lower jaw of cynodonts (Reptilia, Therapsida) and the evolutionary origin of mammal-like adductor jaw musculature." Peabody Museum of Natural History (Yale University) *Postilla* **116**: 1-49.
Barr, Stephen M. 2003. *Modern Physics and Ancient Faith.* Notre Dame, IN: University of Notre Dame Press.
Barrett, Paul M. 2000. "Evolutionary consequences of dating the Yixian Formation." *Trends in Ecology & Evolution* **15** (March): 99-103.
Barrow, John D. 2000. *The Book of Nothing: Vacuums, Voids, and the Latest Ideas about the Origins of the Universe.* New York: Pantheon Books.
Barton, Lynn. 2006. "Why intelligent design will change everything." *WorldNetDaily* (25 March): online text accessed 7/28/2014 (*wnd.com*).
Bates, K. T., & **P. L. Falkingham.** 2012. "Estimating maximum bite performance in *Tyrannosaurus rex* using multi-body dynamics." Royal Society *Biology Letters* **8** (28 August): 20120056.
Batzer, Mark A., & **Prescott L. Deininger.** 2002. "Alu repeats and human genomic diversity." *Nature Reviews Genetics* **3** (May): 370-379.
Bauer, William J. 1979. "Review of Evolution of Living Organisms--By Pierre-Pual Grasse." *Institute for Creation Research* October article posting (online text at *icr.org* accessed 3/8/2016).
Baugh, Carl Edward. 1989. "Academic Justification for Voluntary Inclusion of Scientific Creation in Public Classroom Curricula, Supported by Evidence that Man and dinosaurs were Contemporary." Dissertation for "Pacific International University and Pacific College of Graduate Studies" (online text at *home.texoma.com/~linesden/cem/diss/disv1fr.htm* accessed 5/27/2009).
———. 2009. "Let's Review the Facts." *Creation in the 21st Century* Trinity Broadcasting Network cable TV series (viewed on airing 3/3/2009).
Baumgardner, John R. 1994. "Runaway Subduction as the Driving Mechanism for the Genesis Flood." Creation Science Fellowship *Third International Conference on Creationism* paper (online pdf at *www.icr.org* accessed 10/19/2014).
———. 2003a. "Carbon Dating Undercuts Evolution's Long Ages." *Institute for Creation Research* posting (online text at *icr.org* accessed 2/21/2010).

———. 2003b. "Catastrophic Plate Tectonics: The Physics Behind the Genesis Flood." Creation Science Fellowship *Fifth International Conference on Creationism* paper (online text at *www.globalflood.org/papers/2003ICCcpt.html* accessed 1/11/2011).
———. 2016. "Numerical Modeling of the Large-Scale Erosion, Sediment Transport, and Deposition Processes of the Genesis Flood." Answers in Genesis *Answers Research Journal* **9**: 1-24.
Baumgardner, John R., & **Daniel W. Barnette.** 1994. "Patterns of Ocean Circulation Over the Continents During Noah's Flood." Creation Science Fellowship *Third International Conference on Creationism* paper (online text at *www.icr.org* accessed 1/30/2010).
Baumgardner, John R., Wesley H. Brewer, & **John C. Sanford.** 2013. "Can Synergistic Epistasis Halt Mutation Accumulation? Results from Numerical Simulation," in Marks *et al.* (2013a, 312-337).
Baumgardner, John, John Sanford, Wesley Brewer, Paul Gibson, & **Walter ReMine.** 2008. "Mendel's Accountant: A New Population Genetics Simulation Tool for Studying Mutation and Natural Selection," in Snelling (2008b, 87-98).
Bauval, Robert, & **Adrian Gilbert.** 1994. *The Orion Mystery: Unlocking the Secrets of the Pyramids.* New York: Crown Trade Paperbacks.
Beatty, John. 1982. "Classes and Cladists." *Systematic Zoology* **31** (March): 25-34.
Beck, Felix, Kallayanee Chawengsaksophak, Paul Waring, Raymond J. Playford, & **John B. Furness.** 1999. "Reprogramming of intestinal differentiation and intercalary regeneration in *Cdx2* mutant mice." *Proceedings of the National Academy of Sciences* **96** (22 June): 7318-7323.
Beck, Felix, & **Emma J. Stringer.** 2010. "The role of *Cdx* genes in the gut and in axial development." *Biochemical Society Transactions* **38** (April): 353-357.
Becker, Walt. 1998. *The Link.* New York: Avon Books.
Behe, Michael J. 1996. *Darwin's Black Box: The Biochemical Challenge to Evolution.* New York: The Free Press.
———. 2000. "Irreducible Complexity and the Evolutionary Literature: Response to Critics." *Center for Science & Culture* 31 July posting (online text at *discovery.org* accessed 7/11/2011).
———. 2005a. "APPENDIX II TAB A: Federal Rule of Civil Procedure 26 Disclosure of Expert Testimony" (24 March) re: *Kitzmiller v. Dover Area School District* (online pdf at *ncse.com* accessed 7/4/2014).
———. 2005b. "Rebuttal Analysis of Kenneth Miller's Statement" (15 May) re: *Kitzmiller v. Dover Area School District* (online pdf at *ncse.com* accessed 7/4/2014).
———. 2006. "From Muttering to Mayhem: How Phillip Johnson Got Me Moving," in Dembski (2006a, 37-47).
———. 2007. *The Edge of Evolution: The Search for the Limits of Darwinism.* New York: The Free Press.
———. 2009. "Waiting Longer for Two Mutations." *Genetics* **181** (February): 819-820.
———. 2014a. "From Thornton's Lab, More Strong Experimental Support for a Limit to Darwinian Evolution." Discovery Institute *Evolution News & Views* for 23 June (online text at *evolutionnews.org* accessed 6/25/2014).
———. 2014b. "A Key Inference of *The Edge of Evolution* Has Now Been Experimentally Confirmed." Discovery Institute *Evolution News & Views* for 14 July (online text at *evolutionnews.org* accessed 7/17/2014).
———. 2014c. "It's Tough to Make Predictions, Especially About the Future." Discovery Institute *Evolution News & Views* for 16 July (online text at *evolutionnews.org* accessed 8/13/2014).
———. 2014d. "The Edge of Evolution: Why Darwin's Mechanism Is Self-Limiting." Discovery Institute *Evolution News & Views* for 18 July (online text at *evolutionnews.org* accessed 7/24/2014).
———. 2014e. "An Open Letter to Kenneth Miller and PZ Myers." Discovery Institute *Evolution News & Views* for 21 July (online text at *evolutionnews.org* accessed 8/13/2014).
———. 2015. "Kenneth Miller Resists Chloroquine Resistance." Discovery Institute *Evolution News & Views* for 14 January (online text at *evolutionnews.org* accessed 1/15/2015).
Behe, Michael J., & **David W. Snoke.** 2004. "Simulating evolution by gene duplication of protein features that require multiple amino acid residues." *Protein Science* **13** (October): 2651-2664.
———. 2005. "A response to Michael Lynch." *Protein Science* **14** (September): 2226-2227.
Behrensmeyer, Anna K. 1984. "Taphonomy and the Fossil Record." *American Scientist* **72** (November-December): 558-566.
Beilstein, Mark A., Nathalie S. Nagalingum, Mark D. Clements, Steven R. Manchester, & **Sarah Mathews.** 2010. "Dated molecular phylogenies indicate a Miocene origin for *Arabidopsis*

thaliana." *Proceedings of the National Academy of Sciences* **107** (26 October): 18724-18728.
Beisel, K. W., & **B. Fritzsch.** 2004. "Keeping Sensory Cells and Evolving Neurons to Connect Them to the Brain: Molecular Conservation and Novelties in Vertebrate Ear Development." *Brain, Behavior and Evolution* **64** (3): 182-197.
Belting, Heinz-Georg, Cooduvalli S. Shashikant, & **Frank H. Ruddle.** 1998. "Modification of expression and *cis*-regulation of *Hoxc8* in the evolution of diverged axial morphology." *Proceedings of the National Academy of Sciences* **95** (3 March): 2355-2360.
Ben-Ari, Elia T. 2000. "Hair Today: Untangling the biology of the hair follicle." *BioScience* **50** (April): 303-307.
Bender, Cheryl E., Patrick Fitzgerald, Stephen W. G. Tait, Fabien Llambi, Gavin P. McStay, Douglas O. Tupper, Jason Pellettieri, Alejandro Sánchez Alvarado, Guy S. Salvesen, & **Douglas R. Green.** 2012. "Mitochondrial pathway of apoptosis is ancestral in metazoans." *Proceedings of the National Academy of Sciences* **109** (27 March): 4904-4909.
Bennett, E. Andrew, Heiko Keller, Ryan E. Mills, Steffen Schmidt, John V. Moran, Oliver Weichenrieder, & **Scott E. Devine.** 2008. "Active *Alu* retrotransposons in the human genome." *Genome Research* **18** (December): 1875-1883.
Benoit, Julien, & **Sandra C. Jasinoski.** 2016. "Picking up the pieces: the digital reconstruction of a destroyed holotype from its serial section drawings." *Palaeontologia Electronica* (online @ palaeo-electronica.org) **19** (3): 3T.
Benoit, J., P. R. Manger, & **B. S. Rubidge.** 2016. "Palaeoneurological clues to the evolution of defining mammalian soft tissue traits." *Scientific Reports* (online @ www.nature.com/srep/) **6** (9 May): 25604.
Ben-Shlomo, Herzel, Alexander Levitan, Naomi Editha Shay, Igor Goncharov, & **Shulamit Michaeli.** 1999. "RNA Editing Associated with the Generation of Two Distinct Confirmations of the Trypanosomatid *Leptomonas collosoma* 7SL RNA." *The Journal of Biological Chemistry* **274** (3 September): 25642-25650.
Benson, Roger B. J. 2012. "Interrelationships of basal synapsids: Cranial and postcranial morphological partitions suggest different topologies." *Journal of Systematic Palaeontology* **10** (December): 1-24.
Benton, Michael J. 1990. *Vertebrate Palaeontology: Biology and evolution.* London: Unwin Hyman.
———. 1993a. "Life and Time," in Gould (1993, 22-36).
———. 1993b. "The Rise of the Fishes," in Gould (1993, 79-83).
———. 1993c. "Four Feet on the Ground," in Gould (1993, 111-112).
———. 1997. "Reptiles," in Currie & Padian (1997, 637-642).
———. 2003. *When Life Nearly Died: The Greatest Mass Extinction of All Time.* 2005 pb ed. London: Thames & Hudson.
———. 2005. *Vertebrate Paleontology.* 3rd ed. Oxford: Blackwell Publishing.
———, ed. 2008. *The Seventy Great Mysteries of the Natural World.* London: Thames & Hudson.
———. 2012. "No gap in the Middle Permian record of terrestrial vertebrates." *Geology* **40** (April): 339-342.
Benton, Michael J., & **Philip C. J. Donoghue.** 2007. "Paleontological Evidence to Date the Tree of Life." *Molecular Biology and Evolution* **24** (January): 26-53.
Benton, Michael J., & **Rebecca Hitchin.** 1997. "Congruence between phylogenetic and stratigraphic data on the history of life." *Proceedings of the Royal Society of London* **B** (Biological Sciences) **264** (22 June): 885-890.
Benton, M. J., M. A. Wills, & **R. Hitchin.** 2000. "Quality of the fossil record through time." *Nature* **403** (3 February): 534-537.
Bergman, Jerry. 2006. "Does gene duplication provide the engine for evolution?" *Journal of Creation* (AKA Answers in Genesis *Creation ex Nihilo Technical Journal*) **20** (1): 99-104.
Bergman, Jerry, & **Jeffrey Tomkins.** 2011. "The chromosome 2 fusion of human evolution—part 1: re-evaluating the evidence." *Journal of Creation* (AKA Answers in Genesis *Creation ex Nihilo Technical Journal*) **25** (2): 106-110.
Bergsland, Kristin J., & **Robert Haselkorn.** 1991. "Evolutionary Relationships among Eubacteria, Cyanobacteria, and Chloroplasts: Evidence from the *rpoC1* Gene of *Anabaena* sp. Strain PCC 7120." *Journal of Bacteriology* **173** (June, No. 1`): 3446-3455.
Berlin, Jeremy. 2012. "Sabertooth Vegetarian." *National Geographic* **222** (July): 17.
Berlinski, David. 1996a. "The Soul of Man Under Physics." *Commentary* **101** (January): 38-46.
———. 1996b. "The Deniable Darwin." *Commentary* **101** (June): 19-29.

———. 1997. "The Limits of Darwinism." *Boston Review* **22** (February/March): online text accessed 5/16/2009 (*bostonreview.net*).
———. 1998. "Was There a Big Bang?" *Commentary* **105** (February): 28-38.
———. 2001. "What Brings a World into Being?" *Commentary* **111** (April): 17-23.
———. 2002. "Has Darwin Met His Match?" *Commentary* **114** (December): 31-41.
———. 2006. "The Vampire's Heart." *Discovery Institute* August posting (online text at *www.discovery.org* accessed 8/20/2006, not available as of 4/20/2016).
———. 2010. *The Deniable Darwin and Other Essays*. Seattle, WA: Discovery Institute Press.
Berman, D. S. 1979. "Edaphosaurus (Reptilia, Pelycosauria) from the Lower Permian of Northeastern United States, with description of a new species." *Annals of Carnegie Museum* **48** (11): 185-202.
Berman, David S., Amy C. Henrici, & **Spencer G. Lucas.** 2013. "*Ophiacodon* (Synapsida, Ophiacodontidae) from the Lower Permian Sangre de Cristo Formation of New Mexico." *New Mexico Museum of History and Science* Bulletin 60: 36-41.
Berman, David S., Robert R. Reisz, John R. Bolt, & **Diane Scott.** 1995. "The cranial anatomy and relationships of the synapsid *Varanosaurus* (Eupelycosauria: Ophiacodontidae) from the Early Permian of Texas and Oklahoma." *Annals of Carnegie Museum* **58** (12 May): 99-133.
Bernstein, Peter. 2003. "The ear region of *Latimeria chalumnae*: functional and evolutionary implications." *Zoology* **106** (3): 233-242.
Berry, A., A. Senescau, J. Lelièvre, F. Benoit-Vical, R. Fabre, B. Marchou, & **J. F. Magnaval.** 2006. "Prevalence of *Plasmodium falciparum* cytochrome b gene mutations in isolates imported from Africa, and implications for atovaquone resistance." *Transactions of The Royal Society of Tropical Medicine and Hygiene* **100** (October): 986-988.
Bethell, Tom. 1985. "Agnostic Evolutionists: The Taxonomic Case Against Darwin." *Harper's* **270** (February): 49-52, 56-58, 60-61.
———. 1996. "A New Beginning." *The American Spectator* **29** (September): 16-17.
———. 1999a. "Rethinking Relativity." *The American Spectator* **32** (April): 20-23.
———. 1999b. "The Evolution Wars." *The American Spectator* **32** (December): 18-20.
———. 2000. "No Time for Science." *The American Spectator* **33** (December): 27.
———. 2001a. "Tom Bethell Replies." *The American Spectator* **34** (February): 10.
———. 2001b. "A Map to Nowhere: The genome isn't a code, and we can't read it." *The American Spectator* **34** (April): 51-56.
———. 2002. "That's Life: The Evolution of Edward O. Wilson." *The American Spectator* **35** (March/April): 54-58.
———. 2005. *The Politically Incorrect Guide to Science*. Washington, D.C.: Regnery Publishing, Inc.
Bettencourt-Dias, Mónica, Friedhelm Hildebrandt, David Pellman, Geoff Woods, & **Susana A. Godinho.** 2011. "Centrosomes and cilia in human disease." *Trends in Genetics* **27** (August): 307-315.
Bever, G. S., Tyler R. Lyson, Daniel J. Field, & **Bhart-Anjan S. Bhullar.** 2015. "Evolutionary origin of the turtle skull." *Nature* **525** (10 September): 239-242.
Bhattacharya, Debashish, John M. Archibald, Andreas P. M. Weber, & **Adrian Reyes-Prieto.** 2007. "How do endosymbionts become organelles? Understanding early events in plastid evolution." *BioEssays* **29** (December): 1239-1246.
Bhattacharya, Debashish, Hwan Su Yoon, & **Jeremiah D. Hackett.** 2004. "Photosynthetic eukaryotes unite: endosymbiosis connects the dots." *BioEssays* **26** (January): 50-60.
Bhullar, B. A., J. Marugán-Lobón, F. Racimo, G. S. Bever, T. B. Rowe, M. A. Norell, & **A. Abzhanov.** 2012. "Birds have paedomorphic dinosaur skulls." *Nature* **487** (12 July): 223-226.
Bhullar, Bhart-Anjan S., Zachary S. Morris, Elizabeth M. Sefton, Atalay Tok, Masayoshi Tokita, Bumjin Namkoong. Jasmin Camacho, David A. Burnham, & **Arhat Abzhanov.** 2015. "A Molecular Mechanism for the Origin of a Key Evolutionary Innovation: The Bird Beak and Palate, Revealed by an Integrative Approach to Major Transitions in Vertebrate History." *Evolution* **69** (July): 1665-1677.
Bi, Shundong, Xingsheng Jin, Shuo Li, & **Tianming Du.** 2015. "A new Cretaceous Metatherian mammal from Henan, China." *PeerJ* (online @ peerj.com) **3** (14 April): 896.
Bi, Shundong, Yuanqing Wang, Jian Guan, Xia Sheng, & **Jin Meng.** 2014. "Three new Jurassic euharamiyidan species reinforce early divergence of mammals." *Nature* **514** (30 October): 579-584.
Bininda-Emonds, Olaf R. P., Marcel Cardillo, Kate R. Jones, Ross D. E. MacPhee, Robin M. D. Beck,

Richard Grenyer, Samantha A. Price, Rutger A. Vos, John L. Gittleman, & **Andy Purvis.** 2007. "The delayed rise of present-day mammals." *Nature* **446** (29 March): 507-512.

Bininda-Emonds, Olaf R. P., Jonathan E. Jeffrey, Michael I. Coates, & **Michael K. Richardson.** 2002. "From Haeckel to event-pairing: the evolution of developmental sequences." *Theory in Biosciences* **121** (November): 297-320.

Bininda-Emonds, Olaf R. P., Jonathan E. Jeffrey, & **Michael K. Richardson.** 2003. "Inverting the hourglass: quantitative evidence against the phylotypic stage in vertebrate development." *Proceedings of the Royal Society of London* **B** (Biological Sciences) **270** (22 February): 341-346.

Bininda-Emonds, Olaf R. P., & **Andy Purvis.** 2012. "Comment on 'Impacts of the Cretaceous Terrestrial Revolution and KPg Extinction on Mammal Diversification'." *Science* **337** (6 July): 34.

Bird, Wendell R. 1989. *The Origin of Species Revisited: The Theories of Evolution and of Abrupt Appearance*. 2 vol., 1991 reprint. Nashville, TN: Regency.

Blackburn, Terrence J., Paul E. Olsen, Samuel A. Bowring, Noah M. McLean, Dennis V. Kent, John Puffer, Greg McCone, E. Troy Asbury, & **Mohammed Et-Toumai.** 2013. "Zircon U-Pb Geochronology Links the End-Triassic Extinction with the Central Atlantic Magmatic Province." *Science* **340** (24 May): 941-945.

Blackwell, Gus, Sally Kern, & **Josh Brecheen.** 2014a. "House Bill 1674." Introduced Oklahoma House legislation (online text at *webserver1.lsb.state.ok.us* accessed 2/21/2014).

———. 2014b. "House Bill 1674." Enacted Oklahoma House legislation (online text at *webserver1.lsb.state.ok.us* accessed 3/24/2014).

Blair, Jaime E., & **S. Blair Hedges.** 2005a. "Molecular Clocks Do Not Support the Cambrian Explosion." *Molecular Biology and Evolution* **22** (March): 387-390.

———. 2005b. "Molecular Phylogeny and Divergence Times of Deuterostome Animals." *Molecular Biology and Evolution* **22** (November): 2275-2284.

Blank, Brian E. 2009. "The Calculus Wars." *Notices of the American Mathematical Society* **56** (May): 602-610.

Bock, Walter J. 1965. "The Role of Adaptive Mechanisms in the Origin of Higher Levels of Organization." *Systematic Zoology* **14** (December): 272-287.

———. 2000. "Explanatory history of the origin of feathers." *Integrative and Comparative Biology* (AKA *American Zoologist*) **40** (September): 478-485.

Boistel, Renaud, Thierry Aubin, Peter Cloetens, François Peyrin, Thierry Scotti, Phillippe Herzog, Justin Gerlach, Nicolas Pollet, & **Jean-François Aubry.** 2013. "How minute sooglossid frogs hear without a middle ear." *Proceedings of the National Academy of Sciences* **110** (17 September): 15360-15364.

Boisvert, Catherine A. 2005. "The pelvic fin and girdle of *Panderichthys* and the origin of tetrapod locomotion." *Nature* **438** (22 December): 1145-1147.

Boisvert, Catherine A., Elga Mark-Kurik, & **Per E. Ahlberg.** 2008. "The pectoral fin of *Panderichthys* and the origin of digits." *Nature* **456** (4 December): 636-638.

Bok, Jinwoong, Weise Chang, & **Doris K. Wu.** 2007. "Patterning and morphogenesis of the vertebrate inner ear." *International Journal of Developmental Biology* **51** (6/7): 521-533.

Bok, Jinwoong, Steven Raft, Kyoung-Ah Kong, Soo Kyung Koo, Ursula C. Dräger, & **Doris K. Wu.** 2011. "Transient retinoic acid signaling confers anterior-posterior polarity to the inner ear." *Proceedings of the National Academy of Sciences* **108** (4 January): 161-166.

Bolt, John R., & **R. Eric Lombard.** 1985. "Evolution of the amphibian tympanic ear and the origin of frogs." *Biological Journal of the Linnean Society* **24** (January): 83-89.

Bomphrey, Richard James, Graham K. Taylor, & **Adrian L. R. Thomas.** 2009. "Smoke visualization of free-flying bumblebees indicates independent leading-edge vortices on each wing pair." *Experiments in Fluids* **46** (May): 811-821.

Bonaparte, Jose. 2008. "On the phylogenetic relationships of *Vincelestes neuquenianus*." *Historical Biology* **20** (April): 81-86.

Bonaparte, José, Agustin G. Martinelli, & **Cesar L. Schultz.** 2005. "New information on *Brasilodon* and *Brasilitherium* (Cynodontia, Probainognathia) from the Late Triassic of southern Brazil." *Revista Brasileira de Paleontologia* **8** (January/April): 25-46.

Bond, Mariano, Marcelo F. Tejedor, Kenneth E. Campbell Jr., Laura Chornogubsky, Nelson Novo, & **Francisco Goin.** 2015. "Eocene primates of South America and the African origins of New World monkeys." *Nature* **520** (23 April): 538-541.

Bontrager, Krista Kay. 2011. "Explore Evolution: A High School Textbook Review." *Reasons to Believe* posting 1 November (online text at *www.reasons.org* accessed 8/30/2016).

Boonstra, L. D. 1972. "Discard the names Theriodontia and Anomodontia: a new classification of the Therapsida." *Annals of the South African Museum* **59**: 315-338.

Borczyk, Bartosz, Jan Kusznierz, Łukasz Paśko, & Edyta Turniak. 2014. "Scaling of the sexual size and shape skull dimorphism in the sand lizard (*Lacerta agilis* L.)." *Vertebrate Zoology* **64** (2): 221-227.

Bork, Robert H. 1996. *Slouching Towards Gomorrah: Modern Liberalism and American Decline.* New York: Regan Books.

Bornens, Michel. 2012. "The Centrosome in Cells and Organisms." *Science* **335** (27 January): 422-426.

Bornens, Michel, & Juliette Azimzadeh. 2007. "Origin and Evolution of the Centrosome," in Jékely (2007, 119-129).

Boswell, Rolfe. 1941. *Nostradamus Speaks.* New York: Crowell.

Botelho, João Francisco, Luis Ossa-Fuentes, Sergio Soto Acuña, Daniel Smith-Paredes, Daniel Nuñez-León, Miguel Salinas-Saavedra, Macarena Ruiz-Flores, & Alexander O. Vargas. 2014. "New Developmental Evidence Clarifies the Evolution of Wrist Bones in the Dinosaur-Bird Transition." *PLoS Biology* (online @ plosbiology.org) **12** (September): e1001957.

Botelho, João Francisco, Daniel Smith-Paredes, Sergio Soto Acuña, Jingmai O'Connor, Verónica Palma, & Alexander O. Vargas. 2016. "Molecular Development of Fibular Reduction in Birds and Its Evolution from Dinosaurs." *Evolution* **70** (March): 543-554.

Botha, J., F. Abdala, & R. Smith. 2007. "The oldest cynodont: new clues on the origin and early diversification of the Cynodontia." *Zoological Journal of the Linnean Society* **149** (March): 477-492.

Botha, Jennifer, & Anusuya Chinsamy. 2000. "Growth patterns deduced from the bone histology of the cynodonts *Diademodon* and *Cynognathus*." *Palaeontology* **20** (December): 705-711.

———. 2004. "Growth and life habits of the Triassic cynodont *Trirachodon*, inferred from bone histology." *Acta Palaeontologica Polonica* **49** (4): 619-627.

———. 2005. "Growth patterns of *Thrinaxodon liorhinus*, a non-mammalian cynodont from the Lower Triassic of South Africa." *Journal of Vertebrate Paleontology* **13** (8 June): 171-184.

Botha, Jennifer, Julia Lee-Thorp, & Anusuya Chinsamy. 2005. "The palaeoecology of the non-mammalian cynodonts *Diademodon* and *Cynognathus* from the Karoo Basin of South Africa, using stable light isotope analysis." *Palaeogeography, Palaeoclimatology, Palaeoecology* **223** (1 August): 303-316.

Botha-Brink, Jennifer, & Sean P. Modesto. 2007. "A mixed-age classed 'pelycosaur' aggregation from South Africa: earliest evidence of parental care in amniotes?" *Proceedings of the Royal Society of London* **B** (Biological Sciences) **274** (22 November): 2829-2834.

Boudinot, Brendan E. 2015. "Contributions to the knowledge of Formicidae (Hymenoptera, Aculeata): a new diagnosis of the family, the first global male-basd key to subfamilies, and a treatment of early branching lineages." *European Journal of Taxonomy* Research Monograph 120: 1-62.

Bousema, Teun, & Chris Drakeley. 2011. "Epidemiology and Infectivity of *Plasmodium falciparum* and *Plasmodium vivax* Gametocytes in Relation to Malaria Control and Elimination." *Clinical Microbiology Reviews* **24** (April): 377-410.

Bousquet, François, Tetsuya Nojima, Benjamin Houet, Isabelle Chauvel, Sylvie Chaudy, Stéphane Dupas, Daisuke Yamamoto, & Jean-François Ferveur. 2012. "Expression of a desaturase gene, *desat1*, in neural and nonneural tissues separately affects perception and emission of sex pheromones in *Drosophila*." *Proceedings of the National Academy of Sciences* **109** (3 January): 240-254.

Boyd, Clint A., Timothy P. Cleland, Nico L. Marrero, & Julia A. Clarke. 2011. "Exploring the effects of phylogenetic uncertainty and consensus trees on stratigraphic consistency scores: a new program and a standardized method." *Cladistics* **27** (February): 52-62.

Boyd, Steven W., & Andrew A. Snelling, eds. 2014. *Grappling with the Chronology of the Genesis Flood: Navigating the Flow of Time in Biblical Narrative.* Green Forest, AR: Master Books.

Bradley, Walter. 2001. "Why I Believe the Bible Is Scientifically Reliable," in Geisler & Hoffman (2001, 175-196).

Brady, Seán G., Ted R. Schultz, Brian L. Fisher, & Philip S. Ward. 2006. "Evaluating alternative hypotheses for the early evolution and diversification of ants." *Proceedings of the National Academy of Sciences* **103** (28 November): 18172-18177.

Branch, Glenn, & Eugenie C. Scott. 2013. "Peking, Piltdown, and Paluxy: creationist legends about

paleoanthropology." *Evolution: Education & Outreach* (online @ evolution-outreach.com) **6**: 27.
Braunstein, Evan M., Dennis C. Monks, Vimla S. Aggarwal, Jelena S. Arnold, & **Bernice E. Morrow.** 2009. "*Tbx1* and *Brn4* regulate retinoic acid metabolic genes during cochlear morphogenesis." *BMC Developmental Biology* (online @ ncbi.nlm.nih.gov) **9** (29 May): 31.
Brawand, David, Walter Wahli, & **Henrik Kaessmann.** 2008. "Loss of Egg Yolk Genes in Mammals and the Origin of Lactation and Placentation." *PLoS Biology* (online @ plosbiology.org) **6** (March): e63.
Bray, Patrick G., Rowena E. Martin, Leann Tilley, Stephen A. Ward, Kiaran Kirk, & **David A. Fidock.** 2005. "Defining the role of PfCRT in *Plasmodium falciparum* chloroquine resistance." *Molecular Microbiology* **56** (April): 323-333.
Brazeau, Martin D., & **Per E. Ahlberg.** 2006. "Tetrapod-like middle ear architecture in a Devonian fish." *Nature* **439** (19 January): 318-321.
Brecheen, Josh. 2010a. "Brecheen discusses evolution and Darwinian Theory." *Durant Daily Democrat* (Durant, Oklahoma) posting 18 December: online text accessed 12/29/2010 (*durantdemocrat.com*).
———. 2010b. "Brecheen says the religion of evolution is plagued with falsehoods." *Durant Daily Democrat* (Durant, Oklahoma) posting 24 December: online text accessed 12/29/2010 (*durantdemocrat.com*).
———. 2011. "Senate Bill 554." Introduced Oklahoma Senate legislation (online text at *webserver1.lsb.state.ok.us* accessed 1/24/2011).
———. 2012. "Senate Bill 1742." Introduced Oklahoma Senate legislation (online text at *webserver1.lsb.state.ok.us* accessed 2/3/2012).
———. 2014. "Senate Bill 1765." Introduced Oklahoma Senate legislation (online text at *webserver1.lsb.state.ok.us* accessed 1/29/2014).
———. 2015. "Senate Bill 665." Introduced Oklahoma Senate legislation (online text at *webserver1.lsb.state.ok.us* accessed 1/30/2015).
———. 2016. "Senate Bill 1322." Introduced Oklahoma Senate legislation (online pdf at *webserver1.lsb.state.ok.us* accessed 1/26/2016).
Bremer, Kåre. 1987. "Tribal Interrelationships of the Asteraceae." *Cladistics* **3** (September): 210-253.
———. 2000. "Early Cretaceous lineages of monocot flowering plants." *Proceedings of the National Academy of Sciences* **97** (25 April): 4707-4711.
Bremer, Kåre, & **Mats H. G. Gustafsson.** 1997. "East Gondwana ancestry of the sunflower alliance of families." *Proceedings of the National Academy of Sciences* **94** (19 August): 9188-9190.
Briggs, Derek E. G. 1995. "Experimental taphonomy." *Palaios* **10** (December): 539-550.
———. 2015. "The Cambrian explosion." *Current Biology* **25** (5 October): R864-R868.
Briggs, Derek E. G., Richard A. Fortey, & **Matthew A. Wills.** 1992. "Morphological Disparity in the Cambrian." *Science* **256** (19 June): 1670-1673.
Bright, Kerry L., & **Mark D. Rausher.** 2008. "Natural Selection on a Leaf-Shape Polymorphism in the Ivyleaf Morning Glory." *Evolution* **62** (August): 1978-1990.
Brinster, Ralph L., Howard Y. Chen, Myrna E. Trumbauer, Mary K. Yagle, & **Richard D. Palmiter.** 1985. "Factors affecting the efficiency of introducing foreign DNA into mice by microinjecting eggs." *Proceedings of the National Academy of Sciences* **82** (1 July): 4438-4442.
Britten, Roy J. 2004. "Coding sequences of functioning human genes derived entirely from mobile element sequences." *Proceedings of the National Academy of Sciences* **101** (30 November): 16825-16830.
Bromham, Lindell. 2003. "What can DNA Tell us About the Cambrian Explosion?" *Integrative and Comparative Biology* (AKA *American Zoologist*) **43** (February): 148-156.
Bromham, Lindell, Andrew Rambaut, Richard Fortey, Alan Cooper, & **David Penny.** 1998. "Testing the Cambrian explosion hypothesis by using a molecular dating technique." *Proceedings of the National Academy of Sciences* **95** (31 October): 12386-12389.
Brothers, Denis J. 1999. "Phylogeny and evolution of wasps, ants and bees (Hymenoptera, Chrysidoidea, Vespoidea, and Apoidea)." The Norwegian Academy of Science and Letters *Zoologica Scripta* **28** (1-2): 233-249.
———. 2011. "A new Late Cretaceous family of Hymenoptera, and phylogeny of the Plumariidae and Chrysidoidea (Aculeata)." *ZooKeys* (online @ ncbi.nlm.nih.gov) **130** (24 September): 515-542.
Brower, Andrew V. Z. 2000. "Evolution Is Not a Necessary Assumption of Cladistics." *Cladistics* **16** (March): 143-154.

Brown, Walter T. Jr. 2008. "Old DNA, Bacteria, Proteins, and Soft Tissue?" *In the Beginning: Compelling Evidence for Creation and the Flood* Chapter 68 (online text at *www.creationscience.com* accessed 6/19/2013).

Bruce, L. L. 2007. "Evolution of the Nervous System in Reptiles," in Bullock & Kaas (2007, 125-156).

Bruner, Emiliano, Giorgio Manzi, & **Juan Luis Arsuaga.** 2003. "Encephalization and allometric trajectories in the genus *Homo*: Evidence from the Neandertal and modern lineages." *Proceedings of the National Academy of Sciences* **100** (23 December): 15335-15340.

Brusatte, Stephen L., Graeme T. Lloyd, Steve C. Wang, & **Mark A. Norell.** 2014. "Gradual Assembly of Avian Body Plan Culminated in Rapid Rates of Evolution across the Dinosaur-Bird Transition." *Current Biology* **24** (20 October): 2386-2392.

Brush, Alan H. 1996. "On the origin of feathers." *Journal of Evolutionary Biology* **9** (January): 131-142.

———. 2000. "Evolving a protofeather and feather diversity." *Integrative and Comparative Biology* (AKA *American Zoologist*) **40** (September): 631-639.

———. 2006. "Follicles and the origin of feathers." *Current Zoology* (AKA *Acta Zoologica Sinica*) **52** (Supplement): 122-124.

Buchholtz, Emily A. 2007. "Modular evolution of the Cetacean vertebral column." *Evolution & Development* **9** (May-June): 278-289.

Buchholtz, Emily A., Amy C. Booth, & **Katherine E. Webbink.** 2007. "Vertebral Anatomy in the Florida Manatee, *Trichechus manatus latirostris*: A Developmental and Evolutionary Analysis." *The Anatomical Record: Advances in Integrative Anatomy and Evolutionary Biology* **290** (June): 624-637.

Buchtová, Marcela, Oldřich Zahradníček, Simona Balková, & **Abigail S. Tucker.** 2013. "Odontogenesis in the Veiled Chameleon (*Chamaeleo calyptratus*)." *Archives of Oral Biology* **58** (February): 118-133.

Budd, Graham E. 2003. "The Cambrian Fossil Record and the Origin of Phyla." *Integrative and Comparative Biology* (AKA *American Zoologist*) **43** (February): 157-165.

———. 2008. "The earliest fossil record of the animals and its significance." *Philosophical Transactions of the Royal Society of London* **B** (Biological Sciences) **363** (27 April): 1425-1434.

Budd, Graham E., & **Soren Jensen.** 2000. "A critical reappraisal of the fossil record of the bilaterian phyla." *Biological Reviews of the Cambridge Philosophical Society* **75** (May): 253-295.

Buff, Bruce. 2016. *The Soul of the Matter.* New York: Simon & Schuster/Howard Books.

Bull, J. J., M. R. Badgett, H. A. Wichman, J. P. Huelsenbeck, D. M. Hillis, A. Gulati, & **I. J. Molineaux.** 1997. "Exceptional Convergent Evolution in a Virus." *Genetics* **147** (December): 1497-1507.

Bullock, Theodore H., & **Jon H. Kaas,** eds. 2007. *Evolution of Nervous Systems: A Comprehensive Reference, Vol. 2: Non-Mammalian Vertebrates.* San Diego, CA: Academic Press.

Bunce, Michael, Trevor H. Worthy, Tom Ford, Will Hoppitt, Eske Willerslev, Alexei Drummond, & **Alan Cooper.** 2003. "Extreme reversed sexual size dimorphism in the extinct New Zealand moa *Dinornis*." *Nature* **425** (11 September): 172-175.

Bunce, M., T. H. Worthy, M. J. Phillips, R. N. Holdaway, E. Willerslev, J. Haile, B. Shapiro, R. P. Scofield, A. Drummond, P. J. J. Kamp, & **A. Cooper.** 2009. "The evolutionary history of the extinct ratite moa and New Zealand Neogene paleogeography." *Proceedings of the National Academy of Sciences* **106** (8 December): 20646-20651.

Burggren, Warren. 2000. "And the Beat Goes On: A Brief Guide to the Hearts of Vertebrates." *Natural History* **109** (April): 62-65.

Burkhardt, Pawel. 2015. "The origin and evolution of synaptic proteins—choanoflagellates lead the way." *The Journal of Experimental Biology* **218** (15 February): 506-514.

Burkhardt, Pawel, Christian M. Stegmann, Benjamin Cooper, Tobias H. Kloepper, Cordelia Imig, Frédérique Varoqueaux, Marcus C. Wahl, & **Dirk Fasshauer.** 2011. "Primordial neurosecretory apparatus identified in the choanoflagellate *Monosiga brevicollis*." *Proceedings of the National Academy of Sciences* **108** (13 September): 15264-15269.

Burns, Kevin J., Shannon J. Hackett, & **Nedra K. Klein.** 2002. "Phylogenetic Relationships and Morphological Diversity in Darwin's Finches and Their Relatives." *Evolution* **56** (June): 1240-1252.

Bush, George, Brent A. Vogt, Jennifer Holmes, Anders M. Dale, Douglas Greve, Michael A. Jenike, & **Bruce R. Rosen.** 2002. "Dorsal anterior cingulate cortex: A role in reward-based decision making." *Proceedings of the National Academy of Sciences* **99** (8 January): 523-528.

Bush, Jeffrey O., & **Rulang Jiang.** 2012. "Palatogenesis: morphogenetic and molecular mechanisms of secondary palate development." *Development* (AKA *Journal of Embryology and Experimental*

Morphology) **139** (15 January): 231-243.

Butler, Ann B. 1996. "Levels of organization and the evolution of isocortex." *Trends in Neuroscience* **19** (March): 89.

Butler, P. M., & **W. A. Clemens.** 2001. "Dental morphology of the Jurassic holotherian mammal *Amphitherium*, with a discussion of the evolution of mammalian post-canine dental formulae." *Palaeontology* **44** (February): 1-20.

Butler, Richard J., Paul M. Barrett, & **David J. Gower.** 2012. "Reassessment of the Evidence for Postcranial Skeletal Pneumaticity in Triassic Archosaurs, and the Early Evolution of the Avian Respiratory System." *PLoS ONE* (online @ plosone.org) **7** (March): e34094.

Butterfield, Natalie C., Vicki Metzis, Edwina McGlinn, Stephen J. Bruce, Brandon J. Wainwright, & **Carol Wicking.** 2009. "Patched 1 is a crucial determinant of asymmetry and digit number in the vertebrate limb." *Development* (AKA *Journal of Embryology and Experimental Morphology*) **136** (15 October): 3515-3524.

Cahoon-Metzger, Sharon M., Guoying Wang, & **Sheryl A. Scott.** 2001. "Contribution of BDNF-Mediated Inhibition in Patterning Avian Skin Innervation." *Developmental Biology* **232** (1 April): 246-254.

Cain, Joseph Allen. 1988. "Creationism and Mammal Origins." *Journal of Geological Education* **36**: 94-105.

Calvo, Jorge, Juan Porfiri, Bernardo González Riga, & **Domenica Dos Santos**, eds. 2011. *Paleontologia y Dinosaurios desde America Latina.* Mendoza, Argentina: Universidad Nacional de Cuyo.

Cameron, Chris B., James R. Garey, & **Billie J. Swalla.** 2000. "Evolution of the chordate body plan: New insights from phylogenetic analyses of deuterostome phyla." *Proceedings of the National Academy of Sciences* **97** (25 April): 4469-4474.

Camp, Ashby. 1998. "Reappraising the 'Crown Jewel'." Creation Research Society *Creation Matters* (September/October): 1-5.

Campàs, O., R. Mallarino, A. Herrel, A. Abzhanov, & **M. P. Brenner.** 2010. "Scaling and shear transformations capture beak shape variation in Darwin's finches." *Proceedings of the National Academy of Sciences* **107** (23 February): 3356-3360.

Campbell, John Angus, & **Stephen C. Meyer**, eds. 2003. *Darwinism, Design, & Public Education.* East Lansing, MI: Michigan State University Press.

Campione, Nicolás E., & **Robert R. Reisz.** 2010. "*Varanops brevirostris* (Euoekycosauria: Varanopidae) from the Lower Permian of Texas, with Discussion of Varanopid Morphology and Interrelationships." *Journal of Vertebrate Paleontology* **30** (May): 724-746.

Canudo, José Ignacio, & **Gloria Cuenca-Bescós.** 1996. "Two new mammalian teeth (Multituberculata and Peramura) from the Lower Cretaceous (Barremian) of Spain." *Cretaceous Research* **17** (April): 215-228.

Capuco, Anthony V., & **R. Michael Akers.** 2009. "The origin and evolution of lactation." *Journal of Biology* (online @ jbiol.com) **8** (24 April): 37.

Carlisle, Christopher, with **W. Thomas Smith Jr.** 2006. *The Complete Idiot's Guide® to Understanding Intelligent Design.* New York: Alpha Books (Penguin Group).

Carlisle, David Brez. 1995. *Dinosaurs, Diamonds, and Things from Outer Space: The Great Extinction.* Stanford, CA: Stanford University Press.

Carney, Ryan M., Jakob Vinther, Matthew D. Shawkey, Liliana D'Alba, & **Jörg Ackermann.** 2012. "New evidence on the colour and nature of the isolated *Archaeopteryx* feather." *Nature Communications* **3** (24 January): 637.

Carr, Archie. 1963. *The Reptiles.* New York: Time-Life Books.

Carr, M., B. S. C. Leadbeater, R. Hassan, M. Nelson, & **S. L. Baldauf.** 2008. "Molecular phylogeny of choanoflagellates, the sister group to Metazoa." *Proceedings of the National Academy of Sciences* **103** (28 October): 16641-16646.

Carrano, Matthew T., Timothy J. Gaudin, Richard W. Blob, & **John R. Wible**, eds. 2006. *Amniote Paleobiology: Perspectives on the Evolution of Mammals, Birds, and Reptiles.* Chicago: University of Chicago Press.

Carrington, Richard. 1963. *The Mammals.* New York: Time-Life Books.

Carriveau, Gary W., & **Mark C. Han.** 1976. "Thermoluminescent Dating and the Monsters of Acambaro." *American Antiquity* **41** (October): 497-500.

Carroll, Robert L. 1988. *Vertebrate Paleontology and Evolution.* New York: W. H. Freeman & Co.

———. 2001. "The Origin and Early Radiation of Terrestrial Vertebrates." *Journal of Paleontology* **75** (November): 1202-1213.

Carroll, Sean B. 1994. "Developmental regulatory mechanisms in the evolution of insect diversity." *Development* (AKA *Journal of Embryology and Experimental Morphology*) **120** (January Supplement): 217-223.

———. 1995. "Homeotic genes and the evolution of arthropods and chordates." *Nature* **376** (10 August): 479-485.

———. 2000. "Endless Forms: The Evolution of Gene Regulation and Morphological Diversity." *Cell* **101** (9 June): 577-580.

———. 2005a. *Endless Forms Most Beautiful: The New Science of Evo Devo and the Making of the Animal Kingdom.* New York: W. W. Norton & Co.

———. 2005b. "Evolution at Two Levels: On Genes and Form." *PLoS Biology* (online @ plosbiology.org) **3** (July): e245.

———. 2007. "God as Genetic Engineer." *Science* **316** (8 June): 1427-1428.

Carroll, Sean B., Jennifer K. Grenier, & **Scott D. Weatherbee**. 2001. *From DNA to Diversity: Molecular Genetics and the Evolution of Animal Design.* Oxford: Blackwell Science.

Carroll, Sean B., Scott D. Weatherbee, & **James A. Langeland.** 1995. "Homeotic genes and the regulation and evolution of insect wing number." *Nature* **375** (4 May): 58-61.

Carroll, Sean Michael, Eric A. Ortlund, & **Joseph W. Thornton.** 2011. "Mechanisms for the Evolution of a Derived Function in the Ancestral Glucocorticoid Receptor." *PLoS Genetics* (online @ plosgenetics.org) **7** (June): e1002117.

Carter, Cameron S., Todd S. Braver, Deanna M. Barch, Matthew M. Botvinick, Douglas Noll, & **Jonathan D. Cohen.** 1998. "Anterior Cingulate Cortex, Error Detection, and the Online Monitoring of Performance." *Science* **280** (1 May): 747-749.

Carter, Cameron S., Angus M. Macdonald, Matthew Botvinick, Laura L. Ross, V. Andrew Stenger, Douglas Noll, & **Jonathan D. Cohen.** 2000. "Parsing executive processes: Strategic vs. evaluative functions of the anterior cingulate cortex." *Proceedings of the National Academy of Sciences* **97** (15 February): 1944-1948.

Carter, Robert W. 2008. "Platypus thumbs its nose (or bill) at evolutionary scientists." *Creation Ministries International* posting 23 May (online text at *creation.com* accessed 5/16/2013).

———. 2011. "Five concise responses to atheistic arguments." *Creation Ministries International* posting 26 March (online text at *creation.com* accessed 5/16/2013).

Carvalho-Santos, Zita, Juliette Azimzadeh, José B. Pereira-Leal, & **Mónica Bettencourt-Dias.** 2011. "Tracing the origins of centrioles, cilia, and flagella." *The Journal of Cell Biology* (AKA *The Journal of Biophysics and Biochemical Cytology*) **194** (25 July): 165-175.

Catchpoole, David, Jonathan Sarfati, & **Carl Wieland**, ed. **Don Batten**. 2007. *The Creation Answers Book.* Powder Springs, GA: Creation Books Publishers.

Catuneanu, O., H. Wopfner, P. G. Eriksson, B. Cairncross, B. S. Rubidge, R. M. H. Smith, & **P. J. Hancox.** 2005. "The Karoo basins of south-central Africa." *Journal of African Earth Sciences* **43** (October): 211-253.

Cavalier-Smith, Thomas. 1997. "The blind biochemist." *Trends in Ecology & Evolution* **12** (April): 162-163.

Cavanaugh, David P., & **Todd Charles Wood.** 2002. "A Baraminological Analysis of the Tribe Heliantheae *sensu latu* (Asteraceae) Using Analysis of Pattern (ANOPA)." *Occasional Papers of the Baraminology Study Group* (online @ bryancore.org) No. 1: 1-11.

Cavanaugh, David P., Todd Charles Wood, & **Kurt P. Wise.** 2003. "Fossil Equidae: A Monobaraminic, Stratomorphic Series." Creation Science Fellowship *Fifth International Conference on Creationism* paper (online text at *bryancore.org* accessed 7/27/2011).

Cebra-Thomas, Judith, Fraser Tan, Seeta Sistla, Eileen Estes, Gunes Bender, Christine Kim, Paul Riccio, & **Scott F. Gilbert.** 2005. "How the Turtle Forms its Shell: A Paracrine Hypothesis of Carapace Formation." *Journal of Experimental Zoology* **304** (15 November): 558-569.

Chaffee, Sarah. 2016. "Students, Scientism, and Straw Men." Discovery Institute *Evolution News & Views* 20 August posting (online text at *evolutionnews.org* accessed 9/17/2016).

Chaffey, Tim. 2016. "*New Scientist* Writer Rejects Scientific Evidence for the Ice Age, Fossil Record, and Flooding on Mars." *Answers in Genesis* article for 12 September (online text at *answersingenesis.org* accessed 9/12/2016).

Chaimanee, Yaowalak, Olivier Chavasseau, K. Christopher Beard, Aung Aung Kyaw, Aung Naing Soe, Chit Sein, Vincent Lazzari, Laurent Marivaux, Bernard Marandat, Myat Swe, Mana

Rugbumrung, Thit Lwin, Xavier Valentin, Zin-Maung-Maung-Thein, & Jean-Jacques Jaeger. 2012. "Late Middle Eocene primate from Myanmar and the initial anthropoid colonization of Africa." *Proceedings of the National Academy of Sciences* **109** (26 June): 10293-10297.

Chaimanee, Yaowalak, Dominique Jolly, Mouloud Benammi, Paul Tafforeau, Danielle Duzer, Issam Moussa, & Jean-Jacques Jaeger. 2003. "A Middle Miocene hominoid from Thailand and orangutan origins." *Nature* **422** (6 March): 61-65.

Chaimanee, Yaowalak, Renaud Lebrun, Chotima Yamee, & Jean-Jacques Jaeger. 2011. "A new Middle Miocene tarsier from Thailand and the reconstruction of its orbital morphology using a geometric-morphometric method." *Proceedings of the Royal Society of London* **B** (Biological Sciences) **278** (7 July): 1956-1963.

Chaimanee, Yaowalak, Varavudh Suteethorn, Pratueng Jintasakul, Chavalit Vidthayanon, Bernard Marandat, & Jean-Jacques Jaeger. 2004. "A new orang-utan relative from the Late Miocene of Thailand." *Nature* **427** (29 January): 439-441.

Chang, Cheng, Ping Wu, Ruth E. Baker, Philip K. Maini, Lorenzo Alibardi, & Cheng-Ming Chuong. 2009. "Reptile scale paradigm: Evo-Devo, pattern formation and regeneration." *International Journal of Developmental Biology* **53** (5/6): 813-826.

Chapman, Geoff. 2002. "Factsheet No. 50: Ears to Hear." *Creation Resources Trust* posting (online pdf at *www.c-r-t.co.uk* accessed 8/17/2016).

Chapman, Glen W. 2004. "Satan is Here and Working Hard." *Chapman Research Group* June posting (online pdf at *chapmanresearch.org* accessed 3/18/2015).

Chapman, Matthew. 2007. *40 Days and 40 Nights: DARWIN, Intelligent Design, GOD, OxyContin AND OTHER Oddities ON TRIAL IN Pennsylvania.* 2008 pb ed. New York: Collins (HarperCollins).

Chapman, Susan Caroline. 2011. "Can you hear me now? Understanding vertebrate middle ear development." *Frontiers in Bioscience* **16** (1 January): 1675-1692.

Charles, Cyril, Vincent Lazzari, Paul Tafforeau, Thomas Schimmang, Mustafa Tekin, Ophir Klein, & Laurent Viriot. 2009. "Modulation of *Fgf3* dosage in mouse and men mirrors evolution of mammalian dentition." *Proceedings of the National Academy of Sciences* **106** (29 December): 22364-22368.

Chatterjee, Sankar. 1983. "An Ictidosaur Fossil from North America." *Science* **220** (10 June): 1551-1553.

Chavali, Pavithra L., Monika Pütz, & Fanni Gergely. 2014. "Small organelle, big responsibility: the role of centrosomes in development and disease." *Philosophical Transactions of the Royal Society of London* **B** (Biological Sciences) **369** (5 September): 20130468.

Cheetham, Erika. 1985. *The Further Prophecies of Nostradamus: 1985 and Beyond.* New York: Perigee Books.

Chen, Chia-Wei Janet, Han-Sung Jung, Ting-Xin Jiang, & Cheng-Ming Chuong. 1997. "Asymmetric Expression of Notch/Delta/Serrate Is Associated with the Anterior-Posterior Axis of Feather Buds." *Developmental Biology* **188** (1 August): 181-187.

Chen, Chih-Feng, John Foley, Pin-Chi Tang, Ang Li, Ting Xin Jiang, Ping Wu, Randall B. Widelitz, & Cheng Ming Chuong. 2015. "Development, regeneration, and evolution of feathers." *Annual Review of Animal Biosciences* **3**: 169-195.

Chen, Ju Jiun, Bart-Jan Janssen, Andrina Williams, & Neelima Sinha. 1997. "A Gene Fusion at a Homeobox Locus: Alterations in Leaf Shape and Implications for Morphological Evolution." *The Plant Cell* **9** (August): 1289-1304.

Chen, Liangbiao, Arthur L. DeVries, & Chi-Hing C. Cheng. 1997a. "Evolution of antifreeze glycoprotein gene from a trypsinogen gene in Antarctic notothenioid fish." *Proceedings of the National Academy of Sciences* **94** (15 April): 3811-3816.

———. 1997b. "Convergent evolution of antifreeze glycoproteins in Antarctic notothenioid fish and Arctic cod." *Proceedings of the National Academy of Sciences* **94** (15 April): 3817-3822.

Chen, Pei-ji, Zhi-ming Dong, & Shuo-nan Zhen. 1998. "An exceptionally well-preserved theropod dinosaur from the Yixian Formation of China." *Nature* **391** (8 January): 147-152.

Chen, Sidi, Benjamin H. Krinsky, & Manyuan Long. 2013. "New genes as drivers of phenotypic evolution." *Nature Reviews Genetics* **14** (September): 645-660.

Chen, Sidi, Yong E. Zhang, & Manyuan Long. 2010. "New Genes in *Drosophila* Quickly Become Essential." *Science* **330** (17 December): 1682-1685.

Cheng, Chi-Hing, & Liangbiao Chen. 1999. "Evolution of an antifreeze glycoprotein." *Nature* **401** (30 September): 443-444.

Cheng, Qin, Gregor Lawrence, Carol Reed, Anthony Stowers, Lisa Ranford-Cartwright, Alison

Creasey, Richard Carter, & **Allan Saul.** 1997. "Measurement of *Plasmodium falciparum* growth rates in vivo: A test of malaria vaccines." *The American Journal of Tropical Medicine and Hygiene* **57** (October): 495-500.

Chiappe, M. Eugenia, Andrei S. Kozlov, & **A. J. Hudspeth.** 2007. "The Structural and Functional Differentiation of Hair Cells in a Lizard's Basilar Papilla Suggests an Operational Principle of Amniote Cochleas." *The Journal of Neuroscience* **27** (31 October): 11978-11985.

Chiari, Ylenia, Vincent Cahais, Nicolas Galtier, & **Frédéric Delsuc.** 2012. "Phylogenomic analyses support the position of turtles as the sister group of birds and crocodiles (Archosauria)." *BMC Biology* (online @ biomedcentral.com) **10** (27 July): 65.

Chichinadze, Konstantin, Ann Lazarashvili, & **Jaba Tkemaladze.** 2013. "RNA in centrosomes: Structure and possible functions." *Protoplasma* **250** (February): 397-405.

Chick, Jack T. 1972. "Big Daddy?" *Chick Publications* pamphlet.

Chimento, Nicolás R., Federico L. Agnolin, & **Fernando E. Novas.** 2012. "The Patagonian fossil mammal *Necrolestes*: a Neogene survivor of Dryolestoidea." *Revista del Museo Argentino de Ciencias Naturales* **14** (2): 261-306.

Chinappi, Mauro, Allegra Via, Paolo Marcatili, & **Anna Tramontano.** 2010. "On the Mechanism of Chloroquine Resistance in *Plasmodium falciparum.*" *PLoS ONE* (online @ plosone.org) **5** (November): e14064.

Chinnery, Brenda J., & **David B. Weishampel.** 1998. "*Montanoceratops cerorhynchus* (Dinosauria: Ceratopsia) and relationships among basal neoceratopsians." *Journal of Vertebrate Paleontology* **18** (September): 569-585.

Chinsamy-Turan, Anusuya, ed. 2011. *Forerunners of Mammals: Radiation Histology Biology.* Bloomington, IN: Indiana University Press.

Chirat, Régis, Derek E. Moulton, & **Alain Goriely.** 2013. "Mechanical basis of morphogenesis and convergence evolution of spiny seashells." *Proceedings of the National Academy of Sciences* **110** (9 April): 6015-6020.

Chittick, Donald E. 1984. *The Controversy: Roots of the Creation-Evolution Conflict.* Portland, OR: Multinomah Press.

Chitwood, Daniel H., Aasgish Ranjan, Ciera C. Martinez, Lauren R. Headland, Thinh Thiem, Ravi Kumar, Michael F. Covington, Tommy Hatcher, Daniel T. Naylor, Sharon Zimmerman, Nora Downs, Nataly Raymundo, Edward S. Buckler, Julin N. Maloof, Mallikarjuna Aradhya, Bernard Prins, Lin Li, Sean Myles, & **Neelima R. Sinha.** 2014. "A Modern Ampelography: A Genetic Basis for Leaf Shape and Venation Patterning in Grape." *Plant Physiology* **164** (January): 259-272.

Chitwood, Daniel H., & **Neelima R. Sinha.** 2016. "Evolutionary and Environmental Forces Sculpting Leaf Development." *Current Biology* **26** (4 April): R297-R306.

Chourrout, D., F. Delsuc, P. Chourrout, R. B. Edvardsen, F. Rentzsch, E. Renfer, M. F. Jensen, B. Zhu, P. de Jong, R. E. Steele, & **U. Technau.** 2006. "Minimal ProtoHox cluster inferred from bilaterian and cnidarian Hox components." *Nature* **442** (10 August): 684-687.

Christensen-Dalsgaard, Jakob, Christian Brandt, Maria Wilson, Magnus Wahlberg, & **Peter T. Madsen.** 2011. "Hearing in the African lungfish (*Protopterus annectens*): pre-adaptation to pressure hearing in tetrapods?" Royal Society *Biology Letters* **7** (February): 139-141.

Christensen-Dalsgaard, Jakob, & **Catherine E. Carr.** 2008. "Evolution of a sensory novelty: Tympanic ears and the associated neural processing." *Brain Research Bulletin* **75** (18 March): 365-370.

Christiansen, Per, & **Niels Bonde.** 2002. "Limb proportions and avian terrestrial locomotion." *Journal of Ornithology* **143** (July): 356-371.

Christiansen, Per, & **Stephen Wroe.** 2007. "Bite forces and evolutionary adaptations to feeding ecology in carnivores." *Ecology* **88** (February): 347-358.

Chu, Ka Hou, Ji Qi, Zu-Guo Yu, & **Vo Anh.** 2004. "Origin and Phylogeny of Chloroplasts Revealed by a Simple Correlation Analysis of Complete Genomes." *Molecular Biology and Evolution* **21** (January): 200-206.

Chudinov, Peter K. 1965. "New Facts about the Fauna of the Upper Permian of the U.S.S.R." *The Journal of Geology* **73** (January): 117-130.

Chuong, Cheng-Ming, Rajas Chodankar, Randall B. Widelitz, & **Ting-Xin Jiang.** 2000. "*Evo-Devo* of feathers and scales: building complex epithelial appendages." *Current Opinion in Genetics & Development* **10** (August): 449-456.

Chuong, C.-M., N. Patel, J. Lin, H.-S. Jung, & **R. B. Widelitz.** 2000. "Sonic hedgehog signaling pathway in vertebrate epithelial appendage morphogenesis: perspectives in development and evolution." *Cellular and Molecular Life Sciences* **57** (October): 1672-1681.

Chuong, Cheng-Ming, Sheree A. Ting, Randall B. Widelitz, & Yun-Shain Lee. 1992. "Mechanism of skin morphogenesis. II. Retinoic acid modulates axis orientation and phenotypes of skin appendages." *Development* (AKA *Journal of Embryology and Experimental Morphology*) **115** (July): 839-852.

Chuong Cheng-Ming, Randall B. Widelitz, Sheree Ting-Berreth, & Ting-Xin Jiang. 1996. "Early Events During Avian Skin Appendage Regeneration: Dependence on Epithelial-Mesenchymal Interaction and Order of Molecular Reappearance." *The Journal of Investigative Dermatology* **107** (October): 639–646.

Chuong, Cheng-Ming, Ping Wu, Fu-Cheng Zhang, Xing Xu, Minke Yu, Randall B. Widelitz, Tin-Xin Jiang, & Lianhai Hou. 2003. "Adaptation to the sky: Defining the feather with integumental fossils from mesozoic China and experimental evidence from molecular laboratories." *Journal of Experimental Zoology* **296B** (15 August): 42-56.

Churakov, Gennady, Manoj K. Sadasivuni, Kate R. Rosenbloom, Dorothée Huchon, Jürgen Brosius, & Jürgen Schmitz. 2010. "Rodent Evolution: Back to the Root." *Molecular Biology and Evolution* **27** (June): 1315-1326.

Cifelli, Richard L. 1999. "Tribosphenic mammal from the North American Early Cretaceous." *Nature* **401** (23 September): 363-366.

———. 2000. "Counting premolars in early eutherian mammals." *Acta Palaeontologica Polonica* **45** (2): 195-198.

———. 2001. "Early Mammalian Radiations." *Journal of Paleontology* **75** (November): 1214-1226.

Cifelli, Richard L., & Brian M. Davis. 2003. "Marsupial Origins." *Science* **302** (12 December): 1899-1900.

———. 2013. "Jurassic fossils and mammalian antiquity." *Nature* **500** (8 August): 160-161.

Cisneros, Juan Carlos, Fernando Abdala, Bruce S. Rubidge, Paula Camboim Dentzien-Dias, & Ana de Oliveira Bueno. 2011. "Dental Occlusion in a 260-Million-Year-Old Therapsid with Saber Canines from the Permian of Brazil." *Science* **331** (25 March): 1603-1605.

Civáň, Peter, Peter G. Foster, Martin T. Embley, Ana Séneca, & Cymon J. Cox. 2014. "Analyses of Charophyte Chloroplast Genomes Help Characterize the Ancestral Chloroplast Genome of Land Plants." *Genome Biology and Evolution* **6** (April): 897-911.

Clack, Jennifer A. 1989. "Discovery of the earliest-known tetrapod stapes." *Nature* **342** (23 November): 425-427.

———. 1998. "A new Early Carboniferous tetrapod with a *melange* of crown-group characters." *Nature* **394** (2 July): 66-69.

———. 2002. "An early tetrapod from 'Romer's Gap'." *Nature* **418** (4 July): 72-76.

———. 2004. "From Fins to Fingers." *Science* **304** (2 April): 57-58.

Clack, Jennifer A., Per E. Ahlberg, Henning Blom, & Sarah M. Finney. 2012. "A new genus of Devonian tetrapod from North-East Greenland, with new information onf the lower jaw of *Ichthyostega*." *Palaeontology* **55** (January): 75-86.

Clack, J. A., P. E. Ahlberg, S. M. Finney, P. Dominguez Alonso, J. Robinson, & R. A. Ketcham. 2003. "A uniquely specialized ear in a very early tetrapod." *Nature* **425** (4 September): 65-69.

Clarey, Tim. 2015. *Dinosaurs: Marvels of God's Design—The The Science of the Biblical Account*. Green Forest, AR: Master Books.

Clark, James M., Hans-Dieter Sues, & David Berman. 2000. "A new specimen of *Hesperosuchus agilis* from the upper Triassic of New Mexico and the interrelationships of basal crocodylomorph archosaurs." *Journal of Vertebrate Paleontology* **20** (December): 683-704.

Clarke, Julia A., Zhonghe Zhou, & Fucheng Zhang. 2006. "Insight into the evolution of avian flight from a new clade of Early Cretaceous ornithurines from China and the morphology of *Yixianornis grabaui*." *Journal of Anatomy* **208** (March): 287-308.

Claude, Julien, Peter C. H. Pritchard, Haiyan Tong, Emmanuel Paradis, & Jean-Christophe Auffray. 2004. "Ecological Correlates and Evolutionary Divergence in the Skull of Turtles: A Geometric Morphometric Assessment." *Systematic Biology* **53** (December): 933-948.

Clegg, Michael T., Michael P. Cummings, & Mary L. Durbin. 1997. "The evolution of plant nuclear genes." *Proceedings of the National Academy of Sciences* **94** (22 July): 7791-7798.

Clement, Alice M., & John A. Long. 2010. "Air-breathing adaptation in a marine Devonian lungfish." Royal Society *Biology Letters* **6** (August): 509-512.

Clément, Gaël, Per E. Ahlberg, Alain Blieck, Henning Blom, Jennifer A. Clack, Edouard Poty, Jacques Thorez, & Philippe Janvier. 2004. "Devonian tetrapod from western Europe." *Nature* **427** (29 January): 412-413.

Climate Progress. 2011. "Oklahoma Lawmaker Sally Kern proposes bill that forces teachers to question evolution, climate science." *Climate Progress* posting 29 January (online text at *climateprogress.org* accessed 2/16/2011).

Close, Roger A., Brian M. Davis, Stig Walsh, Andrzei S. Wolniewicz, Matt Friedman, & **Roger B. J. Benson.** 2016. "A lower jaw of *Palaeoxonodon* from the Middle Jurassic of the Isle of Skye, Scotland, sheds new light on the diversity of British stem therians." *Palaeontology* **59** (January): 155-169.

Coates, M. I., & **J. A. Clack.** 1991. "Fish-like gills and breathing in the earliest known tetrapod." *Nature* **352** (18 July): 234-236.

Coates, Michael I., & **Martin J. Cohn.** 1998. "Fins, limbs, and tails: outgrowths and axial patterning in vertebrate evolution." *BioEssays* **20** (May): 371-381.

Coates, Michael I., Neil H. Shubin, & **Edward B. Daeschler.** 2004. "Response to Comment on 'The Early Evolution of the Tetrapod Humerus." *Science* **305** (17 September): 1715.

Codd, Jonathan R., Phillip L. Manning, Mark A. Norell, & **Steven F. Perry.** 2008. "Avian-like breathing mechanics in maniraptoran dinosaurs." *Proceedings of the Royal Society of London* **B** (Biological Sciences) **275** (22 January): 157-161.

Coen, Enrico. 1999. *The Art of the Gene: How Organisms Make Themselves.* Oxford: Oxford University Press.

Cohn, Norman. 1993. *Cosmos, Chaos and the World to Come: The Ancient Roots of Apocalyptic Faith.* New Haven, CT: Yale University Press.

———. 1996. *Noah's Flood: The Genesis Story in Western Thought.* New Haven, CT: Yale University Press.

Colbert, Edwin H., & **Michael Morales.** 1991. *Evolution of the Vertebrates: A History of the Backboned Animals Through Time.* 4th ed. New York: Wiley-Liss.

Combrinck, Jill M., Tebogo E. Mabotha, Kanyile K. Ncokazi, Melvin A. Ambele, Dale Taylor, Peter J. Smith, Heinrich C. Hoppe, & **Timothy J. Egan.** 2013. "Insights into the Role of Heme in the Mechanism of Action of Antimalarials." *ACS Chemical Biology* **8** (January): 133-137.

Comfort, Ray. 2013. "Evolution vs. God: Companion Guide." Living Waters *Evolution vs. God* posting (online pdf at *www.evolutionvsgod.com* accessed 3/28/2014).

Conrad, J., & **C. A. Sidor.** 2001. "Re-evaluation of *Tetraceratops insignis* (Synapsida. Sphenacodontia)." *Journal of Vertebrate Paleontology* **21** (March): 42A.

Conservapedia. 2016. "Platypus." Entry last modified 24 June (online text at *conservapedia.com* accessed 10/7/2016).

Conway Morris, Simon. 1998. *The Crucible of Creation: The Burgess Shale and the Rise of Complexity.* Oxford: Oxford University Press.

Cook, L. M., B. S. Grant, L. J. Saccheri, & **J. Mallet.** 2012. "Selective bird predation on the peppered moth: the last experiment of Michael Majerus." Royal Society *Biology Letters* **8** (August): 609-612.

Cooper, Henry S. F. Jr. 2001. "The Perfect Fossil." *Natural History* **110** (November): 12-13.

Cooper, Kimberly L., Karen E. Sears, Aysu Uygur, Jennifer Maier, Karl-Stephan Baczkowski, Margaret Brosnahan, Doug Antczak, Julian A. Skidmore, & **Clifford J. Tabin.** 2014. "Patterning and post-patterning modes of evolutionary digit loss in mammals." *Nature* **511** (3 July): 41-45.

Cooper, Lisa Noelle, Annalisa Berta, Susan D. Dawson, & **Joy S. Reidenberg.** 2007. "Evolution of Hyperphalangy and Digit Reduction in the Cetacean Manus." *The Anatomical Record: Advances in Integrative Anatomy and Evolutionary Biology* **290** (June): 654-672.

Coppedge, David F. 2005. "Did Old Metamorphic Rocks Form in Just 10 Years?" *Creation-Evolution Headlines* posting 30 June (online text at *creationsafaris.com* accessed 3/27/2013).

———. 2006. "Mature at Birth: Universe Discredits Evolution." Institute for Creation Research *Acts & Facts* (October): 1-4.

———. 2007a. "Missing Link, or Just Jawboning About Ear Evolution?" *Creation-Evolution Headlines* posting 19 March (online text at *creationsafaris.com* accessed 1/21/2014).

———. 2007b. "Geological Truisms Questioned." *Creation-Evolution Headlines* posting 27 March (online text at *creationsafaris.com* accessed 1/21/2014).

———. 2007c. "Did Indians See Jurassic Beast?" *Creation-Evolution Headlines* posting 30 March (online text at *creationsafaris.com* accessed 1/21/2014).

———. 2008a. "Cosmology's Errors Bars." Institute for Creation Research *Acts & Facts* (July): 15.

———. 2008b. "Modeling Just-So Stories for Earth History." *Creation-Evolution Headlines* posting 6 September (online text at *creationsafaris.com* accessed 3/22/2012).

———. 2008c. "Comet Conundrums Resist Bluffing." *Creation-Evolution Headlines* posting 9 September (online text at *creationsafaris.com* accessed 3/22/2012).

———. 2010. "Mt. St. Helens Recalls Overturned Paradigms." *Creation-Evolution Headlines* posting 18 May (online text at *creationsafaris.com* accessed 2/20/2011).

———. 2011. "Baloney Detector." *Creation-Evolution Headlines* undated posting (online text at *creationsafaris.com* accessed 6/18/2011).

Corley, Matt. 2009. "Sally Kern Returns To Blame America's 'Economic Woes' On 'Same-Sex Marriage' and 'Abortion'." *Think Progress* 30 June posting (online text at *thinkprogress.org* accessed 2/16/2011).

Cottrell, Gilles, Lise Musset, Véronique Hubert, Jacques Le Bras, Jérôme Clain, & **the Atovaquone-Proguanil Treatment Failure Study Group.** 2014. "Emergence of Resistance to Atovaquone-Proguanil in Malaria Parasites: Insights from Computational Modeling and Clinical Case Reports." *Antimicrobial Agents and Chemotherapy* **58** (August): 4504-4514.

Coulter, Ann. 2006. *Godless: The Church of Liberalism*. New York: Crown.

Couso, Juan Pablo. 2009. "Segmentation, metamerism and the Cambrian explosion." *International Journal of Developmental Biology* **53** (8-10): 1305-1316.

Cox, C. Barry. 1998. "The jaw function and adaptive radiation of the dicynodont mammal-like reptiles of the Karoo basin of South Africa." *Zoological Journal of the Linnean Society* **122** (January): 349-384.

Coyne, Jerry A. 2009. *Why Evolution Is True*. New York: Viking Press.

———. 2016. "His white suit unsullied by research, Tom Wolfe tries to take down Charles Darwin and Noam Chomsky." *The Washington Post* posting 31 August: online text accessed 9/2/2016 (*www.washingtonpost.com*).

Cracraft, Joel, & **Michael J. Donoghue**, eds. 2004. *Assembling the Tree of Life*. Oxford: Oxford University Press.

Crawford, Nicholas G., Brant C. Faircloth, John E. McCormack, Robb T. Brumfield, Kevin Winker, & **Travis C. Glenn.** 2012. "More than 1000 ultraconserved elements provide evidence that turtles are the sister group of archosaurs." Royal Society *Biology Letters* **8** (October): 783-786.

Creager, Charles Jr. 2003. "So called transition from synapsid reptiles to mammals." *Genesis Mission* undated posting (online text at *genesismission.4t.com* accessed 9/23/2003, not available as of 10/21/2014).

———. 2008. "Transition from synapsid reptiles to mammals (Talk.Origins)." Entry last modified 24 June (online text at *creationwiki.org* accessed 9/23/2016.

Creation Tips. 2008. "The great platypus and echidna mystery." *Creation Tips* 9 November updated posting (online text at *www.creationtips.com* accessed 4/29/2013).

CreationWiki. 2011. "John Woodmorappe." Entry last modified 12 July (online text at *creationwiki.org* accessed 5/27/2012).

———. 2014. "Synapsid." Entry last modified 26 July (online text at *creationwiki.org* accessed 7/25/2016).

———. 2015. "Creation vs. evolution." Entry last modified 29 October (online text at *creationwiki.org* accessed 7/2/2016).

Cremo, Michael A., & **Richard L. Thompson.** 1993. *Forbidden Archaeology: The Hidden History of the Human Race*. San Diego, CA: Bhaktiveda Institute.

Creuzet, Sophie, Gérard Couly, Christine Vincent, & **Nicole M. Le Douarin.** 2002. "Negative effect of Hox gene expression on the development of the neural crest-derived facial skeleton." *Development* (AKA *Journal of Embryology and Experimental Morphology*) **129** (15 September): 4301-4313.

Criswell, Daniel C. 2009. "A Review of Mitoribosome Structure and Function Does not Support the Serial Endosymbiotic Theory." Answers in Genesis *Answers Research Journal* **2**: 107-115.

Crompton, A. W. 1963. "The Evolution of the Mammalian Jaw." *Evolution* **17** (December): 431-439.

———. 1972. "The evolution of the jaw articulation of cynodonts," in Joysey & Kemp (1972, 231-251).

Crompton, A. W., & **F. A. Jenkins Jr.** 1979. "Origin of Mammals," in Lillegraven *et al.* (1979, 59-73).

Crompton, Alfred W., & **Zhe-Xi Luo.** 1993. "Relationships of the Liassic mammals *Sinoconodon*, *Morganucodon oehleri*, and *Dinnetherium*," in Szalay *et al.* (1993, 30-44).

Crompton, A. W., & **Pamela Parker.** 1978. "Evolution of the Mammalian Masticatory Apparatus." *American Scientist* **66** (March-April): 192-201.

Crompton, A. W., & **Ai-Lin Sun.** 1985. "Cranial structure and relationships of the Liassic mammal

Sinoconodon." *Zoological Journal of the Linnean Society* **85** (October): 99-119.
Crompton, A. W., C. Richard Taylor, & James A. Jagger. 1978. "Evolution of homeothermy in mammals." *Nature* **272** (23 March): 333-336.
Crozier, Ross H. 2006. "Charting uncertainty about ant origins." *Proceedings of the National Academy of Sciences* **103** (28 November): 18029-18030.
Cruz, Nicky. 1973. *Satan on the Loose*. Old Tappan, NJ: Fleming H. Revell Company.
Cuenca-Bescós, Gloria, José I. Canudo, José M. Gasca, Miguel Moreno-Azanza, & **Richard L. Cifelli.** 2014. "Spalacotheriid 'symmetrodonts' from the Early Cretaceous of Spain." *Journal of Vertebrate Paleontology* **34** (November): 1427-1436.
Cuozzo, Frank P., Emilienna Rasoazanabary, Laurie R. Godfrey, Michelle L. Sauther, Ibrahim Antho Youssouf, & **Marni M. LaFleur.** 2013. "Biological variation in a large sample of mouse lemurs from Amboasary, Madagascar: Implications for interpreting variation in primate biology and paleobiology." *Journal of Human Evolution* **64** (January): 1-20.
Currie, Philip J., & **Kevin Padian**, eds. 1997. *Encyclopedia of Dinosaurs*. San Diego, CA: Academic Press.
Czerkas, Sylvia J., & **Stephen A. Czerkas.** 1991. *Dinosaurs: A Global View*. New York: Mallard Press.
Czerkas, Stephen A., & **Alan Feduccia.** 2014. "Jurassic archosaur is a non-dinosaurian bird." *Journal of Ornithology* **155** (October): 841-851.
Czerkas, Stephen A., & **Chongxi Yuan.** 2002. "An arboreal maniraptoran from northeast China." Dinosaur Museum (Blanding, Utah) *The Dinosaur Museum Journal* **1**: 63-95.

Daeschler, Edward B., & **Neil Shubin.** 1998. "Fish with fingers?" *Nature* **391** (8 January): 133.
Daeschler, Edward B., Neil H. Shubin, & **Farish A. Jenkins Jr.** 2006. "A Devonian tetrapod-like fish and the evolution of the tetrapod body plan." *Nature* **440** (6 April): 757-763.
Dalton, Rex. 2000a. "Feathers fly over Chinese fossil bird's legality and authenticity." *Nature* **403** (17 February): 689-690.
———. 2000b. "Fake bird fossil highlights the problem of illegal trading." *Nature* **404** (13 April): 696.
———. 2000c. "Chasing the dragons." *Nature* **406** (31 August): 930-932.
Danielsson, Olle, & **Hans Jörnvall.** 1992. "'Enzymogenesis': Classical liver alcohol dehydrogenase origin from the glutathione-dependent formaldehyde dehydrogenase line." *Proceedings of the National Academy of Sciences* **89** (1 October): 9247-9251.
Danowitz, Melinda, Rebecca Domalski, & **Nikos Solounias.** 2015. "The cervical anatomy of *Samotherium*, an intermediate-necked giraffid." *Royal Society Open Science* **2** (November): 150521.
Danowitz, Melinda, Aleksandr Vasilyev, Victoria Kortlandt, & **Nikos Solounias.** 2015. "Fossil evidence and stages of elongation of the *Giraffa camelopardalis* neck." *Royal Society Open Science* **2** (October): 150393.
Dashzeveg, D., & **Z. Kielan-Jaworowska.** 1984. "The lower jaw of an aegialodontid mammal from the Early Cretaceous of Mongolia." *Zoological Journal of the Linnean Society* **82** (September): 217-227.
Dashzeveg, Demberelyn. 1975. "New primitive therian from the early Cretaceous of Mongolia." *Nature* **256** (31 July): 402-403.
———. 1994. "Two Previously Unknown Eupantotheres (Mammalia, Eupanthrotheria)." *American Museum Novitates* **3433** (February): 1-76.
Datavo, Aléssio, & **Richard P. Vari.** 2013. "The Jaw Adductor Muscle Complex in Teleostean Fishes: Evolution, Homologies and Revised Nomenclature (Osteichthyes: Actinopterygii)." *PLoS ONE* (online @ plosone.org) **8** (April): e60846.
Davidson, Eric H., K. J. Peterson, & **R. Andrew Cameron.** 1995. "Origin of Bilaterian Body Plans: Evolution of Developmental Regulatory Mechanisms." *Science* **270** (24 November): 1319-1325.
Davies, Peter L., Jason Baarsnes, Michael J. Kuiper, & **Virginia K. Walker.** 2002. "Structure and function of antifreeze proteins." *Philosophical Transactions of the Royal Society of London* **B** (Biological Sciences) **337** (29 July): 927-935.
Davis, Brian M. 2012. "Micro-computed tomography reveals a diversity of Peramuran mammals from the Purbeck Group (Berriasian) of England." *Palaeontology* **55** (July): 789-817.
Davis, Brian M., & **Richard L. Cifelli.** 2011. "Reappraisal of the tribosphenidan mammals from the Trinity Group (Aptain-Albian) of Texas and Oklahoma." *Acta Palaeontologica Polonica* **56** (3): 441-462.

Davis, John Jefferson. 1999a. "Response to Paul Nelson and John Mark Reynolds," in Moreland & Reynolds (1999, 80-84).

———. 1999b. "Response to Robert C. Newman," in Moreland & Reynolds (1999, 137-141).

Davis, Marcus C., Randall D. Dahn, & **Neil H. Shubin.** 2007. "An autopodial-like pattern of Hox expression in the fins of a basal actinopterygian fish." *Nature* **447** (24 May): 473-476.

Davis, Percival, & **Dean H. Kenyon.** 1993. *Of Pandas and People: The Central Question of Biological Origin.* 2nd ed. Dallas, TX: Haughton Publishing Co.

Davit-Béal, Tiphaine, Abigail S. Tucker, & **Jean-Yves Sire.** 2009. "Loss of teeth and enamel in tetrapods: fossil record, genetic data and morphological adaptations." *Journal of Anatomy* **214** (April): 477-501.

Dawkins, Richard. 1986. *The Blind Watchmaker: Why the Evidence of Evolution Reveals a Universe Without Design.* 1996 pb ed. New York: W. W. Norton & Co.

———. 2009. *The Greatest Show on Earth: The Evidence for Evolution.* 2010 pb ed. New York: The Free Press.

Dayel, Mark J., Rosanna A. Alegado, Stephen R. Fairclough, Tara C. Levin, Scott A. Nichols, Kent McDonald, & **Nicole King.** 2011. "Cell differentiation and morphogenesis in the colony-forming choanoflagellate *Salpingoeca rosetta.*" *Developmental Biology* **357** (1 September): 73-82.

De Bakker, Merijn A. G., Donald A. Fowler, Kelly den Oude, Esther M. Dondorp, M. Carmen Garrido Navas, Jaroslaw O. Horbanczuk, Jean-Yves Sire, Danuta Szczerbińska, & **Michael K. Richardson.** 2013. "Digit loss in archosaur evolution and the interplay between selection and constraints." *Nature* **500** (22 August): 445-448.

Debuysschere, Maxime. 2016. "The Kuehneotheriidae (Mammaliaformes) from Saint-Nicolas-de-Port (Upper Triassic, France): a Systematic Review." *Journal of Mammalian Evolution* (in press, 15 September advanced access at *springer.com*): s10917-06-9335-z.

Debuysschere, M., E. Gheerbrant, & **R. Allain.** 2015. "Earliest known European mammals: a review of the Morganucodonta from Saint-Nicholas-de-Port (Upper Triassic, France)." *Journal of Systematic Palaeontology* **13** (10): 825-855.

De Groote, Isabelle, Linus Girdland Flink, Rizwaan Abbas, Silvia M. Bello, Lucia Burgia, Laura Tabitha Buck, Christopher Dean, Alison Freyne, Thomas Higham, Chris G. Jones, Robert Kruszynski, Adrian Lister, Simon A. Parfitt, Matthew M. Skinner, Karolyn Shindler, & **Chris B. Stringer.** 2016. "New genetic and morpological evidence suggests a single hoaxer created 'Piltdown man'." *Royal Society Open Science* **3** (August): 160328.

DeHaan, Robert F., & **John L. Wiester.** 1999. "The Cambrian Explosion: The Fossil Record & Intelligent Design." *Touchstone* (July/August): 65-69.

De la Monte, Suzanne M., Kasra Ghanbari, William H. Frey, Iraj Beheshti, Paul Averback, Stephen L. Hauser, Hossein A. Ghanbari, & **Jack R. Wands.** 1997. "Characterization of the AD7C-NTP cDNA expression in Alzheimer's disease and measurement of a 41-kD protein in cerbrospinal fluid." *The Journal of Clinical Investigation* **100** (15 December): 3093-3104.

Delgado, Sidney, Didier Casane, Laure Bonnaud, Michel Laurin, Jean-Yves Sire, & **Marc Girondot.** 2001. "Molecular Evidence for Precambrian Origin of Amelogenin, the Major Protein of Vertebrate Enamel." *Molecular Biology and Evolution* **18** (December): 2146-2153.

DeLong, Brad. 1997. "Conservative Fear of Albert Einstein." *Bradford Delong* 16 June posting (online text at *j-bradford-delong.net* accessed 4/16/2016).

Delwiche, Charles F., Robert E. Andersen, Debashish Bhattarcharya, Brent D. Mischler, & **Richard M. McCourt.** 2004. "Algal Evolution and the Early Radiation of Green Plants," in Cracraft & Donoghue (2004, 121-137).

DeMar, R., & **H. R. Barghusen.** 1972. "Mechanics and the Evolution of the Synapsid Jaw." *Evolution* **26** (April): 622-637.

Dembski, William A. 1998a. "The Bible by Numbers." *First Things* (August/September): 61-64.

———. 1998b. *The Design Inference: Eliminating Chance Through Small Probabilities.* Cambridge: Cambridge University Press.

———. 1999. *Intelligent Design: The Bridge Between Science & Theology.* Downers Grove, IL: InterVarsity Press.

———. 2002. *No Free Lunch: Why Specified Complexity Cannot Be Purchased without Intelligence.* Lanham, MD: Rowman & Littlefield Publishers, Inc.

———. 2005. "Intelligent Design's Contribution to the Debate Over Evolution: A Reply to Henry Morris." *Freedom, Technology, Education* posting (online text at *billdembski.com* accessed 1/1/2016).

———. 2006a, ed. *Darwin's Nemesis*. Downers Grove, IL: IVP Books (InterVarsity Press).
———. 2006b. "Ann Coulter weighs in on Darwinism." *Uncommon Descent* posting 26 April (online text at *uncommondescent.com* accessed 7/7/2016).
———. 2006c. "For sheer smarminess, this one is hard to beat...." *Uncommon Descent* posting 10 July (online text at *uncommondescent.com* accessed 7/13/2006).
———. 2006d. "Read my lips: 'I take all responsibility for any errors in those chapters'," *Uncommon Descent* posting 24 July (online text at *uncommondescent.com* accessed 4/5/2011).
Dembski, William A., & **James M. Kushiner**, eds. 2001. *Signs of Intelligence: Understanding Intelligent Design*. Grand Rapids, MI: Brazos Press (Baker Book House).
Dembski, William A., & **Jonathan Wells.** 2008. *The Design of Life: Discovering Signs of Intelligence in Biological Systems*. Dallas, TX: Foundation for Thought and Ethics.
———. 2016a. "The Challenge of Adaptational Packages." Discovery Institute *Evolution News & Views* 22 June posting (online text at *evolutionnews.org* accessed 6/23/2016).
———. 2016b. "Why Fossils Cannot Demonstrate Darwinian Evolution." Discovery Institute *Evolution News & Views* 6 July posting (online text at *evolutionnews.org* accessed 7/7/2016).
Deng, Cheng, C.-H. Christina Cheng, Hua Ye, Ximiao He, & **Liangbiao Chen.** 2010. "Evolution of an antifreeze protein by neofunctionalization under escape from adaptive conflict." *Proceedings of the National Academy of Sciences* **107** (14 December): 21593-21598.
Denhoed, Andrea. 2015. "Tom Wolfe Looks Over His Notes." *The New Yorker* (28 February): online text accessed 1/2/2016 (*newyorker.com*).
Denton, Michael. 1985. *Evolution: A Theory in Crisis*. 1986 US ed. Bethesda, MD: Adler & Adler.
———. 1998. *Nature's Destiny: How the Laws of Biology Reveal Purpose in the Universe*. New York: The Free Press.
———. 2013. "The Types: A Persistent Structuralist Challenge to Darwinian Pan-Selectionism." *BIO-Complexity* (online @ bio-complexity.org) **2013** (3): 1-18.
———. 2016a. *Evolution: Still a Theory in Crisis*. Seattle, WA: Discovery Institute Press.
———. 2016b. "Non-Adaptive Order: An Existential Challenge in Darwinian Evolution." Discovery Institute *Evolution News & Views* 15 February posting (online text at *evolutionnews.org* accessed 4/23/2016).
———. 2016c. "The Types: Why Shared Characteristics Are Bad News for Darwinism." Discovery Institute *Evolution News & Views* 29 February posting (online text at *evolutionnews.org* accessed 4/23/2016).
———. 2016d. "The Evo-Devo Revolution: LEGOs or Transformers?" Discovery Institute *Evolution News & Views* 7 March posting (online text at *evolutionnews.org* accessed 4/24/2016).
———. 2016e. "Human Language: Noam Chomsky, Universal Grammar, and Natural Selection." Discovery Institute *Evolution News & Views* 11 April posting (online text at *evolutionnews.org* accessed 4/24/2016).
———. 2016f. "On the Diversification of Fur, Feathers, and Scales, the Mystery Remains." Discovery Institute *Evolution News & Views* 6 July posting (online text at *evolutionnews.org* accessed 7/7/2016).
Depew, Michael J., Thomas Lufkin, & **John L. R. Rubenstein.** 2002. "Specification of Jaw Subdivisions by *Dlx* Genes." *Science* **298** (11 October): 381-385.
Depew, Michael J., Carol A. Simpson, Maria Morasso, & **John L. R. Rubenstein.** 2005. "Reassessing the *Dlx* code: the genetic regulation of branchial arch skeletal pattern and development." *Journal of Anatomy* **207** (November): 501-561.
De Roos, Albert D. G. 2007. "A critique on the endosymbiotic theory for the origin of mitochondria." *Telic Thoughts* posting 17 October by Mike Gene (online text at *telicthoughts.com* accessed 11/24/2009).
DeWitt, David A. 2013. "What about the Similarity Between Human and Chimp DNA?" *Answers in Genesis* 14 January posting of *The New Answers Book 3* Chapter 10 (online text at *www.answersingenesis.org* accessed 2/17/2015).
DeYoung. Don, & **Jason Lisle.** 2013. "Does Astronomy Confirm a Young Universe?" *Answers in Genesis* 9 June 2014 posting of *The New Answers Book 3* Chapter 19 (online text at *www.answersingenesis.org* accessed 2/17/2015).
Dhouailly, Danielle. 2009. "A new scenario for the evolutionary origin of hair, feather, and avian scales." *Journal of Anatomy* **214** (April): 587-606.
Dhouailly, Danielle, Margaret H. Hardy, & **Philippe Sengel.** 1980. "Formation of feathers on chick foot scales: a stage-dependent morphogenetic response to retinoic acid." *Development* (AKA

Journal of Embryology and Experimental Morphology) **58** (August): 63-78.
Diaz-Horta, Oscar, Clemer Abad, Levent Sennaroglu, Joseph Foster II, Alexander DeSmidt, Guney Bademci, Suna Tokgoz-Yilmaz, Duygu Duman, F. Basak Cengiz, M'hamed Grati, Suat Fitoz, Xue Z. Liu, Amjad Farooq, Faiqa Imtiaz, Benjamin B. Currall, Cynthia Casson Morton, Michiru Nishita, Yasuhiro Minami, Zhongmin Lu, Katherina Walz, & **Mustafa Tekin.** 2016. "ROR1 is essential for proper innervation of auditory hair cells and hearing in humans and mice." *Proceedings of the National Academy of Sciences* **113** (24 May): 5993-5998.
Diego Daza, Juan, Rui Diogo, Peter Johnston, & **Virginia Abdala.** 2011. "Jaw Adductor Muscles Across Lepidosaurs: A Reappraisal." *The Anatomical Record: Advances in Integrative Anatomy and Evolutionary Biology* **294** (October): 1765-1782.
Dines, James P., Erik Otárola-Castillo, Peter Ralph, Jesse Alas, Timothy Daley, Andrew D. Smith, & **Matthew D. Dean.** 2014. "Sexual Selection Targets Cetacean Pelvic Bones." *Evolution* **68** (November): 3296-3306.
Diogo, R., V. Abdala, M. A. Aziz, N. Logergan, & **B. A. Wood.** 2008. "From fish to modern humans—comparative anatomy, homologies and evolution of the head and neck musculature." *Journal of Anatomy* **213** (October): 391-424.
Diogo, Rui, Yaniv Hinits, & **Simon M. Hughes.** 2008. "Development of mandibular, hyoid and hypobranchial muscles in the zebrafish: homologies and evolution of these muscles within body fishes and tetrapods." *BMC Developmental Biology* (online @ ncbi.nlm.nih.gov) **8** (28 February): 24.
Di Peso, Charles C. 1953. "The Clay Figurines of Acambaro, Guanajuato, Mexico." *American Antiquity* **18** (October): 388-389.
Di-Poï, Nicolas, & **Michel C. Milinkovitch.** 2016. "The anatomical placode in reptile scale morphogenesis indicates shared ancestry among skin appendages in amniotes." *Science Advances* **2** (24 June): e1600708 (online at *sciencemag.org*).
Discovery Institute. 2010. "Background on David Coppedge and the Lawsuit Against NASA's Jet Propulsion Laboratory." *Center for Science & Culture* staff posting 19 April (online text at *discovery.org* accessed 9/4/2010).
———. 2011. "Background on David Coppedge and the Lawsuit Against NASA's Jet Propulsion Laboratory Revised." *Center for Science & Culture* staff posting 17 February (online text at *discovery.org* accessed 2/17/2011).
———. 2012. "Background on David Coppedge and the Lawsuit Against NASA's Jet Propulsion Laboratory." *Center for Science & Culture* staff updated posting 2 November (online text at *discovery.org* accessed 6/6/2016).
———. 2014. "A Scientific Dissent From Darwinism." *Dissent from Darwin* posting of April revision (online pdf at *www.dissentfromdarwin.org* accessed 5/26/2014).
Dlussky, G. M. 1999. "The First Find of the Formicoidea (Hymenoptera) in the Lower Cretaceous of the Northern Hemisphere." *Paleontological Journal* **33** (3): 274-277.
———. 2012. "New Fossil Ants of the Subfamily Myrmeciinae (Hymenoptera, Formicidae) from Germany." *Paleontological Journal* **46** (3): 288-292.
Dlussky, G. M., & **K. S. Perfilieva.** 2003. "Paleogene Ants of the Genus *Archimyrmex* Cockerell, 1923 (Hymenoptera, Formicidae, Myrmeciinae)." *Paleontological Journal* **37** (1): 39-47.
Dlussky, Gennady M., & **Alexander G. Radchenko.** 2006. "A new ant genus from the late Eocene European amber." *Acta Palaeontologica Polonica* **51** (3): 561-567.
———. 2009. "Two new primitive ant genera from the late Eocene European ambers." *Acta Palaeontologica Polonica* **54** (3): 435-441.
Dlussky, Gennady, Alexander Radchenko, & **Dmitry Dubovikoff.** 2014. "A new enigmatic ant genus from late Eocene Danish Amber and its evolutionary and zoogeographic significance." *Acta Palaeontologica Polonica* **59** (4): 931-939.
Dobzhansky, Theodosius. 1973. "Nothing in Biology Makes Sense Except in the Light of Evolution." *The American Biology Teacher* **35** (March): 125-129.
Dodson, Peter. 1996. *The Horned Dinosaurs: A Natural History.* Princeton, NJ: Princeton University Press.
———. 1997a. "Ceratopsia," in Currie & Padian (1997, 106).
———. 1997b. "Neoeratopsia," in Currie & Padian (1997, 473-478).
———. 2000. "Origin of birds: The final solution?" *Integrative and Comparative Biology* (AKA *American Zoologist*) **40** (September): 504-512.
Dodson, Peter, & **Philip J. Currie.** 1990. "Neoceratopsia," in Weishampel *et al.* (1990, 593-618).

Doll, Bradley B., Kent E. Hutchison, & **Michael J. Frank.** 2011. "Dopaminergic Genes Predict Individual Differences in Susceptibility to Confirmation Bias." *The Journal of Neuroscience* **31** (20 April): 6188-6198.
Domning, D. P. 2001. "The earliest known fully quadrupedal sirenian." *Nature* **413** (11 October): 625-627.
Donoghue, Michael J. 2005. "Key innovations, convergence, and success: macroevolutionary lessons from plant phylogeny." *Paleobiology* **31** (June): 77-93.
Dorrell, Richard G., & **Christopher J. Howe.** 2012a. "What makes a chloroplast? Reconstructing the establishment of photosynthetic symbioses." *Journal of Cell Science* **125** (15 April): 1865-1875.
———. 2012b. "Functional remodeling of RNA processing in replacement chloroplasts by pathways retained from their predecessors." *Proceedings of the National Academy of Sciences* **109** (13 November): 18879-18884.
———. 2015. "Integration of plastids with their hosts: Lessons learned from dinoflagellates." *Proceedings of the National Academy of Sciences* **112** (18 August): 10247-10254.
Dorrell, Richard G., & **Alison G. Smith.** 2011. "Do Red and Green Make Brown? Perspectives on Plastid Acquisitions with Chromalveolates." *Eukaryotic Cell* **10** (July): 856-868.
Douglas, Susan E. 1999. "Evolutionary History of Plastids." *The Biological Bulletin* **196** (June): 397-399.
Do-While Jones. 2015. "Dino-chickens." *Science Against Evolution* posting June (online text at *www.scienceagainstevolution.info* accessed 8/10/2015).
Downard, James. 2003a. "A Tale of Two Citations." *Talk Reason* posting 30 June (online text at *talkreason.org* accessed 1/10/2010).
———. 2003b. "Three Macroevolutionary Episodes." *Troubles in Paradise* posting (online pdf at *www.tortucan.wordpress.com/* accessed 1/7/2015).
———. 2004a. "'Rules of the Game." *Troubles in Paradise* chapter posting (online pdf at *www.tortucan.wordpress.com/* accessed 1/7/2015).
———. 2004b. "'Dem Bones." *Troubles in Paradise* chapter posting (online pdf at *www.tortucan.wordpress.com/* accessed 1/7/2015).
———. 2004c. "Dinomania." *Troubles in Paradise* chapter posting (online pdf at *www.tortucan.wordpress.com/* accessed 1/7/2015).
———. 2004d. "Creationism Lite." *Troubles in Paradise* chapter posting (online pdf at *www.tortucan.wordpress.com/* accessed 1/7/2015).
———. 2004e. "Planet of the Apes." *Troubles in Paradise* chapter posting (online pdf at *www.tortucan.wordpress.com/* accessed 1/7/2015).
———. 2004f. "'Cuz the Bible Tells Me So." *Troubles in Paradise* chapter posting (online pdf at *www.tortucan.wordpress.com/* accessed 1/7/2015).
———. 2006a. "Secondary Addiction: Ann Coulter on Evolution." *Talk Reason* posting 29 June (online text at *talkreason.org* accessed 1/10/2010).
———. 2006b. "Secondary Addiction Part II: Ann Coulter on Evolution." *Talk Reason* posting 9 July (online text at *talkreason.org* accessed 1/10/2010).
———. 2006c. "The DEMBSKI ALERT." *Talk Reason* posting 23 July (online text at *talkreason.org* accessed 1/10/2010).
———. 2006d. "Secondary Addiction Part III: Ann Coulter on Evolution." *Talk Reason* posting 25 July (online text at *talkreason.org* accessed 1/10/2010).
———. 2014. "Taking 'Teach the Controversy' Critical Analysis Seriously." *Troubles in Paradise* posting (online pdf at *www.tortucan.wordpress.com* accessed 1/7/2015).
———. 2015a. "1.5: Dissing Darwin." *Troubles in Paradise* 4 December updated posting (online pdf at *www.tortucan.wordpress.com* accessed 3/16/2016).
———. 2015b. "1.7: Teach the *Kulturkampf.*" *Troubles in Paradise* 4 December updated posting (online pdf at *www.tortucan.wordpress.com* accessed 3/16/2016).
———. 2015c. "1.6: A Brief History of Creationism." *Troubles in Paradise* 26 December updated posting (online pdf at *www.tortucan.wordpress.com* accessed 5/2/2016).
———. 2016. "1.3: Quote Mining and the Case of Punctuated Equilibrium." *Troubles in Paradise* 11 January updated posting (online pdf at *www.tortucan.wordpress.com* accessed 3/16/2016).
Downs, Jason P., Edward B. Daeschler, Farish A. Jenkins, & **Neil H. Shubin.** 2008. "The cranial endoskeleton of *Tiktaalik roseae.*" *Nature* **455** (16 October): 925-929.
Doyle, Arthur Conan. 1922. *The Coming of the Fairies*. 1972 reprint. New York: Samuel Weiser.
———. 1926. *The History of Spiritualism*. 2 vols. London: Cassell & Co.

———. 1930. *The Edge of the Unknown.* New York: G. P. Putnam's Sons.
Doyle, Shaun. 2011. "'Transitional form' in mammal ear evolution—more cacophony." *Journal of Creation* (AKA Answers in Genesis *Creation ex Nihilo Technical Journal*) **22** (3): 42-45.
Drincovich, María F., Paula Casati, & Carlos S. Andreo. 2001. "NADP-malic enzyme from plants: a ubiquitous enzyme involved in different metabolic pathways." Federation of European Biochemical Societies *FEBS Letters* **490** (9 February): 1-6.
Drincovich, Maria F., Paula Casati, Carlos S. Andreo, Saul J. Chessin, Vincent R. Franceschi, Gerald E. Edwards, & Maurice S. B. Ku. 1998. "Evolution of C_4 Photosynthesis in *Flaveria* Species: Isoforms of NADP-Malic Enzyme." *Plant Physiology* **117** (July): 733-744.
Drosnin, Michael. 1998. *The Bible Code.* New York: Touchstone.
———. 2003. *The Bible Code II: The Countdown.* New York: Penguin Books.
Druzinsky, Robert E., Alison H. Doherty, & Frits L. De Vree. 2011. "Mammalian Masticatory Muscles: Homology, Nomenclature, and Diversification." *Integrative and Comparative Biology* (AKA *American Zoologist*) **51** (August): 224-234.
Duboule, Denis. 1998. "*Hox* is in the hair: a break in colinearity?" *Genes & Development* **12** (1 January): 1-4.
Duman, John G. 2015. "Animal ice-binding (antifreeze) proteins and glycolipds: an overview with emphasis on physiological function." *The Journal of Experimental Biology* **218** (1 June): 1846-1855.
Dunbar, Robin, & Louise Barrett. 2000. *Cousins: Our primate relatives.* London: Dorling Kindersley Publishing, Inc.
Durrett, Richard, & Deena Schmidt. 2007. "Waiting for regulatory sequences to appear." *The Annals of Applied Probability* **17** (1): 1-32.
———. 2008. "Waiting for Two Mutations: With Applications to Regulatory Sequence Evolution and the Limits of Darwinian Evolution." *Genetics* **180** (November): 1501-1509.
———. 2009. "Reply to Michael Behe." *Genetics* **181** (February): 821-822.
Dyall, Sabrina D., Mark T. Brown, & Patricia J. Johnson. 2004. "Ancient Invasion: From Endosymbionts to Organelles." *Science* **304** (9 April): 253-257.
Dyall, Sabrina D., & Patricia J. Johnson. 2000. "Origin of hydrogenosomes and mitochondria: evolution and organelle biogenesis." *Current Opinion in Microbiology* **3** (1 August): 404-411.
Dyke, Gareth, Roeland de Kat, Colin Palmer, Jacques van der Kindere, Darren Naish, & Bharathram Ganapathisubramani. 2013. "Aerodynamic performance of the feathered dinosaur *Microraptor* and the evolution of feathered flight." *Nature Communications* (online @ nature.com) **4** (18 September): 2489.
Dyke, Gareth J., & Mark A. Norell. 2005. "*Caudipteryx* as a non-avialan theropod rather than a flightless bird." *Acta Palaeontologica Polonica* **50** (1): 101-116.

Eames, B. Frank, & Richard A. Schneider. 2008. "The genesis of cartilage size and shape during development and evolution." *Development* (AKA *Journal of Embryology and Experimental Morphology*) **135** (1 December): 3947-3958.
Eaton, Jeffrey G., & Richard L. Cifelli. 1988. "Preliminary report on Late Cretaceous mammals of the Kaiparowits Plateau, southern Utah." *Geology* **26** (December): 45-55.
Eaton, Jeffrey G., Richard L. Cifelli, J. Howard Hutchison, James I. Kirkland, & J. Michael Parrish. 1999. "Cretaceous Vertebrate Faunas from the Kaiparowits Plateau, South-Central Utah," in Gillette (1999, 345-353).
Eberle, Jaelyn J. 1999. "Bridging the Transition between Didelphodonts and Taeniodonts." *Journal of Paleontology* **73** (September): 936-944.
Ecker, Ronald L. 1990. *Dictionary of Science & Creationism.* Buffalo, NY: Prometheus Books.
Edgecombe, Gregory D. 2009. "Palaeontological and Molecular Evidence Linking Arthropods, Onychophorans, and other Ecdysozoa." *Evolution: Education & Outreach* **2** (June): 178-190.
Edwards, D., K. L. Davies, & L. Axe. 1992. "A vascular conducting strand in the early land plant *Cooksonia*." *Nature* **357** (25 June): 683-685.
Egan, Louisa C., Laurie R. Santos, & Paul Bloom. 2007. "The Origins of Cognitive Dissonance: Evidence From Children and Monkeys." *Psychological Science* **21** (November): 978-983.
Ekdale, Eric G., J. David Archibald, & Alexander O. Averianov. 2004. "Petrosal bones of placental mammals from the Late Cretaceous of Uzbekistan." *Acta Palaeontologica Polonica* **49** (1): 161-176.
Eldredge, Niles. 1982. *The Monkey Business: A Scientist Looks at Creationism.* New York:

Washington Square Press.
———. 1991. *Fossils: The Evolution and Extinction of Species*. New York: Harry N. Abrahms.
———. 1995. *Reinventing Darwin: The Great Debate at the High Table of Evolutionary Theory*. New York: John Wiley & Sons.
———. 2000. *The Triumph of Evolution: And the Failure of Creationism*. New York: W. H. Freeman & Co.
———. 2005. *Darwin: Discovering the Tree of Life*. New York: W.W. Norton & Co.
———. 2008. "The Early 'Evolution' of 'Punctuated Equilibria'." *Evolution: Education & Outreach* **1** (April): 107-113.
———. 2015. *Eternal Ephemera: Adaptation and the Origin of Species from the Nineteenth Century Through Punctuated Equilibria and Beyond*. New York: Columbia University Press.
Ellis, Richard. 2001. *Aquagenesis: The Origin and Evolution of Life in the Sea*. New York: Viking Press.
Elsberry, Wesley R. 2007. "The 'Explore Evolution' Companion." *AntiEvolution* 14 July posting (online text at *www.antievolution.org* accessed 7/12/2016).
Ely, Bert, Tracey W. Ely, William B. Crymes Jr., & **Scott A. Minnich.** 2000. "A Family of Six Flagellin Genes Contributes to the *Caulobacter crescentus* Flagellar Filament." *Journal of Bacteriology* **182** (September, No. 17): 5001-5004.
Engel, Michael S., & **David A. Grimaldi.** 2005. "Primitive New Ants in Cretaceous Amber from Myanmar, New Jersey, and Canada (Hymenoptera: Formicidae)." *American Museum Novitates* **3485** (25 July): 1-23.
Enns, Pete. 2010. "Genesis 1 and a Babylonian Creation Story." *The BioLogos Forum* 18 May posting (online text at *biologos.org* accessed 5/19/2011).
Enyart, Bob. 2013. "Triceratops Soft Tissue with Mark Armitage." *Bob Enyart Live* 4 October summary of broadcast topic (online text at *kgov.com* accessed 7/27/2014).
———. 2014. "RSR's List of Scholars Doubting Darwin." *Bob Enyart Live* 29 August summary of broadcast topic (online text at *kgov.com* accessed 6/8/2016).
Erickson, Gregory M., Oliver W. M. Rauhut, Zhonghe Zhou, Alan H. Turner, Brian D. Inouye, Dongyu Hu, & **Mark A. Norell.** 2009. "Was Dinosaurian Physiology Inherited by Birds? Reconciling Slow Growth in *Archaeopteryx*." *PLoS ONE* (online @ plosone.org) **4** (October): e7390.
Erwin, Douglas H. 1998. "The end of the beginning: recoveries from mass extinctions." *Trends in Ecology & Evolution* **13** (September): 344-349.
———. 2001. "Lessons from the past: Biotic recoveries from mass extinctions." *Proceedings of the National Academy of Sciences* **98** (8 May): 5399-5403.
EurekAlert! 2016. "How the mouse outlived the giant." *EurekAlert!* 11 February posting (online text at *www.eurekalert.org* accessed 12/29/2015).
Evans, Jay D., & **Diana E. Wheeler.** 1999. "Differential gene expression between developing queens and workers in the honey bee, *Apismellifera*." *Proceedings of the National Academy of Sciences* **96** (11 May): 5575-5580.
Eve, Raymond A., & **Francis B. Harrold.** 1991. *The Creationist Movement in Modern America*. Boston: Twayne Publishers.
Everett, Vera. 2009. "Soft Tissue Fossilization." *Answers in Genesis* 4 November "Answers in Depth" article (online text at *answersingenesis.org* accessed 2/3/2015).
Evolution Dismantled. 2014. "Transitional Fossil Sequences and the Evolution of Life." *Evolution Dismantled* undated posting (online text at *evolutiondismantled.com* accessed 8/3/2016).
Evolution News & Views. 2012a. "In the Current Issue of Nature, an Argument for Intelligent Design in All but Name." Discovery Institute *Evolution News & Views* for 7 March (online text at *evolutionnews.org* accessed 12/21/2012).
———. 2012b. "Facts of the Coppedge Lawsuit Contradict the Spin from Jet Propulsion Lab and National Center for Science Education." Discovery Institute *Evolution News & Views* for 12 March (online text at *evolutionnews.org* accessed 3/17/2012).
———. 2012c. "On David Coppedge Trial, Darwin Bloggers Groping Their Way in the Dark." Discovery Institute *Evolution News & Views* for 15 March (online text at *evolutionnews.org* accessed 3/17/2012).
———. 2012d. "Why the Coppedge Trial Matters." Discovery Institute *Evolution News & Views* for 6 March (online text at *evolutionnews.org* accessed 3/8/2012).
———. 2012e. "'Reporting' on the David Coppedge Trial, *Time* Magazine Joins the JPL Legal Team." Discovery Institute *Evolution News & Views* for 19 March (online text at *evolutionnews.org*

accessed 3/5/2013).
———. 2012f. "Too Bad David Coppedge Isn't a Peppered Moth." Discovery Institute *Evolution News & Views* for 4 May (online text at *evolutionnews.org* accessed 5/8/2012).
———. 2013. "Our Ears Are Amazing! They Must Have Evolved." Discovery Institute *Evolution News & Views* for 29 March (online text at *evolutionnews.org* accessed 4/4/2013).
———. 2015. "*Privileged Species* with Geneticist Michael Denton Gets Its Online Premiere; See It Now!" Discovery Institute *Evolution News & Views* for 24 March (online text at *evolutionnews.org* accessed 9/29/2016).
———. 2016a. "Denton's *Evolution: Still a Theory in Crisis*, a Fresh and Powerful Challenge to Darwinism." Discovery Institute *Evolution News & Views* 26 January posting (online text at *evolutionnews.org* accessed 4/23/2016).
———. 2016b. "Conversations with Michael Denton: The Galápagos Finches as a 'Two-Edged Sword'." Discovery Institute *Evolution News & Views* 27 January posting (online text at *evolutionnews.org* accessed 4/23/2016).
———. 2016c. "Denton's Challenge: Are Leaf Shapes Adaptive?" Discovery Institute *Evolution News & Views* 13 April posting (online text at *evolutionnews.org* accessed 4/24/2016).
———. 2016d. "Galápagos Finches: A Failed Evolutionary Icon that Won't Go Away." Discovery Institute *Evolution News & Views* 8 May posting (online text at *evolutionnews.org* accessed 5/9/2016).
———. 2016e. "Humans, Chickens, and the Vitellogenin Pseudogene—Suming Up." Discovery Institute *Evolution News & Views* 25 May posting (online text at *evolutionnews.org* accessed 6/19/2016).
———. 2016f. "In *The Design of Life*, Dembski and Wells Offer a Powerful Survey of the Case for Intelligent Design." Discovery Institute *Evolution News & Views* 27 June posting (online text at *evolutionnews.org* accessed 6/29/2016).

Fairclough, Stephen R., Zehua Chen, Eric Kramer, Qiandong Zeng, Sarah Young, Hugh M. Robertson, Emina Begovic, Daniel J. Richter, Carsten Russ, M. Jody Westbrook, Gerard Manning, B. Franz Lang, Brian Haas, Chad Nusbaum, & **Nicole King.** 2013. "Premetazoan genome evolution and the regulation of cell differentiation in the choanoflagellate *Salpingoeca rosetta*." *Genome Biology* (online @ genomebiology.com) **14** (18 February): R15.
Fan, Yuxin, Elena Linardopoulou, Cynthia Friedman, Eleanor Williams, & **Barbara J. Trask.** 2002. "Genomic Structure and Evolution of the Ancestral Chromosome Fusion Site in 2q13-2q14.1 and Paralogous Regions on Other Human Chromosomes." *Genome Research* **12** (November): 1663-1672.
Farmer, Colleen G. 1999. "Evolution of the Vertebrate Cardio-Pulmonary System." *Annual Review of Physiology* **61**: 573-592.
———. 2015. "Similarity of Crocodilian and Avian Lungs Indicates Unidirectional Flow Is Ancestral for Archosaurs." *Integrative and Comparative Biology* (AKA *American Zoologist*) **55** (December): 962-971.
Farmer, C. G., T. J. Uriona, D. B. Olsen, M. Steenblik, & **K. Sanders.** 2008. "The Right-to-Left Shunt of Crocodilians Serves Digestion." *Physiological and Biochemical Zoology* **81** (March/April): 125-137.
Farrell, John. 2000. "Did Einstein cheat?" *Salon.com* 6 July (online text at *salon.com* accessed 4/16/2016).
Fastovsky, David E., & **David B. Weishampel.** 1996. *The Evolution and Extinction of the Dinosaurs*. Cambridge: Cambridge University Press.
Faulkner, Danny R. 2013. "What About Cosmology?" *Answers in Genesis* 20 June 2014 posting of *The New Answers Book 3* Chapter 21 (online text at *www.answersingenesis.org* accessed 2/17/2015).
Fedak, Tim J., Hans-Dieter Sues, & **Paul E. Olsen.** 2015. "First record of the tritylodontid cynodont *Oligokyphus* and cynodont postcranial bones from the McCoy Brook Formation of Nova Scotia, Canada." *Canadian Journal of Earth Sciences* **52** (April): 244-249.
Fedorov, Alexei, Scott Roy, Xiaohong Cao, & **Walter Gilbert.** 2003. "Phylogenetically Older Introns Strongly Correlate With Module Boundaries in Ancient Proteins." *Genome Research* **13** (June 6a): 1155-1157.
Feduccia, Alan. 1985. "On why the dinosaurs lacked feathers," in Hecht *et al.* (1985, 75-79).
———. 1993. "Aerodynamic Model for the Early Evolution of Feathers Provided by *Propithecus*

(Primates, Lemuridae)." *Journal of Theoretical Biology* **160** (21 January): 159-164.
———. 1999a. "1,2,3 = 2,3,4: Accommodating the cladogram." *Proceedings of the National Academy of Sciences* **96** (27 April): 4740-4742.
———. 1999b. *The Origin and Evolution of Birds.* 2nd ed. New Haven, CT: Yale University Press.
———. 2001. "Fossils and avian evolution." *Nature* **414** (29 November): 507-508.
———. 2002. "Birds are Dinosaurs: Simple Answer to a Complex Problem." *The Auk* **119** (October): 1187-1201.
———. 2003. "Bird origins: problem solved, but the debate continues..." *Trends in Ecology & Evolution* **18** (January): 9-10.
———. 2012. *Riddle of the Feathered Dragons: Hidden Birds of China.* New Haven, CT: Yale University Press.
———. 2013. "Bird Origins Anew." *The Auk* **130** (January): 1-12.
Feduccia, A., T. Lingham-Soliar, & J. R. Hinchcliffe. 2005. "Do feathered dinosaurs exist? Testing the hypothesis on neontological and paleontological evidence." *Journal of Morphology* **266** (November): 125-166.
Fekete, Donna M., & Drew M. Noden. 2013. "A Transition in the Middle Ear." *Science* **339** (22 March): 1396-1397.
Fellowes, M. D. E., G. J. Holloway, & J. Rolff, eds. 2005. *Insect Evolutionary Ecology.* Wallingford, UK: Royal Entomological Society.
Fernàndez-Busquets, Xavier, André Körnig, Iwona Bucior, Max M. Burger, & Dario Anselmetti. 2009. "Self-Recognition and Ca^{2+}-Dependent Carbohydrate-Carbohydrate Cell Adhesion Provide Clues to the Cambrian Explosion." *Molecular Biology and Evolution* **26** (November): 2551-2561.
Ferrell, Vance. 2001. *The Evolution Cruncher: Scientific facts which annihilate evolutionary theory.* Altamont, TN: Evolution Facts, Inc.
Ferris, Kathleen G., Tullia Rushton, Anna B. Greenlee, Katherine Toll, Benjamin K. Blackman, & John H. Willis. 2015. "Leaf shape evolution has a similar genetic architecture in three edaphic specialists within the *Mimulus guttatus* species complex." *Annals of Botany* **116** (August): 213-223.
Field, Daniel J., Jacques A. Gauthier, Benjamin L. Krieg, Davide Pisani, Tyler R. Lyson, & Kevin J. Peterson. 2014. "Toward consilience in reptile phylogeny: miRNAs support an archosaur, not lepidosaur, affinity for turtles." *Evolution & Development* **16** (July-August): 189-196.
Finger, Thomas E., Naoyuki Yamamoto, Harvey J. Karten, & Patrick R. Hof. 2013. "Evolution of the Forebrain—Revisiting the Pallium." *The Journal of Comparative Neurology* **521** (November): 3601-3603.
Finn, Roderick Nigel, Jelena Kolarevic, Heidi Kongshaug, & Frank Nilsen. 2009. "Evolution and differential expression of a vertebrate vitellogenin gene cluster." *BMC Evolutionary Biology* (online @ biomedcentral.com) **9** (5 January): 2.
Finn, Roderick Nigel, & Borge A. Kristoffersen. 2007. "Vertebrate Vitellogenin Gene Duplication in Relation to the '3R Hypothesis': Correlation to the Pelagic Egg and the Oceanic Radiation of Teleosts." *PLoS ONE* (online @ plosone.org) **2** (January): e169.
Fisher, Nicholas, Roslaini Abd Majid, Thomas Antoine, Mohammed Al-Helal, Ashley J. Warmen, David J. Johnson, Alexandre S. Lawrenson, Hilary Ranson, Paul M. O'Neill, Stephen A. Ward, & Giancarlo A. Biagini. 2012. "Cytochrome *b* Mutation Y268S Conferring Atovaquone Resistance Phenotype in Malaria Parasite Results in Reduced Parasite bc_1, Catalytic Turnover and Protein Expression." *The Journal of Biological Chemistry* **287** (23 March): 9731-9741.
Fisher, Paul E., Dale E. Russell, Michael K. Stoskopf, Reese E. Barrick, Michael Hammer, & Andrew A. Kuzmitz. 2000. "Cardiovascular Evidence for an Intermediate or Higher Metabolic Rate in an Ornithischian Dinosaur." *Science* **288** (21 April): 503-505.
Flannery, Michael A. 2011. *Alfred Russel Wallace: A Rediscovered Life.* Seattle, WA: Discovery Institute Press.
Flannery, Timothy F., Michael Archer, Thomas H. Rich, & Robert Jones. 1995. "A new family of monotremes from the Cretaceous of Australia." *Nature* **377** (5 October): 418-420.
Flynn, John J., J. Michael Parrish, Berthe Rakotosamimanana, William F. Simpson, & André R. Wyss. 1999. "A Middle Jurassic mammal from Madagascar." *Nature* **401** (2 September): 57-60.
Foitzik, Kerstin, Ralf Paus, Tom Doetschman, & G. Paolo Dotto. 1999. "The TGF-β2 Isoform Is Both a Required and Sufficient Inducer of Murine Hair Follicle Morphogenesis." *Developmental Biology* **212** (15 August): 278-289.
Foley, Jim. 1997. "Creationism and the Platypus." *The TalkOrigins Archive* 18 February updated

posting (online text at *talkorigins.org* accessed 10/7/2016).
Forey, Peter L. 1988. "Golden jubilee for the coelacanth *Latimeria chalumnae*." *Nature* **336** (29 December): 727-732.
Fortunato, Sofia A. V., Marcin Adamski, Olivia Mendivil Ramos, Sven Leininger, Jing Liu, David E. K. Ferrier, & **Maja Adamska.** 2014. "Calcisponges have a ParaHox gene and dynamic expression of dispersed NK homeobox genes." *Nature* **514** (30 October): 620-623.
Foth, Christian. 2014. "Comment on the absence of ossified sternal elements in basal paravian dinosaurs." *Proceedings of the National Academy of Sciences* **111** (16 December): E5334.
Fox, Richard C. 1977. "Notes on the dentition and relationships of the Late Cretaceous insectivore *Gypsonictops* Simpson." *Canadian Journal of Earth Sciences* **14** (August): 1823-1831.
———. 2015. "A revision of the Late Cretaceous-Paleocene eutherian mammal *Cimolestes* Marsh, 1889." *Canadian Journal of Earth Sciences* **52** (December): 1137-1149.
Fox, Richard C., & **Bruce G. Naylor.** 2006. "Stagodontid marsupials from the Late Cretaceous of Canada and their systematic and functional implications." *Acta Palaeontologica Polonica* **51** (1): 13-36.
Francis, Joseph W. 2013. "What About Bacteria?" *Answers in Genesis* 23 January 2015 posting of *The New Answers Book 3* Chapter 31 (online text at *www.answersingenesis.org* accessed 2/17/2015).
Frank, Gerhard, Werner Hemmert, & **Anthony W. Gummer.** 1999. "Limiting dynamics of high-frequency electromechanical transduction of outer hair cells." *Proceedings of the National Academy of Sciences* **96** (13 April): 4420-4425.
Franklin, Craig E., & **Michael Axelsson.** 1994. "The intrinsic properties of an *in situ* perfused crocodile heart." *The Journal of Experimental Biology* **186** (January): 269-288.
———. 2000. "An actively controlled heart valve." *Nature* **406** (24 August): 847-848.
Franklin, Craig, Frank Seebacher, Gordon C. Grigg, & **Michael Axelsson.** 2000. "At the Crocodilian Heart of the Matter," with Response by **Dale A. Russell, Michael K. Stoskop, Paul E. Fisher**, & **Reese E. Barrick**. *Science* **289** (8 September): 1687-1688.
Fraser, Nicholas C., & **Hans-Dieter Sues**, eds. 1994. *In the Shadow of the Dinosaurs—Early Mesozoic Tetrapods*. Cambridge: Cambridge University Press.
Fraser, N. C., K. Padian, G. M. Walkden, & **A. L. M. Davis.** 2002. "Basal Dinosauriform Remains from Britain and the Diagnosis of the Dinosauria." *Palaeontology* **45** (February): 79-95.
Freeland, Joanna R., & **Peter T. Boag.** 1999. "Phylogenetics of Darwin's Finches: Paraphyly in the Tree-Finches, and Two Divergent Lineages in the Warbler Finch." *The Auk* **116** (July): 577-588.
Freeman, Eric F. 1976. "Mammal Teeth from the Forest Marble (Middle Jurassic) of Oxfordshire, England." *Science* **194** (3 December): 1053-1055.
Freitas, Renata, Carlos Gómez-Marín, Jonathan Mark Wilson, Fernando Casares, & **José Luis Gómez-Skarmeta.** 2012. "*Hoxd13* Contribution to the Evolution of Vertebrate Appendages." *Developmental Cell* **23** (11 December): 1219-1229.
Friedman, Robert, & **Austin L. Hughes.** 2001. "Pattern and Timing of Gene Duplication in Animal Genomes." *Genome Research* **11** (November): 1842-1847.
———. 2003. "The Temporal Distribution of Gene Duplication Events in a Set of Highly Conserved Human Gene Families." *Molecular Biology and Evolution* **20** (January): 154-161.
Friel, John P., & **Peter C. Wainwright.** 1997. "A Model System of Structural Duplication: Homologies of Adductor Mandibulae Muscles in Tetraodontiform Fishes." *Systematic Biology* **46** (September): 441-463.
Fritzsch, Bernd. 1987. "Inner ear of the coelacanth fish *Latimeria* has tetrapod affinities." *Nature* **327** (14 May): 153-154.
———. 2003. "The ear of *Latimeria chalumnae* revisited." *Zoology* **106** (3): 243-248.
Fritzsch, Bernd, Kirk W. Beisel, Sarah Pauley, & **Garrett Soukup.** 2007. "Molecular evolution of the vertebrate mechanosensory cell and ear." *International Journal of Developmental Biology* **51** (6/7): 663-678.
Fritzsch, Bernd, Daniel F. Eberl, & **Kirk W. Beisel.** 2010. "The role of bHLH genes in ear development and evolution: revisiting a 10-year-old hypothesis." *Cellular and Molecular Life Sciences* **67** (September): 3089-3099.
Fritzsch, Bernd, Israt Jahan, Ning Pan, Jennifer Kersigo, Jeremy Duncan, & **Benjamin Kopecky.** 2011. "Dissecting the molecular basis of organ of Corti development: Where are we now?" *Hearing Research* **276** (June): 16-26.
Fritzsch, Bernd, Ning Pan, Israt Jahan, Jeremy S. Duncan, Benjamin J. Kopecky, Karen L. Elliott,

Jennifer Kersigo, & **Tian Yang.** 2013. "Evolution and development of the tetrapod auditory system: an organ of Corti-centric perspective." *Evolution & Development* **15** (January): 63-79.
Fröbisch, Jörg. 2011. "On Dental Occlusion and Saber Teeth." *Science* **331** (25 March): 1525-1528.
Fröbisch, Jörg, Rainer R. Schoch, Johannes Muller, Thomas Schindler, & **Dieter Schweiss.** 2011. "A new basal sphenacodontid synapsid from the Late Carboniferous of the Saar-Nahe Basin, Germany." *Acta Palaeontologica Polonica* **56** (1): 113-120.
Fröbisch, Nadia B. 2016. "Teenage tetrapods." *Nature* **537** (15 September): 311-312.
Fundamentalist Journal. 1988. "Satan's Final Assault." *Fundamentalist Journal* **8** (October): 20-22.
Futuyma, Douglas J. 1982. *Science on Trial: The Case for Evolution.* 1983 pb ed. New York: Pantheon Books.
———. 1998. *Evolutionary Biology.* 3rd ed. Sunderland, MA: Sinauer Associates, Inc.

Gabaldón, Toni, & **Martijn A. Huynen.** 2003. "Reconstruction of the Proto-Mitochondrial Metabolism." *Science* **301** (1 August): 609.
———. 2007. "From Endosymbiont to Host-Controlled Organelle: The Hijacking of Mitochondrial Protein Synthesis and Metabolism." *PLoS Computational Biology* (online @ ploscompbiol.org) **3** (November): e219.
Gabryszewski, Stanislaw J., Charin Modchang, Lise Musset, Thanat Chookajorn, & **David A. Fidock.** 2016. "Combinatorial Genetic Modeling of pfcrt-Mediated Drug Resistance Evolution in *Plasmodium falciparum*." *Molecular Biology and Evolution* **33** (June): 1554-1570.
Galis, Frietson. 1999. "Why Do Almost All Mammals Have Seven Cervical Vertebrae? Developmental Constraints, *Hox* Genes, and Cancer." *Journal of Experimental Zoology* **285** (April): 19-26.
———. 2001. "Digit identity and digit number: indirect support for the descent of birds from theropod dinosaurs." *Trends in Ecology & Evolution* **16** (January): 16.
Galis, Frietson, David R. Carrier, Joris van Alphen, Steven D. van der Mije, Tom J. M. Van Dooren, Johan A. J. Metz, & **Clara M. A. ten Broek.** 2014. "Fast running restricts evolutionary change of the vertebral column in mammals." *Proceedings of the National Academy of Sciences* **111** (5 August): 11401-11406.
Galis, Frietson, Martin Kundrát, & **Johan A. J. Metz.** 2010. "Hox Genes, Digit Identities and the Theropod/Bird Transition." International Institute for Applied Systems Analysis *Interim Report* IR-05-065 (online text at *iiasa.ac.at* accessed 10/30/2010).
Galis, Frietson, Martin Kundrát, & **Barry Sinervo.** 2003. "An old controversy solved: bird embryos have five fingers." *Trends in Ecology & Evolution* **18** (January): 7-9.
Galis, Frietson, & **Johan A. J. Metz.** 2007. "Evolutionary novelties: the making and breaking of pleiotropic constraints." *Integrative and Comparative Biology* (AKA *American Zoologist*) **47** (September): 409-419.
Galis, Frietson, Jacques J. M. van Alphen, & **Johan A. J. Metz.** 2001. "Why five fingers? Evolutionary constraints on digit numbers." *Trends in Ecology & Evolution* **16** (November): 637-646.
Galis, Frietson, Tom J. M. Van Dooren, Johan D. Feuth, Johan A. J. Metz, Andrea Witkam, Sebastiaan Ruinard, Marc J. Steigenga, & **Liliane C. D. Wijnaendts.** 2006. "Extreme Selection in Humans Against Homeotic Transformations of Cervical Vertebrae." *Evolution* **60** (December): 2643-2654.
Gal-Mark, Nurit, Schraga Schwartz, & **Gil Ast.** 2008. "Alternative splicing of *Alu* exons—two arms are better than one." *Nucleic Acids Research* **36** (1 April): 2012-2023.
Gamlin, Linda, & **Gail Vines**, eds. 1986. *The Evolution of Life.* 1991 pb ed. New York: Oxford University Press.
Gans, Carl, & **Ernest Glen Wever.** 1972. "The Ear and Hearing in Amphisbaenia (Reptilia)." *Journal of Experimental Zoology* **179** (January): 17-34.
Gao, Xiang, & **Michael Lynch.** 2009. "Ubiquitous internal gene duplication and intron creation in eukaryotes." *Proceedings of the National Academy of Sciences* **106** (8 December): 20818-20823.
García-Fernàndez, Jordi. 2005a. "Hox, ParaHox, ProtoHox: facts and guesses." *Heredity* **94** (February): 145-152.
———. 2005b. "The genesis and evolution of Homeobox gene clusters." *Nature Reviews Genetics* **6** (December): 881-892.
García-Fernàndez, Jordi, Salvatore D'Aniello, & **Hector Escriva.** 2007. "Organizing chordates with an organizer." *BioEssays* **29** (June): 619-624.

García-Fernàndez, Jordi, & **Peter W. H. Holland.** 1994. "Archetypal organization of the amphioxus *Hox* gene cluster." *Nature* **370** (18 August): 563-566.
García-Fernandez, Jordi, Senda Jimenez-Delgado, Juan Pascual-Anaya, Ignacio Maeso, Manuel Irimia, Carolina Minguillon, Elia Benito-Guitierrez, Josep Gardenyes, Stephanie Bertrand, & **Salvatore D'Aniello.** 2009. "From the American to the European amphioxus: towards experimental Evo-Devo at the origin of chordates." *International Journal of Developmental Biology* **53** (8/9/10): 1359-1366.
Gardner, Erle Stanley. 1969a. *Man With the Big Hat.* New York: William Morrow.
———. 1969b. "Acambaro Mystery." *Desert Magazine* **32** (October): 18-21, 36-37.
Gauger, Ann K. 2014. "A Pretty Sharp Edge: Reflecting on Michael Behe's Vindication." Discovery Institute *Evolution News & Views* for 28 July (online text at *evolutionnews.org* accessed 8/14/2014).
———. 2016a. "The Vitellogenin Pseudogene Story: Unequally Yolked." Discovery Institute *Evolution News & Views* 24 May posting (online text at *evolutionnews.org* accessed 6/14/2016).
———. 2016b. "Vincent Torley Thinks I Have Egg on My Face." Discovery Institute *Evolution News & Views* 10 June posting (online text at *evolutionnews.org* accessed 6/14/2016).
Gayon, Jean. 2000. "History of the concept of allometry." *Integrative and Comparative Biology* (AKA *American Zoologist*) **40** (November): 748-758.
Gee, Henry. 1999. *In Search of Deep Time: Beyond the Fossil Record to a New History of Life.* New York: The Free Press.
Gehring, W. J. 1987. "Homeo Boxes in the Study of Development." *Science* **236** (5 June): 1245-1252.
Gehrke, Andrew R., Igor Schneider, Elisa de la Calle-Mustienes, Juan J. Tena, Carlos Gomez-Marin, Mayuri Chandran, Tetsuya Nakamura, Ingo Brassch, John H. Postlethwait, José Gómez-Skarmeta, & **Neil H. Shubin.** 2015. "Deep conservation of wrist and digit enhancers in fish." *Proceedings of the National Academy of Sciences* **112** (20 January): 803-808.
Geisler, Norman L., & **Paul K. Hoffman**, eds. 2001. *Why I Am a Christian: Leading Thinkers Explain Why They Believe.* Grand Rapids, MI: Baker Books.
Gene, Mike. 2005. "Endosymbiotic Theory." *Telic Thoughts* posting 30 August (online text at *telicthoughts.com* accessed 1/25/2010).
Genesis Park. 2011a. "Ancient Dinosaur Depictions." *Genesis Park Room 1: The Dinosaurs* undated posting (online text at *www.genesispark.org* accessed 7/28/2011).
———. 2011b. "Cryptozoology." *Genesis Park Room 1: The Dinosaurs* undated posting (online text at *www.genesispark.org* accessed 7/28/2011).
———. 2011c. "Dragons in History." *Genesis Park Room 1: The Dinosaurs* undated posting (online text at *www.genesispark.org* accessed 7/28/2011).
———. 2011d. "Abrupt Appearance in the Fossil Record." *Genesis Park Room 3: The Story of the Fossils* undated posting (online text at *www.genesispark.org* accessed 7/28/2011).
Gerber, André, Mary A. O'Connell, & **Walter Keller.** 1997. "Two forms of human double-stranded RNA-specific editase 1 (hRED1) generated by the insertion of an Alu cassette." *RNA* **3** (May): 453-463.
Gilbert, Scott F. 2006. *Developmental Biology.* 8th ed. Sunderland, MA: Sinauer Associates, Inc.
Gilbert, Scott F., Grace A. Loredo, Alla Brukman, & **Ann C. Burke.** 2001. "Morphogenesis of the turtle shell: the development of a novel structure in tetrapod evolution." *Evolution & Development* **3** (March): 47-58.
Gilder, George. 2004. "Biocosm." *Wired* October issue (online text at *wired.com* accessed 1/4/2013).
———. 2006. "Evolution and Me: The Darwinian Theory Has Become an All-Purpose Obstacle to Thought Rather than an Enabler of Scientific Advance." Discovery Institute *Evolution News & Views* posting of *National Review* 17 July article (online text at *evolutionnews.org* accessed 4/11/2014).
Gill, Pamela G., Mark A. Purnell, Nick Crumpton, Kate Robson Brown, Neil J. Gosling, M. Stampanoni, & **Emily J. Rayfield.** 2014. "Dietary specializations and diversity in feeding ecology of the earliest stem mammals." *Nature* **512** (21 August): 303-305.
Gillespie, Neal C. 1979. *Charles Darwin and the Problem of Creation.* Chicago: University of Chicago Press.
Gillette, David D., ed. 1999. *Vertebrate Paleontology in Utah.* Miscellaneous Publication 99-1. Salt Lake City, UT: Utah Geological Survey.

Gingerich, Philip D. 1994. "The Whales of Tethys." *Natural History* **103** (April): 86-88.
Gingerich, Philip D., Daryl P. Domning, Caroline E. Blane, & **Mark D. Uhen.** 1994. "Cranial Morphology of *Protosiren fraasi* (Mammalia, Sirenia) from the Middle Eocene of Egypt: A New Study Using Computed Tomography." *Contributions from the Museum of Paleontology, University of Michigan* **29** (30 November): 41-67.
Gingerich, Philip D., Munir ul Haq, Wighart von Koenigswald, William J. Sanders, R. Holly Smith, & **Iyad S. Zalmout.** 2009. "New Protocetid Whale from the Middle Eocene of Pakistan: Birth on Land, Precocial Development, and Sexual Dimorphism." *PLoS ONE* (online @ plosone.org) **4** (February): e4366.
Gingerich, Philip D., Munir ul Haq, Iyad S. Zalmout, Intizar Hussain Khan, & **M. Sadiq Malkani.** 2001. "Origin of Whales from Early Artiodactyls: Hands and Feet of Eocene Protocetidae from Pakistan." *Science* **293** (21 September): 2239-2242.
Gingerich, Philip D., S. Mahmood Raza, Muhammad Arif, Mohammad Anwar, & **Xiaoyuan Zhou.** 1994. "New whale from the Eocene of Pakistan and the origin of cetacean swimming." *Nature* **368** (28 April): 844-847.
Gingerich, Philip D., B. Holly Smith, & **Elwyn L. Simons.** 1990. "Hind Limbs of Eocene *Basilosaurus*: Evidence of Feet in Whales." *Science* **249** (13 July): 154-157.
Gingerich, Philip D., Neil A. Wells, Donald E. Russell, & **S. M. Ibrahim Shah.** 1983. "Origin of Whales in Epicontinental Remnant Seas: New Evidence from the Early Eocene of Pakistan." *Science* **220** (22 April): 403-406.
Giovannoni, Stephen J., Seán Turner, Gary J. Olsen, Susan Barns, David J. Lane, & **Norman R. Pace.** 1999. "Evolutionary Relationships among Cyanobacteria and Green Chloroplasts." *Journal of Bacteriology* **170** (August, No. 8): 3584-3592.
Gish, Duane T. 1978. *Evolution: The Fossils Say NO!* San Diego, CA: Creation-Life Publishers.
———. 1980. "The Origin of Mammals." Institute for Creation Research *Acts & Facts* (September) posting (online text at *icr.org* accessed 3/17/2016).
———. 1981. "The Mammal-Like Reptiles." Institute for Creation Research *Acts & Facts* IMPACT No. 102 (December) posting (online text at *icr.org* accessed 3/17/2016).
———. 1985. *Evolution: The Challenge of the Fossil Record.* Rev. ed. of 1978, *Evolution: The Fossils Say NO!*. El Cajon, CA: Institute for Creation Research.
———. 1990. *The Amazing Story of Creation from science and the Bible.* El Cajon, CA: Institute for Creation Research.
———. 1993a. *Creation Scientists Answer Their Critics.* El Cajon, CA: Institute for Creation Research.
———. 1993b. "An Interview with Dr. Duane Gish." Alpha Omega Institute Institute (Grand Junction, CO) "Spotlight on Science" *Think & Believe* **10** (September/October): 3.
———. 1995. *Evolution: The Fossils STILL Say NO!* Rev. ed. of 1985, *Evolution: The Challenge of the Fossil Record.* El Cajon, CA: Institute for Creation Research.
Gishlick, Alan D. 2006. "Baraminology." *Reports of the National Center for Science Education* **26** (July-August): 17-21.
Godefroit, Pascal, ed. 2012. *Bernissart Dinosaurs and Early Cretaceous Terrestrial Ecosystems.* Bloomington, IN: Indiana University Press.
Godefroit, Pascal, Andrea Cau, Hu Dong-Yu, François Escuillié, Wu Wenhao, & **Gareth Dyke.** 2013. "A Jurassic avialan dinosaur from China resolves the early phylogenetic history of birds." *Nature* **498** (20 June): 359-362.
Godefroit, Pascal, Helena Demuynck, Gareth Dyke, Dongyu Hu, François Escuillié, & **Philippe Claeys.** 2013. "Reduced plumage and flight ability of a new Jurassic paravian theropod from China." *Nature Communications* (online @ nature.com) **4** (22 January): 1394.
Godefroit, Pascal, Sofia M. Sinitsa, Danielle Dhouailly, Yuri L. Bolotsky, Alexander V. Sizov, Maria E. McNamara, Michael J. Benton, & **Paul Spagna.** 2014a. "A Jurassic ornithischian dinosaur from Siberia with both feathers and scales." *Science* **345** (25 July): 451-455.
———. 2014b. "Response to Comment on 'A Jurassic ornithischian dinosaur from Siberia with both feathers and scales'." *Science* **346** (24 October): 434.
Godefroit, Pascal, & **Denise Sogogneau-Russell.** 1999. "Kuehneotheriids from Saint-Nicolas-de-Port (Late Triassic of France)." *Geologica Belgica* **2** (3-4): 181-196.
Godfrey, Laurie R., ed. 1983a. *Scientists Confront Creationism.* 1984 pb ed. New York: W. W. Norton & Co.
———. 1983b. "Creationism and Gaps in the Fossil Record," in Godfrey (1983a, 193-218).
Godwin, Alan R., & **Mario R. Capecchi.** 1998. "*Hoxc13* mutant mice lack external hair." *Genes &*

Development **12** (1 January): 11-20.
Göhlich, Ursula B., Luis M. Chiappe, James M. Clark, & **Hans-Dieter Sues.** 2005. "The systematic position of the Late Jurassic alleged dinosaur *Macelongnathus* (Crocodylomorpha: Sphenosuchia)." *Canadian Journal of Earth Sciences* **42** (March): 307-321.
Gompel, Nicolas, & **Sean B. Carroll.** 2003. "Genetic mechanisms and constraints governing the evolution of correlated traits in drosophilid flies." *Nature* **424** (21 August): 931-935.
Goodman, Jeffrey. 1979. *We Are the Earthquake Generation.* New York: Berkley Books.
Gorbunov, Dmitry, Mattia Sturlese, Florian Nies, Murielle Kluge, Massimo Bellanda, Roberto Battistutta, & **Dominik Oliver.** 2014. "Molecular architecture and the structural basis for anion interaction in prestin and SLC26 transporters." *Nature Communications* (online @ nature.com) **5** (8 April): 3622.
Gordon, Malcolm S., & **Everett C. Olson.** 1995. *Invasions of the Land: The Transitions of Organisms from Aquatic to Terrestrial Life.* New York: Columbia University Press.
Gore, Rick. 2003. "The Rise of the Mammals." *National Geographic* **203** (April): 2-37.
Goswami, Anjali, Guntupalli V. R. Prasad, Paul Upchurch, Dong M. Boyer, Erik R. Seiffert, Omkar Verma, Emmanuel Gheerbrant, & **John J. Flynn.** 2011. "A radiation of arboreal basal eutherian mammals beginning in the Late Cretaceous of India." *Proceedings of the National Academy of Sciences* **108** (27 September): 16333-16338.
Gould, Stephen E., William B. Upholt, & **Robert A. Kosher.** 1995. "Characterization of Chicken Syndecan-3 as a Heparan Sulfate Proteoglycan and Its Expression during Embryogenesis." *Developmental Biology* **168** (1 April): 438-451.
Gould, Stephen Jay. 1966. "Allometry and size ontogeny and phylogeny." *Biological Reviews of the Cambridge Philosophical Society* **41** (December): 587-640.
———. 1981. "Evolution as Fact and Theory." *Discover* **2** (May): 34-37.
———. 1990. "An Earful of Jaw." *Natural History* **99** (March): 12, 15-16, 18, 20, 22-23.
———. 1991. *Bully for Brontosaurus: Reflections in Natural History.* New York: W.W. Norton & Co.
———. 1992. "Impeaching a Self-Appointed Judge." *Scientific American* **267** (July): 118-121.
———, ed. 1993. *The Book of Life.* New York: W. W. Norton & Co.
———. 2002. *The Structure of Evolutionary Theory.* Cambridge, MA: Belknap Press (Harvard University Press).
Gould, Stephen Jay, & **Niles Eldredge.** 1977. "Punctuated equilibria: the tempo and mode of evolution reconsidered." *Paleobiology* **3** (July): 115-151.
———. 1993. "Punctuated equilibrium comes of age." *Nature* **366** (18 November): 223-227.
Gow, C. E. 1981. "*Pachygenelus, Diarthrognathus* and the double jaw articulation." *Palaeontologia Africana* **24**: 15.
———. 2001. "A partial skeleon of the tritheledontid *Pachygenelus* (Therapsida: Cynodontia)." *Palaeontologia Africana* **37**: 93-97.
Gräf, Ralph, Petros Batsios, & **Irene Meyer.** 2015. "Evolution of centrosomes and the nuclear lamina: Amoebozoan assets." *European Journal of Cell Biology* **94** (June): 249-256.
Graham, Jeffrey B., Nicholas C. Wegner, Lauren A. Miller, Corey J. Jew, N Chin Lai, Rachel M. Berquist. Lawrence R. Frank, & **John A. Long.** 2014. "Spiracular air breathing in polypterid fishes and its implications for aerial respiration in stem tetrapods." *Nature Communications* (online @ nature.com) **5** (23 January): 3022.
Grassé, Pierre-P. 1977. *Evolution of Living Organisms: Evidence for a New Theory of Transformation.* English edition of 1973 French version. New York: Academic Press.
Gray, Michael W. 2014. "The Pre-Endosymbiont Hypothesis: A New Perspective on the Origin and Evolution of Mitochondria." *Cold Spring Harbor Perspectives in Biology* **6** (March): a016097.
———. 2015. "Mosaic nature of the mitochondrial proteome: Implications for the origin and evolution of mitochondria." *Proceedings of the National Academy of Sciences* **112** (18 August): 10133-10138.
Gray, Noel-Marie, Kimberly Kainec, Sandra Madar, Lucas Tomko, & **Scott Wolfe.** 2007. "Sink or Swim? Bone Density as a Mechanism for Buoyancy Control in Early Cetaceans." *The Anatomical Record: Advances in Integrative Anatomy and Evolutionary Biology* **290** (June): 638-653.
Green, H. L. H. H. 1937. "VIII—The Development and Morphology of the Teeth of *Ornithorhynchus.*" *Philosophical Transactions of the Royal Society of London* **B** (Biological Sciences) **228** (26 November): 367-420.
Greenspahn, Frederick E. 1983. "Biblical Views of Creation." *Creation/Evolution* **4** (Summer): 30-38.
Gregory, W. K. 1927. "Hesperopithecus Apprently Not an Ape Nor a Man." *Science* **66** (16

December): 579-581.
Griffiths, M., R. T. Wells, & **D. J. Hand.** 1991. "Observations on the skulls of fossil and extant echidnas (Monotremata: Tachyglossidae)." *Australian Mammalology* **14**: 87-101.
Grimaldi, David, & **Donat Agosti.** 2000. "A formicine in New Jersey Cretaceous amber (Hymenoptera: Formicidae) and early evolution of ants." *Proceedings of the National Academy of Sciences* **97** (5 December): 13678-13683.
Groppi, Teno. 2002a. "Dr. Kent Hovind." *Teno's God and Country Center* undated posting (online text at *www.baptistlink.com/godandcountry/* accessed 5/7/2002, not available as of 1/16/2014).
———. 2002b. "E-Z Evolution Refuters." *Teno's God and Country Center* undated posting (online text at *www.baptistlink.com/godandcountry/* accessed 5/7/2002, not available as of 1/16/2014).
———. 2002c. "US Constitution Party." *Teno's God and Country Center* undated posting (online text at *www.baptistlink.com/godandcountry/* accessed 5/7/2002, not available as of 1/16/2014).
Guler, Jennifer L., John White III, Margaret A. Phillips, & **Pradipsinh K. Rathod.** 2015. "Atovaquone Tolerance in *Plasmodium falciparum* Parasites Selected for High-Level Resistance to a Dihydroorotate Dehydrogenase Inhibitor." *Antimicrobial Agents and Chemotherapy* **59** (January): 686-689.
Guliuzza, Randy J. 2010. "Fit & Function: Design in Nature." Institute for Creation Research *Acts & Facts* (February): 10-11.
Gunji, Megu, & **Hideki Endo.** 2016. "Functional cervicothoracic boundary modified by anatomical shifts in the neck of giraffes." *Royal Society Open Science* **3** (February): 150604.
Gunkel, Herrmann, trans. Charles A. Muenchow. 1895. "The influence of Babylonian mythology upon the Biblical creation story," in Anderson (1984, 25-52).
Guo, Baocheng, Ming Zou, & **Andreas Wagner.** 2012. "Pervasive Indels and Their Evolutionary Dynamids after the Fish-Specific Genome Duplication." *Molecular Biology and Evolution* **29** (October): 3005-3022.
Guo, Ting, Kenji Mandai, Brian G. Condie, S. Rasika Wickramasinghe, Mario R. Capecchi, & **David D. Ginty.** 2011. "An evolving NGF-Hoxd1 signaling pathway mediates development of divergent neural circuits in vertebrates." *Nature Neuroscience* **14** (January): 31-36.
Gurdon, J. B., & **Ian Wilmut.** 2011. "Nuclear Transfer to Eggs and Oocytes." *Cold Spring Harbor Perspectives in Biology* **3** (June): a002659.

Haarsma, Loren, Serita Nelesen, Ethan VanAndel, James Lamine, & **Peter VandelHaar.** 2016. "Simulating evolution of protein complexes through gene duplication and co-option." *Journal of Theoretical Biology* **399** (21 June): 22-32.
Hadrys, Thorsten, Rob DeSalle, Sven Sagasser, Nino Fischer, & **Bernd Schierwater.** 2005. "The Trichoplax *PaxB* Gene: A Putative Proto-*PaxA/B/C* Gene Predating the Origin of Nerve and Sensory Cells." *Molecular Biology and Evolution* **22** (July): 1569-1578.
Hagen, Joel B. 2003. "The Statistical Frame of Mind in Systematic Biology from *Quantitative Zoology* to *Biometry*." *Journal of the History of Biology* **36** (Summer): 353-384.
Halder, Georg, Patrick Callaerts, & **Walter J. Gehring.** 1995. "Induction of Ectopic Eyes by Targeted Expression of the *eyeless* Gene in *Drosophila*." *Science* **267** (24 March): 1788-1792.
Hall, Barry G. 2004. "In Vitro Evolution Predicts that the IMP-1 Metallo-β-Lactamase Does Not Have the Potential To Evolve Increased Activity against Imipenem." *Antimicrobial Agents and Chemotherapy* **48** (March): 1032-1033.
Hall, Brian K. 2003. "*Evo-Devo*: evolutionary developmental mechanisms." *International Journal of Developmental Biology* **47** (7/8): 491-495.
———. 2009. *The Neural Crest and Neural Crest Cells in Vertebrate Development and Evolution.* New York: Springer.
Hall, Marshall. 2011. "False Science & The Pharisee Religion: Satan's Ultimate Weapons Against The Bible And God." *The Earth Is Not Moving* undated posting (online text at *www.fixedearth.com* accessed 6/12/2011).
Halliday, Thomas John Dixon, & **Anjali Goswami.** 2016. "Eutherian morphological disparity across the end-Cretaceous mass extinction." *Biological Journal of the Linnean Society* **118** (May): 152-168.
Halliday, Thomas J. D., Paul Upchurch, & **Anjali Goswami.** 2015. "Resolving the relationships of Paleocene placental mammals." *Biological Reviews* (15 December advance posting, online pdf at *onlinelibrary.wiley.com* accessed 10/10/2016): 12242.
Ham, Ken. 1998. *The Great Dinosaur Mystery SOLVED!* Green Forest, AR: Master Books.

———. 2012. "Kentucky Horses That Will Lead You Astray." *Around the World with Ken Ham* blog 28 April posting (online text at *blogs.answersingenesis.org* accessed 4/30/2012).
Ham, Ken, & **Roger Patterson.** 2013. "Should Christians Be Pushing to Have Creation Taught in Government Schools?" *Answers in Genesis* 17 September posting of *The New Answers Book 3* Chapter 3 (online text at *www.answersingenesis.org* accessed 2/17/2015).
Ham, Ken, Jonathan Sarfati, & **Carl Wieland**, ed. Don Batten. 2000. *The Revised & Expanded Answers Book.* 2004 pb ed. (rev. from 1990). Green Forest, AR: Master Books.
Hampl, Vladimir, Jeffrey D. Silberman, Alexandra Stechmann, Sara Diaz-Triviño, Patricia J. Johnson, & **Andrew J. Roger.** 2008. "Genetic Evidence for a Mitochondriate Ancestry in the 'Amitochondriate' Flagellate *Trimastix pyriformis*." *PLoS ONE* (online @ plosone.org) **3** (January): e1383.
Händeler, Katharina, Yvonne P. Grzymbowski, Patrick J. Krug, & **Heike Wägele.** 2009. "Functional chloroplasts in metazoan cells—a unique evolutionary strategy in animal Nlife." *Frontiers in Zoology* (online @ frontiersinzoology.com) **6** (1 December): 28.
Hanegraaff, Hank. 1998. *The Face that Demonstrates the Farce of Evolution.* Nashville, TN: Word Publishing.
———. 2002. *Bible Answer Man* radio 11 November broadcast (heard on airing 11/11/2002).
Hanson, Robert W., ed. 1986. *Science and Creation: Geological, Theological, and Educational Perspectives.* New York: Macmillan.
Hanson, Thor. 2011. *Feathers: The Evolution of a Natural Miracle.* New York: Basic Books (Perseus Books).
Hapgood, Charles H. 1966. *Maps of the Ancient Sea Kings: Evidence of Advanced Civilization in the Ice Age.* Philadelphia: Chilton.
Hapgood, Charles H., with **James H. Campbell.** 1958. *Earth's Shifting Crust.* New York: Pantheon Books.
Hardin, Garrett. 1968. "The Tragedy of the Commons." *Science* **162** (13 December): 1243-1248.
———. 1974. *Mandatory Motherhood: The True Meaning of "Right to Life."* Boston: Beacon Press.
———. 1984. "'Scientific Creationism'--Marketing Deception as Truth," in Montague (1984, 159-166).
Hargreaves, Adam D., Martin T. Swain, Matthew J. Hegarty, Darren W. Logan, & **John F. Mulley.** 2014. "Restriction and Recruitment—Gene Duplication and the Origin and Evolution of Snake Venom Toxins." *Genome Biology and Evolution* **6** (August): 2088-2095.
Harjunmaa, Enni, Aki Kallonen, Maria Voutilainen, Keijo Hämäläinen, Marja L. Mikkola, & **Jukka Jernvall.** 2012. "On the difficulty of increasing dental complexity." *Nature* **483** (15 March): 324-327.
Harjunmaa, Enni, Kerstin Seidel, Teemu Häkkinen, Elodie Renvoisé, Ian J. Corfe, Aki Kallonen, Zhao-Qun Zhang, Alistair R. Evans, Marja L. Mikkola, Isaac Salazar-Cuidad, Ophir D. Klein, & **Jukka Jernvall.** 2014. "Replaying evolutionary transitions from the dental fossil record." *Nature* **512** (7 August): 44-48.
Harms, Michael J., & **Joseph M. Thornton.** 2014. "Historical contingency and its biophysical basis in glucocorticoid receptor evolution." *Nature* **512** (14 August): 203-207.
Harris, Matthew P., John F. Fallon, & **Richard O. Prum.** 2002. "*Shh-Bmp2* Signaling Module and the Evolutionary Origin and Diversification of Feathers." *Journal of Experimental Zoology* **294** (15 August): 160-176.
Harris, Matthew P., Sean M. Hasso, Mark W. J. Ferguson, & **John F. Fallon.** 2006. "The Development of Archosaurian First-Generation Teeth in a Chicken Mutant." *Current Biology* **16** (21 February): 371-377.
Harris, Matthew P., Scott Williamson, John F. Fallon, Hans Meinhardt, & **Richard O. Prum.** 2005. "Molecular evidence for an activator-inhibitor mechanism in development of embryonic feather branching." *Proceedings of the National Academy of Sciences* **102** (16 August): 11734-11739.
Harris, Sam, Sameer A. Sheth, & **Mark S. Cohen.** 2008. "Functional Neuroimaging of Belief, Disbelief, and Uncertainty." *Annals of Neurology* **63** (February): 141-147.
Harrub. Brad. 2015. "A Dinosaur Discovered in a Mammal's Stomach?!" *The Creation Club* 1 June posting (online text at *thecreationclub.com* accessed 4/2/2016).
Harrub, Brad, and **Bert Thompson.** 2001. "Archaeopteryx, Archaeoraptor, and the 'Dinosaurs-To-Birds' Theory (Part I)." *Reason & Revelation* **21** (April): 25-32.
Hartman, Byron H., Thomas A. Reh, & **Olivia Bermingham-McDonogh.** 2010. "Notch signaling specifies prosensory domains via lateral induction in the developing mammalian inner ear."

Proceedings of the National Academy of Sciences **107** (7 September): 15792-15797.

Hartmann, Markus, Thomas R. Schneider, Andrea Pfeil, Gabriele Heinrich, William N. Lipscomb, & **Gerhard H. Braus.** 2003. "Evolution of feedback-inhibited β/α barrel isoenzymes by gene duplication and a single mutation." *Proceedings of the National Academy of Sciences* **100** (4 February): 862-867.

Hartmann, William K., & **Ron Miller.** 1991. *The History of Earth: An Illustrated Chronicle of an Evolving Planet.* New York: Workman Publishing.

Havird, Justin C., Matthew D. Hall, & **Damian K. Dowling.** 2015. "The evolution of sex: A new hypothesis based on mitochondrial mutational erosion." *BioEssays* **37** (September): 951-959.

Havukainen, Heli, Øyvind Halskau, Lars Skjaerven, Bent Smedal, & **Gro V. Amdam.** 2011. "Deconstructing honeybee vitellogenin: novel 40kDa fragment assigned to its N terminus." *The Journal of Experimental Biology* **214** (15 February): 582-592.

Hay, Angela, & **Miltos Tsiantis.** 2006. "The genetic basis for differences in leaf form between *Arabidopsis thaliana* and its wild relative *Cardamine hirsuta*." *Nature Genetics* **38** (August): 942-947.

Hayden, Benjamin Y., John M. Pearson, & **Michael L. Platt.** 2009. "Fictive Reward Signals in the Anterior Cingulate Cortex." *Science* **324** (15 May): 948-950.

Hayward, Alan. 1985. *Creation and Evolution.* 1995 pb ed. Minneapolis, MN: Bethany House Publishers.

Hazen, Robert M. 2012. *The Story of Earth: The First 4.5 Billion Years from Stardust to Living Planet.* New York: Viking Press.

Hecht, Max H., John H. Ostrom, Günter K. Viohl, & **Peter Wellnhofer**, eds. 1985. *The Beginnings of Birds: Proceedings of the International Archaeopteryx Conference, Eichstatt.* Eichstatt, Germany: Freunde des Jura-Museums.

Hedges, S. Blair. 2012. "Amniote phylogeny and the position of turtles." *BMC Biology* (online @ biomedcentral.com) **10** (27 July): 64.

Hedges, S. Blair, & **Laura L. Poling.** 1999. "A Molecular Phylogeny of Reptiles." *Science* **283** (12 February): 998-1001.

Heidel, Alexander. 1951. *The Babylonian Genesis: The Story of Creation.* 2nd ed. Chicago: University of Chicago Press.

Helariutta, Yrjö, Mika Kotilainen, Paula Elomaa, Nisse Kalkkinen, Kåre Bremer, Teemu H. Teeri, & **Victor A. Albert.** 1996. "Duplication and functional divergence in the chalcone synthase gene family of Asteraceae: Evolution with substrate change and catalytic simplification." *Proceedings of the National Academy of Sciences* **93** (20 August): 9033-9038.

Hendey, Q. B. 1973. "Carnivore remains from the Kromdraai Australopithecine site (Mammalia: Carnivora)." *Annals of the Transvaal Museum* **28** (31 January): 99-112.

Hennig, Raoul, Jennifer Heidrich, Michael Saur, Lars Schmüser, Steven J. Roeters, Nadja Hellmann, Sander Woutersen, Mischa Bonn, Tobias Weidner, Jürgen Markl, & **Dirk Schneider.** 2015. "IM30 triggers membrane fusion in cyanobacteria and chloroplasts." *Nature Communications* (online @ nature.com) **6** (8 May): 7018.

Hennig, Willi, trans. D. Dwight Davis & Rainer Zangerl. 1966. *Phylogenetic Systematics.* Urbana, IL: University of Illinois Press.

Hennigan, Tom. 2015. "An Initial Estimate of the Numbers and Identification of Extant Non-Snake/Non-Amphisbaenian Lizard Kinds: Order Squamata." Answers in Genesis *Answers Research Journal* **8**: 171-186.

Higgins, Penny. 2006. "Use and Abuse of the Fossil Record: Defining Terms." *CSI* 9 September posting (online text at *www.csicop.org* accessed 10/29/2016).

Hillenius, Willem J. 1994. "Turbinates in Therapsids: Evidence for Late Permian Origins of Mammalian Endothermy." *Evolution* **48** (April): 207-229.

Hillman, Chris. 2001. "Some Scientifically Inaccurate Claims Concerning Cosmology and Relativity." *University of California* (Riverside) 31 January updated posting (online *math.ucr.edu* text archived at *web.archive.org* accessed 4/16/2016).

Hinchcliffe, Richard. 1990. "Towards a Homology of Process: Evolutionary Implications of Experimental Studies on the Generation of Skeletal Pattern in Avian Limb Development," in Maynard Smith & Vida (1990, 119-131).

———. 1997. "The Forward March of the Bird-Dinosaur Halted?" *Science* **278** (24 October): 596-597.

———. 2008. "Bird wing digits & their homologies: reassessment of developmental evidence for a

2,3,4 identity." *Oryctos* **7**: 7-12.

Hirasawa, Tatsuya, & **Shigeru Kuratani.** 2013. "A new scenario of the evolutionary derivation of the mammalian diaphragm." *Journal of Anatomy* **222** (May): 504-517.

Hitching, Francis. 1982. *The Neck of the Giraffe or Where Darwin Went Wrong.* New Haven, CT: Ticknor & Fields.

Hittinger, Chris Todd, & **Sean B. Carroll.** 2007. "Gene duplication and the adaptive evolution of a classic genetic switch." *Nature* **449** (11 October): 677-681.

Hjort, Karin, Alina V. Goldberg, Anastasios D. Tsaousis, Robert P. Hirt, & **T. Martin Embley.** 2010. "Diversity and reductive evolution of mitochondria among microbial eukaryotes." *Philosophical Transactions of the Royal Society of London* **B** (Biological Sciences) **365** (12 March): 713-727.

Hodge, Bodie. 2010. "Was the Dispersion at Babel a Real Event?" *Answers in Genesis* 19 August posting of *The New Answers Book 2* Chapter 28 (online text at *www.answersingenesis.org* accessed 2/22/2014).

Hodge, Bodie, & **Tim Lovett.** 2013. "What Did Noah's Ark Look Like?" *Answers in Genesis* 3 September posting of *The New Answers Book 3* Chapter 2 (online text at *www.answersingenesis.org* accessed 2/17/2015).

Hodge, Bodie, & **Georgia Purdom.** 2013. "What Are 'Kinds' in Genesis?" *Answers in Genesis* 16 April posting of *The New Answers Book 3* Chapter 4 (online text at *www.answersingenesis.org* accessed 2/17/2015).

Hodge, Bodie, & **Paul F. Taylor.** 2013. "Doesn't the Bible Support Slavery?" *Answers in Genesis* 19 January 2015 posting of *The New Answers Book 3* Chapter 33 (online text at *www.answersingenesis.org* accessed 5/10/2016).

Hodges, Matthew E., Nicole Scheumann, Bill Wickstead, Jane A. Langdale, & **Keith Gull.** 2010. "Reconstructing the evolutionary history of the centriole from protein components." *Journal of Cell Science* **123** (1 May): 407-413.

Hofstadter, Douglas R. 1979. *Godel, Escher, Bach: An Eternal Golden Braid.* 1980 pb ed. New York: Vintage Books.

Hogue, John. 1997. *Nostradamus: The Complete Prophecies.* 1999 pb ed. Boston: Element Books.

———. 2002. *Nostradamus: The New Millennium.* Rev. ed. of 1994 *Nostradamus: The New Revelation.* London: Element.

Holland, Linda Z., Ricard Albalat, Kaoru Azumi, Èlia Benito-Gutiérrez, Matthew J. Blow, Marianne Bronner-Fraser, Frederic Brunet, Thomas Butts, Simona Candiani, Larry J. Dishaw, David E. K. Ferrier, Jordi Garcia-Fernàndez, Jeremy J. Gibson-Brown, Carmela Gissi, Adam Godzik, Finn Hallböök, Dan Hirose, Kazuyoshi Hosomichi, Tetsuro Ikuta, Hidetoshi Inoko, Masanori Kasahara, Jun Kasamatsu, Takeshi Kawashima, Ayuko Kimura, Masaaki Kobayashi, Zbynek Kozmik, Kaoru Kubokawa, Vincent Laudet, Gary W. Litman, Alice C. McHardy, Daniel Meulemans, Masaru Nonaka, Robert P. Olinski, Zeev Pancer, Len A. Pennacchio, Mario Pestarino, Jonathan P. Rast, Isidore Rigoutsos, Marc Robinson-Rechavi, Graeme Roch, Hidetoshi Saiga, Yasunori Sasakura, Masanobu Satake, Yutaka Satou, Michael Schubert, Nancy Sherwood, Takashi Shiina, Naohito Takatori, Javier Tello, Pavel Vopalensky, Shuichi Wada, Anlong Xu, Yuzhen Ye, Keita Yoshida, Fumiko Yoshizaki, Jr-Kai Yu, Qing Zhang, Christian M. Zmasek, Pieter J. de Jong, Kazutoyo Osoegawa, Nicholas H. Putnam, Daniel S. Rokhsar, Noriyuki Satoh, & **Peter W. H. Holland.** 2008. "The amphioxus genome illuminates vertebrate origins and cephalochordate biology." *Genome Research* **18** (July): 1100-1111.

Holland, Peter W. H. 2015. "Did homeobox gene duplications contribute to the Cambrian explosion?" *Zoological Letters* (online @ zoologicalletters.biomedcentral.com) **1** (13 January): 1.

Hölldobler, Bert, & **Hiltrud Engel-Siegel.** 1984. "On the Metapleural Gland of Ants." *Psyche* **91** (3-4): 201-224.

Hölldobler, Bert, Nicola J. R. Plowes, Robert A. Johnson, Upul Nishshanka, Chongming Liu, & **Athula B. Attygalle.** 2013. "Pygidial gland chemistry and potential alarm-recruitment function in column foraging, but not solitary, Nearctic *Messor* harvesting ants (Hymenoptera: Formicidae: Myrmicinae)." *Journal of Insect Physiology* **59** (September): 863-869.

Hölldobler, Bert, & **Edward O. Wilson.** 1990. *The Ants.* Cambridge, MA: Belknap Press (Harvard University Press).

Holliday, Casey M., & **Lawrence M. Witmer.** 2007. "Archosaur Adductor Chamber Evolution: Integration of Musculoskeletal and Topological Criteria in Jaw Muscle Homology." *Journal of Morphology* **268** (June): 457-484.

Holroyd, Patricia A., Takehisa Tsubamoto, Naoko Egi, Russell L. Ciochon, Masanaru Takai, Soe

Thura Tun, Chit Sein, & **Gregg F, Gunnell.** 2006. "A rhinocerotid perissodactyl from the Late Middle Eocene Pondalung Formation, Myanmar." *Journal of Vertebrate Paleontology* **26** (June): 491-494.

Hopson, James A. 1977. "Relative Brain Size and Behavior in Archosaurian Reptiles." *Annual Review of Ecology and Systematics* **8**: 429-448.

———. 1987. "The Mammal-like Reptiles: A study of transitional fossils." *The American Biology Teacher* **49** (January): 16-26.

———. 1995. "Patterns of evolution in the manus and pes of non-mammalian therapsids." *Journal of Vertebrate Paleontology* **15** (September): 615-639.

Hopson, James A., & **James W. Kitching.** 2001. "A probainognathian cynodont from South Africa and the phylogeny of nonmammalian cynodonts." *Bulletin of the Museum of Comparative Zoology* **156** (October): 5-35.

Hopson, James A., & **Hans-Dieter Sues.** 2006. "A traversodont cynodont from the Middle Triassic (Ladinian) of Baden-Wurttemberg (Germany)." *Palaontologische Zeitschrift* **80** (2): 124-129.

Horowitz, Alana. 2012. "Paul Broun: Evolution, Big Bang 'Lies Straight From The Pit Of Hell'." *Huffington Post* posting 6 October (online text at *huffingtonpost.com* accessed 10/6/2012).

Horvath, Julie E., Cassandra L. Gulden, Rhea U. Vallente, Marla Y. Eichler, Mario Ventura, John D. McPherson, Tina A. Graves, Richard K. Wilson, Stuart Schwartz, Mariano Rocchi, & **Evan E. Eichler.** 2005. "Punctuated duplication seeding events during the evolution of human chromosome 2p11." *Genome Research* **15** (July): 914-927.

Hou, Lianhai, Larry D. Martin, Zhonghe Zhou, & **Alan Feduccia.** 1996. "Early Adaptive Radiation of Birds: Evidence from Fossils from Northeastern China." *Science* **274** (15 November): 1164-1167.

Hou, Lianhai, Larry D. Martin, Zhonghe Zhou, Alan Feduccia, & **Fucheng Zhang.** 1999. "A diapsid skull in a new species of the primitive bird *Confuciusornis*." *Nature* **399** (17 June): 679-682.

Hou, Lian-hai, Zhonghe Zhou, Larry D. Martin, & **Alan Feduccia.** 1995. "A beaked bird from the Jurassic of China." *Nature* **377** (19 October): 616-618.

Hovind, Kent. 1996. "Unmasking the False Religion of Evolution." Undated transcript of lecture series (online text at *home1/gte.net/dmadh* accessed 4/5/2002, not available as of 12/5/2009).

———. 1998a. "Part 1a. The Age of the Earth." *Dr. Hovind's Creation Seminars* illustrated 2003 posting of video transcript by **Michel Snoeck** last modified 13 October 2012 (online text at *www.wiseoldgoat.com* accessed 1/8/2014).

———. 1998b. "Part 1b. The Age of the Earth." *Dr. Hovind's Creation Seminars* illustrated 2003 posting of video transcript by **Michel Snoeck** last modified 13 October 2012 (online text at *www.wiseoldgoat.com* accessed 1/8/2014).

———. 1999a. "Part 2. The Garden of Eden." *Dr. Hovind's Creation Seminars* illustrated 2003 posting of video transcript by **Michel Snoeck** last modified 13 October 2012 (online text at *www.wiseoldgoat.com* accessed 1/8/2014).

———. 1999b. "Part 3a. Dinosaurs and the Bible." *Dr. Hovind's Creation Seminars* illustrated 2003 posting of video transcript by **Michel Snoeck** last modified 7 January 2014 (online text at *www.wiseoldgoat.com* accessed 1/8/2014).

———. 1999c. "Part 3b. Dinosaurs Alive Today." *Dr. Hovind's Creation Seminars* illustrated 2003 posting of video transcript by **Michel Snoeck** last modified 13 October 2012 (online text at *www.wiseoldgoat.com* accessed 1/8/2014).

———. 1999d. "Part 4a. Lies in the Textbooks." *Dr. Hovind's Creation Seminars* illustrated 2003 posting of video transcript by **Michel Snoeck** last modified 13 October 2012 (online text at *www.wiseoldgoat.com* accessed 1/8/2014).

———. 1999e. "Part 4b. More Lies in the Textbooks." *Dr. Hovind's Creation Seminars* illustrated 2003 posting of video transcript by **Michel Snoeck** last modified 13 October 2012 (online text at *www.wiseoldgoat.com* accessed 1/8/2014).

———. 1999f. *Koinonia House* audio posting of Chuck Missler interview (August) personal transcription.

———. 2000. "A Battle Plan—Practical Steps to Combat Evolution." *Creation Science Evangelism* undated posting (online text at *www.drdino.com* accessed 1/16/2001).

———. 2002a. "Dr. Hovind's $250,000 Offer." *Creation Science Evangelism* undated posting (online text at *www.drdino.com* accessed 7/23/2002).

———. 2002b. "The Possum, the Redwood Tree, and the Kidney Bean: 'Our Ancestors'." *Creation Today* 3 August 2010 modified posting of undated original text (online text at *www.creationtoday.com* accessed 1/8/2014).

———. 2003a. "Part 5a. The Dangers of Evolution." *Dr. Hovind's Creation Seminars* illustrated 2005 posting of video transcript by **Michel Snoeck** last modified 13 October 2012 (online text at *www.wiseoldgoat.com* accessed 1/8/2014).

———. 2003b. "Part 5b. The Dangers of Evolution, continued…." *Dr. Hovind's Creation Seminars* illustrated 2005 posting of video transcript by **Michel Snoeck** last modified 13 October 2012 (online text at *www.wiseoldgoat.com* accessed 1/8/2014).

———. 2003c. "Part 5c. The Dangers of Evolution, continued…." *Dr. Hovind's Creation Seminars* illustrated 2005 posting of video transcript by **Michel Snoeck** last modified 13 October 2012 (online text at *www.wiseoldgoat.com* accessed 1/8/2014).

———. 2003d. "Part 7a. Questions and Answers Session." *Dr. Hovind's Creation Seminars* illustrated 2005 posting of video transcript by **Michel Snoeck** last modified 13 October 2012 (online text at *www.wiseoldgoat.com* accessed 1/8/2014).

———. 2006. "The Age of the Earth." *Creationism* posting by Paul Abrahmson of Tennessee Seminar 1a transcript (online text at *www.creationism.org* accessed 1/8/2014).

———. 2007a. *Are You Being Brainwashed? Propaganda in Science Textbooks*. Pensacola, FL: Creation Science Evangelism.

———. 2007b. *Seminar Notebook: A resource designed to supplement the* Creation Seminar Series. 9th ed. Pensacola, FL: Creation Science Evangelism.

———. 2010. "10 Questions for Evolutionists." *Creation Today* 29 July posting (online text at *www.creationtoday.org* accessed 8/17/2014).

———. 2012a. "Points to Ponder About the Flood and Noah's Ark." *Answers From the Evidence Bible* undated posting (online text at *www.evidencebible.com* accessed 9/13/2012).

———. 2012b. "Questions for Evolutionists." *Answers From the Evidence Bible* undated posting (online text at *www.evidencebible.com* accessed 9/13/2012).

Howe, C. J., A. C. Barbook, R. E. R. Nisbet, P. J. Lockhart, & **A. W. D. Larkum.** 2008. "The origin of plastids." *Philosophical Transactions of the Royal Society of London* **B** (Biological Sciences) **363** (27 August): 2675-2685.

Hoy, Ronald R. 2012. "Convergent Evolution in Hearing." *Science* **338** (16 November): 894-895.

Hoyle, Fred, & **Chandra Wickramasinghe.** 1993. *Our Place in the Cosmos: The Unfinished Revolution*. London: JM Dent.

Hu, Dongyu, Lianhai Hou, Lijun Zhang, & **Xing Xu.** 2009. "A pre-*Archaeopteryx* troodontid theropod from China with long feathers on the metatarsus." *Nature* **461** (1 October): 640-643.

Hu, Dongyu, Li Li, Lianhai Hou, & **Xing Xu.** 2011. "A new enantiornithine bird from the Lower Cretaceous of western Liaoning, China." *Journal of Vertebrate Paleontology* **31** (January): 154-161.

Hu, Dongyu, Xing Xu, Lianhai Hou, & **Corwin Sullivan.** 2012. "A new enantiornithine bird from the Lower Cretaceous of western Liaoning, China, and its implications for early avian evolution." *Journal of Vertebrate Paleontology* **32** (May): 639-645.

Hu, Yaoming, Jin Meng, & **James M. Clark.** 2009. "A new tritylodontid from the Upper Jurassic of Xinjiang, China." *Acta Palaeontologica Polonica* **54** (3): 385-391.

Hu, Yaoming, Jin Meng, Chuankui Li, & **Yuanqinq Wang.** 2010. "New basal eutherian mammal from the Early Cretaceous Jehol biota, Liaoning, China." *Proceedings of the Royal Society of London* **B** (Biological Sciences) **277** (22 January): 229-236.

Hu, Yaoming, Jin Meng, Yuanqinq Wang, & **Chuankui Li.** 2005. "Large Mesozoic mammals fed on young dinosaurs." *Nature* **433** (13 January): 149-152.

Hu, Yaoming, Yuanqinq Wang, Zhexi Luo, & **Chuankui Li.** 1997. "A new symmetrodont mammal from China and its implications for mammalian evolution." *Nature* **390** (13 November): 137-142.

Huang, Bau-lin, & **Susan Mackem.** 2014. "Use it or lose it." *Nature* **511** (3 July): 34-35.

Huang, Diying, Michael S. Engel, Chenyang Cai, Hao Wu, & **André Nel.** 2012. "Diverse transitional giant fleas from the Mesozoic era of China." *Nature* **483** (8 March): 201-204.

Huang, Ruijin, Qixia Zhi, Ketan Patel, Jörg Wilting, & **Bodo Christ.** 2000. "Dual origin and segmental organisation of the avian scapula." *Development* (AKA *Journal of Embryology and Experimental Morphology*) **127** (1 September): 3789-3794.

Huang, Ruiqi, Frank Hippauf, Diana Rohrbeck, Maria Haustein, Katrin Wenke, Janie Feike, Noah Sorelle, Birgit Piechulla, & **Todd J. Barkman.** 2012. "Enzyme functional evolution through improved catalysis of ancestrally nonpreferred substrates." *Proceedings of the National Academy of Sciences* **109** (21 February): 2966-2971.

Huchon, Dorothée, Ole Madsen, Mark J. J. B. Sibbald, Kai Ament, Michael J. Stanhope, François

Catzeflis, Wilfried W. de Jong, & **Emmanuel J. P. Douzery.** 2002. "Rodent Phylogeny and a Timescale for the Evolution of Glires: Evidence from an Extensive Taxon Sampling Using Three Nuclear Genes." *Molecular Biology and Evolution* **19** (July): 1053-1065.

Hudson, Richard R., & **Jerry A. Coyne.** 2002. "Mathematical Consequences of the Genealogical Species Concept." *Evolution* **56** (August): 1557-1565.

Huey, Raymond B., George W. Gilchrist, Margen L. Carlson, David Berrigan, & **Luís Serra.** 2000. "Rapid Evolution of a Geographic Cline in Size in an Introduced Fly." *Science* **287** (14 January): 308-309.

Hughes, D. P., D. J. C. Kronauer, & **J. J. Boomsma.** 2008. "Extended Phenotype: Nematodes Turn Ants into Bird-Dispersed Fruits." *Current Biology* **18** (8 April): R294-R295.

Hughes, Elizah M., John R. Wible, Michelle Spaulding, & **Zhe-Xi Luo.** 2015. "Mammalian Petrosal from the Upper jurassic Morrison Formation of Fruita, Colorado." *Annals of Carnegie Museum* **83** (15 May): 1-17.

Hughes, William O. H., A. N. M. Bot, & **J. J. Boomsma.** 2010. "Caste-specific expression of genetic variation in the size of antibiotic-producing glands of leaf-cutting ants." *Proceedings of the Royal Society of London* **B** (Biological Sciences) **277** (22 February): 609-615.

Hughes, William O. H., Seirian Sumner, Steven Van Borm, & **Jacobus J. Boomsma.** 2003. "Worker caste polymorphism has a genetic basis in *Acromyrmex* leaf-cutting ants." *Proceedings of the National Academy of Sciences* **100** (5 August): 9894-9397.

Huh, Sung-Ho, Katja Närhi, Päivi H. Lindfors, Otso Häärä, Lu Yang, David M. Ornitz, & **Marja L. Mikkola.** 2013. "Fgf20 governs formation of primary and secondary dermal condensations in developing hair follicles." *Genes & Development* **27** (15 February): 450-458.

Huminiecki, Lukasz, & **Kenneth H. Wolfe.** 2004. "Divergence of Spatial Gene Expression Profiles Following Species-Specific Gene Duplications in Human and Mouse." *Genome Research* **14** (October): 1870-1879.

Hunt, Kathleen. 1997. "Transitional Vertebrate Fossils FAQ." *The TalkOrigins Archive* 17 March updated posting (online text at *talkorigins.org* accessed 9/14/2016).

Hunter, Cornelius G. 2001. *Darwin's God: Evolution and the Problem of Evil.* Grand Rapids, MI: Brazos Press (Baker Book House).

———. 2003. *Darwin's Proof: The Triumph of Religion over Science.* Grand Rapids, MI: Brazos Press (Baker Book House).

Hunter, John P., & **Jukka Jernvall.** 1995. "The hypocone as a key innovation in mammalian evolution." *Proceedings of the National Academy of Sciences* **92** (7 November): 10718-10722.

Hurum, Jørn M., Robert Presley, & **Zofia Kielan-Jaworowka.** 1996. "The middle ear in multituberculate mammals." *Acta Palaeontologica Polonica* **41** (3): 253-275.

Huse, Scott M. 1997. *The Collapse of Evolution.* 3rd ed. Grand Rapids, MI: Baker Books.

Hutchinson, John R., & **Kevin Padian.** 1997. "Carnosauria," in Currie & Padian (1997, 94-97).

Huttenlocker, Adam. 2009. "An investigation into the cladistic relationships and monophyly of therocephalian therapsids (Amniota: Synapsida)." *Zoological Journal of the Linnean Society* **157** (December): 865-891.

Huttenlocker, Adam K., David Mazierski, & **Robert R. Reisz.** 2011. "Comparative Osteohistology of Hyperelongate Neural Spines in the Edaphosauridae (Amniota: Synapsida)." *Palaeontology* **54** (May): 573-590.

Huttenlocker, Adam K., & **Elizabeth Rega.** 2011. "The Paleobiology and Bone Microstructure of Pelycosaurian-Grade Synapsids," in Chinsamy-Turan (2011, 91-119).

Huttenlocker, Adam K., Elizabeth Rega, & **Stuart S. Sumida.** 2010. "Comparative anatomy and osteohistology of hyperelongate neural spines in the sphenacodontids *Sphenacodon* and *Dimetrodon* (Amniota: Synapsida)." *Journal of Morphology* **271** (December): 1407-1421.

Huttenlocker, Adam K., & **Christian A. Sidor.** 2016. "The first karenitid (Therapsida, Therocephalia) from the upper Permian of Gondwana and the biogeography of Permo-Triassic therocephalians." *Journal of Vertebrate Paleontology* **36** (July): e1111897.

Ichihashi, Yasunori, José Antonio Aguilar-Martínez, Moran Farhi, Daniel H. Chitwood, Ravi Kumar, Lee V. Millon, Jie Peng, Julin N. Maloof, & **Neelima R. Sinha.** 2014. "Evolutionary developmental transcriptomics reveals a gene network module regulating interspecific diversity in plant leaf shape." *Proceedings of the National Academy of Sciences* **111** (24 June): E2616-E2621.

IJdo, J. W., A. Baldini, D. C. Ward, S. T. Reeders, & **R. A. Wells.** 1991. "Origin of human chromosome 2: an ancestral telomere-telomere fusion." *Proceedings of the National Academy of*

Sciences **88** (15 October): 9051-9055.

Imada, Katsumi, Tohru Minimino, Yumiko Uchida, Miki Kinoshita, & **Keiichi Namba.** 2016. "Insight into the flagella type III export revealed by the complex structure of the type III ATPase and its regulator." *Proceedings of the National Academy of Sciences* **113** (29 March): 3633-3638.

Ingermanson, Randall. 2004. "A Review of The Bible Code II by Michael Drosnin." *Talk Reason* 6 December posting (online text at *talkreason.org* accessed 12/6/2004).

Ingersoll, Julie. 2013. "Howard Phillips, Founding Father of Religious Right, Has Died." *Religion Dispatches* 21 April (online text at *religiondispatches.org* accessed 4/27/2013).

Institute for Biblical & Scientific Studies. 2008. "Other Views: Carl Baugh." *IBSS* posting last updated 30 April (online text at *www.bibleandscience.com* accessed 5/19/2011).

Iordansky, N. N. 2010. "Pterygoideus Muscles and Other Jaw Adductors in Amphibians and Reptiles." *Biology Bulletin* (Russia) **37** (December): 905-914.

Irmis, Randall B., Sterling J. Nesbitt, Kevin Padian, Nathan D. Smith, Alan H. Turner, Daniel Woody, & **Alex Downs.** 2007. "A Late Triassic Dinosauromorph Assemblage from New Mexico and the Rise of Dinosaurs." *Science* **317** (20 July): 358-361.

Irmis, Randall B., Sterling J. Nesbitt, & **Hans-Dieter Sues.** 2013. "Early Crocodylomorpha." *Geological Society, London, Special Publications* **379**: 257-302.

Irmis, Randall B., & **Jessica H. Whiteside.** 2012. "Delayed recovery of non-marine tetrapods after the end-Permian mass extinction tracks global carbon cycle." *Proceedings of the Royal Society of London* **B** (Biological Sciences) **279** (7 April): 1310-1318.

Isaacs, Darek. 2009. *The Extinction of Evolution.* Alachua, FL: Bridge-Logos.

———. 2010. *Dragons or Dinosaurs? Creation or Evolution?* Alachua, FL: Bridge-Logos.

Ito, Hisashi, & **Ayumi Tanaka.** 2014. "Evolution of a New Chlorophyll Metabolic Pathway Driven by the Dynamic Changes in Enzyme Promiscuous Activity." *Plant Cell Physiology* **55** (March): 593-603.

Ivakhnenko, M. F. 2013. "Cranial morphology of *Dvinia prima* amalitzky (Cynodontia, Theromorpha)." *Paleontological Journal* **47** (March): 210-222.

Izuma, Keise, Madoka Matsumoto, Kou Murayama, Kazuyuki Samejima, Norihiro Sadato, & **Kenji Matsumoto.** 2010. "Neural correlates of cognitive dissonance and choice-induced preference change." *Proceedings of the National Academy of Sciences* **107** (21 December): 22014-22019.

Jablonski, David. 1998. "Geographic Variation in the Molluscan Recovery from the End-Cretaceous Extinction." *Science* **279** (27 February): 1327-1330.

———. 2001. "Lessons from the past: Evolutionary impacts of mass extinctions." *Proceedings of the National Academy of Sciences* **98** (8 May): 5393-5398.

———. 2002. "Survival without recovery after mass extinctions." *Proceedings of the National Academy of Sciences* **99** (11 June): 8139-8144.

Janis, Christine. 1993. "Victors By Default," in Gould (1993, 171-172).

———. 1994. "The Sabertooth's Repeat Performances." *Natural History* **103** (April): 78-83.

Jankowski, Roger. 2013. *The Evo-Devo Origin of the Nose, Anterior Skull Base and Midface.* New York: Springer.

Jarcho, Johanna M., Elliot T. Berkman, & **Matthew D. Lieberman.** 2011. "The neural basis of rationalization: cognitive dissonance reduction during decision-making." *Social Cognitive & Affective Neuroscience* **6** (September): 460-467.

Jasinoski, Sandra C., Fernando Abdala, & **Vincent Fernandez.** 2015. "Ontogeny of the Early Triassic Cynodont *Thrinaxodon liorhinus* (Therapsida): Cranial Morphology." *The Anatomical Record: Advances in Integrative Anatomy and Evolutionary Biology* **298** (August): 1440-1464.

Jeffery, Jonathan E., Olaf R. P. Bininda-Emonds, Michael I. Coates, & **Michael K. Richardson.** 2002. "Analyzing evolutionary patterns in amniote embryonic development." *Evolution & Development* **4** (July-August): 292-302.

Jékely, Gáspár, ed. 2007. *Eukaryotic Membranes and Cytosketon: Origins and Evolution.* Austin, TX: Landes Bioscience.

Jelesko, John G., Ryan Harper, Masaki Furuya, & **Wilhelm Gruissem.** 1999. "Rare germinal unequal crossing-over leading to recombinant gene formation and gene duplication in *Arabidopsis.*" *Proceedings of the National Academy of Sciences* **96** (31 August): 10302-10307.

Jenkins, F. A. Jr., & **F. R. Parrington.** 1976. "The postcranial skeletons of the Triassic mammals *Eozostrodon, Megazostrodon* and *Erythrotherium.*" *Philosophical Transactions of the Royal Society of London* **B** (Biological Sciences) **273** (26 February): 387-431.

Jensen, Bjarke, Tobias Wang, Vincent M. Christoffels, & **Antoon F. M. Moormon.** 2013. "Evolution and development of the building plan of the vertebrate heart." *Biochimica et Biophysica Acta* (Molecular Cell Research) **1833** (April): 783-794.

Jernvall, Jukka. 2000. "Linking development with generation of novelty in mammalian teeth." *Proceedings of the National Academy of Sciences* **97** (14 March): 2641-2645.

Jernvall, Jukka, Soile V. E. Keränen, & **Irma Thesleff.** 2000. "Evolutionary modification of development in mammalian teeth: Quantifying gene expression patterns and topography." *Proceedings of the National Academy of Sciences* **97** (19 December): 14444-14448.

Ji, Qiang, Philip J. Currie, Mark A. Norell, & **Shu-An Ji.** 1998. "Two feathered dinosaurs from northeastern China." *Nature* **393** (25 June): 753-761.

Ji, Qiang, Zhe-Xi Luo, & **Shu-an Ji.** 1999. "A Chinese triconodont mammal and mosaic evolution of the mammalian skeleton." *Nature* **398** (25 March): 326-330.

Ji, Qiang, Zhe-Xi Luo, Chong-Xi Yuan, & **Alan R. Tabrum.** 2006. "A Swimming Mammaliaform from the Middle Jurassic and Ecomorphological Diversification of Early Mammals." *Science* **311** (24 February): 1123-1127.

Ji, Qiang, Zhe-Xi Luo, Chong-Xi Yuan, John R. Wible, Jian-Ping Zhang, & **Justin A. Georgi.** 2002. "The earliest known eutherian mammal." *Nature* **416** (25 April): 816-822.

Ji, Qiang, Zhe-Xi Luo, Xingliao Zhang, Chong-Xi Yuan, & **Li Xu.** 2009. "Evolutionary Development of the Middle Ear in Mesozoic Therian Mammals." *Science* **326** (9 October): 278-281.

Ji, Qiang, Mark A. Norell, Ke-Qin Gao, Shu-An Ji, & **Dong Ren.** 2001. "The distribution of integumentary structures in a feathered dinosaur." *Nature* **410** (26 April): 1084-1088.

Jiang, Hongying, Na Li, Vivek Gopalan, Martine M. Zilversmit, Sudhir Varma, Vijayaraj Nagarajan, Jian Li, Jianbing Mu, Karen Hayton, Bruce Henschen, Ming Yi, Robert Stephens, Gilean McVean, Philip Awadalla, Thomas E. Wellems, & **Xin-zhuan Su.** 2011. "High recombination rates and hotspots in a *Plasmodium falciparum* genetic cross." *Genome Biology* (online @ genomebiology.com) **12** (4 April): R33.

Jiang, Ting-Xin, Han-Sung Jung, Randall B. Widelitz, & **Cheng-Ming Chuong.** 1999. "Self-organization of periodic patterns by dissociated feather mesenchymal cells and the regulation of size, number and spacing of primordia." *Development* (AKA *Journal of Embryology and Experimental Morphology*) **126** (15 November): 4997-5009.

Jiggins, Chris D., Igor Emelianov, & **James Mallet.** 2005. "Assortive Mating and Speciation as Pleiotropic Effects of Ecological Adaptation: Examples in Moths and Butterflies," in Fellowes *et al.* (2005, 451-473).

Johnson, Brian R., Marek L. Borowiec, Joanna C. Chiu, Ernest K. Lee, Joel Atallah, & **Philip S. Ward.** 2013. "Phylogenomics Resolves Evolutionary Relationships among Ants, Bees, and Wasps." *Current Biology* **23** (21 October): 2058-2062.

Johnson, Phillip E. 1991. *Darwin on Trial.* Washington, D.C.: Regnery Gateway.

———. 1993a. *Darwin on Trial.* 2nd ed. Downers Grove, IL: InterVarsity Press.

———. 1993b. "God and Evolution: An Exchange (II)." *First Things* (June/July): 38-41.

———. 1995. *Reason in the Balance: The Case Against NATURALISM in Science, Law & Education.* Downers Grove, IL: InterVarsity Press.

———. 1997. *Defeating Darwinism by Opening Minds.* Downers Grove, IL: InterVarsity Press.

———. 1998. "What Would Newton Do?" *First Things* (November): 25-31.

———. 1999. "The Church of Darwin." *Wall Street Journal* (16 August): A14.

Johnston, Peter. 2014. "Homology of the Jaw Muscles in Lizards and Snakes—A Solution from a Comparative Gnathostome Approach." *The Anatomical Record: Advances in Integrative Anatomy and Evolutionary Biology* **297** (March): 574-585.

Jones, Marc E. H., Cajsa Lisa Anderson, Christy A. Hipsley, Johannes Müller, Susan E. Evans, & **Rainer R. Schoch.** 2013. "Integration of molecules and new fossils supports a Triassic origin for Lepidosauria (lizards, snakes, and tuatara)." *BMC Evolutionary Biology* (online @ biomedcentral.com) **13** (25 September): 208.

Jones, Marc E. H., Alan J. D. Tennyson, Jennifer P. Worthy, Susan E. Evans, & **Trevor H. Worthy.** 2009. "A sphenodontine (Rhynchocephalia) from the Miocene of New Zealand and palaeobiogegraphy of the tuatara (*Sphenodon*)." *Proceedings of the Royal Society of London* **B** (Biological Sciences) **276** (7 April): 1385-1390.

Jones, Terry D., James O. Farlow, John H. Ruben, Donald M. Henderson, & **Willem J. Hillenius.** 2000. "Cursoriality in bipedal archosaurs." *Nature* **406** (17 August): 716-718.

Jones, Terry D., John A. Ruben, Larry D. Martin, Evgeny N. Kurochkin, Alan Feduccia, Paul F. A.

Maderson, Willem J. Hillenius, Nicholas R. Geist, & **Vladimir Alifanov.** 2000. "Nonavian Feathers in a Late Triassic Archosaur." *Science* **288** (23 June): 2202-2205.

Joyce, Walter G., Spencer G. Lucas, Torsten M. Scheyer, Andrew B. Heckert, & **Adrian P. Hunt.** 2009. "A thin-shelled reptile from the Late Triassic of North America and the origin of the turtle shell." *Proceedings of the Royal Society of London* **B** (Biological Sciences) **276** (7 February): 507-513.

Joyce, Walter G., Rainer R. Schoch, & **Tyler R. Lyson.** 2013. "The girdles of the oldest fossil turtle, *Proterochersis robusta*, and the age of the turtle crown." *BMC Evolutionary Biology* (online @ biomedcentral.com) **13** (6 December): 266.

Joysey, Kenneth Alan, & **A. E. Friday**, eds. 1982. *Problems of Phylogenetic Reconstruction*. The Systematics Association Special Volume 21. London: Academic Press.

Joysey, K. A., & **T. S. Kemp.** 1972. *Studies in Vertebrate Evolution: Essays Presented to F. R. Parrington*. Edinburgh: Oliver & Boyd.

Judson, Olivia. 2002. *Dr. Tatiana's Sex Advice to All Creation.* New York: Metropolitan Books (Henry Holt & Co.).

Julian, Glennis E., Jennifer H. Fewell, Jürgen Gadau, Robert A. Johnson, & **Debbie Larrabee.** 2002. "Genetic determination of the queen caste in an ant hybrid zone." *Proceedings of the National Academy of Sciences* **99** (11 June): 8157-8160.

Jung, Han-Sung, Philippa H. Francis-West, Randall B. Widelitz, Ting-Xin Jiang, Sheree Ting-Berreth, Cheryll Tickle, Lewis Wolpert, & **Cheng-Ming Chuong.** 1998. "Local Inhibitory Action of BMPs and Their Relationships with Activators in Feather Formation: Implications for Periodic Patterning." *Developmental Biology* **196** (1 April): 11-23.

Kaas, Jon H. 2013. "The Evolution of Brains from Early Mammals to Humans." *Wiley Interdisciplinary Reviews: Cognitive Science* **4** (January): 35-38.

Kabbany, Jennifer. 2014. "LAWSUIT: University Fired Scientist For Finding Soft Tissue on Dinosaur Horn." *The College Fix* 25 July posting (online text at *www.thecollegefix.com* accessed 7/28/2014).

Kaltenegger, Elisabeth, Eckart Eich, & **Dietrich Ober.** 2013. "Evolution of Homospermidine Synthase in the Convolvulaceae: A Story of Gene Duplication, Gene Loss, and Periods of Various Selection Pressures." *The Plant Cell* **25** (April): 1213-1227.

Kammerer, Christian F., Kenneth D. Angielczyk, & **Jörg Fröbisch.** 2011. "A Comprehensive Taxonomic Revision of Dicynodon (Therapsida, Ammodontia) and Its Implications for Dicynodont Phylogeny." *Journal of Vertebrate Paleontology* **31** (December Supplement 1): 1-158.

———, eds. 2014. *Early Evolutionary History of the Synapsids.* New York: Springer.

Kammerer, Christian F., John J. Flynn, Lovasoa Ranivoharimanana, & **André R. Wyss.** 2012. "Ontogeny in the Malagasy Traversodontid *Dadadon isaloi* and a Reconstruction of its Phylogenetic Relationships." *Fieldiana Life and Earth Sciences* **5** (October): 112-125.

Kangas, Aapo T., Alistair R. Evans, Irma Thesleff, & **Jukka Jernvall.** 2004. "Nonindependence of mammalian dental characters." *Nature* **432** (11 November): 211-214.

Kardong, Kenneth V. 2005. *Vertebrates: Comparative Anatomy, Function, Evolution*. 4th ed. New York: McGraw-Hill College.

Karten, Harvey J. 1997. "Evolutionary developmental biology meets the brain: The origins of mammalian cortex." *Proceedings of the National Academy of Sciences* **94** (1 April): 2800-2804.

———. 2015. "Vertebrate brains and evolutionary connectomics: on the origins of the mammalian 'neocortex'." *Philosophical Transactions of the Royal Society of London* **B** (Biological Sciences) **370** (19 December): 20150060.

Kavanagh, Kathryn D., Alistair R. Evans, & **Jukka Jernvall.** 2007. "Predicting evolutionary patterns of mammalian teeth from development." *Nature* **449** (27 September): 427-432.

Kavanagh, Kathryn D., Oren Shoval, Benjamin B. Winslow, Uri Alon, Brian P. Leary, Akinori Kan, & **Clifford J. Tabin.** 2013. "Developmental bias in the evolution of phalanges." *Proceedings of the National Academy of Sciences* **110** (5 November): 18190-18195.

Kawasaki, Kazuhiko, Anne-Gaelle Lafont, & **Jean-Yves Sire.** 2011. "The Evolution of Milk Casein Genes from Tooth Genes Before the Origin of Mammals." *Molecular Biology and Evolution* **28** (July): 2053-2061.

Keeling, Patrick J. 2010. "The endosymbiotic origin, diversification and fate of plastids." *Philosophical Transactions of the Royal Society of London* **B** (Biological Sciences) **365** (12 March): 729-748.

———. 2014. "The Impact of History on Our Perception of Evolutionary Events: Endosymbiosis and the Origin of Eukaryotic Complexity." *Cold Spring Harbor Perspectives in Biology* **6** (February): a016196.

Keller, G., A. Sahni, & **S. Bajpai.** 2009. "Deccan volcanism, the KT mass extinction and dinosaurs." Indian Academy of Sciences *Journal of Biosciences* **34** (November): 709-728.

Keller, Roberto A., Christian Peters, & **Patricia Beldade.** 2014. "Evolution of thorax architecture in ant castes highlights trade-off between flight and ground behaviors." *eLife* (online @ elifesciences.org) **3** (7 January): e01539.

Kelley, Joanna L., Jan E. Aagaard, Michael J. MacCoss, & **Willie J. Swanson.** 2010. "Functional Diversification and Evolution of Antifreeze Proteins in the Antarctic Fish *Lycodichthys dearborni*." *Journal of Molecular Evolution* **71** (August): 111-118.

Kellner, Alexander W. A., Xiaolin Wang, Helmut Tischlinger, Diogenes de Almeida Campos, David W. E. Hone, & **Xi Meng.** 2010. "The soft tissue of *Jeholopterus* (Pterosauria, Anurognathidae, Batrachognathinae) and the structure of the pterosaur wing membrane." *Proceedings of the Royal Society of London* **B** (Biological Sciences) **277** (22 January): 321-329.

Kemp, Tom S. 1979. "The Primitive Cynodont *Procynosuchus*: Functional Anatomy of the Skull and Relationships." *Philosophical Transactions of the Royal Society of London* **B** (Biological Sciences) **285** (14 February): 73-122.

———. 1980. "The Primitive Cynodont *Procynosuchus*: Structure, Function and Evolution of the Postcranial Skeleton." *Philosophical Transactions of the Royal Society of London* **B** (Biological Sciences) **288** (7 January): 217-258.

———. 1982a. *Mammal-like Reptiles and the Origin of Mammals*. London: Academic Press.

———. 1982b. "The reptiles that became mammals." *New Scientist* **93** (4 March): 581-584.

———. 1986. "The Skeleton of a Baurioid Therocephalian Therapsid from the Lower Triassic (*Lystrosaurus* Zone) of South Africa." *Journal of Vertebrate Paleontology* **6** (September): 215-232.

———. 2005. *The Origin and Evolution of Mammals*. Oxford: Oxford University Press.

———. 2006a. "The origin and early radiation of the therapsid mammal-like reptiles: a palaeobiological hypothesis." *Journal of Evolutionary Biology* **19** (July): 1231-1247.

———. 2006b. "The origin of mammalian endothermy: a paradigm for the evolution of complex biological structure." *Zoological Journal of the Linnean Society* **147** (August): 473-488.

———. 2007a. "The origin of higher taxa: macroevolutionary processes, and the case of the mammals." *Acta Zoologica* (Stockholm) **88** (January): 3-22.

———. 2007b. "The concept of correlated progression as the basis for the evolutionary origin of major new taxa." *Proceedings of the Royal Society of London* **B** (Biological Sciences) **274** (7 July): 1667-1673.

———. 2007c. "Acoustic transformer function of the postdentary bones and quadrate of a nonmammalian cynodont." *Journal of Vertebrate Paleontology* **27** (June): 431-441.

———. 2009a. "The endocranial cavity of a nonmammalian cynodont, *Chiniquodon theotenicus*, and its implications for the origin of the mammalian brain." *Journal of Vertebrate Paleontology* **29** (December): 1188-1198.

———. 2009b. "Phylogenetic interrelationships of therapsids: testing for polytomy." *Palaeontologia africana* **44** (December): 1-12.

———. 2011. "The Origin and Radiation of Therapsids," in Chinsamy-Turan (2011, 3-28).

Kenyon, Cynthia. 1994. "If Birds Can Fly, Why Can't We? Homeotic Genes and Evolution." *Cell* **78** (29 July): 175-180.

Kerby, Carl. 2011. "The 'phylogenetic charts' prove evolution, right!" *Reasons for Hope* 30 May posting (online text at *www.rforh.com*accessed 5/10/2014).

———. 2013. "Phylogenetic Charts: Dinosaurs!" *Kirk Cameron* 21 January posting (online text at *kirkcameron.com* accessed 5/10/2014).

Kermack, Doris M., K. A. Kermack, & **Frances Mussett.** 1968. "The Welsh pantothere *Kuehneotherium praecursoris*." *Zoological Journal of the Linnean Society* **47** (April): 407-423.

Kermack, K. A., Patricia M. Lees, & **Frances Mussett.** 1965. "*Aegialodon dawsoni,* A New Trituberculosectorial Tooth from the Lower Wealdon." *Proceedings of the Royal Society of London* **B** (Biological Sciences) **162** (27 July): 535-554.

Kermack, K. A., Frances Mussett, & **H. W. Rigney.** 1973. "The lower jaw of *Morganucodon*." *Zoological Journal of the Linnean Society* **53** (September): 87-175.

———. 1981. "The skull of *Morganucodon*." *Zoological Journal of the Linnean Society* **71** (January): 1-158.

Kern, Sally. 2008. "House Bill 2211." Introduced Oklahoma House legislation (online text at *webserver1.lsb.state.ok.us* accessed 10/20/2008).

———. 2011. "House Bill 1551." Introduced Oklahoma House legislation (online text at *webserver1.lsb.state.ok.us* accessed 1/24/2011).

———. 2016. "House Bill 3045." Introduced Oklahoma House legislation (online pdf at *webserver1.lsb.state.ok.us* accessed 2/5/2016).

Kessler, Sharon, & **Neelima Sinha.** 2004. "Shaping up: the genetic control of leaf shape." *Current Opinion in Plant Biology* **7** (February): 65-72.

Kherdjemil, Yacine, Robert L. Lalonde, Rushikesh Sheth, Annie Dumouchel, Gemma de Martino, Kyriel M. Pineault, Deneen M. Wellik, H. Scott Stadler, Marie Andrée Akimenko, & **Marie Kmita.** 2016. "Evolution of *Hoxa11* regulation in vertebrates is linked to the pentadactyl state." *Nature* **539** (3 November): 89-92.

Kielan-Jaworowska, Zofia. 1969. "Evolution of the therian mammals in the Late Cretaceous of Asia. Part II. Postcranial skeleton in *Kennalestes* and *Asioryctes*." *Palaeontologia Polonica* **37**: 65-86.

———. 1992. "Interrelationships of Mesozoic Mammals." *Historical Biology* **6** (3): 185-202.

Kielan-Jaworowska, Zofia, Richard L. Cifelli, & **Zhe-Xi Luo.** 2004. *Mammals from the Age of Dinosaurs: Origins, Evolution, and Structure.* New York: Columbia University Press.

Kielan-Jaworowska, Zofia, & **Terry E. Lancaster.** 2004. "A new reconstruction of multitberculate endocranial casts and encephalization quotient of *Kryptobaatar*." *Acta Palaeontologica Polonica* **49** (2): 177-188.

Kim, Hyi-Gyung, Sterling C. Keeley, Peter S. Vroom, & **Robert K. Jansen.** 1998. "Molecular evidence for an African origin of the Hawaiian endemic *Hesperomannia (Asteraceae)*." *Proceedings of the National Academy of Sciences* **95** (22 December): 15440-15445.

Kim, Jung-Woong, Hyun-Jin Yang, Adam Phillip Oel, Matthew John Brooks, Li Jia, David Charles Plachetzki, We Li, William Ted Allison, & **Anand Swaroop.** 2016. "Recruitment of Rod Photoreceptors from Short-Wavelength-Sensitive Cones during the Evolution of Nocturnal Vision in Mammals." *Developmental Cell* **37** (20 June): 520-532.

Kim, Ki-Joong, & **Robert K. Jansen.** 1995. "*ndhF* sequence evolution and the major clades in the sunflower family." *Proceedings of the National Academy of Sciences* **92** (24 October): 10379-10383.

King, Barbara J. 1994. *The Information Continuum: Evolution of Social Information Transfer in Monkeys, Apes, and Hominids.* Santa Fe, NM: School of American Research Press.

———. 2007. *Evolving God: A Provocative View of the Origins of Religion.* New York: Doubleday.

———. 2013. *How Animals Grieve.* Chicago: University of Chicago Press.

———. 2016a. "There's No Controversy: Let's Stop Failing Our Children On Evolution." NPR *13.7 cosmos & culture* blog 28 July posting (online text at *www.npr.org* accessed 9/17/2014).

———. 2016b. "When Science Stands Up To Creationism." NPR *13.7 cosmos & culture* blog 11 August posting (online text at *www.npr.org* accessed 9/17/2014).

King, G. M., B. W. Oelofsen, & **B. S. Rubridge.** 1989. "The evolution of the dicynodont feeding system." *Zoological Journal of the Linnean Society* **96** (June): 185-211.

King, Gillian M., & **Bruce S. Rubidge.** 1993. "A taxonomic revision of small dicynodonts with postcanine teeth." *Zoological Journal of the Linnean Society* **107** (February): 131-154.

King, Nicole, & **Sean B. Carroll.** 2001. "A receptor tyrosine kinase from choanoflagellates: Molecular insights into early animal evolution." *Proceedings of the National Academy of Sciences* **98** (18 December): 15032-15037.

King, Nicole, Christopher T. Hittinger, & **Sean B. Carroll.** 2003. "Evolution of Key Cell Signaling and Adhesion Families Predates Animal Origins." *Science* **301** (18 July): 361-363.

King, Nicole, M. Jody Westbrook, Susan L. Young, Alan Kuo, Monika Abedin, Jarrod Chapman, Stephen Fairclough, Uffe Hellstein, Yoh Isogai, Ivica Letunic, Michael Marr, David Pincus, Nicholas Putnam, Antonis Rokas, Kevin J. Wright, Richard Zuzow, William Dirks, Matthew Good, David Goodstein, Derek Lemons, Wanqing Li, Jessica B. Lyons, Andrea Morris, Scott Nichols, Daniel J. Richter, Asaf Salamov, JGI Sequencing, Peer Bork, Wendell A. Lim, Gerard Manning, W. Todd Miller, William McGinnis, Harris Shapiro, Robert Tjian, Igor V. Grigoriev, & **Daniel Rokhsar.** 2008. "The genome of the choanoflagellate *Monosiga brevicollis* and the origin of the metazoans." *Nature* **451** (14 February): 783-788.

Kirchner, James W. 2002. "Evolutionary speed limits inferred from the fossil record." *Nature* **415** (3 January): 65-68.

Kishimoto, Jiro, Robert E. Burgeson, & **Bruce A. Morgan.** 2000. "Wnt signaling maintains the hair-inducing activity of the dermal papilla." *Genes & Development* **14** (15 May): 1181-1185.

Kitazawa, Taro, Masaki Takechi, Tatsuya Hirasawa, Noritaka Adachi, Nicolas Narboux-Nême, Hideaki Kume, Kazuhiro Maeda, Tamami Hirai, Sachiko Miyagawa-Tomita, Yukiko Kurihara, Jiro Hitomi, Giovanni Levu, Shigeru Kuratani, & **Hiroki Kurihara.** 2015. "Developmental genetic bases behind the independent origin of the tympanic membrane in mammals and diapsids." *Nature Communications* (online @ nature.com) **6** (22 April): 6853.

Kitcher, Philip. 1982. *Abusing Science: The Case Against Creationism.* 1998 pb ed. Cambridge, MA: MIT Press.

Klinghoffer, David. 2012. "'Reporting' on the David Coppedge Trial, *Time* Magazine Joins the JPL Legal Team." Discovery Institute *Evolution News & Views* for 19 March (online text at *evolutionnews.org* accessed 3/5/2013).

———. 2015. "In *The New Yorker*, Tom Wolfe Compares Persecution of Intelligent Design Advocates to the 'Spanish Inquisition'." Discovery Institute *Evolution News & Views* 28 July posting (online text at *evolutionnews.org* accessed 1/2/2016).

———. 2016a. "A 'Nightmarish Scenario' for Darwinism—the Curse of Non-Adaptive Order." Discovery Institute *Evolution News & Views* 16 March posting (online text at *evolutionnews.org* accessed 4/24/2016).

———. 2016b. "*World* Magazine Tells David Coppedge's Powerful Story." Discovery Institute *Evolution News & Views* 15 May posting (online text at *evolutionnews.org* accessed 6/6/2016).

———. 2016c. "Is Intelligent Design a Peasant's Revolt?" Discovery Institute *Evolution News & Views* 5 July posting (online text at *evolutionnews.org* accessed 6/29/2016).

———. 2016d. "Helpful Survey of the Best ID Literature Illuminates Distortions of the Evolution Debate." Discovery Institute *Evolution News & Views* 15 July posting (online text at *evolutionnews.org* accessed 7/23/2016).

———. 2016e. "Darwin Apologists Could Take a Lesson from This Christian Apologist." Discovery Institute *Evolution News & Views* 4 August posting (online text at *evolutionnews.org* accessed 9/2/2016).

———. 2016f. "In *The Kingdom of Speech*, Tom Wolfe Tells the Story of Evolution's Epic Tumble." Discovery Institute *Evolution News & Views* 30 August posting (online text at *evolutionnews.org* accessed 9/2/2016).

———. 2016g. "William Dembski on the 'Science vs. Science' Debate, the 'Two Strands' in ID Research, and More." Discovery Institute *Evolution News & Views* 8 September posting (online text at *evolutionnews.org* accessed 9/17/2016).

———. 2016h. "Intelligent Design Is the Backdrop for a Dan Brown-Style Thriller—Bruce Buff's *The Soul of the Matter*." Discovery Institute *Evolution News & Views* 13 September posting (online text at *evolutionnews.org* accessed 9/17/2016).

———. 2016bi. "London *Spectator* Hails Denton's *Evolution: Still a Theory in Crisis* as a 'Best book' of 2016." Discovery Institute *Evolution News & Views* 21 November posting (online text at *evolutionnews.org* accessed 11/28/2016).

Klotz, Catherine, Marie-Christine Dabauvalle, Michel Paintrand, Thomas Weber, Michel Bornens, & **Eric Karsenti.** 1990. "Parthenogenesis in *Xenopus* Eggs Requires Centrosome Integrity." *The Journal of Cell Biology* (AKA *The Journal of Biophysics and Biochemical Cytology*) **110** (February): 405-415.

Kmita, Marie, Nadine Fraudeau, Yann Hérault, & **Denis Duboule.** 2002. "Serial deletions and duplications suggest a mechanism for the collinearity of *Hoxd* genes in limbs." *Nature* **420** (14 November): 145-150.

Koentges, Georgy, & **Toshiyuki Matsuoka.** 2002. "Jaws of the Fates." *Science* **298** (11 October): 371, 373.

Konstantinidis, P., & **M. P. Harris.** 2011. "Same but Different: Ontogeny and Evolution of the *Musculus adductor mandibulae* in the Tetradontiformes." *Journal of Experimental Zoology* **316** (15 January): 10-20.

Kopplin, Zack. 2016. "School Teaching Creationism With Video From Islamic Sex Cult." *The Daily Beast* 15 May (online text at *www.thedailybeast.com* accessed 5/16/2016).

Korsinczky, Michael, Nanhua Chen, Barbara Kotecka, Allan Saul, Karl Rieckmann, & **Qin Cheng.** 2000. "Mutations in *Plasmodium falciparum* Cytochrome *b* That Are Associated with Atovaquone Resistance Are Located at a Putative Drug-Binding Site." *Antimicrobial Agents and Chemotherapy* **44** (August): 2100-2108.

Koshiba-Takeuchi, Kazuko, Alessandro D. Mori, Bogac L. Kaynak, Judith Cebra-Thomas, Tatyana

Sukonnik, Romain O. Georges, Stephany Latham, Laural Beck, R. Mark Henkelman, Brian L. Black, Eric N. Olson, Juli Wade, Jun K. Takeuchi, Mona Nemer. Scott F. Gilbert, & Benoit G. Bruneau. 2009. "Reptilian heart development and the molecular basis of cardiac chamber evolution." *Nature* **461** (3 September): 95-98.

Kowald, Axel, & Tom B. I. Kirkwood. 2011. "Evolution of the mitochondrial fusion-fission cycle and its role in aging." *Proceedings of the National Academy of Sciences* **108** (21 June): 10237-10242.

Kramerov, D. A., & N. S. Vassetzky. 2011. "Origin and evolution of SINEs in eukaryotic genomes." *Heredity* **107** (December): 487-495.

Krause, David W., Simone Hoffmann, John R. Wible, E. Christopher Kirk, Julia A. Schultz, Wighart von Koenigswald, Joseph R. Groenke, James B. Rossie, Patrick M. O'Connor, Erik R. Seiffert, Elizabeth R. Dumont, Waymon L. Holloway, Raymond R. Rogers, Lydia J. Rahantarisoa, Addison D. Kemp, & Haingoson Andriamialison. 2014. "First cranial remains of a gondwanatherian mammal reveal remarkable mosaicism." *Nature* **515** (27 November): 512-517.

Kriegs, Jan Ole, Gennady Churakov, Jerzy Jurka, Jürgen Brosius, & Jürgen Schmitz. 2007. "Evolutionary history of 7SL RNA-derived SINEs in Supraprimates." *Trends in Genetics* **23** (April): 158-161.

Kriegs, Jan Ole, Jürgen Schmitz, Wojciech Makalowski, & Jürgen Brosius. 2005. "Does the AD7c-NTP locus encode a protein?" *Biochimica et Biophysica Acta* (Gene Structure and Expression) **1727** (21 January): 1-4.

Kromik, Andreas, Reiner Ulrich, Marian Kusenda, Andreas Tipold, Veronika M. Stein, Maren Hellige, Peter Dziallas, Frieder Hadlich, Philipp Widmann, Tom Goldammer, Wolfgang Baumgärtner, Jürgen Rehage, Dierck Segelke, Rosemarie Weikard, & Christa Kühn. 2015. "The Mammalian Cervical Vertebrae Blueprint Depends on the *T (brachyury)* Gene." *Genetics* **199** (March): 873-883.

Kruger, Ashley, Bruce S. Rubidge, Fernando Abdala, Elizabeth Gomani Chindebvu, & Louis L. Jacobs. 2015. "*Lende chiweta*, a New Therapsid from Malawi, and Its Influence on Burnetiamorph Phylogeny and Biogeography." *Journal of Vertebrate Paleontology* **36** (November): e1008698.

Ku, Maurice S. B., Yuriko Kano-Murakami, & Makoto Matsuoka. 1996. "Evolution and Expression of C_4 Photosynthesis Genes." *Plant Physiology* **111** (August): 949-957.

Kuban, Glen J. 2011. "Stegosaurus Carving on a Cambodian Temple?" *Kuban's Paluxy Website* posting (online text at *paleo.cc* accessed 11/16/2011).

Kugler, Charles. 1978. "Pygidial Glands in the Myrmicine Ants (Hymenoptera, Formicidae)." *Insectes Sociaux, Paris* **25** (3): 267-274.

Kuijper, Sanne, Annemiek Beverdam, Carla Kroon, Antje Brouwer, Sophie Candille, Gregory Barsh, & Frits Meijlink. 2005. "Genetics of shoulder girdle formation: roles of Tbx15 and aristaless-like genes." *Development* (AKA *Journal of Embryology and Experimental Morphology*) **132** (1 April): 1601-1610.

Kulessa, Holger, Gail Turk, & Brigid L. M. Hogan. 2000. "Inhibition of Bmp signaling affects growth and differentiation in the anagen hair follicle." *The EMBO Journal* **19** (15 December): 6664-6674.

Kundrát, Martin. 2004. "When Did Theropods Become Feathered?—Evidence for Pre-Archaeopteryx Feathery Appendages." *Journal of Experimental Zoology* **302B** (15 July): 355-364.

———. 2007. "Avian-like attributes of a virtual brain model of the oviraptorid theropod *Conchoraptor gracilis*." *Naturwissenschaften* **94** (June): 499-504.

Kundrát, Martin, Václav Seichert, Anthony P. Russell, & Karel Smetana Jr. 2002. "Pentadactyl Pattern of the Avian Wing Autopodium and Pyramid Reduction Hypothesis." *Journal of Experimental Zoology* **294** (15 August): 152-159.

Kurkin, A. A. 2011. "Permian Amonodonts: Paleobiogeography and Distribution of the Group." *Paleontological Journal* **45** (July): 432-444.

Kusky, T. M., B. F. Windley, & M.-G. Zhai. 2007. "Tectonic evolution of the North China Block: from orogen to craton to orogen." *Geological Society, London, Special Publications* **280**: 1-34.

Kwasniak, Janet. 2014. "Do we have a reptilian brain?" *Neuro-patch* 9 March posting (online text at *dyslectem.info* accessed 3/4/2016).

Laaß, Michael. 2015. "Bone-conduction hearing and seismic sensitivity of the Late Permian anomodont *Kawingasaurus fossilis*." *Journal of Morphology* **276** (February): 121-143.

———. 2016. "The origins of the cochlea and impedance matching hearing in synapsids." *Acta Palaeontologica Polonica* **61** (2): 267-280.

Lai, Lien B., Lin Wang, & Timothy M. Nelson. 2002. "Distinct But Conserved Functions for Two

Chloroplastic NADP-Malic Enzyme Isoforms in C_3 and C_4 *Flaveria* Species." *Plant Physiology* **128** (January): 125-139.

Lakshmanan, Viswanathan, Patrick G. Bray, Dominik Verdier-Pinard, David J. Johnson, Paul Horrocks, Rebecca A. Muhle, George E. Alakpa, Ruth H. Hughes, Steve A. Ward, Donald J. Krogstad, Amar Bir Singh Sidhu, & **David A. Fidock.** 2005. "A critical role for PfCRT K76T in *Plasmodium falciparum* verapamil-reversible chloroquine resistance." *The EMBO Journal* **24** (6 July): 2294-2305.

Lamanna, Matthew C., Hai-Lu You, Jerald D. Harris, Luis M. Chiappe, Shu-An Ji, Jun-Chiang Lü, & **Qiang Ji.** 2006. "A partial skeleton of an enantiornithine bird from the Early Cretaceous of northwestern China." *Acta Palaeontologica Polonica* **51** (3): 423-434.

Lamb, Trip, & **David A. Beamer.** 2012. "Digits Lost Or Gained? Evidence for Pedal Evolution in the Dwarf Salamander Complex (Eurycea, Plethodontidae)." *PLoS ONE* (online @ plosone.org) **7** (May): e37544.

Lambert, David, & **The Diagram Group.** 1985. *The Field Guide to Prehistoric Life*. New York: Facts On File.

———. 1990. *The Dinosaur Data Book*. New York: Avon Books.

Lambert, W. G. 2013. *Babylonian Creation Myths*. Winona Lake, IN: Eisenbrauns.

Lamont, André. 1943. *Nostradamus Sees All*. Rev. ed. Philadelphia: W. Foulsham Co.

Lane, Christopher A. 1999. *Tonopah*. Grand Rapids, MI: Zondervan Publishing House.

Lang, Dietmar, Ralf Thoma, Martina Henn-Sax, Reinhard Sterner, & **Matthias Wilmanns.** 2000. "Structural Evidence for Evolution of the β/α Barrel Scaffold by Gene Duplication and Fusion." *Science* **289** (1 September): 1546-1550.

Langbein, Lutz, Michael A. Rogers, Hermelita Winter, Silke Praetzel, Ulrike Beckhaus, Hans-Richard Rackwitz, & **Jürgen Schweizer.** 1999. "The Catalog of Human Hair Keratins: I. EXPRESSION OF THE NINE TYPE I MEMBERS IN THE HAIR FOLLICLE." *The Journal of Biological Chemistry* **274** (9 July): 19874-19884.

Langbein, Lutz, Michael A. Rogers, Hermelita Winter, Silke Praetzel, & **Jürgen Schweizer.** 2001. "The Catalog of Human Hair Keratins: II. EXPRESSION OF THE SIX TYPE II MEMBERS IN THE HAIR FOLLICLE AND THE COMBINED CATALOG OF HUMAN TYPE I AND II KERATINS." *The Journal of Biological Chemistry* **276** (14 September): 35123-35132.

Langer, Max C. 2014. "The origins of Dinosauria: Much ado about nothing." *Palaeontology* **57** (May): 469-478.

Langer, Max C., & **Jorge Ferigolo.** 2013. "The Late Triassic dinosauromorph *Sacisaurus agudoensis* (Caturrita Formation; Rio Grande do Sul, Brazil): anatomy and affinities." *Geological Society, London, Special Publications* **379**: 353-392.

Langer, Max C., Sterling J. Nesbitt, Jonathas S. Bittencourt, & **Randall B. Irmis.** 2013. "Non-dinosaurian Dinosauromorpha." *Geological Society, London, Special Publications* **379**: 157-186.

Langston, Wann Jr. 1965. "*Oedaleops campi* (Reptilia: Pelycosauria) New Genus and Species from the Lower Permian of New Mexico, and the family *Eothyrididae*." *Bulletin of the Texas Memorial Museum* **9** (15 January): 1-47.

LaPolla, John S., Gennady M. Dlussky, & **Vincent Perrichot.** 2013. "Ants and the Fossil Record." *Annual Review of Entomology* **58**: 609-630.

Larroux, Claire, Bryony Fahey, Danielle Liubicich, Veronica F. Hinman, Marie Gauthier, Milena Gongora, Kathryn Green, Gert Wörheide, Sally P. Leys, & **Bernard M. Degnan.** 2006. "Developmental expression of transcription factor genes in a demosponge: insights into the origin of metazoan multicellularity." *Evolution & Development* **8** (March-April): 150-173.

Larroux, Claire, Graham N. Luke, Peter Koopman, Daniel S. Rokhsar, Sebastian M. Shimeld, & **Bernard M. Degnan.** 2008. "Genesis and Expansion of Metazoan Transcription Factor Gene Classes." *Molecular Biology and Evolution* **25** (May): 980-996.

Larsson, Hans C., Audrey C. Heppleston, & **Ruth M. Elsey.** 2010. "Pentadactyl ground state of the manus of *Alligator mississippiensis* and insights into the evolution of digital reduction in Archosauria." *Journal of Experimental Zoology* **314B** (15 November): 571-579.

Laurens, Kristin R., Elton T. C. Ngan, Alan T. Bates, Kent A. Kiehl, & **Peter F. Liddle.** 2003. "Rostral anterior cingulate cortex dysfunction during error processing in schizophrenia." *Brain* **126** (March): 610-622.

Laurin, Michel. 1993. "Anatomy and Relationships of *Haptodus garnettensis*, a Pennsylvanian Synapsid from Kansas." *Journal of Vertebrate Paleontology* **13** (June): 200-229.

———. 1994. "Re-evaluation of *Cutleria Wilmarthi*, an Early Permian synapsid from Colorado."

Journal of Vertebrate Paleontology **14** (March): 134-138.
Laurin, Michel, & **Vivian de Buffrénil.** 2016. "Microstructural features of the femur in early ophiacodontids: A reappraisal of ancestral habitat use and lifestyle of amniotes." *Comptes Rendus Palevol* **15** (January): 115-127.
Laurin, Michel, Marc Girondot, & **Armand de Ricqlès.** 2000. "Early tetrapod evolution." *Trends in Ecology & Evolution* **15** (March): 118-123.
Laurin, Michel, & **Robert R. Reisz.** 1990. "*Tetraceratops* is the oldest known therapsid." *Nature* **345** (17 May): 249-250.
———. 1996. "The Osteology and Relationships of *Tetraceratops insignis*, the Oldest Known Therapsid." *Journal of Vertebrate Paleontology* **16** (March): 95-102.
Laurin, M., & **R. Soler-Gijón.** 2010. "Osmotic tolerance and habitat of early stegocephalians: indirect evidence from parsimony, taphonomy, palaeobiogeography, physiology and morphology." *Geological Society, London, Special Publications* **339**: 151-179.
Lautenschlager, Stephan. 2016. "Reconstructing the past: methods and techniques for the digital restoration of fossils." *Royal Society Open Science* **3** (October): 160342.
Lebedev, Oleg A. 1997. "Fins made for walking." *Nature* **390** (6 November): 21-22.
Lebo, Lauri. 2008. *The Devil in Dover: An Insider's Story of Dogma v. Darwin in Small-Town America.* New York: The New Press.
Le Bras, Jacques, & **Remy Durand.** 2003. "The mechanisms of resistance to antimalarial drugs in *Plasmodium falciparum*." *Fundamental and Clinical Pharmacology* **17** (May): 147-153.
Lee, Michael S. Y. 1993. "The Origin of the Turtle Body Plan: Bridging a Famous Morphological Gap." *Science* **261** (24 September): 1716-1720.
———. 1994. "The Turtles' Long-Lost Relatives." *Natural History* **103** (June): 63-65.
———. 1996. "Correlated progression and the origin of turtles." *Nature* **379** (29 February): 812-815.
———. 1997. "Pareiasaur phylogeny and the origin of turtles." *Zoological Journal of the Linnean Society* **120** (July): 197-280.
———. 1998. "Phylogenetic Uncertainty, Molecular Sequences, and the Definition of Taxon Names." *Systematic Biology* **47** (December): 710-726.
———. 2013. "Turtle origins: insights from phylogenetic retrofitting." *Journal of Evolutionary Biology* **26** (December): 2729-2738.
Lee, Michael S. Y., Julien Soubrier, & **Gregory D. Edgecombe.** 2013. "Rates of Phenotypic and Genomic Evolution during the Cambrian Explosion." *Current Biology* **23** (7 October): 1-7.
Lee, Sang Hun. 1991. "From Evolution Theory to a New Creation Theory—Errors in Darwinism and a Proposal from Unification Thought." *True Parents* undated posting (online text at *www.tparents.org* accessed 8/2/2016).
Lee, Sang-Hwy, Olivier Bédard, Marcela Buchtova, Katherine Fu, & **Joy M. Richman.** 2004. "A new origin for the maxillary jaw." *Developmental Biology* **276** (1 December): 207-224.
Lemieux, Claude, Christian Otis, & **Monique Turmel.** 2000. "Ancestral chloroplast genome in *Mesostigma viride* reveals an early branch of green plant evolution." *Nature* **403** (10 February): 649-652.
Leoni, Edgar. 1961. *Nostradamus and His Prophecies.* 1982 reprint of orig. *Nostradamus: Life and Literature*. New York: Bell Publishing Co.
Lester, Keith S., & **Alan Boyde.** 1986. "Scanning microscopy of platypus teeth." *Anatomy and Embryology* **174** (April): 15-26.
Lester, K. S., A. Boyde, C. Gilkeson, & **M. Archer.** 1987. "Marsupial and Monotreme Enamel Structure." *Scanning Microscopy* **1** (March): 401-420.
Levin, Malcolm P. 1992. "Life—How It Got Here: A Critique of a View from the Jehovah's Witnesses." *Creation/Evolution* **12** (Summer): 29-34.
Lewin, Roger. 1981. "Bones of Mammals' Ancestors Fleshed Out." *Science* **212** (26 June): 1492.
———. 1987. *Bones of Contention: Controversies in the Search for Human Origins.* New York: Simon & Schuster.
Li, Chun-Li, Xiao-Chun Wu, Olivier Rieppel, Li-Ting Wang, & **Li-Jun Zhao.** 2008. "An ancestral turtle from the Late Triassic of southwestern China." *Nature* **456** (27 November): 497-501.
Li, Ying, Zhen Liu, Peng Shi, & **Jianzhi Zhang.** 2010. "The hearing gene *Prestin* unites echolocating bats and whales." *Current Biology* **20** (26 January): R56-R57.
Liberles, David A., Makayla D. M. Tisdell, & **Johan A. Grahnen.** 2011. "Binding constraints on the evolution of enzymes and signaling proteins: the important role of negative pleiotropy." *Proceedings of the Royal Society of London* **B** (Biological Sciences) **278** (7 July): 1930-1935.

Life Before Man. 1972. New York: Time-Life Books.

Lightner, Jean K. 2010. "Identification of a large sparrow-finch monobaramin in perching birds (Aves: Passeriformes)." *Journal of Creation* (AKA Answers in Genesis *Creation ex Nihilo Technical Journal*) **24** (3): 117-121.

———. 2013. "An Initial Estimate of Avian Ark Kinds." Answers in Genesis *Answers Research Journal* **6**: 409-466.

Lillegraven, Jason A., Zofia Kielan-Jaworowska, & **William A. Clemens.** 1979. *Mesozoic Mammals: The First Two Thirds of Mammalian History*. Los Angeles: University of California Press.

Lin, Chijen R., Chrissa Kioussi, Shawn O'Connell, Paola Briata. Daniel Szeto, Forrest Liu, Juan Carlos Izpisúa Belmonte, & **Michael G. Rosenfeld.** 1999. "Pitx2 regulates lung asymmetry, cardiac positioning and pituitary and tooth morphogenesis." *Nature* **401** (16 September): 279-282.

Lin, Hao, Lifang Niu, Neil A. McHale, Masaru Ohme-Takagi, Kirankumar S. Mysore, & **Million Tadege.** 2013. "Evolutionarily conserved repressive activity of WOX proteins mediates leaf blade outgrowth and floral organ development in plants." *Proceedings of the National Academy of Sciences* **110** (2 January): 366-371.

Lindsey, Hal., with **C. C. Carlson.** 1970. *The Late Great Planet Earth*. 1973 pb ed. New York: Bantam Books.

———. 1972. *Satan is Alive and Well on Planet Earth*. Grand Rapids, MI: Zondervan Publishing House.

Ling, Qihua, & **Paul Jarvis.** 2013. "Dynamic regulation of endosymbiotic organelles by ubiquitination." *Trends in Cell Biology* **23** (August): 399-408.

Lingham-Soliar, Theagarten. 1999. "Rare soft tissue preservation showing fibrous structures in an ichthyosaur from the Lower Lias (Jurassic) of England." *Proceedings of the Royal Society of London* **B** (Biological Sciences) **266** (7 December): 2367-2373.

———. 2003. "Evolution of birds: ichthyosaur integumental fibers conform to dromaeosaur protofeathers." *Naturwissenschaften* **90** (September): 428-432.

———. 2010. "Dinosaur protofeathers: pushing back the origin of feathers into the Middle Triassic?" *Journal of Ornithology* **151** (January): 193-200.

———. 2011. "The evolution of the feather: *Sinosauropteryx*, a colourful tail." *Journal of Ornithology* **152** (July): 567-577.

———. 2013. "The evolution of the feather: scales on the tail of *Sinosauropteryx* and an interpretation of the dinosaur's opisthotonic posture." *Journal of Ornithology* **154** (April): 455-463.

———. 2014. "Comment on 'A Jurassic ornithischian dinosaur from Siberia with both feathers and scales'." *Science* **346** (24 October): 434.

Lingham-Soliar, Theagarten, Alan Feduccia, & **Xiaolin Wang.** 2007. "A new Chinese specimen indicates that 'protofeathers' in the Early Cretaceous theropod dinosaur *Sinosauropteryx* are degraded collagen fibres." *Proceedings of the Royal Society of London* **B** (Biological Sciences) **274** (7 August): 1823-1829.

Linnen, Catherine R., & **Brian D. Farrell.** 2008. "Phylogenetic analysis of nuclear and mitochondrial genes reveals evolutionary relationships and mitochondrial introgression in the *sertifer* species group of the genus *Neodiprion* (Hymenoptera: Diprionidae)." *Molecular Phylogenetics and Evolution* **48** (July): 240-257.

Lisle, Jason. 2010. "Can Creationists Be 'Real' Scientists?" *Answers in Genesis* 13 May posting of *The New Answers Book 2* Chapter 14 (online text at *www.answersingenesis.org* accessed 5/27/2016).

———. 2013. "What Is the Best Argument for the Existence of God?" *Answers in Genesis* 9 September 2014 posting of *The New Answers Book 3* Chapter 27 (online text at *www.answersingenesis.org* accessed 2/17/2015).

Lisle, Jason, & **Mike Riddle.** 2013. "What Are Some Good Questions to Ask an Evolutionist?" *Answers in Genesis* 2 December 2014 posting of *The New Answers Book 3* Chapter 30 (online text at *www.answersingenesis.org* accessed 2/17/2015).

Litingtung, Ying, Randall D. Dahn, Yina L., John F. Fallon, & **Chin Chiang.** 2002. "*Shh* and *Gli3* are dispensable for limb skeleton formation but regulate digit number and identity." *Nature* **418** (29 August): 979-983.

Liu, Jun. 2007. "The taxonomy of the traversodontid cynodonts *Exaeretodon* and *Ischignathus*." *Revista Brasileira de Paleontologia* **10** (May/August): 133-136.

Liu, Jun, & **Fernando Abdala.** 2014. "Phylogeny and Taxonomy of the Traversodontidae," in Kammerer *et al.* (2014, 255-279).

Liu, Jun, & **G. S. Bever.** 2015. "The last diadectomorph sheds light on Late Paleozoic tetrapod biogeography." Royal Society *Biology Letters* **11** (May): 20150100.
Liu, Jun, & **Paul Olsen.** 2010. "The Phylogenetic Relationships of Eucynodontia (Amniota: Synapsida)." *Journal of Mammalian Evolution* **17** (September): 151-176.
Liu, Jun, Bruce Rubidge, & **Jinling Li.** 2009. "New basal synapsid supports Laurasian origin for therapsids." *Acta Palaeontologica Polonica* **54** (3): 393-400.
Liu, Shaofeng, Paul L. Heller, & **Guowei Zhang.** 2003. "Mesozoic basin development and tectonic evolution of the Dabieshan orogenic belt, central China." *Tectonics* **22** (August): 12.
Liu, Yang, James A. Cotton, Bin Shen, Xiuqun Han, Stephen J. Rossiter, & **Shuyi Zhang.** 2010. "Convergent sequence evolution between echolocating bats and dolphins." *Current Biology* **20** (26 January): R53-R54.
Liu, Yong-Qing, Hong-Wei Kuang, Nan Peng, Huan Xu, Peng Zhang, Neng-Sheng Wang, & **Wei An.** 2015. "Mesozoic basins and associated palaeogeographic evolution in North China." *Journal of Palaeogeography* **4** (April): 189-202.
Liu, Zhen, Gong-Hua Li, Jing-Fei Huang, Robert W. Murphy, & **Peng Shi.** 2012. "Hearing Aid for Vertebrates via Multiple Episodic Adaptive Events on Prestin Genes." *Molecular Biology and Evolution* **29** (September): 2187-2198.
Logsdon, John M. Jr., & **W. Ford Doolittle.** 1997. "Origin of antifreeze protein genes: A cool tale in molecular evolution." *Proceedings of the National Academy of Sciences* **94** (April): 3485-3487.
Lombard, R. Eric. 1979. "Evolutionary Principles of the Mammalian Middle Ear." *Evolution* **33** (December): 1230.
Lonfat, Nicolas, Thomas Montavon, Fabrice Darbellay, Sandra Gitto, & **Denis Duboule.** 2014. "Convergent evolution of complex regulatory landscapes and pleiotropy at *Hox* loci." *Science* **346** (21 November): 1004-1006.
Long, John A., & **Malcolm S. Gordon.** 2004. "The Greatest Step in Vertebrate History: A Paleobiological Review of the Fish-Tetrapod Transition." *Physiological and Biochemical Zoology* **77** (September-October): 700-719.
Long, Manyuan, Esther Betrán, Kevin Thornton, & **Wen Wang.** 2003. "The origin of new genes: Glimpses from the young and old." *Nature Reviews Genetics* **4** (November): 865-875.
Long, Manyuan, Michael Deutsch, Wen Wang, Esther Betrán, Frédéric G. Brunet, & **Jianming Zhang.** 2003. "Origin of new genes: evidence from experimental and computational analyses." *Genetica* **118**: 171-182.
Long, M., & **C. H. Langley.** 1993. "Natural selection and the origin of jingwei, a chimeric processed functional gene in Drosophila." *Science* **260** (2 April): 91-95.
Lopatin, Alexey V., & **Alexander O. Averianov.** 2004a. "A new species of *Tribosphenomys* (Mammalia: Rodentiaformes) from the Paleocene of Mongolia." *New Mexico Museum of History and Science* Bulletin 26: 169-176.
———. 2004b. "The Earliest Rodents of the Genus *Tribosphenomys* from the Paleocene of Central Asia." *Doklady Biological Sciences* (Russia) **397** (July-August): 336-337.
———. 2006. "An Aegialodontid Upper Molar and the Evolution of Mammal Dentition." *Science* **313** (25 August): 1092.
———. 2007. "*Kielantherium*, a basal tribosphenic mammal from the Early Cretaceous of Mongolia, with new data on the aegialodontian dentition." *Acta Palaeontologica Polonica* **52** (3): 441-446.
Lopez-Rios, Javier, Amandine Duchesne, Dario Speziale, Guillaume Andrey, Kevin A. Peterson, Philipp Germann, Erkan Ünal, Jing Liu, Sandrine Floriot, Sarah Barbey, Yves Gallard, Magdalena Müller-Gerbl, Andrew D. Courtney, Christophe Klopp, Sabrina Rodriguez, Robert Ivanek, Christian Beisel, Carol Wicking, Dagmar Iber, Benoit Robert, Andrew P. McMahon, Denis Duboule, & **Rolf Zeller.** 2014. "Attenuated sensing of SHH by *Ptch1* underlies evolution of bovine limbs." *Nature* **511** (3 July): 46-51.
Loredo, Grace A., Alla Brukman, Matthew P. Harris, David Kagle, Elizabeth E. Leclair, Rachel Gutman, Erin Denney. Emily Henkelman, B. Patrick Murray, John F. Fallon, Rocky S. Tuan, & **Scott F. Gilbert.** 2001. "Development of an Evolutionarily Novel Structure: Fibroblast Growth Factor Expression in the Carapacial Ridge of Turtle Embryos." *Journal of Experimental Zoology* **291** (15 October): 274-281.
Loxton, Daniel, & **Donald R. Prothero.** 2013. *Abominable Science! Origins of the Yeti, Nessie, and Other Famous Cryptoids.* New York: Columbia University Press.
Lü, Junchang, & **Stephen L. Brusatte.** 2015. "A large, short-armed, winged dromaeosaurid (Dinosauria: Theropoda) from the Early Cretacous of China and its implications for feather

evolution." *Scientific Reports* (online @ www.nature.com/srep/) **5** (16 July): 11775.
Lu, Mei-Fang, Carolyn Pressman, Rex Dyer, Randy L. Johnson, & **James F. Martin.** 1999. "Function of Rieger syndrome gene in left-right asymmetry and craniofacial development." *Nature* **401** (16 September): 276-278.
Lubick, Naomi. 2013. "Biologists Tell Dueling Stories of How Turtles Get Their Shells." *Science* **341** (26 July): 329.
Lucas, Spencer G. 2004. "A global hiatus in the Middle Permian tetrapod fossil record." *Stratigraphy* **1** (1): 47-64.
———. 2005. "Olson's Gap or Olson's Bridge: An Answer." *New Mexico Museum of History and Science* Bulletin 30: 185-186.
———. 2013. "No gap in the Middle Permian record of terrestrial vertebrates: COMMENT." *Geology* **41** (September): e293.
Lucas, Spencer G., Andrew B. Heckert, Jerald D. Harris, Dieter Seegis, & **Rupert Wild.** 2001. "Mammal-like tooth from the Upper Triassic of Germany." *Journal of Vertebrate Paleontology* **21** (2): 397-399.
Lucas, Spencer G., Adrian P. Hunt, Andrew B. Heckert, & **Justin A. Spielmann.** 2007. "Global Triassic tetrapod biostratigraphy and biochronology: 2007 status." *New Mexico Museum of History and Science* Bulletin 41: 229-240.
Lucas, Spencer G., & **Zhexi Luo.** 1993. "*Adelobasileus* from the Upper Triassic of West Texas: the oldest mammal." *Journal of Vertebrate Paleontology* **13** (3): 309-334.
Luo, Shengzhan D., & **Bruce S. Baker.** 2015. "Constraints on the evolution of a *doubkesex* target gene arising from doublesex's pleiotropic deployment." *Proceedings of the National Academy of Sciences* **112** (24 February): E852-E861.
Luo, Zhe-Xi. 1994. "Sister-group relationships of mammals and transformations of diagnostic mammalian characters," in Fraser & Sues (1994, 98-128).
———. 2001. "The inner ear and its bony housing in tritylodontids and implications for evolution of the mammalian ear." *Bulletin Museum of Comparative Zoology* **156** (October): 81-97.
———. 2007. "Transformation and diversification in early mammal evolution." *Nature* **450** (13 December): 1011-1019.
———. 2011. "Developmental Patterns in Mesozoic Evolution of Mammal Ears." *Annual Review of Ecology, Evolution and Systematics* **42**: 355-380.
———. 2014. "Tooth structure re-engineered." *Nature* **512** (7 August): 36-37.
Luo, Zhe-Xi, Peiji Chen, Gang Li, & **Meng Chen.** 2007. "A new eutriconodont mammal and evolutionary development in early mammals." *Nature* **446** (15 March): 288-293.
Luo, Zhe-Xi, Richard L. Cifelli, & **Zofia Kielan-Jaworowska.** 2001. "Dual origin of tribosphenic mammals." *Nature* **409** (4 January): 53-57.
Luo, Zhe-Xi, & **Alfred W. Crompton.** 1994. "Transformation of the Quadrate (Incus) Through the Transition from Non-Mammalian Cynodonts to Mammals." *Journal of Vertebrate Paleontology* **14** (September): 341-374.
Luo, Zhe-Xi, Alfred W. Crompton, & **Spencer T. Lucas.** 1995. "Evolutionary Origins of the Mammalian Promontorium and Cochlea." *Journal of Vertebrate Paleontology* **15** (March): 113-121.
Luo, Zhe-Xi, Alfred W. Crompton, & **Ai-Lin Sun.** 2001. "A New Mammaliaform from the Early Jurassic and Evolution of Mammalian Characteristics." *Science* **292** (25 May): 1535-1540.
Luo, Zhe-Xi, Stephen M. Gatesy, Farish A. Jenkins Jr., William W. Amaral, & **Neil H. Shubin.** 2015. "Mandibular and dental characteristics of Late Triassic mammaliaform *Haramiyavia* and their ramifications for basal mammal evolution." *Proceedings of the National Academy of Sciences* **112** (22 December): E7101-E7109.
Luo, Zhe-Xi, Qiang Ji, John R. Wible, & **Chong-Xi Yuan.** 2003. "An Early Cretaceous Tribosphenic Mammal and Metatherian Evolution." *Science* **302** (12 December): 1934-1940.
Luo, Zhe-Xi, Qiang Ji, & **Chong-Xi Yuan.** 2007. "Convergent dental adaptations in pseudo-tribosphenic and tribosphenic mammals." *Nature* **450** (1 November): 93-97.
Luo, Zhe-Xi, Zofia Kielan-Jaworowska, & **Richard L. Cifelli.** 2002. "In quest for a phylogeny of Mesozoic mammals." *Acta Palaeontologica Polonica* **47** (1): 1-78.
———. 2004. "Evolution of Dental Replacement in Mammals." *Bulletin of Carnegie Museum of Natural History* **36** (December): 159-175.
Luo, Zhe-Xi, Qing-Jin Meng, Qiang Ji, Di Liu, Yu-Guang Zhang, & **April I. Neander.** 2015. "Evolutionary development in basal mammaliaforms as revealed by a docodontan." *Science* **347**

(13 February): 760-764.
Luo, Zhe-Xi, Qing-Jin Meng, Di Liu, Yu-Guang Zhang, & Chong-Xi Yuan. 2016. "Cruro-pedal structure of the paulchoffatiid *Rugosodon eurasiaticus*." *Palaeontologia Polonica* **67**: 149-169.
Luo, Zhe-Xi, Irina Ruf, Julia A. Schultz, & Thomas Martin. 2011. "Fossil evidence on evolution of inner ear cochlea in Jurassic mammals." *Proceedings of the Royal Society of London* **B** (Biological Sciences) **278** (7 January): 28-34.
Luo, Zhe-Xi, Chong-Xi Yuan, Wing-Jin Meng, & Qiang Ji. 2011. "A Jurassic eutherian mammal and divergence of marsupials and placentals." *Nature* **476** (25 August): 442-445.
Luskin, Casey. 2006. "'The Vampire's Heart' – A Response by David Berlinski to James Downard." Discovery Institute *Evolution News & Views* for 16 August (online text at *evolutionnews.org* accessed 8/20/2006).
———. 2007. "Design of Life." Discovery Institute *Evolution News & Views* for 19 November (online text at *evolutionnews.org* accessed 4/20/2016).
———. 2008a. "An 'Ulnare' and an 'Intermedium' a Wrist Do Not Make: A Response to Carl Zimmer." *Center for Science & Culture* 1 August posting (online text at *discovery.org* accessed 2/13/2014).
———. 2008b. "Is the Latest 'Feathered Dinosaur' Actually a Secondarily Flightless Bird?" Discovery Institute *Evolution News & Views* for 12 November (online text at *evolutionnews.org* accessed 5/11/2016).
———. 2014. "So, Michael Behe Was Right After All; What Will the Critics Say Now?" Discovery Institute *Evolution News & Views* for 16 July (online text at *evolutionnews.org* accessed 7/17/2014).
———. 2009a. "Double Standard on Textbook Treatments of Evolution." *Explore Evolution* critical response 2 February (online text at *www2.exploreevolution.com* accessed 1/4/2011).
———. 2009b. "A Mis-Aimed Critique of Inquiry Based Learning." *Explore Evolution* critical response 2 February (online text at *www2.exploreevolution.com* accessed 1/4/2011).
———. 2009c. "When Did 'Neo-Darwinism' Become a Dirty Word?" *Explore Evolution* critical response 23 February (online text at *www2.exploreevolution.com* accessed 1/4/2011).
———. 2009d. "Antibiotic Resistance Revisited." *Explore Evolution* critical response 23 February (online text at *www2.exploreevolution.com* accessed 1/4/2011).
———. 2009e. "MSNBC's Birthday Present to Charles Darwin: Puff-Pieces on Evolution (Part 2)." Discovery Institute *Evolution News & Views* for 23 February (online text at *evolutionnews.org* accessed 12/4/2009).
———. 2009f. "Response to Brian Metscher's Book Review in Evolution & Development." *Explore Evolution* critical response 24 February (online text at *www2.exploreevolution.com* accessed 1/4/2011).
———. 2010a. "The NCSE's Biogeographic Conundrums: A Defense of Explore Evolution's Treatment of Biogeography." *Explore Evolution* critical response 19 January (online text at *www2.exploreevolution.com* accessed 1/4/2011).
———. 2010b. "Sea Monkey Hypotheses Refute the NCSE's Biogeography Objections to *Explore Evolution*." Discovery Institute *Evolution News & Views* for 2 March (online text at *evolutionnews.org* accessed 1/29/2012).
———. 2010c. "ACLU Lawyer and ScienceBloggers Make Off-Base Arguments Against Coppedge Case." Discovery Institute *Evolution News & Views* for 21 April (online text at *evolutionnews.org* accessed 8/11/2010).
———. 2011. "A Fishy Story About AntiFreeze Gene Evolution." Discovery Institute *Evolution News & Views* for 27 January (online text at *evolutionnews.org* accessed 1/27/2011).
Luu, Phan, & Michael I. Posner. 2003. "Anterior cingulate cortex regulation of sympathetic activity." *Brain* **126** (October): 2119-2120.
Lynch, Michael. 1999. "The Age and Relationships of the Major Animal Phyla." *Evolution* **53** (April): 319-325.
———. 2002. "Gene Duplication and Evolution." *Science* **297** (9 August): 945-947.
———. 2005. "Simple evolutionary pathways to complex proteins." *Protein Science* **14** (September): 2217-2225.
Lyson, Tyler R., Gabe S. Bever, Bhart-Anjan S. Bhullar, Walter G. Joyce, & Jacques A. Gauthier. 2010. "Transitional fossils and the origin of turtles." Royal Society *Biology Letters* **6** (December): 830-833.
Lyson, Tyler R., Gabe S. Bever, Torsten M. Scheyer. Allison Y. Hsiang, & Jacques A. Gauthier. 2013.

"Evolutionary Origin of the Turtle Shell." *Current Biology* **23** (17 June): R513-R515.

Lyson, Tyler R., Bhart-Anjan S. Bhullar, Gabe S. Bever, Walter G. Joyce, Kevin de Queiroz, Arhat Abzhanov, & **Jacques A. Gauthier.** 2013. "Homology of the enigmatic muchal bone reveals novel reorganization of the shoulder girdle in the evolution of the turtle shell." *Evolution & Development* **15** (September-October): 317-325.

Lyson, Tyler, & **Scott F. Gilbert.** 2009. "Turtles all the way down: loggerheads at the root of the chelonian tree." *Evolution & Development* **11** (March-April): 133-135.

Lyson, Tyler R., & **Walter G. Joyce.** 2012. "Evolution of the turtle bauplan: the topological relationship to the scapula relative to the ribcage." Royal Society *Biology Letters* **8** (December): 1028-1031.

Lyson, Tyler R., Emma R. Schachner, Jennifer Botha-Brink, Torsten M. Scheyer, Markus Lambertz, G. S. Bever, Bruce S. Rubidge, & **Kevin de Queiroz.** 2014. "Origin of the unique ventilatory apparatus of turtles." *Nature Communications* (online @ nature.com) **5** (7 November): 5211.

MacDonald, Mary E., & **Brian K. Hall.** 2001. "Altered Timing of the Extracellular-Matrix-Mediated Epithelial-Mesenchymal Interaction that Initiates Mandibular Skeletogenesis in Three Inbred Strains of Mice: Development, Heterochrony, and Evolutionary Change in Morphology." *Journal of Experimental Zoology* **291** (15 October): 258-273.

Mace, S. R., & **T. C. Wood.** 2005. "Statistical Evidence for Five Whale Holobaramins (Mammalia: Cetacea)." *Occasional Papers of the Baraminology Study Group* **5**: 15.

MacLean, R. Craig, Graham Bell, & **Paul B. Rainey.** 2004. "The evolution of pleiotropic fitness tradeoff in *Pseudomonas fluorescens*." *Proceedings of the National Academy of Sciences* **101** (25 May): 8072-8077.

Macrini, Thomas E., Guillermo W. Rougier, & **Timothy Rowe.** 2007. "Description of a Cranial Endocast From the Fossil Mammal *Vincelestes neuquenianus* (Theriiformes) and Its Relevance to the Evolution of Endocranial Characters in Therians." *The Anatomical Record: Advances in Integrative Anatomy and Evolutionary Biology* **290** (July): 875-892.

Madar, S. I. 2007. "The Postcranial Skeleton of Early Eocene Pakicetid Cetaceans." *Journal of Paleontology* **81** (January): 176-200.

Madar, S. I., J. G. M. Thewissen, & **S. T. Hussain.** 2002. "Additional holotype remains of *Ambulocetus natans* (Cetacea, Ambulocetidae), and their implications for locomotion in early whales." *Journal of Vertebrate Paleontology* **22** (July): 405-422.

Maddin, Hillary C., David C. Evans, & **Robert R. Reisz.** 2006. "An Early Permian varanodontine: Varanopid (Synapsida: Eupelycosauria) from the Richards Spur Locality, Oklahoma." *Journal of Vertebrate Paleontology* **26** (December): 957-966.

Maddin, Hillary C., Christian A. Sidor, & **Robert R. Reisz.** 2008. "Cranial anatomy of *Ennatosaurus tecton* (Synapsida: Caseidae) from the Middle Permian of Russia and the evolutionary relationships of Caseidae." *Journal of Vertebrate Paleontology* **28** (March): 160-180.

Maderson, Paul F. A., & **Lorenzo Alibardi.** 2000. "The development of the sauropod integument: A contribution to the problem of the origin and evolution of feathers." *Integrative and Comparative Biology* (AKA *American Zoologist*) **40** (September): 513-529.

Maguire, Finlay, & **Thomas A. Richards.** 2014. "Organelle Evolution: A Mosaic of Mitochondrial Functions." *Current Biology* **24** (2 June): R518-R520.

Maier, Wolfgang, & **Irina Ruf.** 2016. "Evolution of the mammalian middle ear: a historical review." *Journal of Anatomy* **228** (February): 270-283.

Majerus, Michael E. N. 2009. "Industrial Melanism in the Peppered Moth, *Biston betularia*: An Excellent Teaching Example of Darwinian Evolution in Action." *Evolution: Education & Outreach* **2** (March): 63-74.

Makarkin, Vladimir N., & **S. Bruce Archibald.** 2014. "An unusual new fossil genus probably belonging to the Psychopsidae (Neuroptera) from the Eocene Okanagan Highlands, western North America." *Zootaxa* **3838** (18 July): 385-391.

Mallarino, Ricardo, Otger Campàs, Joerg A. Fritz, Kevin J. Burns, Olivia G. Weeks, Michael P. Brenner, & **Arhat Abzhanov.** 2012. "Closely related bird species demonstrate flexibility between beak morphology and underlying developmental programs." *Proceedings of the National Academy of Sciences* **109** (2 October): 16222-16227.

Mallarino, Ricardo, Peter R. Grant, B. Rosemary Grant, Anthony Herrel, Winston P. Kuo, & **Arhat Abzhanov.** 2011. "Two developmental modules establish 3D beak-shape variation in Darwin's finches." *Proceedings of the National Academy of Sciences* **108** (8 March): 4057-4062.

Mallo, Moisés. 1998. "Embryological and genetic aspects of middle ear development." *International Journal of Developmental Biology* **42** (January): 11-22.
———. 2001. "Formation of the Middle Ear: Recent Progress on the Developmental and Molecular Mechanisms." *Developmental Biology* **231** (15 March): 410-419.
Manley, Geoffrey A. 2001. "Evidence for an Active Process and a Cochlear Amplifier in Nonmammals." *Journal of Neurophysiology* **86** (August): 541-549.
———. 2012. "Evolutionary Paths to Mammalian Cochleae." *Journal of the Association for Research in Otolaryngology* **13** (December): 733-743.
Manley, Geoffrey A., & **Johanna E. M. Kraus.** 2010. "Exceptional high-frequency hearing and matched vocalizations in Australian pygopod geckos." *The Journal of Experimental Biology* **213** (1 June): 1876-1885.
Mantyla, Kyle. 2012. "Santorum: Satan is Systematically Destroying America." People for the American Way *Right Wing Watch* 16 February posting (online text at *www.rightwingwatch.org* accessed 11/3/2013).
Manuel, Michaël. 2009. "Early evolution of symmetry and polarity in metazoan body plans." *Comptes Rendus Biologies* **332** (February-March): 184-209.
Maor, Eli. 1987. *To Infinity and Beyond: A Cultural History of the Infinite.* Princeton, NJ: Princeton University Press.
Marcus, Amy Dockser. 2000. *The View from Nebo: How Archaeology Is Rewriting the Bible and Reshaping the Middle East.* Boston: Little, Brown & Co.
Margulis, Lynn, & **Karlene V. Schwartz.** 1988. *Five Kingdoms: An Illustrated Guide to the Phyla of Life on Earth.* 2nd ed. New York: W. H. Freeman & Co.
———. 1998. *Five Kingdoms: An Illustrated Guide to the Phyla of Life on Earth.* 3rd ed. New York: W. H. Freeman & Co.
Marin, Birger, Eva C. M. Nowack, & **Michael Melkonian.** 2005. "A Plastid in the Making: Evidence for a Second Primary Endosymbiosis." *Protist* **156** (13 December): 425-432.
Marivaux, Laurent, Jean-Loup Welcomme, Pierre-Olivier Antoine, Gregoire Metais, Ibrahim M. Baloch, Mouloud Benammi, Yaowalak Chaimanee, Stephane Ducrocq, & **Jean-Jacques Jaeger.** 2001. "A Fossil Lemur from the Oligocene of Pakistan." *Science* **294** (19 October): 587-591.
Marivaux, Laurent, Jean-Loup Welcomme, Pierre-Olivier Antoine, Grégoire Métais, Ibrahim M. Baloch, Mouloud Benammi, Yaowalak Chaimanee, Stéphane Ducrocq, & **Jean-Jacques Jaeger.** 2001. "A Fossil Lemur from the Oligocene of Pakistan." *Science* **294** (19 October): 587-591.
Marlétaz, Ferdinand, Ignacio Maeso, Laura Faas, Harry V. Isaacs, & **Peter W. H. Holland.** 2015. "Cdx ParaHox genes acquired distinct developmental roles after gene duplication in vertebrate evolution." *BMC Biology* (online @ biomedcentral.com) **13** (1 August): 56.
Marshall, Charles R. 2006. "Explaining the Cambrian 'Explosion' of Animals." *Annual Review of Earth and Planetary Sciences* **34**: 355-384.
Marshall, Charles R., & **James W. Valentine.** 2010. "The Importance of Preadapted Genomes in the Origin of the Animal Bodyplans and the Cambrian Explosion." *Evolution* **64** (May): 1189-1201.
Marshall, J. S., J. D. Stubbs, & **W. C. Taylor.** 1996. "Two Genes Encode Highly Similar Chloroplastic NADP-Malic Enzymes in *Flaveria* (Implications for the Evolution of C_4 Photosynthesis)." *Plant Physiology* **111** (November): 1251-1261.
Marshall, Wallace F., & **Joel L. Rosenbaum.** 1999. "Are there nucleic acids in the centrosome?" *Current Topics in Developmental Biology* **49**: 187-205.
Marszalek, Joseph R., Pilar Ruiz-Lozano, Elizabeth Roberts, Kenneth R. Chien, & **Lawrence S. B. Goldstein.** 1999. "Situs inversus and embryonic ciliary morphogenesis defects in mouse mutants lacking the KIF3A subunit of kinesin-II." *Proceedings of the National Academy of Sciences* **96** (27 April): 5043-5048.
Martin, Arnaud, Riccardo Papa, Nicola J. Nadeau, Ryan I. Hill, Brian A. Counterman, Georg Halder, Chris D. Jiggins, Marcus R. Kronforst, Anthony D. Long, W. Owen McMillan, & **Robert D. Reed.** 2012. "Diversification of complex butterfly wing patterns by repeated regulatory evolution of a *Wnt* ligand." *Proceedings of the National Academy of Sciences* **109** (31 July): 12632-12637.
Martin, James F., Allan Bradley, & **Eric N. Olson.** 1995. "The paired-like homeo box gene MHox." *Genes & Development* **9** (15 May): 1237-1249.
Martin, Jobe. 2002. *The Evolution of a Creationist: A Laymen's Guide to the Conflict Between the Bible and Evolutionary Theory.* Rev. ed. of 1994 orig. Rockwall, TX: Biblical Discipleship Publishers.
Martin, Thomas. 2005. "Postcranial anatomy of *Haldanodon exspectatus* (Mammalia, Docodonta)

from the Late Jurassic (Kimmeridgian) of Portugal and its bearing for mammalian evolution." *Zoological Journal of the Linnean Society* **145** (October): 219-248.

———. 2006. "Early Mammalian Evolutionary Experiments." *Science* **311** (24 February): 1109-1110.

Martin, Thomas, Jesús Marugán-Lobón, Romain Vullo, Hugo Martin-Abad, Zhe-Xi Luo, & **Angela D. Buscalioni.** 2015. "A Cretaceous eutriconodont and integument evolution in early mammals." *Nature* **526** (15 October): 380-384.

Martin, Thomas, & **Irina Ruf.** 2009. "On the Mammalian Ear." *Science* **326** (9 October): 243-244.

Martindale, Mark Q. 2005. "The evolution of metazoan axial properties." *Nature Reviews Genetics* **6** (December): 917-927.

Martindale, Mark Q., John R. Finnerty, & **Jonathan Q. Henry.** 2002. "The Radiata and the evolutionary origins of the bilaterian body plan." *Molecular Phylogenetics and Evolution* **24** (September): 358-365.

Martindale, Mark Q., Kevin Pang, & **John R. Finnerty.** 2004. "Investigating the origins of triploblasty: 'mesodermal' gene expression in a diploblastic animal, the sea anemone *Nematostella vectensis* (phylum, Cnidaria; class, Anthozoa)." *Development* (AKA *Journal of Embryology and Experimental Morphology*) **131** (15 May): 2463-2474.

Martinelli, Agustin G., & **José F. Bonaparte.** 2011. "Postcanine replacement in *Brasilodon* and *Brasilitherium* (Cynodontia, Probainognathia) and its bearing in cynodont evolution," in Calvo *et al.* (2011, 179-186).

Martinelli, Agustin G., José F. Bonaparte, Cesar L. Schultz, & **Rogerio Rubert.** 2005. "A new tritheledontid (Therapsida, Eucynodontia) from the Late Triassic of Rio Grande do Sul (Brazil) and its phylogenetic relationships among carnivorous non-mammalian eucynodonts." *Ameghiniana* **42** (March): 191-208.

Martinelli, Agustín G., Marcelo de la Fuente, & **Fernando Abdala.** 2009. "*Diademodon tetragonus* Seeley, 1894 (Therapsida: Cynodontia) in the Triassic of South America and its biostratigraphic implications." *Journal of Vertebrate Paleontology* **29** (September): 852-862.

Martinelli, Agustin G., & **Guillermo W. Rougier.** 2007. "On *Chaliminia musteloides* (Eucynodontia: Tritheledontidae) from the Late Triassic of Argentina, and a phylogeny of Ictidosauria." *Journal of Vertebrate Paleontology* **27** (2): 442-460.

Martinelli, Agustín G., Marina Bento Soares, & **Cibele Schwanke.** 2016. "Two New Cynodonts (Therapsida) from the Middle-Early Late Triassic of Brazil and Comments on South American Probainognathans." *PLoS ONE* (online @ plosone.org) **11** (October): e0162945.

Martinez, Ricardo N., Paul C. Sereno, Oscar A. Alcober, Carina E. Colombi, Paul R. Renne, Isabel P. Montañez, & **Brian S. Currie.** 2011. "A Basal Dinosaur from the Dawn of the Dinosaur Era in Southwestern Pangaea." *Science* **331** (14 January): 206-210.

Marx, Felix G., & **Mark D. Uhen.** 2010. "Climate, Critters, and Cetaceans: Cenozoic Drivers of the Evolution of Modern Whales." *Science* **327** (19 February): 993-996.

Maryańska, Teresa, Halszka Osmólska, & **Mieczysław Wolsan.** 2002. "Avialan status for Oviratorosauria." *Acta Palaeontologica Polonica* **47** (1): 97-116.

Marzoli, Andrea, Paul R. Renne, Enzo M. Piccirillo, Marcia Ernesto, Giuliano Bellieni, & **Angelo De Min.** 1999. "Extensive 200-Million-Year-Old Continental Flood Basalts of the Central Atlantic Magmatic Province." *Science* **284** (23 April): 616-618.

Matzke, Nicholas J. 2016. "The evolution of antievolution policies after *Kitzmiller* v. *Dover*." *Science* **351** (1 January): 28-30.

Mayr, Ernst. 1981. "Biological Classification: Toward a Synthesis of Opposing Methodologies." *Science* **214** (30 October): 510-516.

———. 1994. "Recapitulation Reinterpreted: The Somatic Program." *The Quarterly Review of Biology* **69** (June): 223-232.

Mayr, Gerald, D. Stefan Peters, Gerhard Plodowski, & **Olaf Vogel.** 2002. "Bristle-like integumentary structures at the tail of the horned dinosaur *Psittacosaurus*." *Naturwissenschaften* **89** (August): 361-365.

Mazur, Allan. 2010. "Do Americans Believe Modern Earth Science?" *Evolution: Education & Outreach* **3** (December): 629-632.

Mazet, François, & **Sebastian M. Shimeld.** 2002. "Gene duplication and divergence in the early evolution of vertebrates." *Current Opinion in Genetics & Development* **12** (August): 393-396.

Mazierski, David M., & **Robert R. Reisz.** 2010. "Description of a new specimen of *Ianthasaurus hardestiorum* (Eupelycosauria: Edaphosauridae) and a re-evaluation of edaphosaurid phylogeny." *Canadian Journal of Earth Sciences* **47** (August): 901-912.

McDowell, Josh. 1972. *Evidence That Demands A Verdict.* 1979 reprint. San Bernardino, CA: Here's Life.
———. 1975. *More evidence that demands a verdict: Historical evidences for the Christian scriptures.* San Bernardino, CA: Campus Crusade for Christ.
———. 1986. "Where do dinosaurs fit into the biblical story?" *Answers to Tough Questions Skeptics Ask About the Christian Faith* p. 197 (online text at *www.josh.org* accessed 1/7/2011).
———. 1989a. *Christianity: HOAX or HISTORY?* Wheaton, IL: Tyndale House Publishers.
———. 1989b. *Skeptics Who Demand a Verdict.* Wheaton, IL: Tyndale House Publishers.
McDowell, Josh, & **Bob Hostetler.** 1992. *Don't Check Your Brains at the Door.* Dallas, TX: Word Publishing.
McDowell, Josh, & **Don Stewart.** 1981. *Reasons Skeptics Should Consider Christianity.* Wheaton, IL: Tyndale House Publishers.
McDowell, Josh, with **David Stoop.** 1982. *Resurrection Growth Guide.* San Bernardino, CA: Here's Life.
McDowell, Sean. 2016a. "Is Evolution A Theory in Crisis?" *Sean McDowell Blog* 26 July posting (online text at *seanmcdowell.org* accessed 6/26/2016).
———. 2016b. "How is the Intelligent Design Movement Doing? Interview with William Dembski." *Sean McDowell Blog* 8 September posting (online text at *seanmcdowell.org* accessed 9/17/2016).
McFadden, Geoffrey Ian. 1990. "Evidence that cryptomonad chloroplasts evolved from photosynthetic eukaryotic endosymbionts." *Journal of Cell Science* **95** (February): 303-308.
———. 1999. "Endosymbiosis and evolution of the plant cell." *Current Opinion in Plant Biology* **2** (December): 513-519.
———. 2001a. "Chloroplast Origin and Integration." *Plant Physiology* **125** (January): 50-53.
———. 2001b. "Primary and secondary endosymbiosis and the origin of plastids." *Journal of Phycology* **37** (December): 951-959.
———. 2014. "Origin and Evolution of Plastids and Photosynthesis in Eukaryotes." *Cold Spring Harbor Perspectives in Biology* **6** (January): a016121.
McGowan, Christopher. 1984. *In the Beginning ... A Scientist Shows Why the Creationists Are Wrong.* Buffalo, NY: Prometheus Books.
McKay, Bailey D., & **Robert M. Zink.** 2015. "Sisyphean evolution in Darwin's finches." *Biological Reviews of the Cambridge Philosophical Society* **90** (August): 689-698.
McLatchie, Jonathan. 2010. "The Recapitulation Myth." Discovery Institute *Evolution News & Views* 29 June posting (online text at *evolutionnews.org* accessed 7/11/2010).
———. 2012a. "On the Origin of Mitochondria: Reasons for Skepticism on the Endosymbiotic Story." Discovery Institute *Evolution News & Views* **Jonathan M** 10 January posting (online text at *evolutionnews.org* accessed 2/13/2012).
———. 2012b. "On the Evolution of the Mammalian Middle Ear." Discovery Institute *Evolution News & Views* **Jonathan M** 25 July posting (online text at *evolutionnews.org* accessed 3/12/2016).
McLaughlin, Donald. 2016. "National Center for Science Education Misrepresents Oklahoma Academic Freedom Bill." Discovery Institute *Evolution News & Views* for 25 January (online text at *evolutionnews.org* accessed 4/23/2016).
McLaughlin, William I. 1994. "Resolving Zeno's Paradoxes." *Scientific American* **271** (November): 84-89.
McLellan, Tracy. 2005. "Correlated evolution of leaf shape and trichomes in *Begonia dregei* (Begoniaceae)." *American Journal of Botany* **92** (October): 1616-1623.
McLeroy, Don. 2003. "Historical Reality." Posting of 8 September paper (online text at *pages.suddenlink.net/don_mcleroy/* accessed 9/17/2013).
McSweeney, Paul L. H., & **Patrick F. Fox**, eds. 2013. *Advanced Dairy Chemistry Vol. 1A Proteins: Basic Aspects.* 4th ed. New York: Springer.
Meehan, T. J., & **Larry D. Martin.** 2010. "New leptictids (Mammalia: Insectivora) from the Early Oligocene." *Neues jahrbuch fur Geologie und Palaontologie Abhandlungen* **256** (April): 99-107.
Méheust, Raphaël, Ehud Zelzion, Debashish Bhattacharya, Philippe Lopez, & **Eric Bapteste.** 2016. "Protein networks identify novel symbiogenetic genes resulting from plastid endosymbiosis." *Proceedings of the National Academy of Sciences* **113** (29 March): 3579-3584.
Mehlert, A. W. (Bill). 1988. "A Critique of the Alleged Reptile to Mammal Transition." *Creation Research Society Quarterly* **25** (June): 7-15.
———. 1993. "The origin of mammals: a study of some important fossils." Answers in Genesis *Creation ex Nihilo Technical Journal* **7** (2): 122-139.

Meier, Rudolf, Paul Kores, & Steven Darwin. 1991. "Homoplasy Slope Ratio: A Better Measurement of Observed Homoplasy in Cladistic Analyses." *Systematic Zoology* **40** (March): 74-88.

Melin, Amanda D., Yuka Matsushita, Gillian L. Moritz, Nathaniel J. Dominy, & Shoji Kawamura. 2013. "Inferred L/M cone opsin polymorphism of ancestral tarsiers sheds dim light on the origin of anthropoid primates." *Proceedings of the Royal Society of London* **B** (Biological Sciences) **280** (22 May): 20130189.

Méndez-Vilas, A., ed. 2013. *Microbial Pathogens and Strategies for Combating Them: Science, Technology and Education.* Badajoz, Spain: Formatex Research Center.

Mendivil Ramos, Olivia, Daniel Barker, & David E. K. Ferrier. 2012. "Ghost Loci Imply Hox and ParaHox Existence in the Last Common Ancestor of Animals." *Current Biology* **22** (23 October): 1951-1956.

Menegaz, Rachel A., Samantha V. Sublett, Said D. Figueroa, Timothy J. Hoffman, & Matthew J. Ravosa. 2009. "Phenotypic Plasticity and Function of the Hard Palate in Growing Rabbits." *The Anatomical Record: Advances in Integrative Anatomy and Evolutionary Biology* **292** (February): 277-284.

Meng, Jin, Xijun Ni, Chuankui Li, K. Christopher Beard, Daniel L. Gebo, Yuanqing Wang, & Hongjiang Wang. 2007. "New Material of Alagomyidae (Mammalia, Glires) from the Late Paleocene Subeng Locality, Inner Mongolia." *American Museum Novitates* **3597** (December): 1-29.

Meng, Jin, Yuanqing Wang, & Chuankui Li. 2011. "Transitional mammalian middle ear from a new Cretaceous Jehol eutriconodont." *Nature* **472** (14 April): 181-185.

Meng, Jin, & André R. Wyss. 2001. "The Morphology of *Tribosphenomys* (Rodentiaformes, Mammalia): Phylogenetic Implications for Basal Glires." *Journal of Mammalian Evolution* **8** (March): 1-71.

Meng, Qing-Jin, Qiang Ji, Yu-Guang Zhang, Di Liu, David M. Grossnickle, & Zhe-Xi Luo. 2015. "An arboreal docodont from the Jurassic and mammaliaform ecological diversification." *Science* **347** (13 February): 764-768.

Menon, Gopinathan K., & Jaishri Menon. 2000. "Avian epidermal lipids: Functional considerations and relationship to feathering." *Integrative and Comparative Biology* (AKA *American Zoologist*) **40** (September): 540-552.

Menton, David N. 2007. "*Tiktaalik* and the Fishy Story of Walking Fish." *Answers in Genesis* 7 March posting (online text at *answersingensis.org* accessed 2/17/2015).

———. 2013a. "Vestigial Organs—Evidence for Evolution?" *Answers in Genesis* 7 July 2014 posting of *The New Answers Book 3* Chapter 24 (online text at *www.answersingenesis.org* accessed 2/17/2015).

———. 2013b. "Is *Tiktaalik* Evolution's Greatest Missing Link?" *Answers in Genesis* 11 July 2014 posting of *The New Answers Book 3* Chapter 25 (online text at *www.answersingenesis.org* accessed 2/17/2015).

Meredith, Robert W., John Gatesy, Joyce Cheng, & Mark S. Springer. 2011. "Pseudogenization of the tooth gene enamelysin (*MMP20*) in the common ancestor of extant baleen whales." *Proceedings of the Royal Society of London* **B** (Biological Sciences) **278** (7 April): 993-1002.

Meredith, Robert W., John Gatesy, William J. Murphy, Oliver A. Ryder, & Mark S. Springer. 2009. "Molecular Decay of the Tooth Gene Enamelin (*ENAM*) Mirrors the Loss of Enamel in the Fossil Record of Placental Mammals." *PLoS Genetics* (online @ plosgenetics.org) **5** (September): e1000634.

Meredith, Robert W., John Gatesy, & Mark S. Springer. 2013. "Molecular decay of enamel matrix protein genes in turtles and other edentulous amniotes." *BMC Evolutionary Biology* (online @ biomedcentral.com) **13** (23 January): 20.

Meredith, Robert W., Jan E. Janečka, John Gatesy, Oliver A. Ryder, Colleen A. Fisher, Emma C. Teeling, Alisha Goodbla, Eduardo Eizirik, Taiz L. L. Simão, Tanja Stadler, Daniel L. Rabosky, Rodney L. Honeycutt, John J. Flynn, Colleen M. Ingram, Cynthia Steiner, Tiffani L. Williams, Terence J. Robinson, Angela Burk-Herrick, Michael Westerman, Nadia A. Ayoub, Mark S. Springer, & William J. Murphy. 2011. "Impacts of the Cretaceous Terrestrial Revolution and KPg Mammal Diversification." *Science* **334** (28 October): 521-524.

Merker, Stefan, Christine Driller, Dyah Perwitasari-Farajallah, Joko Pamungkas, & Hans Zischler. 2009. "Elucidating geological and biological processes underlying the diversification of Sulawesi tarsiers." *Proceedings of the National Academy of Sciences* **106** (26 May): 8459-8464.

Merrell, Allyson J., & Gabrielle Kardon. 2013. "Development of the diaphragm, a skeletal muscle

essential for mammalian respiration." Federation of European Biochemical Societies *FEBS Journal* **280** (September): 12274.
Merrick, J. R., M. Archer, G. M. Hickey, & **M. S. Y. Lee**, eds. 2006. *Evolution and Biogeography of Australasian Vertebrates*. Oatlands, AU: Auscipub.
Metcalf, S. J., R. F. Vaughan, M. J. Benton, J. Cole, M. J. Simms, & **D. L. Dartnall.** 1992. "A new Bathonian (Middle Jurassic) microvertebrate site, within the Chipping Norton Limestone Formation at Hornsleasow Quarry, Gloucestershire." *Proceedings of the Geologists' Association* **103** (4): 321-342.
Metscher, Brian D. 2009. "Postcards from The Wedge: review and commentary on *Explore Evolution: The Arguments For and Against Neo-Darwinism* by Stephen C. Meyer et al." *Evolution & Development* **11** (January-February): 124-125.
Meulemans Medeiros, Daniel, & **J. Gage Crump.** 2012. "New Perspectives on Pharyngeal Dorsoventral Patterning in Development and Evolution of the Vertebrate Jaw." *Developmental Biology* **371** (15 November): 121-135.
Meyer, Stephen C. 2009. *Signature in the Cell: DNA and the Evidence for Intelligent Design*. New York: HarperOne (HarperCollins).
———. 2013. *Darwin's Doubt: The Explosive Origin of Animal Life and the Case for Intelligent Design*. New York: HarperOne (HarperCollins).
Meyer, Stephen C., Scott Minnich, Jonathan Moneymaker, Paul A. Nelson, & **Ralph Seelke.** 2007. *Explore Evolution: The Arguments For and Against Neo-Darwinism*. Melbourne, Australia: Hill House Publishers.
Meyer, Stephen C., Marcus Ross, Paul Nelson, & **Paul Chien.** 2003. "The Cambrian Explosion: Biology's Big Bang," in Campbell & Meyer (2003, 323-402).
Michell, G. B. 1932. "The So-Called 'Babylonian Epic of Creation'." *Journal of the Transactions of the Victoria Institute* (AKA *Faith and Thought*) **64**: 102-122.
Micol, José Luis, & **Sarah Hake.** 2003. "The Development of Plant Leaves." *Plant Physiology* **131** (February): 389-394.
Milius, Susan. 2000. "Toothy valves control crocodile hearts." *Science News* **158** (26 August): 133.
Milla, Rubén, & **Peter B. Reich.** 2011. "Multi-trait interactions, not phylogeny, fine-tune leaf size reduction with increasing altitude." *Annals of Botany* **107** (March): 455-465.
Millar, Ronald. 1972. *The Piltdown Hoax*. New York: St. Martin's Press.
Millard, A. R. 1967. "New Babylonian 'Genesis' Story." *Tyndale Bulletin* **18**: 3-18.
Miller, Craig T., Deborah Yelon, Didier Y. R. Stainer, & **Charles B. Kimmel.** 2003. "Two *endothelin* 1 effectors, *hand2* and *bapx1*, pattern ventral pharyngeal cartilage and the jaw joint." *Development* (AKA *Journal of Embryology and Experimental Morphology*) **130** (1 April): 1353-1365.
Miller, Kenneth R. 1999. *Finding Darwin's God: A Scientist's Search for Common Ground Between God and Evolution*. New York: Cliff Street Books.
———. 2014. "Edging Towards Irrelevance: A commentary on recent claims by the Discovery Institute on the evolution of drug resistance in malaria." *Miller and Levine* posting (online pdf at *www.millerandlevine.com* accessed 12/23/2014).
Milton, Richard. 1997. *Shattering the Myths of Darwinism*. Rev. ed. of orig. 1992 (Britain) *Facts of Life*. Rochester, VT: Park Street Press.
Mindell, David P., Michael D. Sorenson, & **Derek E. Dimcheff.** 1998. "Multiple independent origins of mitochondrial gene order in birds." *Proceedings of the National Academy of Sciences* **95** (1 September): 10693-10697.
Minguillón, Carolina, & **Jordi García-Fernàndez.** 2003. "Genesis and evolution of the *Evx* and *Mox* genes and the extended Hox and ParaHox gene clusters." *Genome Biology* (online @ genomebiology.com) **4** (23 January): R12.
Minguillón, Carolina, Jeremy J. Gibson-Brown, & **Malcolm P. Logan.** 2009. "*Tbx4/5* gene duplication and the origin of vertebrate paired appendages." *Proceedings of the National Academy of Sciences* **106** (22 December): 21726-21730.
Mitchell, Edward D. 1989. "A New Cetacean from the Late Eocene La Meseta Formation Seymour Island, Antarctic Peninsula." *Canadian Journal of Fisheries and Aquatic Sciences* **46** (12): 2219-2235.
Mitchell, Elizabeth. 2010. "Doesn't Egyptian Chronology Prove That the Bible Is Unreliable?" *Answers in Genesis* 22 July posting of *The New Answers Book 2* Chapter 24 (online text at *www.answersingenesis.org* accessed 2/22/2014).
———. 2013a. "'Primitive' Mammal Reveals Advanced Designs Instead of Evolutionary Beginning."

News to Know 9 September posting (online text at *www.answersingenesis.org* accessed 10/16/2016).
———. 2013b. "Iron Key to Preserving Dinosaur Soft Tissue." *Answers in Genesis* article for 4 December (online text at *www.answersingenesis.org* accessed 4/18/2015).
———. 2014a. "Review: *Your Inner Reptile*." *Answers in Genesis* article for 19 April (online text at *www.answersingenesis.org* accessed 4/3/2015).
———. 2014b. "Recapitulation Repackaged and Re-Applied." *Answers in Genesis* article for 31 July (online text at *www.answersingenesis.org* accessed 6/13/2015).
———. 2014c. "Rodent-like Mammal Shakes the Evolutionary Tree." *Answers in Genesis* article for 29 November (online text at *www.answersingenesis.org* accessed 2/17/2015).
———. 2015. "Rewinding Evolution from Bird Beak to Dinosaur Snout." *News to Know* 11 July posting (online text at *www.answersingenesis.org* accessed 5/11/2016).
———. 2016a. "Can Scientists Rewind Supposed Evolution of Dinosaurs to Birds?" *News to Know* 28 April posting (online text at *www.answersingenesis.org* accessed 4/28/2016).
———. 2016b. "Do Naked Bearded Dragons Reveal Common Ancestry of Scales, Feathers, and Fur?" *News to Know* 3 October posting (online text at *www.answersingenesis.org* accessed 10/3/2016).
Mitgutsch, Christian, Michael K. Richardson, Rafael Jiménez, José E. Martin, Peter Kondrashov, Merijn A. G. de Bakker, & **Marcelo R. Sánchez-Villagra.** 2012. "Circumventing the polydactyly 'constraint': the mole's 'thumb'." Royal Society *Biology Letters* **8** (February): 74-77.
Mivart, St. George. 1871. *On the Genesis of Species.* New York: D. Appleton & Co.
Moazen, Mehran, Neil Curtis, Paul O'Higgins, Susan E. Evans, & **Michael J. Fagan.** 2009. "Biomechanical assessment of evolutionary changes in the lepidosaurian skull." *Proceedings of the National Academy of Sciences* **106** (19 May): 8273-8277.
Moczek, Armin P. 2010. "Phenotypic plasticity and diversity in insects." *Philosophical Transactions of the Royal Society of London* **B** (Biological Sciences) **365** (February): 593-603.
———. 2011. "The origins of novelty." *Nature* **473** (5 May): 34-35.
Moczek, Armin P., & **Debra J. Rose.** 2009. "Differential recruitment of limb patterning genes during development and diversification of beetle horns." *Proceedings of the National Academy of Sciences* **106** (2 June): 8992-8997.
Modesto, Sean P., & **Robert R. Reisz.** 2002. "An enigmatic new diapsid reptile from the Upper Permian of Eastern Europe." *Journal of Vertebrate Paleontology* **22** (December): 851-855.
Modesto, Sean P., Christian A. Sidor, Bruce S. Rubidge, & **Johann Welman.** 2001. "A second varanopsid skull from the Upper Permian of South Africa: implications for Late Permian 'pelycosaur' evolution." *Lethaia* **34** (December): 249-259.
Modesto, Sean P., Roger M. H. Smith, Nicolás E. Campione, & **Robert R. Reisz.** 2011. "The last 'pelycosaur': a varanopid synapsid from the *Pristerognathus* Assemblage Zone, Middle Permian of South Africa." *Naturwissenschaften* **98** (December): 1027-1034.
Molnar, Ralph E., Sergei M. Kurzanov, & **Dong Zhiming.** 1990. "Carnosauria," in Weishampel *et al.* (1990, 169-209).
Monday, Steven R., Scott A. Minnich, & **Peter C. H. Feng.** 2004. "A 12-Base-Pair Deletion in the Flagellar Master Control Gene *flhC* Causes Nonmotility of the Pathogenic German Sorbitol-Fermenting *Escherichia coli* O157:H^- Strains." *Journal of Bacteriology* **186** (April, No. 8): 2319-2327.
Montague, Ashley, ed. 1984. *Science and Creationism.* New York: Oxford University Press.
Montavon, Thomas, Jean-François Le Garrec, Michel Kerszberg, & **Denis Duboule.** 2008. "Modeling *Hox* gene regulation in digits: reverse collinearity and the molecular origin of thumbness." *Genes & Development* **22** (1 February): 346-359.
Montavon, Thomas, Natalia Soshnikova, Bénédicte Mascrez, Elisabeth Joye, Laurie Thevenet, Erik Splinter, Wouter de Laat, François Spitz, & **Denis Duboule.** 2011. "A Regulatory Archipelago Controls *Hox* Genes Transcription in Digits." *Cell* **147** (23 November): 1132-1145.
Montealegre-Z., Fernando, Thorin Jonsson, Kate A. Robson-Brown, Matthew Postles, & **Daniel Robert.** 2012. "Convergent Evolution Between Insect and Mammalian Audition." *Science* **338** (16 November): 968-971.
Moolhuijzen, Paula, Jerzy K. Kulski, David S. Dunn, David Schibeci, Roberto Barrero, Takeshi Gojobori, & **Matthew Bellgard.** 2010. "The transcript repeat element: the human Alu sequence as a component of gene networks influencing cancer." *Functional & Integrative Genomics* **10** (August): 307-319.
Mooney, Chris. 2014. "This Picture Has Creationists Terrified." *Mother Jones* posting 4 February

(online text at *motherjones.com* accessed 3/28/2014).

Morales, Jorge, & **Martin Pickford.** 2005. "Carnivores from the Middle Miocene Ngorora Formation (13-12 Ma), Kenya." *Estudios geológicos* (Madrid) **61** (3-6): 271-284.

———. 2011. "A new paradoxurine carnivore from the Late Miocene Siwaliks of India and a review of the bunodont viverrids of Africa." *Geobios* **44** (March-June): 271-277.

Moran, Laurence A. 2014. "Michael Behe and the edge of evolution." *Sandwalk* posting 21 July (online text at *sandwalk.blogspot.com* accessed 8/14/2014).

Moreland, J. P., ed. 1994. *The Creation Hypothesis: Scientific Evidence for an Intelligent Designer.* Downers Grove, IL: InterVarsity Press.

Moreland, J. P., & **John Mark Reynolds**, eds. 1999. *Three Views on Creation and Evolution.* Grand Rapids, MI: Zondervan Publishing House.

Morgan, Bill. 2005. "Two Tricks by Evolutionists." *Creation vs. Evolution* 17 January posting (online text at *www.fishdontwalk.com* accessed 12/12/2012).

Morgan, Bruce A., Roslyn W. Orkin, Selina Noramly, & **Alejandro Perez.** 1998. "Stage-Specific Effects of *Sonic Hedgehog* Expression in the Epidermis." *Developmental Biology* **201** (1 September): 1-12.

Morgante, Michele, Stephan Brunner, Giorgio Pea, Kevin Fengler, Andrea Zuccolo, & **Antoni Rafalski.** 2005. "Gene duplication and exon shuffling by helitron-like transposons generate intraspecies diversity in maize." *Nature Genetics* **37** (September): 997-1002.

Morris, Henry M. 1963. *The Twilight of Evolution.* 1995 pb ed. Grand Rapids, MI: Baker Book House.

———. 1975. "No. 26—Resolution for Equitable Treatment of both Creation and Evolution." Institute for Creation Research *Impact* (August) posting (online text at *icr.org* accessed 7/2/2016).

———, ed. 1985. *Scientific Creationism.* 1996 pb ed. (rev. from 1974). Green Forest, AR: Master Books.

Morris, Henry M., & **John D. Morris.** 1996. *The Modern Creation Trilogy Volume II: Science and Creation.* Green Forest, AR: Master Books.

Morris, Henry M., & **Gary E. Parker.** 1987. *What Is Creation Science?* 1997 pb ed. (rev. from 1982). Green Forest, AR: Master Books.

Morris, Henry III. 2013a. *The Book of Beginnings.* 3 vol. Dallas, TX: Institute for Creation Research.

———. 2013b. "Satan's Strategic Plan." *Days of Praise* for 28 October (online text via *my.icr.org* accessed 10/28/2013).

Morris, Henry III, John D. Morris, Jason Lisle, James J. S. Johnson, Nathaniel Jeanson, Randy Guliuzza, Jeffrey Tomkins, Jake Hebert, Frank Sherwin, & **Brian Thomas**. 2013. *Creation Basics & Beyond: An In-Depth Look at Science, Origins, and Evolution.* Dallas, TX: Institute for Creation Research.

Morris, John D. 2013. "Do Fossils Show Signs of Rapid Burial?" *Answers in Genesis* 30 December posting of *Answers Book 3* Chapter 9 (online text at *www.answersingenesis.org* accessed 2/12/2014).

Morrison, Aaron R. 2007. "The Intricate and Masterful Design of the Human Ear." *Apologetics Press* posting (online text at *www.apologeticspress.org* accessed 8/22/2016).

Morrow, Tom. 1999. "Review: Forbidden Archaeology's Impact." *Reports of the National Center for Science Education* **19** (May-June): 14-17.

Mortenson, Terry. 2004. *The Great Turning Point: The Church's Catastrophic Mistake on Geology—before Darwin.* Green Forest, AR: Master Books.

Mortenson, Terry, & **Roger Patterson.** 2013. "Do Evolutionists Believe Darwin's Ideas about Evolution?" *Answers in Genesis* 26 January 2015 posting of *The New Answers Book 3* Chapter 28 (online text at *www.answersingenesis.org* accessed 2/17/2015).

Morton, Glenn R. 1996. "Noah's Ark: A Feasibility Study." *The TalkOrigins Archive* posting 22 November (online text at *talkorigins.org* accessed 7/3/2016).

Motani, Ryosuke, Da-Yong Jiang, Guan-Bao Chen, Andrea Tintori, Olivier Rieppel, Cheng Ji, & **Jian-Dong Huang.** 2015. "A basal ichthyosauriform with a short snout from the Lower Triassic of China." *Nature* **517** (22 January): 485-488.

Motani, Ryosuke, Da-yong Jiang, Andrea Tintori, Olivier Rieppel, & **Guan-bao Chen.** 2014. "Terrestrial Origin of Viviparity in Mesozoic Marine Reptiles Indicated by Early Triassic Embryonic Fossils." *PLoS ONE* (online @ plosone.org) **9** (February): e88640.

Motani, Ryosuke, Nachio Minoura, & **Tatsuro Ando.** 1998. "Ichthyosaurian relationships illuminated by new primitive skeletons from Japan." *Nature* **393** (21 May): 255-257.

Mountcastle, Andrew M., & **Stacey A. Combes.** 2013. "Wing flexibility enhances load-lifting capacity in bumblebees." *Proceedings of the Royal Society of London* **B** (Biological Sciences) **280** (22 May): 20120531.

Müller, Gerd B., & **Günter P. Wagner.** 1991. "NOVELTY IN EVOLUTION: Restructuring the Concept." *Annual Review of Ecology, Evolution and Systematics* **22**: 229-256.

Muller, Werner A. 1996. *Developmental Biology.* 1997 English trans. New York: Springer.

Muncaster, Ralph O. 1997. *Creation versus Evolution: New Scientific Discoveries.* Mission Viejo, CA: Strong Basis To Believe.

Murphy, William J., Jan E. Janecka, Tanja Stadler, Eduardo Eizirik, Oliver A. Ryder, John Gatesy, Robert W. Meredith, & **Mark S. Springer.** 2012. "Response to Comment on 'Impacts of the Cretaceous Terrestrial Revolution and KPg Extinction on Mammal Diversification'." *Science* **337** (6 July): 34.

Musser, Anne M. 2003. "Review of the monotreme fossil record and comparison of palaeontological and molecular data." *Comparative Biochemistry and Physiology* **Part A** (Molecular & Integrative Physiology) **136** (December): 927-942.

———. 2006. "Furry Egg-layers: Monotreme Relationships and Radiations," in Merrick *et al.* (2006, 523-550).

Musser, A. M., & **M. Archer.** 1998. "New Information about the skull and dentary of the Miocene platypus *Obdurodon dicksoni*, and a discussion of ornithorhynchid relationships." *Philosophical Transactions of the Royal Society of London* **B** (Biological Sciences) **353** (July): 1063-1079.

Myers, P. Z. 2008. "I guess 'eponymous' wasn't on the LSAT." *Pharyngula* posting 14 July (online text at *scienceblogs.com/pharyngula* accessed 8/8/2016).

———. 2009. "Jonathan Wells' weird notions about development." *Pharyngula* posting 1 February (online text at *scienceblogs.com/pharyngula* accessed 8/5/2016).

Nagashima, Hiroshi, Shigehiro Kuraku, Katsuhisa Uchida, Yoshie Kawashima Ohya, Yuichi Narita, & **Shigeru Kuratani.** 2007. "On the carapacial ridge in turtle embryos: its developmental origin, function and the chelonian body plan." *Development* (AKA *Journal of Embryology and Experimental Morphology*) **134** (June): 2219-2226.

Nagashima, Hiroshi, Fumiaki Sugahara, Masaki Tekechi, Rolf Ericsson, Yoshie Kawashima-Ohya, Yuichi Narita, & **Shigeru Kuratani.** 2009. "Evolution of the Turtle Body Plan by the Folding and Creation of New Muscle Connections." *Science* **325** (10 July): 193-196.

Naish, Darren. 2009. "Publishing with a hidden agenda: why birds simply cannot be dinosaurs." *Tetrapod Zoology* blog for 17 July (online text at *scienceblogs.com/tetrapodzoology* accessed 10/18/2009).

———. 2015. "*Yi qi* Is Neat But Might Not Have Been the Black Screaming Dino-Dragon of Death." *Tetrapod Zoology* blog for 5 May (online text at *blogs.scientificamerican.com* accessed 5/5/2015).

———. 2016a. "The Integrated Maniraptoran, Part 1." *Tetrapod Zoology* blog for 16 May (online text at *blogs.scientificamerican.com* accessed 5/17/2016).

———. 2016b. "The Integrated Maniraptoran, Part 2: Meet the Maniraptorans." *Tetrapod Zoology* blog for 23 May (online text at *blogs.scientificamerican.com* accessed 9/9/2016).

———. 2016c. "The Integrated Maniraptoran, Part 3: Feathers Did Not Evolve in an Aerodynamic Context." *Tetrapod Zoology* blog for 30 May (online text at *blogs.scientificamerican.com* accessed 9/9/2016).

Nakamura, Tetsuya, Andrew R. Gehrke, Justin Lemberg, Julie Szymaszek, & **Neil H. Shubin.** 2016. "Digits and fin rays share common developmental histories." *Nature* **537** (8 September): 225-228.

Nakayama, Takuro, & **John M. Archibald.** 2012. "Evolving a photosynthetic organelle." *BMC Biology* (online @ biomedcentral.com) **10** (24 April): 35.

Nakayama, Takuro, Yuko Ikegami, Takeshi Nakayama, Ken-ichiro Ishida, Yuji Inagaki, & **Isao Inouye.** 2011. "Spheroid bodies in rhopalodiacean diatoms were derived from a single endosymbiotic cyanobacterium." *Journal of Plant Research* **124** (January): 93-97.

National Science Board. 2014. "National Science Board Science and Engineering Indicators 2014." *National Science Foundation* posting (online pdf at *www.nsf.gov/statistics/seind14/* accessed 2/21/2014).

Natural History Museum. 2013. "Reconstructing the *Megatherium* skeleton." London *Natural History Museum* undated posting (online text at *www.nhm.ac.uk* accessed 10/21/2013).

NCSE. 2008. "Sudden Appearance?" *National Center for Science Education* "Critique: Exploring 'Explore Evolution'" 25 September posting (online text via *ncse.com* accessed 1/29/2014).

Near, Thomas J., Alex Dornburg, Kristen L. Kuhn, Joseph T. Eastman, Jillian N. Pennington, Tomaso Patarnello, Lorenzo Zan, Daniel A. Fernández, & Christopher D. Jones. 2012. "Ancient climate change, antifreeze, and the evolutionary diversification of Antarctic fishes." *Proceedings of the National Academy of Sciences* **109** (28 February): 3434-3439.

Nelson, Gareth. 1998. "Colin Patterson (1933-98)." *Nature* **394** (13 August): 626.

Nelson, Paul. 1998. "Colin Patterson Revisits His Famous Question about Evolution" Access Research Network 22 July posting of *Origins & Design* (AKA *Origins Research*) **17** (1): online text accessed 4/7/2016 (*www.arn.org*).

———. 2009a. "Monophyly vs. Polyphyly and Christian Schwabe." *Explore Evolution* critical response 24 February (online text at *www2.exploreevolution.com* accessed 1/4/2011).

———. 2009b. "The Definitions of 'Evolution'." *Explore Evolution* critical response 24 February (online text at *www2.exploreevolution.com* accessed 1/4/2011).

———. 2009c. "The 'Fact' of Evolution." *Explore Evolution* critical response 24 February (online text at *www2.exploreevolution.com* accessed 1/4/2011).

———. 2009d. "Malcolm Gordon and the Origin of Tetrapods." *Explore Evolution* critical response 24 February (online text at *www2.exploreevolution.com* accessed 12/27/2010).

———. 2009e. "Molecular Phylogeny and Phylogenetic Trees." *Explore Evolution* critical response 24 February (online text at *www2.exploreevolution.com* accessed 1/4/2011).

———. 2009f. "Introduction: The Catechism Versus the Data." *Explore Evolution* critical response 24 February (online text at *www2.exploreevolution.com* accessed 1/4/2011).

———. 2009g. "Evolution and Testability." *Explore Evolution* critical response 24 February (online text at *www2.exploreevolution.com* accessed 1/4/2011).

———. 2009h. "The Creationism Gambit." *Explore Evolution* critical response 2 March (online text at *www2.exploreevolution.com* accessed 1/4/2011).

———. 2009i. "Reply to NCSE on Universal Genetic Code." *Explore Evolution* critical response 21 August (online text at *www2.exploreevolution.com* accessed 1/4/2011).

———. 2010. "Seeing Ghosts in the Bushes (Part 2): How Is Common Descent Tested?" Discovery Institute *Evolution News & Views* for 4 February (online text at *evolutionnews.org* accessed 7/11/2012).

Nelson, Paul, & John Mark Reynolds. 1999a. "Young Earth Creationism," in Moreland & Reynolds (1999, 41-75).

———. 1999b. "Conclusion," in Moreland & Reynolds (1999, 95-102).

Nesbitt, Sterling. 2011. "The Early Evolution of Archosaurs: Relationships and the Origin of Major Clades." *Bulletin of the American Museum of Natural History* **352**: 1-292.

Nesbitt, Sterling J., Paul M. Barrett, Sarah Werning, Christian A. Sidor, & Allen J. Charig. 2013. "The oldest dinosaur? A Middle Triassic dinosauriform from Tanzania." Royal Society *Biology Letters* **9** (February): 20120949.

Nesbitt, Sterling J., Richard J. Butler, & David J. Gower. 2013. "A New Archosauriform (Reptilia: Diapsida) from the Manda Beds (Middle Triassic) of Southwestern Tanzania." *PLoS ONE* (online @ plosone.org) **8** (September): e72753.

Neuberger, Michael S., & Brian S. Hartley. 1981. "Structure of an Experimentally Evolved Gene Duplication Encoding Ribitol Dehydrogenase in a Mutant of *Klebsiella aerogenes*." *Journal of General Microbiology* **122**: 181-191.

Newman, C. M., J. E. Cohen, & C. Kipnis. 1985. "Neo-darwinian evolution implies punctuated equilibria." *Nature* **315** (30 May): 400-401.

Newman, Robert C. 1999. "Progressive Creationism (*'Old Earth Creationism'*)," in Moreland & Reynolds (1999, 105-133).

Ni, Xijun, Daniel L. Gebo, Marian Dagosto, Jin Meng, Paul Tafforeau, John J. Flynn, & K. Christopher Beard. 2013. "The oldest known primate skeleton and early haplorhine evolution." *Nature* **498** (6 June): 60-64.

Nicotra, Adrienne B., Andrea Leigh, C. Kevin Boyce, Cynthia S. Jones, Karl J. Niklas, Dana L. Royer., & Hirokazu Tsukaya. 2011. "The evolution and functional significance of leaf shape in the angiosperms." *Functional Plant Biology* **38** (7): 535-552.

Niculita, Hélène. 2006. "Expression of *abdominal-A* homeotic gene in ants with different abdominal morphologies." *Gene Expression Patterns* **6** (January): 141-145.

Niedźwiedzki, Grzegorz, Piotr Szrek, Katarzyna Narkiewicz, Marek Narkiewicz, & Per E. Ahlberg. 2010. "Tetrapod trackways from the early Middle Devonian period of Poland." *Nature* **463** (7 January): 43-48.

Nijhout, H. Frederik. 1990. "Metaphors and the Role of Genes in Development." *BioEssays* **12** (September): 441-446.
———. 1999. "When developmental pathways diverge." *Proceedings of the National Academy of Sciences* **96** (11 May): 5348-5350.
Nijhout, H. F., & **D. J. Emlen.** 1998. "Competition among body parts in the development of insect morphology." *Proceedings of the National Academy of Sciences* **95** (31 March): 3685-3689.
Nilsson, Dan-E. 2003. "Beware of Pseudo-science: A Response to David Berlinski's Attack on My Calculation of How Long it Takes for an Eye to Evolve." *Talk Reason* 13 June posting of 27 March letter (online text at *talkreason.org* accessed 6/15/2003).
———. 2004. "Eye evolution: a question of genetic promiscuity." *Current Opinion in Neurobiology* **14** (August): 407-414.
———. 2009. "The evolution of eyes and visually guided behaviour." *Philosophical Transactions of the Royal Society of London* **B** (Biological Sciences) **364** (12 October): 2833-2847.
Nilsson, Dan-E., & **Susanne Pelger.** 1994. "A pessimistic estimate of the time required for an eye to evolve." *Proceedings of the Royal Society of London* **B** (Biological Sciences) **256** (22 April): 53-58.
Nishimura, José M. Grajales, Esteban Cedillo-Parod, Carmen Rosales-Domínguez, Patricia Padilla-Avila, Antonieta Sánchez-Ríos, Dante J. Morán-Zenteno, Walter Alvarez, Philippe Claeys, José Ruíz-Morales, & **Jesús García-Hernández.** 2000. "Chicxulub impact: The origin of reservoir and seal facies in the southeastern Mexico oil fields." *Geology* **28** (April): 307-310.
Nolan, William A. 1978. *The Baby in the Bottle: An Investigative Review of the Edelin Case and Its Larger Meanings for the Controversy Over Abortion Reform.* New York: Coward, McCann & Geoghegan.
Noramly, Selina, Allison Freeman, & **Bruce A. Morgan.** 1999. "β-catenin signaling can initiate feather bud development." *Development* (AKA *Journal of Embryology and Experimental Morphology*) **126** (15 August): 3509-3521.
Norell, Mark A., & **Julia A. Clarke.** 2001. "Fossil that fills a critical gap in avian evolution." *Nature* **409** (11 January): 181-184.
Norell, Mark A., Qiang Ji, Keqin Gao, Chongxi Yuan, Yibin Zhao, & **Lixia Wang.** 2002. "'Modern' feathers on a non-avian dinosaur." *Nature* **416** (7 March): 36-37.
Norell, Mark A., & **Michael J. Novacek.** 1992a. "The Fossil Record and Evolution: Comparing Cladistic and Paleontologic Evidence for Vertebrate History." *Science* **255** (27 March): 1690-1693.
———. 1992b. "Congruence between superpositional and phylogenetic patterns: Comparing cladistic patterns with fossil records." *Cladistics* **8** (December): 319-337.
Norell, Mark A., & **Xing Xu.** 2005. "Feathered Dinosaurs." *Annual Review of Earth and Planetary Sciences* **33**: 277-299.
Norman, David. 1985. *The Illustrated Encyclopedia of Dinosaurs: An Original and Compelling Insight Into Life in the Dinosaur Kingdom.* New York: Crescent Books.
———. 1994. *Prehistoric Life: The Rise of the Vertebrates.* New York: Macmillan.
Northcutt, R. Glenn. 2011. "Evolving Large and Complex Brains." *Science* **332** (20 May): 926-927.
Northcutt, R. Glenn, & **Jon H. Kaas.** 1995. "The emergence and evolution of mammalian neocortex." *Trends in Neuroscience* **18** (September): 373-379.
———. 1996. "Levels of organization and the evolution of isocortex: Reply." *Trends in Neuroscience* **19** (March): 91-92.
Northrup, Bernard E. 1997. "Taphonomy: A Tool for Studying Earth's Biblical History." Lambert Dolphin's website posting of revised version of 1985 Seattle Creation Conference presentation (online text via *ldolphin.org* accessed 10/7/2003).
Novacek, Michael J., Mark Norell, Malcolm C. McKenna, & **James Clark.** 1994. "Fossils of the Flaming Cliffs." *Scientific American* **271** (December): 60-63, 66-69.
Novacek, Michael J., Guillermo W. Rougier, John R. Wible, Malcolm C. McKenna, Demberelyin Dashzeveg, & **Inés Horovitz.** 1997. "Epipubic bones in eutherian mammals from the Late Cretaceous of Mongolia." *Nature* **389** (2 October): 483-486.
Novas, Fernando E. 1997. "South American Dinosaurs," in Currie & Padian (1997, 678-689).
Noveen, Alexander, Ting-Xin Jiang, & **Cheng-Ming Chuong.** 1995. "Protein Kinase A and Protein Kinase C Modulators Have Reciprocal Effects on Mesenchymal Condensation during Skin Appendage Morphogenesis." *Developmental Biology* **171** (1 October): 677-693.
Novella, Steven. 2008. "Ancient Cambodian Stegosaurus?" *NeuroLogica Blog* posting 19 June (online text at *theness.com* accessed 11/16/2011).
Numbers, Ronald L. 1992. *The Creationists: The Evolution of Scientific Creationism.* 1993 pb ed.

Berkeley, CA: University of California Press.
Nummela, Sirpa, J. G. M. Thewissen, Sunil Bajpai, S. Taseer Hussain, & **Kishor Kumar.** 2004. "Eocene evolution of whale hearing." *Nature* **430** (12 August): 776-778.
———. 2007. "Sound Transmission in Archaeic and Modern Whales: Anatomical Adaptations for Underwater Hearing." *The Anatomical Record: Advances in Integrative Anatomy and Evolutionary Biology* **290** (June): 716-733.
Nutting, Dave . 2007. "Haeckel's Biogenetic Law: Totally False—But Still in the Textbooks!" Alpha Omega Institute Institute (Grand Junction, CO) *Think & Believe* **25** (January/February): 3.
Nutting, Mary Jo. 1997. "Blood Clotting, Mousetraps, and Irreducible Complexity." Alpha Omega Institute Institute (Grand Junction, CO) *Think & Believe* **14** (September/October): 3.
Nygaard, Sanne, Guojie Zhang, Morten Schiott, Cai Li, Yannick Wurm, Haofu Hu, Jiajian Zhou, Lu Ji, Feng Qiu, Morten Rasmussen, Hailin Pan, Frank Hauser, Anders Krogh, Cornelius J. P. Grimmelikhuijzen, Jun Wang, & **Jacobus J. Boomsma.** 2011. "The genome of the leaf-cutting ant *Acrymyrmex echinatior* suggests key adaptations to advance social life and fungus farming." *Genome Research* **21** (August): 1339-1348.

Oard, Michael J. 2013. "Is Man the Cause of Global Warming?" *Answers in Genesis* 17 April posting of *The New Answers Book 3* Chapter 7 (online text at *www.answersingenesis.org* accessed 2/17/2015).
O'Connor, Anne, & **Matthew A. Wills.** 2016. "Measuring Stratigraphic Congruence Across Trees, Higher Taxa, and Time." *Systematic Biology* **65** (September): 792-811.
O'Connor, Jingmai, Luis M. Chiappe, Chunling Gao, & **Bo Zhao.** 2011. "Anatomy of the Early Cretaceous enantiornithine bird *Rapaxavis pani.*" *Acta Palaeontologica Polonica* **56** (3): 463-475.
O'Connor, J. K., Min Wang, Xiaoting Zheng, & **Zhonghe Zhou.** 2014. "Reply to Foth: Preserved cartilage is rare but not absent: Troodontid sternal plates are absent, not rare." *Proceedings of the National Academy of Sciences* **111** (16 December): E5335.
O'Connor, Jingmai K., Yuguang Zhang, Luis M. Chiappe, Qingjin Meng, Li Quanguo, & **Liu Di.** 2013. "A new enantiornithine from the Yixian Formation with the first recognized avian enamel specialization." *Journal of Vertebrate Paleontology* **33** (January): 1-12.
O'Connor, Patrick M., & **Leon P. A. M. Claessens.** 2005. "Basic avian pulmonary design and flow-through ventilation in non-avian theropod dinosaurs." *Nature* **436** (14 July): 253-256.
O'Connor, Patrick M., Joseph J. W. Sertich, Nancy J. Stevens, Eric M. Roberts, Michael D. Gottfried, Tobin L. Hieronymous, Zubair A. Jinnah, Ryan Ridgely, Sifa E. Ngasala, & **Jesuit Temba.** 2010. "The evolution of mammal-like crocodyliforms in the Cretaceous Period of Gondwana." *Nature* **466** (5 August): 748-751.
O'Donnell, Phillip. 2002. "On the Validity of Sea Dragons." *Cryptozoology Research Team* July 2011 posting (online text at *livingdinos.com* accessed 8/17/2014).
———. 2007. "Deadly Cryptids--The Last Living Dragons?" *Cryptozoology Research Team* July 2011 posting (online text at *livingdinos.com* accessed 8/17/2014).
Oftedal, O. T. 2002. "The Mammary Gland and Its Origin During Synapsid Evolution." *Journal of Mammary Gland Biology and Neoplasia* **7** (July): 225-252.
———. 2012. "The evolution of milk secretion and its ancient origins." *Animal* **6** (March): 355-368.
———. 2013. "Origin and Evolution of the Major Constituents of Milk," in McSweeney & Fox (2013, 1-42).
Oftedal, O. T., & **D. Dhouailly.** 2013. "Evo-Devo of the Mammary Gland." *Journal of Mammary Gland Biology and Neoplasia* **18** (June): 105-120.
O'Gorman, Stephen. 2005. "Second Branchial Arch Lineages of the Middle Ear of Wild-Type and *Hoxa2* Mutant Mice." *Developmental Dynamics* **234** (September): 124-131.
Okano, Junko, Shigehiko Suzuki, & **Kohei Shiota.** 2006. "Regional heterogeneity in the developing palate: morphological and molecular evidence for normal and abnormal palatogenesis." *Congenital Anomalies* (Kyoto) **46** (June): 49-54.
Olasky, Marvin N. 1987. "The Real Story of the Trial That 'Disgraced Fundamentalism'." *Fundamentalist Journal* **6** (March): 23-26.
———. 2007. "Darwin slayer." *World Magazine* "When the base cracks" issue posting 21 July (online text at *www.worldmag.com* accessed 5/26/2014).
———. 2011. "Darwin matters: The influence of evolutionary thinking reaches far beyond biology." *World Magazine* posting 2 July (online text at *www.worldmag.com* accessed 8/30/2013).
———. 2012a. "Doubting Darwin." *World Magazine* posting 5 May (online text at

www.worldmag.com accessed 8/30/2013).
———. 2012b. "Museum Visit Reveals a lot of Uncertainties Within Darwinism." *Townhall* 27 July posting (online text at *townhall.com* accessed 8/30/2013).
———. 2015. "A tale of two museums." *World Magazine* posting 7 August (online text at *www.worldmag.com* accessed 8/12/2015).
———. 2016. "Challenging Darwin." *World Magazine* posting 25 June (online text at *world.wng.org* accessed 7/23/2016).
O'Leary, Denyse. 2013a. "The Science Fiction series at your fingertips—human evolution." *Uncommon Descent* posting 19 December (online text at *www.uncommondescent.com* accessed 5/1/2015).
———. 2013b. "The *Science Fictions* series at your fingertips: the human mind." *Uncommon Descent* posting 19 December (online text at *www.uncommondescent.com* accessed 5/1/2015).
———. 2013c. "The *Science Fictions* series at your fingertips—origin of life." *Uncommon Descent* posting 19 December (online text at *www.uncommondescent.com* accessed 5/1/2015).
———. 2014. "Picture that terrifies creationists?" *Uncommon Descent* posting 5 February (online text at *www.uncommondescent.com* accessed 1/2/2016).
———. 2015a. "Evolution: The Fossils Speak, but Hardly with One Voice." Discovery Institute *Evolution News & Views* for 8 July (online text at *evolutionnews.org* accessed 7/8/2015).
———. 2015b. "Evolution Appears to Converge on Goals--But in Darwinian Terms, Is That Possible?" Discovery Institute *Evolution News & Views* for 27 July (online text at *evolutionnews.org* accessed 1/2/2016).
———. 2015c. "Stasis: Life Goes On but Evolution Does Not Happen." Discovery Institute *Evolution News & Views* for 12 October (online text at *evolutionnews.org* accessed 3/18/2016).
———. 2015d. "Conclusions: What the Fossils Told Us in Their Own Words." Discovery Institute *Evolution News & Views* for 20 October (online text at *evolutionnews.org* accessed 2/21/2016).
———. 2015e. "What Can We Hope to Learn About Animal Minds?" Discovery Institute *Evolution News & Views* for 4 December (online text at *evolutionnews.org* accessed 4/22/2016).
———. 2015f. "Furry, Feathery, and Finny Animals Speak Their Minds." Discovery Institute *Evolution News & Views* for 22 December (online text at *evolutionnews.org* accessed 3/4/2016).
O'Leary, Maureen A., Jonathan I. Bloch, John J. Flynn, Timothy J. Gaudin, Andres Giallombardo, Norberto P. Giannini, Suzann L. Goldberg, Brian P. Kraatz, Zhe-Xi Luo, Jin Meng, Xijun Ni, Michael J. Novacek, Fernando A. Perini, Zachary S. Randall, Guillermo W. Rougier, Eric J. Sargis, Mary T. Silcox, Nancy B. Simmons, Michelle Spaulding, Paul M. Velazco, Marcelo Weksler, John R. Wible, & **Andrea L. Cirranello.** 2013a. "The Placental Mammal Ancestor and the Post-K-Pg Radiation of Placentals." *Science* **339** (8 February): 662-667.
———. 2013b. "Response to Comment on 'The Placental Mammal Ancestor and the Post-K-Pg Radiation of Placentals'." *Science* **341** (9 August): 613.
Oliveira, Téo Veiga de, Cesar Leandro Schultz, & **Marina Bento Soares.** 2009. "A partial skeleton of *Chiniquodon* (Cynodontia, Chiniquodontidae) from the Brazilian Middle Triassic." *Revista Brasileira de Paleontologia* **12** (May/August): 113-122.
Olofsson, Peter. 2007. "The Coulter Hoax: How Ann Coulter Exposed the Intelligent Design Movement." *Skeptical Inquirer* **31** (March/April): 48-50.
Olson, Everett C. 1962. "Late Permian terrestrial vertebrates, U.S.A. and U.S.S.R." *Transactions of the American Philosophical Society* **52** (2): 1-224.
———. 1968. "The Family Caseidae." *Fieldiana Geology* **17** (3): 225-349.
Ó Maoiléidigh, Dáibhid, & **A. J. Hudspeth.** 2013. "Effects of cochlear loading on the motility of active outer hair cells." *Proceedings of the National Academy of Sciences* **110** (2 April): 5474-5479.
O'Meara, Rachel N., & **Robert J. Asher.** 2016. "The evolution of growth patterns in mammalian versus nonmammalian cynodonts." *Paleobiology* **42** (August): 439-464.
Ometto, Lino, DeWayne Shoemaker, Kenneth G. Ross, & **Laurent Keller.** 2011. "Evolution of Gene Expression in Fire Ants: The Effects of Developmental Stage, Caste, and Species." *Molecular Biology and Evolution* **28** (April): 1381-1392.
Onoue, Tetsuji, Honami Sato, Tomoki Nakamura, Takaaki Noguchi, Yoshihiro Hidaka, Naoki Shirai, Mitsuru Ebihara, Takahito Osawa, Yuichi Hatsukawa, Yosuke Toh, Mitsuo Koizumi, Hideo Harada, Michael J. Orchard, & **Munetomo Nedachi.** 2012. "Deep-sea record of impact apparently unrelated to mass extinction in the Late Triassic." *Proceedings of the National Academy of Sciences* **109** (20 November): 19134-19139.

O'Reilly, Jill X., Urs Schüffelgen, Steven F. Cuell, Timothy E. J. Behrens, Rogier B. Mars, & Matthew F. S. Rushworth. 2013. "Dissociable effects of surprise and model update in parietal and anterior cingulate cortex." *Proceedings of the National Academy of Sciences* **110** (17 September): E3660-E3669.
Ortiz de Montellano, Bernard. 1996. "The Dogon People Revisited." *Skeptical Inquirer* **20** (November/December): 39-42.
Osanai, Nozomi. 2005. "A Comparison from Secular Historical Records." *Answers in Genesis* 3 August posting of Chapter 7 of *A Comparative Study of the Flood Accounts in the Gilgamesh Epic and Genesis* (online text at *www.answersingenesis.org* accessed 2/14/2013).
Osborne, Henry Field. 1922. "Hesperopithecus, the First Anthropoid Primate Found in America." *Science* **55** (5 May): 463-465.
Osmólska, Halszka. 1990. "Theropoda," in Weishampel *et al.* (1990, 148-150).
Oster, G., & P. Alberch. 1982. "Evolution and Bifurcation of Developmental Programs." *Evolution* **36** (May): 444-459.

Padian, Kevin. 2001. "Cross-testing adaptive hypotheses: Phylogenetic analysis and the origin of bird flight." *Integrative and Comparative Biology* (AKA *American Zoologist*) **41** (June): 598-607.
Padian, Kevin, & Kenneth D. Angielczyk. 2007. "'Transitional Forms' versus Transitional Features," in Petto & Godfrey (2007, 197-230).
Padian, Kevin, & John R. Horner. 2002. "Typology versus transformation in the origin of birds." *Trends in Ecology & Evolution* **17** (March): 120-124.
Painter, K. J., G. S. Hunt, K. L. Wells, J. A. Johansson, & D. J. Headon. 2012. "Towards an integrated experimental-theoretical approach for assessing the mechanistic basis of hair and feather morphogenesis." The Royal Society *Interface Focus* **2** (August): 433-450.
Palatnik, Javier F., Edwards Allen, Xuelin Wu, Carla Schommer, Rebecca Schwab, James C. Carrington, & Detlef Weigel. 2003. "Control of leaf morphogenesis by microRNAs." *Nature* **425** (18 September): 257-263.
Pallen, Mark J., & Nicholas J. Matzke. 2006. "From *The Origin of Species* to the origin of the bacterial flagella." *Nature Reviews Microbiology* **4** (October): 784-790.
Palmer, Douglas. 1999. *Atlas of the Prehistoric World.* New York: Discovery Books.
Panero, Jose, Javier Francisco-Ortega, Robert K. Jansen, & Arnoldo Santos-Guerra. 1999. "Molecular evidence for multiple origins of woodiness and a New World biogeographic connection of the Macaronesian Island endemic *Pericallis* (Asteraceae: Senecioneae)." *Proceedings of the National Academy of Sciences* **96** (23 November): 13886-13891.
Parrish, J. Michael. 1991. "A New Specimen of an Early Crocodylomorph (cf. *Spenosuchus* sp.) from the Upper Triassic Chinle Formation of Petrified Forest National Park, Arizona." *Journal of Vertebrate Paleontology* **11** (June): 193-212.
Parsons, Kevin J., A. Trent Taylor, Kara E. Powder, & R. Craig Albertson. 2014. "Wnt signalling underlies the evolution of new phenotypes and craniofacial variability in Lake Malawi cichlids." *Nature Communications* (online @ nature.com) **5** (3 April): 3629.
Pascual, Rosendo, Michael Archer, Edgardo Ortiz Jaureguizar, José L. Prado, Henk Godthelp, & Suzanne J. Hand. 1992a. "First discovery of monotremes in South America." *Nature* **356** (23 April): 704-706.
———. 1992b. "The first non-Australian monotreme: An Early Paleocene South American platypus (Monotremata, Ornithorhynchidae)," in Augee (1992, 1-14).
Pascual, Rosendo, Francisco J. Goin, Lucia Balarino, & Daniel E. Udrizar Sauthier. 2002. "New data on the Paleocene monotreme *Monotrematum sudamericanum*, and the convergent evolution of triangulate molars." *Acta Palaeontologica Polonica* **47** (3): 487-492.
Paton, R. L., T. R. Smithson, & J. A. Clack. 1999. "An amniote-like skeleton from the Early Carboniferous of Scotland." *Nature* **398** (8 April): 508-513.
Patterson, Colin. 1978. *Evolution.* London: Routledge.
———. 1982. "Morphological Characters and Homology," in Joysey & Friday (1982, 21-74).
———. 1988. "Homology in Classical and Molecular Biology." *Molecular Biology and Evolution* **5** (November): 603-625.
———. 1994. "Null or minimal models," in Scotland *et al.* (1994, 173-92).
———. 1999. *Evolution.* Ithaca, NY: Comstock Publishing Associates (Cornell University Press).
Patterson, Roger. 2009. *Evolution Exposed: Biology.* Petersburg, KY: Answers in Genesis.
Patton, Don. 2008. "Dinosaurs in ancient Cambodian temple." *The Science of Creation* undated

posting (online text at *www.bible.ca* accessed 9/13/2016).
———. 2011. "Mystery of Acambaro, Dinosaur Figurines, Fact or Fraud." *YouTube* 3 February uploading by **JesusLostChildren777** (online video at *www.youtube.com* accessed 8/18/2014).
Paul, Gregory S. 1988. *Predatory Dinosaurs of the World*. 1989 pb ed. New York: Touchstone Books.
Peczkis, Jan. 1993. "Evolving Student Thought: Simulating evolution over many generations." National Science Teachers Association *The Science Teacher* **60** (January): 42-45.
———. 1994. "Implications of body-mass estimates for dinosaurs." *Journal of Vertebrate Paleontology* **14** (December): 520-533.
Pei, Junling, Zhiming Sun, Jing Liu, Jian Liu, Xisheng Wang, Zhenyu Yang, Yue Zhao, & Haibing Li. 2011. "A paleomagnetic study from the Late Jurassic volcanics (155 Ma), North China: Implications for the width of Mongol-Okhotsk Ocean." *Tectonophysics* **510** (4 October): 370-380.
Pelleau, Stéphane, Eli L. Moss, Satish K. Dhingra, Béatrice Volney, Jessica Casteras, Stanislaw J. Gabryszewski, Sarah K. Volkman, Dyann F. Wirth, Eric Legrand, David A. Fidock, Daniel E. Neafsey, & Lise Musset. 2015. "Adaptive evolution of malaria parasites in French Guiana: Reversal of chloroquine resistance by acquisition of a mutation in *pfcri*." *Proceedings of the National Academy of Sciences* **112** (15 September): 11672-11677.
Peng, Yuanyuan, Vladimir Makarkin, Xiaodong Wang, & Dong Ren. 2011. "A new fossil silky lacewing genus (Neuroptera, Psychopsidae) from the Early Cretaceous Yixian Formation of China." *ZooKeys* (online @ ncbi.nlm.nih.gov) **130** (24 September): 217-228.
Pennisi, Elizabeth. 1999. "From Embryos and Fossils, New Clues to Vertebrate Evolution." *Science* **284** (23 April): 575, 577.
———. 2015. "How birds got their beaks." *Science* **348** (15 May): 744.
———. 2016. "Do genomic conflicts drive evolution?" *Science* **353** (22 July): 334-335.
Pennock, Robert T. 1999. *Tower of Babel: The Evidence against the New Creationism*. Cambridge, MA: MIT Press.
Perkins, Sid. 2013a. "Oldest primate skeleton unveiled." *Nature News* (5 June) posting (online text at *nature.com* accessed 3/18/2016).
———. 2013b. "Fossils throw mammalian family tree into disarray." *Nature News* (7 August) posting (online text at *nature.com* accessed 8/8/2013).
———. 2013e. "Fossil reveals features of mammal line that outlived dinosaurs." *Nature News* (15 August) posting (online text at *nature.com* accessed 8/16/2013).
Perloff, James. 1999. *Tornado in a Junkyard: The Relentless Myth of Darwinism*. Arlington, MA: Refuge Books.
Perrichot, Vincent, Loïc Marion, Didier Néraudeau, Romain Vullo, & Paul Tafforeau. 2008. "The early evolution of feathers: fossil evidence from Cretaceous amber of France." *Proceedings of the Royal Society of London* **B** (Biological Sciences) **275** (22 May): 1197-1202.
Perrichot, Vincent, André Nel, Didier Néraudeau, Sébastien Lacau, & Thierry Guyot. 2008. "New fossil ants in French Cretaceous amber (Hymenoptera: Formicidae)." *Naturwissenschaften* **95** (February): 91-97.
Perry, Steven F., & Martin Sander. 2004. "Reconstructing the evolution of the respiratory apparatus in tetrapods." *Respiratory Physiology & Neurobiology* **144** (15 December): 125-139.
Perry, Steven F., Thomas Similowski, Wilfried Klein, & Jonathan R. Codd. 2010. "The evolutionary origin of the mammalian diaphragm." *Respiratory Physiology & Neurobiology* **171** (15 April): 1-16.
Persson, B., J. Hedlund, & H. Jörnvall. 2008. "The MDR superfamily." *Cellular and Molecular Life Sciences* **65** (December #24): 3879-3894.
Petersen, Jörn, René Teich, Burkhard Becker, Rüdiger Cerff, & Henner Brinkmann. 2006. "The *GapA/B* Gene Duplication Marks the Origin of Streptophyta (Charophytes and Land Plants)." *Molecular Biology and Evolution* **23** (June): 1109-1118.
Peterson, Kevin J., & Nicholas J. Butterfield. 2005. "Origin of the Eumetazoa: Testing ecological predictions of molecular clocks against the Proterozoic fossil record." *Proceedings of the National Academy of Sciences* **102** (5 July): 9547-9552.
Peterson, Kevin J., R. Andrew Cameron, & Eric H. Davidson. 2000. "Bilaterian Origins: Significance of New Experimental Observations." *Developmental Biology* **219** (1 March): 1-17.
Peterson, Kevin J., James A. Cotton, James G. Gehling, & Davide Pisani. 2008. "The Ediacaran emergence of bilaterians: congruence between the genetic and the geological fossil records." *Philosophical Transactions of the Royal Society of London* **B** (Biological Sciences) **363** (27 April): 1435-1443.

Peterson, Kevin J., & **Eric H. Davidson.** 2000. "Regulatory evolution and the origin of the bilaterians." *Proceedings of the National Academy of Sciences* **97** (25 April): 4430-4433.

Peterson, Kevin J., Michael R. Dietrich, & **Mark A. McPeek.** 2009. "MicroRNAs and metazoan macroevolution: insights into canalization, complexity, and the Cambrian explosion." *BioEssays* **31** (July): 736-747.

Peterson, Kevin J., Jessica B. Lyons, Kristin S. Nowak, Carter M. Takacs, Matthew J. Wargo, & **Mark A. McPeek.** 2004. "Estimating metazoan divergence times with a molecular clock." *Proceedings of the National Academy of Sciences* **101** (27 April): 6536-6541.

Petto, Andrew J., & **Laurie R. Godfrey**, eds. 2007. *Scientists Confront Intelligent Design and Creationism.* New York: W. W. Norton & Co.

Pian, Rebecca, Michael Archer, & **Suzanne J. Hand.** 2013. "A new, giant platypus, *Obdurodon tharalkooschild*, sp. nov. (Monotremata, Ornithorhynchidae), from the Riversleigh World Heritage Area, Australia." *Journal of Vertebrate Paleontology* **33** (November): 1255-1259.

Piazza, Paolo, C. Donovan Bailey, Marla Cartolano, Jonathan Kriger, Jun Cao, Stephen Ossowski, Korbinian Schneeberger, Fei He, Juliette de Meaux, Neil Hall, Norman MacLeod, Dmitry Filatov, Angela Hay, & **Miltos Tsiantis.** 2010. "*Arabidopsis thaliana* Leaf Form Evolved via Loss of KNOX Expression in Leaves in Association with a Selective Sweep." *Current Biology* **20** (21 December): 2223-2228.

Pickering, Mark, & **James F. X. Jones.** 2002. "The diaphragm: two physiological muscles in one." *Journal of Anatomy* **201** (October): 305-312.

Pigliucci, Massimo. 2008. "What, if Anything, Is an Evolutionary Novelty?" *Philosophy of Science* **75** (December): 887-898.

Pitman, Sean D. 2010. "Ancient Fossils with Preserved Soft Tissues and DNA." *Detecting Design* May 2004 posting revised January (online text at *detectingdesign.com* accessed 6/19/2013).

Pittendrigh, C. S., L. H. Tiffany, & **G. G. Simpson.** 1957. *Life: An Introduction to Biology.* New York: Harcourt, Brace & Co.

Plait, Philip C. 2010a. "Texas congressman uses porn to kill science funding." *Bad Astronomy* blog for 17 May (online text at *blogs.discovermagazine.com/badastronomy* accessed 11/13/2010).

———. 2010b. "Breaking: Republicans derail the COMPETES act." *Bad Astronomy* blog for 19 May (online text at *blogs.discovermagazine.com/badastronomy* accessed 11/13/2010).

———. 2010c. "Rep. Ralph Hall's unbelievable statement on science funding bill." *Bad Astronomy* blog for 2 June (online text at *blogs.discovermagazine.com/badastronomy* accessed 11/13/2010).

———. 2013. "And *These* Are the People Making Laws in Louisiana." *Bad Astronomy* blog posting 23 January (online text at *www.slate.com* accessed 1/26/2013).

———. 2016. "Texas Rep. Louie Gohmert Will Save Us From Gay Space Colonies." *Bad Astronomy* blog posting 1 June (online text at *www.slate.com* accessed 6/18/2016).

Platnick, Norman I. 1982. "Defining Characters and Evolutionary Groups." *Systematic Zoology* **31** (September): 282-284.

Plucinski, Mateusz M., Curtis S. Huber, Sheila Akinyi, Willard Dalton, Mary Eshete, Katherine Grady, Luciana Silva-Flannery, Blaine A. Mathison, Venkatachalam Udhayakumar, Paul M. Arguin, & **John W. Barnwell.** 2014. "Novel Mutation in Cytochrome B of *Plasmodium falciparum* in One of Two Atovaquone-Proguanil Treatment Failures in Travelers Returning From Same Site in Nigeria." *Open Forum Infectious Diseases* **1** (Summer): ofu059.

Pol, Diego, Mark A. Norell, & **Mark E. Siddall.** 2004. "Measures of stratigraphic fit to phylogeny and their sensitivity to tree size, tree shape, and scale." *Cladistics* **20** (February): 64-75.

Pol, Diego, Oliver W. M. Rauhut, Agustina Lecuona, Juan M. Leardi, Xing Xu, & **James M. Clark.** 2013. "A new fossil from the Jurassic of Patagonia reveals the early basicranial evolution and the origins of Crocodyliformes." *Biological Reviews of the Cambridge Philosophical Society* **88** (November): 862-872.

Pollard, Sophie L., & **Peter W. H. Holland.** 2000. "Evidence for 14 homeobox gene clusters in human genome ancestry." *Current Biology* **10** (7 September): 1059-1062.

Polly, P. David. 2000. "Development and evolution of occlude: Evolution of development in mammalian teeth." *Proceedings of the National Academy of Sciences* **97** (19 December): 14019-14021.

———. 2013. "Stuck between the teeth." *Nature* **497** (16 May): 325-326.

Poppe, Kenneth. 2008. *Exposing Darwinism's Weakest Link.* Eugene, OR: Harvest House Publishers.

Posner, Michael I., Mary K. Rothbart, Brad E. Sheese, & **Yiyuan Tang.** 2007. "The anterior cingulate gyrus and the mechanism of self-regulation." *Cognitive, Affective, & Behavioral Neuroscience* **7**

(December): 391-395.
Pratt, David. 2000. "Plate Tectonics: A Paradigm Under Threat." *Journal of Scientific Exploration* **14** (3): 307-352.
———. 2006. "Organized Opposition to Plate Tectonics: The New Concepts in Global Tectonics Group." *Journal of Scientific Exploration* **20** (1): 97-104.
———. 2014. "Evolution and Design." *David Pratt* revised May posting of May 2004 article (online htm at *davidpratt.info* accessed 5/14/2014).
Pray, Leslie A. 2008. "Functions and Utility of *Alu* Jumping Genes." *Nature Education* (online @ nature.com) **1** (1): 93.
Prehistoric Life. 2009. New York: Dorling Kindersley Publishing.
Preuschoft, Holger, & **Ulrich Witzel.** 2005. "Functional Shape of the Skull in Vertebrates: Which Forces Determine Skull Morphology in Lower Primates and Ancestral Synapsids?" *The Anatomical Record (Part A: Discoveries in Molecular, Cellular, and Evolutionary Biology)* **283A** (April): 402-413.
Prihoda, Judit, Atsuko Tanaka, Wilson B. M. de Paula, John F. Allen, Leila Tirichine, & **Chris Bowler.** 2012. "Chloroplast-mitochondria cross-talk in diatoms." *Journal of Experimental Botany* **63** (4 February): 1543-1557.
Prinos, Panagiotis, Suman Joseph, Karen Oh, Barbara I. Meyer, Peter Gruss, & **David Lohnes.** 2001. "Multiple Pathways Governing *Cdx1* Expression during Murine Development." *Developmental Biology* **239** (15 November): 257-269.
Prochazka, Jan, Sophie Pantalacci, Svatava Churava, Michaela Rothova, Anne Lambert, Hervé Lesot, Ophir Klein, Miroslav Peterka, Vincent Laudet, & **Renata Peterkova.** 2010. "Patterning by heritage in mouse molar row development." *Proceedings of the National Academy of Sciences* **107** (31 August): 15497-15502.
Prothero, Donald R. 1992. "Punctuated Equilibrium at Twenty: A Paleontological Perspective." *Skeptic* **1** (Fall): 38-47.
———. 2005. "Evolution: The Fossils say Yes!" *Natural History* **115** (November): 52-56.
———. 2007. *Evolution: What the Fossils Say and Why It Matters.* New York: Columbia University Press.
———. 2008. "What missing link?" *New Scientist* **197** (1 March): 35-41.
Provine, William B. 2004. "Ernst Mayr: Genetics and Speciation." *Genetics* **167** (July): 1041-1046.
Prud'homme, Benjamin, Caroline Minervino, Mélanie Hocine, Jessica D. Cande, Aïcha Aouane, Héloïse D. Dufour, Victoria A. Kassner, & **Nicolas Gompel.** 2011. "Body plan innovation in treehoppers through the evolution of an extra wing-like appendage." *Nature* **473** (5 May): 83-86.
Prum, Richard O. 1999. "Development and Evolutionary Origin of Feathers." *Journal of Experimental Zoology* **285** (15 December): 291-306.
———. 2002. "Why Ornithologists Should Care About the Theropod Origin of Birds." *The Auk* **119** (January): 1-17.
———. 2003. "Are Current Critiques of the Theropod Origin of Birds Science?" *The Auk* **120** (April): 550-561.
———. 2005. "Evolution of the Morphological Innovations of Feathers." *Journal of Experimental Zoology* **304B** (15 November): 570-579.
Prum, Richard O., & **Alan H. Brush.** 2002. "The Evolutionary Origin and Diversification of Feathers." *The Quarterly Review of Biology* **77** (September): 261-295.
———. 2003. "Which Came First, the Feather or the Bird?" *Scientific American* **288** (March): 84-93.
Prum, Richard O., & **Jan Dyck.** 2003. "A hierarchical model of plumage: Morphology, development, and evolution." *Journal of Experimental Zoology* **298B** (15 August): 73-90.
Prum, Richard O., & **Scott Williamson.** 2001. "Theory of the Growth and Evolution of Feather Shape." *Journal of Experimental Zoology* **291** (15/30 April): 30-57.
Pryer, Kathleen M., & **David J. Hearn.** 2009. "Evolution of Leaf Form in Marsileaceous Ferns: Evidence for Heterochrony." *Evolution* **63** (February): 498-513.
Purdom, Georgia. 2006. "'Non-evolution' of the appearance of mitochondria and plastids in eukaryotes: challenges to endosymbiotic theory." *Answers in Genesis* article for 11 October (online text at *www.answersingenesis.org* accessed 1/25/2010).
———. 2008. "Gene Duplication: Evolution Shooting Itself in the Foot." *Answers in Genesis* article for 30 April (online text at *www.answersingenesis.org* accessed 2/23/2014).
———. 2013a. "What about Eugenics and Planned Parenthood?" *Answers in Genesis* 2 June 2014 posting of *The New Answers Book 3* Chapter 17 (online text at *www.answersingenesis.org* accessed 2/17/2015).

———. 2013b. "Did Life Come from Outer Space?" *Answers in Genesis* 27 June 2014 posting of *The New Answers Book 3* Chapter 22 (online text at *www.answersingenesis.org* accessed 2/17/2015).

Purkanti, Ramya, & **Mukund Thattai.** 2015. "Ancient dynamin segments capture early stages of host-mitochondrial integration." *Proceedings of the National Academy of Sciences* **112** (3 March): 2800-2805.

Puthiyaveetil, Sujith, T. Anthony Kavanagh, Peter Cain, James A. Sullivan, Christine A. Newell, John C. Gray, Colin Robinson, Mark van der Giezen, Matthew B. Rogers, & **John F. Allen.** 2008. "The ancestral symbiont sensor kinase CSK links photosynthesis with gene expression in chloroplasts." *Proceedings of the National Academy of Sciences* **105** (22 July): 10061-10066.

Qian, Wenfeng, Di Ma, Che Xiao, Zhi Wang, & **Jianzhi Zhang.** 2012. "The Genomic Landscape and Evolutionary Resolution of Antagonistic Pleiotropy in Yeast." *Cell Reports* **2** (25 October): 1399-1410.

Qian, Wenfeng, & **Jianzhi Zhang.** 2014. "Genomic evidence for adaptation by gene duplication." *Genome Research* **24** (August): 1356-1362.

Qu, Qingming, Tatjana Haitina, Min Zhu, & **Per Erik Ahlberg.** 2015. "New genomic and fossil data illuminate the origin of enamel." *Nature* **526** (1 October): 108-111.

Quiring, Rebecca, Uwe Walldorf, Urs Kloter, & **Walter J. Gehring.** 1994. "Homology of the *Eyeless* Gene of *Drosophila* to the *Small Eye* Gene in Mice and *Aniridia* in Humans." *Science* **265** (5 August): 785–789.

Rabeling, Christian, Jeremy M. Brown, & **Manfred Verhaagh.** 2008. "Newly discovered sister lineage sheds light on early ant evolution." *Proceedings of the National Academy of Sciences* **105** (30 September): 14913-14917.

Rabin-Havt, Ari. 2016. "FLASHBACK: Mike Pence Delivers Entire Speech Denying Evolution." *OXIMITY* 3 August posting (online text at *www.oximity.com* accessed 8/3/2016).

Radford, Benjamin. 2001. "Bogus Nostradamus Prophecies Circulate Following Terrorism." *Skeptical Inquirer* **25** (November/December): 8-9.

Radinsky, Leonard B. 1969. "The Early Evolution of the Perissodactyla." *Evolution* **23** (June): 308-328.

———. 1987. *The Evolution of Vertebrate Design*. Chicago: University of Chicago Press.

Rajagopalan, Lavanya, Nimish Patel, Srinivasan Madabushi, Julie Anne Goddard, Venkat Anjan, Feng Lin, Cindy Shope, Brenda Farrell, Olivier Lichtarge, Amy L. Davidson, William E. Brownell, & **Fred A. Pereira.** 2006. "Essential Helix Interactions in the Anion Transporter Domain of Prestin Revealed by Evolutionary Trace Analysis." *The Journal of Neuroscience* **26** (6 December): 12727-12734.

Ramírez-Chaves, Héctor E., Stephen W. Wroe, Lynne Selwood, Lyn A. Hinds, Chris Leigh, Daisuke Koyabu, Nikolay Kardjilov, & **Vera Weisbecker.** 2016. "Mammalian development does not recapitulate suspected key transformations in the evolutionary detachment of the mammalian middle ear." *Proceedings of the Royal Society of London* **B** (Biological Sciences) **283** (13 January): 20152606.

Rana, Fazale. 2000. "New Challenge to the Bird-Dinosaur Link." *Reasons to Believe* posting 1 April (online text at *www.reasons.org* accessed 4/29/2011).

———. 2011. *Creating Life in the Lab*. Grand Rapids, MI: Baker Books.

———. 2015. "The Creation-Evolution Controversy in 'Jurassic World'." *Reasons to Believe* posting 25 June (online text at *www.reasons.org* accessed 5/11/2016).

Rana, Fazale, & **Hugh Ross.** 2014. *Origins of Life.* Covina, CA: RTB Press.

Randall, Luke. 2013. "Intermediates." *Was Darwin right?* undated posting (online text at *www.wasdarwinright.com* accessed 1/6/2014).

Randi, James, D. J. Grothe, Sadie Crabtree, Rick Adams, Chip Denman, Barnara Drescher, Kylie Sturgess, Matt Lowry, & **Daniel Loxton.** 2012. "The Case of the Cottingley Fairies: Examine the Evidence." *James Randi Educational Foundation* posting (online pdf at *web.randi.org* accessed 10/12/2016).

Rasmussen, D. Tab, & **Kimberley A. Nekaris.** 1998. "Evolutionary History of Lorisiform Primates." *Folia Primatologica* **69** (February Supplement 1): 250-285.

Raspopovic, J., L. Marcon, L. Russo, & **J. Sharpe.** 2014. "Digit patterning is controlled by a Bmp-Sox9-Wnt Turing network modulated by morphogen gradients." *Science* **345** (1 August): 566-570.

RationalWiki. 2015. "A Scientific Dissent From Darwinism." *RationalWiki* posting updated 9 March

(online text at *rationalwiki.org* accessed 6/8/2015).
Rauhut, Oliver W. M., Christian Foth, Helmut Tischlinger, & Mark A. Norell. 2012. "Exceptionally preserved juvenile megalosaurid theropod dinosaur with filamentous integument from the Late Jurassic of Germany." *Proceedings of the National Academy of Sciences* **109** (17 July): 11746-11751.
Rauhut, Oliver W. M., Alexander M. Heyng, Adriana López-Arbarello, & Andreas Hecker. 2012. "A New Rhynchocephalian from the Late Jurassic of Germany with a Dentition That Is Unique amongst Tetrapods." *PLoS ONE* (online @ plosone.org) **7** (October): e46839.
Rauhut, Oliver W. M., Thomas Martin, Edgardo Ortiz-Jaureguizar, & Pablo Puerta. 2002. "A Jurassic mammal from South America." *Nature* **416** (14 March): 165-168.
Raup, David M. 1994. "The role of extinction in evolution." *Proceedings of the National Academy of Sciences* **91** (19 July): 6758-6763.
Raven, John A., & John F. Allen. 2003. "Genomics and chloroplast evolution: what did cyanobacteria do for plants?" *Genome Biology* (online @ genomebiology.com) **4** (3 March): 209.
Raven, Peter H., & George B. Johnson. 1989. *Biology.* 2nd ed. St. Louis, MO: Times Mirror/Mosby College Publishing.
———. 2001. *Biology.* 6th ed. New York: McGraw Hill.
Ravizza, G., & B. Peucker-Ehrenbrink. 2003. "Chemostratigraphic Evidence of Deccan Volcanism from the Marine Osmium Isotope Record." *Science* **302** (21 November): 1392-1395.
Reed, D. A., J. Iriarte-Diaz, & T. G. H. Diekwisch. 2016. "A three dimensional free body analysis describing variation in the musculoskeletal configuration of the cynodont lower jaw." *Evolution & Development* **18** (January-February): 41-53.
Reed, John K. 2008. "Cuvier's analogy and its consequences: forensics vs testimony as historical evidence." *Journal of Creation* (AKA Answers in Genesis *Creation ex Nihilo Technical Journal*) **22** (3): 115-120.
Reed, Robert D., Riccardo Papa, Arnaud Martin, Heather M. Hines, Brian A. Counterman, Carolina Pardo-Diaz, Chris D. Jiggins, Nicola L. Chamberlain, Marcus R. Kronforst, Rui Chen, Georg Halder, H. Frederik Nijhout, & W. Owen McMillan. 2011. "*optix* Drives the Repeated Convergent Evolution of Butterfly Wing Pattern Mimicry." *Science* **333** (26 August): 1137-1141.
Regal, Philip J. 1975. "The Evolutionary Origin of Feathers." *The Quarterly Review of Biology* **50** (March): 35-66.
Reilly, Peter J. 2012. "Young Earth Creationists Whipsawed By IRS." *Forbes* posting 5 October (online text at *www.forbes.com* accessed 1/27/2015).
———. 2013a. "Not Income Tax Evasion – Structuring – That's How They Got Kent Hovind." *Forbes* posting 30 January (online text at *www.forbes.com* accessed 1/15/2014).
———. 2013b. "Is IRS Persecuting Kent Hovind For Creationism?" *Forbes* posting 8 February (online text at *www.forbes.com* accessed 8/5/2013).
———. 2013c. "Wesley Snipes Almost Out—Kent Hovind Remains In Prison." *Forbes* posting 6 April (online text at *www.forbes.com* accessed 1/27/2015).
———. 2013d. "Wesley Snipes Raises Creationist Hopes For Kent Hovind." *Forbes* posting 9 April (online text at *www.forbes.com* accessed 1/27/2015).
———. 2013e. "Has Kent Hovind Given Up Fight Against IRS?" *Forbes* posting 28 September (online text at *www.forbes.com* accessed 1/27/2015).
———. 2014a. "Is Kent Hovind A Tax Protestor?" *Forbes* posting 1 January (online text at *www.forbes.com* accessed 1/27/2015).
———. 2014b. "Time To Let Kent Hovind Go Home." *Forbes* posting 28 July (online text at *www.forbes.com* accessed 1/27/2015).
———. 2014c. "Kent Hovind's Battle With The IRS In Retrospect." *Forbes* posting 29 July (online text at *www.forbes.com* accessed 1/27/2015).
———. 2014d. "Government Coming Down Harder on Kent Hovind." *Forbes* posting 29 October (online text at *www.forbes.com* accessed 1/27/2015).
———. 2014e. "Kent Hovind And Creation Science Evangelism—How Not To Run A Ministry." *Forbes* posting 19 November (online text at *www.forbes.com* accessed 1/27/2015).
———. 2014f. "Was Kent Hovind 2006 Structuring Indicment Flawed?" *Forbes* posting 3 December (online text at *www.forbes.com* accessed 1/27/2015).
———. 2015a. "Kent Hovind Asks Supporters For Noise And Light To Defend Him." *Forbes* posting 12 January (online text at *www.forbes.com* accessed 1/27/2015).
———. 2015b. "Looking At Kent Hovind's Innocence Claims." *Forbes* posting 13 January (online text

at *www.forbes.com* accessed 1/27/2015).
———. 2015c. "Will Kent Hovind Become This Year's Cliven Bundy?" *Forbes* posting 20 January (online text at *www.forbes.com* accessed 1/21/2015).
———. 2015d. "Exclusive—Kent Hovind Claims Congressmen Are Looking Into His Case." *Forbes* posting 22 January (online text at *www.forbes.com* accessed 1/27/2015).
———. 2015e. "A Free Kent Hovind Might Have Backing For A Bigger Better Dinosaur Theme Park." *Forbes* posting 27 January (online text at *www.forbes.com* accessed 1/27/2015).
———. 2015f. "An Awesome Critique Of The Prison Industrial Complex By One Who Knows It Well." *Your Tax Matters Partner* posting 2 February (online text at *ytmp.blogspot.com* accessed 2/2/2015).
———. 2015g. "Will Christian Soldiers Be On The Streets Of Pensacola As Kent Hovind Goes To Trial?" *Forbes* posting 27 February (online text at *www.forbes.com* accessed 2/28/2015).
———. 2015h. "Kent Hovind's Innocence Narrative – Truth Some – While Truth Not So Much." *Your Tax Matters Partner* posting 14 May (online text at *ytmp.blogspot.com* accessed 5/14/2015).
———. 2015i. "The Juror Who Freed Kent Hovind Steps Foward." *Forbes* posting 9 June (online text at *www.forbes.com* accessed 6/22/2015).
———. 2015j. "Troubles About Flat Earth And Other Kent Hovind Developments." *Forbes* posting 26 July (online text at *www.forbes.com* accessed 8/4/2015).
Reiner, Anton. 1996. "Levels of organization and the evolution of isocortex." *Trends in Neuroscience* **11** (March): 89-91.
Reiner, Anton, Kei Yamamoto, & **Harvey J. Karten.** 2005. "Organization and Evolution of the Avian Forebrain." *The Anatomical Record (Part A: Discoveries in Molecular, Cellular, and Evolutionary Biology)* **287A** (November): 1080-1102.
Reisz, Robert R. 1977. "*Petrolacosaurus*, the Oldest Known Diapsid Reptile." *Science* **196** (3 June): 1091-1093.
———. 1993. "Theriodonts." *University of Toronto* undated posting of Biology 356 "Major Features of Vertebrate Evolution" course notes (online pdf at *www.utm.utoronto.ca* accessed 10/3/2016).
———. 2005. "*Oromycter*, A New Caseid from the Lower Permian of Oklahoma." *Journal of Vertebrate Paleontology* **25** (December): 905-910.
Reisz, Robert R., David S. Berman, & **Diane Scott.** 1984. "The Anatomy and Relationships of the Lower Permian Reptile *Araeoscelis*." *Journal of Vertebrate Paleontology* **4** (September): 57-67.
Reisz, Robert R., Stephen J. Godfrey, & **Diane Scott.** 2009. "*Eothyris* and *Oedaleops*: Do These Early Permian Synapsids from Texas and New Mexico form a Clade?" *Journal of Vertebrate Paleontology* **29** (March): 39-47.
Reisz, Robert R., & **Michel Laurin.** 2001. "The reptile *Macroleter*: First vertebrate evidence for correlation of Upper Permian continental strata of North America and Russia." The Geological Society of America *GSA Bulletin* **113** (September): 1229-1233.
Reisz, Robert R., Hillary C. Maddin, Jörg Fröbisch, & **Jocelyn Falconnet.** 2011. "A new large caseid (Synapsida, Caseasauria) from the Permian of Rodez (France), including a reappraisal of '*Casea*' *rutena* Sigogneau-Russell & Russell, 1974." *Geodiversitas* **33** (June): 227-246.
Reisz, Robert R., Sean P. Modesto, & **Diane M. Scott.** 2011. "A new Early Permian reptile and its significance in early diapsid evolution." *Proceedings of the Royal Society of London* **B** (Biological Sciences) **278** (22 December): 3731-3737.
Reisz, Robert R., & **Linda A. Tsuji.** 2006. "An articulated skeleton of *Varanops* with bite marks: The oldest known evidence of scavenging among terrestrial vertebrates." *Journal of Vertebrate Paleontology* **26** (December): 1021-1023.
Reitan, Eric. 2012. "Santorum's War on Satan... er, on Higher Education." *Religion Dispatches* 28 February posting (online text at *religiondispatches.org* accessed 2/29/2012).
ReMine, Walter James. 1993. *The Biotic Message: Evolution versus Message Theory.* St. Paul, MN: St. Paul Science.
Remington, David L., & **Michael D. Purugganan.** 2002. "*GAI* Homologues in the Hawaiian Silversword Alliance (Asteraceae-Madiinae): Molecular Evolution of Growth Regulators in a Rapidly Diversifying Plant Lineage." *Molecular Biology and Evolution* **19** (September): 1563-1574.
Renne, Paul R., Alan L. Deino, Frederik J. Hilgen, Klaudia F. Kuiper, Darren F. Mark, William S. Mitchell III, Leah E. Morgan, Roland Mundil, & **Jan Smit.** 2013. "Time Scales of Critical Events Around the Cretaceous-Paleogene Boundary." *Science* **339** (8 February): 684-687.
Represa, Juan, Dorothy A. Frenz, & **Thomas R. Van De Water.** 2000. "Genetic patterning of embryonic inner ear development." *Acta Oto-laryngologica* **120** (January): 5-10.

Rice, Ritva, Aki Kallonen, Judith Cebra-Thomas, & **Scott F. Gilbert.** 2016. "Development of the turtle plastron, the order-defining skeletal structure." *Proceedings of the National Academy of Sciences* **113** (10 May): 5317-5322.

Rice, Stanley A. 2014. "Confessions of an Oklahoma Evolutionist: The Bad, the Ugly, and the Good." *Reports of the National Center for Science Education* **34** (January-February): 4.1-4.7.

Rice, William R., & **Ellen E. Hostert.** 1993. "Laboratory Experiments on Speciation: What Have We Learned in 40 Years?" *Evolution* **47** (December): 1637-1653.

Rich, Patricia Vickers, Thomas Hewitt Rich, Mildred Adams Fenton, & **Carroll Lane Fenton.** 1996. *The Fossil Book: A Record of Prehistoric Life.* Rev. pb ed. Mineola, NY: Dover Publications, Inc.

Rich, Thomas H. 2008. "The palaeobiogeography of Mesozoic mammals: a review." *Arquivos do Museu Nacional, Rio de Janeiro* **66** (January/March): 231-249.

Rich, Thomas H., James A. Hopson, Pamela G. Gill, Peter Trusler, Sall Rogers-Davidson, Steve Morton, Richard Cifelli, David Pickering, Lesley Kool, Karen Siu, Flame A. Burgmann, Tim Senden, Alistair R. Evans, Barbara W. Wagstaff, Doris Seegets-Villiers, Ian J. Corfe, Timothy F. Flannery, Ken Walker, Anne M. Mussert, Michael Archer, Rebecca Pian, & **Patricia Vickers-Rich.** 2016. "The mandible and dentition of the Early Cretaceous monotreme *Teinolophos trusleri.*" *Alcheringia* (online pdf posting at *www.tandfonline.com* accessed 10/7/2016).

Rich, Thomas H., Patricia Vickers-Rich, Peter Trusler, Timothy F. Flannery, Richard Cifelli, Andrew Constantine, Lesley Kool, & **Nicholas van Klaveren.** 2001. "Monotreme nature of the Australian Early Cretaceous mammal *Teinolophos.*" *Acta Palaeontologica Polonica* **46** (1): 113-118.

Richard, Owain Westmacott, & **Richard Gareth Davies.** 1977. *Imms' General Textbook of Entomology, Vol. 2.* 10th ed. London: Chapman and Hall.

Richards, Thomas A., & **Mark van der Giezen.** 2006. "Evolution of the Isd11-IscS Complex Reveals a Single α-Proteobacterial Endosymbiosis for All Eukaryotes." *Molecular Biology and Evolution* **23** (July): 1341-1344.

Richardson, Michael K., & **Paul M. Brakefield.** 2003. "Hotspots for evolution." *Nature* **424** (21 August): 894-895.

Richardson, M. K., James Hanken, Mayoni L. Gooneratne, Claude Pieau, Albert Raynaud, Lynne Selwood, & **Glenda M. Wright.** 1997. "There is no highly conserved embryonic stage in the vertebrates: implications for current theories of evolution and development." *Anatomy and Embryology* **196** (July): 91-106.

Richardson, Michael K., James Hanken, Lynne Selwood, Glenda M. Wright, Robert J. Richards, Claude Pieau, & **Albert Raynaud.** 1998. "Haeckel, Embryos, and Evolution." *Science* **280** (15 May): 983-985.

Richardson, M. K., Jonathan E. Jeffery, M. I. Coates, & **Olaf R. P. Bininda-Emonds.** 2001. "Comparative methods in developmental biology." *Zoology* **104** (3-4): 278-283.

Richardson, Michael K., & **Gerhard Keuck.** 2001. "A question of intent: when is a 'schematic' illustration a fraud?" *Nature* **410** (8 March): 144.

———. 2002. "Haeckel's ABC of evolution and development." *Biological Reviews of the Cambridge Philosophical Society* **77** (November): 495-528.

Richardson, Paul E., Medha Manchekar, Nassrin Dashti, Martin K. Jones, Anne Beigneux, Stephen G. Young, Stephen C. Harvey, & **Jere P. Segrest.** 2005. "Assembly of Lipoprotein Particles Containing Apolipoprotein-B: Structural Model for the Nascent Lipoprotein Particle." *Biophysical Journal* **88** (April): 2789-2800.

Richmond, Jesse. 2009. "Design and Dissent: Religion, Authority, and the Scientific Spirit of Robert Broom." *Isis* **100** (September): 485-504.

Riddle, Mike. 2013. "Does Evolution Have a ... Chance?" *Answers in Genesis* 30 May 2014 posting of *The New Answers Book 3* Chapter 16 (online text at *www.answersingenesis.org* accessed 2/17/2015).

Ridley, Mark, ed. 1997. *Evolution.* Oxford: Oxford University Press.

Riedl, Rupert. 1977. "A Systems-Analytical Approach to Macro-Evolutionary Phenomena." *The Quarterly Review of Biology* **52** (December): 351-370.

Rieppel, Olivier. 1999. "Turtle Origins." *Science* **283** (12 February): 945-946.

———. 2001. "Turtles as hopeful monsters." *BioEssays* **23** (November): 987-991.

———. 2009. "How Did the Turtle Get Its Shell?" *Science* **325** (10 July): 154-155.

———. 2011. "Willi Hennig's dichotomization of nature." *Cladistics* **27** (February): 103-112.

Rieppel, Olivier, & **Robert R. Reisz.** 1999. "The Origin and Early Evolution of Turtles." *Annual Review of Ecology and Systematics* **30**: 1-22.

Rieseberg, Loren H., Olivier Raymond, David M. Rosenthal, Zhao Lai, Kevin Livingstone, Takuya Nakazato, Jennifer L. Durphy, Andrea E. Schwarzbach, Lisa A. Donovan, & Christian Lexer. 2003. "Major Ecological Transitions in Wild Sunflowers Facilitated by Hybridization." *Science* **301** (29 August): 1211-1216.

Ritvo, Harriet. 1997. *The Platypus and the Mermaid and Other Figments of the Classifying Imagination.* Cambridge, MA: Harvard University Press.

Rivera, Ajna S., M. Sabrina Pankey, David C. Plachetzki, Carlos Villacorta, Anna E. Syme, Jeanne M. Serb, Angela R. Omilian, & Todd H. Oakley. 2010. "Gene duplication and the origins of morphological complexity in pancrustacean eyes, a genomic approach." *BMC Evolutionary Biology* (online @ biomedcentral.com) **10** (30 April): 123.

Robb, Stewart. 1961. *Prophecies on World Events by Nostradamus.* New York: Ace.

Roberts, Henry C. 1949. *The Complete Prophecies of Nostradamus.* New York: Nostradamus, Inc.

Robinson, B. A. 2011. "Comparing two creation stories: from Genesis and Babylonian pagan sources." Ontario Consultants on Religious Tolerance *Religious Tolerance* posting updated 7 February (online text at *www.religioustolerance.org* accessed 5/19/2011).

Robinson, D. Ashley, & David P. Cavanaugh. 1998. "A Quantitative Approach to Baraminology With Examples from the Catarrhine Primates." *Creation Research Society Quarterly* **34** (March): 196-208.

Rocky Mountain Dinosaur Resource Center. 2010. "Didelphodon vorax." *Rocky Mountain Dinosaur Resource Center* 7 December posting (online text at *www.rmdrc.com* accessed 9/15/2016).

Rodríguez-Esteban, Concepción, Javier Capdevila, Aris N. Economides, Jaime Pascual, Ángel Ortiz, & Juan Carlos Izpisúa Belmonte. 1999. "The novel Cer-like protein Caronte mediates the establishment of embryonic left-right asymmetry." *Nature* **401** (16 September): 243-251.

Rogers, Raymond R. 1997. "Ischigualasto Formation," in Currie & Padian (1997, 372-374).

Romano, Marco, & Umberto Nicosia. 2014. "*Alierasaurus ronchii*, gen. et sp. nov., A Caseid from the Permian of Sardinia, Italy." *Journal of Vertebrate Paleontology* **34** (July): 900-913.

———. 2015. "Cladistic Analysis of Caseidae (Caseasauria, Synapsida): Using the Gap-Weighting Method to Include Taxa Based on Incomplete Specimens." *Palaeontology* **58** (November): 1109-1130.

Romer, Alfred Sherwood. 1966. *Vertebrate Paleontology.* 3rd ed. Chicago: University of Chicago Press.

———. 1970. *The Vertebrate Body.* 4th ed. Philadelphia: W. B. Saunders Company.

Romer, A. S., & L. W. Price. 1940. "Review of the Pelycosauria." *The Geological Society of America Special Papers* **28**: 1-534.

Ronchi, Ausonio, Eva Sacchi, Marco Romano, & Umberto Nicosia. 2011. "A huge caseid pelycosaur from north-western Sardinia and its bearing on European Permian stratigraphy and palaeobiogeography." *Acta Palaeontologica Polonica* **56** (4): 723-738.

Rook, Deborah L., & John P. Hunter. 2014. "Rooting Around the Eutherian Family Tree: the Origin and Relations of the Taeniodonta." *Journal of Mammalian Evolution* **21** (March): 75-91.

Rook, Deborah L., John P. Hunter, Dean A. Pearson, & Antoine Bercovici. 2010. "Lower Jaw of the Early Paleocene Mammal *Alveugena* and Its Interpretation as a Transitional Fossil." *Journal of Paleontology* **84** (November): 1217-1225.

Rose, Kenneth D. 2006. "The postcranial skeleton of early Oligocene *Leptictis* (Mammalia: Leptictida), with a preliminary comparison to *Leptictidium* from the middle Eocene of Messel." *Palaeontographica* Abteilung A **278** (October): 37–56.

Rosenau, Josh. 2016a. "School field trips to creationist Ark? Sink that idea right now." *New Scientist* (5 August): online text accessed 8/12/2016 (*newscientist.com*).

———. 2016b. "Creationist Textbooks in Minnesota." *Reports of the National Center for Science Education* **36** (Summer): 10.

Rosenhouse, Jason. 2008. "Review: Darwin Strikes Back." *Reports of the National Center for Science Education* **28** (July-August): 35-36.

Ross, Hugh. 1994. *Creation and Time: A Biblical and Scientific Perspective on the Creation-Date Controversy.* Colorado Springs, CO: Navpress.

———. 1996. *Beyond the Cosmos: The Extra-Dimensionality of God: What Recent Discoveries in Astronomy and Physics Reveal about the Nature of God.* Colorado Springs, CO: Navpress.

———. 1998. *The Genesis Question: Scientific Advances and the Accuracy of Genesis.* Colorado Springs, CO: Navpress.

———. 2008. *Why the Universe Is the Way It Is.* Grand Rapids, MI: Baker Books.

———. 2009. *More Than a Theory: Revealing a Testable Model for Creation*. Grand Rapids, MI: Baker Books.
———. 2014. *Navigating Genesis: A Scientist's Journey through Genesis 1-11*. Covina, CA: RTB Press.
———. 2015a. *A Matter of Days: Resolving a Creation Controversy*. 2nd rev. ed. Covina, CA: RTB Press.
———. 2015b. *Who Was Adam?* Covina, CA: RTB Press.
Ross, L., & B. B. Normark. 2015. "Evolutionary problems in centrosome and centriole biology." *Journal of Evolutionary Biology* **28** (May): 995-1004.
Ross, Marcus R. 2010. "Two: Those Not-So-Dry Bones: Soft Tissue in a *T. Rex* Fossil?" Answers in Genesis *Answers Magazine* (January-March): 43-45.
Rossie, James B., Christopher C. Gilbert, & Andrew Hill. 2013. "Early cercopithecid monkeys from the Tugen Hills, Kenya." *Proceedings of the National Academy of Sciences* **110** (9 April): 5818-5822.
Rossie, James B., Xijun Ni, & K. Christopher Beard. 2006. "Cranial remains of an Eocene tarsier." *Proceedings of the National Academy of Sciences* **103** (21 March): 4381-4385.
Rothschild, Bruce M., & Virginia Naples. 2015. "Decompression syndrome and diving behavior in *Odontochelys*, the first turtle." *Acta Palaeontologica Polonica* **60** (1): 163-167.
Rougier, Guillermo W., Sebastían Apesteguia, & Leandro C. Gaetano. 2011. "Highly specialized mammalian skulls from the Late Cretaceous of South America." *Nature* **479** (3 November): 98-102.
Rougier, Guillermo W., Brian M. Davis, & Michael J. Novacek. 2015. "A deltatheroidan mammal from the Upper Cretaceous Baynshiree Formation, eastern Mongolia." *Cretaceous Research* **52** (January): 167-177.
Rougier, Guillermo W., Amir S. Sheth, Kenneth Carpenter, Lucas Appella-Guiscafre, & Brian M. Davis. 2015. "A New Species of *Docodon* (Mammaliaformes: Docodonta) from the Upper Jurassic Morrison Formation and a Reassessment of Selected Craniodental Characters in Basal Mammaliaformes." *Journal of Mammalian Evolution* **22** (March): 1-16.
Rougier, Guillermo W., John R. Wible, Robin M. D. Beck, & Sebastian Apesteguia. 2012. "The Miocene mammal *Necrolestes* demonstrates the survival of a Mesozoic nontherian lineage into the late Cenozoic of South America." *Proceedings of the National Academy of Sciences* **109** (4 December): 20053-20058.
Rougier, Guillermo W., John R. Wible, & James A. Hopson. 1992. "Reconstruction of the cranial vessels in the Early Cretaceous mammal *Vincelestes neuquenianus*: Implications for the evolution of the mammalian cranial vascular system." *Journal of Vertebrate Paleontology* **12** (June): 188-216.
Rougier, Guillermo W., John R. Wible, & Michael J. Novacek. 1996. "Middle Ear Ossicles of the Multituberculate *Kryptobataar* from the Mongolian Late Cretaceous: Implicaions for Mammaliamorph Relationships and Evolution of the Auditory Apparatus." *American Museum Novitates* **3187** (December): 1-43.
———. 1998. "Implications of *Deltatheridium* specimens for early marsupial history." *Nature* **459** (3 December): 459-463.
Rowe, Timothy. 1993. "Phylogenetic Systematics and the Early History of Mammals," in Szalay *et al.* (1993, 129-145).
———. 1996. "Coevolution of the Mammalian Middle Ear and Neocortex." *Science* **273** (2 August): 651-654.
———. 2004. "Chordate Phylogeny and Development," in Cracraft & Donoghue (2004, 384-409).
Rowe, Timothy B., Thomas E. Macrini, & Zhe-Xi Luo. 2011. "Fossil Evidence on Origin of the Mammalian Brain." *Science* **332** (20 May): 951-957.
Rowe, Timothy, Earle F. McBride, & Paul C. Sereno. 2001. "Dinosaur with a Heart of Stone," with Response by **Dale A. Russell, Paul E. Fisher, Reese E. Barrick**, & **Michael K. Stoskopf.** *Science* **291** (2 February): 783.
Rowe, Timothy, Thomas H. Rich, Patricia Vickers-Rich, Mark Springer, & Michael O. Woodburne. 2008. "The oldest platypus and its bearing on divergence timing of the platypus and echidna clades." *Proceedings of the National Academy of Sciences* **105** (29 January): 1238-1242.
Royer, Dana L., Laura A. Meyerson, Kevin M. Robertson, & Jonathan M. Adams. 2009. "Phenotypic Plasticity of Leaf Shape along a Temperature Gradient in *Acer rubrum*." *PLoS ONE* (online @ plosone.org) **4** (October): e7653.
Royer, Dana L., Daniel J. Peppe, Elisabeth A. Wheeler, & Ülo Niinemets. 2012. "Roles of Climate

and Functional Traits in Controlling Toothed vs. Untoothed Leaf Margins." *American Journal of Botany* **99** (May): 915-922.

Royer, Dana L., Peter Wilf, David A. Kanesko, Elizabeth A. Kowalski, & **David L. Dilcher.** 2005. "Correlations of climate and plant ecology to leaf size and shape: Potential proxies for the fossil record." *American Journal of Botany* **92** (October): 1141-1151.

Ruben, John A., Albert F. Bennett, & **Frederick L. Hisaw.** 1987. "Selective factors in the origin of the mammalian diaphragm." *Paleobiology* **13** (Winter): 54-59.

Ruben, John A., & **Terry D. Jones.** 2000. "Selective factors associated with the origin of fur and feathers." *Integrative and Comparative Biology* (AKA *American Zoologist*) **40** (September): 585-596.

Rubidge, Bruce S. 1995. *Did Mammals Originate in Africa? South African Fossils and the Russian Connection.* Sidney Haughton Memorial Lecture 4.

———. 2005. "Re-uniting lost continents—Fossil reptiles from the ancient Karoo and their wanderlust." *South African Journal of Geology* **108** (March): 135-172.

Rubidge, Bruce S., & **Christian A. Sidor.** 2001. "Evolutionary Patterns Among Permo-Triassic Therapsids." *Annual Review of Ecology and Systematics* **32**: 449-480.

Rudebeck, P. H., M. J. Buckley, M. E. Walton, & **M. F. S. Rushworth.** 2006. "A Role for the Macaque Anterior Cingulate Gyrus in Social Valuation." *Science* **313** (1 September): 1310-1312.

Ruf, Irina, Zhe-Xi Luo, & **Thomas Martin.** 2013. "Reinvestigation of the Basicranium of *Haldanodon exspectatus* (Mammaliaformes, Docodonta)." *Journal of Vertebrate Paleontology* **33** (March): 382-400.

Ruf, Irina, Wolfgang Maier, Pablo G. Rodrigues, & **Cesar L. Schultz.** 2014. "Nasal Anatomy of the Non-mammaliaform Cynodont *Brasilitherium riograndensis* (Eucynodontia, Therapsida) Reveals New Inisght into Mammalian Evolution." *The Anatomical Record: Advances in Integrative Anatomy and Evolutionary Biology* **297** (November): 2018-2030.

Ruta, Marcello, Kenneth D. Angielczyk, Jörg Fröbisch, & **Michael J. Benton.** 2013. "Decoupling of morphological disparity and taxic diversity during the adaptive radiation of anomodont therapsids." *Proceedings of the Royal Society of London* **B** (Biological Sciences) **280** (7 October): 20131071.

Ryan, Aimee K., Bruce Blumberg, Concepción Rodriguez-Esteban, Sayuri Yonei-Tamura, Koji Tamura, Tohru Tsukui, Jennifer de la Peña, Walid Sabbagh, Jason Greenwald, Senyon Choe, Dominic P. Norris, Elizabeth J. Robertson, Ronald M. Evans, Michael G. Rosenfeld, & **Juan Carlos Izpisúa Belmonte.** 1998. "Pitx2 determines left-right asymmetry of internal organs in vertebrates." *Nature* **394** (6 August): 545-551.

Ryan, Joseph F., & **Andreas D. Baxevanis.** 2007. "Hox, Wnt, and the evolution of the primary body axis: insights from the early-divergent phyla." *Biology Direct* (online @ biomedcentral.com) **2** (13 December): 37.

Ryan, Joseph F., Patrick M. Burton, Maureen E. Mazza, Grace K. Kwong, James C. Mullikin, & **John R. Finnerty.** 2006. "The cnidarian-bilaterian ancestor possessed at least 56 homeoboxes: evidence from the starlet sea anemone, *Nematostella vectensis*." *Genome Biology* (online @ genomebiology.com) **7** (24 July): R64.

Ryan, Joseph F., Maureen E. Mazza, Kevin Pang, David Q. Matus, Andreas D. Baxevanis, Mark Q. Martindale, & **John R. Finnerty.** 2007. "Pre-Bilaterian Origins of the Hox Cluster and the Hox Code: Evidence from the Sea Anemone, *Nematostella vectensis*." *PLoS ONE* (online @ plosone.org) **2** (January): e153.

Ryan, Joseph F., Kevin Pang, NISC Comparative Sequencing Program, James C. Mullikin, Mark Q. Martindale, & **Andreas D. Baxevanis.** 2010. "The homeodomain complement of the ctenophore *Mnemiopsis leidyi* suggests that Ctenophora and Porifera diverged prior to the ParaHoxozoa." *EvoDevo* (online @ evodevojournal.com) **1** (October): 9.

Ryan, William, & **Walter Pitman.** 1998. *Noah's Flood: The New Scientific Discoveries About the Event that Changed History.* New York: Simon & Schuster.

Rybczynski, Natalia, Mary R. Dawson, & **Richard H. Telford.** 2009. "A semi-aquatic Arctic mammalian carnivore from the Miocene epoch and the origin of Pinnipedia." *Nature* **458** (23 April): 1021-1024.

Saier, Milton H. Jr. 2004. "Evolution of the bacterial type III protein secretion systems." *Trends in Microbiology* **12** (March): 113-115.

Salazar-Ciudad, Isaac, & **Jukka Jernvall.** 2002. "A gene network model accounting for development and evolution of mammalian teeth." *Proceedings of the National Academy of Sciences* **99** (11 June): 8116-8120.

———. 2010. "A computational model of teeth and the developmental origins of morphological variation." *Nature* **464** (25 March): 583-586.

Salazar-Ciudad, Isaac, & **Miquel Marin-Riera.** 2013. "Adaptive dynamics under development-based genotype-phenotype maps." *Nature* **497** (16 May): 361-364.

San Antonio Bible-Based Science Associaton. 2007. San Antonio Bible-Based Science Association *Communiqué* (March) posting (online text at *www.sabbsa.org* accessed 12/16/2008).

———. 2008a. "The Controversial Dr. Carl Baugh." SABBSA *Communiqué* (June) posting (online text at *www.sabbsa.org* accessed 12/16/2008).

———. 2008b. "Dr. Baugh's Human and Dinosaur Tracks." SABBSA *Communiqué* (July) posting (online text at *www.sabbsa.org* accessed 12/16/2008).

———. 2010. "'Creation by Evolution'—Satan's Newest Attack!" San Antonio Bible-Based Science Association *Communiqué* (June) posting (online text at *www.sabbsa.org* accessed 6/5/2014).

Sanchez, S., P. Tafforeau, & P. E. Ahlberg. 2014. "The humerus of *Eusthenopteron*: a puzzling organization presaging the establishment of tetrapod limb bone marrow." *Proceedings of the Royal Society of London* **B** (Biological Sciences) **281** (19 March): 20140299.

Sanchez, Sophie, Paul Tafforeau, Jennifer A. Clack, & **Per E. Ahlberg.** 2016. "Life history of the stem tetrapod *Acanthostega* revealed by synchrotron microtomography." *Nature* **537** (15 September): 408-411.

Sánchez-Villagra, Marcelo R. 2010. "Developmental palaeontology in synapsids: the fossil record of ontogeny in mammals and their closest relatives." *Proceedings of the Royal Society of London* **B** (Biological Sciences) **277** (22 April): 1139-1147.

Sánchez-Villagra, Marcelo R., Sven Gemballa, Sirpa Nummela, Kathleen K. Smith, & **Wolfgang Maier.** 2002. "Ontogenetic and Phylogenetic Transformations of the Ear Ossicles in Marsupial Mammals." *Journal of Morphology* **251** (March): 219-238.

Sanz, José L., Luis M. Chiappe, Bernardino P. Pérez-Moreno, Angela D. Buscalioni, José L. Moratalla, Francisco Ortega, & **Francisco J. Poyato-Ariza.** 1996. "An Early Cretaceous bird from Spain and its implications for the evolution of avian flight." *Nature* **382** (1 August): 442-445.

Sarfati, Jonathan D. 1999. *Refuting Evolution: A Handbook for Students, Parents, and Teachers Countering the Latest Arguments for Evolution.* Acacia Ridge, Australia: Answers in Genesis.

———. 2000a. "*Archaeoraptor*—Phony 'feathered' fossil." *Creation Ministries International* posting 3 February (online text at *creation.com* accessed 9/8/2016).

———. 2000b. "Ross-Hovind Debate, John Ankerberg Show, October 2000." *Answers in Genesis* article for 21 December (online text at *www.answersingenesis.org* accessed 1/22/2003).

———. 2006. "Noah's Flood and the Gilgamesh Epic." Answers in Genesis *Creation* **28** (September): 12-17.

———. 2008. *Refuting Evolution: A handbook for students, parents, and teachers countering the latest arguments for evolution. Creation Ministries International* posting of 2nd ed. (online text at *creation.com* accessed 4/29/2013).

———. 2014. "When did animals become carnivorous?" *Creation Ministries International* posting 31 August (online text at *creation.com* accessed 10/15/2016).

Sarfati, Jonathan, & **Carl Wieland.** 2002. "Speaking the truth in love." *Answers in Genesis* 2 December response to reader letters (online text at *www.answersingenesis.org* accessed 1/23/2003).

Sassera, Davide, Nathan Lo, Sara Epis, Giuseppe D'Auria, Matteo Montagna, Francesco Comandatore, David Horner, Juli Peretó, Alberto Maria Luciano, Federica Franciosi, Emanuele Ferri, Elena Crotti, Chiara Bazzocchi, Daniele Daffonchio, Luciano Sacchi, Andres Moya, Amparo Latorre, & **Claudio Bandi.** 2011. "Phylogenomic Evidence for the Presence of a Flagellum and *cbb*$_3$ Oxidase in the Free-Living Mitochondrial Ancestor." *Molecular Biology and Evolution* **28** (December): 3285-3296.

Sato, Akie, Colm O'hUigin, Felipe Figueroa, Peter R. Grant, B. Rosemary Grant, Herbert Tichy, & **Jan Klein.** 1999. "Phylogeny of Darwin's finches as revealed by mtDNA sequences." *Proceedings of the National Academy of Sciences* **96** (27 April): 5101-5106.

Sato, Akie, Herbert Tichy, Colm O'hUigin, Peter R. Grant, B. Rosemary Grant, & **Jan Klein.** 2001. "On the Origin of Darwin's Finches." *Molecular Biology and Evolution* **18** (March): 299-311.

Sauka-Spengler, Tatjana, Daniel Meulemans, Matthew Jones, & **Marianne Bronner-Fraser.** 2007. "Ancient Evolutionary Origin of the Neural Crest Gene Regulatory Network." *Developmental Cell*

13 (4 September): 405-420.

Sawyer, Roger H., Travis Glenn, Jeffrey O. French, Brooks Mays, Rose B. Shames, George L. Barnes Jr., Walter Rhodes, & **Yoshinori Ishikawa.** 2000. "The expression of Beta (β) keratins in the epidermal appendages of reptiles and birds." *Integrative and Comparative Biology* (AKA *American Zoologist*) **40** (September): 530-539.

Sawyer, Roger H., & **Loren W. Knapp.** 2003. "Avian skin development and the evolutionary origin of feathers." *Journal of Experimental Zoology* **298B** (15 August): 57-72.

Sawyer, Roger H., Loren Rogers, Lynette Washington, Travis C. Glenn, & **Loren W. Knapp.** 2005. "Evolutionary Origin of the Feather Epidermis." *Developmental Dynamics* **232** (February): 256-267.

Sawyer, Roger H., Brian A. Salvatore, Ta-Tanisha F. Potylicki, Jeffrey O. French, Travis C. Glenn, & **Loren W. Knapp.** 2003. "Origin of Feathers: Feather beta (β) keratins are expressed in discrete epidermal cell populations of embryonic scutate scales." *Journal of Experimental Zoology* **295B** (15 February): 12-24.

Scanlon, John D., & **Michael S. Y. Lee.** 2011. "The Major Clades of Living Snakes: Morphological Evolution, Molecular Phylogeny, and Divergence Dates," in Aldredge & Sever (2011, 55-95).

Schachat, Sandra R., & **George W. Gibbs.** 2016. "Variable wing venation in Agathiphaga (Lepidoptera: Agathiphagidae) is key to understanding the evolution of basal moths." *Royal Society Open Science* **3** (October): 160453.

Schachner, Emma R., Robert L. Cieri, James P. Butler, & **C. G. Farmer.** 2014. "Unidirectional pulmonary airflow patterns in the savannah monitor lizard." *Nature* **506** (20 February): 367-370.

Schachner, Emma R., John R. Hutchinson, & **C. G. Farmer.** 2013. "Pulmonary anatomy in the Nile crocodile and the evolution of unidirectional airflow in Archosauria." *PeerJ* (online @ peerj.com) **1** (26 March): 60.

Schachner, Emma R., Tyler R. Lyson, & **Peter Dodson.** 2009. "Evolution of the Respiratory System in Nonavian Theropods: Evidence from Rib and Vertebral Morphology." *The Anatomical Record: Advances in Integrative Anatomy and Evolutionary Biology* **292** (September): 1501-1513.

Schaefer, Henry, Fred Sigworth, Philip S. Skell, Frank Tipler, Robert Kaita, Michael Behe, Walter Hearn, Tony Mega, Dean Kenyon, Marco Horb, Daniel Kubler, David Keller, James Keesling, Roland F. Hirsch, Robert Newman, Carl Koval, Tony Jelsma, William Dembski, George Lebo, Timothy G. Standish, James Keener, Robert J. Marks, Carl Poppe, Siegfried Scherer, Gregory Shearer, Joseph Atkinson, Lawrence H. Johnston, Scott Minnich, David A. DeWitt, Theodor Liss, Braxton Alfred, Walter Bradley, Paul D. Brown, Marvin Fritzler, Theodore Saito, Muzaffar Iqbal, William S. Pelletier, Keith Delaplane, Ken Smith, Clarence Fouche, Thomas Milner, Brian J. Miller, Paul Nesselroade, Donald F. Calbreath, William P. Purcell, Wesley Allen, Jeanne Drisko, Chris Grace, Wolfgang Smith, Rosalind Picard, Garrick Little, John L. Omdahl, Martin Poenie, Russell W. Carlson, Hugh Nutley, David Berlinski, Neil Broom, John Bloom, James Graham, John Baumgardner, Fred Skiff, Paul Kuld, Yongsoon Park, Moorad Alexanian, Donald Ewert, Joseph W. Francis, Thomas Saleska, Ralph W. Seelke, James G. Harman, Lennart Moller, Raymond C. Bohlin, Fazale R. Rana, Michael Atchison, William S. Harris, Rebecca W. Keller, Terry Morrison, Robert F. DeHaan, Matti Lesola, Bruce Evans, Jim Gibson, David Ness, Bijan Nemati, Edward T. Peltzer, E. Stan Lennard, Rafe Payne, Phillip Savage, Pattle Pun, Jed Macosko, Daniel Dix, Ed Karlow, James Harbrecht, Robert W. Smith, Robert DiSilvestro, David Prentice, Walter Stangl, Jonathan Wells, James Tour, Todd Watson, Robert Waltzer, Vincent Villa, Richard Sternberg, James Tumlin, & **Charles Thaxton.** 2002. "A Scientific Dissent on Darwinism." *National Association for Objectivity in Science* posting of *Discovery Institute* list of 100 (online text at *www.objectivityinscience.org* accessed 3/17/2004).

Schierwater, Bernd, & **Rob Desalle.** 2001. "Current Problems with the Zootype and the Early Evolution of Hox Genes." *Journal of Experimental Zoology* **291** (15 August): 169-174.

Schilthuizen, Menno. 2001. *Frogs, flies, and dandelions: Speciation—The Evolution of New Species.* Oxford: Oxford University Press.

Schlafly, Phyllis. 2001a. "Eagle Forum Collegians Weekly Internet Newsletter." *Eagle Forum* 3 October posting (online text at *www.eagleforum.org* accessed 12/4/2001).

———. 2001b. "Monthly Update." *Eagle Forum* November posting (online text at *www.eagleforum.org* accessed 12/4/2001).

Schlange, Thomas, Hans-Henning Arnold, & **Thomas Brand.** 2002. "BMP2 is a positive regulator of Nodal signaling during left-right axis formation in the chicken embryo." *Development* (AKA *Journal of Embryology and Experimental Morphology*) **129** (July): 3421-3429.

Schmelzle, Thomas, Sirpa Nummela, & **Marcelo R. Sánchez-Villagra.** 2005. "Phylogenetic Transformations of the Ear Ossicles in Marsupial Mammals, with Special Reference to Diprotodontians: A Character Analysis." *Annals of Carnegie Museum* **74** (September): 189-200.
Schmelzle, Thomas, Marcelo R. Sánchez-Villagra, & **Wolfgang Maier.** 2007. "Vestibular labyrinth diversity in diprotodontian marsupial mammals." The Mammalogical Society of Japan *Mammal Study* **32** (June): 83-97.
Schmerler, Samuel B., Wendy L. Clement, Jeremy M. Beaulieu, David S. Chatelet, Lawren Sack, Michael J. Donoghue, & **Erika J. Edwards.** 2012. "Evolution of leaf form correlates with tropical-temperate transitions in *Viburnum* (Adoxaceae)." *Proceedings of the Royal Society of London* **B** (Biological Sciences) **279** (7 October): 20121110.
Schmieder, Martin, Elmar Buchner, Winfried H. Schwarz, Mario Trieloff, & **Philippe Lambert.** 2010. "A Rhaetian $^{40}Ar/^{39}Ar$ age for the Rochechouart impact structure (France) and implications for the latest Triassic sedimentary record." *Meteoritics & Planetary Science* **45** (August): 1225-1242.
Schneider, Igor, Ivy Aneas, Andrew R. Gehrke, Randall D. Dahn, Marcelo A. Nobrega, & **Neil H. Shubin.** 2011. "Appendage expression driven by the *Hoxd* Global Control Region is an ancient gnathostome feature." *Proceedings of the National Academy of Sciences* **108** (2 August): 12782-12786.
Schoch, Rainer R. 2010. "Heterochrony: the interplay between development and ecology exemplified by a Paleozoic amphibian clade." *Paleobiology* **36** (March): 318-334.
———. 2013. "The evolution of major temnospondyl clades: An inclusive phylogenetic analysis." *Journal of Systematic Palaeontology* **6** (August): 673-705.
———. 2014. "Life cycles, plasticity and palaeoecology in temnospondyl amphibians." *Paleobiology* **57** (May): 517-529.
Schoch, Rainer R., & **Hans-Dieter Sues.** 2015. "A Middle Triassic stem-turtle and the evolution of the turtle body plan." *Nature* **523** (30 July): 584-587.
Schoene, Blair, Jean Guex, Annachiara Bartolini, Urs Schaltegger, & **Terrence J. Blackburn.** 2010. "Correlating the end-Triassic mass extinction and flood basalt volcanism at the 100 ka level." *Geology* **38** (May): 387-390.
Schoene, Blair, Kyle M. Samperton, Michael P. Eddy, Gerta Keller, Thierry Adatte, Samuel A. Bowring, Syed F. R. Khadri, & **Brian Gertsch.** 2015. "U-Pb geochronology of the Deccan Traps and relation to the end Cretaceous mass extinction." *Science* **347** (9 January): 182-184.
Schroeder, Gerald. 1997. *The Science of God: The Convergence of Scientific and Biblical Wisdom.* New York: The Free Press.
Schulmeister, Susanne. 2003. "Genitalia and terminal abdominal segments of male basal Hymenoptera (Insecta): morphology and evolution." *Organisms Diversity & Evolution* **3** (3): 253-279.
Schultz, Ted R. 2000. "In search of ant ancestors." *Proceedings of the National Academy of Sciences* **97** (19 December): 14028-14029.
Schwartz, Jeffrey H. 1999. *Sudden Origins: Fossils, Genes, and the Emergence of Species.* New York: John Wiley & Sons.
Schwartz, Robert M., & **Margaret O. Dayhoff.** 1978. "Origins of Prokaryotes, Eukaryotes, Mitochondria, and Chloroplasts." *Science* **199** (27 January): 395-403.
Schweitzer, Mary Higby, & **Cynthia Lee Marshall.** 2001. "A Molecular Model for the Evolution of Endothermy in the Theropod-Bird Lineage." *Journal of Experimental Zoology* **291** (15 December): 317-338.
Schweitzer, Mary H., Mark Marshall, Keith Carron, D. Scott Bohle, Scott C. Busse, Ernst V. Arnold, Darlene Barnard, J. R. Hornery, & **Jean R. Starkey.** 1997. "Heme compounds in dinosaur trabecular bone." *Proceedings of the National Academy of Sciences* **94** (10 June): 6291-6296.
Schweitzer, Mary Higby, Zhiyong Suo, Recep Avci, John M. Asara, Mark A. Allen, Fernando Teran Arce, & **John R. Horner.** 2007. "Analyses of Soft Tissue from *Tyrannosaurus rex* Suggest the Presence of Protein." *Science* **316** (13 April): 277-280.
Schweitzer, Mary Higby, Jennifer L. Wittmeyer, & **John R. Horner.** 2007. "Soft tissue and cellular preservation in vertebrate skeletal elements from the Cretaceous to the present." *Proceedings of the Royal Society of London* **B** (Biological Sciences) **274** (22 January): 183-197.
Schweitzer, Mary H., Jennifer L. Wittmeyer, John R. Horner, & **Jan K. Toporski.** 2005. "Soft-Tissue Vessels and Cellular Preservation in *Tyrannosaurus rex*." *Science* **307** (25 March): 1952-1955.
Schweitzer, Mary Higby, Wenxia Zheng, Timothy P. Cleland, & **Marshall Bern.** 2013. "Molecular analyses of dinosaur osteocytes support the presence of endogenous molecules." *Bone* **52**

(January): 414-423.
Schweitzer, Mary H., Wenxia Zheng, Timothy P. Cleland, Mark B. Goodwin, Elizabeth Boatman, Elizabeth Theil, Matthew A. Marcus, & **Sirine C. Fakra.** 2014. "A role for iron and oxygen chemistry in preserving soft tissues, cells and molecules from deep time." *Proceedings of the Royal Society of London* **B** (Biological Sciences) **281** (22 January): 20132741.
ScienceDaily. 2011. "A new species of fossil silky lacewing insects that lived more than 120 million years ago." *ScienceDaily* 7 October posting (online text at *www.sciencedaily.com* accessed 3/19/2016).
Scotland, Robert W., Darrell J. Siebert, & **David M. Williams**, eds. 1994. *Models in Phylogeny Reconstruction.* Oxford: Clarendon Press.
Senter, Phil. 2010. "Using creation science to demonstrate evolution: application of a creationist method for visualizing gaps in the fossil record to a phylogenetic study of coelurosaurian dinosaurs." *Journal of Evolutionary Biology* **23** (August): 1732-1743.
———. 2011. "Using creation science to demonstrate evolution 2: morphological continuity within Dinosauria." *Journal of Evolutionary Biology* **24** (October): 2197-2216.
Serdobova, Irina M., & **Dmitri A. Kramerov.** 1998. "Short Retroposons of the B2 Superfamily: Evolution and Application for the Study of Rodent Phylogeny." *Journal of Molecular Evolution* **46** (February): 202-214.
Sereno, Paul C. 1990. "Psittacosauridae," in Weishampel *et al.* (1990, 579-592).
———. 1997. "Psittacosauridae," in Currie & Padian (1997, 611-613).
———. 1999a. "Definitions in Phylogenetic Taxonomy: Critique and Rationale." *Systematic Biology* **48** (June): 329-351.
———. 1999b. "The Evolution of Dinosaurs." *Science* **284** (25 June): 2137-2147.
Sereno, Paul C., & **Andrea B. Arcucci.** 1993. "Dinosaurian precursors from the Middle Triassic of Argentina: *Lagerpeton chanarensis*." *Journal of Vertebrate Paleontology* **13** (December): 385-399.
———. 1994. "Dinosaurian precursors from the Middle Triassic of Argentina: *Marasuchus lilloensis, gen. nov.*" *Journal of Vertebrate Paleontology* **14** (March): 53-73.
Sereno, Paul C., Ricardo N. Martinez, Jeffrey A. Wilson, David J. Varricchio, Oscar A. Alcober, & **Hans C. E. Larsson.** 2008. "Evidence for Avian Intrathoracic Air Sacs in a New Predatory Dinosaur from Argentina." *PLoS ONE* (online @ plosone.org) **3** (September): e3303.
Serres, Margrethe H., Alastair R. W. Kerr, Thomas J. McCormack, & **Monica Riley.** 2009. "Evolution by leaps: gene duplication in bacteria." *Biology Direct* (online @ biomedcentral.com) **4** (23 November): 46.
Seymour, R. S., C. L. Bennett-Stamper, S. D. Johnston, D. R. Carrier, & **G. C. Grigg.** 2004. "Evidence for endothermic ancestors of crocodiles at the stem of archosaur evolution." *Physiological and Biochemical Zoology* **77** (November-December): 1051-1067.
Shapiro, Michael D., James Hanken, & **Nadia Rosenthal.** 2003. "Developmental Basis of Evolutionary Digit Loss in the Australian Lizard *Hemiergis*." *Journal of Experimental Zoology* **297B** (15 June): 48-56.
Sharpe, P. T. 2007. "Homeobox genes in ititiation and shape of teeth during development in mammalian embryos," in Teaford *et al.* (2007, 3-11).
Shaver, Mike, Dave Nutting, & **Mary Jo Nutting.** 1996. "Stop that Panda!" Alpha Omega Institute Institute (Grand Junction, CO) *Think & Believe* **13** (March/April): 1.
Shaw, Kerry L., Christopher K. Ellison, Kevin P. Oh, & **Chris Wiley.** 2011. "Pleiotropy, 'sexy' traits, and speciation." *Behavioral Ecology* **22** (November-December): 1154-1155.
Shcherbakov, Dmitry E. 2013. "Permian ancestors of Hymenoptera and Raphidioptera." *ZooKeys* (online @ ncbi.nlm.nih.gov) **358** (4 December): 45-6781.
Sheng, Guojun, & **Ann C. Foley.** 2012. "Diversification and conservation of the extraembryonic tissues in mediating nutrient uptake during amniote development." *Annals of the New York Academy of Sciences* **1271** (October): 97-103.
Shermer, Michael. 2001. *The Borderlands of Science: Where Sense Meets Nonsense.* Oxford: Oxford University Press.
Sherwin, Frank. 2010. "Darwinism's Rubber Ruler." Institute for Creation Research *Acts & Facts* (February): 17.
Sheth, Rushikesh, Luciano Marcon, M. Félix Bastida, Marisa Junco, Laura Quintana, Randall Dahn, Marie Kmita, James Sharpe, & **Maria A. Ros.** 2012. "Hox Genes Regulate Digit Patterning by Controlling the Wavelength of a Turing-Type Mechanism." *Science* **338** (14 December): 1476-1480.

Shigetani, Yasuyo, Fumiaki Sugahara, Yayoi Kawakami, Yasunori Murakami, Shigeki Hirano, & Sigeru Kuratani. 2002. "Heterotopic Shift of Epithelial-Mesenchymal Interactions in Vertebrate Jaw Evolution." *Science* **296** (17 May): 1316-1319.

Shin, Jeong-Oh, Jong-Min Lee, Kyoung-Won Cho, Sungwook Kwak, Hyuk-Jae Kwon, Min-Jung Lee, Sung-Won Cho, Kye-Seong Kim, & Han-Sung Jung. 2012. "MiR-200b is involved in Tgf-β signaling to regulate mammalian palate development." *Histochemistry and Cell Biology* **137** (January): 67-78.

Shubin, Neil. 1998. "Evolutionary cut and paste." *Nature* **394** (2 July): 12-13.

———. 2008. *Your Inner Fish: A Journey Into the 3.5-Billion-Year History of the Human Body.* New York: Pantheon Books.

Shubin, Neil H., A. W. Crompton, Hans-Dieter Sues, & Paul E. Olsen. 1991. "New Fossil Evidence on the Sister-Group of Mammals and Early Mesozoic Faunal Distribution." *Science* **251** (1 March): 1063-1065.

Shubin, Neil H., Edward B. Daeschler, & Michael I. Coates. 2004. "The Early Evolution of the Tetrapod Humerus." *Science* **304** (2 April): 90-93.

Shubin, Neil H., Edward B. Daeschler, & Farish A. Jenkins Jr. 2006. "The pectoral fin of *Tiktaalik roseae* and the origin of the tetrapod limb." *Nature* **440** (6 April): 764-771.

———. 2014. "Pelvic girdle and fin of *Tiktaalik roseae*." *Proceedings of the National Academy of Sciences* **111** (21 January): 893-899.

Shubin, Neil, Cliff Tabin, & Sean Carroll. 1997. "Fossils, genes and the evolution of animal limbs." *Nature* **388** (14 August): 639-647.

Sidor, Christian A. 2001. "Simplification as a Trend in Synapsid Cranial Evolution." *Evolution* **55** (July): 1419-1442.

———. 2003a. "The naris and palate of *Lycaenodon longiceps* (Therapsida: Biarmosuchia) with comments on their early evolution in the therapsida." *Journal of Paleontology* **77** (July): 977-984.

———. 2003b. "Evolutionary trends and the origin of the mammalian lower jaw." *Paleobiology* **24** (Autumn): 605-640.

Sidor, C. A., & P. J. Hancox. 2006. "*Elliotherium kersteni*, a new tritheledontid from the Lower Elliot Formation (Upper Triassic) of South Africa." *Journal of Paleontology* **80** (March): 333-342.

Sidor, Christian A., & James A. Hopson. 1998. "Ghost lineages and 'mammalness': assessing the temporal pattern of character acquisition in the Synapsida." *Paleobiology* **24** (April): 254-273.

Simionato, Elena, Valérie Ledent, Gemma Richards, Morgane Thomas-Chollier, Pierre Kerner, David Coornaert, Bernard M. Degnan, & Michel Vervoort. 2007. "Origin and diversification of the basic helix-loop-helix gene family in metazoans: insights from comparative genomics." *BMC Evolutionary Biology* (online @ biomedcentral.com) **7** (2 March): 33.

Simmons, Geoffrey. 2004. *What Darwin Didn't Know.* Eugene, OR: Harvest House Publishers.

———. 2007. *Billions of Missing Links.* Eugene, OR: Harvest House Publishers.

Simmons, Nancy B., Kevin L. Seymour, Jörg Habersetzer, & Gregg F. Gunnell. 2008. "Primitive Early Eocene bat from Wyoming and the evolution of flight and echolocation." *Nature* **451** (14 February): 818-821.

Simons, Lewis M. 2000. "Archaeoraptor Fossil Trail." *National Geographic* **198** (October): 128-132.

Simpson, George Gaylord. 1929. "The Dentition of *Ornithorhynchus* as Evidence of Its Affinities." *American Museum Novitates* **390** (December): 1-15.

———. 1944. *Tempo and Mode in Evolution.* New York: Columbia University Press.

———. 1953. *The Major Features of Evolution.* New York: Columbia University Press.

———. 1960. "The History of Life," in Tax (1960, 117-180).

———. 1983. *Fossils and the History of Life.* New York: Scientific American Library.

Singh, Nadia D., & Kerry L. Shaw. 2012. "On the scent of pleiotropy." *Proceedings of the National Academy of Sciences* **109** (3 January): 5-6.

Sinha, Satrajit, & Elaine Fuchs. 2001. "Identification and dissection of an enhancer controlling epithelial gene expression in skin." *Proceedings of the National Academy of Sciences* **98** (27 February): 2445-2460.

Slatkin, Montgomery. 1982. "Pleiotropy and Parapatric Speciation." *Evolution* **36** (March): 263-270.

Slaughter, Bob H. 1965. "A therian from the Lower Cretaceous (Albian) of Texas." Peabody Museum of Natural History (Yale University) *Postilla* **93**: 1-18.

Slick, Matt. 2011. "Cretaceous Period." *Christian Apologetics & Research Ministry* 18 September posting (online text at *carm.org* accessed 8/2/2016).

Sloan, Christopher P. 1999. "Feathers for *T. rex*?" *National Geographic* **196** (November): 98-107.

Sloan, Robert E. 1983. "The Transition between Reptiles and Mammals," in Zetterberg (1983, 263-277).

Sloan, Robert E., & **Leigh Van Valen.** 1965. "Cretaceous Mammals from Montana." *Science* **148** (9 April): 220-227.

Slotten, Ross A. 2004. *The Heretic in Darwin's Court: The Life of Alfred Russel Wallace.* New York: Columbia University Press.

Smit, Jan. 2008. "Why did the dinosaurs die out?" in Benton (2008, 42-45).

Smith, Anika. 2010. "Suit Filed in Case of Pro-ID JPL Employee Demotion." *Discovery Institute* direct mailing (2 July).

Smith, Caroline L., Frédérique Varoqueaux, Maike Kittelmann, Rita N, Azzam, Benjamin Cooper, Christine A, Winters, Michael Eitel, Dirk Fasshauer, & **Thomas S. Reese.** 2014. "Novel Cell Types, Neurosecretory Cells, and Body Plan of the Early-Diverging Metazoan *Trichoplax adhaerens*." *Current Biology* **24** (21 July): 1565-1572.

Smith, Roger M. H., & **Jennifer Botha-Brink.** 2011. "Morphology and composition of bone-bearing coprolites from the Late Permian Beaufort Group, Karoo Basin, South Africa." *Palaeogeography, Palaeoclimatology, Palaeoecology* **312** (1 December): 40-53.

Smith, Roger M. H., Bruce S. Rubidge, & **Christian A. Sidor.** 2006. "A New Burnetiid (Therapsida: Biarmosuchia) from the Upper Permian of South Africa and its Biogeographic Implications." *Journal of Vertebrate Paleontology* **26** (June): 331-343.

Smithson, Timothy R., Stanley P. Wood, John E. A. Marshall, & **Jennifer A. Clack.** 2012. "Earliest Carboniferous tetrapod and arthropod faunas from Scotland populate Romer's Gap." *Proceedings of the National Academy of Sciences* **109** (20 March): 4532-4537.

Smithsonian National Museum of Natural History. 2015. "Reconstructing Extinct Animals." *Smithsonian Institution* undated posting (online text at *naturalhistory.si.edu* accessed 4/8/2016).

Snelling, Andrew A., ed. 2008. *Proceedings of the Sixth International Conference on Creationism.* Pittsburgh: Creation Science Fellowship.

———. 2010. "Doesn't the Order of Fossils in the Rock Record Favor Long Ages?" *Answers in Genesis* 9 September posting of *The New Answers Book 2* Chapter 31 (online text at *www.answersingenesis.org* accessed 2/22/2014).

———. 2013a. "How Could Fish Survive the Genesis Flood?" *Answers in Genesis* 16 June 2014 posting of *The New Answers Book 3* Chapter 20 (online text at *www.answersingenesis.org* accessed 2/17/2015).

———. 2013b. "What Are Some of the Best Flood Evidences?" *Answers in Genesis* 13 February posting of *The New Answers Book 3* Chapter 29 (online text at *www.answersingenesis.org* accessed 3/5/2015).

———. 2014. *Earth's Catastrophic Past: Geology, Creation & the Flood.* Green Forest, AR: Master Books.

Snelling, Andrew, & **Bodie Hodge.** 2013. "Did the Continents Split Apart in the Days of Peleg?" *Answers in Genesis* 4 July 2014 posting of *The New Answers Book 3* Chapter 23 (online text at *www.answersingenesis.org* accessed 2/17/2015).

Snelling, Andrew, & **Tom Vail.** 2013. "When and How Did the Grand Canyon Form?" *Answers in Genesis* 6 June 2014 posting of *The New Answers Book 3* Chapter 18 (online text at *www.answersingenesis.org* accessed 2/17/2015).

Sonleitner, Frank J. 1987. "The Origin of Species by Punctuated Equilibrium." *Creation/Evolution* **7** (Spring): 25-30.

———. 1996. "'*The Mysterious Origins of Man*' broadcast on NBC at 7 p.m. EST on February 25, 1996." *Creation/Evolution* **15** (Winter): 30-32.

Soares, Marina B., Cesar L. Schultz, & **Bruno L. D. Horn.** 2011. "New information on *Riograndia guaibensis* Bonaparte, Ferigolo & Ribeiro, 2001 (Eucynodontia, Tritheledontidae) from the Late Triassic of southern Brazil: anatomical and biostratigraphic implications." *Anais da Academia Brasileira de Ciencias* **83** (March): 329-354.

Sordino, Paulo, Frank van der Hoeven, & **Denis Duboule.** 1995. "*Hox* gene expression in teleost fins and the origin of vertebrate digits." *Nature* **375** (22 June): 678-681.

Spectator. 2016. "The best and worst books of 2016, chosen by some of our regular contributors." *The Spectator* (12 November): online text accessed 11/28/2016 (*spectator.co.uk*).

Spencer, Frank. 1990. *Piltdown: A Scientific Forgery.* London: Oxford University Press.

Spindler, Frederik, Jocelyn Falconnet, & **Jörg Fröbisch.** 2016. "*Callibrachion* and *Datheosaurus*, two historical and previously mistaken basal caseasaurian synapsids." *Acta Palaeontologica Polonica*

61 (3): 597-616.

Spindler, F., D. Scott, & **R. R. Reisz.** 2015. "New information on the cranial and postcranial anatomy of the early synapsid *Ianthodon schultzei* (Spehacomorpha: Sphenacodontia), and its evolutionary significance." *Fossil Record* **18** (1): 17-30.

Spoor, F., S. Bajpai, S. T. Hussain, K. Kumar, & **J. G. M. Thewissen.** 2002. "Vestibular evidence for the evolution of aquatic behaviour in early cetaceans." *Nature* **417** (9 May): 163-166.

Springer, Mark S., Robert W. Meredith, Emma C. Teeling, & **William J. Murphy.** 2013. "Technical Comment on 'The Placental Mammal Ancestor and the Post-K-Pg Radiation of Placentals'." *Science* **341** (9 August): 613.

Srivastava, Mansi, Claire Larroux, Daniel R. Lu, Kareshma Mohanty, Jarrod Chapman, Bernard M. Degnan, & **Daniel S. Rokhsar.** 2010. "Early evolution of the LIM homeobox gene family." *BMC Biology* (online @ biomedcentral.com) **8** (18 January): 4.

Srivastava, Mansi, Oleg Simakov, Jarrod Chapman, Bryony Fahey, Marie E. A. Gauthier, Therese Mitros, Gemma S. Richards, Cecilia Conaco, Michael Dacre, Uffe Hellsten, Claire Larroux, Nicholas H. Putnam, Mario Stanke, Maja Adamska, Aaron Darling, Sandie M. Degnan, Todd H. Oakley, David C. Plachetzki, Yufeng Zhai, Marcin Adamski, Andrew Calcino, Scott F. Cummins, David M. Goodstein, Christina Harris, & **Daniel J. Jackson.** 2010. "The *Amphimedon queenslandica* genome and the evolution of animal complexity." *Nature* **466** (5 August): 720-726.

Stadler, Peter F., Claudia Fried, Sonja J. Prohaska, Wendy J. Bailey, Bernhard Y. Misof, Frank H. Ruddle, & **Günter P. Wagner.** 2004. "Evidence for Independent *Hox* gene duplications in the hagfish lineage: A PCR-based gene inventory of *Eptatretus stoutii*." *Molecular Phylogenetics and Evolution* **32** (September): 686-694.

Stahl, Barbara J. 1974. *Vertebrate History: Problems in Evolution.* New York: McGraw-Hill.

———. 1985. *Vertebrate History: Problems in Evolution.* Rev. pb ed. New York: Dover Publications, Inc.

Stebbins, G. Ledyard. 1971. *Processes of Organic Evolution.* 2nd ed. Englewood Cliffs, NJ: Prentice-Hall.

———. 1977. *Processes of Organic Evolution.* 3rd ed. Englewood Cliffs, NJ: Prentice-Hall.

Stettenheim, Peter R. 2000. "The integumentary morphology of modern birds—an overview." *Integrative and Comparative Biology* (AKA *American Zoologist*) **40** (September): 461-477.

Stokstad, Erik. 2002. "'Fantastic' Fossil Helps Narrow Data Gap." *Science* **296** (26 April): 637, 639.

Stovall, J. Willis, Llewellyn I. Price, & **Alfred Sherwood Romer.** 1966. "The Postcranial Skeleton of the Giant Permian Pelycosaur *Cotylorhynchus romeri*." *Bulletin of the Museum of Comparative Zoology* **135** (September): 1-30.

Strahler, Arthur N. 1987. *Science and Earth History—The Evolution/Creation Controversy.* Buffalo, NY: Prometheus Books.

Striedter, G. F. 1997. "The telecephlaon of tetrapods in evolution." *Brain, Behavior and Evolution* **49** (4): 179-213.

Stuckwish, Dale. 2012. "What are the biblical cryptids of cryptozoology." *Pittsburgh Creationism Examiner* 26 December posting (online text at *www.examiner.com* accessed 8/19/2014).

Su, Xin-zhuan, Laura A. Kirkman, Hisashi Fujioka, & **Thomas E. Wellems.** 1997. "Complex Polymorphisms in an ~330 kDa Protein Are Linked to Chloroquine-Resistant P. falciparum in Southeast Asia and Africa." *Cell* **91** (28 November): 593-603.

Sucena, Elio, Isabelle Delon, Isaac Jones, François Payre, & **David L. Stern.** 2003. "Regulatory evolution of *shavenbaby/ovo* underlies multiple cases of morphological parallelism." *Nature* **424** (21 August): 935-938.

Sues, Hans-Dieter. 1985. "The relationships of the Tritylodontidae (Synapsida)." *Zoological Journal of the Linnean Society* **85** (November): 205-217.

———. 2001. "On *Microconodon*, a Late Triassic cynodont from the Newark Supergroup of eastern North America." *Bulletin of the Museum of Comparative Zoology* **156** (October): 37-48.

Sues, Hans-Dieter, & **Farish A. Jenkins Jr.** 2006. "The Postcranial Skeleton of *Kayentatherium wellesi* from the Lower Jurassic Kayenta Formation of Arizona and the Phylogenetic Significance of Postcranial Features in Tritylodontid Cynodonts," in Carrano *et al.* (2006, 114-152).

Sues, Hans-Dieter, Paul E. Olsen, & **Joseph G. Carter.** 1999. "A Late Triassic traversodont cynodont from the Newark Supergroup of North Carolina." *Journal of Vertebrate Paleontology* **19** (June): 351-354.

Sulej, Tomasz, Andrzej Wolniewicz, Niels Bonde, Błażej Błażejowski, Grzegorz Niedźwiedski, & **Mateusz Tałanda.** 2014. "New perspectives on the Late Triassic vertebrates of East Greenland:

preliminary results of a Polish-Danish palaeontological expedition." *Polish Polar Research* **35** (4): 541-552.

Summers, Robert L., Anurag Dave, Tegan J. Dolstra, Sebastiano Bellanca, Rosa V. Marchetti, Megan N. Nash, Sashika N. Richards, Valerie Goh, Robyn L. Schenk, Wilfred D. Stein, Kiaran Kirk, Cecilia P. Sanchez, Michael Lanzer, & **Rowena E. Martin.** 2014. "Diverse mutational pathways converge on saturable chloroquine transport via the malaria parasite's chloroquine resistance transporter." *Proceedings of the National Academy of Sciences* **111** (29 April): E1759-E1767.

Sun, Chen, & **Shicui Zhang.** 2015. "Immune-Relevant and Antioxidant Activities of Vitellogenin and Yolk Proteins in Fish." *Nutrients* **7** (10): 8818-8829.

Sun, Tianjun, Feng-Hsu Lin, Robert L. Campbell, John S. Allingham, & **Peter L. Davies.** 2014. "An Antifreeze Protein Folds with an Interior Network of More Than Semi-Clathrate Waters." *Science* **343** (14 February): 795-798.

Sunderland, Luther D. 1988. *Darwin's Enigma: Fossils and Other Problems.* 1998 pb ed. subtitled as *Ebbing the Tide of Naturalism* (*nota bene*: though the text is unchanged, pagination differs from the original printing). Green Forest, AR: Master Books.

Supp, Dorothy M., Martina Bruckner, Michael P. Kuehn, David P. Witte, Linda A. Lowe. James McGrath, JoMichelle Corrales, & **S. Steven Potter.** 1999. "Targeted deletion of the ATP binding domain of left-right dynein confirms its role in specifying development of left-right asymmetries." *Development* (AKA *Journal of Embryology and Experimental Morphology*) **126** (December): 5495-5504.

Suzuki, Kentaro, Yuji Yamaguchi, Mylah Villacorte, Kenichiro Mihara, Masashi Akiyama, Hiroshi Shimizu, Makoto M. Taketo, Naomi Nakagata, Tadasuke Tsukiyama, Terry P. Yamaguchi, Walter Birchmeier, Shigeaki Kato, & **Gen Yamada.** 2009. "Embryonic hair follicle fate change by augmented β-catenin through Shh and Bmp signaling." *Development* (AKA *Journal of Embryology and Experimental Morphology*) **136** (1 February): 367-372.

Suzuki, Satoshi, Mikiko Abe, & **Masaharu Motokawa.** 2011. "Allometric Comparison of Skulls from Two Closely Related Weasels, *Mustela itatsi* and *M. sibirica*." *Zoological Science* **28** (September): 676-688.

Suzuki, Takayuki. 2013. "How Is Digit Identity Determined During Limb Development?" *Development, Growth & Differentiation* **55** (January): 130-138.

Swift, Dennis. 1994. "The Dinosaurs Of Acambaro: Initial Report." *The Science of Creation* undated posting (online text at *www.bible.ca* accessed 8/18/2014).

———. 1999. "The Dinosaurs Of Acambaro: Preliminary Report From Second Expedition." *The Science of Creation* undated posting (online text at *www.bible.ca* accessed 8/18/2014).

Świło, Marlena, Grzegorz Niedźwiedski, & **Tomasz Sulej.** 2014. "Mammal-like tooth from the Upper Triassic of Poland." *Acta Palaeontologica Polonica* **59** (4): 815-820.

Swisher, Carl C. III, Yuan-qing Wang, Xiao-lin Wang, Xing Xu, & **Yuan Wang.** 1999. "Cretaceous age for the feathered dinosaurs of Liaoning, China." *Nature* **400** (1 July): 58-61.

Switek, Brian. 2007. "Feduccia is at it again." *Laelaps* blog for 6 April (online text at *scienceblogs.com/laelaps* accessed 5/7/2014).

Szalay, Frederick S., Michael J. Novacek, & **Malcolm C. McKenna**, eds. 1993. *Mammal Phylogeny: Mesozoic: Differentiation, Multituberculates, Monotremes, Early Therians, and Marsupials.* New York: Springer.

Tabuce, Rodolphe, Laurent Marivaux, Renaud Lebrun, Mohammed Adaci, Mustapha Bensalah, Pierre-Henri Fabre, Emmanuel Fara, Helder Gomes Rodrigues, Lionel Hautier, Jean-Jacques Jaeger, Vincent Lazzari, Fateh Mebrouk, Stéphane Peigné, Jean Sudre, Paul Tafforeau, Xavier Valentin, & **Mahammed Mahboubi.** 2009. "Anthropoid *versus* strepsirhine status of the African Eocene primates *Algeripithecus* and *Azibius*: craniodental evidence." *Proceedings of the Royal Society of London* **B** (Biological Sciences) **276** (7 December): 4087-4094.

Takechi, Masaki, & **Shigeru Kuratani.** 2010. "History of Studies on Mammalian Middle Ear Evolution: A Comparative Morphological and Developmental Biology Perspective." *Journal of Experimental Zoology* **314B** (15 September): 417-433.

Talk.Origins Archive. 1988. "Creationist Whoppers." *The TalkOrigins Archive* undated posting (online text at *talkorigins.org* accessed 3/21/2016).

Tamura, Koji, Naoki Nomura, Ryohei Seki, Sayuri Yonei-Tamura, & **Hitoshi Yokoyama.** 2011. "Embryological Evidence Identifies Wing Digits in Birds as Digits 1, 2 and 3." *Science* **331** (11 February): 753-757.

Tan, Xiaodong, Jason L. Pecka, Jie Tang, Sandor Lovas, Kirk W. Beisel, & **David Z. Z. He.** 2012. "A motif of eleven amino acids is a structural adaptation that facilitates motor capability of eutherian prestin." *Journal of Cell Science* **125** (15 February): 1039-1047.

Tan, Xiaodong, Jason L. Pecka, Jie Tang, Oseremen E. Okoruwa, Qian Zhang, Kirk W. Beisel, & **David Z. Z. He.** 2011. "From Zebrafish to Mammal: Functional Evolution of Prestin, the Motor Protein of Cochlear Outer Hair Cells." *Journal of Neurophysiology* **105** (January): 36-44.

Tarzia, Wade. 1994. "Forbidden Archaeology: Antievolutionism Outside the Christian Arena." *Creation/Evolution* **14** (Summer): 13-25.

Tashman, Brian. 2013a. "Is Satan Behind the Campaign to Let Gays Join the Boy Scouts?" People for the American Way *Right Wing Watch* 4 February (online text at *www.rightwingwatch.org* accessed 8/9/2013).

———. 2013b. "Stanton: Same-Sex Marriage Is a 'Pernicious Lie of Satan' that Imperils Society and Humanity." People for the American Way *Right Wing Watch* 7 February posting (online text at *www.rightwingwatch.org* accessed 11/27/2013).

———. 2013c. "Santorum: Satan Controls The Film Industry." People for the American Way *Right Wing Watch* 24 October posting (online text at *www.rightwingwatch.org* accessed 11/3/2013).

———. 2016a. "Anne Graham Lotz: Satan Behind Gay Marriage Decision, End Times Looming." People for the American Way *Right Wing Watch* 4 January posting (online text at *www.rightwingwatch.org* accessed 1/8/2016).

———. 2016b. "Darrell Scott: Trump Is Under 'Concentrated Satanic Attack'." People for the American Way *Right Wing Watch* 21 September posting (online text at *www.rightwingwatch.org* accessed 9/22/2016).

Tattersall, Ian. 1995. *The Last Neanderthal: The Rise, Success, and Mysterious Extinction of Our Closest Human Relatives*. New York: Macmillan (Peter N. Névraumont).

Tausta, S. Lorraine, Heather Miller Coyle, Beverly Rothermel, Virginia Stiefel, & **Timothy Nelson.** 2002. "Maize C4 and non-C4 NADP-dependent malic enzymes are encoded by distinct genes derived from a plastid-localized ancestor." *Plant Molecular Biology* **50** (November): 635-652.

Tax, Sol, ed. 1960. *Evolution After Darwin: The University of Chicago Centennial. Volume 1. The Evolution of Life: Its Origin, History and Future*. Chicago: University of Chicago Press.

Taylor, Paul S. 1995. *The Illustrated Origins Answer Book: Concise, Easy-to-Understand Facts about the Origin of Life, Man, and the Cosmos*. 5th ed. Gilbert, AZ: Eden Communications.

Teaford, Mark F., Moya Meredith Smith, & **Mark W. J. Ferguson**, eds. 2007. *Development, Function and Evolution of Teeth*. Cambridge: Cambridge University Press.

Teichmann, Sarah A., Jong Park, & **Cyrus Chothia.** 1998. "Structural assignments to the *Mycoplasma genitalium* proteins show extensive gene duplications and domain rearrangements." *Proceedings of the National Academy of Sciences* **95** (8 December): 14658-14663.

Temple, Robert K. G. 1976. *The Sirius Mystery*. New York: St. Martin's Press.

Testaz, Sandrine, Artem Jarov, Kevin P. Williams, Leona E. Ling, Victor E. Koteliansky, Claire Fournier-Thibault, & **Jean-Loup Duband.** 2001. "Sonic hedgehog restricts adhesion and migration of neural crest cells independently of the Patched-Smoothened-Gli signaling pathway." *Proceedings of the National Academy of Sciences* **98** (23 October): 12521-12526.

Texas Freedom Network. 2009a. "What Does Don McLeroy Really Want to Teach?" *Texas Freedom Network* 18 March posting (online text at *www.tfninsider.org* accessed 3/23/2009).

———. 2009b. "Senate Takes Up McLeroy Nomination!" *Texas Freedom Network* 28 May posting (online text at *www.tfn.wordpress.com* accessed 5/28/2009).

———. 2013a. "Don McLeroy's Strange Testimony on Texas Science Textbooks: 'Support the Bible, and Adopt These Books'." *Texas Freedom Network* 13 September posting by **Dan** (online text at *www.tfninsider.org* accessed 1/25/2014).

———. 2013b. "Don McLeroy's Strange Testimony on Texas Science Textbooks: 'Support the Bible, and Adopt These Books'." *Texas Freedom Network* 17 September posting (online text at *www.tfninsider.org* accessed 1/25/2014).

Thanukos, Anastasia. 2008. "Views from Understanding Evolution: Parsimonious Explanations for Punctuated Patterns." *Evolution: Education & Outreach* **1** (April): 138-146.

Thaxton, Charles B., Walter L. Bradley, & **Roger L. Olsen.** 1984. *The Mystery of Life's Origin: Reassessing Current Theories*. New York: Philosophical Library.

Theobald, Douglas L. 2011. "29+ Evidences for Macroevolution: The Scientific Case for Common Descent." *The TalkOrigins Archive* posting updated 30 September (online text at *talkorigins.org* accessed 11/9/2011).

Theunissen, Lionel. 1997. "Patterson Misquoted: A Tale of Two 'Cites'." *The TalkOrigins Archive* 24 June updated posting (online text at *talkorigins.org* accessed 4/7/2016).
Thewissen, J. G. M., & **Sunil Bajpai.** 2001. "Whale Origins as a Poster Child for Macroevolution." *BioScience* **51** (December): 1037-1049.
Thewissen, J. G. M., M. J. Cohn, L. S. Stevens, S. Bajpai, J. Heyning, & **W. E. Horton Jr.** 2006. "Developmental basis for hind-limb loss in dolphins and origin of the cetacean bodyplan." *Proceedings of the National Academy of Sciences* **103** (30 May): 8414-8418.
Thewissen, J. G. M., Lisa Noelle Cooper, Mark T. Clementz, Sunil Bajpai, & **B. N. Tiwari.** 2007. "Whales originated from aquatic artiodactyls in the Eocene epoch of India." *Nature* **450** (20 December): 1190-1194.
Thewissen, J. G. M., Lisa Noelle Cooper, John C. George, &, **Sunil Bajpai.** 2009. "From Land to Water: the Origin of Whales, Dolphins, and Porpoises." *Evolution: Education & Outreach* **2** (June): 272-288.
Thewissen, J. G. M., & **S. T. Hussain.** 1993. "Origin of underwater hearing in whales." *Nature* **361** (4 February): 444-445.
———. 2000. "*Attockicetus praecursor*, A New Remingtonocetid Cetacean from Marine Eocene Sediments of Pakistan." *Journal of Mammalian Evolution* **7** (September): 133-146.
Thewissen, J. G. M., S. T. Hussain, & **M. Arif.** 1994. "Fossil Evidence for the Origin of Aquatic Locomotion in Archaeocete Whales." *Science* **263** (14 January): 210-212.
Thewissen, J. G. M., & **S. I. Madar.** 1999. "Ankle Morphology of the Earliest Cetaceans and Its implications for the Phylogenetic Relations among Ungulates." *Systematic Biology* **48** (March): 21-30.
Thewissen, J. G. M., S. I. Madar, & **S. T. Hussain.** 1996. *Ambulocetus natans*, an Eocene cetacean (Mammalia) from Pakistan (*Courier Forschungsinstitut Senckenberg*, Band 191). Stuttgart, Germany: E. Schweizbart.
———. 1998. "Whale ankles and evolutionary relationships." *Nature* **395** (1 October): 452.
Thewissen, J. G. M., L. J. Roe, J. R. O'Neil, S. T. Hussain, A. Sahni, & **S. Bajpai.** 1996. "Evolution of cetacean osmoregulation." *Nature* **381** (30 May): 379-380.
Thewissen, J. G. M., & **E. M. Williams.** 2002. "THE EARLY RADIATIONS OF CETACEA (MAMMALIA): Evolutionary Pattern and Developmental Correlations." *Annual Review of Ecology and Systematics* **33**: 73-90.
Thewissen, J. G. M., E. M. Williams, L. J. Roe, & **S. T. Hussain.** 2001. "Skeletons of terrestrial cetaceans and the relationship of whales to artiodactyls." *Nature* **413** (20 September): 277-281.
Think & Believe. 1990a. "MAMMALS FROM REPTILES?" Alpha Omega Institute Institute (Grand Junction, CO) "Spotlight on Science" *Think & Believe* **7** (September/October): 3.
———. 1990b. "FOSSIL ALTERS HISTORY OF MAMMALS?" Alpha Omega Institute Institute (Grand Junction, CO) "Spotlight on Science" *Think & Believe* **7** (September/October): 3.
———. 1997. "Notes & Quotes." Alpha Omega Institute Institute (Grand Junction, CO) *Think & Believe* **14** (July/August): 2.
———. 1998. "The Incredible Immune System." Alpha Omega Institute Institute (Grand Junction, CO) "Spotlight on Science" *Think & Believe* **15** (January/February): 3.
Thomas, Brian. 2011a. "Latest Soft Tissue Study Skirts the Issues." *Institute for Creation Research* posting 5 July (online text at *www.icr.org* accessed 11/20/2014).
———. 2011b. "Published Reports of Original Soft Tissue Fossils." *Institute for Creation Research* posting 21 July (online text at *www.icr.org* accessed 7/27/2014).
———. 2013a. "Triceratops Horn Soft Tissue Foils 'Biofilm' Explanation." *Institute for Creation Research* posting 18 March (online text at *www.icr.org* accessed 7/27/2014).
———. 2013b. "Dinosaur Soft Tissue Preserved by Blood?" *Institute for Creation Research* updated posting 11 December (online text at *www.icr.org* accessed 4/18/2015).
———. 2014. "Second Look Causes Scientist to Reverse Dino-Bird Claim." *Institute for Creation Research* posting 18 July (online text at *www.icr.org* accessed 3/10/2015).
Thomas, David E. 1997. "Hidden Messages and the Bible Code." *Skeptical Inquirer* **21** (November/December): 30-36.
Thomas-Chollier, Morgane, Valérie Ledent, Luc Leyns, & **Michel Vervoort.** 2010. "A non-tree-based comprehensive study of metazoan Hox and ParaHox genes prompts new insights into their origin and evolution." *BMC Evolutionary Biology* (online @ biomedcentral.com) **10** (11 March): 73.
Thompson, Bert. 1995. *Creation Compromises.* Montgomery, AL: Apologetics Press.
Thompson, Bert, & **Brad Harrub.** 2001. "Lesson 9: Creation vs. Evolution—Part I." *Apologetics Press*

Advanced Christian Evidences Correspondence Course posting (online pdf at *www.apologeticspress.org* accessed 3/6/2013).

———. 2002. "15 Answers to John Rennie and *Scientific American's* Nonsense." *Apologetics Press* posting (online pdf at *www.apologeticspress.org* accessed 11/16/2011).

———. 2003. "Lesson 1: Origins: Random Chance or Intelligent Design?" *Apologetics Press Advanced Christian Evidences Correspondence Course* posting (online pdf at *www.apologeticspress.org* accessed 3/6/2013).

Thompson, Hannah, & **Abigail S. Tucker.** 2013. "Dual Origin of the Epithelium of the Mammalian Middle Ear." *Science* **339** (22 March): 1453-1456.

Thompson, Helen. 2014. "Paleoartist Brings Human Evolution to Life." *The Smithsonian Magazine* 7 May posting (online text at *www.smithsonianmag.com* accessed 4/8/2016).

Thomson, J. Michael, Eric A. Gaucher, Michelle F. Burgan, Danny W. De Kee, Tang Li, John P. Aris, & **Steven A. Benner.** 2005. "Resurrecting ancestral alcohol dehydrogenases from yeast." *Nature Genetics* **37** (June): 630-635.

Thulborn, Tony, & **Susan Turner.** 2003. "The last dicynodont: an Australian Cretaceous relict." *Proceedings of the Royal Society of London* **B** (Biological Sciences) **270** (7 May): 985-993.

Ting-Berreth, Sheree A., & **Cheng-Ming Chuong.** 1996. "Local Delivery of TGF β2 Can Substitute for Placode Epithelium to Induce Mesenchymal Condensation during Skin Appendage Morphogenesis." *Developmental Biology* **179** (1 November): 347-359.

Tobin, Paul N. 2000. "The Babylonian Origins of the Creation Myths." *The Rejection of Pascal's Wager* posting (online text at *webspace.webring.com/people/np/paul_tobin* accessed 5/19/2011).

Tokita, Masayoshi, Tomoki Nakayama, Richard A. Schneider, & **Kiyokazu Agata.** 2013. "Molecular and cellular changes associated with the evolution of novel jaw muscles in parrots." *Proceedings of the Royal Society of London* **B** (Biological Sciences) **280** (12 December): 20122319.

Tomescu, Alexandru M. F. 2009. "Megaphylls, microphylls and the evolution of leaf development." *Trends in Plant Science* **14** (January): 5-12.

Tomitani, Akiko. 2006. "Origin and early evolution of chloroplasts." *Paleontological Research* **10** (December): 283-297.

Tomitani, Akiko, Kiyotaka Okada, Hideaki Miyashita, Hans C. P. Matthijs, Terufumi Ohno, & **Ayumi Tanaka.** 1999. "Chlorophyll *b* and phycobilins in the common ancestor of cyanobacteria and chloroplasts." *Nature* **400** (8 July): 159-162.

Tomkins, Jeffrey P. 2011a. "How Genomes are Sequenced and Why It Matters: Implications for Studies in Comparative Genomics of Humans and Chimpanzees." Answers in Genesis *Answers Research Journal* **4**: 81-88.

———. 2011b. "Response to Comments on 'How Genomes are Sequenced and Why It Matters: Implications for Studies in Comparative Genomics of Humans and Chimpanzees'." Answers in Genesis *Answers Research Journal* **4**: 161-162.

———. 2011c. "Genome-Wide DNA Alignment Similarity (Identity) for 40,000 Chimpanzee DNA Sequences Queried against the Human Genome is 86-89%." Answers in Genesis *Answers Research Journal* **4**: 233-241.

———. 2011d. "New Research Undermines Key Argument for Human Evolution." Institute for Creation Research *Acts & Facts* (June): 6.

———. 2011e. "Evaluating the Human-Chimp DNA Myth—New Research Data." Institute for Creation Research *Acts & Facts* (October): 6.

———. 2012a. "Gorilla Genome Is Bad News for Evolution." *Institute for Creation Research* posting 9 March (online text at *www.icr.org* accessed 3/11/2012).

———. 2012b. "Journal Reports Bias in Human-Chimp Studies." Institute for Creation Research *Acts & Facts* (June): 6.

———. 2012c. *More Than a Monkey: The Human-Chimp DNA Similarity Myth*. Create/Space Independent Publishing Platform 8 July publication.

———. 2013a. "Comprehensive Analysis of Chimpanzee and Human Chromosomes Reveals Average DNA Similarity of 70%." Answers in Genesis *Answers Research Journal* **6**: 63-69.

———. 2013b. "Genetic Recombination Study Defies Human-Chimp Evolution." *Institute for Creation Research* posting 31 May (online text at *www.icr.org* accessed 6/18/2013).

———. 2013c "Alleged Human Chromosome 2 'Fusion Site' Encodes an Active DNA Binding Domain Inside a Complex and Highly Expressed Gene—Negating Fusion." Answers in Genesis *Answers Research Journal* **6**: 367-375.

———. 2013d. "Alleged Human Chromsome 2 'Fusion Site' Encodes an Active DNA Binding Domain Inside a Complex and Highly Expressed Gene." Answers in Genesis *Answers Research Journal* **6**: 367-375.

———. 2013e. "New Research Debunks Human Chromosome Fusion." Institute for Creation Research *Acts & Facts* (December): 9.

———. 2015a. "Challenging the BioLogos Claim that a Vitellogenin (Egg-Laying) Pseudogene Exists in the Human Genome." Answers in Genesis *Answers Research Journal* **8**: 403-411.

———. 2015b. "Endosymbiosis: A Theory in Crisis." Institute for Creation Research *Acts & Facts* (November): 13.

———. 2015c. "Vitellogenin Pseudogene Debunked." Institute for Creation Research *Acts & Facts* (January): 8.

Tomkins, Jeffrey, & **Jerry Bergman.** 2011. "The chromosome 2 fusion of human evolution—part 2: re-analysis of the genomic data." *Journal of Creation* (AKA Answers in Genesis *Creation ex Nihilo Technical Journal*) **25** (2): 111-117.

Toumey, Christopher P. 1994. *God's Own Scientists: Creationists in a Secular World.* New Brunswick, NJ: Rutgers University Press.

Tournier, Frédéric, & **Michel Bornens.** 2001. "Centrosomes and parthenogenesis." *Methods in Cell Biology* **67**: 213-224.

Tournier, Frederic, Eric Karsenti, & **Michel Bornens.** 1989. "Parthenogenesis in *Xenopus* eggs injected with centrosomes from synchronized human lymphoid cells." *Developmental Biology* **136** (December): 321-329.

Towers, Matthew, Jason Signolet, Adrian Sherman, Helen Sang, & **Cheryll Tickle.** 2011. "Insights into bird wing evolution and digit specification from polarizing region fate maps." *Nature Communications* (online @ nature.com) **2** (9 August): 426.

Travisano, Michael. 1997. "Long-Term Experimental Evolution in Escherichia coli. VI. Environmental Constraints on Adaptation and Divergence." *Genetics* **146** (June): 471-479.

Truman, Royal, & **Peter Borger.** 2008a. "Genome truncation vs. mutational opportunity: can new genes arise via gene duplication?—Part 1." *Journal of Creation* (AKA Answers in Genesis *Creation ex Nihilo Technical Journal*) **22** (1): 99-110.

———. 2008b. "Genome truncation vs. mutational opportunity: can new genes arise via gene duplication?—Part 2." *Journal of Creation* (AKA Answers in Genesis *Creation ex Nihilo Technical Journal*) **22** (1): 111-119.

Truth in Science. 2008a. "Platypus: a Darwinian Cautionary Tale." *Truth in Science* posting (online text at *www.truthinscience.org.uk* accessed 12/18/2011).

———. 2008b. "Synapsids and the Evolution of Mammals." *Truth in Science* posting (online text at *www.truthinscience.org.uk* accessed 12/18/2011).

———. 2009. "Press Release regarding the Textbook 'Explore Evolution'." *Truth in Science* posting (online text at *www.truthinscience.org.uk* accessed 12/18/2011).

Tsirigos, Aristotelis, & **Isidore Rigoutsos.** 2009. "Alu and B1 Repeats Have Been Selectively Retained in the Upstream and Intronic Regions of Genes of Specific Functional Classes." *PLoS Computational Biology* (online @ ploscompbiol.org) **5** (December): e1000610.

Tsuji, Linda A. 2006. "Cranial anatomy and phylogenetic affinities of the Permian parareptile *Macroleter poezicus*." *Journal of Vertebrate Paleontology* **26** (December): 849-865.

Tsuji, Linda A., Johannes Müller, & **Robert R. Reisz.** 2010. "*Microleter mckinzieorum* gen. et sp. nov. from the Lower Permian of Oklahoma: the basalmost pararetile from Laurasia." *Journal of Systematic Palaeontology* **8** (June): 245-255.

Tsukaya, Hirokazu. 2005. "Leaf shape: genetic controls and environmental factors." *International Journal of Developmental Biology* **49** (5/6): 547-555.

Tsukui, Tohru, Javier Capdevila, Koji Tamura, Pilar Ruiz-Lozano, Concepción Rodriguez-Esteban, Sayuri Yonei-Tamura, Jorge Magallón, Roshantha A. S. Chandraratna, Kenneth Chien, Bruce Blumberg, Ronald M. Evans, & **Juan Carlos Izpisúa Belmonte.** 1999. "Multiple left-right asymmetry defects in *Shh*$^{-/-}$ mutant mice unveil a convergence of the Shh and retinoic acid pathways in the control of *Lefty-1*." *Proceedings of the National Academy of Sciences* **96** (28 September): 11376-11381.

Tucker, Abigail S., Karen L. Matthews, & **Paul T. Sharpe.** 1998. "Transformation of Tooth Type Induced by Inhibition of BMP Signaling." *Science* **282** (6 November): 1136-1138.

Tucker, Abigail, & **Paul Sharpe.** 2004. "The cutting-edge of mammalian development; how the embryo makes teeth." *Nature Reviews Genetics* **5** (July): 499-508.

Tucker, Abigail S., Robert P. Watson, Laura A. Lettice, Gen Yamada, & Robert E. Hill. 2004. "Bapx1 regulates patterning in the middle ear: altered regulatory role in the transition from the proximal jaw during vertebrate evolution." *Development* (AKA *Journal of Embryology and Experimental Morphology*) **131** (15 March): 1235-1245.

Tudge, Colin. 2000. *The Variety of Life: A Survey and a Celebration of all the Creatures that Have Ever Lived.* Oxford: Oxford University Press.

Turmel, Monique, Marie-Christine Gagnon, Charley J. O'Kelly, Christian Otis, & Claude Lemieux. 2008. "The Chloroplast Genomes if the Green Algae *Pyramimonas, Monomastix*, and *Pycnococcus* Shed New Light on the Evolutionary History of Prasinophytes and the Origin of the Secondary Chloroplasts of Euglenids." *Molecular Biology and Evolution* **26** (March): 631-648.

Turner, Alan H., Peter J. Makovicky, & Mark A. Norell. 2007. "Feather Quill Knobs in the Dinosaur *Velociraptor*." *Science* **317** (21 September): 1721.

———. 2012. "A review of dromaeosaurid systematics and paravian phylogeny." *Bulletin of the American Museum of Natural History* **371**: 1-206.

Turner, Alan H., Diego Pol, Julia A. Clarke, Gregory M. Erickson, & Mark A. Norell. 2007. "A Basal Dromaeosaurid and Size Evolution Preceding Avian Flight." *Science* **317** (7 September): 1378-1381.

Uncommon Descent. 2011a. "Backgrounder: Some challenges offered for Lynn Margulis's endosymbiosis theory." *Uncommon Descent* anonymous posting 11 April (online text at *www.uncommondescent.com* accessed 4/19/2011).

———. 2011b. "Older than thought: Lacewings at 120 million years ago." *Uncommon Descent* anonymous posting 8 October (online text at *www.uncommondescent.com* accessed 3/19/2016).

———. 2013. "Genomics scientist Jeffrey Tompkins takes issue with BioLogos' we are 98% chimpanzee claim." *Uncommon Descent* anonymous posting 14 September (online text at *www.uncommondescent.com* accessed 1/2/2016).

———. 2016. "Sudden gene change helped create mammals?" *Uncommon Descent* posting 23 June (online text at *www.uncommondescent.com* accessed 8/31/2016).

Ungar, Peter S. 2010. *Mammal Teeth: Origin, Evolution, and Diversity.* Baltimore, MD: Johns Hopkins University Press.

Unger, Kelley. 2015. "The Book that Inspired Michael Behe." Discovery Institute *Center for Science & Culture* 12 November email (onlin text via *discovery.org* accessed 4/22/2016).

Unwin, David M. 2006. *Pterosaurs From Deep Time.* New York: Pi Press (Peter N. Névraumont).

Urashima, Tadasu, Hiroaki Inamori, Kenji Fukuda, Tadao Saito, Michael Messer, & Olav T. Oftedal. 2015. "4-*O*-Acetyl-sialic acid (Neu4,5Ac_2) in acidic milk ologosaccharides of the platypus (*Ornithorhynchus anatinus*) and its evolutionary significance." *Glycobiology* **25** (June): 683-697.

Uyeda, Josef C., Thomas F. Hansen, Stevan J. Arnold, & Jason Pienaar. 2011. "The million-year wait for macroevolutionary bursts." *Proceedings of the National Academy of Sciences* **108** (20 September): 15908-15913.

Vaidya, Akhil B., & Michael W. Mather. 2000. "Atovaquone resistance in malaria parasites." *Drug Resistance Updates* **3** (November): 283-287.

Valentine, James W. 1989. "Bilaterians of the Precambrian-Cambrian transition and the annelid-arthropod relationship." *Proceedings of the National Academy of Sciences* **86** (1 April): 2272-2275.

———. 1994. "Late Precambrian bilaterians: Grades and clades." *Proceedings of the National Academy of Sciences* **91** (19 July): 6751-6757.

———. 1995. "Why No New Phyla after the Cambrian? Genome and Ecospace Hypotheses Revisited." *Palaios* **10** (April): 190-194.

———. 1997. "Cleavage patterns and the topology of the metazoan tree of life." *Proceedings of the National Academy of Sciences* **94** (22 July): 8001-8005.

———. 2001. "How were the vendobiont bodies patterned?" *Paleobiology* **27** (September): 425-428.

———. 2002. "Prelude to the Cambrian Explosion." *Annual Review of Earth and Planetary Sciences* **30**: 285-306.

———. 2003. "Architectures of Biological Complexity." *Integrative and Comparative Biology* (AKA *American Zoologist*) **43** (February): 99-103.

———. 2004. *On the Origin of Phyla.* 2006 pb ed. Chicago: University of Chicago Press.

———. 2007. "Seeing ghosts: Neoproterozoic bilaterian body plans." *Geological Society, London,*

Special Publications **286**: 369-375.
Valentine, James W., David Jablonski, & Douglas H. Erwin. 1999. "Fossils, molecules and embryos: new perspectives on the Cambrian explosion." *Development* (AKA *Journal of Embryology and Experimental Morphology*) **126** (1 March): 851-859.
Van't Hof, Arjen E., Pascal Campagne, Daniel J. Rigden, Carl J. Yung, Jessica Lingley, Michael A. Quail, Neil Hall, Alistair C. Darby, & Ilik J. Saccheri. 2016. "The industrial melanism mutation in British peppered moths is a transposable element." *Nature* **534** (2 June): 102-105.
Van't Hof, Arjen E., Nicola Edmonds, Martina Daliková, František Marec, & Ilik J. Saccheri. 2011. "Industrial Melanism in British Peppered Moths Has a Singular and Recent Mutational Origin." *Science* **332** (20 May): 958-960.
Van Till, Howard J. 1993. "God and Evolution: An Exchange (I)." *First Things* (June/July): 32-38.
Van Valen, Leigh. 1978. "Why not to be a cladist." *Evolutionary Theory* **3** (August): 285-299.
———. 2002. "How did rodents and lagomorphs (Mammalia) originate?" *Evolutionary Theory* **12** (December): 101-128.
Van Valkenburgh, Blaire, & Ian Jenkins. 2002. "Evolutionary Patterns in the History of Permo-Triassic and Cenozoic Synapsid Predators." *Paleontological Society Papers* **8**: 267-288.
Van Veen, Vincent, & Cameron S. Carter. 2002. "The anterior cingulate as a conflict monitor: fMRI and ERP studies." *Physiology & Behavior* **77** (December): 477-482.
Van Wyhe, John. 2008. *The Darwin Experience*. Washington, DC: National Geographic.
Varela-Lasheras, Irma, Alexander J. Bakker, Steven D. van der Mije, Johan A. J. Metz, Joris van Alphen, & Frietson Galis. 2011. "Breaking evolutionary and pleiotropic constraints in mammals: On sloths, manatees and homeotic mutations." *EvoDevo* (online @ evodevojournal.com) **2** (May): 11.
Vargas, Alexander O., & John F. Fallon. 2005. "Birds have dinosaur wings: The molecular evidence." *Journal of Experimental Zoology* **304B** (15 January): 86-90.
Vargas, Alexander O., Tiana Kohlsdorf, John F. Fallon, John VandenBrooks, & Günter P. Wagner. 2008. "The Evolution of *HoxD-11* Expression in the Bird Wing: Insights from *Alligator mississippiensis*." *PLoS ONE* (online @ plosone.org) **3** (October): e3325.
Vargas, Alexander O., & Günter P. Wagner. 2009. "Frame-shifts of digit identity in bird evolution and Cyclopamine-treated wings." *Evolution & Development* **11** (March-April): 163-169.
Vargas, Alexander O., Günter P. Wagner, & Jacques A. Gauthier. 2009. "*Limusaurus* and bird digit identity." *Nature Precedings* (online @ precedings.nature.com) 6 October: 2009.3828.1.
Varricchio, David J., Frankie D. Jackson, John J. Borkowski, & John R. Horner. 1997. "Nest and egg clutches of the dinosaur *Troodon formosus* and the evolution of avian reproductive traits." *Nature* **385** (16 January): 247-250.
Varricchio, David J., Jason R. Moore, Gregory M. Erickson, Mark A. Norell, Frankie D. Jackson, & John J. Borkowski. 2008. "Avian Paternal Care Had Dinosaur Origin." *Science* **322** (19 December): 1826-1828.
Venema, Dennis R. 2016a. "Vitellogenin and Common Ancestry: Does BioLogos have egg on its face?" *The BioLogos Forum* posting 11 February (online text at *biologos.org* accessed 6/14/2016).
———. 2016b. "Vitellogenin and Common Ancestry: Understanding synteny." *The BioLogos Forum* posting 25 February (online text at *biologos.org* accessed 6/14/2016).
———. 2016c. "Vitellogenin and Common Ancestry: Reading Tomkins." *The BioLogos Forum* posting 11 March (online text at *biologos.org* accessed 6/14/2016).
———. 2016d. "Vitellogenin and Common Ancestry: Tomkins' false dichotomy." *The BioLogos Forum* posting 24 March (online text at *biologos.org* accessed 6/14/2016).
———. 2016e. "Vitellogenin and Common Ancestry: From Egg to Placenta." *The BioLogos Forum* posting 21 April (online text at *biologos.org* accessed 6/14/2016).
Ventura, Mario, Claudia R. Catacchio, Saba Sajjadian, Laura Vives, Peter H. Sudmant, Tomas Marques-Bonet, Tina A. Graves, Richard K. Wilson, & Evan E. Eichler. 2012. "The evolution of African great ape subtelomeric heterochromatin and the fusion of human chromosome 2." *Genome Research* **22** (June): 1036-1049.
Vitaliano, Dorothy B. 1973. *Legends of the Earth: Their Geologic Origins*. Bloomington, IN: Indiana University Press.
Vlad, Daniela, Daniel Kierzkowski, Madlen I. Rast, Francesco Vuolo, Raffaele Dello Ioio, Carla Galinha, Xiangchao Gan, Mohsen Hajheidari, Angela Hay, Richard S. Smith, Peter Huijser, C. Donovan Bailey, & Miltos Tsiantis. 2014. "Leaf Shape Evolution Through Duplication, Regulatory Diversification, and Loss of a Homeobox Gene.'" *Science* **343** (14 February): 780-783.

Vogel, Gretchen. 2012. "Turing Pattern Fingered for Digit Formation." *Science* **338** (14 December): 1406.

Voordeckers, Karin, Chris A. Brown, Kevin Vanneste, Elisa van der Zande, Arnout Voet, Steven Maere, & **Kevin J. Verstrepen.** 2012. "Reconstruction of Ancestral Metabolic Enzymes Reveals Molecular Mechanisms Underlying Evolutionary Innovation through Gene Duplication." *PLoS Biology* (online @ plosbiology.org) **10** (December): e1001446.

Voss, S. Randal, & **H. Bradley Shaffer.** 1997. "Adaptive evolution via a major gene effect: Paedomorphosis in the Mexican axolotl." *Proceedings of the National Academy of Sciences* **94** (9 December): 14185-14189.

Vullo, Romain, Emmanuel Gheerbrant, Christian de Muizon, & **Didier Néraudeau.** 2009. "The oldest modern Therian mammal from Europe and its bearing on stem marsupial paleobiogeography." *Proceedings of the National Academy of Sciences* **106** (24 November): 19910-19915.

Vullo, Romain, Vincent Girard, Dany Azar, & **Didier Néraudeau.** 2010. "Mammalian hairs in Early Cretaceous amber." *Naturwissenschaften* **97** (July): 683-687.

Wade, Nicholas, ed. 1998. *The Science Times Book of Fossils and Evolution.* New York: The Lyons Press.

Wagner, Andreas. 2000a. "The Role of Population Size, Pleiotropy and Fitness Effects of Mutation in the Evolution of Overlapping Gene Functions." *Genetics* **154** (March): 1389-1401.

———. 2000b. "Inferring Lifestyle from Gene Expression Patterns." *Molecular Biology and Evolution* **17** (December): 1985-1987.

Wagner, Günter P. 2005. "The developmental evolution of avian digit homology: An update." *Theory in Biosciences* **124** (November): 165-183.

Wagner, Günter P., & **Chi-Hua Chiu.** 2001. "The Tetrapod Limb: A Hypothesis on Its Origin." *Journal of Experimental Zoology* **291** (15 October): 226-240.

Wagner, Günter P., & **Jacques A. Gauthier.** 1999. "1,2,3 = 2,3,4: A solution to the problem of the homology of the digits in the avian hand." *Proceedings of the National Academy of Sciences* **96** (27 April): 5111-5116.

Wagner, Günter P., & **Manfred D. Laubichler.** 2004. "Rupert Riedl and the Re-Synthesis of Evolutionary and Developmental Biology: Body Plans and Evolvability." *Journal of Experimental Zoology* **302** (15 December): 92-102.

Wagner, Peter J., & **Christian A. Sidor.** 2000. "Age Rank/Clade Rank Metrics--Sampling, Taxonomy, and the Meaning of 'Stratigraphic Consistency'." *Systematic Biology* **49** (September): 463-479.

Walkden, Gordon, Julian Parker, & **Simon Kelley.** 2002. "A Late Triassic Impact Ejecta Layer in Southwestern Britain." *Science* **298** (13 December): 2185-2188.

Walker, A. D. 1990. "A Revision of Sphenosuchus acutus Haughton, a Crocodylomorph Reptile from the Elliott Formation (Late Triassic or Early Jurassic) of South Africa." *Philosophical Transactions of the Royal Society of London* **B** (Biological Sciences) **330** (29 October): 1-120.

Walker, C. A. 1981. "New subclass of birds from the Cretaceous of South America." *Nature* **292** (2 July): 51-53.

Wallace, Alfred Russel. 1910. *The World of Life: A Manifestation of Creative Power, Directive Mind and Ultimate Purpose.* London: Chapman and Hall.

Walsh, John Evangelist. 1996. *Unraveling Piltdown: The Science Fraud of the Century and Its Solution.* New York: Random House.

Wang, Mei, Alexandr P. Rasnitsyn, Chungkun Shih, & **Dong Ren.** 2014. "A new Cretaceous genus of xyelydid sawfly illuminating nygmata evolution of Hymenoptera." *BMC Evolutionary Biology* (online @ biomedcentral.com) **14** (17 June): 131.

Wang, Min, Xiaoting Zheng, Jingmai K. O'Connor, Graeme T. Lloyd, Xiaoli Wang, Yan Wang, Xiaomei Zhang, & **Zhonghe Zhou.** 2015. "The oldest record of ornithuromorpha from the early cretaceous of China." *Nature Communications* (online @ nature.com) **6** (5 May): 6987.

Wang, Min, & **Zhonghe Zhou.** 2016. "A new adult specimen of the basalmost ornithuromorph bird *Archaeorhynchus spathula* (Aves: Ornithuromorpha) and its implications for early avian ontogeny." *Journal of Systematic Palaeontology* (online pdf at *www.tandfonline.com* accessed 11/6/2015) **14**: DOI: 10.1080/14772019.2015.1136968.

Wang, Wen, Jianming Zhang, Carlos Alvarez, Ana Llopart, & **Manyuan Long.** 2000. "The Origin of the *Jingwei* Gene and the Complex Modular Structure of Its Parental Gene, *Yellow Emperor*, in *Drosophila melanogaster*." *Molecular Biology and Evolution* **17** (September): 1294-1301.

Wang, Xia, Gareth J. Dyke, Vlad Codrea, Pascal Godefroit, & Thierry Smith. 2011. "A euenantiornithine bird from the Late Cretaceous Haţeg Basin of Romania ." *Acta Palaeontologica Polonica* **56** (4): 853-857.

Wang, Xiaolin, Zhonghe Zhou, Fucheng Zhang, & Xing Xu. 2002. "A nearly completely articulated rhamphorhynchoid pterosaur with exceptionally well-preserved wing membranes and 'hairs' from Inner Mongolia, northeast China." *Chinese Science Bulletin* **47** (February): 226-230.

Wang, Yuanqing, Yaoming Hu, Jin Meng, & Chuankui Li. 2001. "An Ossified Meckel's Cartilage in Two Cretaceous Mammals and Origin of the Mammalian Inner Ear." *Science* **294** (12 October): 357-361.

Wang, Zhe, Rebecca L. Young, Huiling Xue, & Günter P. Wagner. 2011. "Transcriptomic analysis of avian digits reveals conserved and derived digit identities in birds." *Nature* **477** (29 September): 583-586.

Wang, Zhe, Lihong Yuan, Stephen J. Rossiter, Xueguo Zuo, Binghua Ru, Hui Zhong, Naijian Han, Gareth Jones, Paul D. Jepson, & Shuyi Zhang. 2009. "Adaptive Evolution of *5'HoxD* Genes in the Origin and Diversification of the Cetacean Flipper." *Molecular Biology and Evolution* **26** (March): 613-622.

Wang, Zhi, Ben-Yang Liao, & Jianzhi Zhang. 2010. "Genomic patterns of pleiotropy and the evolution of complexity." *Proceedings of the National Academy of Sciences* **107** (19 October): 18034-18039.

Wang, Zhuo, Juan Pascual-Anaya, Amonida Zadissa, Wenqi Li, Yoshihito Niimura, Zhiyong Huang, Chunyi Li, Simon White, Zhiqiang Xiong, Dongming Fang, Bo Wang, Yao Ming, Yan Chen, Yuan Zheng, Shigehiro Kuraku, Miguel Pignatelli, Javier Herrero, Kathryn Beal, Masafumi Nozawa, Qiye Li, Juan Wang, Hongyan Zhang, Lili Yu, Shuji Shigenobu, Junyi Wang, Jiannan Liu, Paul Flicek, Steve Searle, Jun Wang, Shigeru Kuratani, Ye Yin, Bronwen Aken, Guojie Zhang, & Naoki Irie. 2013. "The draft genomes of soft-shell turtle and green sea turtle yield insights into the development and evolution of the turtle-specific body plan." *Nature Genetics* **45** (June): 701-706.

Ward, Charles A. 1940. *Oracles of Nostradamus.* New York: Charles Scribner's Sons.

Ward, Peter, Conrad Labandeira, Michel Laurin, & Robert A. Berner. 2006. "Confirmation of Romer's Gap as a low oxygen interval constraining the timing of initial arthropod and vertebrate terrestrialization." *Proceedings of the National Academy of Sciences* **103** (7 November): 16818-16822.

Ward, Philip S. 2007. "Phylogeny, classification, and species-level taxonomy of ants (Hymenoptera: Formicidae)." *Zootaxa* **1668**: 549-563.

Ward, Philip S., & Seán G. Brady. 2003. "Phylogeny amd biogeography of the ant subfamily Myrmeciinae (Hymenoptera: Formicidae)." *Invertebrate Systematics* **17** (3): 361-386.

Warren, Wesley C., LaDeana W. Hillier, Jennifer A. Marshall Graves, Ewan Birney, Chris P. Ponting, Frank Grützner, Katherine Belov, Webb Miller, Laura Clarke, Asif T. Chinwalla, Shiaw-Pyng Yang, Andreas Heger, Devin P. Locke, Pat Miethke, Paul D. Waters, Frédéric Veyrunes, Lucinda Fulton, Bob Fulton, Tina Graves, John Wallis, Xose S. Puente, Carlos López-Otin, Gonzalo R. Ordóñez, Evan E. Eichler, Lin Chen, Ze Cheng, Janine E. Deakin, Amber Alsop, Katherine Thompson, Patrick Kirby, Anthony T. Paperfuss, Matthew J. Wakefield, Tsviya Olender, Doren Lancet, Gavin A. Huttley, Arian F. A. Smit, Andrew Pask, Peter Temple-Smith, Mark A. Batzer, Jerilyn A. Walker, Miriam K. Konkel, Robert S. Harris, Camilla M. Whittington, Emily S. W. Wong, Neil J. Gemmell, Emmanuel Buschiazzo, Iris M. Vargas Jentzsch, Angelika Merkel, Juergen Schmitz, Anja Zemann, Gennady Churakov, Jan Ole Kriegs, Juergen Brosius, Elizabeth P. Murchison, Ravi Sachidanandam, Carly Smith, Gregory J. Hannon, Enkhjargal Tsend-Ayush, Daniel McMillan, Rosalind Attenborough, Willem Rens, Malcolm Ferguson-Smith, Christophe M. Lefèvre, Julie A. Sharp, Kevin R. Nicholas, David A. Ray, Michael Kube, Richard Reinhardt, Thomas H. Pringle, James Taylor, Russell C. Jones, Brett Nixon, Jean-Louis Dacheux, Hitoshi Niwa, Yoko Sekita, Xiaoqiu Huang, Alexander Stark, Pouya Kheradpour, Manolis Kellis, Paul Flicek, Yuan Chen, Caleb Webber, Ross Hardison, Joanne Nelson, Kym Hallsworth-Pepin, Kim Delehaunty, Chris Markovic, Pat Minx, Yuocheng Feng, Colin Kremitzki, Makedonka Mitreva, Jarret Glasscock, Todd Wylie, Patricia Wohldmann, Prathapan Thiru, Michael N. Nhan, Craig S. Pohl, Scott M. Smith, Shunfeng Hou, Marilyn B. Renfree, Elaine R. Mardis, & Richard K. Wilson. 2008. "Genome analysis of the platypus reveals unique signatures of evolution." *Nature* **453** (8 May): 175-183.

Watchtower. 1985. *Life—How did it get here? By evolution or by creation?* Watch Tower Bible and Tract Society of Pennsylvania International Bible Students Association (Jehovah's Witness)

anonymous book. New York: Watchtower Bible and Tract Society of New York.
———. 2010. "Has All Life Descended From a Common Ancestor?" Anonymous article in Watch Tower Bible and Tract Society of Pennsylvania *The Origin of Life: Five Questions Worth Asking* pamphlet. New York: Watchtower Bible and Tract Society of New York.
Watson, D. M. S. 1929. "Adaptation." *Nature* **124** (10 August): 231-233.
———. 1931. "55. On the Skeleton of a Bauriamorph Reptile." *Journal of Zoology* (London) **101** (September): 1163-1205.
Weatherbee, Scott D., H. Frederik Nijhout, Laura W. Grunert, Georg Halder, Ron Galant, Jayne Selegue, & **Sean Carroll.** 1999. "*Ultrabithorax* function in butterfly wings and the evolution of insect wing patterns." *Current Biology* **9** (11 February): 109-115.
Webster, Douglas B., Arthur N. Popper, & **Richard R. Fay**, ed. 1992. *The Evolutionary Biology of Hearing.* New York: Springer.
Weil, Anne. 2002. "Upwards and onwards." *Nature* **416** (25 April): 798-799.
———. 2005. "Living large in the Cretaceous." *Nature* **433** (13 January): 116-117.
———. 2011. "A jaw-dropping ear." *Nature* **472** (14 April): 174-176.
Weinberg, Samantha. 2000. *A Fish Caught in Time: The Search for the Coelacanth.* New York: HarperCollins Publishers.
Weinberg, Stanley L. 1980. "Reactions to Creationism in Iowa." *Creation/Evolution* **1** (Fall): 1-8.
Weiner, Jonathan. 1994. *The Beak of the Finch: A Story of Evolution in Our Time.* 1995 pb ed. New York: Vintage Books.
Weishampel, David B. 1990. "Dinosaur Distributions," in Weishampel *et al.* (1990, 63-139).
Weishampel, David B., Peter Dodson, & **Halszka Osmólska**, eds. 1990. *The Dinosauria.* 1992 pb ed. Berkeley, CA: University of California Press.
Welch, John J., Eric Fontanillas, & **Lindell Bromham.** 2005. "Molecular Dates for the 'Cambrian Explosion': The Influence of Prior Assumptions." *Systematic Biology* **54** (August): 672-678.
Wells, Jonathan. 1994. "Darwinism: Why I Went for a Second Ph.D." *True Parents* undated posting (online text at *www.tparents.org* accessed 8/2/2016).
———. 2000a. *Icons of Evolution: Science or Myth? Why Much of What We Teach About Evolution Is Wrong.* 2002 pb ed. Washington, D.C.: Regnery Publishing, Inc.
———. 2000b. "Survival of the Fakest." *The American Spectator* **10** (December): 18-24, 26-27.
———. 2006. *The Politically Incorrect Guide to Darwinism and Intelligent Design.* Washington, D.C.: Regnery Publishing, Inc.
———. 2008. "An Evolutionary Origin of the Centrosome?" Discovery Institute *Evolution News & Views* for 7 May (online text at *evolutionnews.org* accessed 8/1/2016).
———. 2009a. "Anatomical Homology and Circular Definitions." *Explore Evolution* critical response 23 February (online text at *www2.exploreevolution.com* accessed 1/4/2011).
———. 2009b. "Haeckel, Darwin, and Textbooks." *Explore Evolution* critical response 23 February (online text at *www2.exploreevolution.com* accessed 1/4/2011).
———. 2009c. "Fact and Fiction about the Peppered Moth." *Explore Evolution* critical response 23 February (online text at *www2.exploreevolution.com* accessed 1/4/2011).
———. 2009d. "Genetic Toolkits." *Explore Evolution* critical response 23 February (online text at *www2.exploreevolution.com* accessed 1/4/2011).
———. 2009e. "Misrepresenting the Galapagos Finches." *Explore Evolution* critical response 23 February (online text at *www2.exploreevolution.com* accessed 1/4/2011).
———. 2011a. *The Myth of Junk DNA.* Seattle, WA: Discovery Institute Press.
———. 2011b. "Here's Jonathan Wells on destroying Darwinism – and responding to attacks on his character and motives." *Uncommon Descent* posting 5 November (online text at *www.uncommondescent.com* accessed 3/11/2013).
Welten, Monique C. M., Fons J. Verbeek, Annemarie H. Meijir, & **Michael K. Richardson.** 2005. "Gene expression and digit homology in the chicken embryo wing." *Evolution & Development* **7** (January-February): 18-28.
West, Geoffrey B., James H. Brown, & **Brian J. Enquist.** 1997. "A General Model for the Origin of Allometric Scaling Laws in Biology." *Science* **276** (4 April): 122-126.
———. 1999. "The Fourth Dimension of Life: Fractal Geometry and Allometric Scaling of Organisms." *Science* **284** (4 June): 1677-1679.
West, Geoffrey B., William H. Woodruff, & **James H. Brown.** 2002. "Allometric scaling of metabolic rate from molecules and mitochondria to cells and mammals." *Proceedings of the National Academy of Sciences* **99** (19 February): 2473-2478.

West, John Anthony. 1993. *Serpent in the Sky: The High Wisdom of Ancient Egypt*. Wheaton, IL: Quest Books.

West, John G. 2016. "Debating Common Ancestry." Discovery Institute *Evolution News & Views* 14 May posting (online text at *evolutionnews.org* accessed 5/17/2016).

Westenberg, Kerri. 1999. "From Fins to Feet." *National Geographic* **195** (May): 114-126.

Whitcomb, John C., & **Henry M. Morris.** 1961. *The Genesis Flood: The Biblical Record and Its Scientific Implications*. 1966 ed. Philadelphia: The Presbyterian and Reformed Publishing Co.

White, Nicholas J. 1999a. "Antimalarial drug resistance and combination chemotherapy." *Philosophical Transactions of the Royal Society of London* **B** (Biological Sciences) **354** (29 April): 739-749.

———. 1999b. "Delaying antimalarial drug resistance with combination chemotherapy." *Parassitologia* **41** (September): 301-308.

———. 2004. "Antimalarial drug resistance." *The Journal of Clinical Investigation* **113** (15 April): 1084-1092.

Whiteside, Jessica H., Danielle S. Gregson, Paul E. Olsen, & **Dennis V. Kent.** 2011. "Climatically-driven biogeographic provinces of Late Triassic tropical Pangea." *Proceedings of the National Academy of Sciences* **108** (31 May): 8972-8977.

Whiteside, Jessica H., Paul E. Olsen, Timothy Eglinton, Michael E. Brookfield, & **Raymond N. Sambrotto.** 2010. "Compound-specific carbon isotopes from Earth's largest flood basalt eruptions directly linked to the end-Triassic mass extinction." *Proceedings of the National Academy of Sciences* **107** (13 April): 6721-6725.

Whitfield, Philip. 1993. *From So Simple A Beginning: The Book of Evolution*. New York: Macmillan.

Whiting, Michael F. 2004. "Phylogeny of the Holometabolous Insects: The Most Successful Group of Terrestrial Organisms," in Cracraft & Donoghue (2004, 345-361).

Whitlock, John A., & **Joy A. Richman.** 2013. "Biology of tooth replacement in amniotes." *International Journal of Oral Science* **5** (June): 66-70.

Whitney, Megan R., & **Christian A. Sidor.** 2016. "A new therapsid from the Permian Madumabisa Mudstone Formation (Mid-Zambezi Basin) of Southern Zambia." *Journal of Vertebrate Paleontology* **36** (July): e1150767.

Wible, John R., Desui Miao, & **James A. Hopson.** 1990. "The septomaxilla of fossil and recent synapsids and the problem of the septomaxilla of monotremes and armadillos." *Zoological Journal of the Linnean Society* **98** (March): 203-228.

Wible, John R., Michael J. Novacek, & **Guillermo W. Rougier.** 2004. "New Date on the Skull and Dentition in the Mongolian Late Cretaceous Eutherian Mammal *Zalambdalestes*." *Bulletin of the American Museum of Natural History* **281**: 1-144.

Wible, J. R., G. W. Rougier, M. J. Novacek, & **R. J. Asher.** 2007. "Cretaceous eutherians and Laurasian origin for placental mammals near the K/T boundary." *Nature* **447** (21 June): 1003-1006.

———. 2009. "The Eutherian Mammal *Maelestes gobiensis* from the Late Cretaceous of Mongolia and the Phylogeny of Cretaceous Eutheria." *Bulletin of the American Museum of Natural History* **327**: 1-123.

Widelitz, Randall B., Ting-Xin Jiang, Chia-Wei Janet Chen, N. Susan Stott, Han-Sung Jung, Cheng-Ming Chuong. 1999. "Wnt-7a in feather morphogenesis: involvement of anterior-posterior asymmetry and proximal-distal elongation demonstrated with an in vitro reconstitution model." *Development* (AKA *Journal of Embryology and Experimental Morphology*) **126** (15 June): 2577–2587.

Widelitz, Randall B., Ting-Xin Jiang, Jianfen Lu, & **Cheng-Ming Chuong.** 2000. "*β-catenin* in Epithelial Morphogenesis: Conversion of Part of Avian Foot Scales into Feather Buds with a Mutated β-Catenin." *Developmental Biology* **219** (1 March): 98-114.

Widelitz, Randall B., Ting Xin Jiang, Mingke Yu, Ted Shen, Jen-Yee Shen, Ping Wu, Zhicao Yu, & **Cheng-Ming Chuong.** 2003. "Molecular biology of feather morphogenesis: A testable model for evo-devo." *Journal of Experimental Zoology* **298B** (15 August): 109-122.

Widelitz, Randall B., Jacqueline M. Veltmaat, Julie Ann Mayer, John Foley, & **Cheng-Ming Chuong.** 2007. "Mammary glands and feathers: Comparing two skin appendages which help define novel classes during vertebrate evolution." *Seminars in Cell & Developmental Biology* **18** (April): 255-266.

Wieland, Carl, Ken Ham, & **Jonathan Sarfati.** 2002. "Maintaining Creationist Integrity." *Answers in Genesis* article updated 16 December (online text at *www.answersingenesis.org* accessed

1/22/2003).
Wilford, John Noble. 1998. "Early Amphibian Fossil Hints of a Trip Ashore Earlier Than Thought," in Wade (1998, 111-114).
Williams, Blythe A., Richard F. Kay, & **E. Christopher Kirk.** 2010. "New perspectives on anthropoid origins." *Proceedings of the National Academy of Sciences* **107** (16 March): 4797-4804.
Williams, Devon. 2007. "Friday Five: William A. Dembski." Focus on the Family *Citizen Magazine* 14 December posting (online text at *www.citizenlink.org* accessed 2/6/2014).
Willis, Tom. 2000. "More Great Proofs of Evolution: *'The Laws of Cause and Effect, and the 1st and 2nd Laws of Thermodynamics have been invalidated by modern science'* Part 2." *The CSA News* March-April posting (online text at *www.csama.org* accessed 2/20/2001).
Wills, M. A. 2007. "Fossil ghost ranges are most common in some of the oldest and some of the youngest strata." *Proceedings of the Royal Society of London* **B** (Biological Sciences) **274** (7 October): 2421-2427.
Wilson, David B., ed. 1983. *Did the Devil Make Darwin Do It? Modern Perspectives on the Creation-Evolution Controversy.* Ames, IA: Iowa State University Press.
Wilson, Edward O., Frank M. Carpenter, & **William L. Brown Jr.** 1967. "The First Mesozoic Ants." *Science* **157** (1 September): 1038-1040.
Wilson, Gregory P., & **Jeremy A. Riedel.** 2010. "New Specimen Reveals Deltatheroidan Affinities of the North American Late Cretaceous Mammal *Nanocuris*." *Journal of Vertebrate Paleontology* **30** (May): 872-884.
Wilson, Joanne, & **Abigail S. Tucker.** 2004. "Fgf and Bmp signals repress the expression of *Bapx1* in the mandibular mesenchyme and control the position of the developing jaw joint." *Developmental Biology* **266** (1 February): 138-150.
Wise, Kurt P. 1994. "The Origin of Life's Major Groups," in Moreland (1994, 211-234).
———. 2005. "The Flores Skeleton and Human Baraminology." *Occasional Papers of the Baraminology Study Group* (online @ bryancore.org) No. 6: 1-13.
———. 2006. "Swimming with the Dinosaurs." *Answers in Genesis* article for 8 March (online text at *www.answersingenesis.org* accessed 8/11/2016).
Witmer, Lawrence M. 1990. "The craniofacial air sac system of Mesozoic birds (Aves)." *Zoological Journal of the Linnean Society* **100** (December): 327-378.
Wolf, John, & **James S. Mellett.** 1985. "The Role of 'Nebraska Man' in the Creation-Evolution Debate." *Creation/Evolution* **5** (Summer): 31-43.
Wolfe, G. R., F. X. Cunningham, D. Durnfordt, B. R. Green, & **E. Gantt.** 1994. "Evidence for a common origin of chloroplasts with light-harvesting complexes of different pigmentation." *Nature* **367** (10 February): 566-568.
Woltering, Joost M., & **Denis Duboule.** 2010. "The Origin of Digits: Expression Patterns versus Regulatory Mechanisms." *Developmental Cell* **18** (20 April): 526-532.
Woltering, Joost M., Daan Noordermeer, Marion Leleu, & **Denis Duboule.** 2014. "Conservation and Divergence of Regulatory Strategies at *Hox* Loci and the Origin of Tetrapod Digits." *PLoS Biology* (online @ plosbiology.org) **12** (January): e1001773.
Womelsdorf, Thilo, Kevin Johnston, Martin Vinck, & **Stefan Everling.** 2010. "Theta-activity in anterior cingulate cortex predicts task rules and their adjustments following errors." *Proceedings of the National Academy of Sciences* **107** (16 March): 5248-5253.
Wood, Todd Charles. 2002a. "The AGEing Process: Rapid Post-Flood Intrabaraminic Diversification Caused by Altruistic Genetic Elements (AGES)." Geoscience Research Institute *Origins* **54** (1): 5-34.
———. 2002b. "A baraminology tutorial with examples from the grasses (Poaceae)." *Journal of Creation* (AKA Answers in Genesis *Creation ex Nihilo Technical Journal*) **16** (1): 15-25.
———. 2003. "No. 363—Mediated Design." Institute for Creation Research *Impact* (September): i-iv.
———. 2005. "Visualizing Baraminic Distances Using Classical Multidimensional Scaling." Geoscience Research Institute *Origins* **57**: 9-29.
———. 2006. "The Current Status of Baraminology." *Creation Research Society Quarterly* **43** (December): 149-158.
———. 2009. "The truth about evolution." *Todd's Blog* posting 30 September (online text at *toddcwood.blogspot.com* accessed 2/26/2010).
———. 2010a. "Baraminological Analysis Places *Homo habilis, Homo rudolfensis*, and *Australopithecus sediba* in the Human Holobaramin." Answers in Genesis *Answers Research*

Journal **3**: 71-90.
———. 2010b. "Baraminology in Journal of Evolutionary Biology." *Todd's Blog* posting 18 June (online text at *toddcwood.blogspot.com* accessed 3/5/2013).
———. 2011a. "Responding to Senter: Baraminology in JEB again." *Todd's Blog* posting 26 January (online text at *toddcwood.blogspot.com* accessed 7/9/2011).
———. 2011b. "Yet another helpful paper from Phil Senter." *Todd's Blog* posting 9 March (online text at *toddcwood.blogspot.com* accessed 3/5/2013).
———. 2011c. "Using creation science to demonstrate evolution? Senter's strategy revisited." *Journal of Evolutionary Biology* **24** (April): 914-918.
Wood, Todd Charles, & **David P. Cavanaugh.** 2001. "A Baraminological Analysis of Subtribe Flaveriinae (Asteraceae: Helenieae) and the Origin of Biological Complexity." Geoscience Research Institute *Origins* **52** (1): 7-27.
———. 2003. "An Evaluation of Lineages and Trajectories as Baraminological Membership Criteria." *Occasional Papers of the Baraminology Study Group* (online @ bryancore.org) No. 2: 1-6.
Wood, Todd Charles, & **Megan J. Murray.** 2003. *Understanding the Pattern of Life: Origins and Organization of the Species.* Nashville, TN: Broadman & Holman Publishers.
Woodburne, Michael O., Thomas H. Rich, & **Mark S. Springer.** 2003. "The evolution of tribospheny and the antiquity of mammalian clades." *Molecular Phylogenetics and Evolution* **28** (August): 360-385.
Woodburne, Michael O., & **Richard H. Tedford**. 1975. "The First Tertiary Monotreme from Australia." *American Museum Novitates* **2588** (October): 1-11.
Woodmorappe, John. 1996. *Noah's Ark: A Feasibility Study.* Dallas, TX: Institute for Creation Research.
———. 1997a. "An Attempted Atheistic Snow Job on Believers: The Merits of Creationism: Comments on the PBS Debate on Firing Line." *Revolution Against Evolution* undated posting (online text at *www.rae.org* accessed 8/5/2002).
———. 1997b. "John Woodmorappe's refutation of Glen Morton's review of NOAH'S ARK: A FEASIBILITY STUDY." *Revolution Against Evolution* undated posting (online pdf at *www.rae.org* accessed 4/17/2016).
———. 1999. "New Educational Activities for Home Schooling Science: A Hands-on Science Activity that Demonstrates the Atheism and Nihilism of Evolution." *Revolution Against Evolution* 30 June updated posting (online text at *www.rae.org* accessed 5/27/2012).
———. 2001. "Mammal-like reptiles: major trait reversals and discontinuities." *Journal of Creation* (AKA Answers in Genesis *Creation ex Nihilo Technical Journal*) **15** (1): 44-52.
———. 2002. "Walking whales, nested hierarchies, and chimeras: do they exist?" *Journal of Creation* (AKA Answers in Genesis *Creation ex Nihilo Technical Journal*) **16** (1): 111-119.
———. 2013. "How Could Noah Fit the Animals on the Ark and Care for Them?" *Answers in Genesis* 15 October posting of *The New Answers Book 3* Chapter 5 (online text at *www.answersingenesis.org* accessed 2/23/2014).
Woodward, Thomas. 1991. "A Professor Takes Darwin to Court." *Christianity Today* **35** (19 August): 33-35.
———. 2003. *Doubts about Darwin: A History of Intelligent Design.* Grand Rapids, MI: Baker Books.
———. 2006. *Darwin Strikes Back: Defending the Science of Intelligent Design.* Grand Rapids, MI: Baker Books.
Wray, Gregory A., Jeffrey S. Levinton, & **Leo H. Shapiro.** 1996. "Molecular Evidence for Deep Precambrian Divergences Among Metazoan Phyla." *Science* **274** (25 October): 568-573.
Wu, Ping, Lianhai Hou, Maksim Plikus, Michael Hughes, Jeffrey Scehnet, Sanong Suksaweang, Randall B. Widelitz, Ting-Xin Jiang, & **Cheng-Ming Chuong.** 2004. "*Evo-Devo* of amniote integuments and appendages." *International Journal of Developmental Biology* **48** (2-3): 249-270.
Wu, Ping, Ting-Xin Jiang, Jen-Yee Shen, Randall Bruce Widelitz, & **Cheng-Ming Chuong.** 2006. "Morphoregulation of Avian Beaks: Comparative Mapping of Growth Zone Activities and Morphological Evolution." *Developmental Dynamics* **235** (May): 1400-1412.
Wu, Ping, Ting-Xin Jiang, Sanong Suksaweang, Randall Bruce Widelitz, & **Cheng-Ming Chuong.** 2004. "Molecular Shaping of the Beak." *Science* **305** (3 September): 1465-1466.
Wu, Ping, Chen Siang Ng, Jie Yan, Yung-Chih Lai, Chih-Kuan Chen, Yu-Ting Lai, Siao-Man Wu, Jiun-Jie Chen, Weiqi Luo, Randall B. Widelitz, Wen-Hsiung Li, & **Cheng-Ming Chuong.** 2015. "Topographical mapping of α- and β-keratins on developing chick skin integuments: Functional interaction and evolutionary perspectives." *Proceedings of the National Academy of Sciences* **112**

(8 December): E6770-E6779.
Wu, Xiao-Chun, Zhan Li, Bao-Chun Zhou, & Zhi-Ming Dong. 2003. "A polydactylous amniote from the Triassic period." *Nature* **426** (4 December): 516.
Wyss, André. 2001. "Digging Up Fresh Clues About the Origin of Mammals." *Science* **292** (25 May): 1496-1497.

Xing, Lida, Hendrick Klein, Martin G. Lockley, Shi-Li Wang, Wei Chen, Yong Ye, Masaki Matsukawa, & Jian-Ping Zhang. 2013. "Earliest records of theropod and mammal-like tetrapod footprints in the Upper Triassic of Sichuan Basin, China." *Vertebrata PalAsiatica* **51** (3): 184-198.
Xing, Lida, W. Scott Persons IV, Phil R. Bell, Xing Xu, Jianping Zhang, Tetsuto Miyashita, Fengping Wang, & Philip J. Currie. 2013. "Piscivory in the Feathered Dinosaur *Microraptor*." *Evolution* **67** (August): 2441-2445.
Xu, Xing, James Clark, Jonah Choiniere, David Hone, & Corwin Sullivan. 2011. "Reply to '*Limusaurus* and bird digit identity'." *Nature Precedings* (online @ precedings.nature.com) 9 September: 2011.6375.1.
Xu, Xing, James M. Clark, Jinyou Mo, Jonah Choiniere, Catherine A. Forster, Gregory M. Erickson, David W. E. Hone, Corwin Sullivan, David A. Eberth, Sterling Nesbitt, Qi Zhao, Rene Hernandez, Cheng-kai Jia, Feng-lu Han, & Yu Guo. 2009. "A Jurassic ceratosaur from China helps clarify avian digital homologies." *Nature* **459** (18 June): 940-944.
Xu, Xing, Catherine A. Forster, James M. Clark, & Jinyou Mo. 2006. "A basal ceratopsian with transitional features from the Late Jurassic of northwestern China." *Proceedings of the Royal Society of London* **B** (Biological Sciences) **273** (7 September): 2135-2140.
Xu, Xing, & Susan Mackem. 2013. "Tracing the Evolution of Avian Wing Digits." *Current Biology* **23** (17 June): R538-R544.
Xu, Xing, Peter J. Makovicky, Xiao-lin Wang, Mark A. Norell, & Hai-lu You. 2002. "A ceratopsian dinosaur from China and the early evolution of Ceratopsia." *Nature* **416** (21 March): 314-317.
Xu, Xing, Mark A. Norell, Xuewen Kuang, Xiaolin Wang, Qi Zhao, & Chengkai Jia. 2004. "Basal tyrannosaurids from China and evidence for protofeathers in tyrannosaurids." *Nature* **431** (7 October): 680-684.
Xu, Xing, Zhi-lu Tang, & Xiao-lin Wang. 1999. "A therizinosaurid dinosaur with integumentary structures from China." *Nature* **399** (27 May): 350-354.
Xu, Xing, Kebai Wang, Ke Zhang, Qingyu Ma, Lida Xing, Corwin Sullivan, Dongyu Hu, Shuqing Cheng, & Shuo Wang. 2012. "A gigantic feathered dinosaur from the Lower Cretaceous of China." *Nature* **484** (5 April): 92-95.
Xu, Xing, Xiao-lin Wang, & Xiao-chun Wu. 1999. "A dromaeosaurid dinosaur with a filamentous integument from the Yixian Formation of China." *Nature* **401** (16 September): 262-266.
Xu, Xing, Hailu You, Kai Du, & Fenglu Han. 2011. "An *Archaeopteryx*-like theropod from China and the origin of Avialae." *Nature* **475** (28 July): 465-470.
Xu, Xing, & Gou Yu. 2009. "The Origin and Early Evolution of Feathers: Insights from Recent Paleontological and Neontological Data." *Vertebrata PalAsiatica* **47** (4): 311-329.
Xu, Xing, Qi Zhao, Mark Norell, Corwin Sullivan, David Hone, Gregory Erickson, Xiao-Lin Wang, Feng-Lu Han, & Yu Guo. 2009. "A new feathered maniraptoran dinosaur fossil that fills a morphological gap in avian origin." *Chinese Science Bulletin* **54** (February): 430-435.
Xu, Xing, Xiaoting Zheng, Corwin Sullivan, Xiaoli Wang, Lida Xing, Yan Wang, Xiaomei Zhang, Jingmai K. O'Connor, Fucheng Zhang, & Yanhong Pan. 2015. "A bizarre Jurassic maniraptoran theropod with preserved evidence of membranous wings." *Nature* **521** (7 May): 70-73.
Xu, Xing, Xiaoting Zheng, & Hailu You. 2009. "A new feather type in a nonavian theropod and the early evolution of feathers." *Proceedings of the National Academy of Sciences* **106** (20 January): 832-834.
———. 2010. "Exceptional dinosaur fossils show ontogenetic development of early feathers." *Nature* **464** (29 April): 1338-1341.
Xu, Xing, Zhong-he Zhou, & Richard O. Prum. 2001. "Branched integumental structures in *Sinornithosaurus* and the origin of feathers." *Nature* **410** (8 March): 200-204.
Xu, Xing, Zhong-he Zhou, Xiao-lin Wang, Xuewen Kuang, Fucheng Zhang, & Xiangke Du. 2003. "Four-winged dinosaurs from China." *Nature* **421** (23 January): 335-340.

Yahya, Harun, ed. David Livingstone, trans. Ron Evans. 2002. *Romanticism: A Weapon of Satan.* New Delhi: Millat Centre.

———. 2004a. "The Origin of Mammals." *Signs of Allah* posting (online text at *www.evidencesofcreation.com* accessed 4/5/2016).

———, ed. David Livingstone, trans. Carl Nino Rossini. 2004b. *Satan: The Sworn Enemy of Mankind*. Istanbul: Global Publishing.

———. 2009. "Reconstruction (Imaginary Pictures)." *Harun Yahya* 17 August posting (online text at *www.harunyahya.com* accessed 4/8/2016).

Yamamichi, Masato, & **Akira Sasaki.** 2013. "Single-Gene Speciation with Pleiotropy: Effects of Allele Dominance, Population Size, and Delayed Inheritance." *Evolution* **67** (July): 2011-2023.

Yamamoto, Daisuke S., Megumi Sumitani, Koji Tojo, Jae Min Lee, & **Masatsugu Hatakeyama.** 2004. "Cloning of a *decapentaplegic* orthologue from the sawfly, *Athalia rosae* (Hymenoptera), and its expression in the embryonic appendages." *Development Genes and Evolution* **214** (March): 128-133.

Yamato, Maya, & **Nicholas D. Pyenson.** 2015. "Early Development and Orientation of the Acoustic Funnel Provides Insight into the Evolution of Sound Reception Pathways in Cetaceans." *PLoS ONE* (online @ plosone.org) **10** (March): e0118582.

Yang, Ji, Hongya Gu, & **Ziheng Yang.** 2004. "Likelihood Analysis of the Chalcone Synthase Genes Suggests the Role of Positive Selection in Morning Glories (*Ipomoea*)." *Journal of Molecular Evolution* **58** (January): 54-63.

Yang, Ji, Jinxia Huang, Hongya Gu, Yang Zhong, & **Ziheng Yang.** 2002. "Duplication and Adaptive Evolution of the Chalcone Synthase Genes of Dendranthema (Asteraceae)." *Molecular Biology and Evolution* **19** (October): 1752-1759.

Yano, Tohru, & **Koji Tamura.** 2013. "The making of differences between fins and limbs." *Journal of Anatomy* **222** (January): 100-113.

Yates, Tony B., & **Edmund A. Marek.** 2013. "Is Oklahoma really OK? A regional study of the prevalence of biological evolution-related misconceptions held by introductory biology teachers." *Evolution: Education & Outreach* (online @ evolution-outreach.com) **6**: 6.

Yavachev, L. P., O. I. Georgiev, E. A. Braga, T. A. Avdonina, A. E. Bogomolova, V. B. Zhurkin, V. V. Nosikov, & **A. A. Hadziolov.** 1986. "Nucleotide sequence analysis of the spacer regions flanking the rat tRNA transcription unit and identification of repetitive elements." *Nucleic Acids Research* **14** (25 March): 2799-2810.

Yek, Sze Huei, & **Ulrich G. Mueller.** 2011. "The metapleural gland of ants." *Biological Reviews of the Cambridge Philosophical Society* **86** (November): 774-791.

Yokoyama, Ken Daigoro, & **David D. Pollock.** 2012. "SP Transcription Factor Paralogs and DNA-Binding Sites Coevolve and Adaptively Converge in Mammals and Birds." *Genome Biology and Evolution* **4** (11): 1102-1117.

Yoon, Hwan Su, Jeremiah D. Hackett, & **Debashish Bhattacharya.** 2002. "A single origin of the peridinin- and fucoxanthin-containing plastids in dinoflagellates through tertiary endosymbiosis." *Proceedings of the National Academy of Sciences* **99** (3 September): 117248-11729.

Yoon, Hwan Su, Jeremiah D. Hackett, Gabriele Pinto, & **Debashish Bhattacharya.** 2002. "The single, ancient origin of chromist plastids." *Proceedings of the National Academy of Sciences* **99** (26 November): 15507-15512.

Yoon, Hwan Su, Jeremiah D. Hackett, Frances M. Van Dolah, Tetyana Nosenko, Kristy L. Lidie, & **Debashish Bhattacharya.** 2005. "Tertiary Endosymbiosis Driven Genome Evolution in Dinoflagellate Algae." *Molecular Biology and Evolution* **22** (May): 1299-1308.

You, Hai-Lu, Jessie Atterholt, Jingmai K. O'Connor, Jerald D. Harris, Matthew C. Lamanna, & **Da-Qing Li.** 2010. "A second Cretaceous ornithuromorph bird from the Changma Basin, Gansu Province, northwestern China." *Acta Palaeontologica Polonica* **55** (4): 617-625.

Younce, Max D. 2009. *The Truth About Evolution Or; Don't Let Satan Make A Monkey Out of You!* Kearney, NE: Morris Publishing (online pdf at *www.jesus-is-savior.com* accessed 6/24/2013).

Young, Glenn M., Michael J. Smith, Scott A. Minnich, & **Virginia L. Miller.** 1999. "The *Yersinia enterocolitica* Motility Master Regulatory Operon, *flhDC*, Is Required for Flagellin Production, Swimming Motility, and Swarming Motility." *Journal of Bacteriology* **181** (May, No. 9): 2823-2833.

Young, Matt. 1998. "The Bible Code." *Rocky Mountain Skeptic* (March-April): 1, 4-6 (online posting at *inside.mines.edu/~mmyoung/* accessed 7/8/2014).

———. 2013. "JPL finally wins wrongful termination lawsuit." *Panda's Thumb* posting 18 January (online text at *pandasthumb.org* accessed 1/22/2013).

Young, Nathan M., Diane Hu, Alexis J. Lainoff, Francis J. Smith, Raul Diaz, Abigail S. Tucker, Paul A. Trainor, Richard A. Schneider, Benedikt Haligrimsson, & **Ralph S. Marcucio.** 2014. "Embryonic bauplans and the developmental origins of facial diversity and constraint." *Development* (AKA *Journal of Embryology and Experimental Morphology*) **141** (March): 1059-1063.

Young, Rebecca L., Gabe S. Bever, Zhe Wang, & Günter P. Wagner. 2011. "Identity of the Avian Wing Digits: Problems Resolved and Unsolved." *Developmental Dynamics* **240** (May): 1042-1053.

Young, Rebecca L., Vincenzo Caputo, Massimo Giovannotti, Tiana Kohlsdorf, Alexander O. Vargas, Gemma E. May, & Günter P. Wagner. 2009. "Evolution of digit identity in the three-toed Italian skink *Chalcides chalcides*: a new case of digit identify frame shift." *Evolution & Development* **11** (November-December): 647-658.

Young, Rebecca L., & Günter P. Wagner. 2011. "Why Ontogenetic Homology Criteria Can Be Misleading: Lessons From Digit Identity Transformations." *Journal of Experimental Zoology* **314B** (15 May): 165-170.

Youngstown City Schools. 2015. "SCIENCE: BIOLOGY UNIT #4: DIVERSITY OF LIFE (4 WEEKS)." *Youngstown (Ohio) City Schools* 22 May curriculum posting (online pdf at *www.youngstownk12.oh.us* accessed 5/16/2016).

Yu, Mingke, Ping Wu, Randall B. Widelitz, & Cheng-Ming Chuong. 2002. "The morphogenesis of feathers." *Nature* **420** (21 November): 308-312.

Yu, Mingke, Zhicao Yue, Ping Wu, Da-Yu Wu, Julie-Ann Mayer, Marcus Medina, Randall B. Widelitz, Ting-Xin Jiang, & Cheng-Ming Chuong. 2004. "The developmental biology of feather follicles." *International Journal of Developmental Biology* **48** (2-3): 181-191.

Yuan, Chong-Xi, Qiang Ji, Qing-Jin Meng, Alan R. Tabrum, & Zhe-Xi Luo. 2013. "Earliest Evolution of Multituberculate Mammals Revealed by a New Jurassic Fossil." *Science* **341** (16 August): 779-783.

Zákány, József, Catherine Fromental-Ramain, Xavier Warot, & Denis Duboule. 1997. "Regulation of number and size of digits by posterior *Hox* genes: A dose-dependent mechanism with potential evolutionary implications." *Proceedings of the National Academy of Sciences* **94** (9 December): 13695-13700.

Zalmout, Iyad S., Munir Ul-Haq, & Philip D. Gingerich. 2003. "New Species of *Protosiren* (Mammalia, Sirenia) from the Early Middle Eocene of Balochistan (Pakistan)." *Contributions from the Museum of Paleontology, University of Michigan* **31** (15 August): 79-87.

Zardoya, Rafael, & Axel Meyer. 1998. "Complete mitochondrial genome suggests diapsid affinities of turtles." *Proceedings of the National Academy of Sciences* **95** (24 November): 14226-14231.

———. 2001. "The evolutionary position of turtles revised." *Naturwissenschaften* **88** (May): 193-200.

Zelenitsky, Darla K., François Therrien, Gregory M. Erickson, Christopher L. DeBuhr, Yoshitsugu Kobayashi, David A. Eberth, & Frank Hadfield. 2012. "Feathered Non-Avian Dinosaurs from North America Provide Insight into Wing Origins." *Science* **338** (26 October): 510-514.

Zeller, Ulrich. 1989. *"Die Entwicklung und Morphologie des Schadels von* Ornithorhynchus anatinus *(Mammalia: Prototheria: Monotremata)." Abhandlungen der Senckenbergischen Naturforschenden Gesellschaft* **545**: 1-188.

Zeller, U., J. R. Wible, & M. Elsner. 1993. "New ontogenetic evidence on the septomaxilla of *Tamandua* and *Choloepus* (Mammalia, Xenarthra), with a reevaluation of the homology of the mammalian septomaxilla." *Journal of Mammalian Evolution* **1** (March): 31-46.

Zetterberg, Peter, ed. 1983. *Evolution versus Creationism: The Public Education Controversy.* Phoenix, AZ: Oryx Press.

Zhang, Fai-Kui. 1984. "The fossil record of Mesozoic mammals in China." Translated by Will Downs (pdf accessed *paleoglot.org* 9/15/2016) *Vertebrata PalAsiatica* **22** (1): 29-38.

Zhang, Fai-Kui, A. W. Crompton, Z.X. Luo, & C. R. Schaff. 1998. "Pattern of dental replacement of *Sinoconodon* and its implications for evolution of mammals." *Vertebrata PalAsiatica* 36 (3): 197-217.

Zhang, Fucheng, Zhonghe Zhou, Xing Xu, Xiaolin Wang, & Corwin Sullivan. 2008. "A bizarre Jurassic maniraptoran from China with elongate ribbon-like feathers." *Nature* **455** (23 October): 1105-1108.

Zhang, Jianming, Antony M. Dean, Frédéric Brunet, & Manyuan Long. 2004. "Evolving protein functional diversity in new genes of *Drosophila*." *Proceedings of the National Academy of Sciences* **101** (16 November): 16246-16250.

Zhang, Jianming, Manyuan Long, & Liming Li. 2005. "Translational effects of differential codon usage among intragenic domains of new genes in *Drosophila*." *Biochimica et Biophysica Acta* (Gene Structure and Expression) **1728** (May): 135-142.

Zhang, Shicui, Shaohui Wang, Hongyan Li, & Li Li. 2011. "Vitellogenin, a multivalent sensor and an antimicrobial effector." *The International Journal of Biochemistry & Cell Biology* **43** (March): 303-305.

Zhang, Zihui, Defeng Chen, Huitao Zhang, & Lianhai Hou. 2014. "A large enantiornithine bird from the Lower Cretaceous of China and its implications for lung ventilation." *Zoological Journal of the Linnean Society* **113** (December): 820-827.

Zhang, Zihui, Chunling Gao, Qingjin Meng, Jinyuan Liu, Lianhai Hou, & Guangmei Zheng. 2009. "Diversification in an Early Cretaceous avian genus: evidence from a new species of *Confuciusornis* from China." *Journal of Ornithology* **150** (October): 783-790.

Zhao, Yuanxiang, & Steven S. Potter. 2001. "Functional specificity of the *Hoxa13* homeobox." *Development* (AKA *Journal of Embryology and Experimental Morphology*) **128** (15 August): 3197-3207.

Zheng, Wenjie, Xingsheng Jin, & Xing Xu. 2015. "A psittacosaurid-like basal neoceratopsian from the Upper Cretaceous of central China and its implications for basal ceratopsian evolution." *Scientific Reports* (online @ www.nature.com/srep/) **5** (21 September): 14190.

Zheng, Xiaoting, Shundong Bi, Xiaoli Wang, & Jin Meng. 2013. "A new arboreal haramiyid shows the diversity of crown mammals in the Jurassic period." *Nature* **500** (8 August): 199-202.

Zheng, Xiaoting, Jingmai O'Connor, Xiaoli Wang, Min Wang, Xiaomei Zhang, & Zhonghe Zhou. 2014. "On the absence of sternal elements in *Anchiornis* (Paraves) and *Sapeornis* (Aves) and the complex early evolution of the avian sternum." *Proceedings of the National Academy of Sciences* **111** (23 September): 13900-13905.

Zheng, Xiaoting, Xing Xu, Hailu You, Qi Zhao, & Zhiming Dong. 2010. "A short-armed dromaeosaurid from the Jehol Group of China with implications for early dromaeosaurid evolution." *Proceedings of the Royal Society of London* **B** (Biological Sciences) **277** (22 January): 211-217.

Zhou, Chang-Fu, Shaoyuan Wu, Thomas Martin, & Zhe-Xi Luo. 2013. "A Jurassic mammaliaform and the earliest mammalian evolutionary adaptations." *Nature* **500** (8 August): 163-167.

Zhou, Zhonghe, & Fucheng Zhang. 2005. "Discovery of an ornithurine bird and its implication for Early Cretaceous avian radiation." *Proceedings of the National Academy of Sciences* **102** (27 December): 18998-19002.

Zhou, Zhonghe, Fu-Cheng Zhang, & Zhi-Heng Li. 2009. "A New Basal Ornithurine Bird (*Jianchangornis microdonta* gen. et sp. nov.) from the Lower Cretaceous of China." *Vertebrata PalAsiatica* **47** (4): 299-310.

Zhou, Zhonghe, Fucheng Zhang, & Zhiheng Li. 2010. "A new Lower Cretaceous bird from China and tooth reduction in early avian evolution." *Proceedings of the Royal Society of London* **B** (Biological Sciences) **277** (22 January): 219-227.

Zhu, Min, Per E. Ahlberg, Wenjin Zhao, & Liantao Jia. 2002. "First Devonian tetrapod from Asia." *Nature* **420** (19/26 December): 760-761.

Zillmer, Hans J., trans. Tracey J. Evans & Richard Demarest. 1998. *Darwin's Mistake: Antediluvian discoveries prove: dinosaurs and humans coexisted.* 2002 English ed. Munich: Langer Muller.

Zimmer, Carl. 1995. "Back to the Sea." *Discover* **16** (January): 82-84.

———. 1998. *At the Water's Edge: Macroevolution and the Transformation of Life.* New York: The Free Press.

———. 2000a. "The Hidden Unity of Hearts." *Natural History* **109** (April): 56-61.

———. 2000b. *Parasite Rex: Inside the Bizarre World of Nature's Most Dangerous Creatures.* New York: The Free Press.

———. 2001. "Prepared for the Past." *Natural History* **110** (April): 28-29.

———. 2008. "Missing the Wrist." *The Loom* blog for 14 July (online text at *blogs.discovermagazine.com/loom* accessed 8/8/2016).

———. 2009. "The Blind Locksmith Continued: An Update from Joe Thornton." *The Loom* blog for 15 October (online text at *blogs.discovermagazine.com/loom* accessed 11/5/2009).

Index

Introduction

Usually professionals who are preparing for the PMP Certification Exam find two types of questions difficult to solve in their exam.

The first type of question is based on input, tool & technique, and output (ITTO).

With ITTO based questions, you will be given four choices and you have to identify the correct input, tool & technique or the output for the given process in the question.

It is almost impossible to remember each ITTO for the exam. However, if you study properly and understand the project management process flow, these questions will be a piece of cake for you.

The second type of challenging question is the mathematical questions.

These questions involve mathematical calculations and then interpretation. Many people find these types of questions very difficult and tend to avoid them.

Believe me, mathematical questions are very easy to solve and can help you score better in the exam. You only need to understand the concept and logic behind each formula and give them a little practice.

In this eBook, I have explained all the formulas mentioned in the PMBOK Guide. I have also included some other formulas, which are not available in the guide, but questions based on these formulas may appear on the exam.

I strongly believe that once you have gone through this guide, you won't have any problem facing any mathematical question in your PMP Exam.

I hope this guide will help you in your endeavor; however, if you have any suggestions or comments, you are welcome to contact me.

Thanks.

Regards,
M Fahad Usmani, PMP, PMI-RMP
fahad@pmstudycircle.com

Schedule Management

The schedule management knowledge area is one of the most important chapters in the PMBOK Guide from a PMP exam point of view.

You are going to see a lot of questions from this chapter in the exam; therefore, be sure to study this chapter thoroughly and make sure you are able to solve all practice questions given in this chapter.

In the exam, you can see questions from following given topics:

- Triangular Distribution
- Beta Distribution (PERT)
- Standard Deviation, Variance
- Critical Path Method
- Type of Dependency
- Total Float and Free Float
- Schedule Compressions Technique
- Crashing

Now let's discuss each of them in detail.

Triangular and beta distribution are used to calculate the activity duration. To calculate triangular or beta estimate you use three estimates for each activity:

1) Optimistic Estimate
2) Pessimistic Estimate
3) Most Likely Estimate

The optimistic estimate is the least possible time in which the activity can be finished. It is denoted by *to*.

The pessimistic estimate is the worst case scenario. Here you assume the worst case for the activity to complete. Simply put, this is the maximum time required to complete an activity. It is denoted by *tp*.

The most likely estimate is the most probable time the activity will take to be completed. It is denoted by *tm*.

Once you get these three estimates, you can calculate the triangular or beta estimate very easily.

Triangular Distribution

In triangular distribution, the mean is equal to the sum of these three, and divided by three.

Mean = (Optimistic + Most-likely + Pessimist)/3

Or,

Expected Duration (Triangular Distribution) te= (to + tm + tp)/3

Example:

Your team member told you that in your project, the painting work may take 20 days to complete. However, if the weather is not favorable it will take 26 days. He also told you that if all conditions are good, the painting can be completed in 18 days.

Using triangular distribution, calculate the expected duration for this activity.

Solution:

According to triangular estimate,

Expected Duration te= (to + tm + tp)/3
Given data:
tm = 20 days
tp = 26 days
to = 18 days

Therefore, the expected duration = (20 + 26 +18)/3
= 21.33 days

***

Practice Question: 1

You have been given a task which might take 60 days to complete; however, if the conditions are favorable it might be completed in 55 days. In the worst-case scenario, it may take 70 days.

Find the average estimate for this task using triangular estimate.

***

Beta Distribution (PERT)

In triangular distribution each estimate is given equal weightage, added to each other, and then divided by three.

However, in beta distribution which is also known as PERT (Program Evaluation and Review Technique), the most likely estimate is given more weightage, i.e. four times. Afterwards, all estimates are added and divided by six.

Expected Duration (Beta Distribution) te= (to + 4tm + tp)/6

Standard Deviation and Variance

Standard deviation is the amount of variation in a data set. It is the measure of how the numbers are spread.

A low standard deviation means that data points are close to the mean of the set, and a high standard deviation means that the data points are spread out over a wider range.

Standard Deviation (SD) = (tp – to)/6

Variance is the square of the standard deviation.

Variance = $(SD)^2$

Example:

For any activity, you estimate that it can be completed within 10 days. If the weather conditions are bad, this activity can take about 11 days. However, if all conditions are favorable, this activity can be completed in 7 days.

Using PERT, calculate the three point estimate, standard deviation, and variance for this activity.

Solution:

We know the PERT estimate formula:

te= (to + 4tm + tp)/6

Here:

te = PERT Estimate= to be calculated
to = Optimistic Time = 7 days
tm = Most Likely Time = 10 days
tp = Pessimistic Time = 11 days

Substituting these values in the PERT formula:

te= (to + 4tm + tp)/6
= (7 + 4*10 + 11)/6
= (7+40+11)/6
= 58/6
= 9.6 days.

Hence the PERT estimate for this activity is 9.6 days.

Standard Deviation = (tp – to)/6

= (11 – 7)/6
= 4/6
= 0.67

Now,

Variance = (Standard Deviation)^2

= 0.67*0.67
= 0.45

***

Practice Question: 2

You have been given a task which may take 60 days to complete; however, if the conditions are favorable, it may be completed in 55 days. In the worst-case scenario, it may take 70 days.

Find the PERT estimate, standard deviation, and variance for this task.

***

Critical Path Method

You can define the critical path in many ways, such as:

- The critical path is the longest path in the network diagram,

or

- The critical path is the shortest duration in which the project can be completed.

So you can say that in any network diagram, the duration of the critical path is the duration of the project, and no other path can be longer than the critical path.

The critical path has a zero float and so the activities on it will also have the zero float. The activities on the critical path are known as critical activities.

Terms Used in Network Diagram Method

Before we start identifying the critical path and calculating total float, free float, early start, early finish, etc., let's take a quick look at the definitions of these terms

Total Float: The amount of time that the activities or tasks on a non-critical path can be delayed by without affecting the project schedule. The total float is also known as float or slack.

Total Float = Late Finish – Early Finish

On a critical path, float or slack is zero.

Free Float: Free float is associated with activities. Free float is the amount of time that an activity is delayed by without affecting the early start of the next activity.

Free float of Activity X = Early Start of next Activity – Early Finish of Activity X – 1

Early Start: The earliest time when an activity can be started.

Early Finish: The earliest time when an activity can be completed.

Late Start: The latest time that an activity can be started.

Late Finish: The latest time that an activity can be finished.

Activity Duration: The amount of time taken by an activity on a network diagram to complete itself.

Activity Duration = EF – ES + 1 or LF – LS + 1

Type of Dependencies

In a precedence diagram, activities can have four types of dependencies:

1) Finish to Start (FS)
2) Start to Start (SS)
3) Finish to Finish (FF)
4) Start to Finish (SF)

Finish to Start

In a Finish to Start relationship the next activity cannot be started until the previous activity is completed. This relationship is the most commonly used in a network diagram.

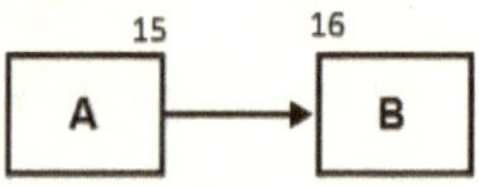

Start to Start

Here the second activity cannot be started until the first activity starts; both activities start simultaneously.

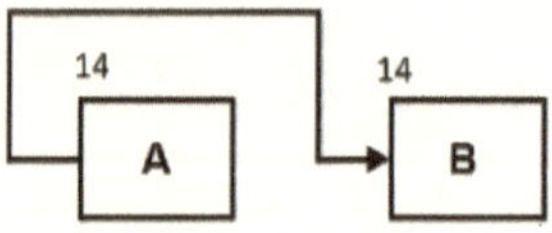

Finish to Finish

Here the second activity cannot be finished until the first activity finishes; simply put, both activities should finish simultaneously.

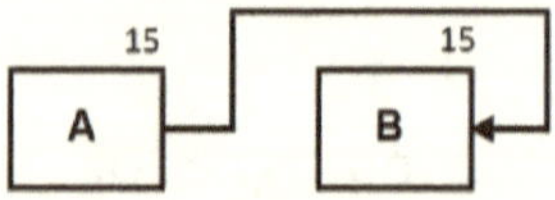

Start to Finish

In this type of dependency, the second activity cannot be finished until the first activity starts. This relationship is the least commonly used in a network diagram.

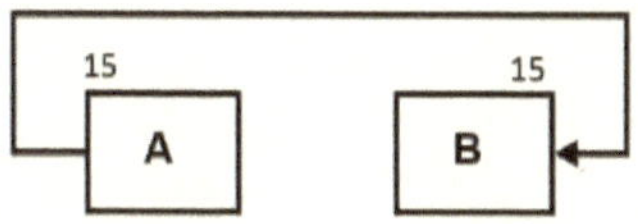

Example 1:

Based on the below diagram, identify the critical path, its duration, float for other paths, Early Start, Early Finish of all activities (considering durations are given in months).

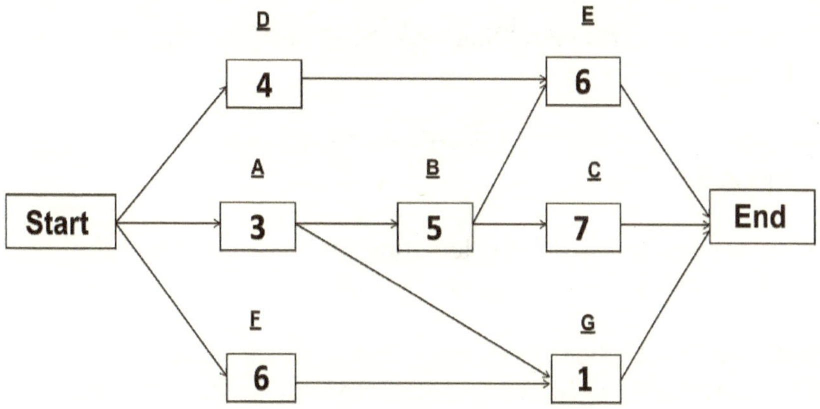

Solution:

Before identifying the Critical Path, we will list out all paths in the network diagram with their duration, and then we will identify the Critical Path.

The above diagram has the following paths:

1. Start->D->E->End (10 months)
2. Start->A->B->C->End (15 months)
3. Start->A->B->E->End (14 months)
4. Start->A->G->End (4 months)
5. Start->F->G->End (7 months)

We can see that the second path "Start->A->B->C->End" is the path with the longest duration; hence, it is the Critical Path.

Float for the first path "Start->D->E->End" is = 15 – 10
= 5 months

Float for the third path "Start->A->B->E->End" is = 15 -14
= 1 month

Float for the fourth path "Start->A->G->End" is = 15 – 4
= 11 months

Float for the fifth path "Start->F->G->End" is = 15 – 7
= 8 months

Now, we will move on to calculating the Early Start and Early Finish of all activities.

Note: For the Critical Path, Early Dates are the same as Late Dates. For example, Early Start and Late Start will be the same, and Early Finish and Late Finish Date will also be the same.

Early Start and Early Finish Dates are calculated with forward pass, and Late Start and Late Finish dates will be calculated with backward pass.

First, we will identify the Early Start and Early Finish of activities on the Critical Path.

Path Start->A->B->C->End

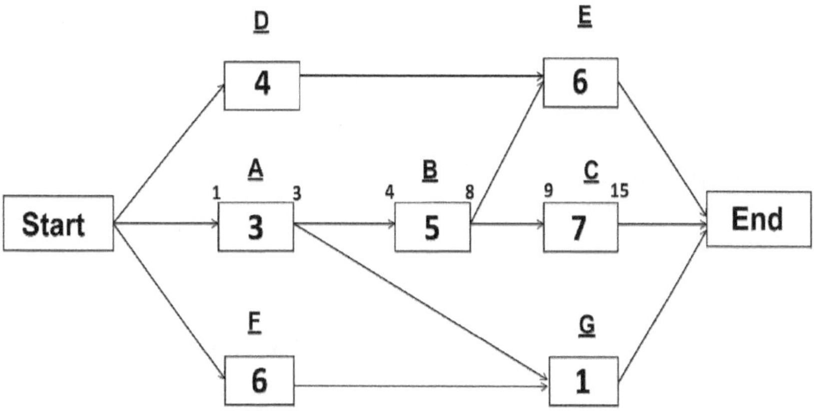

Activity A:

A is the first activity of the path, and let us assume that it starts on day 1.

Hence, Early Start of activity A is 1.

Now we have to calculate the Early Finish of A.

Activity Duration = EF – ES + 1

EF of A = Activity Duration + ES of A – 1

EF = 3 + 1 – 1

EF = 3

Activity B:

Early Start of B = Early Finish of predecessor activity + 1
= Early Finish of Activity A + 1

= 3 + 1
= 4

Early Finish of B = Early Start of B + Activity duration – 1
= 4 + 5 – 1
= 8

Activity C:

Early Start of Activity C = Early Finish of predecessor activity + 1
= Early Finish of Activity B +1
= 8 + 1
= 9

Early Finish of Activity C = Early Start of Activity C + duration of activity C – 1
= 9 + 7 – 1
= 15

Path Start->D->E->End

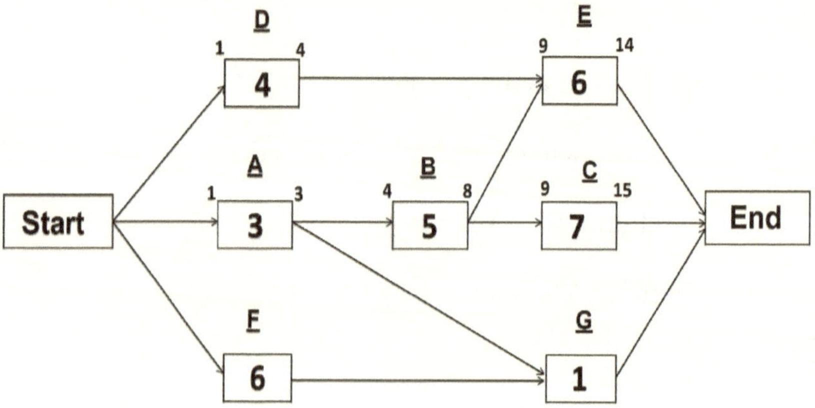

Activity D:

Early Start of Activity D = 1 (since it is the first activity on the path)

Early Finish of Activity D = Early Start of Activity D + Activity duration – 1
= 1 + 4 – 1
= 4

Activity E:

Early Start of Activity E = Early Finish of predecessor activity +1

Activity E has two predecessors, which one will we chose here?

We will choose the activity with the greater Early Finish Date. In this case, Activity B has a greater Early Finish Date; therefore, we will choose Activity B.

Early Start of Activity E = Early Finish of predecessor activity +1
= Early Finish of Activity B + 1
= 8 + 1
= 9

Early Finish of Activity E = Early Start of Activity E + Activity duration – 1
= 9 + 6 – 1
= 14

Path Start->F->G->End

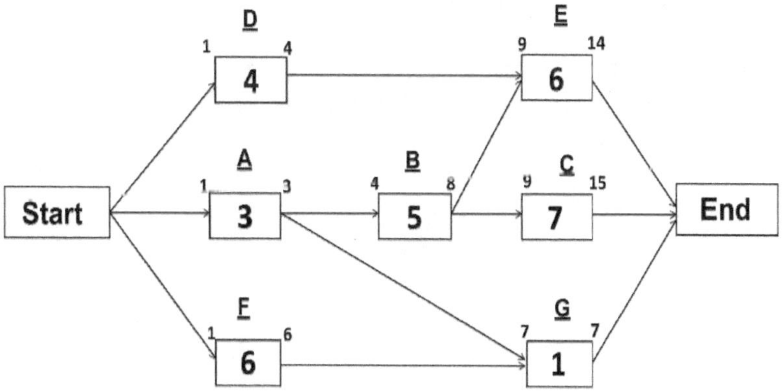

Activity F:

Early Start of Activity F = 1 (since it is the first activity on the path)
Early Finish of Activity F = Early Start of Activity F + Activity duration – 1
= 1 + 6 – 1
= 6

Activity G:

Early Start of Activity G = Early Finish of predecessor activity +1

Here again, activity G has two predecessor activities; therefore, we will choose the activity with the greater Early Finish Date. In this case, activity F has the greater Early Finish Date.

Early Start of Activity G = Early Finish of Activity F + 1
= 6 + 1
= 7

Early Finish of Activity G = Early Start of Activity G + Activity duration – 1
= 7 + 1 – 1
= 7

We have calculated the Early Start and Early Finish of activities in the network diagram. Now it is time to calculate the Late Start and Late Finish of all activities.

In Early Start and Early Finish calculations, we used the forward pass. To calculate Late Start and Late Finish, we will use the backward pass.

Path Start->A->B->C->End

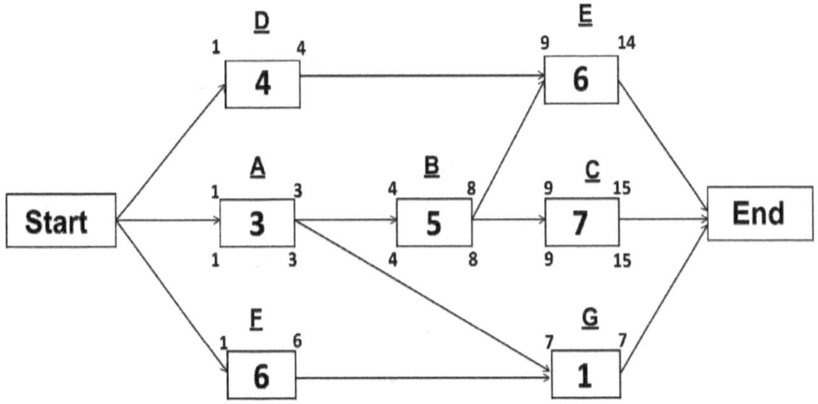

On a Critical Path, Early Start and Late Start Dates are the same. Likewise, Early Finish and Late Finish Dates will also be the same.

Since the Path Start->A->B->C->End is a critical path, Early Start and Early Finish will be the same as Late Start and Late Finish.

Hence,

Activity A

Late Start = 1
Late Finish = 3

Activity B

Late Start = 4
Late Finish = 8

Activity C

Late Start = 9
Late Finish = 15

Path Start->D->E->End

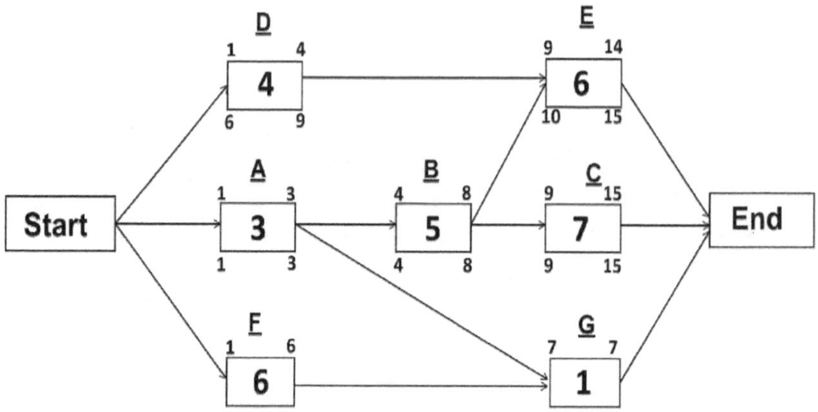

Activity E

The Late Finish of all last activities in any path will be same as the Last Finish of the last activity on the Critical Path because you cannot extend your project beyond this point.

Therefore, Late Finish of Activity E = 15

Late Start of Activity E = Late Finish of Activity E – Activity duration + 1
= 15 – 6 +1
= 10

Activity D

Late Finish of Activity D = Late Start of Successor Activity – 1
= Late Start of Activity E – 1
= 10 – 1
= 9

Late Start of Activity D = Late Finish of Activity D – Activity

duration + 1
= 9 – 4 + 1
= 6

Path Start->F->G->End

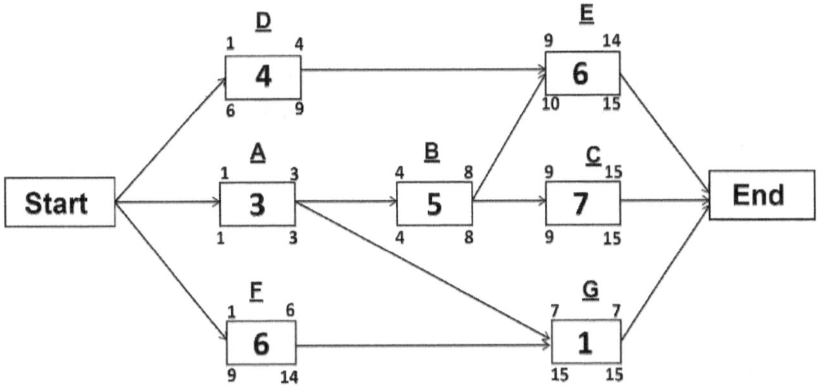

Activity G

The Late Finish of all last activities in any path will be the same as the Last Finish of the last activity on the Critical Path because you cannot cross it.

Therefore, Late Finish of Activity G = 15

Late Start of Activity G = Late Finish of Activity G – Activity duration + 1
= 15 – 1 +1
= 15

Activity F

Late Finish of Activity F = Late Start of Successor Activity – 1
= Late Start of Activity G – 1
= 15 – 1
= 14

Late Start of Activity F = Late Finish of Activity F – Activity

duration + 1
= 14 – 6 + 1
= 9

In the above example, we have calculated the Critical Path, Total Float, Early Start, Early Finish, Late Start, and Late Finish Date.

In this example, I showed you how to calculate Early Start and Early Finish dates when an activity has two predecessors; however, how do we calculate the Late Start and Late Finish dates when an activity has two successors?

Let's look at another example.

***

Example 2:

Based on the below diagram, identify the critical path, its duration, float for other paths, Early Start, and Early Finish of all activities considering the durations are given in months.

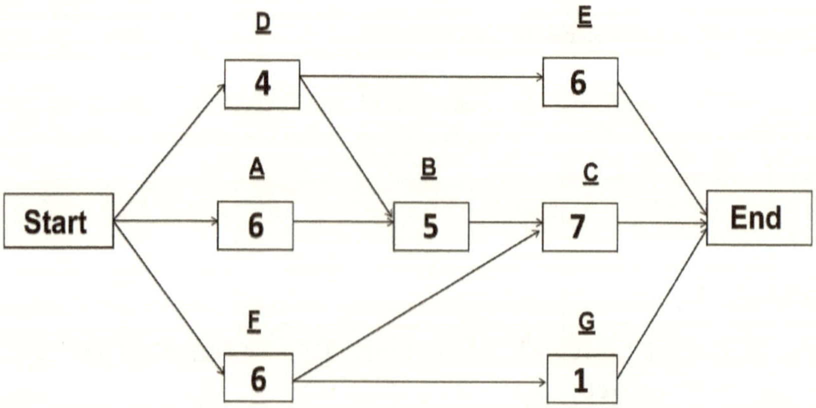

Solution:

Let's first identify the paths in the network diagram and their durations.

The above diagram has five paths:

1) Start->D->E->End (10 months)
2) Start->D->B->C->End (16 months)
3) Start->A->B->C->End (18 months)
4) Start->F->G->End (7 months)
5) Start->F->C->End (13 months)

Path "Start->A->B->C->End" is the Critical Path because it has the longest duration.

Float for the first path "Start->D->E->End" = 18 – 10 = 8 months

Float for the second path "Start->D->B->C->End" = 18 – 16 = 2 months

Float for the fourth path "Start->F->G->End" = 18 – 7 = 11 months

Float for the fifth path "Start->F->C->End" = 18 – 13 = 5 months

In the previous example, we have first calculated the Early Start and Early Finish. In this example, we will start with calculating the Late Start and Late Finish Dates.

Late Start and Late Finish dates are calculated through back pass, which means we will start from the last activity and move back towards the first activity.

Since the path "Start->A->B->C->End" is the Critical Path, we will begin from this path.

Path Start->A->B->C->End

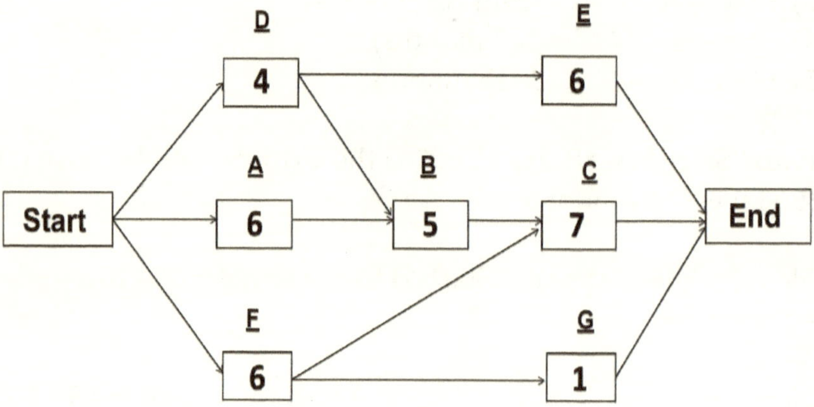

Activity C

The Late Finish will be 18 because the project is supposed to end on the 18th of the month. Also note that the Late Finish for the last activity in any path will be 18, because no activity can go further than this point once the project is completed.

Hence, the Late Finish of Activity C = 18

Late Start of Activity C = Late Finish of Activity C – duration of activity + 1
=18 – 7 + 1
= 12

Activity B

Late Finish of Activity B = Late Start of Successor Activity – 1
= Late Start of Activity C – 1
= 12 – 1
= 11

Late Start of Activity B = Late Finish of Activity B – duration of activity + 1
= 11 – 5 + 1
= 7

Activity A

Late Finish of Activity A = Late Start of Successor Activity – 1
= Late Start of Activity B – 1
= 7 – 1
= 6

Late Start of Activity A = Late Finish of Activity A – duration of activity + 1
= 6 – 6 + 1
= 1

Path Start->D->E->End

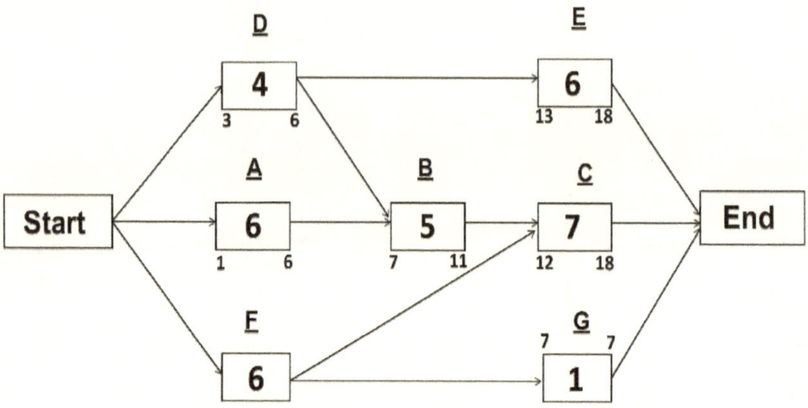

Activity E

Late Finish of Activity E = 18

Late Start of Activity E = Late Finish of Activity E – duration of activity + 1
= 18 – 6 + 1
= 13

Activity D

Late Finish of Activity D = Late Start of successor activity – 1

Here, Activity D has two successor activities, B and E. Which one will we choose?

We will choose the activity with the smaller Late Start date (please note it).

Therefore,

Late Finish of Activity D = Late Start of Activity B – 1
= 7 – 1
= 6

Late Start of Activity D = Late Finish of Activity D – duration of activity + 1
= 6 – 4 + 1
= 3

Path Start->F->G->End

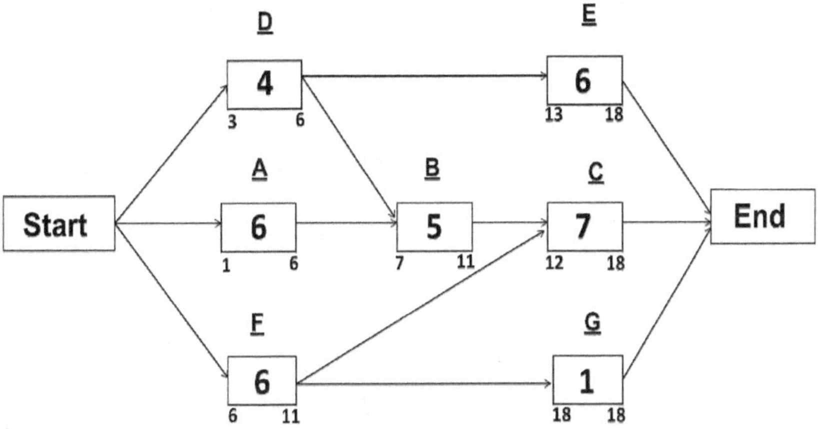

Activity G

Late Finish of Activity G = 18

Late Start of Activity G = Late Finish of Activity G – duration of activity + 1
= 18 – 1 + 1
= 18

Activity F

Late Finish of Activity F = Late Start of successor activity - 1

Here, Activity F has two successor activities C and G. Which one we will choose?

We will choose the activity with the smaller Late Start date (please note it).

Therefore,

Late Finish of Activity F = Late Start of activity C - 1
= 12 - 1
= 11

Late Start of Activity F = Late Finish of Activity F - duration of activity + 1
= 11 - 6 + 1
= 6

Early Start and Early Finish Dates

As we did in our previous example, we can calculate the Early Start and Early Finish Dates for all activities in the given network diagram as follows:

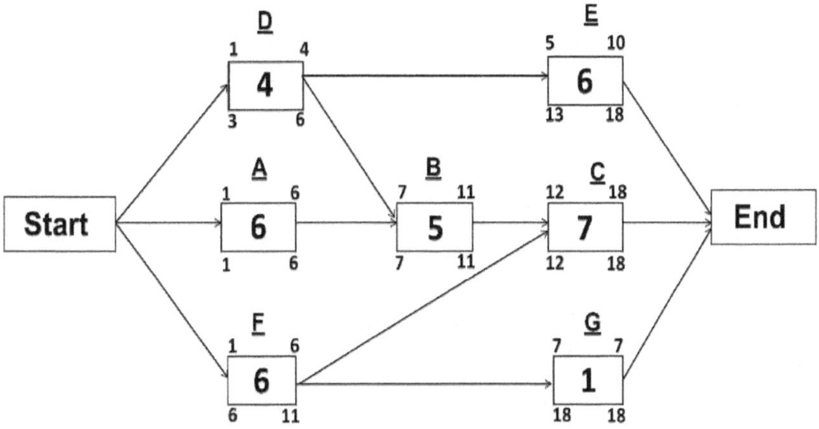

***

Practice Question: 3

Find the Critical Path, Early Start, Early Finish, Late Start, and Late Finish Date of the below given network diagram.

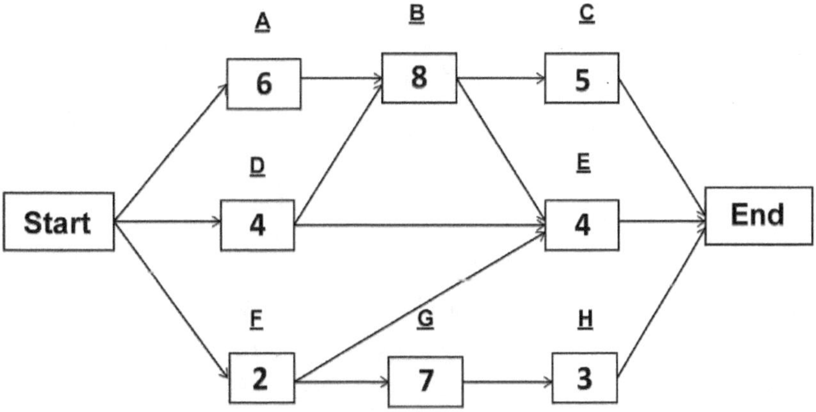

***

Free Float

Free float is the amount of time that an activity can be delayed without delaying the early start of the next activity.

Please note that if two activities are having a common successor single activity, one of these two activities may have the free float.

Free float of Activity X = Early Start of next Activity - Early Finish of Activity X - 1

Example:

For the given below network diagram, find which activity has the free float and calculate the free float.

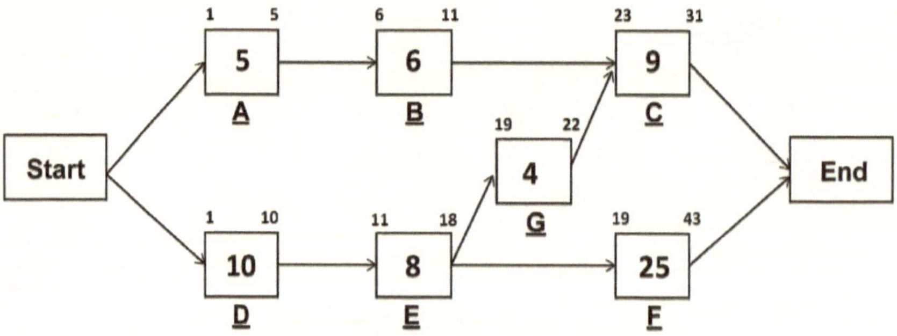

Solution:

In the given diagram only two activities are converging to a single activity, i.e. activity B and activity G. Therefore, one of these two activities can have a free float.

Let's start with activity B.

Activity B is finishing on the 12th and activity C is starting on the 23rd so this activity will have a free float.

Now let's examine activity G.

Activity G is finishing on the 22nd and activity C is starting on the 23rd so this activity cannot have a free float.

We know the formula to calculate the free float:

Free float of Activity X = Early Start of next Activity - Early Finish of Activity X - 1

So in our case,

Free float of Activity B = Early Start of Activity C - Early Finish of Activity B - 1
= 23 - 11 - 1
= 11

Hence, the free float for activity B is 11 days.

***

Practice Question 4:

For the given below network diagram, find which activity has the free float and calculate the free float.

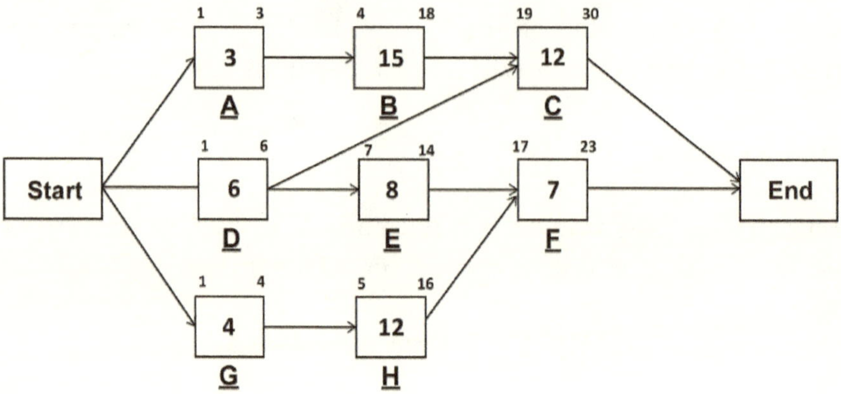

***

Schedule Compression

If your project is delayed and you have to bring it back on schedule, you use schedule compression techniques to do so.

There are other cases as well when you will wish to reduce the duration of the project, such as customer demand or an opportunity for a new project.

There are two schedule compression techniques you can use: fast tracking and crashing.

Fast Tracking

In fast tracking, you review the critical path to find out which sequential activities can be performed parallel or partially parallel to each other.

Once you find out which activities can be fast tracked, you will start working on them to reduce the schedule.

Crashing

Crashing is another schedule compression technique where you add extra resources to the project to compress the schedule.

In crashing, you will review the critical path and see which activities can be completed by adding extra resources. You try to find the activities that can be reduced by a maximum duration by adding the least cost. Once you find those activities, you will apply the crashing technique to them.

The difference between fast tracking and crashing is as follows:

In fast tracking, activities are re-planned to be performed in parallel or partially parallel, while in crashing you add additional resources to activities to finish them early.

In fast tracking, you do not spend extra money while in crashing you do everything, including spending money to shorten the duration of the project.

In the PMP exam, you will see questions based on crashing only (from a math point of view). You may see one or two questions from this topic on your exam.

Example:

Based on the following network diagram and table, find which activities will be crashed and identify the crashing sequence.

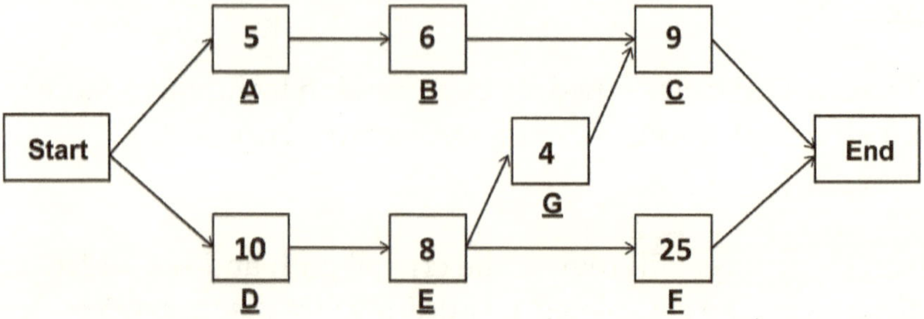

Activity	Duration	Crashed Duration	Cost (USD)	Crashed Cost (USD)
A	5	3	5,000	6,000
B	6	4	6,000	7,500
C	9	6	8,000	9,500
D	10	5	8,500	12,000
E	8	4	7,000	8,500
F	25	14	18,000	25,000
G	4	3	3,000	3,500

Solution:

In the given network diagram path Start-D-F-F-End is the critical path; therefore you will focus on this path to reduce the duration of the project.

To find the crashing cost per day you will use the following formula:

Crashing cost per day = (Crash Cost - Normal Cost)/(Normal Duration - Crash Duration)

The crashing cost per day for each activity is as follows:

Crashing cost per day for activity A = (6,000 – 5,000)/(5 – 3) = 500

Crashing cost per day for activity B = (7,500 – 6,000)/(6 – 4) = 750

Crashing cost per day for activity C = (9,500 – 8,000)/(9 – 6) = 500

Crashing cost per day for activity D = (12,000 – 8,500)/(10 – 5) = 700

Crashing cost per day for activity E = (8,500 – 7,000)/(8 – 4) = 375

Crashing cost per day for activity F = (25,000 – 18,000)/(25 – 14) = 636

Crashing cost per day for activity G = (3,500 – 3,000)/(4 – 3) = 500

Activity	Duration	Crashed Duration	Cost (USD)	Crashed Cost (USD)	Cost Saving Per Day (USD)
A	5	3	5,000	6,000	500
B	6	4	6,000	7,500	750
C	9	6	8,000	9,500	500
D	10	5	8,500	12,000	700
E	8	4	7,000	8,500	375
F	25	14	18,000	25,000	636
G	4	3	3,000	3,500	500

Now if you read the table you will see that activity E has the lowest cost of crashing per day followed by F and D.

Therefore, the sequence of crashing will be E-F-D.

***

Practice Question: 5

Based on the following network diagram and table, find which activities will be crashed and identify the crashing sequence.

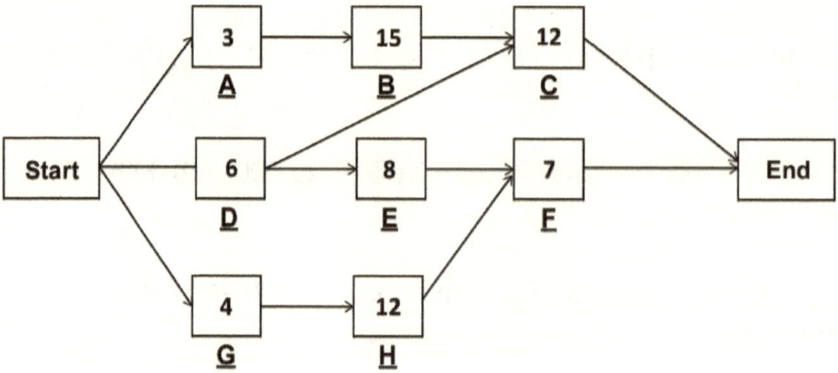

Activity	Duration	Crashed Duration	Cost (USD)	Crashed Cost (USD)
A	3	2	3,000	4,500
B	15	10	18,000	24,000
C	12	9	11,000	13,000
D	6	4	4,000	5,000
E	8	3	6,500	8,000
F	7	4	9,000	11,000
G	4	2	5,000	6,500
H	12	7	13,000	19,000

Miscelleneous Questions

Practice Question: 6

The early start of an activity is 20 and the late start is 32. If the late finish is 51, find the activity duration.

Practice Question: 7

The early start of an activity is 15 and the late start is 26. If the late finish is 45, find the total float of this activity.

Solutions

Practice Question: 1

You have been given a task which might take 60 days to complete; however if the conditions are favorable, it might be completed in 55 days. In the worst-case scenario, it may take 70 days.
Find the average estimate for this task.

Solution:

According to Triangular estimate,

Expected Duration $t_E = (t_O + t_M + t_P)/3$

Given Data:
$t_M = 60$ days
$t_P = 70$ days
$t_O = 55$ days

Therefore, Expected Duration = (60 + 70 +55)/3
= 61.67 days

***

Practice Question: 2

You have been given a task which may take 60 days to complete; however if the conditions are favorable it may be completed in 55 days. In the worst case scenario, it may take 70 days.

Find the PERT estimate, standard deviation, and variance for this task.

Solution:

We know the PERT estimate formula:

$t_E = (t_O + 4t_M + t_P)/6$

Here:
t_O = Optimistic Time = 55 days
t_M = Most Likely Time = 60 days
t_P = Pessimistic Time = 70 days
t_E = PERT Estimate= to be calculated.

Substituting these values in the PERT formula:

$t_E = (t_O + 4t_M + t_P)/6$
= (55 + 4*60 + 70)/6
= (55 + 240 + 70)/6
= 365/6
= 60.83 days.

Hence, the PERT estimate for this activity is 60.83 days.

Standard Deviation = $(t_P - t_O)/6$
= (70 – 55)/6
= 15/6
= 2.5

We know that:

Variance = $(SD)^2$

Variance = 2.5^2
= 6.25

***

Practice Question: 3

Find the Critical Path, Early Start, Early Finish, Late Start, and Late Finish Date of the below given network diagram.

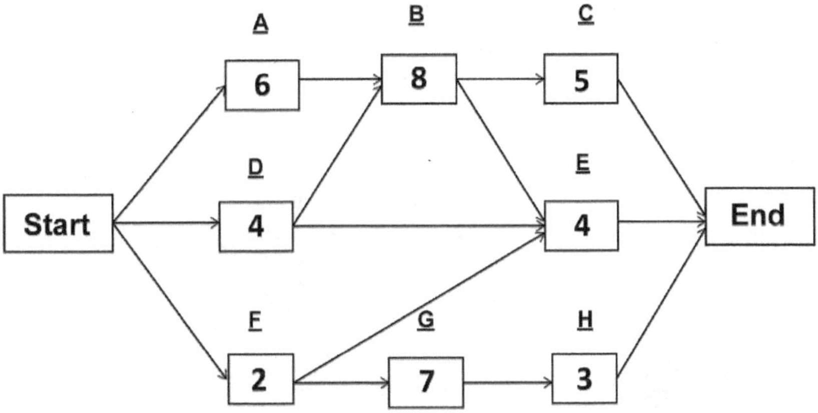

Solution:

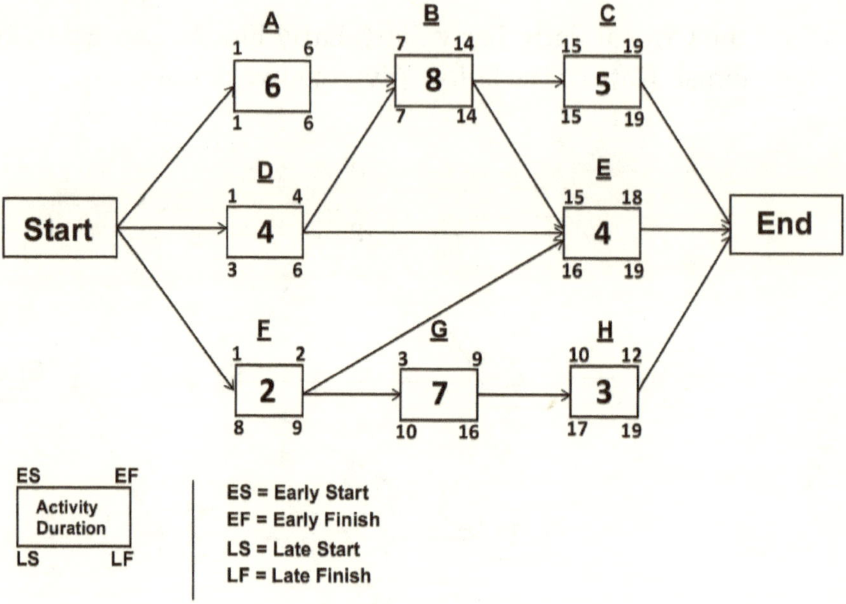

***

Practice Question: 4

For the given below network diagram, find which activity has the free float and calculate the free float.

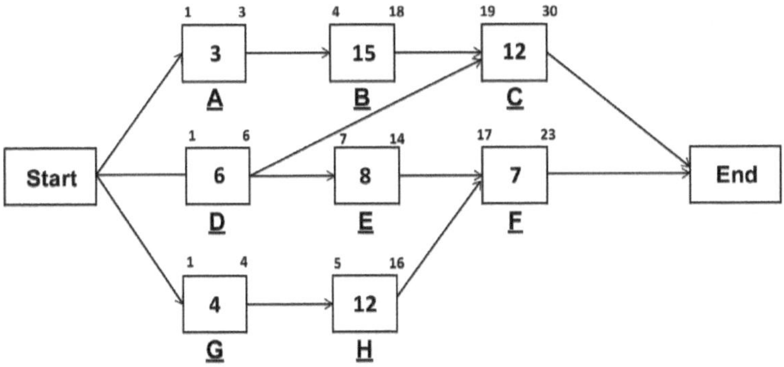

Solution:

In the given diagram only two activities are converging to a single activity, i.e. activity E and activity H. Therefore, one of these two activities can have a free float.

Let's start with activity E.

Activity E is finishing on the 14th and activity F is starting on the 17th, so this activity will have a free float.

Now let's examine activity H.

Activity H is finishing on the 16th and activity F is starting on the 17th, so this activity cannot have a free float.

We know the formula to calculate the free float:

Free float of Activity X = Early Start of next Activity – Early Finish of Activity X – 1

Hence,

Free float of Activity E = Early Start of Activity F - Early Finish of Activity E – 1
= 17 – 14 – 1
= 2

Hence, the free float for activity F is 2 days.

Please note that Activity B and D are also converging to a single activity but they will not have a free float because activity B is on critical path and activity D is followed by another activity which is starting the next day after completing the activity D.

***

Practice Question: 5

Based on the following network diagram and table, find which activities will be crashed and identify the crashing sequence.

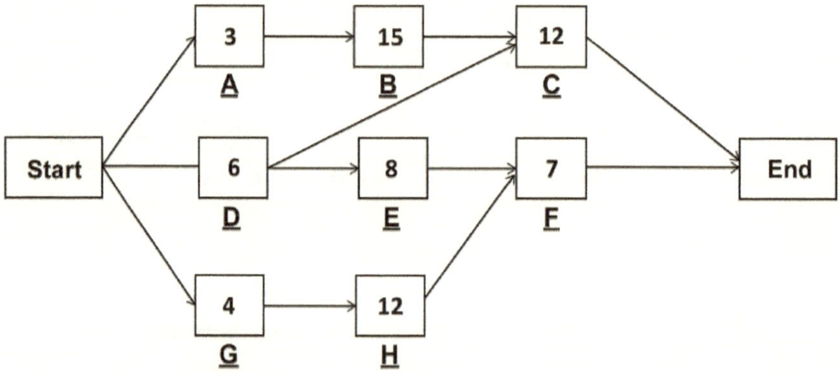

Activity	Duration	Crashed Duration	Cost (USD)	Crashed Cost (USD)
A	3	2	3,000	4,500
B	15	10	18,000	24,000
C	12	9	11,000	13,000
D	6	4	4,000	5,000
E	8	3	6,500	8,000
F	7	4	9,000	11,000
G	4	2	5,000	6,500
H	12	7	13,000	19,000

Solution:

To find the crashing cost per day, you will use the following formula:

Crashing cost per day = (Crash Cost – Normal Cost)/(Normal Duration – Crash Duration)

The crashing cost per day for each activity is as follows:

Crashing cost per day for activity A = (4,500 – 3,000)/(3 – 2) = 1,500

Crashing cost per day for activity B = (24,000 – 18,000)/(15 – 10) = 1,200

Crashing cost per day for activity C = (13,000 – 11,000)/(12 – 9) = 2,000/3 = 666

Crashing cost per day for activity D = (5,000 – 4,000)/(6 – 4) = 500

Crashing cost per day for activity E = (8,000 – 6,500)/(8 – 3) = 300

Crashing cost per day for activity F = (11,000 – 9,000)/(7 – 4) = 666

Crashing cost per day for activity G = (6,500 – 5,000)/(4 – 2) = 750

Crashing cost per day for activity H = (19,000 – 13,000)/(12 – 7) = 1,200

For the given diagram path Start-A-B-C-End is the critical path; therefore, you will try to reduce the duration of activities on this path only.

Activity	Duration	Crashed Duration	Cost (USD)	Crashed Cost (USD)	Cost Saving Per Day (USD)
A	3	2	3,000	4,500	1,500
B	15	10	18,000	24,000	1,200
C	12	9	11,000	13,000	666
D	6	4	4,000	5,000	500
E	8	3	6,500	8,000	300
F	7	4	9,000	11,000	666
G	4	2	5,000	6,500	750
H	12	7	13,000	19,000	1,200

Comparing the cost of reduction per day, the sequence will be as follows:

Activity-C, Activity-B, Activity-A

***

Practice Question: 6

The early start of an activity is 20 and the late start is 32. If the late finish is 51, find the activity duration.

Solution:

We know that:

Activity duration = Late Finish - Late Start + 1
= 51 - 32 + 1
= 20

***

Practice Question: 7

The early start of an activity is 15 and the late start is 26. If the late finish is 45, find the total float of this activity.

Solution:

Given data:
ES = 15
LS = 26
LF = 45

Total Float = LS - ES
= 26 - 15
= 11

Cost Management

PMP aspirants are often afraid of this knowledge area because it requires mathematical calculations.

They try to ignore it and think if they invest this time in some other knowledge area, they will perform better there to cover this loss.

This is the wrong approach. You should invest your time and effort equally in all parts of the PMBOK Guide and not neglect any concept even if it is a little difficult.

This knowledge area has many mathematical formulas; however, believe me: as you start to understand them, it will be a piece of cake for you.

In this chapter, I am going to explain the following topics:

- Project Cost Estimation
- Project Budget
- Earned Value Management
- Elements of Earned Value Management
- Variances
- Indexes
- Forecasting Tools
- Estimate at Completion
- Estimate to Complete
- To Complete Performance Index

Cost Estimation

In the estimate costs process, the project cost is estimated by using the following three techniques:

- Analogous Estimating
- Parametric Estimating
- Bottom-up Estimating

Analogous Estimating

You use this technique when very little detail about the project is available. Therefore, this estimation is not very reliable. The benefits of this technique are: it is very fast, less costly, and it provides a quick result.

In analogous estimating the cost of the project is predicted by comparing it with any similar previously completed project. You will select a project, which is closest to your project from your historical records, and using expert judgment you will determine the cost estimate of your current project.

Analogous estimating is also known as top-down estimating.

Parametric Estimation

This technique also uses historical information to calculate the cost estimates. However, there is a difference between this technique and the analogous estimation technique.

Parametric estimation uses historical information along with statistical data. It takes variables from similar projects and applies them to the current project.

For example, in the previous project you can see what the cost of concrete per cubic meter was. Then you will calculate the concrete requirement for your project and multiply it by the cost obtained from the previous project to get the total cost of concrete for your current project.

In the same way you can calculate the cost of other parts as well.

The accuracy of this technique is better than the analogous estimation.

Bottom-up Estimating

The bottom-up estimating technique is also known as the "definitive technique". This technique is the most accurate and time-consuming technique. Here, the cost of each single activity is determined with the greatest level of detail and then rolls up to calculate the total project cost.

In other words, here the total project work is broken down into the smallest work components. The cost of each component is estimated and then finally, it is aggregated to get the cost estimate of the project.

Project Budget

The project budget is calculated by adding the cost baseline with the management reserve.
Project Budget = Cost Baseline + Management Reserve

The cost baseline is determined by adding the contingency reserve to the cost estimate.

Cost Baseline = Project Cost + Contingency Reserve

Now let's look at some mathematical examples on this topic.

The project budget is also known as the Budget at Completion and denoted by BAC.

Example:

For your project you have calculated the cost of all activities as 150,000 USD. The estimated contingency reserve is 25,000 and the management reserve is 5% of the cost baseline.

Calculate the project budget.

Solution:

We know that

Project Budget = Project Cost + Contingency Reserve + Management Reserve

And the cost baseline:

Cost Baseline = Project Cost + Contingency Reserve

Management Reserve = 5% of cost baseline
= 5% of (150,000 + 25,000)
= 5% of 175,000
= 8,750 USD

Now the project budget = 150,000 + 25,000 + 8,750
= 183,750 USD.

***

Practice Question: 1

You have been assigned as a project manager for a new project and are asked to calculate the budget for it.

You have calculated the cost of all work packages as 200,000 USD. The estimated contingency reserve is 40,000 and the management reserve is 6% of the cost baseline.

Calculate the project budget.

Example:

The cost of your new project is 350,000 USD. The contingency reserve is 75,000 USD and the management reserve is 7% of the cost baseline. What is the rough order of magnitude of the project?

Solution:

Project Budget = Project Cost + Contingency Reserve + Management Reserve
= 350,000 + 75,000 + 7% of (350,000 + 75,000)
= 425,000 + 29,750
= 454,750 USD

We know that the rough order of magnitude is -25% to 75% of the estimated cost.

So, the rough order estimate will be 341,062.5 USD to 795,812.5 USD.

***

Practice Question: 2

The cost of renovation work of an old home is 100,000 USD.

The calculated contingency reserve is 10,000 USD and the management reserve is 4% of the cost baseline. What is the rough order magnitude for this project?

***

Example:

The estimated cost baseline is 150,000 USD. The contingency reserve is 25,000 USD and management reserve is 10,000 USD.

What is the definitive estimate?

Solution:

This question is a little tricky. In the question you have already been given the cost baseline which means the contingency reserve is not needed to get the project budget.

To get the project budget you will simply add the management reserve to the cost baseline.

Project Budget = Cost Baseline + Management Reserve
= 150,000 + 10,000
= 160,000 USD

The definitive estimate is -5% to 10% of the estimated budget.

So the definitive estimate would be 152,000 to 176,000 USD.

***

Practice Question: 3

The estimated project budget is 250,000 USD.

The contingency reserve is 25,000 USD and the management reserve is 10,000 USD. What is the definitive estimate?

***

Earned Value Management

As per the PMBOK Guide, "Earned Value Management (EVM) in its various forms is a commonly used method of performance measurements. It integrates project scope, cost, and schedule measures to help the project management team assess and measure the project performance and progress."

Simply put, earned value management helps you assess the performance of your project.

Once you get the performance data you can take corrective or preventive action as needed.

Elements of Earned Value Management

Earned value management has three basic elements:

1) Planned Value
2) Actual Cost
3) Earned Value

Planned Value

This is the first element of earned value management. Planned value is the approved value of the work to be completed in a given time, or you can say that it is the money that you should have earned as per the schedule.

In other words, planned value is the authorized value of work that has to be completed in a given time period as per the schedule.

Planned value is denoted by PV.

Actual Cost

This is the second element of earned value management. Actual cost is the cost incurred on work completed to date; or simply put, it is the amount of money you have spent to date.

Actual cost is denoted by AC.

Earned Value

This is the last and third element of earned value management. Earned value is the value of the actual completed work to date, or you can say that if the project is terminated today, earned value is the value that the project has produced.

Earned value is denoted by EV.

Variances

In earned value management you have two variances: schedule variance and cost variance.
These variances help you analyze the project's progress, i.e. how you are performing in terms of schedule and cost.

Schedule Variance

Schedule variance tells you whether you are behind or ahead of schedule, and it can be calculated by subtracting planned value from earned value.

Schedule variance is denoted by SV.

Schedule Variance = Earned Value – Planned Value

SV = EV - PV

From the above formula we can conclude that:

- If schedule variance is positive, you are ahead of schedule.
- If schedule variance is negative, you are behind schedule.
- If schedule variance is zero, you are on schedule.

Note: When the project is completed schedule variance becomes zero, because at the end of the project you have earned all planned value.

Cost Variance

Cost variance tells you whether you are under budget or over budget. It can be calculated by subtracting the actual cost from earned value.

Cost variance is denoted by CV.

Cost Variance = Earned Value – Actual Cost

CV = EV - AC

From the above formula we can conclude that:

- If cost variance is positive, you are under budget.
- If cost variance is negative, you are over budget.
- If cost variance is zero, you are on budget.

Indexes

Like variances, indexes also help you analyze the performance of the project.

They help you analyze the efficiency of schedule performance and cost performance of the project.

In earned value management, you have two indexes: schedule performance index and cost performance index.

Schedule Performance Index

The schedule performance index tells you how efficiently you are actually progressing compared to the planned progress.

It can be calculated by dividing earned value by planned value.

The schedule performance index is denoted by SPI.

Schedule Performance Index = (Earned Value)/(Planned Value)

SPI = EV/PV

From the above formula you can conclude that:

- If the schedule performance index is greater than one, you are ahead of schedule.

- If the schedule performance index is less than one, you are behind schedule.

- If the schedule performance index is equal to one, all work is completed.

Cost Performance Index

The cost performance index helps you analyze the cost performance of the project. It is the ratio of the work completed with the actual cost spent on the project.

It can be calculate by dividing earned value by actual cost

Cost Performance Index = (Earned Value)/(Actual Cost)

CPI = EV/AC

- If the cost performance index is one, you are on budget.
- If it is greater than one, you are under budget.
- If it is less than one, you are over budget.

Forecasting Tools

In cost management you have three project forecasting tools:

1) Estimate at Completion
2) Estimate to Complete
3) To Complete Performance Index

Estimate at Completion

The estimate at completion gives you the forecasted value of the project when it is completed.

It tells you how much you may have to spend to complete the project. In other words, you can say that it is the amount of money the project will cost you at the end.

The estimate at completion can be determined by three methods depending on the way the project is performing. However, from a PMP Certification exam point of view, the first method is more important, and there is less chance you will see questions based on the other cases.

Anyway, I'm going to explain all cases mentioned in the PMBOK Guide, so don't worry.

The estimate at completion is denoted by EAC.

Case-I

In this case, you assume that the project will continue to perform to the end as it was performing up until now.

In other words, your future cost performance will be the same as the past cost performance.

Here,

Estimate at Completion = (Budget at Completion) / (Cost Performance Index)

Or,

EAC = BAC/CPI

Case-II

Here you say that until now you have deviated from your budget estimate; however, from now onwards you can complete the remaining work as planned.

Usually this happens when due to unforeseen conditions, an incident happens and your cost elevates. However in this case you are sure that this will not happen again and you can continue with the planned cost estimate.

That is why in this formula, to calculate the EAC you will simply add the money spent to date (i.e. AC) to the budgeted cost for the remaining work.

Here,

EAC = AC + (BAC - EV)

Case-III

In this case, you're over budget, behind schedule, and the client is insisting you complete the project on time.

Here not only the cost but the schedule also has to be taken into consideration.

Here,

EAC = AC + (BAC - EV)/(CPI*SPI)

Estimate to Complete

This is the second forecasting tool.

It is the expected amount of money you will have to spend to complete the remaining work.

It is denoted by ETC.

Estimate to Complete = Estimate at Completion - Actual Cost

ETC = EAC - AC

To Complete Performance Index

The To Complete Performance Index gives you the future cost performance index that you must follow for the remaining work if you want to complete it within the given budget.

It is denoted by TCPI.

You can calculate the TCPI by dividing the remaining work by the remaining funds:

TCPI = (Remaining Work)/(Remaining Funds)

You can calculate the remaining work by subtracting the earned value from the total budget, i.e.

(BAC - EV).

However, there are two cases to determine the remaining funds on hand. In the first case you determine the remaining funds when you are under budget, and in the second case when you are over budget.

In these cases, the formula to calculate the To Complete Performance Index formula will be different.

Case-I: You're Under Budget

In this case, the remaining funds will be calculated by subtracting the "actual cost incurred to date" from the "initial budget", i.e. (BAC - AC).

So, the formula will be:

TCPI = (BAC - EV)/(BAC - AC)

Case-II: You're Over Budget

In this case, the remaining funds will be calculated by subtracting the actual cost incurred to date from the estimate at completion, i.e. (EAC - AC).

Here the TCPI will show you the required cost performance to complete the project with the newly calculated budget.

TCPI = (BAC - EV)/(EAC - AC)

Example:

You have a project which is worth 100,000 USD. Four months have passed, and 60,000 USD has been spent. Upon closer review you find that only 55% of the work is completed.

Determine the Budget at Completion (BAC).

Solution:

The Budget at Completion is the total budget allotted to the project.

In the question, the cost of the project is 100,000 USD.

Hence the BAC is 100,000 USD.

***

Practice Question: 4

You have recently joined an ongoing project. The cost of the project is 350,000 USD. However, 150,000 USD has already been spent and 35% of the work is actually completed.

Determine the Budget at Completion (BAC).

***

Example:

You have a project which is worth 100,000 USD. Four months have passed and 60,000 USD has been spent. Upon closer review you find that only 55% of the work is completed.

Determine the Actual Cost.

Solution:

Actual Cost = Total Cost Spent
= 60,000

Hence the Actual Cost is 60,000 USD.

***

Practice Question: 5

You have recently joined an ongoing project. The cost of the project is 350,000 USD. However, 150,000 USD has already been spent and 35% of the work is actually completed.

Determine the Actual Cost (AC).

***

Example:

You have a project which is worth 100,000 USD. Six months have passed and your schedule says that by now you should have completed 60% of the work. What is the Planned Value of the project?

Solution:

Planned Value = BAC * % of Planned Completed Work
= 100,000 * 0.60
= 60,000 USD

Hence, the Planned Value is 60,000 USD.

***

Practice Question: 6

You have recently joined an ongoing project. The cost of the project is 350,000 USD. However, 150,000 USD has already been spent and 35% of work is actually completed. However, according to the schedule, 43% of the work should have been completed to date.

Determine the Planned Value (PV).

***

Example:

You have a project which is worth 100,000 USD. Six months have passed and upon closer inspection, you find that only 55% of the work is completed so far. Determine the Earned Value for the project.

Solution:

Earned Value = BAC * % of Actual Work Completed
= 100,000 * 0.55
= 55,000 USD

Hence, the Earned Value is 55,000 USD.

***

Practice Question: 7

You have recently joined an ongoing project. The cost of the project is 350,000 USD. However, 150,000 USD has already been spent and 35% of work is actually completed. However, according to the schedule, 43% of the work should have been completed to date.

Determine the Earned Value (EV).

***

Example:

You have a project to be completed in 12 months and the total cost of the project is 200,000 USD.

Six months have passed and 110,000 USD has been spent, but on closer review, you find that only 40% of the work is completed so far; however, as per the schedule 45% should have been completed.

Calculate the Budget at Completion (BAC), Earned Value (EV), Planned Value (PV), and Actual Cost (AC).

Solution:

The Budget at Completion (BAC) is the total budget allocated to the project, and in the question 200,000 USD is the total cost of the project; therefore, Budget at Completion (BAC) is 200,000 USD.

Therefore, the Budget at Completion (BAC) is 200,000 USD.

From the above question you can clearly see that only 40% of the work is actually completed.

The definition of Earned Value says that it is the value of the project that has been earned.

In this case only 40% of the work has been completed.

Hence,

Earned Value = 40% of value of total work
= 40 % of BAC
= 40% of 200,000
= 0.4 * 200,000
= 80,000 USD

Therefore, the Earned Value (EV) is 80,000 USD.

The definition of Planned Value says that Planned Value is the value of the work that should have been completed so far (as per the schedule).

Therefore, in this case we should have completed 45% of the total work.

Hence,

Planned Value = 45% of value of the total work
= 45% of BAC
= 45% of 200,000
= 0.45 * 200,000
= 90,000

Therefore, the Planned Value (PV) is 90,000 USD.

Finding the Actual Cost (AC) is quite simple.

As per the definition of Actual Cost, it is the amount of money you have spent so far.

And in our question, you have spent 110,000 on the project so far.

Hence, the Actual Cost is 110,000 USD.

***

Practice Question: 8

You have been assigned as project manager for an ongoing project. The cost of this project is 600,000 USD, and the duration is two years.

Upon closer review of the project, you find that 250,000 USD has been spent and 32% of the work is completed; although, according to the schedule 37% should have been completed by now.

Determine the Budget at Completion (BAC), Actual Cost (AC), Planned Value (PV), and Earned Value (EV).

***

Example:

You have a project which is worth 100,000 USD. Four months have passed and 50,000 USD has been spent. Upon closer review, you find that only 40% of the work is completed.

Determine the Cost Variance (CV), Cost Performance Index (CPI), and find out whether you are under or over budget.

Solution:

Cost Variance = Earned Value - Actual Cost

Cost Performance Index = (Earned Value)/(Actual Cost)

Earned Value = 40% of total cost of the project
= 40% of 100,000
= 40,000 USD

Actual Cost = 50,000 USD

Hence, Cost Variance = Earned Value – Actual Cost
= 40,000 – 50,000
= –10,000 USD

Cost Performance Index = (Earned Value)/(Actual Cost)
= 40,000/50,000
= 0.80

Since the Cost Variance is negative and Cost Performance Index is less than one, you're over budget.

***

Practice Question: 9

You have recently joined an ongoing project. The cost of the project is 350,000 USD, 150,000 USD has already been spent and 45% of the work is completed. However, according to the schedule, 43% of the work should have been completed to date.

Determine the Cost Variance (CV) and Cost Performance Index (CPI) and deduce whether you are over budget or under budget.

***

Example:

You have a project which is worth 250,000 USD. Four months have passed and 90,000 USD has been spent. Upon closer review, you find that only 30% of the work is completed; however, as per the schedule, 35% of the work should have been completed.

Determine the Schedule Variance (SV), Schedule Performance Index (SPI), and whether you are ahead or behind schedule.

Solution:

Schedule Variance = Earned Value - Planned Value

Schedule Performance Index = (Earned Value)/(Planned Value)

Planned Value = 35% of total work
= 35% of 250,000
= 0.35 * 250,000
= 87,500 USD

Earned Value = 30% of total work
= 30% of 250,000
= 75,000 USD

Schedule Variance = Earned Value - Planned Value
= 87,500 - 75,000
= -12,500 USD

Hence, the Schedule Variance (SV) is -12,500 USD.

Schedule Performance Index = (Earned Value)/(Planned Value)
= 75,000/87,500
= 0.86

Hence, the Schedule Performance Index (SPI) is 0.86.

Since the Schedule Variance is negative and Schedule Performance Index is less than one, you are behind schedule.

***

Practice Question: 10

You have recently joined an ongoing project. The cost of the project is 350,000 USD. However, 150,000 USD has already been spent and 45% of the work is actually completed. However, according to the schedule, 43% of the work should have been completed to date.

Determine the Schedule Variance (SV) and Schedule Performance Index (SPI) and deduce whether you are ahead of schedule or behind schedule.

***

Example (EAC) Case-I:

You have a project which is worth 500,000 USD. Seven months have passed and 350,000 USD has been spent. Upon closer review, you find that only 60% of the work is completed; however, as per the schedule, 65% of the work should have been completed.

Determine the Estimate at Completion (EAC).

Solution:

Estimate At Completion = (Budget At Completion)/(Cost Performance Index)

EAC = BAC/CPI

And,

Cost Performance Index = (Earned Value)/(Actual Cost)

CPI = EV/AC

Earned Value = 60% of BAC
= 0.6 * 500,000
= 300,000 USD

Actual Cost = 350,000 USD

Hence,

CPI = (Earned Value)/(Actual Cost)
= 300,000/350,000
= 0.86

Hence,

EAC = 500,000/0.86

= 581,395 USD

Hence the Estimate at Completion (EAC) is 581,395 USD.

***

Practice Question: 11

You have recently joined an ongoing project. The cost of the project is 350,000 USD, and 150,000 USD has already been spent and 40% of the work is actually completed. However, according to the schedule, 43% of the work should have been completed to date.

Determine the EAC.

***

Example (EAC) Case-II:

You have a project with a budget of 600,000 USD. During the execution phase, an incident happens which costs you a lot of money. However, you are sure that this will not happen again, and you can continue with your calculated performance for the rest of the project.

To date you have spent 300,000 USD, and the value of the completed work is 275,000 USD.

Calculate the Estimate at Completion (EAC).

Solution:

Since the cost elevation is temporary in nature and the rest of the project can be completed as planned, in this case you will use the formula:

EAC = AC + (BAC – EV)

Given in the question,

Actual Cost (AC) = 300,000 USD

Budget at Completion (BAC) = 600,000 USD

Earned Value (EV) = 275,000 USD

Hence,

EAC = 300,000 + (600,000 – 275,000)
= 300,000 + 325,000
= 625,000 USD

Hence, the Estimate at Completion is 625,000 USD.

***

Practice Question: 12

You have a project with a budget of 400,000 USD. During execution phase, one machine fails which costs you a lot of money. However, you are sure that this will not happen again, and you can continue with your calculated performance for the rest of the project.

To date you have spent 200,000 USD, and the value of the completed work is 180,000 USD.
Calculate the Estimate at Completion (EAC).

***

Example (EAC) Case-III:

You have a fixed deadline project with a budgeted cost of 600,000 USD. So far you have spent 300,000 USD and the value of the completed work is 275,000 USD. However, as per the schedule you should have earned 325,000 USD to date.

Calculate the Estimate at Completion (EAC).

Solution:

Given in the question:

Budget at Completion (BAC) = 600,000 USD
Actual Cost (AC) = 300,000 USD
Earned Value (EV) = 275,000 USD

Planned Value (PV) = 325,000 USD

To calculate the EAC, first you have to calculate the CPI and SPI.

SPI = EV/PV
= 275,000/325,000
= 0.85

CPI = EV/AC
= 275,000/300,000
= 0.92

Now, you can use the formula

EAC = AC + (BAC – EV)/(CPI*SPI)
= 300,000 + (600,000 – 275,000)/(0.85*0.92)
= 300,000 + 325,000/0.78
= 300,000 + 416,000
= 716,000 USD

Hence, the Estimate at Completion is 716,000 USD.

***

Practice Question: 13

You have a project with a budgeted cost of 300,000 USD. So far you have spent 100,000 USD and the value of the completed work is 80,000 USD. However, as per the schedule you should have earned 90,000 USD to date.

The client is insisting to complete the project on time.

Calculate the Estimate at Completion (EAC).

***

Example:

You have a project which is worth 500,000 USD. Seven months have passed and 350,000 USD has been spent. Upon closer review, you find that only 60% of the work is completed; however, as per the schedule, 65% of the work should have been completed.

Determine the Estimate to Complete (ETC).

Solution:

ETC = EAC - AC

Estimate At Completion = (Budget At Completion)/(Cost Performance Index)

EAC = BAC/CPI

And,

Cost Performance Index = (Earned Value)/(Actual Cost)

CPI = EV/AC

Earned Value = 60% of BAC
= 0.6 * 500,000
= 300,000 USD

Actual Cost = 350,000 USD

Hence,

Cost Performance Index = (Earned Value)/(Actual Cost)
= 300,000/350,000
= 0.86

EAC = 500,000/0.86
= 581,395 USD

Estimate to Complete = Estimate at Completion - Actual Cost

ETC = EAC - AC
= 581,395 - 350,000
= 231,395 USD

Hence, the Estimate to Complete (ETC) is 231,395 USD.

***

Practice Question: 14

You have recently joined an ongoing project. The cost of the project is 350,000 USD, and 150,000 USD has already been spent and 40% of the work is actually completed. However, according to the schedule, 43% of the work should have been completed to date.

Determine the Estimate to Complete (ETC).

***

Example:

You have a project which is worth 500,000 USD. Seven months have passed and 350,000 USD has been spent.

Upon closer review, you find that only 60% of the work is completed; however, as per the schedule, 65% of the work should have been completed.

Determine the To Complete Performance Index (TCPI).

Solution:

EAC = BAC/CPI

And,

Cost Performance Index = (Earned Value)/(Actual Cost)

CPI = EV/AC

Earned Value = 60% of BAC
= 0.6 * 500,000
= 300,000 USD

Actual Cost = 350,000 USD

Hence,

Cost Performance Index = (Earned Value)/(Actual Cost)
= 300,000/350,000
= 0.86

Hence, the Cost Performance Index (CPI) is 0.86.

It means the project is over budget.

To calculate TCPI, we have to calculate the Estimate at Completion (EAC)

EAC = BAC/CPI
= 500,000/0.86
= 581,395 USD

Since the CPI is less than one, it means the project is over budget; therefore, we will use the formula:

TCPI = (BAC ¬ EV) / (EAC - AC)
= (500,000 - 300,000)/(581,395 - 350,000)
= 200,000/231,395
= 0.86

Hence, the To Complete Performance Index (TCPI) is 0.86

***

Practice Question: 15

You have recently joined an ongoing project. The cost of the project is 350,000 USD, 150,000 USD has already been spent and 45% of the work is actually completed. However, according to the schedule, 43% should have been completed to date.

Determine the To Complete Performance Index (TCPI).

***

Practice Question: 16

You have a project to be completed in 12 months and the total cost of the project is 100,000 USD. Six months have passed and 60,000 has been spent, but upon closer review, you find that only 40% of the work is completed so far; however, as per the schedule 50% should have been completed.

Calculate the following:

(a) Schedule Variance (SV)
(b) Cost Variance (CV)
(c) Schedule Performance Index (SPI)
(d) Cost Performance Index (CPI)
(e) Variance at Completion (VAC)
(f) Estimate at Completion (EAC), ignoring schedule constraint
(g) Estimate to Complete (ETC)
(h) To Complete Performance Index (TCPI)

Miscelleneous Questions

Practice Question:17

The John is a project manager of a project which is a construction of a university building. The BAC of this project is 5,000,000 USD. The duration of this project is four years with equal work planned to be completed each year. Two years have passed and 3,000,000 USD has been spent, though the completed work is only 40%. What is the variance of the project schedule?

Practice Question:18

You have been appointed as project manager of an ongoing project which is running late by 10%.The BAC of this project is 200,000 USD and you have spent 60,000 USD. If the completed work is 25%, what will the TCPI be?

Practice Question:19

The budget of your project is 250,000 USD, you have spent 100,000 and 30% is complete, though as per the schedule 35% should have been completed. The management is concerned and wants to know if you can complete the project within the budget or if a new budget is required. Is a new budget required, and if so what would it be?

Practice Question:20

Your project is 70 percent complete and you have spent 400,000 USD. According to the schedule, you should have completed 80%, but because of some unwanted issues the project is late. The BAC of the project is 500,000 USD. Considering the current cost performance is the same as in the future, what will the project budget be at the end of the project?

Practice Question:21

The cost of your project is 200,000 USD and you have already spent 40% of it. The CPI is 0.70 and you believe your future cost performance will be the same as the current cost performance. Your customer wants to know how much more money is needed to complete the project. What value will you provide him with?

Solutions

Practice Question: 1

You have been assigned as a project manager for a new project and are asked to calculate the budget for it. You have calculated the cost of all work packages as 200,000 USD. The estimated contingency reserve is 40,000 and the management reserve is 6% of the cost baseline.

Calculate the project budget.

Solution:

We know that

Project Budget = Project Cost + Contingency Reserve + Management Reserve

Management Reserve = 6% of cost baseline
= 6% of (200,000 + 40,000)
= 6% of 240,000
= 14,400 USD

Now the project budget = 200,000 + 40,000 + 14,400
= 254,400 USD.

***

Practice Question: 2

The cost of renovation work of an old home is 100,000 USD. The calculated contingency reserve is 10,000 USD and the management reserve is 4% of the cost baseline.

What is the rough order magnitude for this project?

Solution:

Project Budget = Project Cost + Contingency Reserve + Management Reserve
= 100,000 + 10,000 + 4% of (100,000 + 10,000)
= 100,000 + 10,000 + 4,400
= 114,400 USD

We know that rough order of magnitude is -25% to 75% of the estimated cost.

So, the rough order estimate will be 85,800 USD to 200,200 USD.

***

Practice Question: 3

The estimated project budget is 250,000 USD, the contingency reserve is 25,000 USD and the management reserve is 10,000 USD.

What is the definitive estimate?

Solution:

The definitive estimate is -5% to 10% of the estimated budget.

The estimated project budget is 250,000 USD.

So the definitive estimate would be 237,500 USD to 275,000 USD.

In this question, you have already been given the project budget. The contingency and management reserve are given to confuse you.

***

Practice Question: 4

You have recently joined an ongoing project. The cost of the project is 350,000 USD. However, 150,000 USD has already been spent and 35% of the work is actually completed.

Determine the Budget at Completion (BAC).

Solution:

The Budget at Completion is the total budget allotted to the project. In the question, the cost of the project is 350,000 USD.

Hence, the BAC is 350,000 USD.

***

Practice Question: 5

You have recently joined an ongoing project. The cost of the project is 350,000 USD. However, 150,000 USD is has already been spent and 35% of the work is actually completed.

Determine the Actual Cost (AC).

Solution:

The Actual Cost is the amount of money actually spent on the project.

In the given question, you have already spent 150,000 USD.

Hence, the Actual Cost (AC) is 150,000 USD.

***

Practice Question: 6

You have recently joined an ongoing project. The cost of the project is 350,000 USD. However, 150,000 USD has already been spent and 35% of work is actually completed. However, according to the schedule, 43% of the work should have been completed to date.

Determine the Planned Value (PV).

Solution:

Planned Value (PV) is the amount of work completed as per the schedule.
As per the given question,

Planned Value (PV) = 43% of BAC
= 0.43 * 350,000
= 150,500 USD

Hence, the Planned Value is 150,500 USD.

***

Practice Question: 7

You have recently joined an ongoing project. The cost of the project is 350,000 USD.

However, 150,000 USD has already been spent and 35% of work is actually completed. However, according to the schedule, 43% of the work should have been completed to date.

Determine the Earned Value (EV).

Solution:

Earned Value is the monetary value of the work completed.

As per the given question, 35% of the actual work has been completed.

Earned Value = 35% of BAC
= 0.35 * 350,000
= 122,500 USD

Hence, the Earned Value (EV) is 122,500 USD.

***

Practice Question: 8

You have been assigned as project manager for an ongoing project. The cost of this project is 600,000 USD, and the duration is two years.

Upon closer review of the project, you find that 250,000 USD has been spent and 32% of the work is completed; although, according to the schedule 37% should have been completed by now.

Determine the Budget at Completion (BAC), Actual Cost (AC), Planned Value (PV), and Earned Value (EV).

Solution:

The Budget at Completion is the cost of the budget, which is 600,000 USD.

Hence, the Budget at Completion is 600,000 USD.

The Actual Cost is the money that has been spent on the project.

In the given case, 250,000 USD has been spent to date.

Hence, the Actual Cost is 250,000 USD.

The Planned Value is the Scheduled Cost of the project that should have been completed to date.
In the given case, it is 37% of the cost of the project.

Planned Value = 37% of cost of the project
= 37% of 600,000
= 0.37 * 600,000
= 222,000 USD

Hence, the Planned Value is 222,000 USD.

The Earned Value is the amount of the project that has been completed.

In the given question, 32% of the work has been completed.

Earned Value = 32% of the project
= 32% of 600,000
= 0.32 * 600,000
= 192,000 USD

Hence, the Earned Value is 192,000 USD.

***

Practice Question: 9

You have recently joined an ongoing project. The cost of the project is 350,000 USD, 150,000 USD has already been spent and 45% of the work is actually completed. However, according to the schedule, 43% of the work should have been completed to date.

Determine the Cost Variance (CV) and Cost Performance Index (CPI) and deduce whether you are over budget or under budget.

Solution:

Cost Variance = Earned Value - Actual Cost

In the given question,

Earned Value = 0.45 * X 350,000
= 157,500 USD

And,

Actual Cost = 150,000 USD

Hence,

CV = EV - AC
= 157,000 - 150,000
= 7,500 USD

Hence, the Cost Variance (CV) is 7,500 USD.

CPI = EV/AC
= 157,500/150,000
= 1.05

Hence, the Cost Performance Index (CPI) is 1.05.

Since the Cost Variance is positive and Cost Performance Index is greater than one, you are under budget.

***

Practice Question: 10

You have recently joined an ongoing project. The cost of the project is 350,000 USD. However, 150,000 USD has already been spent and 45% of the work is actually completed. However, according to the schedule, 43% of the work should have been completed to date.

Determine the Schedule Variance (SV) and Schedule Performance Index (SPI) and deduce whether you are ahead of schedule or behind schedule.

Solution:

Schedule Variance = Earned Value - Planned Value

SV = EV - PV

In the given question,

Earned Value = 45% of the work
= 0.45 * 350,000
= 157,500 USD

Planned Value = 43% of the work
= 0.43 * 350,000
= 150,500 USD

SV = EV - PV
= 157,500 – 150,500
= 7,000 USD

Hence, the Schedule Variance (SV) is 7,000 USD.

Schedule Performance Index = (Earned Value)/(Planned Value)

SPI = EV/PV
= 157,500/150,500
= 1.05

Hence, the Schedule Performance Index (SPI) is 1.05

Since the SV is positive and SPI is greater than one, you're ahead of schedule.

***

Practice Question: 11

You have recently joined an ongoing project. The cost of the project is 350,000 USD. 150,000 USD has already been spent and 40% of the work is actually completed. However, according to the schedule, 43% of the work should have been completed to date.

Determine the EAC.

Solution:

We know that,

EAC = BAC/CPI

And, CPI = EV/AC

According to the question,

Earned Value = 40% of the work
= 0.40 * 350,000
= 140,000

AC = Actual Cost spent
= 150,000 USD

CPI = EV/AC
= 140,000/150,000
= 0.933

EAC = BAC/CPI

BAC = 350,000 USD

EAC = 350,000/0.933
= 375,133.976 USD

Hence, the Estimate at Completion (EAC) = 375,133.976 USD.

***

Practice Question: 12

You have a project with a budget of 400,000 USD. During the execution phase, one machine fails which costs you a lot of money. However, you are sure that this will not happen again, and you can continue with your calculated performance for the rest of the project.

To date, you have spent 200,000 USD, and the value of the completed work is 180,000 USD.

Calculate the Estimate at Completion (EAC).

Solution:

Since you know that this cost elevation is temporary in nature and the rest of the project can be completed as planned, in this case you will use the formula:

EAC = AC + (BAC – EV)

Given in the question,

Actual Cost (AC) = 200,000 USD

Budget at Completion (BAC) = 400,000 USD

Earned Value (EV) = 180,000 USD

Hence,

EAC = 200,000 + (400,000 – 180,000)
= 200,000 + 220,000
= 420,000 USD

Hence, the Estimate at Completion is 420,000 USD.

***

Practice Question: 13

You have a project with a budgeted cost of 300,000 USD. So far you have spent 100,000 USD, and the value of the completed work is 80,000 USD. However, as per the schedule you should have earned 90,000 USD to date.

The client is insisting to complete the project on time.

Calculate the Estimate at Completion (EAC).

Solution:

Given in the question:

Budget at Completion (BAC) = 300,000 USD
Actual Cost (AC) = 100,000 USD
Earned Value (EV) = 80,000 USD
Planned Value (PV) = 90,000 USD

To calculate the EAC, first you have to calculate the CPI and SPI.

SPI = EV/PV
= 80,000/90,000
= 0.89

CPI = EV/AC
= 80,000/100,000
= 0.80

Now, you can use the formula

EAC = AC + (BAC – EV)/(CPI*SPI)
= 100,000 + (300,000 - 80,000)/(0.80*0.89)
= 100,000 + 220,000/0.71
= 100,000 + 310,000
= 410,000 USD

Hence, the Estimate at Completion is 410,000 USD.

***

Practice Question: 14

You have recently joined an ongoing project. The cost of the project is 350,000 USD, and 150,000 USD has already been spent and 40% of the work is actually completed. However, according to the schedule, 43% of the work should have been completed to date.

Determine the Estimate to Complete (ETC).

Solution:

We know that,

EAC = BAC/CPI

And, CPI = EV/AC

According to the question,

Earned Value = 40% of the work
= 0.40 * 350,000
= 140,000 USD

AC = Actual Cost spent
= 150,000 USD

CPI = EV/AC
= 140,000/150,000
= 0.933

EAC = BAC/CPI

BAC = 350,000 USD

EAC = 350,000/0.933
= 375,133.976

ETC = EAC - AC
= 375,133.976 - 150,000
= 225,133.976 USD

Hence, the Estimate to Complete is 225,133 USD.

***

Practice Question: 15

You have recently joined an ongoing project. The cost of the project is 350,000 USD. 150,000 USD is already spent and 45% of the work is actually completed. However, according to the schedule, 43% should have been completed to date.

Determine the To Complete Performance Index (TCPI).

Solution:

We know that,

CPI = EV/AC

EV = 45% of the work
= 0.45 * 350,000
= 157,500

AC = 150,000 USD

CPI = 157,500/150,000
= 1.05

Hence, the Cost Performance Index is 1.05.

Since the CPI is greater than one, it means the project is under budget; therefore, we will use the formula:

TCPI = (BAC ¬ EV) / (BAC - AC)
= (350,000 – 157,500)/(350,000 – 150,000)
= 195,500/200,000
= 0.96

Hence, the To Complete Performance Index is 0.96.

***

Practice Question: 16

You have a project to be completed in 12 months and the total cost of the project is 100,000 USD. Six months have passed and 60,000 has been spent, but upon closer review you find that only 40% of the work is completed so far; however, as per the schedule 50% should have been completed.

Calculate the following:

(a) Schedule Variance (SV)
(b) Cost Variance (CV)
(c) Schedule Performance Index (SPI)
(d) Cost Performance Index (CPI)
(e) Variance at Completion (VAC)
(f) Estimate at Completion (EAC), ignoring schedule constraint
(g) Estimate to Complete (ETC)
(h) To Complete Performance Index (TCPI)

Solution:

Schedule Variance = Earned Value - Planned Value
SV = EV - PV
SV = 40,000 – 50,000
= - 10,000 USD

Hence, the Schedule Variance is –10,000 USD.

Cost Variance = Earned Value – Actual Cost

CV = EV - AC
CV = 40,000 – 60,000
= -20,000 USD

Hence, the Cost Variance is -20,000 USD.

Schedule Performance Index = (Earned Value)/(Planned Value)

SPI = EV/PV
= 40,000/50,000
= 0.8

Hence, the Schedule Performance Index is 0.8.

Cost Performance Index = (Earned Value)/(Actual Cost)
CPI = EV/AC
= 40,000/60,000
= 0.67

Hence, the Cost Performance Index is 0.67.

Since there are no schedule constraints,

Estimate at Completion = (Budget at Completion)/(Cost performance Index)

EAC = BAC/CPI
= 100,000/0.67
= 149,253 USD

Hence, the Estimate at Completion is 149,253 USD.

Variance at Completion = Budget at Completion - Estimate at Completion

VAC = BAC - EAC
= 100,000 – 149,253
= - 49,253 USD

Hence, the Variance at Completion is –49,253 USD.

Estimate to Complete = EAC –AC
= 149,253 – 60,000
= 89,253 USD

Hence, the Estimate to Complete is 89,253 USD

To Complete Performance Index = (Budget at Completion - Earned Value)/ (Budget at Completion – Actual Cost)

TCPI = (BAC - EV) / (EAC - AC)
= (100,000 - 40,000)/(149,253 - 60,000)
= 60,000/89,253
= 0.672

Hence, the To Complete Performance Index is 0.672

***

Practice Question:17

The John is a project manager of a project which is a construction of a university building. The BAC of this project is 5,000,000 USD. The duration of this project is four years with equal work planned to be completed each year. Two years have passed and 3,000,000 USD has been spent, though the completed work is only 40%. What is the variance of the project schedule?

Solution:

Given data:

BAC = 5,000,000 USD
AC = 3,000,000 USD
EV = 40% of BAC
= 0.40*5,000,000
= 2,000,000 USD

As the project is to have an equal amount completed in each year, the duration of the project is four years; in two years 50% of the work should be completed.

PV = 50% of 5,000,000
= 2,500,000

We know that,

Schedule Variance = Earned Value – Planned Value

SV = EV – PV
= 2,000,000 - 2,500,000
= -500,000 USD

The schedule variance is –500,000 USD.

***

Practice Question:18

You have been appointed as project manager of an ongoing project which is running late by 10%.The BAC of this project is 200,000 USD and you have spent 60,000 USD. If the completed work is 25%, what will the TCPI be?

To calculate the TCPI we will first calculate the EAC, then using the TCPI formula based on EAC, we will calculate the TCPI.

Solution:

Given data:

The project is running late by 10%, which means the cost performance index is 90%
CPI = 0.9

BAC = 200,000 USD
AC = 60,000 USD
EV = 25% of BAC
=0.25*200,000
= 50,000 USD

To calculate the EAC we will assume that the future cost performance will be the same as the current cost performance.

EAC = BAC/CPI
= 200,000/0.9
= 222,222 USD

We know that,

TCPI = (BAC - EV)/(EAC - AC)

= (200,000 - 50,000)/(222,222 - 60,000)
= 150,000/162,222
= 0.93

The TCPI is 0.93

***

Practice Question:19

The budget of your project is 250,000 USD, you have spent 100,000 and 30% is complete.
Though as per the schedule 35% should have been completed. The management is concerned and wants to know if you can complete the project within the budget or if a new budget is required. Is a new budget required, and if so what would it be?

First of all we will calculate the CPI to see if the project is under budget or over budget. If it is over budget, we will then find out the EAC.

Solution:

Given data:

BAC = 250,000 USD
AC = 100,000 USD
EV = 30% of 250,000 USD
= 75,000 USD
PV = 35% of 250,000
= 87,500 USD

Cost Performance Index = (Earned Value)/(Actual Cost)

CPI = EV/AC
= 75,000/87,500
=0.86

Therefore, you cannot complete the project within the given budget, so you will need a new cost estimate, i.e.:

EAC = BAC/CPI
= 250,000/0.86
= 290,697 USD

Hence the new budget will be approximately 290,697 USD.

***

Practice Question:20

Your project is 70 percent complete and you have spent 400,000 USD. According to the schedule, you should have completed 80%, but because of some unwanted issues the project is late. The BAC of the project is 500,000 USD. Considering the current cost performance is the same as in the future, what will the project budget be at the end of the project?

The question is asking you to find the EAC. First of all we will calculate the CPI to see if the project is under budget or over budget. If it is over budget, we will then calculate the EAC.

Solution:

Given data:

BAC = 500,000 USD
EV = 70% of 500,000 = 350,000 USD
PV = 80% of 500,000 USD = 400,000 USD

CPI = EV/AC
= 350,000/400,000
= 0.875

EAC = 500,000/0.875
= 571,428

The project budget at the end will be approximately 570,000 USD.

***

Practice Question:21

The cost of your project is 200,000 USD and you have already spent 40% of it. The CPI is 0.70 and you believe your future cost performance will be the same as the current cost performance. Your customer wants to know how much more money is needed to complete the project. What value will you provide him with?

Here the customer wants to know the ETC.

Solution:

Given data:

CPI = 0.70
BAC = 200,000 USD

EAC = BAC/CPI
= 200,000/0.7
= 285,000 USD

Now we can calculate the ETC.

ETC = 285,000 - 80,000
= 205,000 USD

Therefore, you will need 205,000 USD to complete the remaining part of the project.

Communications Management

In this knowledge area there is only one formula. Questions on this knowledge are relatively easier than other knowledge areas.

The formula used in this knowledge area is:

Communication Channels = $n*(n-1)/2$

Where n is the number of stakeholder.

Example: 1

You are managing a project with seven team members. How many lines of communication do you have on your project?

Solution:

Number of people = 7+1 (including the project manager)
= 8

Number of Communication Channels = $n*(n-1)/2$
$= 8*(8-1)/2$
$= (8*7)/2$
$= 56/2$
$= 28$

Hence, the project has 28 lines of communication.

***

Practice Question: 1

Your project team consists of 19 members excluding you. Calculate the lines of communication in your project.

***

Example: 2

Your team consists of 10 team members including you. During the project execution, you recruit two new personnel.

How many lines of communications have been added?

Solution:

Number of Communication Channels = n*(n–1)/2

Communication channels before adding new members =
10*(10–1)/2
= 10*9/2
= 45

Communication channels after adding new members =
12*(12–1)/2
= 12*11/2
= 66

Lines added = 66 – 45
= 21

Therefore, 21 new communication lines have been added.

***

Practice Question: 2

You have 9 members in your team including you. Three members left the team and you are only able to recruit two new members matching the same skills.

Determine the number of communication lines, and deduce whether it has increased or decreased.

Miscelleneous Questions

Practice Question: 3

The John John's project has a total of 15 stakeholders including him, and at the end of the project 4 people have left the project. What is the number communication channels John has now?

Practice Question: 4

In your project you have 25 stakeholders excluding the project manager. How many communication channels are there?

Practice Question: 5

John's project has a total of 15 stakeholders including him, and at the end of the project 4 people have left the project. What is the number communication channels John has now?

Practice Question: 6

In your project you have 25 stakeholders excluding the project manager. How many communication channels are there?

Solutions

Practice Question: 1

Your project team consists of 19 members excluding you. Calculate the lines of communication in your project.

Solution:

Number of Communication Channels = n*(n–1)/2
Total number of people = 19 + 1
= 20

Hence,

Numbers of Communication Channels = 20 (20 – 1)/2
= 20 X 19/2
= 190

Hence, the project has 190 lines of communication.

***

Practice Question: 2

You have 9 members in your team including you. Three members left the team and you are only able to recruit two new members matching the same skills.

Determine the number of communication lines, and deduce whether it has increased or decreased.

Solution:

Total number of Communication Channels before the members left = $n*(n-1)/2$
$= 9 * (9-1)/2$
$= 9 * 8/2$
$= 36$

Current number of members = $9 - 3 + 2$
$= 8$

Current Communication Channels = $8 * (8-1)/2$
$= 8 * 7/2$
$= 28$

Hence, the communication lines have decreased by 8.

***

Practice Question: 3

The John John's project has a total of 15 stakeholders including him, and at the end of the project 4 people have left the project. What is the number communication channels John has now?

Solution:

The question is asking about John's communication channels, not projects. The total number of staff is 10, so John will have 10 communication channels.

***

Practice Question: 4

In your project you have 25 stakeholders excluding the project manager. How many communication channels are there?

Solution:

In your project you have 25 stakeholders excluding the project manager. How many communication channels are there?

***

Practice Question: 5

John's project has a total of 15 stakeholders including him, and at the end of the project 4 people have left the project. What is the number communication channels John has now?

Solution:

The question is asking about John's communication channels, not projects. The total number of staff is 10, so John will have 10 communication channels.

***

Practice Question: 6

In your project you have 25 stakeholders excluding the project manager. How many communication channels are there?

Solution:

Total numbers of stakeholders (N) = 25 + 1 = 26

Number of communication channels = $N*(N - 1)/2$
$= 26*25/2$
$= 325$

Therefore, the total number of communication channels is 325

Risk Management

A risk is an unplanned event that, if it occurs, has a positive or negative effect on one or more project objectives such as scope, schedule, cost, quality, etc.

In risk management, you take measures to manage these surprises. Here your objective will be to minimize the impact of negative risks and maximize the impact of positive risks.

Risk management is very important for your project. It helps you manage risks proactively and complete the project successfully.

In the risk management knowledge area, you have some mathematical formulas. The calculations involved here are fairly simple.

In the exam, you will see some mathematical questions from the risk management knowledge area. The questions may belong to the following topics:

- Probability
- Expected Monetary Value
- Decision Tree Method

Though these concepts may look a little difficult, the calculations involved are basic in nature.

Probability

Probability can be defined as the chance of a particular event happening.

For example, if you toss a coin, the probability of showing heads is 0.5 or 50%.

Probability is a mathematical term, and it always lies between 0 and 1.

If the probability of any event is 0, this means that this event will never occur; if it is 1, the event is certain.

Mathematically,

Probability = (Nos of Favorable Event)/(Total Nos of Event)

Also,

(Probability of an Event Happening) + (Probability of the Event Not Happening) = 1

Example:

If you throw a die, what is the probability of a 5 showing up?

Solution:

If you throw the die then you may get any of following results: 1, 2, 3, 4, 5, 6.

So the total number of events = 6
But you want the die to come up 5.
So the number of favorable events = 1

We know the formula,

Probability = (Nos of Favorable Event)/(Total Nos of Event)

So,

Probability of 5 = 1/6

***

Example:

If you throw a die, what is the probability of a 5 not showing up?

Solution:

We have already calculated the probability of 5 = 1/6

We know that:

Probability of an Event Happening + Probability of an Event Not Happening = 1

Therefore,

Probability of 5 not showing= 1 – Probability of 5 showing
= 1 - 1/6
= 5/6

***

Practice Question: 1

After running the Monte Carlo simulation, you come to the conclusion that you have an 85% chance of completing the project within the approved budget.

What is the probability of not completing the project within the approved budget?

Solution:

The probability of completing the project within the approved budget = 0.85

We know that,

Probability of an Event Happening + Probability of an Event Not Happening = 1

Therefore,

The probability of not completing the project within the approved budget = 1 – 0.85
= 0.15

***

Practice Question: 2

In a basket you have three blue and six white balls. You remove one ball from the basket randomly. What is the probability this ball will be blue?

Solution:

Since you have three blue balls, the chance of a favorable result = 3

Total number of balls is 9, therefore, the total number of events = 9

So the probability of picking a blue ball = 3/9
= 1/3

***

Expected Monetary Value

Expected monetary value is a statistical technique in risk management that is used to quantify the risks, which in turn assists the project manager in calculating the contingency reserve.

It is denoted by EMV.

This technique is also used in decision tree method where you decide whether to go with a particular decision or not.

Expected monetary value can be positive or negative.

If the risk is positive, it will have a positive impact on the project, and therefore, the expected monetary value will be positive.

Positive EMV is denoted by a + sign.

However, if the risk is negative, it will have a negative impact on the project, and therefore, the expected monetary value will be negative.

Negative EMV is denoted by a - sign.

The formula to calculate EMV for a single event is as follows:

EMV = Probability * Impact
= P*I

Where P = Probability

I = Impact

For multiple risks, you will calculate the EMV of each risk and then add them all.

Example: 1

For your project, there is a 40% chance that a team member may leave the project, causing a 10,000 USD loss. What is the Expected Monetary Value of this risk event?

Solution:

Expected Monetary Value (EMV) = Probability * Impact
= 0.4 * 10000
= 4,000 USD

Hence, the Expected Monetary Value of this risk event is 4,000 USD.

***

Example: 2

You have a 100,000 USD project in your hands. There is 30% chance you may lose an employee, which may cost you 10,000 USD. Moreover, there is a 40% chance that if you buy certain materials in bulk, you may get a 25% discount, saving you 5,000 USD.

What is the Expected Monetary Value of this project?

Solution:

Expected Monetary Value of the events = - 0.3 * 10,000 + 0.4 * 5000
= -3,000 + 2,000
= - 1,000 USD

The expected monetary value of these two events is -1,000 USD, therefore to get the expected monetary value of the project you will add this value to the project cost.

Expected Monetary Value of the project = 100,000 + 1,000
= 101,000 USD

Hence the Expected Monetary Value of the project is 101,000 USD.

***

Practice Question: 3

Your team member suggests you buy new equipment. With this equipment, there is a 30% chance to save 20,000 USD. However, there is a 10% chance that due to lack of experience, efficiency may decrease, which may cost you 10,000 USD.

Find the Expected Monetary Value (EMV) of these two events.

***

Decision Tree Analysis

You use decision tree analysis when you have multiple choices and you want to select the best option.

In this technique, no special formula is required. You draw the diagram, one branch for each option, and using the expected monetary value formula you calculate the EMV of each choice and make the decision based on this calculation.

Example:

Based on the below decision tree, find the EMV of choice A, and determine that which choice is better than the rest.

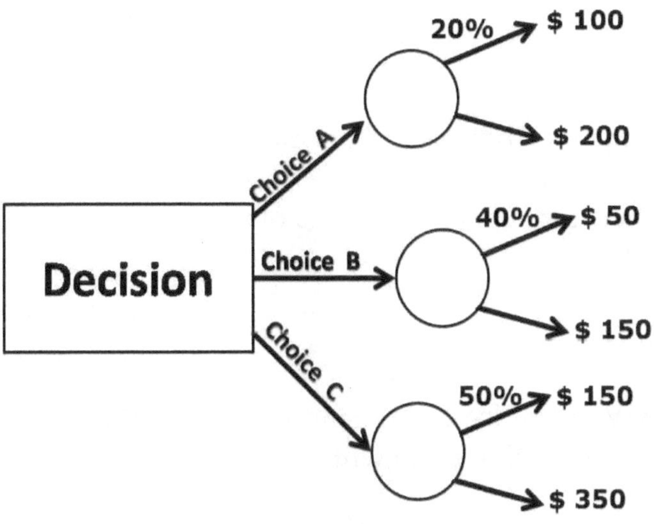

Solution:

For Choice A, there is a 20% chance of spending 100, and 80% (*100 – 20*) chance of spending 200.

Hence, the EMV of Choice A = 20% of 100 + 80% of 200.
= 0.2*100 +0.8*200
= 20 + 160
= 180 USD

Hence, the EMV of Choice A is 180 USD.

Now to decide on the best option, we will calculate the EMV of other choices and select the option with the highest EMV because all risks are opportunities.

Since we have calculated the EMV of A, we will go for B, and C.

EMV of Choice B = 40% of 50 + 60% of 150
= 0.4*50 + 0.6*150
= 20 + 90
= 110 USD

Hence, the EMV of Choice B is 110 USD

EMV of Choice C = 50% of 150 + 50% of 350
= 0.5*150 +0.5*350
= 75 + 175
= 250 USD

Since the EMV of Choice C is the highest in all choices, Choice C will be the best option to go with.

***

Practice Question: 4

Based on the below decision tree, find the EMV of choice A, B, C, and determine which choice is better than the rest.

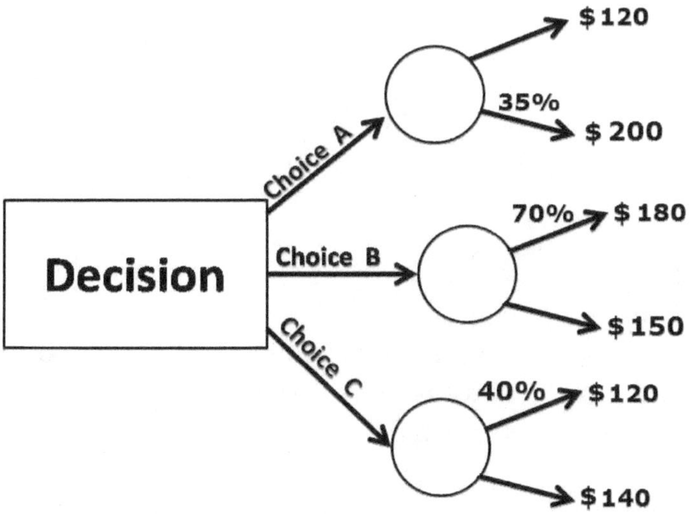

***

Example:

Based on the decision tree given below, deduce which choice is better and its expected monetary value.

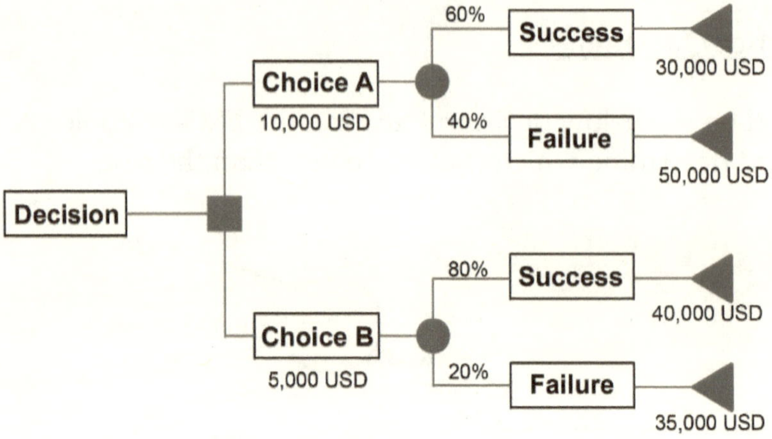

Solution:

To solve this question, you will calculate the EMV of both choices, find which choice is more economically beneficial to the project, and then make a decision based on these findings.

EMV of choice A = 10,000 + 0.6*30,000 – 0.4*50,000
= 10,000 + 18,000 - 20,000
= 8,000 USD

EMV of choice B = 5,000 + 0.8*40,000 – 0.2*35,000
= 5,000 + 32,000 - 7,000
= 30,000 USD

Both choices are representing opportunities, therefore, you will go with the choice which gives you the highest opportunity.

In the given question, choice B is giving you an opportunity of 32,000 USD which is the highest, so you will go with choice B.

***

Practice Question: 5

Based on the decision tree given below, deduce which choice is better and its expected monetary value.

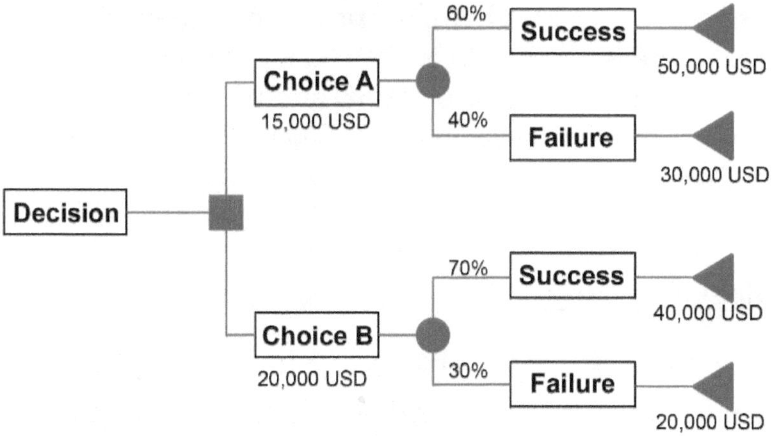

Miscelleneous Questions

Practice Question: 6

You are managing a project to build a part of a big refinery. During the project execution, a team member comes and informs you about a high impact risk. The team member told you that there is a risk with a 50% chance of occurring, and if it occurs it will cost you 50,000 USD and because of this risk your company may lose another 25,000 USD. What is the EMV of this event?

Practice Question: 7

The BAC of your project is 150,000 USD, it is 45% completed, the value of the completed work is 48% and you are about to reach your second milestone. A team member identifies a risk event with a 60% chance of happening and if it occurs, it will cost you 8,000 USD. What is the risk exposure of the event?

Practice Question: 8

In your project, you have one opportunity, which has a 55% chance of occurring, and if it occurs, it will benefit you by 35,000 USD. What is the EMV of this event?

Practice Question: 9

In your project, you have one threat, which has a 35% chance of occurring, and if it occurs it will cost you 2,500 USD. What is the EMV of this event?

Practice Question: 10

A risk has a 40% chance of happening in the sixth month of the project. The duration of the project is one year and three months have passed. What is the chance of the event occurring now?

Practice Question: 11

You have decided to use linear values of 0.1, 0.5 and 0.9 to analyze the probability and impact of risks. These values are assigned as low, moderate and high, respectively. You have identified a risk with a moderate impact but high probability, what will the risk rating of this risk be?

Solutions

Practice Question: 1

After running the Monte Carlo simulation, you come to the conclusion that you have an 85% chance of completing the project within the approved budget.

What is the probability of not completing the project within the approved budget?

Solution:

The probability of completing the project within the approved budget = 0.85

We know that:

Probability of an Event Happening + Probability of an Event Not Happening = 1

Therefore,

The probability of not completing the project within the approved budget = 1 - 0.85
= 0.15

***

Practice Question: 2

In a basket you have three blue and six white balls. You remove one ball from the basket randomly.

What is the probability this ball will be blue?

Solution:

Since you have three blue balls, the chance of a favorable result = 3

Total number of balls is 9, therefore, the total number of events = 9

So the probability of picking a blue ball = 3/9
= 1/3

***

Practice Question: 3

You team member suggests you buy new equipment. With this equipment, there is a 30% chance to save 20,000 USD. However, there is a 10% chance that due to lack of experience, efficiency may decrease, which may cost you 10,000 USD.

Find the Expected Monetary Value (EMV) of these two events.

Solution:

EMV = 0.3 * 20,000 – 0.1 * 10,000
= 6,000 – 1,000
= 5,000 USD

Hence, the EMV for these events is 5,000 USD.

***

Practice Question: 4

Based on the below decision tree, find the EMV of choice A, B, C, and determine which choice is better than the rest.

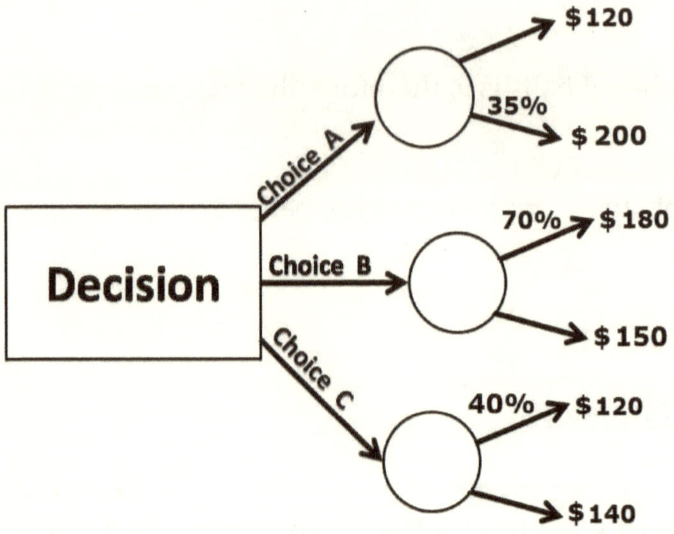

Solution:

EMV of Choice A = 0.65 * 120 + 0.35 * 200
= 78 + 70
= 148 USD

EMV of Choice B = 0.7 * 180 + 0.3 * 150
= 126 + 45
= 171 USD

EMV of Choice C = 0.4 * 120 + 0.6 * 140
= 48 + 84
= 132 USD

Since the EMV of Choice B is the highest, Choice B will be the best choice.

***

Practice Question: 5

Based on the decision tree given below, deduce which choice is better and its expected monetary value.

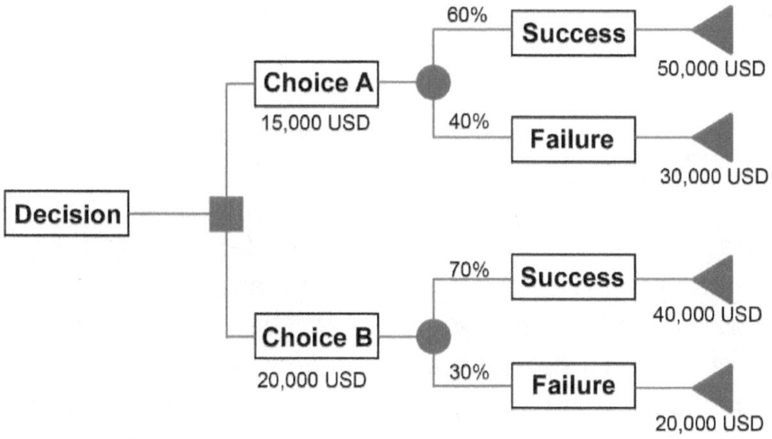

Solution:

As you can see, this diagram has two choices and you have to select one of them.

Therefore first of all you will find the EMV of both choices and then make the best decision.

EMV of choice A = 15,000 + 0.6*50,000 – 0.4*30,000
= 15,000 + 30,000 – 12,000
= 33,000 USD

EMV of choice B = 20,000 + 0.7*40,000 – 0.3*20,000
= 20,000 + 28,000 – 6,000
= 42,000 USD

Both choices are representing opportunities; therefore, you will go with the choice which gives you the highest.

In the given question, choice A is offering you 33,000 and choice B is offering you 42,000; therefore, you will go with choice B.

***

Practice Question: 6

You are managing a project to build a part of a big refinery. During the project execution, a team member comes and informs you about a high impact risk. The team member told you that there is a risk with a 50% chance of occurring, and if it occurs it will cost you 50,000 USD and because of this risk your company may lose another 25,000 USD. What is the EMV of this event?

Solution:

EMV of the event = probability * total impact
= 0.5 *(-50,000 - 25,000)
= 0.5 x -75,000
= -37,000 USD

The expected monetary value of the risk event is -37,500 USD.

***

Practice Question: 7

The BAC of your project is 150,000 USD, it is 45% completed, the value of the completed work is 48% and you are about to reach your second milestone. A team member identifies a risk event with a 60% chance of happening and if it occurs, it will cost you 8,000 USD. What is the risk exposure of the event?

Solution:

Risk exposure = probability of event * Impact
= 0.6 * 8,000
= 4,800

The risk exposure of the event is 4,800 USD.

***

Practice Question: 8

In your project, you have one opportunity, which has a 55% chance of occurring, and if it occurs, it will benefit you by 35,000 USD. What is the EMV of this event?

Solution:

EMV of the event = probability * impact
= 0.55 * 35,000
= 19,250 USD

The EMV of the risk event is 19,250 USD

***

Practice Question: 9

In your project, you have one threat, which has a 35% chance of occurring, and if it occurs it will cost you 2,500 USD. What is the EMV of this event?

Solution:

EMV of the event = probability * impact
= 0.35 * (-2,500)
= - 875 USD

***

Practice Question: 10

A risk has a 40% chance of happening in the sixth month of the project. The duration of the project is one year and three months have passed. What is the chance of the event occurring now?

Solution:

40%. The probability will remain the same until the risk occurs or the event time has passed.

***

Practice Question: 11

You have decided to use linear values of 0.1, 0.5 and 0.9 to analyze the probability and impact of risks. These values are assigned as low, moderate and high, respectively. You have identified a risk with a moderate impact but high probability, what will the risk rating of this risk be?

Solution:

Risk rating = probability rating x impact rating
= 0.9 * 0.5
= 0.45

The risk rating of the risk is 0.45

Procurement Management

Procurement management helps you plan your activities such as what to get done from outside, how to identify a seller, and what kind of contract should be used.

Procurement management involves lot of activities and some of them involve mathematical calculations as well.

As a project manager, it is your responsibility to understand these processes well.

In this chapter I will discuss the procurement concepts that require mathematical calculations. I hope after completing this chapter you will be able to solve mathematical questions on the procurement management knowledge area.

Types of Procurement Contracts

You can divide the procurement contract into three categories:

1) Fixed Price Contract
2) Cost Reimbursable Contract
3) Time and Material

Fixed Price Contract

In a fixed price contract, once the price of the contract is decided and the contract is signed, it cannot be changed. Now the seller is legally bound to complete the task within the agreed amount.

If the cost escalates, it will be on the seller's account.

Fixed price contracts are considered to be risky for sellers.

Fixed price contracts can further be divided into three types:

1) Firm Fixed Price (FFP)
2) Fixed Price Incentive Fee (FPIF)
3) Fixed Price with Economic Price Adjustment (FP-EPA)

Firm Fixed Price Contract (FFP)

In this case, once the contract is signed for a definite fee, the seller has to complete the job for this agreed amount of money. Any cost overrun will be borne by the seller.

This is the most risky contract for the seller and the least risky contract for the buyer.

Fixed Price Incentive Fee (FPIF)

Here the contract price is fixed; however, the seller will be given an additional incentive fee based on their performance.

Although this is a type of fixed price contract, the final price of the contract is not fixed. It depends on the performance of the seller and agreed terms.

The terms used in this type of contract:

Target Cost: This is the anticipated cost of the contract.

Target Profit: This is the agreed profit at the target cost.

Target Price: This is equal to target cost plus target profit.

Ceiling Price: This is the maximum price the buyer will pay. Once it is reached, the seller will bear the cost.

Share Ratio: This is expressed as (buyer ratio)/(seller ratio); for example 80/20. This ratio tells us that if the contract is completed below the ceiling price, the buyer will keep 80% of it and the seller will get 20%.

If the price crosses the target cost and reaches the ceiling price, the buyer will bear 80% and the seller will bear 20%.

Once the ceiling price is crossed, the seller will bear all costs.

Example:

You sign a FPIF contract with a contractor with the following conditions:

Target Cost = 400,000 USD
Target Profit = 50,000 USD
Target Price = 450,000 USD
Ceiling Price = 500,000 USD
Share Ratio = 75/25

The contractor completes the contract for 375,000 USD.

Calculate the actual profit received by the seller, and what is the actual price of the contract.

Solution:

The target cost was 400,000 USD and the contractor has completed it for 375,000 USD, therefore, there is an underrun of 25,000 USD.

Since the ratio of the seller is 25%, he will get the 25% of this 25,000 USD.

Contractor's ratio on saving = 25% of 25,000
= 6,250 USD

The target profit is 50,000 USD and the contractor's share on saving is 6,250 USD, so the total profit will be equal to the sum of these two.

Total Actual Profit received by the seller = 50,000 + 6,250
= 56,250 USD

Actual price to contractor = Actual cost + actual profit
= 375,000 + 56,625
= 431,625 USD

Please note that since the actual price is lower than the ceiling price, you will get exactly 431,625 USD.

However, if this would have been more than the ceiling price, you will get the ceiling price only, i.e. 500,000 USD.

***

Practice Question: 1

You sign a FPIF contract with a seller with the following parameters:
Target Cost = 420,000 USD
Target Profit = 40,000 USD
Target Price = 460,000 USD
Ceiling Price = 525,000 USD
Share Ratio = 60/40

The contractor completes the contract for 440,000 USD.

Calculate the actual profit received by the seller, and what is the actual price of the contract?

***

Point of Total Assumption (PTA)

This concept is used in Fixed Price Incentive Fee (FPIF) contract types. The Point of Total Assumption is the point above which cost is borne by the seller.

In this type of contract, there is a ceiling cost, which considers the worst-case scenario. If the seller crosses this cost, it is assumed that this cost has overrun due to mismanagement on the part of the seller, and it is his responsibility to bear this cost overrun.

Here is the formula to calculate the Point of Total Assumption:

Point of Total Assumption (PTA) = ((Ceiling Price - Target Price)/Buyer's Share Ratio) + Target Cost

Example:

With the following information on hand, calculate the Point of Total Assumption.

Target Cost = 100,000 USD
Target Profit = 10,000 USD
Target Price = 115,000 USD
Ceiling Price = 120,000 USD
Share Ratio = 80% buyer, and 20% seller

Point of Total Assumption (PTA) = ((Ceiling Price – Target Price)/Buyer's Share Ratio) + Target Cost

Point of Total Assumption = (120,000 – 115,000)/0.8 + 100,000
= 6,250 + 100,000
= 106,250 USD

Hence, the Point of Total Assumption is 106,250 USD.

***

Practice Question: 2

With the following information on hand, calculate the Point of Total Assumption.

Target Cost = 375,000 USD
Target Profit = 25,000 USD
Target Price = 400,000 USD
Ceiling Price = 450,000 USD
Share Ratio = 75% buyer, and 25% seller

***

Fixed Price with Economic Price Adjustment Contracts (FP-EPA)

If the duration of the contract is very long, a Fixed Price with Economic Price Adjustment Contract will be used. Here you include a special provision in a clause which protects the seller from inflation.

Cost Reimbursable Contract

Here the seller is reimbursed for their work plus a fee representing their profit. Sometimes this fee will be paid if the seller meets or exceeds the selected project objectives.

This contract can be further divided into four categories:

1) Cost Plus Fixed Fee Contract (CPFF)
2) Cost Plus Incentive Fee Contract (CPIF)
3) Cost Plus Award Fee (CPAF)
4) Cost Plus Percentage of Cost (CPPC)

Note: Keep in mind that in a cost reimbursable contract there is no ceiling price because the seller is going to be reimbursed for all of their costs.

Cost Plus Fixed Fee Contract (CPFF)

Here the seller is paid for all their costs incurred plus a fixed fee (which will not change), regardless of their performance. In this case, the buyer bears the risk.

This contract is used in situations when risk is high and no one is interested in bidding. This type of contract is selected to keep the seller safe from risks.

Cost Plus Incentive Fee Contract (CPIF)

Here the seller will be reimbursed for all costs plus an incentive fee based upon achieving certain performance objectives mentioned in the contract. This incentive will be calculated by using an agreed predetermined formula.

In this contract the risk lies with the buyer; however, this risk is lower than the Cost Plus Fixed Fee where the buyer has to pay a fixed fee along with the cost incurred.

The terms used in this type of contract are:

Target Cost: This is the agreed anticipated cost of the project.

Target Fee: This is the agreed fee as mentioned in the contract.

Share Ratio: This is expressed as (buyer ratio)/(seller ratio). For example 80/20.

Please note that in this contract the minimum and maximum fee is mentioned. So in any case you will not get less than the minimum fee and more than the maximum fee.

Example:

You sign a CPIF contract with a seller with the following parameters:

Target Cost = 500,000 USD
Target Fee = 50,000 USD
Share Ratio = 80/20
Maximum Fee = 75,000 USD
Minimum Fee = 25,000 USD

The seller completes the work for 550,000 USD. How much will the seller receive?

Solution:

The target cost is 500,000 USD and the seller has completed the work for 550,000 USD, therefore there is a cost overrun of 50,000 USD.

Since your share ratio is 20%, you will have to bear 20% of this 50,000 USD, and this amount will be deducted from your target fee.

Your share on overrun = 20% of 50,000
= 0.2*50,000
= 10,000 USD

Actual fee given to you = Target Fee – Your share on overrun
= 50,000 – 10,000
= 40,000 USD

Since this fee is between the maximum and minimum fee, you will get 40,000 USD.

Actual price to the seller = Actual cost + actual fee
= 550,000 + 40,000
= 590,000 USD

Therefore, the actual price paid to the seller is 590,000 USD.

Please note that in this type of contract your fee will never be lower than the minimum fee or more than the maximum fee.

If the calculated fee is less than the lower fee, you will get the lower fee, and if the calculated fee is higher than the calculated fee, you will get the maximum fee.

***

Practice Question: 3

You sign a CPIF contract with a seller with the following parameters:

Target Cost = 200,000 USD
Target Fee = 20,000 USD
Share Ratio = 70/30
Maximum Fee = 30,000 USD
Minimum Fee = 10,000 USD

The seller completes the work for 180,000 USD. How much will the seller receive?

***

Cost Plus Award Fee (CPAF)

In this type of contract, the seller is paid for all costs plus an award fee. This award fee will be based on achieving satisfaction on performance objectives described in the contract.

The evaluation of performance is a subjective matter and you cannot appeal against it.

There is a difference between an incentive fee and an award fee. An incentive fee is calculated based on a formula defined in the contract, and is an objective evaluation. An award fee is dependent on the satisfaction of the client and is evaluated subjectively. An award fee is not subjected to an appeal.

The terms used in this type of contract are:

Estimated Cost: This is the estimated cost of the contract.

Base Fee: This is the minimum profit that the seller will receive.

Award Fee: This is the award fee that the contractor will receive after achieving certain performance objectives.

Cost Plus Percentage of Cost (CPPC)

Here the seller is paid for all costs plus a percentage of these costs. This type of contract is not preferred, because the seller might artificially increase the cost to earn a higher profit.

Time and Materials Contract

This contract is generally used when the deliverable is "labor hours." Here the project manager or the organization will provide the qualifications or experience to the contractor to provide the staff.

A time and materials contract is used to hire experts or any support services.

This contract is a hybrid of fixed price and cost reimbursable contracts. Here the risk is distributed to both parties.

Example: 1

You are planning to rent drilling equipment at a rental cost of 500 USD per day. This is costly equipment to purchase at 100,000 USD. Since you are going to use this equipment for a very long time, you are analyzing whether to buy this equipment or not.

Calculate after how many days the rent paid will be equal to the cost of the equipment.

Solution:

Let's say that after "N" days the cost of rent will be equal to the cost of the equipment.
Therefore,

500*N = 100,000
N = 100,000/500
= 200

Therefore, after 200 days the cost of rent will be equal to the cost of the equipment.

***

Example: 2

You are planning to rent a transport vehicle for a few days. The daily rent for this vehicle is 60 USD and the daily operating cost is 40 USD. This is costly equipment to purchase at 70,000 USD. Since you are going to use this vehicle for a very long time, you are analyzing whether to buy it or not.

Calculate after how many days the rent paid will be equal to the cost of the vehicle.

Solution:

Let's say that after "N" days the cost of rent will be equal to the cost of the vehicle.

Therefore,

60*N = 70,000 + 40*N

20N = 70,000
N = 70,000/20
= 3,500 days

Therefore, after 3,500 days the cost of rent will be equal to the cost of the vehicle.

***

Practice Question: 4

For your office work you need a high end photocopy machine which is available to rent for 150 USD per day. The operating cost is 25 USD. The price of this machine is 125,000 USD. You will need this machine for several months, therefore, you start calculating whether you should purchase it or not.

Calculate the number of days after which it would be advisable to buy this machine.

***

Practice Question: 5

Your organization has signed a cost plus incentive fee contract with a seller. The cost of the contract is 100,000 USD and the target profit is 10% of the target cost. The seller's share ratio is 25% and has spent 90,000 USD to complete the task.

Calculate the profit and actual price paid to the seller.

***

Practice Question: 6

Your company has signed a fixed price incentive fee contract with a seller with the following details:

Target Cost = 300,000 USD
Target Incentive Fee = 50,000 USD
Share Ratio = 70/30
Ceiling Price = 400,000 USD
Actual Cost = 320,000 USD

Calculate the actual price paid to the seller.

Miscelleneous Questions

Practice Question: 7

You have signed a contract to construct a structure attached to the main office building of the client. As per the contract, you will have to complete the task for 150,000 USD. This price includes the cost and a fee. The contract also has a clause, which says that if the price goes above 200,000 USD, the client will not pay for the overage. Now the project is about 90% complete. You have spent more than 150,000 USD and are reaching towards 200,000 USD and you started bearing the cost. How will you describe the situation?

Practice Question: 8

As there is some ambiguity with the scope of work, the client signs a contract with you to pay for all costs involved and an additional remuneration based on an agreed ratio such as 75/25. The remuneration will be paid when you achieve some performance objectives based on objective criteria as mentioned in the contract. What kind of contract is this?

Practice Question: 9

You have signed a contract with a seller that you will pay them for all costs incurred and an extra fee which will be based on some subjective performance criteria defined and incorporated into the contract. The determination of this fee will be solely based on the subjective determination of the seller's performance by you. What kind of contract is this?

Practice Question: 10

You have signed a CPIF contract with a seller with an estimated cost of 300,000 USD. The predetermined fee is 35,000 USD and the share ratio is 70/30. Calculate the profit earned by the seller if he spends 275,000 USD on completing the tasks.

Practice Question: 11

You sign a FPIF contract with a buyer with a target cost of 200,000 USD. The target price and profit are 225,000 USD and 25,000 USD. The ceiling is 250,000 USD and the project is completed for 235,000 USD. Considering the share ratio of 70/30, calculate how much profit you have made?

Practice Question: 12

You are currently in the process of deciding whether to buy or make your own simulation program to help train new employees in your organization. The cost of this software is 100,000 USD and the integration and installation cost is 2,500 USD. However, if you develop this software on your own, it will take 5 engineers 4 months with a 6,000 USD salary per month. Another indirect cost of this process is 5,000 USD. Determine the option selected by you.

Practice Question: 13

In your project, you have a risk with a high probability of occurrence.

You are currently reviewing if you should manage it on your own or entrust it to a third party. If you manage it on your own it will cost you 100,000 USD and then 3,000 USD monthly. If this is managed by a third party, then they will charge you 65,000 USD and 6,000 per month. After how many months will the cost paid to the third party be equal to if you managed it your own?

Solutions

Practice Question: 1

You sign a FPIF contract with a seller with the following parameters:

Target Cost = 420,000 USD
Target Profit = 40,000 USD
Target Price = 460,000 USD
Ceiling Price = 525,000 USD
Share Ratio = 60/40

The contractor completes the contract for 440,000 USD.

Calculate the actual profit received by the seller, and what is the actual price of the contract.

Solution:

The target cost was 420,000 USD and the contractor has completed it for 440,000 USD, therefore, there is an overrun of 20,000 USD.

Since the ratio of the seller is 40%, he will have to bear 40% of this 20,000 USD.

Contractor's ratio on overrun = 40% of 20,000
= 8,000 USD

Since the target profit is 40,000 USD and the contractor's loss is 8,000 USD, the total profit will be equal to the target profit minus contractor's share on overrun.

Total Actual Profit received by the seller = 40,000 – 8,000
= 32,000 USD

Actual price to contractor = Actual cost + actual profit
= 440,000 + 32,000
= 472,000 USD

***

Practice Question: 2

With the following information on hand, calculate the Point of Total Assumption.

Target Cost = 375,000 USD
Target Profit = 25,000 USD
Target Price = 400,000 USD
Ceiling Price = 450,000 USD
Share Ratio = 75% buyer, and 25% seller

Solution:

We know that,

The Point of Total Assumption (PTA) = ((Ceiling Price – Target Price)/ Buyer's Share Ratio) + Target Cost

PTA = ((450,000 – 400,000)/0.75) + 375,000
= 50,000/0.75 + 375,000
= 66,666 + 375,000
= 441,666 USD

Hence, the Point of Total Assumption is 441,666 USD.

***

Practice Question: 3

You sign a CPIF contract with a seller with the following parameters:

Target Cost = 200,000 USD
Target Fee = 20,000 USD
Share Ratio = 70/30
Maximum Fee = 30,000 USD
Minimum Fee = 10,000 USD

The seller completes the work for 180,000 USD.

How much will the seller receive?

Solution:

The target cost is 200,000 USD and the seller has completed the work for 180,000 USD, therefore, there is an underrun of 20,000 USD.

Since your share ratio is 30%, you will get 30% of this 20,000 USD. This amount will be added to your target fee.

Your share on underrun = 30% of 20,000
= 0.3*20,000
= 6,000 USD

Actual fee given to you = Target Fee + Your share on underrun
= 20,000 + 6,000
= 26,000 USD

Since this fee is between the maximum and minimum fee, you will earn a 26,000 USD profit.

Actual price to the seller = Actual cost + actual fee
= 180,000 + 26,000
= 206,000 USD

***

Practice Question: 4

For your office work you need a high-end photocopy machine which is available to rent for 150 USD per day. The operating cost is 25 USD. The price of this machine is 125,000 USD. You will need this machine for several months, therefore, you start calculating whether you should purchase it or not.

Calculate the number of days after which it would be advisable to buy this machine.

Solution:

Let's say that after "N" days the cost of rent will be equal to the cost of the machine.

Therefore,

150*N = 125,000 + 25*N
125*N = 125,000
N = 125,000/125
= 1,000 days

Therefore, after 1,000 days the cost of rent will be equal to the cost of the machine.

***

Practice Question: 5

Your organization has signed a cost plus incentive fee contract with a seller. The cost of the contract is 100,000 USD and the target profit is 10% of the target cost. The seller's share ratio is 25% and has spent 90,000 USD to complete the task.

Calculate the profit and actual price paid to the seller.

Solution:

Given in the question:

Target Cost: 100,000 USD
Target Profit: 10% of 100,000
= 10,000 USD
Share Ratio: 75/25
Actual Cost: 90,000 USD

Since there is a cost underrun, the seller will get his 25% incentive.

Seller's incentive = 25% of 10,000
= 2,500 USD

Actual Profit = Target Profit + Incentive
= 10,000 + 2,500
= 12,500 USD

Actual Cost = Actual Cost + Actual Profit
= 90,000 + 12,500
= 102,500 USD

Hence, the actual price paid to the contractor is 102,500 USD.

***

Practice Question: 6

Your company has signed a fixed price incentive fee contract with a seller with the following details:

Target Cost = 300,000 USD
Target Incentive Fee = 50,000 USD
Share Ratio = 70/30
Ceiling Price = 400,000 USD
Actual Cost = 320,000 USD

Calculate the actual price paid to the seller.

Solution:

The target cost was 300,000 USD and the contractor has completed the work for 320,000 USD. Therefore, there is an overrun of 20,000 USD.

Since the seller's ratio is 30%, he will bear 30% of this 20,000 USD.

Seller's ratio on overrun = 30% of 20,000
= 6,000 USD

The target incentive fee is 50,000 USD and the seller has to bear 6,000 USD.

Total profit to the seller = 50,000 – 6,000
= 44,000 USD

Now,

Actual price paid to the seller = Actual Cost + Actual Profit
= 320,000 + 44,000
= 364,000 USD

Hence, the actual price paid to the contractor is 364,000 USD.

***

Practice Question: 7

You have signed a contract to construct a structure attached to the main office building of the client. As per the contract, you will have to complete the task for 150,000 USD. This price includes the cost and a fee. The contract also has a clause, which says that if the price goes above 200,000 USD, the client will not pay for the overage. Now the project is about 90% complete. You have spent more than 150,000 USD and are reaching towards 200,000 USD and you started bearing the cost. How will you describe the situation?

Solution:

You have crossed the PTA. Once you cross the PTA, you start bearing all costs incurred.

***

Practice Question: 8

As there is some ambiguity with the scope of work, the client signs a contract with you to pay for all costs involved and an additional remuneration based on an agreed ratio such as 75/25. The remuneration will be paid when you achieve some performance objectives based on objective criteria as mentioned in the contract. What kind of contract is this?

Solution:

CPIF, as there are objective performance criteria, it is a CPIF contract.

***

Practice Question: 9

You have signed a contract with a seller that you will pay them for all costs incurred and an extra fee which will be based on some subjective performance criteria defined and incorporated into the contract. The determination of this fee will be solely based on the subjective determination of the seller's performance by you. What kind of contract is this?

Solution:

CPAF, as there are subjective performance criteria, it is a CPAF contract.

***

Practice Question: 10

You have signed a CPIF contract with a seller with an estimated cost of 300,000 USD. The predetermined fee is 35,000 USD and the share ratio is 70/30. Calculate the profit earned by the seller if he spends 275,000 USD on completing the tasks.

Solution:

The target cost is 300,000 USD and the seller completes the work for 275,000 USD. This mean the seller is under budget by 25,000 USD.

As the share ratio is 70:30, the seller's share will be 30%.

Seller ratio = 0.3 * 25,000 = 7,500 USD

Total profit earned by seller = 35,000 + 7,500 = 42,500 USD

***

Practice Question: 11

You sign a FPIF contract with a buyer with a target cost of 200,000 USD. The target price and profit are 225,000 USD and 25,000 USD. The ceiling is 250,000 USD and the project is completed for 235,000 USD. Considering the share ratio of 70/30, calculate how much profit you have made?

Solution:

The target cost is 200,000 USD and you completed it for 210,000 USD, which means there is an overrun of 10,000 USD.

Your share on overrun = 30% of 10,000 = 3,000 USD

As this is your share of loss, this money will be deducted from the profit.

Net profit = 25,000 - 3,000 = 22,000 USD

The actual profit received by you is 22,000 USD

***

Practice Question: 12

You are currently in the process of deciding whether to buy or make your own simulation program to help train new employees in your organization. The cost of this software is 100,000 USD and the integration and installation cost is 2,500 USD. However, if you develop this software on your own, it will take 5 engineers 4 months with a 6,000 USD salary per month. Another indirect cost of this process is 5,000 USD. Determine the option selected by you.

Solution:

You will buy it.

We will calculate the cost of buying and then building and whichever option is lower will be selected.

Cost of buying = 100,000 + 2,500 = 102,500 USD

Cost of DIY = 5 * 6,000 * 4 + 5,000 = 125,000 USD

***

Practice Question: 13

In your project, you have a risk with a high probability of occurrence. You are currently reviewing if you should manage it on your own or entrust it to a third party. If you manage it on your own it will cost you 100,000 USD and then 3,000 USD monthly. If this is managed by a third party, then they will charge you 65,000 USD and 6,000 per month. After how many months will the cost paid to the third party be equal to if you managed it your own?

Solution:

Let us say after "N" months the cost paid to the third party will be equal to the buying price.

100,000 + 3,000N = 65,000 + 6,000N
35,000 = 3,000 N
N = 35,000/3,000
= 11.6

Hence, after 12 months the buying cost will be equal to the rent cost.

Miscellaneous Formulas

In this chapter, we are going to discuss a few concepts which are not mentioned in the PMBOK Guide, but questions based on them have been seen on the exam.

It is possible that from each topic discussed in this chapter, you may see one question. Calculations involved in this chapter are very basic in nature, and with very little practice you can get a good command over them.

Here we are going to discuss the following topics:

- Mean, Variance, and Standard Deviation
- Discounted Cash Flows
- Net Present Value (NPV)
- Internal Rate of Return (IRR)
- Payback Period
- Opportunity Cost
- Cost Benefit Ratio
- Sunk Cost

Mean, Variance, and Standard Deviation

Although the calculations for the Mean, Variance, and Standard Deviation are very simple, the formulas used for these calculations are bit intimidating in appearance. Therefore, I will start directly from the concept and walk you through to the end.

Mean: Mean is an average of all numbers. It is determined by adding all numbers and dividing by the number count.

Variance: Variance is the average of the squared difference from the mean. It is determined by adding all squaring difference (difference between number and mean, and then square) and dividing by the number count.

Standard Deviation: Standard Deviation is the square root of the Variance.

Steps to calculate the Standard Deviation:

(a) Calculate the Mean
(b) Calculate the Variance
(c) Calculate the Standard Deviation

Now, let's look at an example:

Example:

Assuming in your class you have five students, and the height of each student is as follows: 160 cm, 170 cm, 180 cm, 175 cm, and 165 cm.

Calculate the Mean, Variance, and Standard Deviation for these students.

Solution:

Mean = (160+170+180+175+165)/5
= 170 cm

To find the Variance, subtract this 'mean height' from the height of each student, square it, add them all together, and then take the average.

Variance = $[(160-170)^2 + (170-170)^2 + (180-170)^2 + (175-170)^2 + (165-170)^2]/5$
$= [100+0+100+25+25]/5$
$= 250/5$
$= 50$

Hence, the Variance is 50

And, Standard Deviation = Square Root of Variance

Standard Deviation = Square Root of 50
= 7.07

Hence, the Standard Deviation is 7.07 cm.

Now, you might be thinking what is the use of these data?

These data are very important because they give you following information:

(a) The average height of students is 170 cm (Mean)
(b) The lowest height of students = 170 – 7.07 = 162.93 cm
(c) The maximum height of students = 170 + 7.07 = 177.07 cm
(d) The height of most of the students varies from 162.93 cm to 177.07 cm

Let's revise the whole procedure once again:

(1) Calculate the average height of all students
(2) Then subtract the average height from the height of each student, and square it
(3) Add all of them together and take the average
(4) Take the square root

Note: In this example I used population-based data; in this example there were only five students in the class.

However, if you select Sample Data, this means you select a few random numbers from a large data pool; in this case you would have to use (N-1), where N is the number of sample data. In other words, if there was a class of five hundred students in our example, you would have to use (5 – 1) in the denominator of variance formula, i.e. 4.

You also might be thinking that since we have taken the square of the difference, and then take the square root of it, why are we squaring a number if we are going to take the square root of it?

There is a reason for this calculation: if we simply add the difference, positive and negative numbers will cancel each other out.

***

Practice Question: 1

In your school, you select 5 random students and weigh them. Their weights are: 100 Kg, 80 Kg, 75 Kg, 60 Kg, and 50 Kg.

Calculate the Mean, Variance, and Standard Deviation for these students.

***

Discounted Cash Flows

We all know that the amount of money you have today will not have the same value after a few years; likewise money earned after a few years will not have the same value today.

For example, if you have 100 USD in your hand, you buy something from it. Now, there is no guarantee that you will be able to buy the same stuff for 100 USD after five years; you may end up paying a little more money.

In other words, you can say that money in hand in the future is worth less than money in hand today.

Here is the formula to calculate the Future Value of the money:

$$FV = PV (1 + r)^n$$

Where,

FV = Future Value
PV = Present Value
r = Interest Rate
n = Time Period

Example:

You have 10,000 USD in your hand. Find the Future Value of this money after 5 years, considering a 5% interest rate.

Solution:

Given data:

PV = 10,000 USD
FV = ?
r = 5%
n = 5

$FV = PV (1 + r)^n$
$= 10,000 (1+.05)^5$
$= 10,000 * (1.05)^5$
$= 10,000 * 1.2762815$
$= 12,762.815$ USD

Hence, the Future Value of 10,000 USD after five years is 12,762 USD.

***

Practice Question 2:

Find the Future Value of 5,000 USD after 10 years, considering a 6% interest rate.

***

Likewise, you can also calculate the Present Value from the Future Value.

Let's look at another example with reverse calculation.

Example:

You have to select one project out of two. One project will give you 20,000 USD profit in three years while the second project may give you 25,000 USD in profit in five years.

Which project will you select, considering a 5% inflation?

Solution:

In this type of question, we will calculate the Present Values of profit earned by all projects and select the project which gives you the higher profit.

Formula for Future Value:

$FV = PV (1 + r)^n$

Hence, the formula for Present Value will be:

$FV = PV (1 + r)^n$
$PV = FV/[(1 + r)^n]$

The Present Value of the first project:

$PV1 = 20,000/[(1+0.05)^3]$
$= 20,000/(1.05)^3$
$= 20,000/1.158$
= 17,271 USD

Hence, the present value of 20,000 USD is 17,271 USD.

The Present Value of the second project:

$PV2 = 25,000/[(1+0.05)^5]$
$= 25,000/(1.05)^5$
$= 25,000/1.276$
= 19,592 USD

Hence, the Present Value of 25,000 USD is 19,592 USD.

Since the second project will give more profit, you will select the second project.

***

Practice Question 3:

You organization has to select one project out of two projects. The first project, which may last for 4 years, will give you a 50,000 USD profit, and the second project will give a 55,000 USD profit. The duration of the second project is 7 years.

In this scenario, which project will you suggest your organization choose considering a yearly inflation of 4%?

***

Net Present Value

The Net Present Value (NPV) is the total cash generated by the project in today's dollar minus the initial investment.

The formula to calculate the Net Present Value is complicated and is not required for the PMP exam.

However, you need to remember two things regarding the Net Present Value:

(1) If the NPV is negative, don't select the project. Select the project with the positive NPV.
(2) If you have many positive NPVs given in questions, select the highest one.

Example:

You have three projects with net present values of 300,000 USD, 450,000 USD, and 275,000 USD. Due to resource limitations you can complete only one project.

Considering all other parameters are equal, which project will you select?

Solution:

If all other parameters are equal and you have to choose the project based on the net present value, you will select the project with the highest net present value.

In the question the second project has the highest net present value of 450,000 USD; hence you will select this project.

***

Practice Question: 4

Your organization has four projects to choose from but due to resource constraints you are able to select only one of these projects. All parameters are calculated and they are almost equal. The NPV of these four projects are as follows: 150,000 USD, 95,000 USD, 175,000 USD, and 120,000 USD.

Based on the given information, which of the projects will you select?

***

Internal Rate of Return

The Internal Rate of Return (IRR) is the interest rate that makes the Net Present Value zero.
In other words, you can say that at IRR, cash inflows are equal to the money invested into the project.

For the exam, you don't need to do the calculations. Just remember these points:

(1) At IRR, NPV will be zero.
(2) If you have been given many options to select, chose the value with highest IRR.

Example

You have three projects with the following internal rates of return: 12%, 15%, and 11%.

If you have been asked to select one project and consider all other parameters to be equal, which project will you select?

Solution:

When all other parameters are equal, you will choose the project which has the highest IRR.
In this case, the second project has highest IRR at 15%, so you will select this project.

***

Practice Question: 5

Your organization has given you three projects with IRRs of 13%, 9%, and 17%.

You have been asked to select the project which provides the most benefits. Considering all other parameters are equal, which project will you select?

***

Payback Period

It is the ratio of "total cash out" with "average per period cash in".

i.e.

Payback Period = (Total Cash out)/(Average per Period Cash in)

In other words, you can say that it is the time required to recover the cost invested on the project.

If all other parameters are same, the project with minimum payback period will be selected.

Example

You have three projects and have to select one of them. The payback periods for these three projects are 5 years, 3 years, and 7 years. Considering all other parameter are the same, which of the projects will you select?

Solution:

If all other parameters are the same, you will select the project with the shortest payback period. In the question the second project has the shortest payback period of three years, so you will select this project.

***

Practice Question: 6

You have four projects with payback periods of two years, three years, four years and five years.
If all other parameters are equal, which project will you select?

***

Opportunity Cost

It often happens that an organization has many projects but due to lack of resources it cannot take all of them on. Therefore they have to select the best project and forgo the profit of other projects.

This unrealized profit is known as the opportunity cost.

Example:

You have two projects, A and B. The NPV of project A is 500,000 USD and NPV of project B is 550,000 USD. Due to some constraints you have to select only one project, so you go with project B. What is the opportunity cost of project B?

Solution:

Since you have selected project B and did not select project A with an NPV of 550,000 USD, the opportunity cost will be 500,000 USD.

***

Practice Question: 7

Your organization has two projects but they can only manage one, so they have decided to go with the best option.
The first project has an NPV of 200,000 USD and the second has an NPV of 150,000 USD. All other parameters are equal, so your organization selects the best project. What is the opportunity cost?

Solution:

Your organization will select the project with the highest NPV, i.e. the project at 200,000 USD. Since your organization has selected the first project and did not select the second project with an NPV of 150,000 USD, the opportunity cost will be 150,000 USD.

***

Cost Benefit Ratio

Cost benefit ratio compares the cost with the revenue. If the ratio is less than one, it is profitable to choose the project.

You may also see this term referred to as "Benefit Cost Ratio", as it is the ratio between the benefits and the cost. In this case if the ratio is greater than one, the project is profitable.

Example

You are working on a project that is worth 500,000 USD which may give you a revenue of 650,000 USD. What is the cost benefit ratio?

Solution:

Cost Benefit Ratio = 500,000/650,000
= 10/13

***

Practice Question: 8

The cost of a project is 350,000 USD and you estimated that it will earn your organization 475,000 USD in revenue. What is the cost benefit ratio?

***

Sunk Cost

Sunk cost is the cost that you have spent on the project and now it is not recoverable.

For example, let's say that your project is 50% completed and you have spent 100,000 USD on it. However the situation changes, and the project no longer seems viable and is terminated.

In this case the 100,000 USD will be the sunk cost.

You should not make your decision based on the sunk cost. For example, you may think that you have spent a lot of money on it and now you must complete the project.

The objective of the sunk cost is that you should never base your future decisions on the sunk cost.

Example

The cost of your project is 500,000 USD. 40% of it is completed and you have spent 200,000 USD. The value of the complete work is 250,000 USD. Due to some circumstances the project is cancelled. What is the sunk cost?

Solution:

Since you have spent 200,000 USD on the project and it is not recoverable, the sunk cost is 200,000 USD.

***

Practice Question: 9

The budget for your project is 200,000 USD and you have spent 30% of it. The value of the completed work is 50,000 USD. Due to changes in market condition, your company cancels the project.

What is the sunk cost?

***

Miscelleneous Questions

Practice Question: 10

You are producing bearings for automobile companies and during an inspection; you see that more than 96% of your bearings are defect-free. What level of Sigma is this?

Practice Question: 11

What is the future value of an annual income cash flow of 25,000 USD for 5 years at a rate of 8%?

Practice Question: 12

The BAC of your project is 650,000 USD and it is expected to end in two years. The expected inflow for the first year is 3000 USD per month, for the second year 4,000 USD per month and then 5,000 USD per month. Calculate the payback period.

Practice Question: 13

The first project has an NPS of 350,000 USD and a payback period of 5 years and the second project has an NPS of 250,000 USD and a payback period as 4 years. What is the opportunity cost here?

Practice Question: 14

Your project will cost you 850,000 USD and it will be completed in two years.

After the completion of the project, the cash inflow for the first year is 15,000 USD per month and then 20,000 per month. What is the payback period?

Practice Question: 15

The IRR of the first project is 25% and the payback period is 20 months, and for the second project the IRR is 20% and the payback period is 15 months. Which project will you select?

Practice Question: 16

Your company is supplying bearings to automobile companies. To check the quality of deliverables, you check every 50th bearing of the batch. What kind of sampling method is that?

Practice Question: 17

Your company is supplying tires to automobile companies. To check the quality of these tires, you select a few from the batch and check their dimensions and for any manufacturing defects. What kind of sampling method is this?

Practice Question: 18

You are manufacturing gaskets to be used in high-pressure valves in cross-country gas pipelines. As you are producing a large number of gaskets, to check the quality of gaskets you group them in batches of 1000 gaskets and select three gaskets randomly from each batch. What kind of sampling method is this?

Practice Question: 19

Based on the table given below, find the payback period of the project.

Year	Outflow (USD)	Inflow (USD)
1st Year	150,000	-
2nd Year	50,000	50,000
3rd Year	75,000	100,000
4th Year	25,000	200,000
5th Year	20,000	150,000

Practice Question: 20

You are collecting data for defect frequencies in your deliverable. You have identified four defects, A, B, C and D, with the following frequencies: 50, 40, 75, and 10. To draw a Pareto diagram, in what order will you plot the defects?

Practice Question: 21

What do you mean by a benefit cost ratio of 1.5?

Practice Question: 22

You have an opportunity to select from four projects but due to resource constraints, you cannot select more than one project. So you decide to go with one project based on the benefit cost ratio. Which of the following projects will you select?

Practice Question: 23

You have three projects with the following details: Project A with an NPV of 500,000 USD, project B with an NPV of 350,000 USD and a benefit cost ratio of 7:3 and project C with an NPV of 650,000 USD. Out of these three projects, which one will you select?

Practice Question: 24

You purchased a photocopy machine five years ago at 25,000 USD and are now planning to purchase a new one because the quality of the old machine is diminishing. The cost of a new photocopy machine will be 45,000 USD and the present cost of the old machine is 5,000 USD. This 5,000 USD is known as:

Practice Question: 25

A process has three sources of variations, which are not dependent on each other. The standard deviations for of these sources are 5, 7, and 9. Find the standard deviation of the process.

Practice Question: 26

You are the project manager of a project whose budget at completion is 350,000 USD. It is 30% complete and now the client does not see any foreseeable use for this project. What will you recommend to him?

Solutions

Practice Question: 1

In your school, you select 5 random students and weigh them. Their weights are: 100 Kg, 80 Kg, 75 Kg, 60 Kg, and 50 Kg.

Calculate the Mean, Variance, and Standard Deviation for these students.

Solution:

Mean = (sum of weight of students)/(number of students)
= (100 + 80 + 75 + 60 + 50)/5
= 365/5
= 73

Variance = $[(100 - 73)^2 + (80 - 73)^2 + (75 - 73)^2 + (60 - 73)^2 + (50 - 73)^2]/(5 - 1)$
= (729 + 49 + 4 + 169 + 529)/4
= 1480/4
= 370

Standard Deviation = Square Root of Variance
= square root of 370
= 19.23

[Please note, since we are using random data here, we are using (N–1), instead of N]

Therefore, the maximum weight = Mean + Standard Deviation
= 73 + 19.23
= 92.23 Kg

Minimum weight = 73 – 19.23
= 53.77 Kg

Hence the weight of the students varies from 53.77 Kg to 92.23 Kg.

***

Practice Question 2:

Find the Future Value of 5,000 USD after 10 years, considering a 6% interest rate.

Solution:

Given data:

PV = 5,000 USD
FV =?
r = 6%
n = 10

$FV = PV (1 + r)^n$
$= 5000 * (1+.06)^{10}$
$= 5000 * (1.06)^{10}$
$= 5000 * 1.79$
= 8,950 USD

Hence, the Future Value of 5,000 after ten years is 8,950 USD.

***

Practice Question 3:

You organization has to select one project out of two projects. The first project, which may last for 4 years, will give you 50,000 USD profit. The second project will give 55,000 USD profit. The duration of the second project is 7 years.

In this scenario, which project will you suggest your organization choose considering a yearly inflation of 4%?

Solution:

We will calculate the Present Value of profits given by both projects, and the project with the higher value will be selected.

The formula for Present Value is:
$FV = PV (1 + r)^n$
$PV = FV / [(1 + r)^n]$

The Present Value of profit earned by the first project:

$PV1 = FV / [(1 + r)^n]$
$= 50{,}000 / [(1+0.04)^4]$
$= 50{,}000 / (1.04)^4$
$= 50{,}000 / 1.17$
$= 42{,}735$ USD

Hence, the Present Value of profit from the first project is 42,735 USD.

The Present Value of profit earned by the second project:

PV2 = FV/[(1 +r)^n]
= 55,000/[(1+0.04)^7]
= 55,000/[(1.04)^7]
= 55,000/1.32
= 41,666

Hence, the Present Value of profit from the second project is 41,666 USD.

Since the profit earned by the first project is greater than the second project, you will select the first project.

***

Practice Question: 4

Your organization has four projects to choose from but due to resource constraints you are able to select only one of these projects. All parameters are calculated and they are almost equal. The NPV of these four projects are as follows: 150,000 USD, 95,000 USD, 175,000 USD, and 120,000 USD.

Based on the given information, which of the projects will you select?

Solution:

If all other parameters are equal and you have to choose the project based on the NPV, you will select the project with the highest NPV.

In this question the third project has the highest NPV of 175,000 USD; hence you will select this project.

***

Practice Question: 5

Your organization has given you three projects with IRRs of 13%, 9%, and 17%. You have been asked to select the project which provides the most benefits.

Considering all other parameters are equal, which project will you select?

Solution:

When all other parameters are equal, you will choose the project which has the highest IRR.
In this case the third project has the highest IRR at 17%, so you will select the third project.

***

Practice Question: 6

You have four projects with payback periods of two years, three years, four years and five years. If all other parameters are equal, which project will you select?

Solution:

If all other parameters are the same, you will select the project with the shortest payback period. In this question the first project has the shortest payback period of two years, so you will select this project.

***

Practice Question: 7

Your organization has two projects but they can only manage one, so they have decided to go with the best option.
The first project has an NPV of 200,000 USD and the second has an NPV of 150,000 USD. All others parameters are equal, so your organization selects the best project.

What is the opportunity cost?

Solution:

Your organization will select the project with the highest NPV, i.e. the project at 200,000 USD. Since your organization has selected the first project and did not select the second project with an NPV of 150,000 USD, the opportunity cost will be 150,000 USD.

***

Practice Question: 8

The cost of a project is 350,000 USD and you estimated that it will give your organization 475,000 USD in revenue. What is the cost benefit ratio?

Solution:

Cost Benefit Ratio = 350,000/475,000
= 14/19

***

Practice Question: 9

The budget for your project is 200,000 USD and you have spent 30% of it. The value of the completed work is 50,000 USD. Due to changes in market condition, your company cancels the project. What is the sunk cost?

Solution:

Since you have spent 30% of 200,000 USD (i.e. 60,000 USD) on the project and the project has been cancelled, the sunk cost is 60,000 USD.

***

Practice Question: 10

You are producing bearings for automobile companies and during an inspection; you see that more than 96% of your bearings are defect-free. What level of Sigma is this?

Solution:

This is Two Sigma.

Two Sigma = 95.46%

Three Sigma =99.73%

Six Sigma = 99.99966%.

***

Practice Question: 11

What is the future value of an annual income cash flow of 25,000 USD for 5 years at a rate of 8%?

Solution:

Let us say FV is the future value, PV is the present value, "r" is the rate and "t" is the time.

$FV = PV * (1+r/100)^t$
$= 25,000 * (1+8/100)^5$
$= 25,000 * (1.08)^5$
$= 25,000 * 1.47$
$= 36,750$

The future value is 36,750 USD.

***

Practice Question: 12

The BAC of your project is 650,000 USD and it is expected to end in two years. The expected inflow for the first year is 3000 USD per month, for the second year 4,000 USD per month and then 5,000 USD per month. Calculate the payback period.

Solution:

Total cash inflow in the first year = 3,000 * 12 = 36,000 USD

Total cash inflow in the second year = 4,000 * 12 = 48,000 USD

Total cash inflow = 36,000 + 48,000 = 84,000 USD

Balance = BAC - cash inflow in two years
= 650,000 - 84,000
= 566,000 USD

Months needed to cover balance amount = 566,000/5,000
= 113.2
= 114

Total months = 12 + 12 + 114

= 138 months

***

Practice Question: 13

The first project has an NPS of 350,000 USD and a payback period of 5 years and the second project has an NPS of 250,000 USD and a payback period as 4 years. What is the opportunity cost here?

Solution:

Here you will select the second project because of the lower payback period; hence, the opportunity cost will be 350,000 USD.

***

Practice Question: 14

Your project will cost you 850,000 USD and it will be completed in two years.

After the completion of the project, the cash inflow for the first year is 15,000 USD per month and then 20,000 per month. What is the payback period?

Solution:

Cash inflow at the end of the first year = 15,000 * 12 = 180,000 USD

Balance amount = BAC - cash inflow in the first year
= 850,000 - 180,000 = 670,000 USD

Months taken to recover balance amount = 670,000/20,000 = 33.5
= 34

Total months 12 + 34 = 46 months

***

Practice Question: 15

The IRR of the first project is 25% and the payback period is 20 months, and for the second project the IRR is 20% and the payback period is 15 months. Which project will you select?

Solution:

The first project. A higher IRR is preferred over a lower IRR.

***

Practice Question: 16

Your company is supplying bearings to automobile companies. To check the quality of deliverables, you check every 50th bearing of the batch. What kind of sampling method is that?

Solution:

Statistical sampling is a sampling method where you select a sample from a population according to a fixed interval.

***

Practice Question: 17

Your company is supplying tires to automobile companies. To check the quality of these tires, you select a few from the batch and check their dimensions and for any manufacturing defects. What kind of sampling method is this?

Solution:

Random sampling. This is an example of random sampling where you select the test object randomly from the population.

***

Practice Question: 18

You are manufacturing gaskets to be used in high-pressure valves in cross-country gas pipelines. As you are producing a large number of gaskets, to check the quality of gaskets you group them in batches of 1000 gaskets and select three gaskets randomly from each batch. What kind of sampling method is this?

Solution:

This is an example of stratified sampling where you divide the population into "strata" and then choose a random sample from each stratum.

***

Practice Question: 19

Based on the table given below, find the payback period of the project.

Year	Outflow (USD)	Inflow (USD)
1st Year	150,000	-
2nd Year	50,000	50,000
3rd Year	75,000	100,000
4th Year	25,000	200,000
5th Year	20,000	150,000

Solution:

Total outflow = 150,000 + 50,000 + 75,000 + 25,000 + 20,000 = 320,000 USD

Inflow at the end of fourth year = 50,000 + 100,000 + 200,000 = 350,000 USD

***

Practice Question: 20

You are collecting data for defect frequencies in your deliverable. You have identified four defects, A, B, C and D, with the following frequencies: 50, 40, 75, and 10. To draw a Pareto diagram, in what order will you plot the defects?

Solution:

CABD. In a Pareto diagram you keep defects in a descending order of occurrence.

***

Practice Question: 21

What do you mean by a benefit cost ratio of 1.5?

Solution:

A benefit cost ratio of 1.5 shows that you are earning 1.5 USD for each dollar spent.

***

Practice Question: 22

You have an opportunity to select from four projects but due to resource constraints, you cannot select more than one project.

So you decide to go with one project based on the benefit cost ratio. Which of the following projects will you select?

Solution:

Let us compare the BCR of each project:

BCR of Project A = 5/3 = 1.67
BCR of Project B = 4/3 = 1.33
BCR of Project C = 5/4 = 1.25
BCR of Project D = 7/4 = 1.75

You will select the project with the highest BCR ratio (Project D BCR = 1.75).

***

Practice Question: 23

You have three projects with the following details: Project A with an NPV of 500,000 USD, project B with an NPV of 350,000 USD and a benefit cost ratio of 7:3 and project C with an NPV of 650,000 USD. Out of these three projects, which one will you select?

Solution:

Project C. The project with the highest NPV will be selected.

***

Practice Question: 24

You purchased a photocopy machine five years ago at 25,000 USD and are now planning to purchase a new one because the quality of the old machine is diminishing.

The cost of a new photocopy machine will be 45,000 USD and the present cost of the old machine is 5,000 USD. This 5,000 USD is known as:

Solution:

Sunk Cost. Sunk cost is a cost that has already been incurred and you cannot recover it.

***

Practice Question: 25

A process has three sources of variations, which are not dependent on each other. The standard deviations for of these sources are 5, 7, and 9. Find the standard deviation of the process.

Solution:

Variance of first source = 5 x 5 = 25

Variance of second source = 7 x 7 = 49

Variance of third source = 9 x 9 = 81

Variance for the process = 25 + 49 + 81 = 155

Standard deviation = Square root of 155 = 12.5

***

Practice Question: 26

You are the project manager of a project whose budget at completion is 350,000 USD. It is 30% complete and now the client does not see any foreseeable use for this project. What will you recommend to him?

Solution:

You ask him to make a decision without considering the money spent. This is an example of a sunk cost where in decision-making you do not consider the cost that has been spent.

A Cheat Sheet on Formulas

In this chapter I am going to list all the formulas mentioned in this formula guide for your quick reference. I suggest you review these formulas as many times as you can, and if you wish, you can make a printout of it and refer to it whenever you have free time.

All of these formulas are explained in their respective chapters, which you can refer to if needed.

Time Management

Formulas used in this knowledge area:

(1) Expected Duration (Triangular Distribution) $t_E = (t_O + t_M + t_P)/3$
(2) Expected Duration (Beta Distribution) $t_E = (t_O + 4t_M + t_P)/6$
(3) Standard Deviation = $(t_P - t_O)/6$

Beta Distribution is also known as PERT (Program Evaluation and Review Technique) estimate.

Here,

t_P = Pessimistic Estimate
t_M = Most-likely Estimate
t_O = Optimistic Estimate

Activity Duration = EF – ES + 1 or LF – LS + 1

Float = LS - ES or LF – EF

Float is also known as total float.

Here,

EF = Early Finish
ES = Early Start
LF = Late Finish
EF = Early Finish

Free float of activity X = ES of next activity – EF of activity X – 1

Cost Management

Budget at Completion: The total budget allotted to the project. It is denoted by BAC.

Earned Value: The value of work completed to date. It is denoted by EV.

Planned Value: The authorized value of work that has to be completed in a given time period as per the schedule. It is denoted by PV.

Actual Cost: The amount of money that you have spent to date. It is denoted by AC.

Schedule Variance: The difference between the earned value and the planned value. It is denoted by SV.

Schedule Variance = Earned Value – Planned Value
SV = EV – PV

We can say that:

(a) If schedule variance is negative, you're behind schedule.
(b) If schedule variance is positive, you're ahead of schedule.

Cost Variance: The difference between the earned value and the actual cost. It is denoted by CV.

Cost Variance = Earned Value - Actual Cost
CV = EV - AC

We can say that:

(a) If cost variance is negative, you're over budget.

(b) If cost variance is positive, you're under budget.

Schedule Performance Index: The ratio between the earned value and the planned value. It is denoted by SPI.

Schedule Performance Index = (Earned Value)/(Planned Value)
SPI = EV/PV

We can say that:

(a) If schedule performance index is greater than one, you're ahead of schedule.

(b) If schedule performance is less than one, you're behind schedule.

Cost Performance Index: The ratio between the earned value and the actual cost. It is denoted by CPI.

Cost Performance Index = (Earned Value)/(Actual Cost)
CPI = EV/AC

We can say that:

(a) If cost performance index is less than one, you're over budget.

(b) If cost performance index is greater than one, you're under budget.

Estimate at Completion: The total amount of money that the project will cost you in the end, in case your original budget is no longer valid. It is denoted by EAC.

As per the PMBOK Guide, you can calculate the EAC in three different cases.

Case-I

In this case you assume that the project will continue to perform to the end as it was performing up until now.

In other words, your future performance will be the same as the past performance.

Formula to calculate the EAC in Case-I

Estimate at Completion = (Budget at Completion) / (Cost Performance Index)

Or,

EAC = BAC/CPI

Case-II

Here you say that until now you have deviated from your budget estimate; however, from now onwards you can complete the remaining work as planned.

That is why in this formula, to calculate the EAC you will simply add the amount spent to date (i.e. AC) to the budgeted cost for the remaining work.

Formula to calculate the EAC in Case-II

EAC = AC + (BAC – EV)

Case-III

In this case, you're over budget, behind schedule, and the client is insisting that you complete the project on time.

Here not only the cost but the schedule also has to be taken into consideration.

Formula to calculate the EAC in Case-III

EAC = AC + (BAC – EV)/(CPI*SPI)

Estimate to Complete: The expected amount of money that you will have to spend to complete the remaining work. It is denoted by ETC.

Estimate to Complete = Estimate at Completion – Actual Cost

ETC = EAC - AC

To-Complete Performance Index: The estimated cost performance for the project to meet the project's budget goal. It is denoted by TCPI.

There are two formulas to calculate TCPI:

(a) If you're under budget: TCPI = (BAC–EV)/(BAC–AC)
(b) If you're over budget: TCPI = (BAC–EV)/(EAC–AC)

Communications Management

The formula used in this knowledge area:

Communication Channels = n*(n -1)/2

Where n is the number of stakeholders.

Risk Management

Expected Monetary Value = Probability * Impact

EMV = P*I

Where "P" is the probability and "I" is the impact.

Procurement Management

The main formula used in this knowledge area is the formula for the Point of Total Assumption. Others calculations require only basic mathematical functions.

The formula for the Point of Total Assumption is:

Point of Total Assumption (PTA) = ((Ceiling Price - Target Price)/Buyer's Share Ratio) + Target Cost

Miscellaneous Formulas

In the exam you will see many formula-based questions which are not mentioned in the PMBOK Guide.

So here I'm going to list those formulas.

Mean

Mean = (Sum of Data Points/(Nos of data points)

Variance

Variance = [(Mean-First Data)^2 + (Mean-First Data)^2 + (Mean-First Data)^2 + (Mean-First Data)^2 + (Mean-First Data)^2 + ...] /n

Where n is the number of data points.

Standard Deviation

Standard Deviation = Square Root of Variance

Discounted Cash Flows

Here is the formula to calculate the future value of the money:

$FV = PV (1 + r)^n$

Where,
FV = Future Value
PV = Present Value
r = Interest Rate
n = Time Period

###

Thank you for reading my eBook. If you enjoyed it, please take a moment to send me your comments or questions?

Thanks!

Regards,
M Fahad Usmani, PMP, PMI-RMP
fahad@pmstudycircle.com

###

Other eBooks by Me

I have authored three other eBooks for professionals preparing for the PMP and PMI-RMP exam.

These eBooks are: PMP Question Bank, Earned Value Management for the PMP Exam, and PMI-RMP Question Bank.

The first two eBooks are for PMP aspirants and the last one is for PMI-RMP aspirants.

PMP Question Bank will help you test your preparedness before the exam.

Earned Value Management for the PMP Certification exam is for professionals who want more in-depth knowledge on Earned Value Management.

PMI-RMP Question Bank will give you a glimpse of a real exam.

PMP Question Bank

This is my first eBook for PMP aspirants. In this eBook you will find 400 PMP exam sample questions based on the sixth edition of the PMBOK Guide, and aligned with the exam format after March 26, 2018.

All questions are unique, have detailed answers and are supported with references to the PMBOK Guide, whenever applicable.

You can use this question bank before your exam to check your preparedness. It gives you an opportunity to test yourself twice.

https://pmstudycircle.com/pmp-question-bank/

EVM for the PMP Certification Exam

This is my third eBook for PMP aspirants.

If you want more in-depth knowledge on Earned Value Management, this eBook is for you.

Based on the fifth edition of the PMBOK Guide, this eBook has more than 75 examples and questions with detailed solutions, which help you understand the formulas, analyze the logical interpretations, and elevate your confidence in solving earned value management questions.

https://pmstudycircle.com/earned-value-management-for-the-pmp-certification-exam/

PMI-RMP Question Bank

This PMI-RMP Question Bank is for professionals preparing for the PMI-RMP certification exam.

This eBook has 170 PMI-RMP sample exam questions covering all exam areas, and is based on the 5th edition of the PMBOK Guide.

In this eBook I have explained all questions in detail and given cross-references to the PMBOK Guide, whenever applicable.

Although the exam syllabus is short and in the real exam you are going to see many similar types of questions, in this eBook I have tried to avoid duplicating questions. So every question will be a new opportunity for you learn a new concept.

https://pmstudycircle.com/pmi-rmp-question-bank/

Discount Coupon

I have prepared a course "The PMP Exam Preparation Tool" for professionals preparing for the PMP exam. This program has all the necessary tools you will need during your PMP exam preparation.

The course includes study notes, true/false statements, flashcards, practice questions, etc.

The course access is for one year.

The course is available at www.pmsprout.com

Since, you have purchased my PMP Formula Guide, you are entitled to a discount for this course.

Please send me an email at fahad@pmstudycircle.com with proof of purchase to receive the coupon code.

###

www.ingramcontent.com/pod-product-compliance
Lightning Source LLC
Chambersburg PA
CBHW020903180526
45163CB00007B/2612
* 9 7 8 1 5 2 0 5 3 1 9 7 7 *